ON4UN's
Low—Band
DXing

Antennas, Equipment and Techniques for DXcitement on 160, 80 and 40 Meters

JOHN DEVOLDERE, ON4UN

Published By

The American Radio Relay League
225 Main Street
Newington CT 06111

Contents

Foreword

We at ARRL are pleased to publish a significant revision to what has become an essential guidebook for low-banders: **ON4UN's Low-Band DXing** by John Devoldere. This book is a *must* for DXers, contesters and casual operators who even occasionally venture onto the low bands.

As with the author's two previous ARRL books, *Low Band DXing* and *Antennas and Techniques for Low Band DXing*, you will find a wealth of information that springs from the author's rich experience. Newcomers and old-timers alike will benefit from the plenitude of practical information in this extensive revision. John Devoldere, ON4UN, has long been recognized both for his outstanding achievements on the low bands and as a highly successful on-the-air contester and DXer. For this third edition, the author has consulted extensively with leading low-band operators. The extensively revised chapter on receiving antennas is but one of the highlights of this new edition. Of the nearly 650 figures, photos and tables, more than half are new. Whatever your band and antenna preferences, you will find a great deal between the covers of this book to put to practical use at your station.

ON4UN's Low Band DXing covers everything you would want to know about 160, 80 and 40 meters—with the emphasis on antennas. The author presents solid ideas for the ham with limited space as well as those who have the budget and real estate to erect those "monster" antennas. Two new chapters cover low-band DXing from a small lot, and ON4UN's personal odyssey from DXer to contester.

As always, we want to hear from you. Help us make the next edition even better than this one. We've provided a handy form at the back of this book that you can use to share your ideas with us.

David Sumner, K1ZZ
Executive Vice President

Newington, Connecticut
March 1999

Preface

Writing the third edition of *ON4UN's Low-Band DXing* has been a very different experience than the two previous editions. In general, to write a book like this requires one to read everything that's been written on the subject. In the past that was limited to reading and studying (with a certain technical standing) lots of books and going through the Amateur Radio magazines. That normally takes almost as much time as writing the book. For this new edition I had to read six years of magazines (a huge pile), but that was nothing compared to when I wrote the first edition.

The big difference from six years ago is the worldwide change from the industrial age to the information age. The Internet has revolutionized the sharing of information. I subscribe to a number of e-mail reflectors on the Internet, and frequently find interesting discussions on them. For years, I've saved all of this electronic correspondence, for possible later use and reference. This includes literally thousands of messages and tens of Megabytes of data on a hard disk. These Internet messages give a good insight into what radio amateurs are interested in, and what topics need to be addressed in a new book. I also find that they often provide good references and excellent food for thought. The problem is to read and comprehend all this material, and place it in the proper perspective.

I have found many exclusive testimonies on the Internet to highlight certain topics, and I have used them throughout the book. This means that many of the low-band DXers who have shared their views or knowledge or engaged in discussions on one of the relevant Internet reflectors (the Top-Band Reflector, for example) have contributed to this new book, and I thank them for their participation.

The publication of *DXing on the Edge*, written by an eminent Top-Band DXer, has made life a little easier for me. My friend Jeff Briggs, K1ZM, expertly covered the history and the personalities associated with Top Band.

The chapter that took longest in rewriting was Chapter 1, the chapter on propagation. This book is not intended to be a scientific thesis and everybody is invited to prove or disprove my findings, just as scientists are invited to turn their scientific research into helpful tools for the low-band enthusiast.

Some of the chapters have been largely rewritten, others not, as I did not see any need for it. The new material has mainly been the product of "evolution" over the past six years: more information available and faster information exchange via the Internet, more powerful and better antenna modeling programs and faster PCs to handle them. The mystery of 160-meter propagation has not yet been unveiled. I guess that barely the top 5% of the proverbial iceberg has

been discovered so far. I still see a lot of explaining going on *after* it all happened, and I have yet to see reliable low-band propagation predictions. All of this appears to be mostly black magic. But that's probably what makes low-band DXing so attractive to many of us: It is not all exact science, and there still is adventure in it!

I have kept the same chapter names, and have added two new chapters. The chapter dealing with antennas for small lots came by popular request. Indeed, low-band DXing is not a hobby exclusively reserved for farmers and ranchers, as many have proven. In "From Low Band DXing to Contesting," I try to explain the challenges of contesting—why contesting is making such an enormous contribution to the Amateur Radio state of the art, and the major differences between a successful DXer's station and a successful contester's station.

One of my great experiences in writing this new edition of *ON4UN's Low-Band DXing* has been to work with several mentors. Friends and experts alike have been found willing to coach, support, advise and help me with nearly all of the chapters. I am indebted a great deal to these fine gentlemen and true friends of mine. They were my perfect critics, coaches, counselors and godfathers during the many, many months of hard work. Thank you Peter (ON6TT), George (W2VJN), Nathan (NW3Z), Frank (W3LPL), Lew (K4VX), Klaus (DJ4AX), Uli (DJ2YA), John (WØUN), Tim (W3LR), George (K2UO) and Frank (DL2CC). Without you it would have been so much harder, and there would have been only half of the pleasure and satisfaction!

I also would like to thank the hundreds of true-blue

low-band DXers who sent in the questionnaire that made it possible to take a fairly accurate snapshot of the active low-banders. And all of you who sent me information, pictures and descriptions of your new antennas, as well as those readers of the previous edition who pointed out errors and typos. Finally a word of thanks to the various authors and publishers who let me quote from their work or use figures from their publications.

This book is not a bible, despite what some have said. The Bible is supposed to relate the only and absolute truth. This is a book about our hobby, about our technical hobby, about applied science, about engineering. It's been written *by* a human and *for* humans, for hams eager to learn. There is no *absolute truth* in this book. But there is certainly a lot in here from which many of us can learn. This book does not tell you what's *the* best antenna. There simply is no best antenna. That's also what makes our hobby so much fun and so exciting! Think how boring it would be if we all had the same (best) antenna, would live in the same QTH without any noise (of course), would all enjoy the same propagation, and finally all have the same skills. Variety, evolution, change, progress, education, improvement—that's what makes life a challenge. That's what makes our hobby, and more particularly low-band DXing, such a treat.

As a closing word, let me quote Carl R. Stevenson, who wrote: "I have little respect for know-nothings of radio and don't care for yackers, beepers, gamers, DXers and others who only use the radio medium as a playing field for their testosterone-induced macho electronic-paintball-war endeavors. Those things are *not* the heart of ham radio. Ham radio has always been intended as a place to experiment, tinker, invent and advance the state of the art, not as an arcade for mindless games."

During the 20th century, Amateur Radio has been a hobby full of magic. The sense of discovery, of adventure, was dominant. Unfortunately today, at the advent of the 21st century, this sense of magic is largely vanishing. When young kids see you operate your ham station, they may actually comment, "Well, I can do just as well with my cellular phone." The world is indeed at the fingertips of everyone these days. And for discovering the world, for global communications, it's hard to beat the Internet. We're living in an ever faster-changing world now. Amateur Radio is no longer in the first place about communications. It is about technology, about education, about friendship, about experimentation. This is what it has always intended to be. Undoubtedly, Amateur Radio will survive for its one and only valid reason.

I think we have a responsibility: we must show youngsters visiting our stations the magic that once thrilled us, and share with them the fascination of our hobby. One or two decades ago, we were all fascinated by the new world of digital electronics. Let's show the young kids the fascinating world of RF electronics. Today, RF engineers are a dying species. We can help preserve this species. We have the responsibility to bring Amateur Radio into the 21st century in the right way and to introduce new people into the hobby.

I hope this new edition will help advance the state of the art of Amateur Radio and be instrumental in giving you increased satisfaction and joy in this wonderful hobby.

I dedicate this book to Frida, Marleen and Stefan.

John Devoldere, ON4UN
September 1998

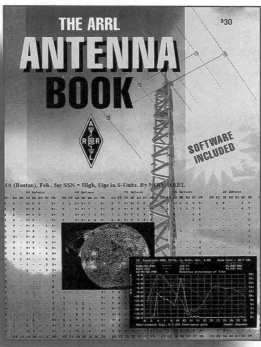

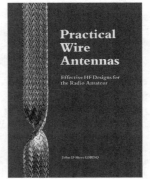

Propagation

1 PROPAGATION

THANKS

I would like to say a word of thanks to Rod Graves, VE7FPT, Ted Cohen, N4XX, and Cary Oler for answering questions and providing some excellent food for thought on the subject of radio propagation on the low bands. This chapter, however, does not necessarily reflect the opinions of any of these propagation experts.

There are two areas related to Amateur Radio where we have seen more publications in the past 5 years than ever before: propagation and (modeled) antennas. It will come as no surprise when I say that the availability of more powerful computers, not only in industry and research, but also in ham shacks is a major reason for this evolution. In recent years some excellent articles dealing in detail with very specific propagation phenomena on 160 meters have been published (Ref. 140, 141, 142). This is to be expected, since 160 meters really excels in terms of still-unexplained propagation phenomena.

There are numerous publications (Ref. 101, 103, 104, 105 and 167) that cover the basic principles of radio propagation by ionospheric refraction. Let me recommend in particular Robert Brown's, NM7M, excellent book *The Little Pistol's Guide to HF Propagation* (Ref. 167). This book is a must for anyone who wants to have a more than a casual understanding of propagation.

The existence of books of this caliber makes it easy for me, since I will not have to explain the basics. A lot of it is covered in this book, although it does not cover or explain all our observations.

This chapter on low-band propagation is not written as a general study book on propagation. This chapter is written for the more advanced ham, one who is a dedicated low-band DXer and who tries to understand propagation to help work an elusive country or maybe to generate a better contest score.

You will soon find out that the general rules that govern propagation on the low bands are rather simple. I will explain that you need a path in darkness and that you (most often) have sunrise and sunset peaks. These are the simple but very important ground rules. Today, however, it is still totally impossible to predict exactly when you will enjoy a good night for low-band propagation. Researchers have tried (and are still trying) to correlate propagation with measurable data. So far, they mainly seem to be able to explain a lot of what has happened, after the fact. Predicting low-band propagation is still a bit like witchcraft, almost like long-term weather prediction!

But, don't let this scare you off. This chapter is to a large extent based on observations, your observations and mine. I am using the testimonies of literally hundreds of low-band DXers to identify interesting propagation phenomena and to quantify possible mechanisms that govern these facts. These can be either long-time and widely accepted ones, or in some cases more speculative and unproved mechanisms. It is important that you recognize "odd" propagation phenomena and circumstances and that you know how to take advantage of them. That's what this chapter is all about.

In the age of ever-faster and more-powerful computers, numerical modeling has produced a number of models that simulate the behavior of the extremely complex ionosphere. Specialists in ionospheric propagation come up with plausible explanations for many of our observations, and the computer models they produce seem to match real-life observations better and better each day. But all the mysteries are far from being cleared up!

I will try to cover propagation on the three low bands (40, 80 and 160 meters), hopefully without confusing my readers! I will explain similarities, but also pinpoint important differences. Understanding the basic mechanics is very essential. If you realize that on 160 meters the opening over an 18,000-km path will occur maybe one day a week—and then only during a specific time of the year—and that the opening will last maybe three to five minutes, you will begin to realize how important it is to know when to try to make that contact.

A spectacular example to illustrate this was my QSO on 160 meters with ZL7DK in early March 1998. I knew I only had a three to five-minute window, both in the morning and the evening. After observing these two windows for almost two weeks, one morning the weak signals from Chatham Island finally came through and I was able to work them. That morning "I had the skip." Other mornings, the signals made it into either England, France or Germany, without any spillover into Belgium. Why? In addition to just noting these facts, I will try later to describe a mechanism that fits these facts.

In recent years numerous HF-propagation programs have popped up. While these have proven their use for predicting propagation on the higher bands (the MUF-related bands), I must admit that I have never had any use for them on the low bands, certainly not on 160 meters.

In the worlds of commercial HF broadcasting and HF point-to-point communications, the challenge consists in finding the optimum frequency or maybe the best angle of radiation (to select the right transmitting antenna) that will give the most reliable propagation, as a function of the time of day. In our hobby of low-band DXing, the problem is quite different. The challenge is to determine the best time (month, day, and hour) to make a contact on a given (low-band) frequency, with a given antenna setup, between two specific locations.

Cary Oler and Ted Cohen, N4XX, wrote in their excellent article "The 160-Meter Band: An Enigma Shrouded in a Mystery" (Ref. 142): "Top Band is one of the last frontiers for radio propagation enthusiasts. It involves regions of the Earth's environment that are very difficult to explore and are poorly understood. These factors have led to our failure to predict propagation conditions with any level of accuracy. They also account for our inability to explain some of the puzzling mixtures of conditions that make this one of the most interesting and volatile bands available to the Amateur service."

Bill Tippet, W4ZV, said it very nicely: "If 160 were perfectly predictable, we would all become bored with it and take up another hobby. Let's just enjoy it as it is, because we'll never be able to figure it out!"

So, don't expect this chapter to predict all kind of exotic openings on 160 meters for you!

■ 1. PROPAGATION VERSUS TIME

Let's have a look at the following time cycles:
The Sunspot Cycle
The 27-Day Sun Rotation Cycle
The Seasonal Cycle
The Time of Day

1.1. The Sunspot Cycle

It is well known that HF radio propagation by ionospheric refraction can be greatly influenced by the sunspot cycle. This is simply because ionization is caused mainly by ultraviolet radiation (UV) from the sun, and UV is highly dependent on solar activity.

Solar activity can influence HF propagation in three major areas:
• MUF (Maximum Usable Frequency).
• D-layer activity (absorption).
• The occurrence of magnetic disturbances.

1.1.1. The MUF

The *critical frequency* is the highest frequency at which a signal transmitted straight up at a 90° elevation angle is returned to Earth. The critical frequency is continuously measured in several hundred places around the world by devices called *ionosondes*. At frequencies higher than the critical frequency, all energy will travel through the ionosphere and be lost in space (**Fig 1-1**). The critical frequency varies with sunspot cycle, time of year and day, as well as geographical location. Typical values are 9 MHz at noon and 5 MHz at night. During periods with low sunspot activity the

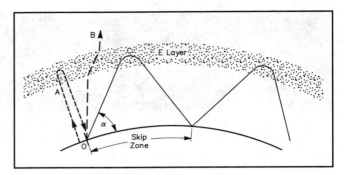

Fig 1-1—Ionospheric propagation. At A, we see refraction of a vertically transmitted wave—this means that the frequency is *below* the critical frequency. At B, the angle is too high or the frequency too high and the refraction is insufficient to return the wave to Earth. At C, we have the highest angle at which the refracted wave will return to Earth. The higher the frequency the lower the angle a will become. Note the skip zone where there is no signal at all.

critical frequency can be as low as 3 MHz. During those times we can even witness dead zones on 80 meters at night.

Fig 1-2A shows a world map created by the program *Proplab Pro* (a sophisticated ionospheric modeling and ray-tracing program), with contours for the critical frequencies. These are abbreviated , standing for the critical frequency of the "ordinary" wave at the layer. Fig 1-2A is for midwinter (January 1) and a low Smoothed Sunspot Number (SSN = 020) at 0000 UTC. Fig 1-2B shows a similar map for mid-summer (July 1) with a very high sunspot number (SSN = 240) at 1200 UTC.

At frequencies slightly higher than the critical frequency, *refraction* (that is, bending in the ionosphere) will occur for a relatively high wave angle and for all lower elevation angles. As we increase the frequency, the maximum elevation angle at which we have ionospheric refraction will become lower and lower. At 30 MHz during periods of high solar activity, such angles can be less than 10°.

The relation between the MUF, the critical frequency and the wave elevation angle is:

$$MUF = F_{crit} / \sin(\alpha)$$

where α = wave angle of elevation in degrees.

Table 1-1 gives an overview of the multiplication factor $1/\sin \alpha$ (also called secant α) for a number of elevation angles (α). For the situation where the critical frequency is as low as 2 MHz, any 3.8-MHz energy radiated at angles higher than 30° will be lost into space. This is one reason for using an antenna with a low radiation angle for the low bands.

The MUF is the highest frequency at which reliable radio communications by ionospheric propagation can be maintained over a given path—this is called the "classical MUF." Another common use of the term MUF refers to a median statistical value. Fifty percent of the time, the actual MUF observed on any given day will be higher than the median (and 50% of the time it will be lower). If you take

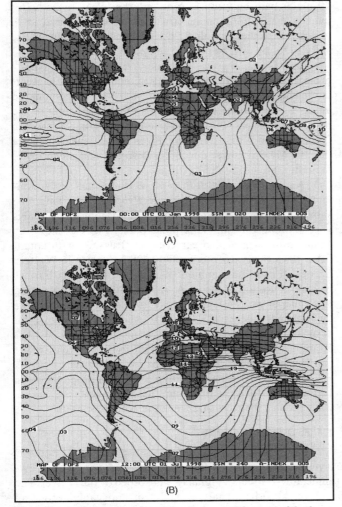

Fig 1-2—At A, Mercator-projection world map with the critical f_0F_2 frequencies for midwinter and a low Smoothed Sunspot Number (SSN=20) at 0000 UTC. At B, map for midsummer with a high sunspot number (SSN=240) at 1200 UTC. (*Maps generated by Proplab Pro software program.*)

85% of the (median) MUF, the path should support signal propagation 90% of the time. This is also called FOT, standing for Frequency of Optimum Traffic. At 115% of the (median) MUF, it should only support signal propagation 10% of the time. This is called the Highest Possible Frequency, or HPF.

Finally, just because signal propagation is supported does not mean that we will be able to communicate. For that, you have to consider the (S+N)/N, which is a function of the modulation type used (this determines the minimum receiver bandwidth) and the noise field at the receiver, among other things.

The MUF changes with time and with specific locations on the Earth, or to be more exact, with the geographic location of the ionospheric refraction points. The MUF for a given path with multiple refraction points will be equal to the lowest MUF along the path. **Fig 1-3A** shows a typical

3000-km MUF chart on a Mercator projection map produced by *Proplab Pro*. (See Section 2.2.7.2.)

Each point of the MUF map is the midpoint of a 3000-km path. If an MUF of 7 MHz is shown for a given location, then you can expect an MUF of 7 MHz for any 3000-km path for which the mid-point is the given location. In other words, the MUF for a signal transmitted 1500 km away from the given location and for which the path goes through the ionosphere above that location is the value shown at that location on the map.

It is generally accepted that the optimum communication frequency (FOT) is about 80% to 85% of the MUF. On much lower frequencies, the situation is less than optimal, as the absorption in the ionosphere increases and atmospheric noise generally increases. We now have computer programs available that will accurately predict MUF and FOT for a given path and a given level of solar activity. These programs are very useful to predict propagation on the higher bands, as well as for 40 meters. Their usefulness is limited on 80 meters and they are even less useful on 160 meters, since propagation on that band is not ruled by MUF.

A high MUF means good conditions on 10 and 15 meters. On the low bands though, only slightly higher absorption can also be expected during sunspot maxima. During low sunspot cycle years, the MUF is often below values sustaining 7 MHz long-distance propagation. For example, during the low-sunspot years, the 7-MHz path between Europe and the USA very frequently closes down during the night and contacts are only possible near sunset (western end) or near sunrise (eastern end of the path).

Even 80 meters can sometimes suffer from this phenomenon. The low sunspot years are therefore not always the best years for low-band propagation—despite the widely held belief of many low-banders.

Fig 1-3B shows a so-called *frequency map*, generated using the *Miniprop Plus* program. This map (which is really an MUF map) allows you to quickly assess the frequencies you can use into a given area of the world. Fig 1-3B is for a very low sunspot number (SN = 11.9) on January 1 at 0800 UTC (Europe sunrise). Note that the map says there is 3.6-MHz, but no 7.1-MHz, propagation between the USA and Europe, confirming what I showed earlier. Fig 1-3C is for a SN of 100, which guarantees enough ionization for a

Table 1-1

α (degrees)	$1/sin(\alpha)$
10	5.8
20	2.9
30	2.0
40	1.6
50	1.3
60	1.2
70	1.1
80	1.0
90	1.0

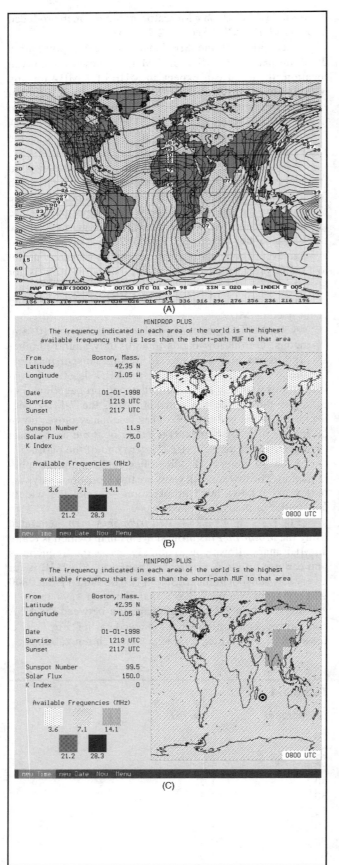

Fig 1-3—At A, Mercator projection of the world showing the 3000-km equal-MUF lines for January 1 (0000 UTC) for an SSN of 20 (A index=5). Note that there are areas in the Northern Hemisphere where during the night the MUF is below 7 MHz. This confirms the experience that during low-sunspot years 40 meters may produce only weak signals during the night on some paths (for example, Europe to the US). Note the great-circle path between Western Europe and California. (Map generated by *Proplab Pro* program.) At B, a frequency map for January 1, 0800 UTC (SSN=11.9, K=0).The MUF is below 7 MHz on the path between Europe and the USA. At C, a similar frequency map; the only difference is that the Sunspot Number is 99.5. This time the path between Europe and the USA is open on 40 meters. (B and C maps were generated by W6EL's *Miniprop Plus* program.)

40-meter path between Europe and the USA. In both frequency maps the geomagnetic A index was entered as 0, which means there would be no solar-induced geomagnetic activity to disturb the path. (See Section 2.2.5.)

Fig 1-4 shows an "Oblique Azimuthal Equidistant" projection from *Proplab Pro* (commonly called a great-circle map) that allows us to see the MUF values encountered along a great circle path between a QTH (the center of the map) and a target QTH. In this example, the great-circle map is centered on Western Europe. From these maps we can see that the MUF is lower during local winter and much lower at night than during the daytime.

Another very useful map projection is the "Polar Azimuthal Equidistant" projection in *Proplab Pro*. This map shows either the Northern or the Southern Hemisphere and is very suitable for analyzing polar paths. **Fig 1-5** shows an example of such a map centered on Brussels, with equal-MUF lines for midwinter at 0800 UTC (sunrise in Western Europe) for a sunspot number of 20. Note the low MUF zones between Europe and the USA. Again, this shows that during low sunspot years 40 meters may go dead at night, just like the higher-frequency bands. **Fig 1-6** shows the same map for a sunspot number of 200, indicating that 40 as well as 30 meters will remain open all night long between Europe and North America at this level of solar activity.

Finally, MUFs have nothing to do with propagation on 160 meters, since the maximum usable frequencies are always greater than 1.8 MHz, even at solar minimum.

1.1.2. D-layer activity

During the day, the lowest ionospheric layer is the D layer, at an altitude of 60 to 90 km. **Fig 1-7** shows how low-angle, low-frequency signals are absorbed by the D layer. The D layer absorbs signals, rather than noticeably refracting them, because it is much more dense than the other higher ionospheric layers. The density of neutral, nonionized particles, which make up the bulk of the mass in this region, is 1000 times greater in the D layer than in the E layer. (See Ref. 121.) For a low-frequency signal to propagate through any layer without large losses, the number of neutral atoms should be small. Statistically speaking, a free electron in the

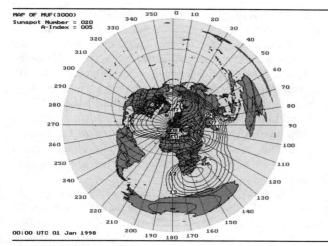

Fig 1-4—Map showing 3000-km equal-MUF lines, displayed on a Great-Circle Projection (also known commonly as an Azimuthal Projection). In this example only the MUF lines up to 10 MHz are shown, in order not to clutter the map. All the radial lines departing from the center QTH (Belgium, in Western Europe) are great-circle lines, indicating the straight-line beam headings to all target locations.This same map projection (without the MUF lines) is used to show beam headings from a particular QTH to DX around the world. (Map generated by Proplab Pro program.)

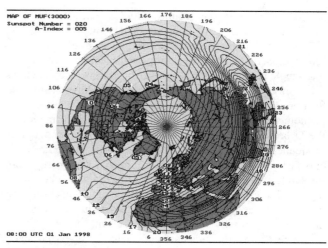

Fig 1-5—The same 3000-km equal-MUF information as shown in Fig 1-4, but now displayed on a Polar Azimuthal Equidistant projection. This example is for 0800 UTC (sunrise in W Europe). Notice the low MUF over the North Atlantic and North America. This shows why 40 meters can often go dead between Europe and the US during low sunspot years. (Map created by Proplab Pro software.)

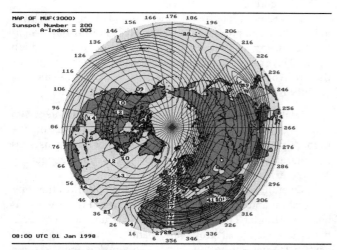

Fig 1-6—The same 3000-km equal-MUF map as in Fig 1-5, but for a very high Sunspot Number of 200. Note that even 10 MHz will remain open all night long between the US and Europe under these solar conditions. (Map created by Proplab Pro software.)

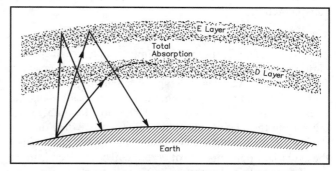

Fig 1-7—D-layer absorption. The higher-angle signals pass through the D-layer and are reflected by the E/F layer. Low-angle signals are absorbed. This explains the need for a high-angle radiator to work short distances in the daytime. This may also be the mechanism explaining why high-angle antennas often work better near sunrise and sunset since they cut through the D-layer better on the low bands.

D layer during the day would collide with nearby neutral atoms about 10 million times per second! Electrons are thus not given much of a chance to refract signals in the D layer and absorption occurs instead. The "collision frequency" is high, resulting in high levels of signal absorption.

During a typical day, the level of ionization of the D layer follows the solar zenith angle, but it is greatly influenced by the level of solar x-ray flares. During the night, the ionization level of the D layer drops dramatically but some very small level persists. It is the degree of this remaining ionization of the D layer that determines the attenuation on the lower bands during the night. This effect is of course, mostly pronounced on 160 meters. Small variations in D-layer ionization can cause large fluctuations in signal absorption on Top Band. This is especially important in the multi-hop propagation modes where the signal has to traverse the D layer twice for each hop.

The absorption level is inversely proportional to the arrival angle of the signal, so high-angle signals pass through the D layer relatively unattenuated. This is one reason our high-angle (low-to-the-ground) dipoles work so well for local traffic on 80 meters during the daytime.

I think that this mechanism also plays a role in the often-reported phenomenon where, for periods shortly after sunrise, high-angle antennas often take over from low-angle antennas for working very long (eg, long-path) distances. (See also Section 1.4.4.4). How might the sunspot cycle affect this phenomenon? When sunspot activity is low, the formation of the D layer is slower; D layer build-up before noon is less pronounced, while the evening disintegration of the layer occurs faster. This is because there is generally less energy from the sun to create and sustain the high ionization level of the D layer. This means, in turn, that at a sunspot minimum absorption in the D layer will be somewhat less than at a sunspot maximum, especially around dusk and dawn.

The absorption mechanism of the D layer has been studied repeatedly during solar eclipses. Reports (Ref. 121 and 125) show that during an eclipse, D-layer attenuation is greatly reduced and propagation similar to nighttime conditions occurs on short-range paths. This means that propagation between two stations that are fully in darkness will only be influenced by the variation in the remaining D-layer density to a very small degree. Things are different when operating in the gray line, where signals have to punch through an already much-ionized D layer. For that reason I believe that attenuation from the D layer when operating gray line will be more pronounced during years of high sunspot activity.

1.1.3. Magnetic disturbances

Auroral activity is one of the important low-band propagation characteristics and is still largely a field of research for scientists. Amateurs living within a radius of a few thousand miles from the magnetic poles know all about the consequences of the phenomenon! The most favorable periods during the 11-year sunspot cycle (that is, when there is the least geomagnetic activity) occur during the up-phase of the cycle. The phenomenon of aurora will be covered in more detail later in this chapter.

1.1.4. Low-band propagation during high sunspot years

For years, the generally accepted notion was that low-band DXing was not favored during high-sunspot years, but not everyone shares this opinion now. Since there must be enough ionization to sustain some sort of propagation mechanism (refraction, ducting, etc), higher levels of solar activity should in theory be advantageous for low-band DXing, as well as for DXing in general. On the low bands this is especially true for the 40-meter band.

In the past there were few DX signals on the low bands during the high-sunspot years. To a large degree this was due to the absence of other DXers on those bands. The relative lack of specific interest in low-band DXing kept the run-of-the-mill DXers away from the low bands when 20 meters was open day and night. Multi-band and specific low-band awards increased emphasis on low-band operating during major contests. Further, the tremendous growth of the elite group of Top-Band

DXers has been very instrumental in raising the activity on the low bands all through the cycle and all through the year.

Around 20 years ago the average DXer "discovered" the low bands and added them to the list of what were considered DX bands. Nowadays, every DXpedition includes 40, 80, and even 160-meter work in their operating schedules. Many DXpeditions even set out to work only the low bands! Even in the middle of the summer during high sunspot years we can now hear several stations calling CQ DX on Top Band!

On 160 meters, propagation doesn't change much through the solar cycle. The mechanisms that are influenced by the cycle are:

1. Auroral mechanism: It appears that we enjoy the geomagnetically quietest years during the minimum and the rising phase of the sunspot cycle (Ref. 142). (See Section 2.2.)

2. Ionization levels in the E and F layers during night: Ted Cohen and Cary Oler (Ref. 142) state that correlation between sunspot numbers and signal strength is only about 5% as strong on 160 meters as the correlation on the high bands. This is because on the low bands reflection during night time in the ionosphere happens in the E and F layers, which have dropped to minimal levels of ionization during the night. And those nighttime minimum levels depend very little on the levels during daytime (when the influence by the cycle is very important, as witnessed by the behavior of the high HF bands).

3. D layer: The remaining ionization of the D layer during the night plays a role, especially in multi-hop propagation. During the twilight periods (gray line) the absorption of the D layer is more substantial than during the night, hence the influence of the sun spots in this mechanism—there is slightly more attenuation during high sunspot years.

In short, the sunspot cycle does have some influence on low-band propagation, but by far not as much as on the higher bands. Low sunspot numbers are by themselves no guarantee at all for good low-band (and especially 160-meter) conditions.

Over the years, I have seen that at the bottom of the solar cycle more and more DXers have discovered the low bands and in particular Top Band. They build better antennas, gain better knowledge of propagation and are increasingly successful DXing on the low bands. This makes them stay on these challenging bands, even after 10 and 15 meters have opened up.

1.2. The 27-Day Solar Cycle

The sun rotates around its own axis in approximately 27 Earth days. Sunspots and other phenomena on the sun can last several solar rotations. This means that we can expect similar radiation conditions from the sun to return every 27 days. DXers (both on the low and the higher HF bands) look forward to a repeat of very good conditions 27 days after the last very good ones have occurred—and their expectation are often met. This is probably the only somewhat reliable propagation prediction system for 160 meters that we have at this time! If conditions are good today, the lack of *bad things* is why propagation may possibly be good 27 days from today.

However, if conditions today are very bad (maybe due to a solar flare), there's no telling whether conditions will be bad in 27 days, since a solar flare certainly does not repeat every 27 days. A recurring coronal hole, however, would most likely repeat in the next 27-day period, since these tend to hang around for at least several solar rotations. Unfortunately, during the early years of a solar cycle this 27-day predicting system may not be very reliable because the sunspots don't hang around for even one full revolution most of the time. As the cycle matures, the 27-day recurrence becomes more important.

1.3. The Seasonal Cycle

We all know the mechanism that originates our seasons: the declination of the sun relative to the equator. This tilt reaches a maximum of 23.5° around December 21 and June 21 (**Fig 1-8**). This coincides with the middle of the Northern Hemisphere winter propagation season and the middle of the summer propagation season. At those times the days are longest or shortest and the sun rises to the highest or lowest point at local noon in the non-equatorial zones.

On the equator, the sun will rise to its highest point at local noon twice a year, at the equinoxes around September 21 and March 21. These are the times of the year when the sun-Earth axis is perpendicular to the Earth's axis (sun declination is zero), and when nights and days are equally long at any place on Earth (equi = equal, nox = night). On December 21 and June 21, the sun is still very high at the equator (90° − 23.5° = 66.5°). The maximum height of the sun at any latitude on Earth is given by the expression:

Height = 90° − north latitude + 23.5° (with a maximum of 90°)

In other words, the sun never rises higher than 23.5° at the poles, and never higher than 53.5° where the latitude is 60°. Because of this mechanism, it is clear that the seasonal influence of the sun on low-band propagation will be complementary in the Northern and Southern Hemispheres. Any influence will be most prominent near the poles, and less pronounced in the equatorial zones (±23.5° of the equator). But how do the changing seasons influence propagation?

1. The longer the sun's rays can create and activate the D layer, the more absorption there will be during the dusk and dawn periods. During local winter, the sun will rise to a much lower apex and the rate of sunrise will be much lower. Accordingly, D-layer ionization will build up more slowly.

2. When the sun rises quickly (local summer in areas away from the equator) the configuration of the D, E and F layers necessary to set up a possible wave-ducting mode will last for a much shorter time than in winter, when the sun rises more slowly. *Gray-line propagation* will thus last longer in the winter than it will during summer.

3. Many thunderstorms are generated in the summer (in non-equatorial areas). Electrical noise (QRN) will easily mask weaker DX signals and discourage even the most ardent DX operator. Take the north-south path (US-to-South America or Europe-to-Africa): The Northern Hemisphere summer may be the most advantageous season, since QRN is likely to be of less intensity than the QRN during the Southern Hemisphere summer.

4. When the nights are longest (in local winter), you will have the greatest possible time for DX openings. Indeed, you must be in darkness or twilight not to suffer from excessive D-layer absorption and have acceptable conditions for long-distance propagation on the low bands.

5. The occurrence of the auroral phenomenon is most pronounced around equinox (March-April and September-October).

1.3.1. Winter (15 October to 15 February in the Northern Hemisphere)

Winter is characterized by low MUFs, shorter days, lots of darkness, sun rising slowly, longer gray-line duration (see Section 1.4.4.1) and no electrostatic discharges (QRN) from local thunderstorms. This period is best for all stations located

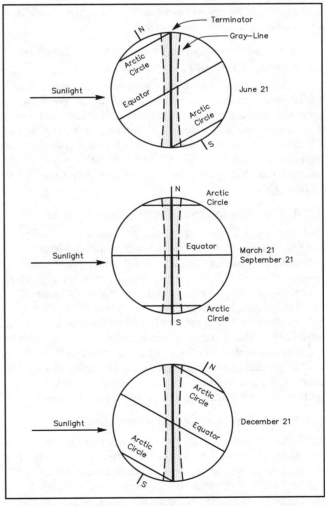

Fig 1-8—These drawings show the declination of the sun at the different positions (solid vertical lines) at different times of the year. The gray line is represented as a zone of variable width (shaded area) to emphasize that its behavior near the poles differs from that near the equator.

in the Northern Hemisphere during the winter. Conversely, this condition will not exist in the Southern Hemisphere. Therefore the winter period in the Northern Hemisphere is ideal for east-to-west and west-to-east propagation between two stations both located in the Northern Hemisphere. Typical paths are US-to-Europe, US-to-Japan, US-to-Asia, etc.

1.3.2. Summer (15 April to 15 August in the Northern Hemisphere)

Summer is characterized by higher MUFs, longer days, faster rising sun, increased D-layer activity at dusk and dawn, and higher probabilities of QRN due to local thunderstorms. These factors create the worst conditions one can expect for east-to-west or west-to-east propagation in the Northern Hemisphere. What we should realize, however, is that while the large majority of amateurs may be fighting the local QRN in the Northern Hemisphere in summertime, our friends down under are enjoying excellent winter conditions. This means that summertime is the best time for transequatorial propagation (for example, from Europe to Southern Africa or North America to the southern part of South America).

1.3.3. Equinox Period (15 August to 15 October and 15 February to 15 April)

During these periods the ionospheric conditions are fairly similar in both the Northern and the Southern Hemispheres: similar MUF values, days and nights approximately 12 hours long on both sides of the equator, reduced QRN, etc. Clearly this is the ideal season for "oblique transequatorial propagation" on the NE-SW and NW-SE paths. Typical examples are Europe to New Zealand and West Coast US to Indian Ocean.

1.3.4. Propagation into the equatorial zones

In principle, all seasons can produce good conditions for propagation from the Northern or Southern Hemisphere into, but not across, the equatorial zone. On 80 and 40 meters the only real limiting factors can be the MUF distribution along the path and especially the amount of QRN in the equatorial zone itself. Unfortunately, there is no rule of thumb to tell us everything about the electrical storm activities in these zones. From Europe, we work African stations and stations in the southern part of South America on 160 meters mainly during the months of June, July and August. A similar situation exists between North America and the southern parts of Africa and South America.

1.3.5. Low bands are open year-round

I have concluded that it is not true that DX on 80 and 160 meters can only be worked during the local winter, a popular belief not so very long ago. The equinox period is the best time of the year for equatorial and transequatorial propagation, while in the middle of our Northern Hemisphere summer (if QRN is acceptably low for us), rare DX stations from down under or from the equatorial zones are often worked. Even good east-west openings can occur in the middle of the local summer, so long as there is a darkness path and the QRN level makes listening for the

signals at all possible. Of course the operators on both sides must be willing to try and not take for granted that it won't work. Over the years I have worked quite a few rare ones over an east-west path in summertime. Here are a few examples:

XYØRR (Burma) was worked on 160 meters on Sep 3, 1991, shortly before his sunrise. After a QSO on 80-meter SSB, we moved to 160 meters, where a 579 was exchanged. After the QSO XYØRR called CQ a few times, but nobody else came back. Those convinced that summer time is not a good time for 160 meters were wrong once again.

A little story behind this contact is that I normally don't have a "long" Beverage up for that direction during the summer. Some 2.5-m (8-ft) tall corn was growing in the field in that direction. That day I spent a memorable couple of hours putting an insulated wire right on top of the corn field. You must try that sometime for fun!

Another striking example is what happened during the DXpedition of Rudi, DK7PE, to S21ZC in early August, 1992. The first night he was on 160 meters I was in the middle of a local thunderstorm (S9 + 40 static crashes) and there was no chance for a contact. The next day, the QRN was down to S7 and a perfect QSO (579) was made over quite a long path (comparable with a path from Sudan (ST) to the US East Coast or from the West Coast of Africa to California).

1.4. The Daily Cycle

We know how the Earth's rotation around its axis creates the mechanism of day and night. The transition from day to night is very abrupt in equatorial zones. The sun rises and sets very quickly; the opposite is true in the polar zones. Let us, for convenience, subdivide the day into three periods:

1. Daytime: from after sunrise (dawn) until before sunset (dusk)

2. Nighttime: from after sunset (dusk) until before sunrise (dawn)

3. Dawn/dusk: sunrise and sunset (twilight periods)

1.4.1. Daytime

Around local sunrise, the D layer builds up under the influence of radiation from the sun. Maximum D-layer ionization and activity is reached shortly after local noon. This means that from minimum absorption (due to the D layer) before sunrise, the absorption will gradually increase until a maximum is reached just after local noon. The degree of absorption will depend on the height of the sun at any given time.

For example, near the poles, such as in northern Scandinavia, the sun rises late and sets early in local winter. The consequence will be a late and slow buildup of the D layer. In the middle of the winter the sun may be just above the horizon (for regions just below the Arctic Circle, situated at $90° - 23.5° = 66.5°$ above the equator), or actually below the horizon all day long for locations above the Arctic Circle. Absorption in the D layer will be minimal or nonexistent under these circumstances. This is why stations located in the polar regions can actually work 80-meter DX almost 24 hours a day in winter (provided there are quiet geomagnetic conditions—see Section 2.2). Contacts between Finland or Sweden and the Pacific or the US West Coast are not uncommon around local noon in northern Sweden and northern Finland at that time of year on 40 and 80 meters.

I suppose this is not a good example of "typical" daytime conditions, as in those polar regions we never actually have typical daytime conditions in midwinter but remain in dusk and dawn periods all day long!

I've mentioned before that during typical daytime conditions, when the D-layer ionization is very intense, low-angle signals will be totally absorbed while high-angle signals can get through and be refracted in the E layer (160 and 80 meters). Only at peak ionization, just after noon, may the absorption be noticeable on high-angle signals. The signal strength of local stations, received through ionospheric refraction, will dip to a minimum just after local noon. As stated before, in order to obtain good local coverage on 80 meters during daytime, you must have an antenna with a high vertical angle of radiation. We will later see that this can very easily be obtained using a low dipole. On 160 meters, middle-of-the-day propagation is essentially limited to ground-wave signals. On the opposite end of the low-band spectrum, 40 meters basically stays open for DX almost 24 hours per day in winter time, albeit with much-attenuated signals around local noon.

1.4.2. Nighttime (black-line propagation)

After sunset, the D layer dissipates and almost completely disappears. Consequently, good propagation conditions on the low bands can be expected if both ends of the path, plus the area in between, are in darkness. The greatest distances can be covered if both ends of the path are at the opposite ends of the darkness zone (both located near the terminator, which is the dividing line between day and night). During nighttime in a period of low solar activity, the critical frequency may descend to values below 3.7 MHz and dead zones (skip zones) will show up regularly. Skip zones are also common on 40 meters during nighttime. Skip zones due to MUF do not occur on 160 meters since the MUF is always higher than 1.8 MHz.

In contrast to gray-line propagation, Brown, NM7M, calls propagation with one of the stations at the terminator "dark-line propagation" (Ref. 140)

1.4.3. Midway midnight peak

North-South (± 30° headings) paths exhibit a clear propagation peak at local midnight halfway on the path, both on 80 and 160 meters. When I make skeds on these bands with African or Indian Ocean stations, I will always try to have them at "midway midnight."

Although it has been generally believed that an east-west path only exhibits a sunrise and a sunset peak, many critical observers have witnessed (at least on 160 meters) a similar midway midnight peak. This peak certainly does not exist on 40 meters, where the signal peaks are only there before and around sunset and around or after sunrise. Although I had observed the same phenomenon several times, it was Peter, DJ8WL, who raised the question on the Internet. The exact

mechanism may not be understood but it probably is connected to the fact that, at that time, the sun is exactly "on the other side" of the Earth, creating the most favorable ionization conditions in the E and F regions of the ionosphere on the dark side of the Earth. Later, I will explain how to calculate these peak times, using sunrise-sunset times.

1.4.4. Dusk and dawn: twilight periods

As mentioned before, the terminator is the dividing line between one half of the Earth in daylight and the other half in darkness. The visual transition from day to night and vice versa happens quite abruptly in the equatorial zones and much more slowly in the polar zones. (See Section 1.4.1.) The so-called *gray line* is a gray band between day and night, usually referred to as the *twilight zone*. Actually, "gray zone" might have been a more appropriate term than gray line. Dusk and dawn periods produce very interesting propagation conditions that are not limited to the low bands. However, the mechanisms involved can differ very substantially between the high bands (10, 15 and 20 meters) and the low bands (40, 80 and 160 meters).

For low-band operators in particular, it is extremely important that they be able to visualize the situation, using maps or globes (see Section 2.2.7.2) that show the terminator, the great-circle lines and the auroral oval. For a long time many have speculated about what actually produces the enhanced propagation condition we all experience almost daily on the low bands at either dusk or dawn. It has become widely accepted that these twilight effects are due to the onset of specific propagation mechanisms—ones that are characterized by lower loss than the standard multi-hop model. The mechanisms involved will be discussed in more detail later.

But besides the role of the specific propagation mechanisms at dusk and dawn, there is another reason why we are able to work DX much better during these twilight periods. When the sun is rising in the morning, all signals coming from the east (which can often cause a great deal of QRM during the night) are greatly attenuated by the D layer existing in the east. The net result is often a much quieter band from one direction (east in the morning and west in the evening), resulting in much better signal-to-noise ratios on weak signals from the opposite direction. If only one end of the path is in twilight, this creates a one-way enhancement only.

It is also important to know how long these special propagation conditions exist—in other words, to know how long the effects of the radio-twilight periods last. You should understand that the rate of change from darkness to daylight (and vice versa) depends on the rate of sunrise (or sunset). There are two factors that determine this rate: the season (the sun rises faster in summer than in winter) and the latitude of your location (the sun rises very high near the equator, and peaks in the sky at low angles near the poles).

Another important point—I have often observed that propagation paths perpendicular to the terminator seem to enjoy the greatest signal enhancement. I believe it is because these paths travel the shortest distance through the D layer.

1.4.4.1. Definition of a low-band gray line

When both ends of a path are in the twilight zone, one at sunrise and the other at sunset, then we have a gray-line situation. The effect of advantageous propagation conditions at sunrise and sunset has been recognized since the early days of low-band DXing. Dale Hoppe, K6UA, and Peter Dalton, W6NLZ, first called the zone in which the special propagation condition exists the gray line (Ref. 108). The gray line is a zone centered around the geographical terminator.

In the past, some authors have shown the gray-line zone as a zone of equal width all along the terminator. This is incorrect so far as the related radio-propagation phenomenon is concerned. R. Linkous, W7OM, recognized this varying zone width and accordingly emphasized its importance in his excellent article, "Navigating To 80-meter DX" (Ref. 109).

The mechanism that determines the width of the gray lines means that we have a narrow gray line near the equator and a wider gray line near the poles. The time span during which we will benefit from typical gray-line conditions will accordingly be shorter near the equator and longer in the polar regions. Therefore the gray-line phenomenon is of less importance to low-band DXers living in equatorial regions than to their colleagues close to the polar circles. This does not mean that there is less enhancement near the equator at sunrise or sunset; it just means that the duration of the enhanced period is shorter.

Some authors (Ref. 108 and 118) have stated that gray-line propagation always occurs along the terminator. On the low bands there has been only occasional proof of such propagation. From the following examples of gray-line propagation, it should be clear that propagation does not happen along the gray line but rather through the dark zone, on a path that (in most cases) is nearly perpendicular to the terminator. Gray-line propagation on the low bands is a different affair from what often is called gray-line propagation on the HF bands, where the propagation path does follow the direction of the gray line.

A well-known example has been described by K9LA (Ref. 152). Many of us remember the "great" low-band conditions between Heard Island and the US East Coast in January, 1997. This had been considered to be an extremely difficult path. **Fig 1-9A** shows the theoretical great-circle path between Heard Island and New York (mid January at 2300 UTC). Note that the path (heading of 250°) makes an angle of approximately 30° with the terminator on Heard Island. Fig 1-9B shows the same path, this time with New York as the center of the azimuthal projection map. This path (heading of 130°) also makes a sharp angle (25°) with the terminator at the US-end of the path. Most, if not all US East Coast stations who worked VK0IR and who had access to a variety of directive receiving antennas (such as Beverages) noted that the VK0IR signals arrived at a heading of approximately 60°, right across Europe.

Fig 1-9C shows the path between Heard Island and Spain. The path (beaming 300°) now makes a perfect 90° angle with the terminator on Heard Island. Similarly, the path between the US East Coast and Spain (beaming 65°) also makes a perfect 90° angle with the terminator at the US end of the path (Fig 1-9D). I am convinced that the signals at Heard Island traveled across Europe to the US East Coast. This is supported by testimonies from US stations and it again confirms my belief that enhanced gray-line conditions most often go together with a signal azimuth that is nearly perpendicular to the terminator.

K1GE confirmed that this has occurred with other stations from the Indian Ocean as well. I was listening every day during the Heard Island DXpedition and witnessed that the signals faded out completely in Europe at exactly the same time they faded out in North America. This seems to confirm that, indeed, the path to the US was right across Europe. What makes this path skew to more northerly regions is explained later in this chapter.

To be fully correct, I must admit that the QSO between the East Coast and Heard Island at 2344 UTC is only half a gray-line QSO. However, stations a little further inland in the USA who worked VK0IR just prior to that time did it on a double-sided gray-line path.

The VK0IR expedition was a living testimony to the fact that the width of the gray line depends on the latitude of the station involved. Many remember how VK0IR (located at 53° south latitude) was worked almost every day on 160 meters until more than 30 minutes after local sunrise, while 80-meter QSOs were made as late as 0050 UTC, 1½ hours after sunrise on Heard Island during that DXpedition. This is a remarkable circumstance where significant signal enhancement occurred for a signal direction perpendicular to the terminator at both ends of the path.

1.4.4.2. Other examples of remarkable gray-line propagation

Another striking example of gray-line enhancement involved a QSO I had on 80 meters with Kingman Reef, which is a particularly hard path late in the Northern Hemisphere winter season from Europe. I made QSOs with Kingman Reef and Palmyra around May 1, 1988. If we analyze the sunset and sunrise times for that date, we see that sunset is roughly 40 minutes after sunrise at my location. This means that there can be theoretically no opening, but we can of course force things a little and take advantage of the gray line. How shall we do that? Split the 40 minutes in half and try a QSO 20 minutes before sunset in Palmyra (or Kingman Reef) and 20 minutes after sunrise here in Belgium.

Does this sound like a nice 50/50 deal? Certainly not! Those Pacific islands are situated only about 6° north of the equator, while Belgium is 51° north. This means that the gray line lasts just seconds on KH5 and maybe 40 minutes in Belgium. The skeds were made right at Pacific sunset time. On Kingman the QSO was made 5 minutes after Kingman sunset, on Palmyra 4 minutes after sunset, and in both cases 40 to 45 minutes after sunrise in Belgium (where the gray line is fairly "wide"). This is a striking example of how knowledge of the mechanism of propagation can help you realize a very difficult QSO. As proof of how marginal a situation it really was, only one QSO was made with Europe

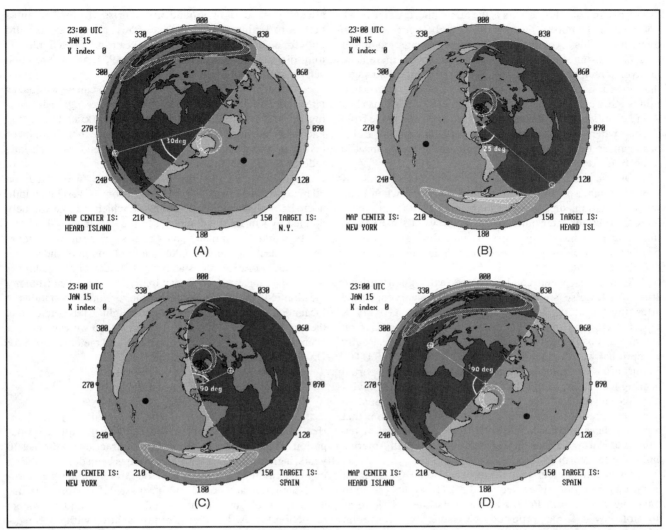

Fig 1-9—At A, the great-circle, short path from Heard Island to New York. The angle compared with the terminator at Heard is quite sharp. At B, the other end of the same path, you see also a sharp angle (25°) compared to the terminator. At C, the path from Heard Island to Spain in Europe makes for a perfect 90° angle with the equator, as it does for the bent path from the US East Coast to Europe. Most, if not all East Coast stations reported that the Heard Island signals came in on a path over Europe. Similarly, the crooked path across Europe met the critical conditions for maximum signal enhancement at sunrise or sunset. This was a perfect example of a bent path, witnessed by many observers. (*Figures created using* DX-AID *software, with additions by ON4UN.*)

from Palmyra, and only two from Kingman Reef.

Over shorter paths, stations can often be worked on 80 meters until hours after sunrise (or hours before sunset). Contacts between the US East Coast and Europe are quite common in midwinter up to 2 hours after sunrise in Europe on 80 meters.

The width of the gray line on 160 meters is much more restricted than on 80 meters. Even in the middle of the winter, I have seldom worked real long-haul QSOs more than 15 minutes after sunrise on Top Band. I often copy stations quite late after sunrise, but with rather weak signals on a very quiet band. This is possible mainly because of the absence of noise from the east. This phenomenon has been confirmed many times by Jack, VE1ZZ, who can copy European stations on 160 meters up to 3 hours before his sunset. This does not

mean he was able to contact the many European stations he heard. For that he had to wait almost two hours! Anyhow, at that time of the day the European stations are probably listening to the east. VE1ZZ further reported that the signals were quite weak and only copiable because there was absolutely no noise whatsoever when this happened.

At this time in the afternoon toward the daylight area, the D-layer acts as a shield to ionospheric propagated noise. This provides a much lower noise floor for the station in sunlight. The stations at the eastern end of the path are subject to high noise levels because they do not have the advantage of the D-layer absorption attenuating the atmospheric noise propagated into their area. Just watch your S meter on 160 meters at 2 PM during winter and then again after sunset and you will probably see 20 or 25 dB difference in noise level. This is the atmospheric noise level, not man-made noise. Twenty dB is a difficult spread to overcome. This explains the one-way propagation under such circumstances. At the same time, in Europe there is no D-layer screen and signals from the east are 50 dB stronger than Jack's signal before his sunset. This one-way propagation is quite common during quiet magnetic conditions (low A and K indices). Similar observations have been made from Europe where UA9/Ø stations are heard long before European sunset, but again one has to wait until almost sunset in order to make contact.

On 160 meters there are rare times when I can work DX well after sunrise from my location in Europe—when I can work ZL stations on a genuine long path (21,500 km) about 30 to 45 minutes after our sunrise. The few ZLs worked on this long path had to be worked right through a wall of English stations, who were enjoying their sunrise peak at exactly that time.

Sometimes, when conditions are really good with no atmospheric noise, signals can also be heard a very long time after sunrise at the eastern end of the path. GW3YDX reported copying many W6/7 stations as well as KL7 on 160 meters until more than an hour after local sunrise. Really exceptional was the fact that he copied K6SE at 1130 UTC, 3 hours after local sunrise! That same day, GM3POI reported hearing KL7RA at 1230 UTC, also on Top Band (in midwinter)! We have, of course, to be careful and not extrapolate these observations to the whole of Europe—the stations that reported these extraordinary conditions are located at the fringes of Europe, almost at the back door of North America and quite far North (52° to 53° N).

To me, it is clear that these exceptional QSOs must occur on a *bent path*. (See Section 3.3.) The signals seem to travel near the North Pole, staying in darkness as long as possible. W8LT has reported the same experience on 80 meters, working EI and G stations between 1000 and 1100 UTC. He confirmed that the best antenna for this propagation was a half-square antenna, broadside to N/S. To him, this indicated that a crooked path was involved. I believe that this kind of propagation can only occur when magnetic conditions are exceptionally quiet.

Similar conditions are quite common on 40 meters during the European winter. With a good Yagi antenna I can work North America 24 hours a day on 40 meters, when there is no magnetic disturbance. At local noon, when the sun is highest, I hear W8s and W9s quite often, followed somewhat later by the W6 and W7/VE7 stations, all on a direct polar path. The West Coast will keep coming through with the beam pointed approximately 350°, until at 1430 UTC the band will also open on the long path. Shortly later, the bent short path will close.

These specific propagation paths and times are applicable only to moderate-distance DX when it comes to 160 meters. Real long-haul propagation on 160 meters seems to follow the rule of enhanced propagation only occurring at dawn/dusk. During a long period of tests (in November and December) on 160 meters between New Caledonia (FK8CP) and Europe I found that the signals always peaked right around my sunrise (from 3 minutes before, to 3 minutes after sunrise). This "short peak" is valid only for a very long path to Western Europe. FK8CP reports long-distance openings into Asia (UA9, UAØ) from much earlier until a little later after his sunrise.

Close-in DX (2000 km or 1500 miles) can be worked as late as 45 minutes after sunrise on 160 meters, again depending on one's latitude. FK8CP reports working DX in the Pacific as late as 50 minutes after his sunrise in the middle of his local summer (which is quite late, considering the latitude of New Caledonia).

On 40 meters, the gray line is of course "wider" than on 80 meters (remember the gray line is a zone of variable width, not to be confused with the terminator, which is a line uniquely defined by sun-Earth geometry). In the winter, long-haul DX can be worked until many hours after sunrise (or many hours before sunset), again depending on the latitude of the station concerned. For example, stations at latitudes of 55° or higher will find 40 meters open all day long in winter. Even from my location (51° north), I have been able to work W6 stations at local noon time, about 3 hours into daylight. At the same time, the band sometimes opens up to the east, so we can say that even for my "modest" latitude of 51° N, 40 meters is open for DX 24 hours a day on better days.

1.4.4.3. Ionospheric ducting and chordal-hop propagation

Multi-hop propagation with intermediate ground reflections has long been the traditional way to explain propagation of radio waves by ionospheric refraction. In the last 20 years, a great deal of scientific work has been done, enabling us to calculate exact path losses due to ionospheric absorption (deviative and non-deviative losses), free-space attenuation (path-distance related) and earth (ground or water) reflection losses. The theory of propagation with intermediate ground reflections is satisfactory to explain short and medium-range contacts.

Some experts are skeptical, however, that the same theories can adequately explain the high level of signals often experienced over very long distances. This is especially so when gray-line propagation and genuine long-path distances are involved, since the old theory would require many intermediate ground reflections and these would necessarily involve very high losses.

I favor the concept that when an entire very long-distance path is well into darkness on 160 and 80 meters, propagation is primarily by multi-hop propagation, with little or no D-layer absorption. This means that only a few intermediate ground reflections must be involved. This sort of model seems to fit observations well when neither end of the path is in the twilight zone.

Even this model, however, does not explain precisely why there is generally a significant signal enhancement where either or both ends of a very long-distance path are in the twilight zone. In the twilight zone, there should be more D-layer absorption compared to the full-darkness situation. Thus, there must be another mechanism involved that compensates for this additional twilight-zone D-layer loss and actually ends up with a loss balance that is lower than the full-darkness multi-F-hop model, with its low level of D-layer absorption.

Signal ducting could be the answer to this question. Recently propagation-prediction software tools (Ref. 153) have been developed that include three-dimensional ray-tracing models, including geomagnetic effects. *Proplab Pro* is one such program that produces 80 and 160-meter predictions that explicitly compute ducting modes for many long-distance paths. The results from this program correspond better than other programs to what we hams have actually observed for many years. (See Section 2.2.7.2.)

I first came across a description of the phenomenon of ionospheric ducting in an article by Y. Blanarovich in 1980 (Ref. 110). Almost 20 years later, there appear to be several schools of thought concerning the existence and the mechanism of ionospheric ducts.

Arguments in favor of multiple ionospheric/ground hops

Some hams maintain that ionospheric ducts do not exist and that all long-distance propagation must be by multiple-hop Earth-ionosphere propagation. In its favor, this multi-hop-only view does help explain why paths that are across saltwater generally produce much stronger signals, apparently due to the minimal reflection losses at saltwater reflection points. The all-duct theory cannot adequately explain this over-saltwater phenomenon. Robbins, KL7Y, who's been working in the field of HF radar, states categorically that HF signals really do bounce off the Earth and that the losses on Earth reflections (especially from saltwater) are mostly insignificant—usually much less than the losses due to an ionospheric reflection, at least for frequencies below the MUF.

Textbooks, including *Ionospheric Radio* by Davies, provide charts that indicate that sea-water reflection loss is a fraction of a dB on 160 meters for all but the lowest angles (3° or less). A land reflection might typically average several dB, so it is easy to see how long paths over saltwater could produce tremendous signals on the low bands compared to paths that require multiple reflections over poor ground. This is one area where I believe many propagation programs fail. Since most programs do not know the geography at the reflection point, they plug in an "average value," something like 2 or 3 dB. On an all-water long path with multiple ground reflections, the program could be off by 10 to 20 dB. One of the exceptions is *Proplab Pro*, which contains an earth/water/ice geographical data base that is used to compute realistic reflection losses.

Robbins, KL7Y, continues by saying that you actually can hear for yourself the different bounces of your signal: "*If you have a quiet QTH, try fast QSK and a quick dit at full power on 160. I usually can hear my backscattered signal quite well when the band is open. Sometimes the echo is discernible as a very quick string of echoes, with each echo representing another bounce. The phenomenon seems to work much better on a high horizontal radiator, although it can be heard on a vertical antenna at times. The backscatter is sometimes surprisingly strong, S9 or better. Sometimes*

you can make the backscatter more noticeable by using the RIT so the echo comes back at a different note than your sidetone."

Others, however, based on experimental observations (Ref. 100) and theoretical studies (Refs. 131 and 151), have come to the conclusion that very specific propagation modes were causing exceptionally strong signals to be heard over very long paths. Signal ducting and chordal-hop propagation are forms of ionospheric propagation without intermediate ground reflections and appears to offer reasonable explanations for long-distance propagation.

Signal ducting

Due to the Earth's round shape and the layered structure of the ionosphere, waveguide-like channels (ducts) can appear, in which radio waves can propagate over long distances. **Fig 1-10** shows a nighttime electron-density distribution over the North Atlantic Ocean near Iceland, a likely reflection point for a propagation path from North America to Europe. This is for 0000 UTC on January 1 for a smoothed sunspot number of 20 and an A Index of 5. There is a dip in density above the E-layer peak. Under the proper conditions, this valley can be responsible for setting up a waveguide-like 160-meter E-F layer duct, bounded by the F layer at the top and the top of the E layer on the bottom (Ref. 151). The trick is to get the signal to enter this kind of duct.

Cary Oler and Ted Cohen (Ref. 142) point out that this kind of ducting is most typical for 160 meters because Top Band signals can be refracted more effectively at higher wave angles than signals at higher frequencies can be.

From the launching point of such a duct path, refraction occurs when the signal travels through the E layer, resulting in bending of the waves into a lower angle (in other words, there is not a complete reflection). The wave is propagated further at the required angle to start the ducting. Relatively high launch angles from the Earth are required to punch though the D and E layer, so that the wave can be finally reflected up into the E-F region. This may be another explanation why higher wave-angle antennas often beat out very low-angle antennas, especially on 160 meters.

In addition, horizontal ionospheric tilting is often required to be able to work into a duct. This condition typically exists at the terminator, provided that the angle between the propagation path and the terminator is close to 90° (Ref. 151).

At the receiver end of the duct, a disturbance of some sort, perhaps a horizontal tilt, causes the signal to break out of the duct and refract back to Earth. The exit of the signal from the duct often produces a spotlight-like illumination of the Earth, making signals very strong in a specific location, while being inaudible only a few hundred miles away.

This kind of propagation is typical for 160. Unusually strong signals can be explained by the fact that the signal only crosses through the lossy D region twice: once when leaving the transmitting site and once again at the end of the duct. We know that it is the D layer that causes the absorption, even during the night, when the D layer is "thin" and only slightly ionized. In this scenario we consider no other

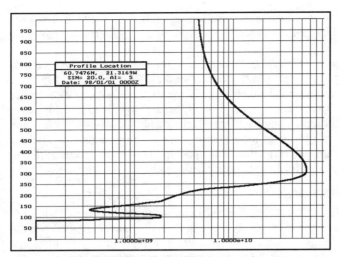

Fig 1-10—Electron-density profile for a point near Iceland, roughly half-way between Europe and North America on the North Atlantic path. Notice the dip in electron density between the E and F layer. This dip is responsible for waveguide-like propagation, with the bottom side of the F layer and the top side of the E layer serving as "duct walls."

causes of absorption; for example, aurora.

Signal ducting can be combined with multi-hop, either at one end or at both ends. The *Proplab Pro* program often indicates a combination of the above mechanisms and also graphically shows sometimes-considerable horizontal path bending away from the great-circle heading.

In a genuine double-ended gray-line case, both ends of the path can enjoy the signal boost originated by the special launching conditions in the dusk and dawn areas. This also can occur when just one end of the path goes through twilight. On 160 meters this enhancement is very typical on the Europe-to-Western US path. Provided the geomagnetic situation is quiet, in Western Europe we can work the US West Coast both at their sunset and then again at our sunrise, while during the period in-between, we rarely reach as far west. The paths close because we no longer receive the benefit of refraction into lower effective launch angles at the E-layer penetration, and/or from loss of sufficient E-layer ionization to allow bouncing off the topside of the E-layer boundary of the duct.

An observation mentioned previously that seems to support the existence of ducts is that they are often very location-selective, at both ends of the duct. I have not only experienced this "spotlight propagation" on 160 meters but also repeatedly on 80 meters. The long path between the US West Coast and Europe often produces very strong signals, but in small areas. I remember several occasions when N7UA was S9+ at my QTH (approximately 20 minutes before his local sunset), while he was 4 to 5 S units weaker 200 km to the east of me (in the direction from which his signal arrives). One explanation for this phenomenon is that a particular exit condition from a duct existed at that time, creating this spotlight-like situation, illuminating only a very specific geographic area. This may also

explain why a DXpedition can often manage to copy calls from what we hear as a massive, unruly pileup without too much difficulty.

A very remarkable example of such propagation was experienced during the recent ZL7DK DXpedition. At the end of February and early March, 1998, the sunrise and sunset at my QTH and on Chatham Island occur within minutes of each another. This means we have a genuine double-ended gray line but with a very small window of time. ZL7DK made QSOs into Europe on 160 meters on most days, both at their sunrise and sunset. I listened every day, during both windows, and could hear Gs, DLs, etc working him without the slightest trace of a signal here. Then one day I heard ZL7DK coming out of the noise at exactly my sunrise and he remained copiable for about 10 minutes. During that time frame only I worked him—nobody else even called him. Apparently that day the duct-exit configuration put only Belgium in the spotlight.

Another observation that seems to support the existence of such ionospheric ducts is that on such paths usually only the two ends of the path have propagation, and except for stations within a few thousand km (which is really a short distance compared to the length of the total path) no other stations in-between can copy the signal. This was also the case with the Chatham Island station. While the great-circle bearing is approximately 5° (West of North) from Belgium, I copied ZL7DK best on a Beverage pointing at 310°, right across the USA. At that time there was not a single US station calling ZL7DK though.

During the 10 minutes that the path lasted between my QTH and ZL7DK, I witnessed what seemed like waves of "galactic" noise on the band. I mentioned this to Cary Oler and he explained this as follows: "*Ducting could explain the waves of noise you heard as well. Since storms in North America can produce static that is in the Top-Band frequencies and since static can be broadcast in almost any direction relative to a thunderstorm, it makes perfect sense that the static signals could enter the same duct as valid transmitter signals. This would effectively transmit the atmospheric storm noise from North America to your location—a real potential problem world-wide. So the noise you may be hearing during those undulating wave-like periods could very well be originating from some very distant location. I hope this helps perhaps explain what you heard. I can't say conclusively, of course, but it makes sense that this might be happening to you. This same explanation can also be applied to instances where one station is plagued by noise and another station not too far away is noiseless. Since noise is a signal that can be ducted as well, on Top Band, many people may think they're hearing local noise when in fact they're just the poor unlucky recipients of distantly ducted atmospheric radio noise.*"

Ionospheric signal ducting often seems to go hand-in-hand with signal path skewing. Cary Oler and Ted Cohen, N4XX, mention that signals do not always follow the exact great circle path, but deviate in various directions according to changes in the shape of the ionospheric layers and the orientation of the signal to the Earth's magnetic field (Ref.

142). It is not clear to me, however, whether these models support the degree of path skewing (often > 45°) that are often seen, especially on 160 meters.

Chordal hops

Other authors have pictured a very specific way of signal ducting, called *chordal hop propagation*. In this system the waves are guided along the concave bottom of the ionospheric F_2 layer acting as a single-walled duct.

The flat angles of incidence necessary for chordal-hop propagation are possible through refraction in the E layer and because of the tilt in the E layer at both ends of the path. Chordal-hop propagation modes over long distances are estimated by some to account for up to 12 dB of gain due to the omission of the ground reflection losses. Y. Blanarovich, VE3BMV, described a very similar theory (Ref. 110). Long-delayed echoes, or "around-the-world echoes" witnessed by several amateurs on frequencies as low as 80 meters, can only be explained by propagation mechanisms excluding intermediate ground reflections.

As it stands today, it seems to me that propagation mechanisms without intermediate ground reflection described by different authors are somewhat speculative, since it has not and cannot be measured. Just because it works on a model doesn't mean that it necessarily works like that in the real world! The reason, of course, that these mechanisms have been widely accepted as fact is that they fit a lot of observations better than other theories or explanations.

1.4.4.4. High wave angles at sunrise/sunset

Over the years, many have observed that enhanced gray-line propagation goes hand-in-hand with high-angle radiation. Hams generally accept the notion that low-radiation angles are required for DX work on the low bands. Those that have the choice between low-angle antennas and a high-angle radiator (a very low dipole) confirm that 95 to 99% of the time the low-angle antenna is the better one, but there are the occasions when the low dipole will be the winner. This only seems to happen during the gray-line period (dusk or dawn) though.

In the 1960s, Stew Perry, W1BB, speculated that at sunrise and sunset the ionosphere acted like a big wall behind the receiving or transmitting location and it focused weak 160-meter signals like a giant poorly reflective dish on one area at a time just ahead of the densely ionized region in sunlight. This seems like an acceptable explanation, since it appears that losses at low incident (grazing) angles are very high near the LUF (Lowest Usable Frequency) of a path. This may also explain why very often at sunrise and at sunset high-angle antennas seem to perform better than low-angle antennas.

Another possible way of explaining the fact that high-angle antennas often have an edge at sunrise/sunset is that the high-angle signal on its way to the F layer can punch right through the E layer. The lower-angle signal spends too much time passing through the D and the E layer and it is absorbed more. In other words, over the same distance a two-hop, high-angle signal can be considerably stronger than a single-hop, low-angle signal due to the effects of the D and E layer.

G3PQA notes that his dipole (high-angle antenna) always outperforms his low Beverage (which has very little off-the-side, high-angle radiation) when working ZLs on 80 meters on long path at equinox, when there is a daylight gap and when conditions peak after sunrise. This also might confirm that a high angle is required to pierce through the D and E layer in order to get into a ducting mechanism.

Y. Blanarovich, K3BU, testified that during one of his recent Top-Band operations from VE1ZZ's QTH: "*I had an inverted V at 70 ft and four square L-vertical array, and I was able to crack the 'one way afternoon' Europeans with the Inverted V almost two hours before the four square was heard. These verticals have an ocean of radials under them and sitting at the ocean shore on a small hill.*" This means they produce extremely low radiation angles, which evidently is not what's needed under these circumstances.

1.4.4.5. Antipodal focusing

Most low-band DXers know that it is relatively easy to work into regions near the antipodes (points directly opposite one's QTH on the globe). This is despite the fact that those are the longest distances one can encounter—one would expect weak signals as a result. The phenomenon of ray-focusing in near-antipodal regions must be used to explain the high field strengths encountered at those long distances. This is in addition to the gray-line phenomenon. Antipodal focusing is based on the fact that all great circles from a given QTH intersect at the antipode of that QTH. Therefore, radio waves radiated by an antenna in a range of azimuthal directions and propagating around the Earth along great-circle paths are being focused at the antipodal point. Exact focusing can occur only under ideal conditions—that is, if the refracting properties of the ionosphere are ideal and perfectly homogeneous all over the globe. As these conditions do not exist in the real world (patchy clouds, MUF variation, etc), antipodal focusing will exist only over a limited range of propagation paths (great-circle directions) at a given time.

The smaller the section of the azimuthal shell involved in the focusing (ie, the narrower the beamwidth), the closer actual properties will approximate ideal conditions. This means that in order to gain maximum benefit from antipodal focusing, the optimum azimuth (yielding the lowest attenuation and the lowest noise level) has to be known.

Fixed, highly directive antennas (fixed on the geographical great-circle direction) may not be ideal, however, since the optimum azimuth is changing all the time (winter vs equinox, vs summer). Rotatable or switchable arrays are the ideal answer, but omnidirectional antennas or antennas with a wide forward pattern perform very well for paths near the antipodes (by summing all paths, as with "diversity" antennas). The focusing gain can be as high as 30 dB at the antipodes and will range in the order of 15 dB at distances a few thousand kilometers away from the exact antipode.

While the effect on 80 meters seems to spread quite a distance from the theoretical antipode, on 160 meters the focusing appears to be more localized. On 160 meters, the Gs can benefit from this effect into ZL, while on the European continent the effect seems to be all but non-existent. This is very different from 80 meters. Since the antipodal focusing example given above obviously coincides with gray-line propagation, and since the gray-line period is much more restricted in time on 160 meters compared to 80 meters, it should be clear that the focusing phenomenon applies to a much narrower region on Top Band.

■ 2. PROPAGATION VS LOCATION

Working the low bands is very different, depending on whether you live near or in the arctic regions, at average latitudes or near the equator. Let's analyze what causes these differences.

Previously I have referred to the geographical location of the station. There is a close relationship between the time and the location when considering the influence of solar activity. Location is the determining factor for five different aspects of low-band propagation:
1. Latitude of your station vs rate of sunrise/sunset
2. Magnetic disturbances
3. Local atmospheric noise (QRN)
4. Effects caused by the electron gyrofrequency
5. Polarization and Power coupling

2.1. Latitude of Your Location Vs Solar Activity

This aspect has already been dealt with in detail previously. The latitude of the QTH will influence the MUF, the best season for a particular path and the width of the gray-line zone.

2.2. Magnetic disturbances (aurora)

F. R. Bond, in his book *Aurora Australis* wrote "The aurora (Southern and Northern Lights) is mankind's only visible marker of the interactions taking place in the vast and complicated region of the Earth's magnetosphere." Brown, NM7M, recently added: "... and Top Band Propagation is another aspect of those interactions." (Ref. 140)

Auroral absorption, most often evidenced by the aurora at high latitudes, is a very important factor in long-distance propagation mechanisms on the low bands. It is certainly the most important one for those living at geomagnetic latitudes of 60° or more, as well as for all of us living in more southerly regions when we are trying to work stations on paths that cross areas affected by the aurora. We are interested in what effect this phenomenon (which we hams mostly refer to simply as *aurora*) has on radio propagation, particularly on the low bands.

2.2.1. Auroral absorption

Auroral absorption (AA) is very frequent and takes place due to the influx of auroral electrons. The ionization density of the affected areas in the ionosphere is very high and absorption of signals on 1.8 MHz can exceed 35 dB.

Auroral absorption is relatively brief in duration, occurring during the times of auroral displays. Absorption regions tend to be elongated in longitude and narrow in latitude, just like the aurora display itself.

AA events are always accompanied by geomagnetic activity due to ionospheric current systems. Hence, there is much interest in the records of auroral-zone magnetometers for predicting times of low magnetic activity (or conversely, periods of high auroral absorption).

2.2.2. Coronal mass ejections (CMEs)

Sporadic outbursts of plasma, called *Coronal Mass Ejections* (CMEs), represent the release of considerable matter/mass from the solar corona. They are the sources of blasts of solar wind that can disrupt the geomagnetic field, give rise to auroral ionization and shut down propagation on the low bands.

Only the plasma from a CME that goes out of the sun in the direction of the Earth may possibly hit the geomagnetic field to cause a magnetic disturbance. CMEs off the back side of the sun do not bother us, since they represent material ejected into space in directions that never can result in an encounter with the Earth.

2.2.3. Aurora

The plasma coming from the solar corona during a CME is called *interplanetary* plasma. *Magnetospheric* plasma is plasma that is trapped within the Earth's magnetic field. The solar wind consists mainly of protons and electrons, the plasma generated during CMEs. Magnetic activity here on Earth results from the impact of the solar wind on the Earth's magnetosphere.

The solar wind blowing by the Earth's magnetic field acts like a huge dynamo, where huge electrical currents are generated. This energy is often pent up in the Earth's magnetosphere. At times the energy is violently released, accelerating electrons in the tail regions of the Earth's magnetosphere. These electrons, since they are charged particles, are constrained to follow the magnetic field lines of the Earth. And since many of these field lines penetrate the Earth in the high-latitude regions, these electrons end up with trajectories that take them into the high-latitude ionosphere, where they collide with constituent particles and ionize the lower regions of the ionosphere. This process also releases photons of light, which we see as auroral activity. The increased electron density and disturbed ionization patterns contribute to increases in auroral absorption and can cause signals to begin experiencing multipathing and fading.

As far as low-band propagation is concerned, the auroral belt at a height of approximately 65 miles (100 km) acts much like the D layer does during the day—it absorbs all low-band signals trying to go through the belt. Sustained periods of low auroral activity appears to be most common during the rising phase of the solar cycle. **Fig 1-11** shows the relation between the solar flux and the A index over a typical solar cycle.

The auroral belt is centered around the magnetic poles. The magnetic North Pole lies about 11° south of the geographic North Pole and 71° west of Greenwich. The magnetic South Pole is situated 12° north of the geographic South Pole and 111° east of Greenwich. The intensity of the auroral phenomenon determines the diameter, the width and the ionization level of the auroral belt. At very low activity the oval retracts to a major-axis dimension of approximately 3500 km, with a belt width of only a few hundred km. During a very heavy aurora the belt can grow to a major axis dimension of more than 8000 km, with a belt width of more than 3000 km. Ionization in the auroral oval is usually not constant all the way around. The ionization is—as a rule—minimum at the local noon meridian and maximum at local midnight. Of course, local noon is of very little interest to us low banders since our signals typically propagate only in darkness.

The Earth rotates around the axis going through its geographic poles, while the aurora oval is centered around the magnetic poles. This means that the position of the often irregular shaped oval changes position continuously with respect to the Earth underneath.

2.2.4. Effects caused by the auroral oval on low-band propagation

The auroral belts (also called auroral ovals or even auroral doughnuts) have a profound impact on propagation. If the low-band path over which you are communicating goes along or through the auroral oval, the result is usually degraded propagation caused by the strong signal absorption. On the higher bands (20 meters and up) fast selective fading (multipathing) is a common sign of aurora. I have very seldom heard this on Top Band, and only infrequently on 80 meters, where these episodes always seem to be of short duration.

During exceptionally quiet geomagnetic conditions (K-indices of zero for at least 8 hours), the auroral zone might shrink to a major-axis dimension of about 40% compared to when geomagnetic conditions are heavily disturbed, while its width might be reduced to a few hundred kilometers. The ionization levels in the shrunken oval can be extremely low during extended fully quiet conditions. Under such circumstances most polar paths will suffer hardly any degradation.

During disturbed conditions, however, the auroral oval can very rapidly grow to an average size of some 8000 km.

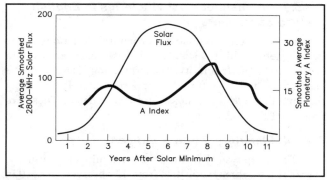

Fig 1-11—This graph shows the geomagnetic activity (measured A index) as a function of the solar cycle. It appears that the geomagnetic activity is lowest during the upswing part of the solar cycle.

Under such conditions all paths that cross or touch this extended oval will be affected by severe absorption in the D and E regions and through other instabilities of the auroral ionosphere.

Cary Oler and Ted Cohen (Ref. 142) state that when the auroral zone is contracted, it is possible for Top Band signals to pass through the auroral zone without suffering heavy absorption by skirting underneath the aurora oval. During periods of very quiet geomagnetic activity, the width (not the diameter!) of the aurora belt is only approximately 500 km. On the other hand, radio signals reflected from the E layer can travel over distances of 500 to 2000 km through the stratosphere and atmosphere, on their way from or to Earth for a propagation hop. This means that with proper geometry low-band signals can literally skip underneath and through the auroral zone into the polar ionosphere inside the auroral belt, where the ionosphere is more stable. They then continue from the polar ionosphere back into the ionosphere at latitudes below the aurora belt, without ever coming in contact with the lossy region of the belt itself.

I have often found that propagation into or through polar regions favors the use of low-angle antennas much more than propagation into or across equatorial zones. I suppose this is so because low-angle hopping has more chances to skip underneath the aurora oval, hence suffering less attenuation than would be the case with higher-angle hopping.

Besides "undershooting" the auroral doughnuts, a common way stations outside the oval deal with the auroral phenomenon is to launch their signals in a non-great-circle route, called a "bent path" or a "crooked path" away from the aurora oval. A signal launched directly at the auroral oval will bend away from it because the enhanced ionization in the auroral zone creates horizontal ionization gradients. These horizontal gradients refract signals in the horizontal plane.

For stations inside the auroral belt, propagation to the world outside the belt is all but impossible once a geomagnetic disturbance has set in. It has been reported though that stations inside the aurora oval can often hear quite well, but do not seem to get out at all. For example, VY1JA experienced much frustration during the 1998 November Sweepstakes contest hearing strong stations that couldn't hear him. I suspect that the launch angle to skip under the oval was not right at VY1JA's end.

For stations just outside the aurora belt near the North Pole, usually the only (marginal) opening is directly to south. Frequently, these stations enjoy better propagation towards the equator than stations 1000 or 2000 km further south when the aurora is on.

On at least one occasion on 160 meters, I experienced propagation conditions similar to those on VHF during an extremely heavy aurora. Around 1600 to 1800 UTC on February 8, 1986, I heard and worked KL7 and KH6 stations on 80 meters, at the same time when auroral reflection was very dominant on VHF and 28 MHz. From Europe, this was on a path straight across the North Pole and the signals had the buzzy sound typical for auroral reflection. This seems to me to indicate that under exceptional conditions (the aurora was extremely intense), aurora can be beneficial to low-band DXing. This particular aurora generated an A index of 238. K-index values were reported between 8 and 9—this was one of the largest geomagnetic storms since 1960. A similar situation existed in January 1987, when in Europe we could work KL7 stations for several days on 160 meters.

Will, DJ7AA, recently reported a similar occurrence (February 18, 1998): *"Around 0130 I heard K1UO with a very big signal out from nothing working a SP3 station. I went on 1835 and one CQ brought me a huge pileup with really big signals banging in here, even from call areas like W5 or W0. I wonder what NA0Y was running, I think he was the loudest ever heard W0 here in my place. Interesting, all signals having a little flutter on it sounds like aurora, and they all coming in over my Beverage to South America, about 3-4 S-units stronger than on my big 500-m beverage to 320 degrees... At 0300 the band died. When checking the NOAA home page I saw a very big auroral zone over the Northern Hemisphere at this time, while WWW said: Major Storm..."*

There is almost always a temporary enhancement of conditions right after a sudden rise in the K index. Except for polar paths, it seems that a low K index doesn't help for most 160-meter propagation.

Enhanced propagation conditions shortly after a major aurora appear quite regularly. I witnessed a striking example on 80 meters in November 1986, when only nine hours after a major disturbance, N7AU produced S9 signals on the long path for more than 30 minutes, just before sunset in Belgium. Normally, long-path openings occur to the US West Coast from Belgium only between the middle of December and the middle of January, and even then the openings are extremely rare. During the November opening, I heard N7UA calling CQ EUROPE with signals between S6 and S9 for almost an hour. The propagation was very selective, since only Belgian stations were returning his calls! A few days earlier DJ4AX was heard working the West Coast and giving 57 reports while the W6/W7 stations were completely inaudible in Belgium, only 200 miles to the Northwest!

I presume that an ionospheric-ducting phenomenon was responsible for such propagation. This means that very specific launching conditions had to be present at both sides of the path. Recent thought is that duct "exit" conditions are very critical and area-selective—more so for longer path lengths. It also may be that auroral disturbances can create and enhance such critical conditions.

2.2.5. A and K indices

The most common way to quantify the level of geomagnetic activity is through the A and K indices.

2.2.5.1. The local K index

The K index indicates the magnitude of irregular variations in the magnetic field over a 3-hour period. The index is calculated from the actual measured values at each observatory station. There are a number of these observatories worldwide. Since magnetic-field measurements vary greatly depending on location, the raw measurements are normalized to produce a K index specific to each observatory.

The K-index scale is quasi-logarithmic, increasing as the geomagnetic field becomes more disturbed. K indices range in value from 0 to 9 (0 = dead quiet, to 9 = extremely disturbed). The K index that we monitor on radio station WWV is an index derived from magnetometer measurements made at the Table Mountain Observatory located just north of Boulder, Colorado, and hence is referred to as the "Boulder K index." Every 3 hours new K indices are determined and the broadcasts are updated.

2.2.5.2. The local A index

The underlying concept of the A index is to provide a longer-term picture of geomagnetic activity using measurements averaged over some time frame. The A index is the mathematical average of the raw *a* indices over the last 24 hours.

The overall A index is an averaged measure of geomagnetic activity derived from the 3-hour K-index measurements. For each 3-hour K index, a conversion is made to the *a* index using a conversion table (see **Table 1-2**). The A index is the average of the last 8 a indices.

A indices are always linked to a specific day. Therefore, estimated A indices are issued during the day itself. For example, the Boulder A index (in the WWV announcement) is the 24-hour A index derived from the eight 3-hour K indices recorded at Boulder. The first estimate of the Boulder A index is at 1800 UTC. This estimate is made using the six observed Boulder K indices available at that time (0000 to 1800 UTC) and the best available prediction for the remaining two K indices. At 2100 UTC, the next observed Boulder K index is measured and the estimated A index is reevaluated and updated if necessary. At 0000 UTC, the eighth and last Boulder K index is measured and the actual Boulder A index is produced. For the 0000 UTC announcement and all subsequent announcements the word "estimated" is dropped and the actual Boulder A index is broadcast.

A and a-indices range in value from 0 to 400 and are derived from K indices based on the table of equivalents (see **Table 1-3**). Both A and K indices (for Boulder, Colorado) are broadcast by WWV (on 2.5, 5, 10, 15 and 20 MHz) every hour at 18 minutes past the hour.

2.2.5.3. Geomagnetic activity terms in English instead of numbers

For an overall assessment of natural variations in the geomagnetic field, six standard English terms are used to report geomagnetic activity. The terminology is based on the estimated A index for the 24-hour period directly preceding the time the WWV broadcast was last updated. These are listed in Table 1-3.

2.2.5.4. Planetary A and K indices

The Geophysical Institute in Goetteningen, Germany, averages the geomagnetic data from 12 observatories (10 in the Northern Hemisphere and 2 in the Southern Hemisphere) to give planetary values, A_p and K_p (the subscript p stands for Planetary).

Table 1-4 shows an example of K, A, A_p and K_p-

Table 1-2

A_p index	Corresponding K_p
0-2	0
3-5	1
6-10	2
11-20	3
21-35	4
36-61	6
62-102	6
103-166	7
167-268	8
>269	9

Table 1-3

Category	A index range
0-7	Quiet
8-15	Unsettled
16-29	Active
30-49	Minor Storm
50-99	Major Storm
100-400	Severe Storm

indices from a Boulder report from June 24, 1997. It lists the Daily Geomagnetic Data from Fredricksburg, Virginia; College, Alaska; and the Estimated Planetary values from NOAA. You will see differences between the observations (at Fredricksburg and College) and the estimated A_p and K_p values. Both A and K-indices are available from various sources on the Internet. Listings of the A_p and K_p indices are available at: **ftp://ftp.ngdc.noaa.gov/STP/GEOMAG-NETIC_DATA/INDICES/KP_AP/** (Note: the upper-case letters are important.) There is a file for each year.

2.2.5.5. Converting K values to auroral oval average diameter

For the radio amateur it is important to assess the size and the width of the auroral oval in order to be able to evaluate (on a map or on a globe) whether a given path will touch or pass through the auroral oval. Simplifying somewhat, we can say that the oval is a circular ring, of which the statistical equivalent average radius (at midnight) is given in **Table 1-5**.

If you only have K values, these data allow you to manufacture oval disks of various diameters, which can be used as overlays on maps or on a globe, to help visualize possible crossings of great circle paths with the auroral oval. Of course, the oval is a statistical description and does not describe how the ionization is distributed or how energetic it may be. In other words, the local intensity of the aurora is not the same at all points of the oval and at all times of the day.

2.2.6. Viewing the aurora from satellites

The only source of really reliable information is to use real-time maps that now are available from various sources on the Internet. Today we have satellites that produce almost real-time pictures. These can give us much more information than what we've had before.

2.2.6.1. The NOAA/POES satellite (formerly TIROS)

Views of both North and South Poles and the aurora oval are available on the following Internet Web pages: **http://www.sel.noaa.gov/pmap/**. These are updated when the NOAA Polar-Orbiting Operational Environmental Satellite (POES) satellite passes by about every hour. The satellite maps out the auroral zone for that moment. The maps are continuously updated and represent the best available picture for that moment. The POES images are based on particle-sensor readings the spacecraft makes as it passes over the polar regions. Instruments on board continually monitor the power flux of the protons and electrons that could produce aurora in the atmosphere. These readings are valid only for those longitudes where the spacecraft passes overhead. The readings may be considerably different at other positions along the auroral oval. This is why SEC must examine the results of 100,000 other polar passes in order to form a statistical picture of what is most likely happening elsewhere. This means that what the pictures show are based on data obtained from previous passes. It is not a true real-time

Table 1-4

Date		Fredricksburg Local		College, AK Local		Estimated Planetary
June	A	K indices	A	K indices	A_p	K_p indices
16	8	1-2-1-2-2-3-3-1	3	1-0-0-3-0-1-1-0	5	1-1-0-2-2-2-3-1
17	6	1-2-2-1-1-2-2-2	1	0-1-1-0-0-0-1-0	5	0-2-2-1-1-2-2-2
18	3	1-1-1-1-1-1-1-1	0	0-0-0-0-0-0-0-0	4	0-1-1-1-1-2-1-2
19	11	2-2-4-2-2-2-2-3	6	1-2-3-3-2-0-0-1	10	3-2-4-3-2-2-2-2
20	6	2-1-2-2-2-2-1-2	2	0-1-2-0-2-0-0-0	5	2-1-1-1-2-2-1-2
21	2	0-0-0-0-0-1-1-2	2	0-0-0-3-0-0-0-0	3	0-0-0-1-1-2-1-1
22	15	2-4-3-3-3-3-3-2	4	1-2-3-1-1-0-1-0	9	1-3-3-2-3-2-2-2

Table 1-5

K Index	Oval Average Radius (km)	Oval Average Width (km)
0	1800	500
1	2050	800
2	2300	1100
3	2550	1400
4	2800	1700
5	3050	2000
6	3300	2300
7	3550	2600
8	3800	2900
9	4050	3200

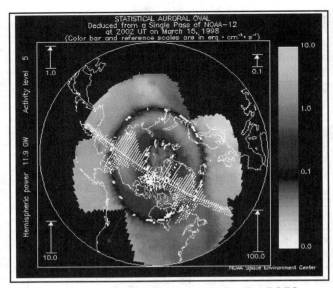

Fig 1-12—North Polar view generated by the POES satellite, during a magnetically relatively quiet period (K=2). Inside the bright, light-shaded auroral-zone "doughnut" dark areas are areas of higher ionization. Darker areas outside the doughnut indicate lower levels. The white line shows the orbit of the satellite making the measurements, and the dots on either side represent the measurements done in the directions of the stacked black dots. (*Source: NOAA Web page.*)

real picture, but a combination of real-time data and best-fit extrapolations taken from a huge data base.

Table 1-6 lists the conversion from Total Hemisphere Power, as reported by NOAA, to the more familiar K values. **Fig 1-12** shows a typical POES-generated polar view during a very quiet geomagnetic spell. The black line shows the orbit of the satellite making the measurements and the dots on either side represent the measurements done in the direction of the stacked black dots. The arrow in the lower-left quadrant shows the local noon meridian, where the width of the oval is usually smallest. Note that the local noon meridian would not suggest any propagation on the low bands, since propagation at local noon is impossible anyhow due to D-layer absorption. **Fig 1-13** shows a typical POES-generated polar view during a geomagnetically upset period.

2.2.6.2. The POLAR satellite

The POLAR spacecraft images, which shows images of the auroral oval taken at various wavelengths, can be viewed at: **http://solar.uleth.ca/solar/www/aurora.html**. These images give you an almost real-time picture of the polar

Table 1-6

Power (Gigawatts)	K Index
1-2	0
2-4	1−
4-6	1
6-10	2−
10-16	2
16-24	2+
24-39	3
39-61	3+
61-96	4
>96	5
>200	8
>500	9

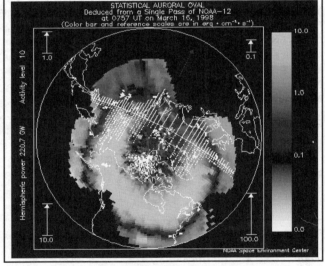

Fig 1-13—North Polar view generated by the POES satellite, during a magnetically upset period (K=8). Very dark areas inside the auroral doughnut areas are areas of high ionization, while the lighter tones outside the doughnut show much less ionization. The width of the oval is generally smallest at the local noon meridian. This particular view is for 0757 UTC. Note that the auroral oval is at its widest around local midnight (across the USA) while the activity is minimal around local noon. The white arrow in the upper quadrant shows the local noon meridian. (*Source: NOAA Web page.*)

region, including details of isolated regions of enhanced activity. This can be particularly useful when studying a particular radio-propagation path, since very often a radio signal may be able to follow a path through a quiet region of the auroral ionosphere, while other slightly different paths may cause the signal to pass through a region experiencing a local substorm.

The POLAR spacecraft provides data when it is in the proper orientation for taking pictures. It has quite an elongated orbit, which means that it can skim by the Earth so closely at times that its sensors are only able to see small regions. I suspect this may also be why sometimes no pictures are released. The above-mentioned Web page normally lists three types of pictures:

The visible-light picture

Fig 1-14 shows visible-light pictures of the Earth showing the auroral oval. It is obvious that the aurora can only be seen over the dark side of the Earth.

The UV picture

Fig 1-15 shows an ultraviolet image for the northern polar auroral oval, taken by the POLAR spacecraft. Lines of latitude and longitude are denoted by the dashed lines. Continental outlines are also shown for reference.

The polar view map

Fig 1-16 shows the current position of the auroral oval determined by the polar-orbiting Defense Meteorological Satellite Program (DMSP) satellites. The dark-gray oval defines the approximate shape of the diffuse auroral oval as determined by the electron-precipitation sensors on-board the DMSP spacecraft. The auroral oval line represents the equatorward boundary of the diffuse auroral oval and is therefore useful for radio communicators and others to determine precise auroral-oval locations. The map also indicates an *effective Q Index*, which is similar to the geomagnetic K index. It is a quasi-logarithmic value and is based on

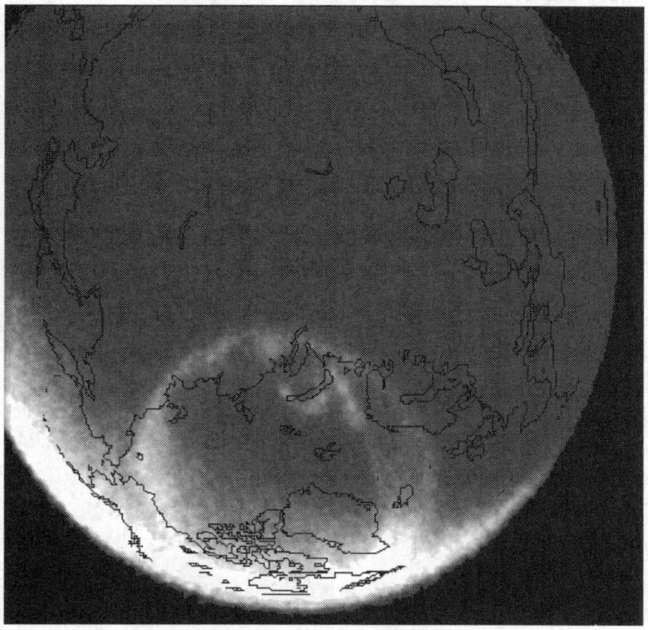

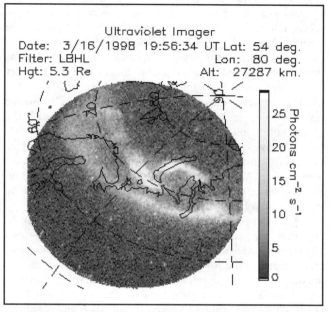

Fig 1-15—UV picture, equivalent to the visible-light picture shown in Fig 1-14. (*Source: STD at* http://solar.uleth.ca)

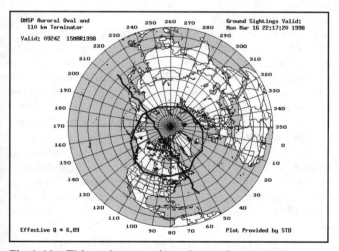

Fig 1-16—This polar-cap view shows the current position of the auroral oval determined by the polar orbiting Defense Meteorological Satellite Program (DMSP) satellites and overlaid onto the map. This oval represents the equatorward boundary of the diffuse auroral region and is useful for radio communicators and others to determine propagation paths compared to the auroral oval. The somewhat jagged straight line is the terminator. (*Source: STD at* http://solar.uleth.ca)

Fig 1-14—Visible-light picture taken by the POLAR Spacecraft. One can see the terminator and the auroral oval blending into the daylight zone. (*Source: Solar Terrestrial Dispatch at* http://solar.uleth.ca)

geomagnetic data from several high-latitude (auroral-zone) magnetic observatories. A Q index of 8 is really bad news for low banders using high-latitude paths. A Q index of 0 or 1 is best and a Q index of 3 or 4 is so-so.

2.2.7. Putting it into practice

2.2.7.1. Getting geomagnetic data

As a general rule, low A_p and K_p-numbers acquired at stations near the polar regions for a sustained period are prerequisites for good conditions on paths that go near or through these polar regions. It is the K index that is the most important one, since it gives you a more differentiated status than the A index. "Near the poles" means that the Boulder figures are not the most suitable ones! K indices obtained from the observatories in Inuvik, Baker Lake and Cambridge Bay in Canada are ideal because they are located within the aurora belt, when it is active.

We have to realize that K and A indices are measurements derived from what *has already* occurred. If these indices have been zero for 8 hours (or longer) and provided that there is no abrupt change, the chances are good that the low bands will be in fair-to-possibly-good shape on polar paths.

Because of the sun's 27-day rotation cycle, it is probable that low geomagnetic activity will be recurrent with a 27-day period, especially during the declining and the minimum phases of the solar cycle. During the ascending and the maximum phases though the recurrent trend often becomes very unreliable. It is a good idea for the serious low-band DXer to make a continuous log of broadcast A and K values. Such logged A indices are especially useful to predict the level of magnetic activity in another 27 days.

I guess most Top Banders have come to grips with the fact that K and A indices are there to confirm what they have already witnessed—good or bad conditions. Recently N6TR, well known Top Band DXer from the West Coast complained: "... *I am very skeptical that any of the numbers mean much. I have had good openings with high K numbers, and no openings with long standing low numbers. About the only thing I can count on is that interesting things seem to happen just as the K starts to rise.*"

Since auroral absorption is often initiated by CMEs on the sun, we should be able to predict auroras 2 to 4 days before they hit us (at least for the CME-induced auroras). Before satellite technology was available, we had no detailed information on CMEs, and forecasting was based only on the use of recurrence tendencies, extrapolating conditions only from logged A and K-index values from 27 and 54 days earlier.

Today we have a number of satellites that keep a constant eye on the sun and send a continuous flow of data to the Earth, data that is being converted into "readable" reports that are available in abundance on various Web sites on the Internet. You can subscribe to a very useful daily summary of auroral activity, which will be sent to you by e-mail, on: **http://solar.uleth.ca/solar/www/sublists.html**.

These reports forecast magnetic storms based on sun-surface and solar-wind observations, done from satellites.

Such reports—together with viewing the NOAA generated images themselves—are helping the low-band DXers to better understand what makes it all tick, and to better plan their activities.

Since mid March 1998, Cary Oler has also made available a 160-meter Web site, which can be reached at: **http://solar.uleth.ca/solar/www/topband.html**. This Web site, called "The Top Band radio propagation section," contains a variety of tools for the 160-meter DXer. In addition to high latitude K values, which are updated every hour, Cary Oler has created a table where he shows the probability of DX contacts for a large number of polar paths. It also contains the latest auroral-zone pictures and maps (visible and UV) as shown in Figs 1-14, 1-15 and 1-16. These predictions are limited in that they only take into account the influence of magnetic disturbances. This is somewhat of a one-way information—When conditions are magnetically disturbed, we know the paths going through the polar areas will be dead. But low K indices, even for a relatively long spell are no full guarantee that the path will be okay.

There are other mechanisms that enter into the picture on 160 meters and that determine the overall attenuation on a given path. These mechanisms are still largely unknown or, at least, are open to speculation.

SWARM is another software program that was developed by the Solar Terrestrial Dispatch, headed by Cary Oler. Check out the Web page at **http://solar.uleth.ca/solar/www/swarm.html** for details. *SWARM* (Solar Warning And Real-time Monitor) monitors everything from geomagnetic and ionospheric conditions to solar activity and solar wind conditions, all in *real-time*. It is particularly valuable for the prediction of quiet geomagnetic intervals and the arrival of interplanetary disturbances. This breakthrough is made possible by the software's ability to read, process, graph, print and interpret real-time solar wind data from the newly launched Advanced Composition Explorer (ACE) spacecraft.

In January 1998, the ACE (Advanced Composition Explorer) spacecraft began sending nearly continuous measurements of the solar wind from its vantage point outside of the Earth's magnetosphere (about a million kilometers upstream of the Earth, between the Earth and the Sun). This distance makes it possible for the satellite to detect the arrival of interplanetary disturbances up to an hour before they impact the Earth. The spacecraft is currently providing about 23 hours of continuous solar-wind measurements each day. The data are received by ground stations in Japan and the US and then immediately transmitted to the Space Environment Center, where they are made available over the Internet for the *SWARM* computer program to read and use.

The one hour lead-time gives the users of *SWARM* a headstart, since they can take advantage of the fleeting propagation enhancements that can occur on the low bands (and other bands) shortly after the arrival of these disturbances. The software also audibly alerts you when geomagnetic activity surpasses certain threshold levels. You need never again spend valuable time calling CQ on a polar path when the K index is up to 4 or higher. Low-band DXers, using the *SWARM* software determine precisely when geomagnetic storming is likely to commence. They will also be informed beforehand about the potential intensity of geomagnetic storming. The software also reports all possible related data, such as solar-flux values and sunspot numbers. *SWARM* lets you look at real pictures of the aurora (visible and UV) and you can download and print up to 19 different types of daily, weekly and monthly reports from forecast centers around the world.

SWARM will even produce a plot showing the solar-wind arrival times at the Earth, based on solar wind velocities observed at either the WIND or ACE spacecraft and the distances of these spacecraft from the Earth at any given moment. *SWARM* users can predict, with an accuracy of several minutes, precisely when a disturbance will impact Earth. The software package was written by people at STD (Solar Terrestrial Dispatch, in Canada), headed by Cary Oler (see also Ref. 142), who is president. It requires Windows 95 or 98 (or Windows NT 4.0 or later) and a connection to the Internet (either hard-wired or through a standard dial-up Internet Service Provider).

2.2.7.2. Viewing the paths

If we wish to know whether certain paths will be disturbed, we must visualize the path as well as the auroral oval, which will immediately reveal what's going on for that given path for a given auroral intensity. Maps in various projections, as well as globes can be used. You may have to fabricate your own aurora ovals to use with the map or globe. (See Section 2.2.5.5.)

Nowadays, when every ham has at least one computer in the shack, computer programs do just what we want, with much less hassle. There are numerous propagation-prediction programs around, but only a few address the auroral phenomenon. The following three programs all have their own merits and shortcomings, but at least they address geomagnetic conditions: *DX-AID*, *Miniprop Plus* and *Proplab Pro*. These three programs were extensively used to create the figures in this chapter.

I find *DX-AID* extremely useful for generating great-circle and Mercator-projection maps, including variable-sized auroral ovals, based on the K index, which you have to enter. The program runs perfectly well in a Windows 95 environment, where you can call it up with a single keystroke when needed. The program is written by Peter Oldfield. He can be contacted at **poldfield@compuserve.com**. *DX-AID* is not only a mapping program, it also does classic HF-propagation forecasting, which is really of little use to us low-banders (except on 40 meters).

Miniprop Plus, by Sheldon Shallon ,W6EL, is another user-friendly program that has some excellent mapping possibilities. The program is no longer commercially available unfortunately.

Proplab Pro (formerly *Skylab Pro*) is a professional ray-tracing program. Cary Oler calls it a "High-Frequency Propagation Laboratory." It is a full-fledged propagation-prediction program that also generates a range of maps. It is probably the most sophisticated and most advanced program that is available, but it is possibly too professional for the average ham or low-bander, even a very dedicated one. But if you really want to study propagation in fine detail, I can highly recommend this program. It actually does three-dimensional ray tracing and will predict and plot skewed paths!

There are a number of other tools available (*Geoclock*, *Geochron*, *The DX-Edge*, *Gray-Line Globe*, *Dx4WIN*) but I have found few that address the issue of the auroral oval, which means they are less than ideal for visualizing transpolar propagation paths on the low bands.

2.2.7.3. Correlating geomagnetic data with conditions

Statistical analysis has been done on a representative group of long-haul DX QSOs from the US West Coast on 160 meters (over a two-month period). The occurrences were checked against the K index: 62% of all QSOs were made on days with a K index of zero; 30% with an index of 1 and not one QSO with a K index above 3.

In another study, the Top Band Monitor did a survey and tried to correlate A-index figures with days of good conditions on Top Band during the winter of 1993/1994. The author tried to link upward swings in A index with good Top Band conditions, and downward swings with bad conditions. My conclusion from studying the data was that only 10% of the good openings on 160 meters were correlated to a downward swing in the A index.

But we should not forget that magnetic activity is *not* the only phenomenon that rules conditions on the low band, and more specifically on 160 meters. There are still many unknown mechanisms that makes the residual attenuation on 160 meters vary significantly, even when the geomagnetic activity is low.

Sometimes we have weeks of really good conditions (for example at the end of November and in December 1997) followed again by weeks of fairly "flat" propagation. During both periods there were both up and down-swings of geomagnetic activity. Recently, it has been suggested that atmospheric conditions may have to do with 160-meter propagation, but that could be pure speculation. The lower bands—and more particularly Top Band—are still areas where many things still have to be "discovered." That's what makes these bands so interesting and appealing to many!

2.3. Local Atmospheric Noise

Most local atmospheric noise (static or QRN) is generated by electrical storms or thunderstorms. We know that during the summer QRN is the major limiting factor for copying weak signals on the low bands, at least for those regions where thunderstorm activities are serious. To give you an idea of the frightening power involved, a thunderstorm has up to 50 times more potential energy than an atomic bomb! There are an estimated 1800 thunderstorms in progress over the Earth's surface at any given time throughout the year. The map in **Fig 1-17** shows the high degree of variation in frequency of thunderstorms in the US. On the average there is a lightning strike somewhere on the Earth every 10 milliseconds, generating a tremendous amount of

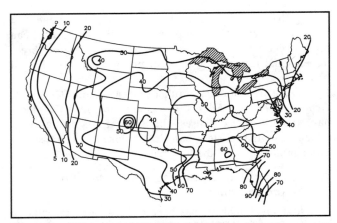

Fig 1-17—This map shows the mean number of thunderstorm days in the US. The figure is related to both mountainous terrain and seasonal weather patterns.

radio-frequency energy.

In the Northern Hemisphere above 35° latitude, QRN is almost nonexistent from November until March. In the middle of the summer, when an electrical storm is near, static crashes can produce signals up to 40 dB over S9, and make even local QSOs impossible (and dangerous). It is obvious that in equatorial zones, where electrical storms are very common all year long, this phenomenon will be the limiting factor for low-band DXing. This is why we cannot generally speak of an ideal season for DXing into the equatorial zones, since QRN is a random possibility all year long.

Using highly directive receiving antennas such as Beverage antennas or small loops can be helpful to reduce QRN from electrical storms by producing a null in the direction of the storm. Unless directly overhead, electrical storms in general have a fairly sharp directional pattern.

Rain, hail or snow are often electrically charged and can cause a continuous QRN hash when they come into contact with antennas. Some antennas are more susceptible to this *precipitation noise* than others; vertical antennas seem to be worst in this respect. Closed-loop antennas generally behave better than open-ended antennas (such as dipoles), while Beverage receiving antennas are almost totally insensitive to this phenomenon.

2.4. Effects caused by the Electron Gyrofrequency

Scientists specializing in ionospheric propagation have recently made Top Band DXers aware of mechanisms that seem to determine propagation there. The theory concerning gyrofrequencies on 160 meters is covered in detail in literature (Ref. 142).

The gyrofrequency is a measure of the interaction between an electron in the Earth's atmosphere and the Earth's magnetic field. The closer a transmitted signal's frequency is to the gyrofrequency, the more energy is absorbed from the signal by the electron. This is particularly true for radio

waves traveling *perpendicular* to the magnetic field. Gyrofrequencies are not influenced by the sun but change with location and vary between 700 and 1600 kHz. A map of the D/E-region electron gyrofrequencies is shown in **Fig 1-18**.

You should remember that Top Band signals will be less strongly absorbed and behave more like a conventional signal is expected to behave the farther the frequency is removed from the electron gyrofrequency. Check the map in Fig 1-18 to determine the values of gyrofrequency your signals will encounter for a given path.

Absorption is higher along paths where the signal frequency is closer to the electron gyrofrequency and *particularly* on paths that are normal to the magnetic field. In other words, north-south paths are less affected than mainly east-west paths, such as from the US East Coast to Europe, or US East Coast to Japan. Similar paths in other parts of the world may not be as sensitive because gyrofrequencies are lower.

If I had to quantify the impact of the gyrofrequency on Top Band propagation and compare it to the impact of the auroral oval, then I'd say that the auroral oval is the proverbial elephant while the gyrofrequency is the mere mouse.

2.5. Polarization and Power Coupling on 160 Meters

Power coupling has to do with the way waves generated by the transmit antenna "couple" into the ionosphere. It appears that the polarization of the antenna plays an important role in achieving optimal coupling (minimum losses). In certain areas of the world vertical polarization will produce strongest signals, while in other areas horizontal polarization will.

Fortunately, in the US as well as in Europe, vertical polarization is the way to go. There are areas of the world, however, where horizontal polarization is the more suitable polarization. This is true for large parts of Asia, Africa and parts of Australia. The geomagnetic latitude of the location is an important factor in this mechanism. **Fig 1-19** is a map of the Earth's magnetic latitude compared to the geographic latitude and longitude.

Most top-banders use transmitting antennas with vertical polarization, which fortunately seems to be the right choice from a power coupling point of view, at least if your QTH is at an average or higher-than-average latitude. Things are more complicated when it comes to stations with low magnetic latitudes (near the equator). The challenge there is to have an antenna with horizontal polarization for good coupling, but at the same time be able to produce low takeoff angles. For dipoles we are talking of heights of at least a half wavelength, something that is not often feasible on 160 meters.

■ 3. PROPAGATION PATHS

This section discusses the following items to help increase our understanding of low-band propagation paths:
1. Great-circle short path
2. Great-circle long path
3. Particular non great-circle paths

3.1. Great-Circle Short Path

Great circles are all circles obtained by cutting the globe with a plane going through the center of the Earth. All great circles are 40,000 km (24,860 miles) long. The equator is a particular great circle, the cutting plane being perpendicular to the Earth's axis. Meridians are other great circles, going through both poles.

When we speak about a great-circle map we usually mean an azimuthal-equidistant projection map. This map, when covering the entire world, has the unique property of showing the great circles as straight lines, as well as showing distances to any point on the map from the center point. On such a projection, the antipode of the center location will be represented by the outer circle of the map. Great-circle maps are specific to a particular location. They are most commonly used for determining rotary beam headings for DX work from a particular location.

There are various sources on the Internet where you can download great-circle maps or programs to make such maps.

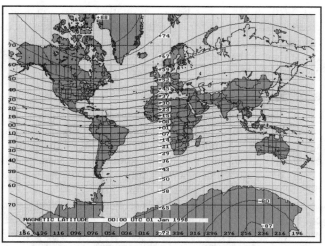

Fig 1-19—Mercator-projection world map showing the geomagnetic latitudes. These are not the same as the geographical latitudes, since the magnetic North and South Poles do not coincide with the geographic poles.

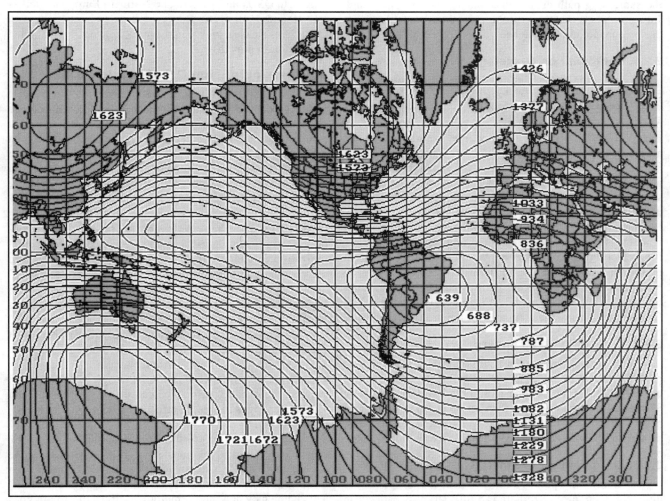

Fig 1-18—This grid shows the worldwide distribution of the gyro frequencies. These values are determined by the Earth's intrinsic magnetic field and are not influenced by the solar cycle. (*Courtesy C. Oler, STD.*)

Fig 1-20 shows a number of azimuthal-projection maps created using *DX-Aid* for Boston, Omaha, San Francisco, Brussels, Moscow and Tokyo. The advantage of the great-circle map is that headings are straight lines. The negative point is the extreme distortion near the antipodes.

3.2. Great-Circle Long Path

A *long-path* condition exists when the station at the eastern side of the path is having *sunset* at approximately the same time as the station at the western end of the path is experiencing *sunrise*. A second condition is that propagation occurs on a path with a beam heading that is 180° opposite to the short-path great-circle direction.

We will see further how a "crooked path" propagation can satisfy the first condition, but is not a genuine long-path propagation. One example is the path from Western Europe to Japan at 0745 UTC in mid-winter. This involves a path over Northern Siberia and not across South America, as it would if it were a true long path.

John Kaufman, W1FV, in Massachusetts, has noticed that his sunrise long-path openings on 80 meters are best during the sunspot cycle peak. During the low sunspot years the same long path all but disappears. On the other hand, John comments that the sunset long paths seem to dominate during sunspot cycle troughs.

3.2.1. Long path on 40 meters

Long-path QSOs are quite common on 40 meters. From Europe we have a "genuine" long path to the US West Coast around 1500 to 1600 UTC in midwinter. A very similar long path exists between Japan and Europe around sunrise time in Europe, especially around equinox time. In midwinter, when all the darkness is in the Northern Hemisphere, there still is "some" long path between Europe and Japan, but there is a generally much-stronger path that is a somewhat crooked short-path across Northern Siberia. In general, the signal direction is about the same as the usual short path direction. During midwinter, both long and crooked paths exist simultaneously for about 10 or 15 minutes, around 0745 UTC. (See **Fig 1-21**.) This often makes copy very difficult, because of multipath propagation due to the different time delays on both paths. During the recent JA low-band contest (in January 1998), I had to ask several JA stations to slow down their CW for me to copy them because of the multipath echoes.

3.2.2. Long path on 80 meters

You can only be "nearly" sure that it is a genuine long path when at both sides of the path the strongest signals come from the genuine long-path direction. Long paths on 80 meters are less common than on 40 meters. Very often paths we call "long paths" are actually crooked or bent paths, somewhere between long and short path. (See Section 3.3.1.3.)

3.2.3. Long path on 160 meters

Genuine long-path QSOs very near the antipodes are quite common, provided there is a full-darkness path. I can hear the G stations working ZL long path on 160 meters

approximately 30 minutes after my sunrise, but only on very rare occasions have I been able to work ZL on long path myself. There have been other near-antipode long-path QSOs between VK6HD (Perth) and the US East Coast in midwinter (for example, the QSO between K1ZM and VK6HD at 2115 UTC on January 27, 1985).

Real long-path QSOs on 160 meters only seem to occur during a period centered around the one or two years at the minimum of the sunspot cycle. WØZV remembers a few genuine long-path QSOs made from Colorado; for example, with UA9UCO and JJ1VKL/4S7. Another one that made history was between PY1RO and several JA stations at JA sunrise. Other long-path contacts were made between US East Coast stations and well known calls, such as 9M2AX, VK6HD and VS6DO.

During the 1987-1988 season, my first winter on 160 meters, I tried for weeks to make a so-called long-path QSO with N7UA, but neither of us heard each other (see also Section 3.3.1.4). During December 1992, I ran a daily test with FK8CP on the long path (his sunset is within minutes of my sunrise), but we never made a QSO either.

During December 1997, N7UA had numerous so-called long-path QSOs into Eastern and Northern Europe as well as UK around 1510 UTC. But were these genuine long-path QSOs? Let's have a look at non-great-circle paths, also called crooked or bent paths.

3.3. Particular Non Great-Circle Paths

Many paths on 40, 80 and 160 meters are great-circle paths. It is obvious that paths over relatively short distances are more or less straight-line great-circle paths. Let us assume that paths are basically of the great-circle type, unless there is a good reason for them not to be. This is true even for north-south transequatorial paths (eg, Europe-Africa, North America-South America). Deviations from great-circle headings will rarely be observed on such paths and over such long distances.

According to general knowledge about the mechanisms involved in ionospheric reflection (refraction), a horizontally bent propagation path can only be induced by horizontal ionization gradients in the ionosphere. Signals traveling into a layer with a higher degree of ionization will be refracted. This means that signals will bend away from areas with higher ionization. This is the theoretical approach and is most definitely correct.

On the other hand, there is what we actually observe. It is also generally accepted that we observe bent propagation paths that match two sets of circumstances:
1. Bent paths avoiding propagation through the auroral oval
2. Other bent paths (probably caused by significant and abrupt changes in free-electron density in the horizontal plane).

Before we go into further details on this subject, let's recall a few typical circumstances. The first case is the classic one, where we can identify signals coming from headings bent away from the normal great-circle directions. This happens when the normal great-circle short path goes through the auroral oval during periods of high geomagnetic

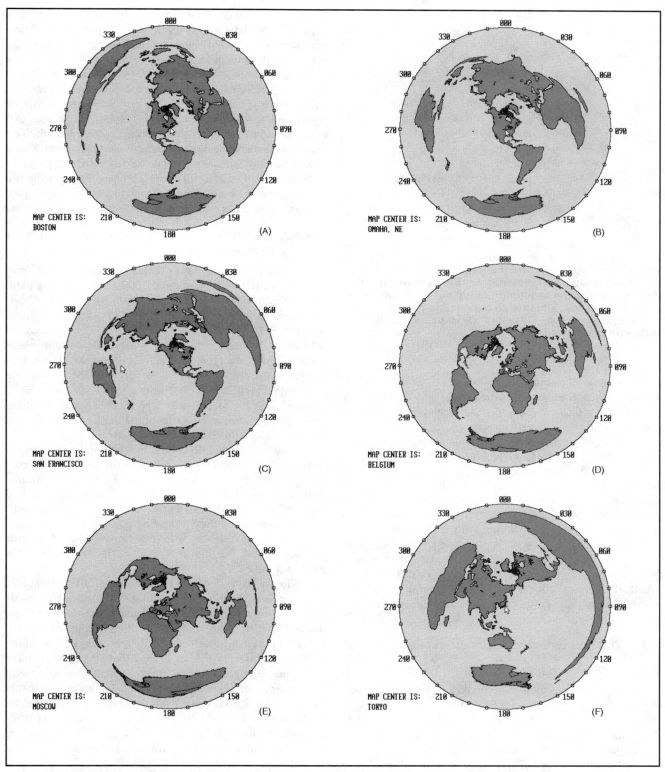

Fig 1-20—Azimuthal projections for Boston, Omaha, San Francisco, Brussels, Moscow and Tokyo. (*Maps generated by* DX-AID.)

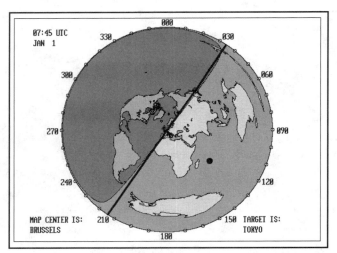

Fig 1-21—The great-circle path for midwinter (0745 UTC) shows the short and the long path that exist simultaneously on 40 meters from Belgium to Japan. Both run along the terminator, a typical situation for the higher bands, which also is possible on our highest low-frequency band, 40 meters. (*Map generated by DX-AID, with additions by ON4UN.*)

activity. Many of us have witnessed occasions when signals seem to be bending away from the auroral belt. In Europe we work US West Coast stations by beaming to Central or South America under such circumstances.

A second class of bent paths was witnessed by many who live on the US East Coast and in the Midwest. During the exciting month of January 1997, they will remember how the VKØIR signals, both on 80 and especially on 160 meters, always peaked right across Europe (60°) instead of by the direct short path, which is about 110°. This could *not* have been a case of seemingly bending away from the aurora oval because of refraction away from higher ionization—it almost was like the signals were being *attracted* to it! It must be said, however, that during the stay of VKØIR on Heard Island, the geomagnetic conditions were generally very quiet. The reason for these kind of bent paths must be different. Let me try to analyze the second case first.

3.3.1. The non-heterogeneous ionosphere

3.3.1.1. The Mechanism

The general mechanism that makes signals deviate from the great-circle line is horizontal ionization gradients in the ionosphere. What causes these gradients apparently is not agreed to by all scientists, and I have not seen any model that explains the actual observations.

We often think of radio waves as a single ray of energy sent in a specific direction, refracted in the ionosphere and reflected from a perfectly flat reflecting surface on the Earth. This has been the standard, simplified method of visualizing radio propagation. HF energy is, however, in most practical cases radiated over a range of azimuths and over a range of elevation angles.

The ionosphere is not a perfect mirror, but should rather

be thought of as a cloudy and patchy non-heterogeneous region, with different layers of varying ionization. Radio waves are more heavily attenuated traveling through certain regions than through others.

Cary Oler and Ted Cohen (Ref. 142) point out that *"weak sporadic-E clouds, that might not affect the higher frequencies, can achieve a substantial impact on 160 meter signals by increasing absorption or refracting signals."* I suspect that such sporadic-E clouds could induce a waveguide-like hop between the F layer and the sporadic-E cloud. Sporadic E thus should be considered as a possible cause for skewed paths.

There still is a lot of discussion ongoing in scientific circles about the precise mechanisms that could trigger path skewing. Here are some regular, well-documented skew paths.

3.3.1.2. The classic example: ZL propagation from Europe

New Zealand is about 19,000 km on the short path from my QTH in Belgium, or about 21,000 km on the long path, very close to being the antipode. From Belgium the short-path heading to New Zealand is 25 to 75° and the long path heading between 205 and 255°. When I work ZLs on 80 meters on long path during the Northern Hemisphere winter, signals always arrive via North America, at a heading of approximately 300°—this is 90° off from the great-circle long-path direction. The path is not a great-circle path, but is inclined as if the signal were trying to leave the Southern Hemisphere as fast as possible (both the ZLs and Europeans beam across North America in the winter).

As we continue into spring, the optimum path between Western Europe and New Zealand will move from across North America to across Central America (February-March). Eventually, beaming across South America will yield the best signals later in the year (from April onward). Somewhere around the Spring Equinox all three paths produce equally good signals, when the strongest signal strengths are heard (see the discussion of antipodal focusing in Section 1.4.4.5).

Theory says that there are an indefinite number of great-circle paths to the antipode. As low-band DX signals travel only over the dark side of the globe, however, the usable number of great-circle headings is limited to 180° (assuming there is no aurora activity screening off part of the aperture). This very seldom means that signals will arrive with equal strength over 180°, however, not to mention with the proper phase! There is another well-known principle that the strongest signals come from those directions where the attenuation is least. Although according to scientists there seems to be no correlation, I have observed that these are directions where the signals travel through areas with the lowest MUF.

These New Zealand-to-Western Europe QSOs are examples of gray-line propagation. None of these favored propagation paths ever coincide with the terminator itself. I have found that the actual path is more or less perpendicular to the terminator at all times of the year.

To summarize, I have observed for over 30 years that on

80-meter long paths and paths to areas near the antipodes, the signal paths are skewed in such a way that the signals apparently travel the longest possible distance in the hemisphere where it is winter, as witnessed from the direction of arrival of the signals.

3.3.1.3. South America across North America in (Northern Hemisphere) winter

A similar path bending is also quite common over "shorter" paths. During the European winter, signals from Argentina and Chile regularly arrive in Europe at beam headings pointed directly at North America, up to 90° from the expected great-circle direction. The signals from South America appear to travel straight north in order to "escape" the summer conditions in the Southern Hemisphere, and are then propagated toward Europe. One striking example was when I worked 3Y1EE (Peter Is) on 80 meters (January 28 1987). The signals were totally inaudible from the great-circle direction (190°) but were solid Q5 from 310° (coming across North America). Similarly, when I worked CE0Y/SM0AGD on 160 meters (October 1992), signals were only readable on a Beverage beaming 290°, while the great-circle direction to Easter Island is approximately 250°.

3.3.1.4. The pseudo-long path between Europe and US West Coast

The so-called long path on 80 meters between Scandinavia and Eastern Europe to the US West Coast is actually another example of a crooked path. Looking at the darkness distribution on Earth, (**Fig 1-22** and **Fig 1-23**) it is clear that a genuine (reciprocal) long path is out of the question, as signals would travel for nearly 20,000 km in daylight (across Africa, South Indian Ocean, Antarctica and South Pacific) around 1300 UTC, or along the entire path in the twilight zone around 1600 UTC. In Europe the beam headings generally indicate an optimum azimuthal angle of approximately 90°, which again is almost perpendicular to the terminator. Along their way, the signals will be least attenuated in those areas of the ionosphere where the MUF is lowest. They thus seem to travel along a crooked path, avoiding areas of higher absorption. OZ8BV reports a 90° to 100° direction when working the West Coast on 80-meter long path from southern Denmark. Ben is using a 3-element Yagi at 54 m (180 ft) and is well placed to confirm this path (the genuine long path would be 150 to 160°).

D. Riggs, N7AM, who is using a rotary quad for 80 meters, wrote: "*We have learned that the 80-meter long path between the Pacific Northwest and Scandinavia is following the LUF (lowest usable frequency). I have always believed that the long path to Europe was not across the equator but leaves us at 240° and since the MUF is highest at the equator it cannot continue at 240° but it bends westerly going under the Hawaiian islands, across the Philippines under Japan and across the Asian continent to Scandinavia. The MUF charts prove this fact. The fact that the long path to Europe lies north of the equator is proven by the northern Europeans and after the West Coast peak.*"

During the 1996-1998 seasons a number of QSOs were made between the US West Coast and Scandinavian stations on 160 meters. SM4CAN, SM4HCM and SM3CVM all confirmed that the signals were coming from due east (90°), instead of on the short path (335°) or long path (145°). N7UA stated that he was using his JA Beverage, as the stations were *not* audible on the over-the-pole European Beverage. SM4HCM called it "*a skewed path, somewhere between long path and short path*" (see **Fig 1-24**). N7UA added that "*... the lower the frequency, the more the long path moves toward North, away from the true reciprocal heading.*"

These observations seem to be related to the principle that the lower the frequency, the more easily a horizontal ionization gradient can cause path skewing (see also Section 3.3.1.1 and Ref. 147). There must be horizontal ionization gradients present for such skew paths to occur.

3.3.1.5 Europe to Alaska

Besides the so-called short path between Europe and KL7 we can often work Alaska on a so-called long path (that is *not* actually a long path) before their sunrise and just after our sunset on 80 meters. At that time (around 1600 UTC in midwinter) signals usually arrive in Europe from east/northeast. This is clearly a bent path across Siberia to Alaska, thus avoiding the auroral belt (**Fig 1-25**) and certainly is not a real (reciprocal) long path. If the path were a genuine long path, it would go right across the South Pole, which is in continuous daylight at that time of the year. KL7Y notes the signals came in from the southwest, approximately 45° from the genuine long path.

But when geomagnetic conditions have been quiet for a long period, it sometimes *is* possible for signals to travel

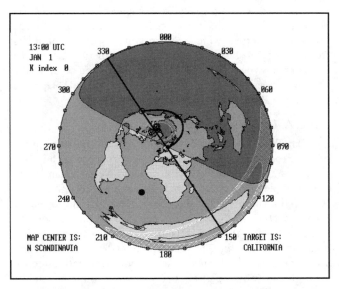

Fig 1-22—The 80-meter long path between Europe and the US West Coast is neither a short path (the lighter line right through the auroral belt), nor a genuine long path (20,000 km in daylight). Instead, it is a crooked path (the curved darker path). In Europe signals generally arrive at headings of 70° to 110° from True North. (*Map generated by* DX-AID, *with additions by* ON4UN.)

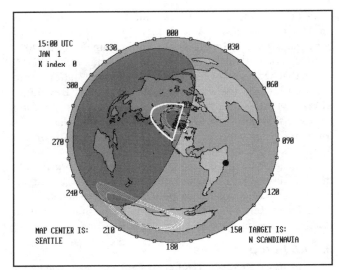

Fig 1-23—Great-circle map centered on Seattle, showing the 160-meter path for contacts into northern Scandinavia around 1500 UTC in midwinter. The actual path is a crooked one (white curved line). At both ends of the path, the direction is again perpendicular to the nearby terminator. Propagation along the terminator in the twilight zone is impossible because there is too much D-layer absorption in that zone. (*Map generated by DX-AID, with additions by ON4UN.*)

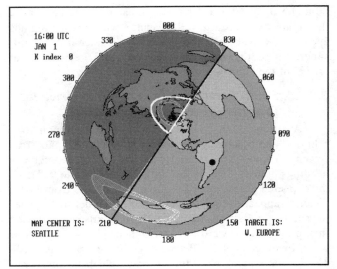

Fig 1-24—Great-circle map centered on Seattle, showing the so-called 80-meter midwinter long path to Western Europe (light curved line) around 1600 UTC, which is clearly a crooked path that skirts the auroral oval. The real long path, the path directly along the termintor, is open on 40 meters, however. (*Map generated by DX-AID, with modifications by ON4UN.*)

through the heart of the auroral zones. I remember an amazing such QSO in the 1970s with KL7U on 80 meters at about 1600 UTC, hearing him only when listening at 350°, which is the direct short path right across the magnetic North Pole (the straight line short path in Fig 1-25). Going only by the time of the contact, this would easily be defined as a long-path QSO; however, it was not, since the signals did not come from the real long-path direction (which is approximately 160°) but almost from the regular short-path direction. It is obvious that this can happen only when there is no auroral absorption at all, as this short path goes right across the magnetic North Pole. This path is the equivalent of the JA-to-Europe path.

3.3.1.7. The Heard Island case

The VKØIR example has been covered in detail previously. From the viewpoint of path skewing, the path is very similar to the path between Europe and New Zealand. This means to me that signals seem to travel best through areas of low MUF. Therefore the path from Heard Island to the US East Coast (and Midwest) seems to travel as much as possible through the Northern Hemisphere, avoiding higher MUF areas in the South. The bent path also meets the most advantageous launching conditions when the path direction (at both ends of the path) is perpendicular to the terminator at those points. (See Section 1.4.4.1.) This explains why these signals were received on the US East Coast via a bent path across Europe in January 1997 (see Fig 1-9C).

3.3.2. Avoiding the auroral zones

The second reason for path deviation is to avoid the auroral oval. In the Northern Hemisphere, during those times

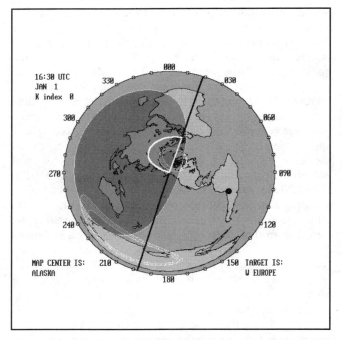

Fig 1-25—The Alaska-to-Western-Europe 80-meter path around 1630 UTC in midwinter. The real long path is totally impossible because it travels for some 25,000 km in daylight. The genuine short path travels in the twilight zone, hence there is a high degree of attenuation. The actual path takes off from Alaska in a westerly direction and arrives in Europe east of the short-path bearing. Note again that the two real signal directions are perpendicular to the terminator. (*Map generated by DX-AID, with additions by ON4UN.*)

when there is aurora, signals on the low bands will often appear to arrive from a more southerly direction than one would expect from great-circle considerations. The path between Western North America and Europe is greatly affected by this phenomenon, because the magnetic North Pole lies right in that path. Between Japan and Europe there is much less influence.

The aurora was described in detail earlier, as well as its effect on radio propagation. Let us analyze a few paths that suffer frequently from the effects of aurora. To view these paths, get your globe, maps or switch on your favorite Mapping program that can show the auroral oval. With *DX-AID* (see also Section 2.2.7.2) you can plot the great-circle paths and enter various values of the K index, and watch what evil the auroral oval is doing.

The short path between the West Coast of the US and Western Europe has always been a "critical" path, because of the interference of the auroral oval with the great-circle path. For the same reason, the short path is very rare between the West Coast and Northern Scandinavia (Northern Scandinavia being inside the auroral oval).

Fig 1-26A shows the path between Western Europe and Seattle (in midwinter) at sunrise in Europe, for a K index of 0. The auroral oval has retracted to its minimum size and width, and although the great-circle path goes through the auroral oval, attenuation will probably be minimal, either by the signal skirting under the ionosphere at the narrow oval or by apparently slightly bending around the oval. Fig 1-26B shows the same for a K index of 9 (heavily disturbed magnetic conditions). Note the extreme width and major-axis size of the oval. Under such conditions, since Seattle actually lies on the border of the oval there may be little escape from the aurora. As a rule, stations further south (for example, Southern California stations) may possibly make it into Europe, beaming across South America.

I remember one striking case like this where I made a good QSO with W6RJ on 80 meters where Bob was using a KLM-Yagi beaming to South America, while I was using my South America Beverage to copy him. The direct path, across the aurora oval was totally dead at that time. Obviously the attenuation on this crooked path is far greater than on a straight path when there is no auroral absorption, so only stations with good antennas and some power may regularly experience this path.

WØZV reported similar path skewing on 160 meters from Colorado. In one instance (April 1988) Bill worked SM6CPY in the middle of a severe ionospheric disturbance, beaming 110°, while the direct bearing is 27°!

An even more striking example of path skewing due to aurora is the path between Europe and Alaska on 80 meters. Looking at the map or globe, there is a great-circle path that is only about 7500 km long, but it beams right across the magnetic North Pole. The distance is similar to the distance between Western Europe and Florida. Straight short-path openings between Europe and Alaska are rare exceptions, happening only a few times a year when the K index has been at zero for some period of time. The main difference between the Seattle and the

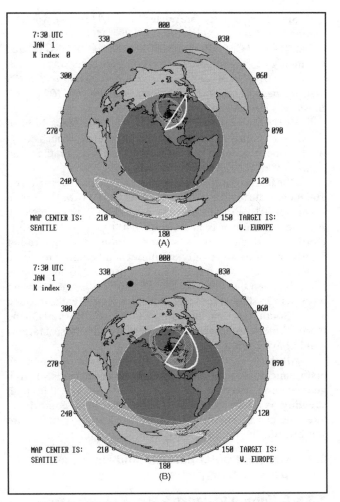

Fig 1-26—The West-Coast-to-Europe "short" path on 80 meters. At A, we have minimum geomagnetic activity (K=0) and the signals either travel the genuine direct path or may be very slightly bent southward. At B, the situation for a K index of 9 (major disturbance). Note the extent of the aurora zone. The only way to make it into Europe is for stations at both ends of the path to beam toward South America. (*Map generated by DX-AID, with additions by ON4UN.*)

Alaska case is that from Seattle it takes much more path bending to go around the oval. Because of its more northerly location, Alaska actually lies *inside* the auroral oval.

W4ZV, who operated for many years from Colorado as WØZV points out that he never saw skew paths from Colorado to JA, since the JA bearing (315° from Colorado) is nowhere close to Magnetic North (13° from True North). He also added that he worked Europe *much* more frequently from Colorado peaking on his 70° Beverage than on his 40° Beverage, which was the true great-circle bearing from there. Under *very* severe geomagnetic conditions (K index = 6), signals would even peak on his 110° Beverage! W4ZV, now in North Carolina, confirms that a similar path exists from the US East Coast to the Far East. Signals from JA quite frequently skew to the south during geomagnetic disturbances.

Similar experiences are related by N5JA (Texas), who said that he found the best direction for UU4JMG changing from 40° at 0145 UTC, to 60° at 0215 UTC, 90° at 0230 UTC and eventually 120°! He reported shifts going the other way as well: *"I've seen ON4UN in years past become first audible on the 90°, then be best on the 60°, and later best on the 40° or 20° Beverages"* (what goes up, must come down!)

KØHA, top notch Top Bander from the black hole of the USA (Nebraska) has reported similar experiences copying European stations best on his 140° Beverage rather than on his 43° or 86° antennas.

One last striking example of a crooked path occurred on 160 meters during the first night of the CQ-160-M Contest in January 1991: With the exception of VE1ZZ, not one North American station was heard until 0400 UTC. At that time North American stations started coming through rather faintly, but they were only audible when beaming to South America (240°). No signals from the "usual" 290-320° direction! Between 0400 and 0700 UTC, 80 W/VE stations were worked in 25 states/provinces. All of the signals came through across South America, including K6RK in California! On the North American Beverage, only a few of those stations would have been worked.

All these skewed paths are likely caused by magnetic disturbances, as they all happened during periods of high geomagnetic activity. In general, one can say that path skewing is quite common during disturbed conditions. Skewing can be anything from 30° to almost 90° off (more southward) the normal great-circle direction.

3.3.3. Crooked polar paths in midwinter, or pseudo-long paths

3.3.3.1. Europe to Japan

In Northern Europe we can work Japan in midwinter, just after our sunrise (0745 UTC) on 80 meters. The most-common opening to Japan on 80 meters is at JA sunrise, around 2200 UTC. At first you might be tempted to call the other 0745 UTC opening a long-path opening. Careful analysis using directive receiving antennas has shown that the signals come from a direction slightly east of north rather than their true long-path heading of 210°. The 0745-UTC opening is rather short at my QTH (typically 15 to a maximum of 30 minutes in midwinter). **Fig 1-27** shows the great-circle path *along* the terminator, along the gray-line zone, and is *not* a typical example of gray-line propagation, where low-band signals usually travel more or less perpendicular to the terminator. These are certainly far from ideal conditions for low attenuation, and the signals on the JA path are substantially weaker than the signals normally heard from Japan from the same heading, but at JA sunrise. Low-band propagation over long distances along the terminator is clearly not the rule. If present, signals are weaker than what can normally be expected over similar distances at paths that do not follow the terminator.

If we look at the darkness/daylight distribution across the world at that time (0745 UTC in midwinter), we see that we have indeed more than one path possibility: a range of paths ranging from true short path (30°) to alternative crooked paths bent slightly east or even west of the magnetic North Pole, all across areas in darkness. These alternative bent paths go right through the North Pole auroral zones, and hence will very seldom produce stronger signals than the path along the terminator.

A number of years ago Hoppe and others (Ref. 108 and Ref. 118) considered that low band gray-line propagation went *along* the terminator, as is the case on the higher bands. It has, however, been proven over and over again that this is *not* the rule on the low bands, where the most spectacular propagation enhancements always occur when the propagation paths are perpendicular to the terminator. (See also Section 1.3 and 1.3.1.) The Europe-to-Japan short path at 0745 UTC in midwinter is a remarkable exception to this general rule.

Often such very specific propagation paths (more or less along the terminator) are very area selective, probably because ducting phenomena are involved, occurring only when the necessary launching and exiting conditions exist. In several cases I was able to work several JA stations with signals up to S9, while the same stations were reported to be undetectable in Germany only a few hundred kilometers away. In all cases the signals were loudest when they were coming in about 10° east of north. In Japan the openings seem to be very area-selective as well, as can be judged from the call areas worked. Northern Japan (JA7 and JA8) leads the

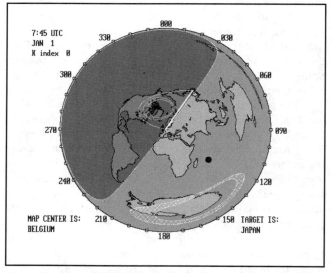

Fig 1-27—The direct short path between Europe and Japan at Europe sunrise in midwinter is an example of quite exceptional propagation. Adding the auroral oval to the great-circle map from Fig 1-21, we see that the alternative path (the light curved line) would skirt around the auroral belt. It would be at right angles to the terminator at both path extremes, and because of the consideration would be the most-favored path. In actual practice, however, I have never seen JA signals come out of the northwest along that path at that time of the day. Instead, the signals come on the short path from Japan. (*Map generated by DX-AID, with additions by ON4UN.*)

opening. Central Japan follows 30 minutes later, and southern Japan often is too late for this kind of opening at my QTH. In midwinter there is a 1.5-hour spread in sunset time between northern and southern Japan.

On 40 meters the situation is somewhat similar, and yet very different at the same time. At 0745 UTC in midwinter, JA 40-meter signals arrive from north or northeast (as on 80 meters), but when that path fades about 15 to 30 minutes later, it is immediately replaced by a genuine long path, where the signals now come in across South America, along the terminator as shown in Fig 1-21. I have never observed this genuine long path across South America on 80 meters, let alone 160 meters.

At JA sunrise time at 2200 UTC, the short-path direction is almost at right angles to the terminator (see **Fig 1-28B**). A similar good launching angle (with respect with the terminator) occurs at sunset in Europe around 1600 UTC (Fig 1-28A). From that point of view the 1600 UTC and the 2200 UTC openings are almost identical. In real life, however, we in Europe find the 2200 UTC opening much better. This may also be due to the fact that 1600 UTC is in the middle of the night in Japan!

3.3.3.2. New England to Japan

Brown, NM7M, analyzed a similar path (Ref. 140) on 160 meters: W1 to JA. He also found two openings, one he calls a "gray-line" path (the opening around 2140 UTC in midwinter) and another one that he called a "black-line" path. I consider the former path as a most atypical gray-line path for low frequencies, even though both ends are in twilight (a double gray-line situation). My understanding is that the few signals heard or worked on that path came out of the northeast direction, which is a crooked path across Europe. See **Fig 1-29A**.

The theoretical great-circle path between W1 and Japan at 2140 UTC also goes parallel with the terminator; hence there should be no enhancement. The alternative crooked path travels across Europe, and once more enjoys good launching conditions (path direction perpendicular to terminator) at both ends of the path.

The great-circle path at 1200 UTC makes an angle of about 50° with the terminator, and it is common knowledge (many US East Coast stations claim paths from 270° to 205°) that many of these paths are bent southward (making the angle with the terminator more like 90°). The 1200-UTC opening is a one-sided gray-line opening, since it occurs at sunrise on the East Coast (Fig 1-29B).

QSOs have, of course, been made much earlier (0945 UTC). At that time we have a most typical "black-line" path occurring at midnight for the point half-way between the path ends (Fig 1-29C). Here too, slight skewing around the polar regions is very common. At this time of the day there is no signal enhancement by the gray-line phenomena, as both ends of the line are well into darkness.

3.3.3.3. Other paths

Similar paths exist on 80 meters in midwinter between

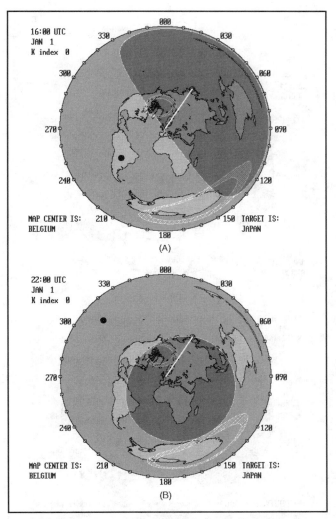

Fig 1-28—The classic paths between western Europe and Japan. At A, the situation at sunset in Europe, and at B, for sunrise in Japan. At both times, the path direction is almost perpendicular to the terminator. (*Map generated by DX-AID, with additions by ON4UN.*)

California and Central Asia (Mongolia) around 0030 UTC, between Eastern Europe (Moscow) and the northern Pacific (Wake Island) around 0615 UTC, and between the US East Coast (and the northern part of the US Midwest) and northern Scandinavia around 1230 UTC. All of those polar-region paths are east of the North Pole and should not be influenced by aurora as much as paths going west of it. Use your globe, map or mapping program (eg, *DX-AID*) to visualize the paths that appear to avoid the auroral belt, if necessary.

3.3.4. Selective paths/areas

I have often wondered how a DX station, or even more so, a DXpedition can make contacts on the low bands while hoards of stations keep calling after a QSO has been started. We've all seen cases where 80% of the pileup just keeps calling (especially in Europe)! For an observer located in the middle of all these callers this is pure chaos.

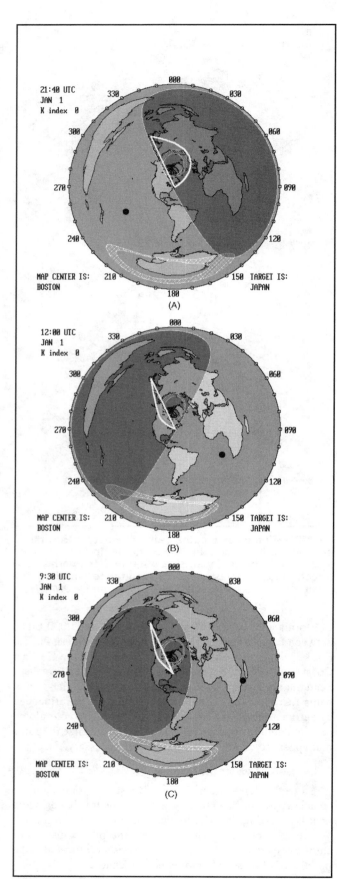

I am convinced that selective skip is the reason why the DX station can work stations in this seeming chaos. ZS6EZ, eminent low-bander and DXpeditioner stated: "*On my DXpedition, it was very often like a searchlight, where only one very small area is audible at a time, and that area moves around with time. There is no regular movement either. You might have W1, W8, W9, W0, W3, W4, W5, W0, W7, W0, W9, etc in a single opening. The pileup is never audible; just a few stations at a time.*" He adds: "*This is why I contend that the argument of the East Coast Wall doesn't hold in most sunrise openings to Africa. If the 'spotlight' is on the West Coast, the other coast is not even audible.*"

John, K4TO, commented to me on the same subject: "*I have also observed that given propagation into a particular area, one signal might be 10 dB stronger than another, even though both stations were running the same power and comparable antennas. Given a few seconds' time lapse, and the situation reverses. The station who was stronger before is now the weaker of the two. Having observed this first hand, I find that my own anxiety level about working a particular DX station is now reduced to near zero. I feel that sooner or later, the propagation will come to favor me. I wish more DXers would have this same experience. I believe everyone would relax and enjoy the hobby more. Operating manners and techniques would improve greatly.*"

I think that area-selectivity may be partly due to polarization rotation of signals on 160 meters. This can also explain the slow QSB we often witness on Top Band. Sometimes we must wait a few minutes for polarization to "come back" in order to hear a marginal signal again. On Top Band this is why, where weak signals are involved, it is essential to get a full call at the first try. Often a second try at a partially received call brings dead silence, until a minute or so later, when the same signal slowly comes out of the noise again.

Signal ducting between the E and F layers provides another possible explanation for the "searchlight effect." The landfall of the searchlight would relate to finding stations with appropriate launch angles such that the RF could enter or exit the duct.

Dan Robbins, KL7Y, came up with another theory that might help explain selective propagation. He concluded from his work with HF radars that the ionosphere can act as a "filter" for the angle of radiation: "*Near the MUF, high angles are 'filtered' out. But below the MUF, low angles may*

Fig 1-29—The two openings (for both 160 and 80 meters) from New England to Japan. At A, the theoretical 2140-UTC straight-line short path, which goes along the terminator and right through the aurora oval. This is not the way the signals actually travel. A bent path across Europe (curved lighter line) arrives on both ends of the path at right angles to the terminator. At B, the morning 1200-UTC opening. Actual signals arrive in New England from beam headings somewhat further west than the short-path heading in order to skirt the auroral oval. At C, the geometry for an all-night ("black-line") path (shown in white) at 0930 UTC. Even at this time, the beam heading is often bent further south than the direct path. (*Map generated by DX-AID, with additions by ON4UN.*)

also be filtered out. Once angles are filtered out they are gone, and the same narrow range of angles will propagate over and over again, even if the filtering conditions disappear on later hops. If the range of allowable angles is narrow, the propagation will occur in narrow distance bands from the transmitter, since distance is a function of angle of radiation. This is why one guy works the S9 DX and his buddy 300 miles away hears nothing. This is why the band appears dead, except for that loud 3B8 or whatever."

In other words, for any given path there will be an optimum frequency and an optimum range of angles of radiation that produce the maximum signal. If we are restricted to one frequency, then there will be one range of optimum radiation angles for that path. According to Dan Robbins the range of optimum angles may be surprisingly narrow at times—narrow enough to account for almost all instances of selective propagation.

Fortunately, most of the antennas we use on the low bands have rather broad vertical lobes, nothing like the arrays used for OTH radar. This effect of "angle selectivity" is thus largely smoothed by the broad lobe of our transmit antennas.

3.3.5. Conclusion

On the low bands we continuously witness bent propagation paths. By that, I mean that at one end (or more likely at both ends of the path) a signal appears to arrive from a direction that is vastly different from the great-circle heading. Documentation of these facts are so overwhelming that there is no doubt about these bent paths.

For the low-band operator it is important to know that these bent pads are quite common and to be able to anticipate them. Many of the observations are clearly linked with high geomagnetic activity. In these cases signals seem to come from areas away from the auroral zones, which seems logical enough.

There are, however, many documented cases of signal-path bending on paths that do not go through high-latitude areas. These cases of apparent path bending cannot be related to enhanced geomagnetic activity.

■ 4. TOOLS FOR SUCCESSFUL DXING ON THE LOW BANDS

4.1. Sunrise/Sunset Information

Now that you have read through the foregoing paragraphs, let's try to be practical. The most important tool is still, whatever one may say, sunrise/sunset information. I use tables in a booklet form—it's faster than starting up a computer program, although a lot of logging programs now have that information built in.

4.1.1. The ON4UN sunrise/sunset tables

The booklet of sunrise/sunset tables that I created several years ago shows sunrise and sunset times for over 500 different locations in the world (including 100 different locations in the US). Increments are given per half month. **Fig 1-30** shows an example of a printout for one location.

I have tried all the propagation aids that are described in this book, and many others as well. The only aid that I regularly

use is the sunrise/sunset tables. Why? You can grab the tables anytime and look up the required information in seconds. Just keep the little booklet within reach on your operating desk! The tables never get outdated, since sunrise/sunset times hardly change over the years. All the graphical systems (globe, slide-rule or computer-screen world maps) are far too inaccurate to be useful for 80 and especially 160 meters.

There are some copies of this handy sunrise/sunset booklet (100 pages) still available. Send $10 plus $5 for worldwide airmail postage to John Devoldere, ON4UN, Poelstraat 215, B9820, Merelbeke, Belgium.

4.1.2. General rules for using sunrise/sunset times

For all E-W, W-E, NW-SE and NE-SW paths there are normally two propagation peaks that can be expected (for short path):

1. The first peak will occur around sunrise of the station at the eastern end of the path.
2. The second peak occurs around sunset for the station at the western end of the path.

For N-S paths there are no pronounced peaks around either sunset or sunrise. Often the peak seems to occur near midnight.

The use of the tables can best be explained with a few examples.

4.1.3. Example 1

What are the peak propagation times between Belgium and Japan on February 15? From the tables:

Belgium: 15 Feb: SRW = 0656 SSW = 1659
Japan: 15 Feb: SRE = 2130 SSE = 0824

where

SRE = sunrise, eastern end
SRW = sunrise, western end
SSE = sunset, eastern end
SSW = sunset, western end

The first peak is around sunrise in Japan or SRE = 2130

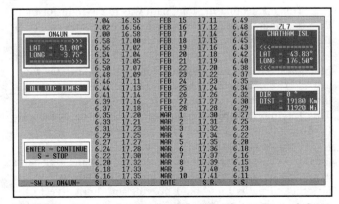

Fig 1-30—Screen dump of the sunrise/sunset module of the ON4UN LOW BAND SOFTWARE. This example shows the sunrise/sunset times for Belgium and ZL7, in 1 day increments from February 15 to March 10. These precise data are invaluable when short windows are available, such as in this case.

UTC. This is after sunset in Belgium (SSW = 1659), so the path is in darkness. Always check this.

The second peak is around sunset in Belgium or SSW = 1659 UTC. This too is after sunset in Japan (0824 UTC) so the path is in darkness.

Is there a possibility for a long-path opening on the lower bands? The definition of a long-path opening says we must have sunset at the eastern end before sunrise at the western end of the path. In the example above this is not true, because SRW at 0656 UTC is not earlier than SSE at 0824 UTC.

4.1.4. Example 2. Is there a long-path opening from Japan to Belgium on January 1?

Belgium: 1 Jan SRW = 0744 UTC, SSW = 1549 UTC
Japan: 1 Jan SRE = 2152 UTC, SSE = 0740 UTC

Here, SRW (0744 UTC) is later than SSE (0740 UTC). This is indeed a valid condition for a long-path opening. It will be of short duration and will be centered around 0746 UTC. (See Section 3.2.3.)

Being a so-called long path does not mean that the direction of signal arrival is the opposite of the short path, however! In Section 3.3.3 I explained that this so-called long path to Japan is a typical example of a midwinter crooked path across the pole.

4.1.5. Pre-sunset and post-sunrise QSOs

In practice, long-path openings are possible even when the paths are partially in daylight. Near the terminator we are in the gray-line zone and can take advantage of the enhanced propagation in these zones. The width of the gray-line zone has been discussed earlier. A striking example of an excellent genuine long-path QSO was a contact I made with Arie, VK2AVA, on March 19 1976, at 0700 UTC on 80 meters. The long-path distance is 22,500 km. Note that the QSO was made almost right at equinox (March 21) and the path is a textbook example of a NE-SW path. On that day we had the following conditions:

Sunrise west (Belgium) = 0555 UTC
Sunset east (Sydney, Australia) = 0812 UTC.

This means that the long path was in daylight for more than two hours. The QSO was made one hour after sunrise in Belgium and more than one hour before sunset in Australia.

Another similar example was a QSO with VK0GC from Macquarie Island (long-path distance 21,500 km). On January 21, 1985, I made a long-path contact on 80 meters that lasted from 0800 until 0830 UTC, with excellent signals. This was more than one hour before sunset on Macquarie (0950) and almost one hour after sunrise in Belgium (0731).

Because the locations of these stations (VK2 and VK0) are fairly close to the antipodes from Belgium, the long paths can safely be considered genuine long paths. Indeed there are no crooked paths that could provide an alternative to the genuine long paths. The gray-line globe is a unique tool to help you visualize a particular path like this.

4.1.6. Calculating the half-way local midnight peak

For East-West paths ($\pm$ 45°) in addition to the usual sunrise and sunset peak, there is often a so-called halfway midnight peak. (See Section 1.4.3.) The time of this peak can be calculated as follows:

First use the sunset/sunset tables or a computer program to determine both sunrise and sunset for both ends of the path. Let's use as an example a path between Denver (Colorado) and Belgium on January 15. The sunrise/sunset data are:

Colorado sunset: 0001 UTC, sunrise: 1419 UTC
Belgium sunset: 1606 UTC, sunrise: 0738 UTC
Midnight in Colorado : 0001 + (1419 - 0001)/2 = 0710 UTC
Midnight in Belgium: 1606 + (2400 + 0738 − 1606)/2 = 2352 UTC

The halfway midnight peak time is calculated as the mathematical average between the two midnight times:

Local halfway midnight is: (2400 − 2352 + 0710) / 2 = 0339 UTC

For North-South paths ($\pm$ 30°) there is a distinct propagation peak at local midnight at the halfway spot. This is true for 80 and 160 meters. This peak is commonly called the midnight peak. How do we calculate the exact time of this midnight peak?

Example: path between New York and Paraguay on June 15
New York sunset: 0028 UTC, sunrise 0925 UTC
Paraguay sunset: 2108 UTC, sunrise: 1034 UTC
Calculate the two local midnight (sun) times as follows
Midnight New York: (0925 − 0028) / 2 = 0429 UTC
Midnight Paraguay: (1034 + 2400 − 2108) / 2 = 0643 UTC
Halfway midnight time: (0429 + 0643) / 2 = 0536 UTC.

4.1.7. Calculating sunrise/sunset times

Table 1.7 lists a BASIC program developed by Van Heddegem, ON4HW, to calculate sunrise and sunset times. It is based on classical astronomy and calculates precise sunrise and sunset times. The program does not use any arc sin or arc cos functions as they are not available in all BASIC dialects.

4.2. Propagation Predicting Computer Programs

Earl, K6SE, in a message on the Internet, wrote: "*I gave up long ago on trying to predict DX conditions on 160 meters. One major observation I've made over the years is that on a night when conditions are good to EU from here (northwest), conditions to JA (northeast) that same morning are not exceptionally good. And, if conditions are good to JA in the morning, that evening conditions to EU are not good. The only 'prediction' I use now is: if conditions are exceptionally good in any particular direction on a given night (ie, to EU 28 December 97 in the Stew Perry contest), I hope there will be a repeat one solar rotation later (27 days). Generally, I've come to the conclusion that 160-meter conditions are unpredictable; hi, so I just check the band every night to see what's happening.*"

To be perfectly honest, in my 37 years of DXing on 80 meters and in my 12 years on 160 meters, I have never used a propagation "prediction" program. Usually when

Table 1-7

```
10 CLEAR
20 PI = 3.1415927
30 SW = -0.97460409
40 CW = 0.22393492
50 SE = 0.39777700
60 CE = 0.91748213
70 K1 = -0.014834754
80 PRINT "ENTER WEST LONGITUDE IN DECIMAL DEGREES"
90 INPUT LO
100 PRINT "ENTER NORTH LATITUDE IN DECIMAL DEGREES"
110 INPUT LA
120 LO = LO*PI/180
130 LA=LA*PI/180
140 PRINT "ENTER DAY NUMBER(1 TO 365)"
150 INPUT D
160 M = (2 * PI * D * LO) / 365.24219 - 0.05347
170 L = M - 1.3449463
180 C1 = 1 - 0.03343 * COS (M)
190 C2 = .99944 * SIN (M) / C1
200 C3 = (COS (M) - .03343) / C1
210 C4 = SW * C3 + CW * C2
220 C5 = CW * C3 - SW * C2
230 C6 = SE * C4 : REM SINE OF SUN DECLINATION
240 B1 = K1 - C6 * SIN(LA)
250 B2 = (COS (LA)^2 * (1-C6^2) - B1^2
260 IF B2 <= O THEN R$ = "NO SR" : S$ = "NO SS" : GOTO 340
270 B 3 = ATN(B1 / SQR (B2) - PI/2
280 B4 = ATN ((COS (L) * CE * C4 - SIN (L) * C5) / SIN (L) * CE * C4 + COS (L) *
C5))
290 GOSUB 370
300 R$=STR$ (B6)
310 B3=-B3
320 GOSUB 370
330 S$ = STR$ (B6)
340 PRINT "SUNRISE: ";R$
350 PRINT "SUNSET: ";S$
360 GOTO 140
370 B5 = B4 + B3 + LO + PI
380 IF B5 < 0 THEN B5 = B5 + 2*PI
390 B5 = INT (B5 * 720 / PI +0.5) : REM MINUTES PAST 0000 UTC
400 IF B5 > 14139 THEN B5 = B5 - 1440
410 B6 = 0.4 * INT (B5 / 60) + B5 / 100 : REM TIME IN HH.MM
420 RETURN
```

operating one of the major CQ WW contests, one of my friends would run the propagation predictions on his computer and bring them along. I have never found anything in there that I did not know. The easy paths, the well-known conditions were there, clearly on paper, but the marginal ones, the exotic ones were not. Never. And by the way, these prediction programs usually do not cover 160 meters. Rightfully so, because basically propagation on TopBand relies on different mechanisms than the higher bands (since 160-meter propagation is not MUF related).

Tom, N4KG, who says that he "lives" on the low bands, recently wrote: "*I find very little correlation between solar flux and low-band propagation, particularly on 160 and 80 meters. Each of the low bands has a distinct characteristic. Forty meters does follow solar flux in that the MUF can drop below 7 MHz when the flux levels are very low. During these times, 40 meters will close shortly after sunset on Northern paths and there will be no European sunrise opening to the USA. With slightly higher solar activity, 40 meters will stay open all night. On 80 meters, solar flux is not so much of a factor. The BEST times are just before local sunset and after local sunrise. There is a significant enhancement at the terminator, which seems to be a daily event.*

There seems to be some validity to the theory that propagation is enhanced at the very start of a solar disturbance, but then the band goes flat for several days until the ionosphere stabilizes again. 160 meters does not correlate with anything! *You may see nice enhancements at sunset and sunrise, or you may see nothing at sunset and sunrise but find good openings between 0200 and 0500 to Europe (from USA). On signals from the east, I have seen peaks at their sunrise and I have seen them peak a full hour before sunrise (before any daylight) and then vanish. Propagation-prediction programs seem to be almost useless and often misguiding on the low bands because they fail to model the focusing effects at sunrise and sunset and they do NOT look at non-great circle paths. Most over the pole paths to the opposite side of the world come via skewed LP, at least here in the eastern USA...*"

I could not have said it better! This being said, the bulk of the current propagation prediction programs *are* useful for predicting 40-meter propagation, while some of them include very useful viewing and mapping facilities. (See Section 2.2.7.2.)

However, there may be light at the end of the tunnel. Rod Graves, VE7FTP, an active low-band DXer, has written a program that attempts some new approaches in predicting the more "odd" openings, the ones that other programs don't seem to recognize. The concept of the program is quite novel and employs a zone method developed by the author. It supports (E-F region) ducted paths and also directly addresses skewed paths. These are not apparent skews due to magnetic disturbances, but real skews due to the structure of the ionosphere at a given time. The program seems to predict very-long-distance and long-path openings on 80 and 160 meters much better than any other. The limitation of the software, at this time, is that it does not take into account the auroral absorption. The software is still in a development phase but may become available at a later date. VE7FTP also admits that signals from locations near the antipode may travel paths other than the ones his program checks.

Rod continues to say: "*Like most others, I do not use a program to tell me when the bands might be open to unknown locations. I just turn on the receiver to see what, if anything, is coming in from wherever and enjoy the discovery of the unexpected DX. However, when I want to make a sked with someone or when there is a new DXpedition, I run the program to determine when the band might be open to that specific location and when the predicted optimum times are. I use this as my guide for when to be sure to be listening or for when to make the sked. Sometimes the program shows only what is common lore, or just provides another way of determining what could have been done with sunrise/sunset tables, or gray-line devices but the program does it very conveniently for me. Sometimes, however, the program reveals times of openings, or times of peak signal strength, that are not obvious. This is especially true for paths over 10,000 km, particularly on the low bands (40-160) where refraction and ducting phenomena are more prominent.*"

For me, the final "acid test" that would make me a firm believer in propagation forecasting programs is when such a program would successfully forecast the odd 80 and 160-meter openings *and paths* (directions)—such as the Europe to ZL path described before. I also would like to see a quantitative confirmation of the bent paths that we so often experience as described in Section 3.3.1.2. In order to make that possible and in order for anyone to be able to model the entire system that determines attenuation on 80 and 160 meters, we will have to know *all* the mechanisms. Today I have the impression that we see only the top of the iceberg, but the issue is being addressed.

In the meantime, sunrise/sunset tables and a good mapping program (such as *DX-AID*) are in my opinion your best companions in your search for a new one on 80 or 160 meters.

4.3. ON4UN Low Band Software

4.3.1. ON4UN Propagation Programs

When writing the original *Low-Band DXing* book, I developed a number of computer programs as tools for the active DXer. The programs were fully color compatible, and available only for MS-DOS on a 3½-inch diskette (see order form in the back of this book). The software runs perfectly under Windows (also 95).

While the majority of programs are technical programs related to antenna design (and are covered in the antenna chapters of this book), a group of programs deals with the propagation aspects of low-band DXing. The propagation software contains the following modules:

1. SUNRISE/SUNSET TIMES.

This program lists the sunrise and sunset times in half-month increments for the user's QTH. The user's QTH can be preprogrammed and saved to disk. It can be changed at any time, however. On the screen, the sunset and sunrise times of a "target QTH" are listed side-by-side with your own time. The target QTH can be specified either by coordinates or by name. On-screen you also see the great-circle direction as well as distance.

The software works in miles as well as kilometers. You can also modify the display increments and can also list the times in single-day increments if you wish, which is very handy for following the gray line on DXpeditions. **Fig 1-31** shows a screen print giving relevant data for the ZL7 (Chatham) expedition from February 15 to March 10, 1998.

2. DATABASE.

When specified by name, the coordinates are looked up in a database containing over 550 locations worldwide. This database is accessible by the user for updating or adding more locations. The database can contain data for up to 750 locations and can be sorted in alphabetical order of the country name or the radio prefix. You can also print the data on paper.

3. LISTING SUNRISE/SUNSET TIMES.

This program also allows you to list (scroll on-screen) or make a full printout of the sunrise and sunset times (plus directions and distances from your QTH) for a given day of the year. This is a really nice feature for DXpeditioners and for contesting!

4. GRAY-LINE PROGRAM.

This program uses a unique algorithm that adapts the effective radio width of the gray-line zone to the location and the time of the year. This width is also different for 80 and 160 meters. In addition, the user can specify a minimum distance under which he is not interested in gray-line information. The printout (on screen or paper) lists the distance to the target QTH, the beginning and ending times of the gray-line window, as well as the effective width of the gray line at the target QTH. Fig 1-31 shows a printout of a gray-line run for Belgium on February 27. Notice the short Chatham Island opening predicted between 0614 and 0623 UTC. I made the QSO on 160 meters at 0623 UTC and copied ZL7DK until 0634 UTC!

The cost of the *LOW BAND SOFTWARE* is $50. It can be ordered from either of the following sources: John Devoldere, ON4UN, 215 Poelstraat, B9820 Merelbeke, Belgium, or George Oliva, K2UO, 5 Windsor Dr, Eatontown, NJ 07724 USA. See also the order form in the back of this publication.

The antenna software modules are the subject of Chapter 4 in this book.

■ 5. TYPICAL CONDITIONS ON 40, 80 AND 160 METERS

5.1. 40 Meters

- Forty meters is like an HF band that works at night time (it's almost like VHF for a Top-Bander!)
- Propagation prediction can be done by classic MUF-based programs.
- Gray-line propagation occurs *along* the terminator, like on the higher HF bands.
- Gray-line zones can be very wide (many hours even at medium latitudes).
- Skewed paths are not as common as on 80 and 160 meters.
- Allows you to work any distance, if properly equipped.

5.2. 80 Meters

- With well-equipped stations at both ends, just about any distance and path can be covered at the right time of the year.
- During low sunspot years, 80-meter propagation may be influenced by MUF.
- Gray-line enhancement always appears to happen on paths perpendicular to the terminator.

5.3. 160 Meters

- Propagation is not at all dictated by MUF and only marginally by the solar cycle.
- Besides the auroral phenomenon we still do *not* know exactly what makes a good DX night or a bad one. Mystery is still a big part of Top Band!
- Auroral absorption is most pronounced on 160 meters.
- Skewed paths occur most frequently on Top Band.
- Gray-line enhancement always appear to happen on paths perpendicular to the terminator.
- One-sixty meters has a distinct geographic area in which working DX is like a piece of cake—anything in a circle of approximately 5000 km around one's own QTH. For instance, from Europe, working the East Coast of the USA, can be done daily. The "light-gray" zone is W8 and W9 land. WØ land is "dark gray" and for anything beyond that conditions must be well above normal. This is quite different from 80 meters, where longer distances are possible every day and where the transition between "easy" and "difficult" seems to be much more vague.
- Long path on 160 meters is rare, except for stations very near the antipodes.
- If 80 meters is swinging, there is no

PRF	COUNTRY	CITY	DIST.	DIR.	START	END	WIDTH
--	ANTARCTICA	MCMURDO	16915	172	17.18	17.31	38
--	ANTARCTICA	GEN.BELGRADO	14537	190	6.31	6.34	37
3D2	ROTUMA		15660	10	6.25	6.31	5
3D2	FIJI		16278	9	6.24	6.30	5
5W	WESTERN SAMOA	APIA	15829	352	17.30	17.36	5
KH3	JOHNSTON ISL		12423	353	17.33	17.38	5
KH8	AMERICAN SAMOA		15868	350	17.23	17.29	5
T2	TUVALU	FUNAFUTI	15239	6	6.15	6.21	5
T31	CENTRAL KIRIBATI	CANTON ISL	14609	354	17.34	17.38	5
ZD9	GOUGH ISL		10235	190	6.27	6.34	9
ZK2	NIUE		16384	348	17.20	17.26	5
ZK3	TOKELAU		15294	353	17.32	17.38	5
ZL7	CHATHAM ISL		19180	0	6.14	6.23	10
ZL7	CHATHAM ISL		19180	0	17.31	17.38	10
ZL8	KERMADEC ISL		17618	4	6.14	6.20	6

END OF RUN   PRESS ANY KEY TO CONTINUE

Fig 1-31—The ON4UN LOW BAND SOFTWARE gray-line calculates all possible gray-line openings for a given day. This example is for March 6, 1998, the day ON4UN worked ZL7DK on 160 meters. Gray line was predicted between 0626 and 0636 UTC and the QSO was made at 0633 UTC!

guarantee that 160 meters will be any good, and vice versa. The same is true when you compare 40 meters with 80 meters—40 meters can be poor, yet 80 meters is swinging. So don't extrapolate from the lower or the higher band. This simply (almost) never works.

- Very typical for 160 meters is a slow and deep QSB, especially on the very marginal paths. It's advisable to get a call right the first time, for there may not be a second time, or it may be minutes later! I have seldom seen this happen on 80 meters.
- The 160-meter band usually has very pronounced peaks at sunrise (sometimes also at halfway midnight), especially for the really long-haul paths. For really long-haul distances, the sharp peak is usually within minutes of sunrise. You can almost set your watch by it. The sunset peak on 160 meters is much less pronounced. There seems to be a broad "peak" within an hour or so after sunset.
- On 160 meters the skip is often very selective (for various reasons).

■ 6. FUTURE WORK

6.1. Beacons

Chris Burger, ZS6EZ, recently was complaining on the Internet about the fact that from his QTH much 160-meter DX is hidden behind the (Australis) auroral donut: *"Based on our past experience unusual openings do occur, however, and probably more often than we are aware of because either no one is listening or transmitting from the other end of the path."*

One solution to assessing band conditions is a beacon system. Ten-meter enthusiasts figured this out years ago, and the 14.100-MHz worldwide beacon system has been useful to many DXers—remember how we all checked these beacons before VKØIR hit the airwaves?

Until the end of the 1980s, the 160-meter band was home to numerous commercial stations with whom we shared the band. Old-timers will remember how OSN, a military station in Belgium (on 1831.5), served as a perfect beacon, except for the Belgians, where it rendered a good portion of the DX window useless.

Maybe we need an initiative to start agreeing on and then reserving a single frequency for 160-meter beacons. These could transmit for a prescribed one-minute interval, similar to the current 14.100-MHz system.

6.2. Coordinated Amateur Radio Observation System (CAROS)

Cary Oler, president of Solar Terrestrial Dispatch is currently studying 160-meter propagation in greater depth, hoping to isolate more factors that could lead to improved models for propagation analysis and prediction. STD solicits the involvement of all individuals who communicate or regularly listen on 160 meters. Cary appreciates receiving as much input as possible regarding observed contacts and propagation conditions on Top Band. This includes reports throughout the Northern Hemisphere's summer as well as during the usual wintertime DXing months.

In support of this and other radio communicators on

higher frequencies, Cary Oler has developed CAROS, which can be accessed through the STD on the World Wide Web at: **http://solar.uleth.ca/solar/www/caros.html**. All reports submitted are archived. The reports are analyzed in detail and studied in combination with ionospheric data. Through a collection effort such as this, Cary hopes to be able to pry loose some of the secrets of 160-meter propagation. It is obvious that the success of this project is highly dependent on the number of reliable reports that are received. You can also submit your observations directly to: **http://solar.uleth.ca/solar/www/subcaros.html**.

DXpedition logs on 160 and 80 meters also hold valuable information on propagation for these bands. In recent years we have seen more and more highly specialized low-band DXpeditions who rigorously stayed on the low bands during every possible long-haul opening. It has become common practice to do a thorough analysis of these logs. The knowledge that we gather from these analyses will undoubtedly further our understanding of propagation on our mysterious low-frequency bands.

■ 7. THE 160-METER MYSTERY

Understanding and predicting propagation on 40 meters is pretty straightforward and 80 meters is well understood as well. With the right equipment and knowledge on both ends, you can probably work 300 countries in a year on 80 meters.

One-sixty is a totally different ball game. The more I have been active on 160 meters, the more I am convinced of how little we really know about propagation on that band. True, we know a few of the parameters that influence propagation, but far from all. For a long time I have kept daily records of the K and A indices, sunspot numbers, etc together with my observations of conditions on 160 meters, in order to try to find a correlation between the data and the propagation. But I have found very little or none; only negative correlations. We know more or less when it definitely will not work, but not for sure when it will work!

Of course, we must realize that on Top Band we are in a gray area where things are sometimes possible, but often not. There are dozens of parameters that make things happen or not happen. They all seem to influence a delicate mechanism that makes really long-haul propagation on 160 meters work every now and then. Understanding all of the parameters and being able to quantify them and feed them into a computer that will tell exactly when we can work that evasive DX station halfway around the globe will probably be an illusion forever.

There is no interest from the broadcasters in this subject. Broadcasters and utility traffic operators are interested in knowing the frequency that will give them best, most-reliable propagation. They are not interested in studying the subject of "marginal propagation," just on the edge of what is possible. Therefore, long-haul DXing on 160 meters will probably always remain a real hunting game, where limited understanding, feeling, expertise, and luck will be determining factors for success. Don't forget your hunting weapons—the antennas and the equipment.

DX-OPERATING ON THE LOW BANDS

2. DX-OPERATING ON THE LOW BANDS

THANKS TO PETER, ON6TT

My friend and neighbor Peter Casier, ON6TT, well-known DXpeditioner and Africa traveler, has accepted doing the proofreading of this chapter. Despite the fact that Peter has only been a ham for less than 10 years, he is certainly well respected by the DX and Contest community for his various achievements. It was Peter who went directly to see our minister of PTT back in 1992 and addressed the issue of high-power licenses for contesting, and pushed until he got what he wanted. From an avid contester Peter evolved into an adventurous DXpeditioner. His first trip was to Clipperton (1992), followed by Howland (1993), Peter I (1994), and Heard Island (1997). Since 1994 Peter has worked for various aid organizations in Africa, which turned out to provide the opportunity to work, travel and work DX from rare places. All his DX work, much of it on the low bands, made Peter, in my eyes, the ideal critic and proofreader for this chapter. And he did it thoroughly, as with all his undertakings. Thank you, Peter.

Trey, N6TR, considers 160-meter DXing a disease. But the symptoms he described apply to the other low bands as well:

• Desire to be on the radio at sunrise.
• Desire to be on the radio at sunset.
• Desire to be on the radio at all times in between sunset and sunrise.
• Desire to struggle for months to work a single station in a new country.
• Never being satisfied with the antenna system and constantly trying new ones.
• Only comes down to see the family after working a new country (to gloat). During one of those fantastic openings, will come down after each new country and hold up fingers indicating how many new countries were worked so far. These events are rare, and occur about once or twice in a century.
• Drinks lots of water before going to bed with the sole purpose of waking up in the wee hours of the morning to see if a new country can be found.
• Has problems getting to work on time during the winter months.
• Sends equipment and wire to people in unworked countries, hoping that the end result will be their QSL card on the wall.
• Spends thousand of dollars going to rare countries just so other people can work it.

And these are only some of the better-known symptoms. According to Rush Drake, W7RM, it's a painful disease, "To work DX on 160 you've got to love pain." Earl, K6SE, changed that to, "You've got to love torture."

Who am I to disagree with such eminent low band DXers?

One-sixty meters is usually referred to as Top Band, the band at the top of the (wavelength) spectrum, the band with top-notch operators, the band that's a top challenge and that gives you top excitement and satisfaction. Gary, NI6T, says "One sixty? Not a band, but an obsession."

All kidding aside, low-band DXing is a highly competitive technical hobby. It is certainly not a communications sport for the appliance operators. It is one area of Amateur Radio where it really helps to be knowledgeable. This is not a "plug and play" hobby!

■ 1. MYTHS

Gerry, VE6LB, who is a successful low-band DXer from an urban QTH, using simple antennas, summed up a few myths:

1. There is no (or little) DX on the low bands.
2. You need a big antenna and high power (it's only for the big guns) to work DX on the low bands.
3. DX is so scarce that you need to spend many hours (mostly late at night) to find DX on the low bands.
4. Any DX to be found on the low bands is on CW.
5. There is no low-band DX during the summer.
6. The low bands are too noisy to work DX.

■ 2. REALITY

Let's look at some facts:

1. All countries have been available on 40 meters and there are quite a few DXers that have all of them on 40 except for P5. At this time, all countries (with the exception of P5, BV9P and BS7H) have been available on 80 meters, and probably not more than a handful of countries have not—so far—been available on Top Band. Every year several Top Band DXers work DXCC in less than a year (as reported in the *Low Band Monitor*).
2. You will probably never win the CQ Worldwide

160-Meter Contest from a suburban lot with a 50-ft antenna-height restriction. But you can work DXCC on the low bands, even with 100 W from a typical suburban lot. I have friends who have never run power (more than 100 W) and have over 100 countries on Top Band! It is true—of course—that the better the means, the more you'll be sitting in the front row when the show is on.

3. Most of the DX on the low bands can be worked around sunset or sunrise. This is a better arrangement than on 10 meters, where the DX shows up in the middle of the day when most of us are at work.

4. Too bad not all the low-band DX is on CW. (That's a personal note. I love CW so much better than phone!) Seriously, there are countries that are only available on Phone and others only on CW. That's the name of the game. When it comes to Top Band though, CW is the name of the game! It's Top Band, and CW, that separate the players.

5. Ever thought that when it's summer here, it's winter on the other side of the equator?

6. Noise, whatever its origin, is one of the main challenges for the low-band DXer, but it certainly does not stop real hams from DXing. This is not a broadcast hobby, or a communicator's hobby. In this (low-band) hobby we are driven to move the boundaries of what is possible.

7. Well, all of this does *not* mean that working DX on the low bands is just a piece of cake, a nice pastime for the appliance type operator. But what makes so many love the low bands for chasing DX?

■ 3. WHAT MAKES PEOPLE CHASE DX ON THE LOW BANDS?

I included this question in my questionnaire that I sent out early 1998 via the Internet. The answer is the same for literally everyone who works the low bands: It's the challenge, the sense of fulfillment and of having done something difficult (see Section 16.5).

Low-band DXers always wander near the edge of what is possible. The most successful low-band DXers are the pioneers that keep moving this edge. Improved understanding of propagation, together with better equipment and most of all, better antennas, make it possible to dig deeper and deeper into the noise to catch the previously elusive layer of buried signals. The Top Band DXers are those balancing themselves on the cutting edge of the DXing sword.

If you are looking for an easy pastime, stay away from low-band DXing. Maybe one of the many lists that are abundant on the higher bands is something for you. K1ZM, who now has over 290 countries on 160 meters, wrote on his survey reply: "160 is truly a MAN's as well as a GENTLEMAN's band. You want a challenge? Get on 160."

On 80 and 40 meters you do not need to have a genuine "antenna farm" to work DXCC within one year. Even on 160, urban QTHs with small and low antennas regularly produce DXCCs on Top Band. I have included a (small) chapter on "Working 160-Meter DX from a Small Suburban Lot." There are many examples of rather modest stations on a small suburban lot that have done extremely

well. My friend George, K2UO, worked over 200 countries on 160 from a 1/2-acre suburban lot. To be so successful from an average QTH requires a better than average knowledge of propagation as well as a substantial dose of perseverance, however.

■ 4. THE FREQUENCIES

The frequencies used for DXing on the low bands are not the same in all countries. Therefore it is important that you know where to look for the DX.

4.1. 160 Meters

The frequency allocations on 160 meters vary widely all over the world, but almost all countries with the exception of Japan have a CW section in the 1820 to 1840-kHz window.

In some areas of the world, 160 meters is a shared band. This means there can be QRM from non-amateur users. In Europe just about all ship-to-shore and other commercial or military stations are systematically being moved outside the 1820 to 1850-kHz segment. But it is a general trend that a lot of the commercial stations that operated on these frequencies have moved to satellite operations, which has led the authorities to freeing more spectrum for Amateur Radio. For a while the situation has been too ludicrous in Europe. Germany would only allow 1832 to 1835 for SSB, Belgium would only allow 1830 to 1850, with 10 W, while in England they had 1820 to 2000, and that's less than 100 km away. Politicians and administrators may love borders, but radio waves ignore them!

In Europe it is still generally accepted that 1840 is the bottom end of the phone band. But it appears many are not aware that on LSB their sidebands spread 2.5 to 3.0 kHz down, and that they are taking out 40 percent of the primary DX CW window in Europe. In addition, we hear those SSB nets and rag-chewers often operating as low as 1838, carrier frequency. With this, I would beg and urge all SSB stations not to transmit on SSB below 1842.5 kHz.

Over the past 5 years, 1820 to 1840 kHz has been the de facto DX window for 160-meter CW, with 1840 to

Fig 2-1—Pierre, HB9AMO, was the first to collect all 40 zones on Top Band back in 1987.

Fig 2-2—Sunset over HB9AMO's quarter-wave vertical for 160 meters.

1850 kHz serving as phone DX band for all countries including the ex-USSR countries (1850 to 1930 kHz). From 1907.5 kHz through 1912.5 kHz is the Japanese 160-meter DX window. From 1830 to 1840 kHz is generally considered the European CW transmit segment, while 1820 to 1830 kHz is considered the DX window in Europe. (That's where the DX is, and European, as well as the US, stations should stay out of there.)

DXpeditions seem to use the 1823 to 1828-kHz window most of the time. More recently they have made the wise decision to work on the so-called half-frequencies (eg, 1823.5 kHz). This avoids the spurs and birdies on even kHz that are present in some receivers. In addition, avoid 1818, the W1AW (ARRL) broadcast frequency used for code practice and bulletins.

While it is true that frequency assignments are not the same all over, it seems that the minor differences are not a real problem. Over the years it seems that the different administrations are aligning themselves. With the exception of our JA friends, we all play on the same playgrounds— 1820 to 1850 being the boundaries of the DX field.

Several requests have been made by JA amateurs to their national authorities to obtain frequencies in the regular window, but so far nothing positive has been forthcoming.

In Russia, as well as in the CIS (former USSR) countries, the official band plan is 1830 to 1840 for CW, and 1840 to 1930 for SSB and CW. During contests they are also allowed to use SSB between 1830 and 1840.

4.2. 80 Meters

Although the 80-meter band is not allocated uniformly in all continents and countries, this does not really represent a problem for the DXer. On CW all countries have an allocation starting at 3500 kHz. The DX window for CW is the same all over the world: 3500 to 3510 kHz. A secondary window exists between 3525 and 3530 kHz, which is the lower limit for General and Advanced Class amateurs in the US.

The 80-meter SSB DX window is 3775 to 3800 kHz. The IARU has internationally recognized both the 3500 to 3510 window as well as the one at 3575 to 3800 as DX

windows. This means that it is recommended that all amateurs stay out of these windows for local contacts, during the times that the bands are open for intercontinental contacts. Although not enforced by law, this is a gentleman's agreement that we should all obey.

In the middle of the day, the DX segments can be used for local work, although one should be aware that local QSOs could cause great QRM to a DXer (at, say, 500 miles) who is already in the gray-line zone, and who could just enjoy peak propagation conditions at his QTH. In Europe situations like this occur almost daily in the winter, when northern Scandinavian stations can work the Pacific and the West Coast of the US from 1300 to 1400 UTC, while western Europe is in bright daylight and does not hear the DX at all. Western Europeans can hear the Scandinavians quite well, and consequently the Scandinavians can hear western Europe certainly well enough to get QRMed by other hams there. The DXer must be aware of these situations so as not to interfere with counterparts in other areas.

Band plan for Russia and CIS countries (former USSR)
- 3500 - 3580 CW
- 3580 - 3600 RTTY, Packet, CW
- 3600 - 3800 SSB, CW

(This is generally the same as in all Western European countries.)

Band plan for Australia
- 3500 - 3700 CW
- 3535 - 3625 Novice segment
- 3535 - 3620 SSB/AM
- 3620 - 3640 FSK
- 3640 - 3700 SSB/AM
- 3795 - 3800 DX window SSB/CW (lowest SSB carrier frequency 3798)

Band plan for Japan

The 80-meter band in Japan has three windows:
- 3500 - 3575 kHz
- 3747 - 3754 kHz
- 3791 - 3805 kHz

Band plan for USA

Since the FCC decided to expand SSB privileges in the US, first down to 3775 and later to 3750 kHz for Amateur Extra amateurs, the DX window has de facto expanded from below 3750 to 3800 kHz during openings to the US. However, the top 10 kHz is still the focal area.

Recommendations

Many amateurs are unaware that 80 meters is a shared band in many parts of the world. In the US, 80 meters sounds like a VHF band compared to what it sounds like in Europe. Because of the many commercial stations on the band, the 25-kHz DX window can often hold only five QSOs in between the extremely strong commercial stations in the local evening hours. If you are fortunate enough to live in a region where 80 meters is an exclusive band, please be aware of this, and bear with those who must continuously fight the commercial QRM.

US DXers generally complain about the narrow segment that has been officially reserved for DX on 80 meters. Most suggest the US should follow the European band plan (3775 to 3800 kHz). They also complain bitterly about the little cooperation from certain rag-chewers ("pig farmers" as they're sometimes called in the US), who have another 200 kHz, which could be used for their local contacts.

The increased popularity of 80-meter DXing, together with the few DX channels available in the phone DX window, have created a new problem where certain individuals would sit on a frequency in the DX window for hours (seems like days) on end, without giving anyone else a chance. This problem is nonexistent on CW, where you have an abundance of DX channels in the DX window.

4.3. 40 Meters

Forty meters is pretty straightforward. The CW DX QSOs happen between 7000 and 7010 kHz, and with rare exceptions around 7025 kHz.

With the European, or non-US, phone band being as narrow as it is, you can find DX anywhere between 7040 and 7100 kHz, with 7045 to 7080 kHz as a focal area.

■ 5. SPLIT-FREQUENCY OPERATION

The split-frequency technique is highly recommended for the rare DX station or DXpedition working the low bands. It is the most effective way of making as many QSOs as possible during the short low-band openings, because the marginal conditions often encountered are conducive to great chaos if stations are calling the DX on his frequency. It also gives a fair chance to the stations that have the best propagation to the DX station. With list operations this is not necessarily so, and stations having peak propagation can bite off their fingernails while the MC is passing along stations who barely make contact and have to fight to get a 33 report. With split operation the DXer with a good antenna and with good operating practice is bound to have a lead over the modest station, which is only fair. Why else would we build a station that performs better than the average?

There are two good reasons for the DX station to work split frequency:

1. First he must realize that when he stops his CQ, there are likely to be many stations calling him. Though he might pick out a good strong signal, others may still call him, and his reply to a particular station may be lost in the QRM. This will be evidenced by a slow QSO rate, even though the DX station hears the callers well. If he works split, the callers will have more chance to get the DXer's reply right the first time. The reason here is obviously that the callers cannot cope with the QRM they are creating themselves on the DX's frequency. In this case the DX station should simply specify a single frequency (eg, up 5) where he will be listening.
2. Another reason is that there are such large numbers of stations calling the DX station that the DX cannot

Fig 2-3

discriminate the callers. In this case it is the DX station who will not be able to handle the situation without going split. In this case he will specify a frequency range where he will be listening, in order to spread out the callers, and make the layer less thick.

A few general rules apply for split-frequency operation:

1. If possible, the DX station should operate in a part of the band where the stations from the area he is working cannot operate, or in a section of the band that is generally considered the DX section.
2. The DX station should indicate his listening frequency at least every minute. It only takes a second to do so, and it goes a long way toward keeping order.
3. The listening frequency should be well *outside* the DX window. Too often I hear a DX station on 3503 listening 5 up, ruining a major part of the DX window. There is no reason why he should not listen 10 or 20 up. The same applies to phone operation where the DX station transmitting in the window should listen outside the DX window for replies.
4. If the DX station is working by call areas, he should impose enough authority to reject those calling from areas other than those specified. He should not stay with a particular call area too long. At five stations from each area, at a rate of three QSOs a minute (that's fast!), it still takes almost 20 minutes to get through the 10 US call areas!
5. It is a good idea for the DX station to check his own part of the world to make sure the frequency is clear. This can be done periodically, especially if there is a sudden unexplained drop in QSO rate. Changing the transmit frequency a few kHz may bring relief.
6. If the DX station's listening frequency is being jammed, he should specify a frequency range instead of a single listening frequency.
7. On CW the split should be at least 5 kHz. Splits of less than 5 kHz should be avoided, because the pileup's clicks are likely to spread onto the DX station's frequency.

Depending on the band plan in the country of the DX station, split-frequency operation may be unavoidable. This

Fig 2-4—Seated is Jack, VE1ZZ, in his North American 160-meter lighthouse facing Europe. Yuri, K3BU (standing), has been operating various Top Band contests from Jack's unique QTH.

is the case when working the US from Europe on 40-meter phone. Under such circumstances, always make it a point to indicate your receiving frequency accurately, and make it a single frequency, or, if the pileup is too big, make it a reasonable range (10 kHz is usually sufficient). There is really no need to take more of the frequency spectrum than is absolutely necessary. One valid reason to use a wider spectrum than normally necessary is to elude deliberate jammers.

5.1. 160 Meters

Rare DX stations should as a rule operate split frequency. The generally accepted transmit window for the DX stations is 1820 to 1830, with 1823 to 1828 as the most popular range. It is good practice to use "half" frequencies, eg, 1823.5, 1824.5 (see also Section 4.1).

In Japan, Top Band operation is restricted to 1907.5 to 1912.5 kHz (only CW). The JA stations usually listen between 1820 and 1835 (see also Section 4.1).

If you have a frequency allocation around 1910 kHz, stay away from the JA windows around sunrise and sunset time in Japan. Never call JA stations on their frequency. Force them to go split frequency. They will be happy to listen around 1830 and it will greatly improve the QSO rate.

If you are a Japanese station, and as long as your transmit frequencies are restricted to 1907.5 to 1912.5, always work split frequency when the band is open into Europe. If not, you will attract many stations from Eastern Europe on your frequency, and they will cover up your signal at farther away DX stations. If you work split,

indicate your QSX frequency at every call. I have witnessed time after time JA stations working European stations (or trying to) without indicating their listening frequency. Send something like "QSX 25," or "QSX 32," repeatedly in every CQ and QRZ.

5.2. 80 Meters

On CW the main reason to go split is that the response to the DX station becomes too great (too big a pileup). Another nice reason would be for the DX station to listen "up 25" for the General class stations in the US.

On SSB I can think of many good reasons to go split: in the first place, not to occupy the DX window more than necessary. Therefore the DX station should always indicate a listening frequency outside the DX window (below 3750 kHz). US stations wanting to work Europe should transmit above 3800 and listen below 3750 in order to keep the DX window as uncongested as possible.

Middle East stations should transmit on a frequency below 3750 kHz when working North America in order to avoid the QRM for European stations. Stations in the Pacific working Europe should transmit above 3800 kHz and listen below 3800 kHz to avoid US QRM.

It is not reasonable for a European to transmit inside the US phone band (3780 kHz, for example) and listen on 3805 kHz. If this is done, two windows inside the US subband are occupied for one QSO, and the potential for QRM and confusion is increased. The inverse situation is equally undesirable.

5.3. 40 Meters

The nature of the frequency allocations in the different regions makes split-frequency operation a very common practice on 40-meter phone.

The US stations in the 7150 to 7300-kHz window should be aware of the fact that they operate in the middle of very strong broadcast stations in Europe. These broadcast stations are not on the same frequency 24 hours a day, and what may be a clear frequency one minute can be totally covered by a 60-over-S9 BC station the next minute. These BC stations usually appear on the hour or on the half-hour. Especially in contests, make sure your "clear" transmit frequency remains clear! During contests various DX stations may be using the same listening frequency when working split. Therefore it is essential that the caller not only give his own call, but also the call of the station being called. Going just by timing does not always guarantee a "real" QSO.

Not only the BC stations cause problems, but also non-Amateur Radio phone traffic between 7000 and 7100, much of which comes from Mexico and South America. These pirate stations usually transmit on channels that are organized in 5 kHz steps. Therefore it might be a good idea for the DX station working into the USA to try multiples of 2.5 kHz to avoid this kind of QRM.

■ 6. THE CLARIFIER

Zero beat is a term indicating that the two stations in contact are transmitting on exactly the same frequency. It is common practice to zero beat on phone. The RIT (Receiver Incremental Tuning) or RX Clarifier on present-day transceivers has created a problem where stations in QSO drift apart and use RIT to compensate instead of making sure that they stay on the same frequency. Fortunately, modern equipment is practically immune to frequency drift, so this is not as much of a problem as it used to be, especially with some of the home-built equipment.

As an example of where this is a problem, let us assume station A does not have a stable VFO. If station A and station B start a QSO at zero beat (station A and station B on exactly the same frequency), there are three sequences of events that a monitoring station might observe:

1. Neither station uses RIT: One station always follows the other. The QSO may wander all over, but at least there will be no sudden frequency jumps when passing the microphone, and the QSO will be on one drifting frequency.
2. One station uses RIT, the other does not: If we are still listening on the same frequency that the QSO began on, there will be a frequency jump at the start of transmission of one of the stations, but not for the other station. The QSO will still drift.
3. Both stations use RIT: Again, if we are listening on the original frequency, there will be a frequency jump at the start of transmission for both stations, and the two stations may drift away from one another. The QSO will take up more space on the band, and it will be very annoying to listen to. Should RIT be used in such a case? Decide for yourself.

Some of the recent transceivers not only have an RIT control but also an XIT control (TX clarifier). This one makes things even more complicated. Be careful when using it. There are some instances where RIT could be a welcome feature:

1. Some operators like to listen to SSB signals that sound very high-pitched, like Donald Duck. That means that they tune in too high on LSB. To the other operator, their transmission will sound too low-pitched, because they are no longer zero beat. Tuning in a station using RIT will allow one to listen to the voice pitch he prefers, while staying zero beat with the other station(s) on frequency.
2. When trying to beat a pileup, it can be advantageous to sound a little high in frequency. Adjusting the RIT slightly in such a case will yield that result.
3. Let us assume our transceiver is designed for working CW at an 800-Hz beat note, and it is only when listening at this note that the transmit frequency will be exactly the same as the receiving frequency. One can use RIT to offset the transmit frequency (by, say, 300 Hz) to bring the note down to 500 Hz, and still transmit on the receiving frequency. So you can see that RIT can be a useful feature without creating unnecessary QRM at the same time.

As previously stated, most DX QSOs on 80 meters are on only one frequency. There is no need to waste space on the band by working a station slightly off your frequency. It can, however, sometimes be necessary to work split-frequency, such as when someone causes deliberate QRM, when working rare DX stations or DXpeditions or when working into areas where a different band plan exists.

By the way, I never use the RIT on my transceiver. As far as I am concerned, the modern transceiver with dual VFOs simply has no need for RIT or XIT.

■ 7. ZERO BEAT

The terminology zero beating stems from the AM days. On AM one used to really zero-beat. When the transmitter VFO is tuned to the receiving frequency, a beat note is produced. This note is the mixing product of the two signals, and can be heard as an audio note. When the tone becomes 0 Hz, the transmitter and receiver are on the same frequency. Then we say the operator has zero beat the received signal.

On CW, most transceivers are designed so you are transmitting on the same frequency as the station you are working *only if the beat note is some specific frequency*. With the older variety of transceivers this was a fixed beat note, usually 800 Hz. This beat-note frequency is usually specified in the operating manual. Because many hams do not care for the specified 800-Hz beat note, they just listen to what pleases them (450 Hz is my preference). As a result of this, those operators are always off frequency by 350 Hz or so on CW. This is not a problem if the receiving station uses a 2-kHz filter, but it could be a real problem if he uses a 250 or 500-Hz filter. Also, think of all the wasted space on the band. This is the reason that in the past I advocated the use of a separate receiver and transmitter on CW. Then, at least one could really listen to one's own frequency!

The more modern transceivers now have the provision for operating right on frequency on CW. A good transceiver should at least have an adjustable beat note. The CW monitor note should shift accordingly. Continuously adjustable down to 200 Hz is the best. Some people like to listen at very low pitches. (W4ZV likes 250 Hz!) The only precaution here is to tune in the station you want to work at exactly the same beat note as your CW monitor note. That's all there is

Fig 2-5—Collection of 50 QSLs that made the first ever 160-meter WAS from Japan. The proud owner of these cards and the 160-meter WAS award is Yasuo, JA3ONB.

to it. This way, one can get easily within 50 Hz of the other station and still listen to the preferred beat note.

There is a lot of personal preference involved in choosing the beat note itself. It is generally accepted that it is very tiring to listen to a beat note over 700 Hz for extended periods of time. Also, the ability of the normal ear to discriminate signals very close in frequency is best at lower frequencies. For example, listen to a station with a beat note of 1000 Hz. Assume a second station of very similar signal strength and keying characteristics starts transmitting 50-Hz off frequency (at a 950-Hz or 1050-Hz beat note). Separating these two signals with IF or audio filters would be very difficult. Let us assume we have to rely on the "filters" in our ears to do the discrimination. The relative frequency difference is

$$\left(\frac{1050 - 1000}{1000}\right) \times 100 = 5\%$$

If you were using a 400-Hz beat note, the offender would have been at 450 Hz (or 350 Hz), which is a 13% relative frequency difference. This is much more easily discernible to the ear.

It's a good idea to do some checks with a local station (or on a second receiver) to make sure you are "zero beat" on CW. This will save you lots of frustration. In contests it has often happened to me that I had cleaned up my QRG and worked even the very weak signals only to find out that there was a guy with an S9 signal calling me 400 Hz up. He was strong, but I never heard him, while I easily worked stations that were 40 or 50 dB weaker than he…

■ 8. BEING THE RARE DX

You don't have to be on a DXpedition to be a rare one. There are still dozens of countries where the number of licensed radio amateurs can be counted on one hand. Operating as a resident or temporary resident from a much-wanted country is very similar to working from a DX-pedition. The required expertise to make low-band DXing a success is the same as required from top notch DXpeditioners.

8.1. DXpeditions and the Low Bands

Twenty-five years ago it was a rarity to have a DXpedition show up on 80 meters. One sixty was out of the question. That was just the gentlemen's band for daytime rag-chews. Eighty meters was thought by most to be only for local QSOs as well. The majority of DXpeditioners did not seem to know better.

Fortunately there has been a positive evolution over the years, and for most expeditions 80 or 160 meters has been just another band. And during the lower parts of the sunspot cycle they are definitely capable of bringing in a lot more DX than 21 or 28 MHz! The 5-Band DXCC, 5-Band WAS and 5-Band WAZ awards have greatly promoted low-band DXing. So have the single-band scores and record listings in the DX contests.

Until a few years ago some DXpeditioners would only appear on the low bands the last day or the last two days of

Fig 2-6—Five of Japan's best known 160-meter operators (left to right): JA4CQS, JA5AUC, JA4DND, JA3ONB and JA3EMU.

their operation (and even brag about their willingness to operate the low bands at all). Nowadays, we see quite a few DXpeditions aimed primarily at working the low bands, especially emphasizing CW, which is, without any doubt, the ideal mode for working DX on the low bands.

For a DXpedition, making it a policy to stay all the time on the band that guarantees the best QSO rates, will of course not result in many Top-Band QSOs, but it will guarantee a better cash return rate with the QSLs. If you care about the low bands at all, pure logic tells you to start tackling 160 and 80 meters from the first day, as there may not be openings every day. Waiting for the last day to find out that there is no propagation just proves that one does not understand low-band propagation at all, or that one simply does not care about low-band contacts. W0CD writes in his survey reply, "DXpeditions going to new countries should give more time to 160 to be sure there is decent propagation. Not just a few hours the last night."

It is very important that a DXpedition prepares itself well for the low bands. It is not a bad idea to ask an experienced low-band DXer to calculate band openings for the low bands. But fortunately, most of the well-organized DXpeditions now include at least one low-band expert. The DXpeditions should be able to discern the parts of the world to which propagation is peaking at a given time on a given low band. Nothing is more frustrating than to hear a Far East station working Europe on 80 or 160 during the 10-minute window that this path is open to the US East Coast.

Joerg, YB1AQS, from the famous ZL7DK team said it so well, "As we've found all the years—the 160-meter antenna has, if possible, to be the first one up and the last one down."

8.2. DXpedition Frequencies

On 80 and 40 meters, the typical DXpedition frequencies are within the bottom 10 kHz of the bands for CW, usually listening 5 to 10 kHz up, and sometimes around the 25-kHz mark. On phone the DXpedition playgrounds are usually in the 3795 to 3805 window for 80, while on 40 meters you can find them anywhere between 7040 and 7100. Please specify a listening frequency *outside* the DX windows!

On 160 CW, 1823 to 1826 kHz (with 1825 as a focal point) seems to be the most widely used DXpedition spot, with QSX 1830 to 1835 in areas where these frequencies are available.

It is important that a DXpedition announces its frequencies well beforehand. The Internet is the ideal tool for doing this. It's not a bad idea to publish an escape freq-uency in case of QRM. Stick to the published frequencies; otherwise all your credibility is gone. If you anticipate problems in this respect, publish 2 or 3 frequencies plus or minus.

When operating the low bands from a DXpedition, when leaving one band, always announce where you are going, and repeat the information a few times (not too fast on CW!).

8.3. Split Frequency

DXpeditions usually operate split frequency, both on CW and SSB. The late Gus Browning, W4BPD, was one of the first to include the low bands in his DXpeditions in the early 1960s. The advantage of split-frequency operation on the low bands is even more outstanding than on the higher bands, because the openings are much shorter and signals can be much weaker than on the higher bands. Working split makes it easier for calling stations to hear the DX. Otherwise, the strong pileup of callers will inevitably cover up the DX station, resulting in a very low QSO rate.

Sometimes we hear DXpeditions spreading the pileups over too wide a portion of the band, which is not generally advantageous for the QSO rate, and most of all very inconsiderate toward the other users of the band. It is also not unheard of for two DXpeditions to be on at the same time, both listening in the same part of the band. The net result of this is maximum confusion and frustration for everyone involved. There will inevitably be many "not in log" QSOs, where people ended up in the wrong log.

Calling by call areas seems to have become the standard approach to handle a pileup that's become too big to be handled without instructions. This is a fine procedure, provided one does not stay with the same call area for say more than 5 QSOs. Otherwise one might lose propagation to certain areas before going around the 10 call areas. Even at a 2-QSO-per-minute rate (which is high for the low bands), it takes almost half an hour to go through the 10 US call areas! When working the US on 160 meters it makes no sense doing this, as the propagation usually is very area-selective.

An alternative I have heard is calling by country. This inevitably leads to frustration. Why did he call for Holland and not for Belgium. Holland is only 20 km from here, why do they get a chance and not me? I think this alternative is to be avoided.

One sixty is a little special. Often on long haul paths, the copy is not 100% at all times, and skip is very often area selective and moving around. The secret to success here is to keep things simple. Don't try to send any complicated instructions. Half of your audience won't copy it anyhow! Simple instructions like, USA 5/10 UP or, EU 7 UP are okay. More complicated instructions will inevitably lead to chaos on 160.

On the other hand, if your pileup grows too big, you can eliminate those that copy you from those that "pretend" to copy you by suddenly changing your QSX and then quickly work the ones that really copy you!

If the pile is not too big, an excellent system is to use just one exact frequency on which the DXpedition will listen. This system is very friendly and works extremely well when the pileup is not too big. It at least keeps the callers from the DXer's frequency. With modern equipment having digital readouts to 10 Hz and an abundance of memories, this approach seems very reasonable.

Do not just send UP. This inevitably will attract people calling less than 1 kHz from the DXpedition's frequency. Instead, specify QSX 5, or UP 5/10. In Europe, 3500 to 3505 kHz has often been covered with a lot of QRM in recent years.

It is a good idea during the preparation phase of a DXpedition to ask DXers in several parts of the world what they consider as being the best transmit frequencies for the low bands. This way they can avoid showing up on a frequency where there is always a carrier or where every few minutes a commercial station pops up.

8.4. Controlling the Pileup

Sometimes, you hear a beautifully smooth pileup. A dream! A pleasure to listen to! Pure music! Sometimes, it's pure chaos. In the first instance it is the DXpeditioner who's responsible for this. Therefore a few hints on how to control a pileup:

- Avoid frustrating your public.
- Avoid sounding frustrated; inspire confidence.
- Show authority but not temper.
- Keep your instructions simple.
- Stick to your instructions yourself. Never make any "out of turn" QSOs.
- Change the QSX frequency if the pileup grows too big. Those that copy you well will immediately follow.
- Avoid copying half calls; this just slows down the QSO rate (especially on 160 where slow and deep fading is commonplace). Working with half-calls *only* works with JA stations, but certainly *not* with a European pileup.
- Always repeat the full call to tell the DX station that he's in the log, so he won't call you again.
- On 160 and 80 meters when paths are very marginal, send the call of the station you are replying to several times. After sending the report, send his call again and use a standard way of ending each QSO (TU, 73, etc). This is the sign for the crowd to start calling.
- Do not change your way of operating. Have a well thought-out strategy and pattern (rhythm), and stick to it. This will induce confidence with your public and help them avoid being frustrated.

Fig 2-7—View of the fantastic QTH of Domenico, I8UDB, top notch 80-meter SSB DXer. Not only is his house built on the top of a small island, it provides views of the Mediterranean in all directions. The tower is designed by Domenico, and stands 135 feet tall, unguyed. On top of the tower is a husky Rohn 40-like rotating mast. The antennas are a 3-el M² 80-meter Yagi, a 4-el KLM 40-meter Yagi and a 6-el 15-meter Yagi. The 160-meter antennas are a delta loop and a sloping ¹/₂-wave dipole.

- Ask from time to time if your frequency is still okay (especially if your rate suddenly drops).
- Do not hesitate to ask your audience to look for a new transmit frequency for you.

8.5. Calling CQ on a Seemingly Dead Band

We frequently hear or read that every respectable and wise DXer spends all his time listening (see Section 19), and only transmits when he's sure to make a contact. He never calls CQ DX, just listens all the time and grabs the DX before someone else does. This rule for sure applies to the DXer, to the hunter.

However, this rule does *not* apply to the hunted, to the DX! If we want the golden rule for the DXers of "listen, listen, listen" to work, then the golden rule for the DX should be "CALL, CALL, and KEEP CALLING!" And please, don't give up after just a few minutes. I have already explained in

Section 8.2 why DXpeditions should have the patience and persistence to call CQ after CQ on seemingly dead bands, at the times they published. This is the only way to catch the short, and otherwise elusive, openings. You can be assured that there are hundreds of faithful low-band DXers digging for the DXpedition's signal.

Recently I heard a DXpedition in the Pacific (with not so-experienced low-band operators) state that they "went to 160, and listened for 10 minutes, but heard nothing." No, indeed, it is hard to hear all the stations listening. It is the DXpeditioner who should call CQ. Another statement I read on the DX reflector, "We will look again but if the past couple of days are evidence of what we will find on 160, we will be on 80." Again, "looking" is not a way to do it. Calling CQ is the *only* way for the rare DX station.

Also, don't go away after just one or two contacts, even if there are no replies for a moment. You probably will be announced in the DX Cluster, but it takes some time before the news gets out on the Cluster! It also takes a few minutes before you get your buddies out of bed and on the band with their amplifiers warmed up.

The problem sometimes is, "who is the DXer and who is the DX?" For European stations, a W6 or a W7 is good DX on Top Band. So we expect them to call CQ. Same the other way around: Europe is good DX for the US West Coast, so they expect us Europeans to call CQ. Fortunately we don't have ambiguity when a DXpedition is involved.

The ZL7DK guys said it so well, " During our stay we got at least one good opening into all possible directions, but in average not more than two per destination. The openings into the critical directions (mostly the polar paths) have to have absolute priority. The paths are open maybe 5 or 10 minutes a day, if they are open. If you are dedicated to work stations on these difficult bands and difficult paths, you must be there every day (to call CQ) in order not to miss any opening."

8.6. Information Support for DXpeditions

After a rather shy attempt at the pilot control concept during the AH1A expedition, it was the famous VKØIR expedition in January 1997 that set the example on how perfect logistics can help a difficult DXpedition to be a huge success.

With DXpeditions it is important that there is an "off-line" link between the expedition and the customer, the audience. In the past, feedback and information had to be forwarded during prime operation time, which really was a shame. As a consequence the information flow was minimal.

Well-organized expeditions can use Pacsat (packet radio via satellite) or HF digital communications to establish a solid link between the rare spot and the home base. In the worst case, they can use Internet e-mail via a satellite telephone system. This always works.

The DXpedition pilot takes care of all the information flow to and from the DXpedition via one of these links. He organizes himself to have a maximum of information from the "public" and to feed a maximum of information from the rare spot back to the public. He is the DXpedition's spokesman, the public relations man. Nowadays, e-mail via Internet seems to have become the standard method of communications for such information.

The most important information flow concerns:

• What does the DXpedition hear during the low-band openings, what are the problems, what are the schedules (times and frequencies)?
• What, and when, is the public hearing the DXpedition, and what could be improved (suggestions)?
• Make the log available in (almost) real-time.

The first two items are there to optimize the results and to create confidence that all is being done to "make" it. The real-time logs are important to avoid stations from making a "back-up" contact (not 100% sure my "first" QSO was a good one). I once missed a country (Malpelo) on 160 by not making a backup QSO, so I really cannot blame anyone for doing so if not 100% sure about the first try. Having the logs available on the Internet avoids this situation. It also avoids utter frustration when you're not in the log while you thought you were!

Fig 2-8—The author, ON4UN, and Peter, DJ8WL, good for a combined total of nearly 550 countries on Top Band, at the Ham Radio Convention in Friedrichshafen, Germany.

■ 9. SENDING CW ON THESE NOISY LOW BANDS

There's no doubt about it. CW is far superior to Phone when it comes to marginal conditions. It also separates the Men from the others.

One of the situations that makes copying signals very difficult is QRN. It appears there are two families of QRN: high latitude QRN and tropical QRN. The difference is that crashes of tropical QRN generally last much longer than those generated by high-latitude QRN. Also with higher latitude QRN, the pauses between crashes as a rule last longer.

Thus, if you want your call to make it through high latitude QRN, high speed CW can sometimes be a solution. Dan, K8RN, who operated VK9LX on Top Band said, "QRN was very bad, even with Beverages for receive. It seemed to me that if the stations calling sent their call fast, they had a better chance of making it through (between) the static crashes. If the speed was too low nothing made it through." But high speed CW is no good at all to pierce through tropical QRN.

Rolf, SM5MX, XV7SW, recently commented on the Top Band reflector, "In this kind of tropical QRN, each QRN bang often lasts long enough to mask a call sign completely. From the DX end you may just understand that somebody is there and call QRZ?, but the same thing will happen again at the next bang, the next one, and the next, and so on, if the speed is too high. So I found it tremendously helpful when people reduced the speed. Once you are able to pick out a letter here and there, you may be able to paste together a full call sign and eventually make it."

Referring to another issue regarding high speed CW on the low bands, Tom, N4KG, commented, "High speed CW on the low bands by DX stations contributes to confusion and disorderly conduct in the pileups. Half of the callers can't copy anything but their own call signs, even with a good signal on a quiet band."

In fact, I think it is the DX station that determines the CW speed. His sending speed should be the speed he expects the stations to use for replying. Tom, N4KG, added, "DX stations sending above 30 WPM on the low bands actually reduce their rate and promote more broken calls. 25 to 28 WPM seems to work well for most cases. On long polar routes, with weak signals, QSB and QRN, high speed is counterproductive. Sending a call twice at 25 WPM takes less time than three times at 30 WPM and is more readily copied."

Joergen, YB1AQS, formulated it as follows: "Even if you can hear everybody crystal clear—don't shoot at them in CW with 35 WPM! 22 WPM on 160 meters and 28 WPM on 80 meters are enough. Repeat their call sign 2 times before the report and at least once at the end."

■ 10. NETS AND LIST OPERATIONS

List operations, which occur daily on the HF bands, stem from net operations such as the Pacific DX net, the P29JS/VK9NS net, and others. In these nets, a "master of ceremonies" (MC) will check in both the DX and the non-DX stations, usually by area. After completing the check-in procedure, the MC directs the non-DX stations (one at a time, in turn) to call and work the DX station. In most cases the non-DX station has indeed worked the DX station, but there was no competition, no challenge, no know-how involved. Some even call the MC on the telephone to get on his list. What satisfaction can one derive from such a QSO? Yes, it gives the QRP operator a better chance to work the DX station, and the only thing you have to do is copy your report. The MC will often QSP your call. And if the DX station is a DXpedition, there is a good chance that he will give everyone a 59 report, so it becomes even simpler. Just like shooting fish in a barrel, in my opinion.

In the late 1970s, list operations run by organized nets suddenly started showing up in great numbers all over the bands, run by "voluntary" MCs. This list system soon spread to 80 and 40-meter phone. Fortunately, lists have never made it on CW. This is yet another reason why real DXers love CW (Ref. 505). Many prominent low-band DXers dislike the list system, but most have learned to live with it.

The list system cannot be used if the DX station refuses to take part in it. Fortunately, we see more and more such operations, and they have all proved that DXpeditions can work much better without lists. I have witnessed stations asking Carl, WB4ZNH, operating as 3C1BG, if they could run a list for him. Carl was insulted by the proposition, so he asked the station to announce on his transmitting frequency (he was working split, of course) that he would not work anyone whose call sign was passed along by another station.

If the DX station is involved in a list operation, it generally means he cannot cope with the situation. The ability to cope with a pileup is part of "the game" for rare DX stations. There should definitely be no excuse for such things to happen to DXpeditions. If you are not a good enough operator to handle the situation yourself, you should not go on a DXpedition.

A number of years ago I worked a DXpedition in the Caribbean on 80, and the operator asked me to make a list of approximately 100 (!) European stations for him, which he would then work the next (!) morning. I flatly refused to do so with the comment that it was ridiculous to ask this, as he was S9 plus in Europe for hours every day, and he could probably work two stations per minute if he worked split frequency. He was offered the frequency so he could find another European station to "work" for him. An HB9 station took a long list for the next hour or more. The confusion the next morning was worse than anything I'd ever heard. Half of those on the list were not there, and more wanted to get on the list. The same operator on that DXpedition must have recalled my refusal to make the list for him, as years later he refused to work me on 160 meters from a later DXpedition. But this time I did not hear him working off lists, so he must have learned something after all.

Don't ask a DX station to make a list for him. It will simply offend him if he is a good operator.

If the inexperienced DX station unfortunately chooses to work from a list, here are the "11 Commandments" of list operation that the DX station and the MC should stick to:

1. The MC station should have absolute Q5 copy of the DX station.
2. The lists should be taken at the time of operation. Only short lists should be taken. Ideally they should contain no more than 10 stations.
3. The MC should try to be as objective as possible when picking calls in the pileup. He should not ask the pileup "only to give the suffix (last two letters) of their call." This is even illegal in most countries. (W3BGN calls it "the last 2 letter syndrome.")
4. The MC should make use of a second station far enough away to cover different areas. The second station could also take short lists on a different frequency.

5. The MC should never pick up stations that continually break out of turn, or keep calling when no list is being taken. Tail ending is a good way to get one's call in, provided it really is tail ending. It is extremely frustrating to hear so-called tail enders calling right when the DX station is giving the report and a second later give a Q5 report to the DX station.
6. The MC should never make mention of deliberate jamming on the frequency. If there is deliberate jamming and many comments, shouldn't he suspect that he is doing something wrong after all?
7. The MC should listen for other DX when taking a list (sometimes DX likes to work other DX). The MC should be aware of propagation conditions to the DX station as well as the gray line conditions at the DX station's QTH. When taking the lists, the MC should use selective calling, always giving priority to stations that are near their sunset or sunrise and are about to lose propagation to the DX station. In other words, the MC should be very knowledgeable about low-band DXing.
8. The MC should never relay a report.
9. In order to speed up the operation, a calling station should be given no more than two tries by the MC to get his report across. If the station can't make it with two tries, chances are that he would not even be answered by the DX if he was on his own. If it is clear that guesswork is going on, the MC should continue with the next station.
10. It is up to the MC, but I strongly suggest that the MC check the exchange of reports and make sure that the correct reports are confirmed at both ends to ensure that a valid QSO took place. He should make sure that the report is confirmed at both ends. If no exchange can be made, he should advise the DX station that no QSO took place, and if anyone has relayed a report, the MC might advise the DX station to change the report and try another exchange.
11. The list system should only be used as a last resort.

There are also a few rules for the "mere participant" in the list game:

1. The QSO should consist of a fully exchanged and confirmed report. The caller should confirm the report with the DX station so that both the MC and the DX station can make sure a valid QSO was made.
2. The caller should not repeat the DX station's call sign. The exchange length should be kept to a minimum. Unless you want to make a fool of yourself don't say "last heard you were 55." Was that yesterday you heard him last?
3. The caller should not get on the list if he cannot copy the DX station reasonably well. If the caller does not come back when the DX station turns it over to him, and if this situation repeats itself, he is just making a fool of himself.
4. If the caller cannot hear the MC, but hears the DX very well, he can try to ask the DX station to ask the MC to

put him on the list. This happened to me when trying to work a station in Africa when the MC was in Germany and skip prevented my getting on the list in the normal way.

In almost all cases, list operation can be avoided by working split frequency.

The fact that I give so much attention to list operations does not in any way mean that I agree with this modus operandi. I think it is always a poor solution. Because we are confronted with a very limited bandwidth on 80-meter SSB (maybe 5 or 6 channels between 3775 and 3800 kHz), and because there always will be a number of poor operators as well as newcomers, it is likely that we will have to accept lists every now and then. If you really hate lists as much as I have come to, stay away from 80-meter SSB, and try CW. I have never seen a list on CW. Better yet, try 160 meters for a real challenge!

I am convinced though, that probably more than 50% of the QSOs made on nets and in lists would never take place without the "help" of the MC. Such QSOs are immensely deflated in value, and one can ask what kind of satisfaction one can get from faking a contact! It happens all the time, though, on SSB. K4ZW complained bitterly in his survey reply: "Last year's episodes with stateside stations making phantom QSO with JT1BG using the help of a European MC was terrible. Even though these people clearly could not hear Bator, they would count 3×3 and then QSL the same report. Very disappointing to hear that on the low bands."

To close the subject of list working, let me quote this story by Don Newlands, VE3HGN, originally published in January 1985 *Radiosporting*.

My granddaddy used to fish for food, my dad fished for sport, and I don't fish at all. I buy fish at the fish store, along with most other folks. No fuss, no muss. Evidently we three had something in common: we all wanted fish. Granddad would stand in the icy water, casting and reeling until his limbs went numb; one man in harmony with nature. My memory of him is a bit misty. He didn't talk much, but seemed confident and content.

My dad, on the other hand, had an expensive yacht, filled with gear. He bragged about his catch and had a lot of big ones stuffed and varnished which he would point to with pride.

Now I haven't time for all that crab. I'm prepared to stand a few minutes in line at a fish market and I can either eat it or mount it, but for anybody to freeze his butt in water? Or lay out big bucks for a boat just to fish from? Today, he's hopelessly out of touch.

Now you may ask what this has to do with ham radio? Simple! My grandpappy was an original DXer, wire antennas, 20 watts and a spark that could set a house on fire. My dad was a DXer with stacked antennas, 12 Beverages, phased verticals, transmission lines as thick as a wrist, 2-meter

spotting nets, DX Cluster and all. And he kept the family up all night with his yelling and screaming.

From my perspective this all sounds primitive and disorganized. I prefer a simple lil' tribander at 40 ft and a low-band trap dipole with the apex at 35 ft. All I do is give my call (or just the suffix—I know it's not legal, but it's in) to the list-taker, and in a minute it's all over: "last heard (whenever that was) you were 55, rifle shot, bang bang." All I then do is listen to the MC's "Good contact." No fuss, no muss. Oh, my certificates (all framed, of course) are the same as theirs. Now that's what I call progress!

CQ DX, CQ DX is a call of the endangered species. Now it's "put me on the list." Now, how do you get on the list? Ya 'phone 'em, that's how! And while you're on the phone, ask 'em if they have any fish.

■ 11. ARRANGING SKEDS FOR THE LOW BANDS

Once you work your way up the DXCC ladder, you will inevitably come to a point where you will start asking stations on the "higher" bands (if you work the higher bands) for skeds on the low bands. Years ago if you asked someone on 15 or 20 meters for a sked on the low bands, the typical reply would have been either, "Sorry, I only work DX," or, "Yes, I work 40 meters." In the last 20 years, many high-band DXers have extended their horizons, and 80 meters has been added to their vocabulary and low-band definition. The answer has now become, "Yes, I am QRV on 40 and 80." But for most, 160 meters is still very much unknown and unexplored territory.

Once in a while, the high-band DXer will say, "Well, you're not the first one to ask; maybe I should do something about it. Can you tell me what antenna I should use." So, don't hesitate to ask; at least it will help some of the others realize that our HF spectrum extends all the way down to 160 meters.

If you are asked to help with antenna suggestions for 160 meters, don't give the candidate newcomer the idea that a wet string will do wonders. I always advocate an inverted L for transmitting, and if the inverted L is too noisy for receive, I tell him to try his 80-meter dipole for receiving. If he has an amplifier that works on 160, tell him to use it. Power is helpful on 160. Don't fool yourself, and don't fool the candidate newcomer.

If you get a positive answer, it will be your turn to indicate the best time. Take into consideration that you are the asking party, and try an opening that is not in the middle of his night. Rather, get yourself up in the middle of the night! Also, don't go by a single sked. Arrange a minimum of three skeds, or maybe a week's skeds, in order to hit the day with the right propagation. Tell the other party that the band may be okay only one day out of 3 or 5. Find out how much power he runs on 160 and what antenna he is using, so that you know what signal to expect.

Don't forget to have your sunrise and sunset information ready at all times. Most computer logging programs nowadays include it (eg, *DXbase*), and it is likely that the information is on line for you.

Tell your sked not to call you. Rather, you should call him (don't give his full call; just his suffix!), or just call CQ DX at the sked time exactly on the agreed frequency. Make the others (the listeners) a little nervous if they don't know about the sked, and do everything that's necessary not to give away "your" sked to strangers. If you work QSX, don't give away the listening frequency before you've worked him! If you think your chances are not the best (maybe because your setup is not as good as your friend's), call your friend who's got a better station and inform him about the sked. He will probably help you, and you still have a good chance of being number 2 to work the new one. By the way, once you have worked the new one, sign properly, give the full call of the station you've just worked, and clear the frequency (CL). If you have packet radio and are connected into a DX Cluster, announce the station you've just worked, so that others have all the info and may work him as well. Don't be so altruistic as to put the sked in the Cluster, unless you're a bit of a masochist. It is not advisable to announce any station you are trying to work before you work them. This will only make it more difficult for you—remember who comes first!

It's also a good idea to arrange an "escape" frequency: 40 meters is in most cases the most suitable band to go to after the sked, to exchange information about the test on 160.

■ 12. GETTING THE RARE ONES

Working the first 100 countries on 80 or 40 meters is fairly easy. Well equipped stations have done it in one contest weekend. Anyone with a good station should be able to do it easily within a year. A growing number of stations have achieved DXCC on 160 meters. The major DX contests (CQ World-Wide DX, ARRL International DX, WAE, All Asia, CQ World-Wide 160-Meter, ARRL 160-Meter, etc) are excellent opportunities to increase low-band scores. All DXpeditions worth that name now also include a fair amount of low-band operating in their operating scheme.

■ 13. GETTING THE LATEST INFORMATION

In the old days (almost sounds like pre-history) we had dozens of DX bulletins, all over the world, to inform us. This was the Geoff Watts era.

Then came packet radio, and along with it the DX Clusters. Information was much "fresher" then, and within a day or so a message sent from the US would arrive in Europe.

Local DX-information nets, mainly on 2-meter FM, which were thriving 10 years ago, have all but gone.

Today just about all dedicated DXers use the Internet, e-mail and the Web to get the latest information. The information transfer is lightning fast. During the VKØIR DXpedition there were up to three bulletins a day sent on the Internet (various reflectors on e-mail plus a dedicated Web page), so that news from the island was in all the

Fig 2-9—Bob Eshleman, W4DR, top all-band DXer, has 350 countries on 40, 338 on 80 and 283 on 160, an amazing total. On top of that Bob is active on all other bands, including the 30, 17 and 12-meter bands, as well.

homes within hours of it being released on Heard Island. DXers quickly got used to the Internet and its powerful possibilities. The Internet is certainly changing the way of DXing for many of us. For the better? It looks like the days of literally chasing the DX are gone. It's almost become "shooting" the DX. Especially on the higher bands, where it's now almost like shooting fish in a barrel.

DX Clusters consist of a network of packet radio links, where DX announcements are circulated to all stations that have checked into the Cluster. The same stations also provide the real-time DX information that is circulated to all other stations that have checked in with one of the nodes of the network. DX Clusters have been with us for quite a few years now. DX Clusters are thriving better than ever. There are now gateways via TCP/IP to the Internet, which allows DX Clusters to be linked from all over the world.

While packet radio DX Clusters are generally used to "spot" DX stations, the bulk of general information on DXpeditions, etc is nowadays generally distributed by e-mail on the Internet.

It is true that DX Clusters and the Internet have changed DXing in general. There have been publications where the DX Clusters have been pictured as the greatest evil in Amateur Radio. It is said to undermine the art of listening. Scott, W4PA, has a very strong view about this issue, "Shut off packet radio, and do it like a man." It is true that the DX Clusters are changing the face of DXing. The "little" station with the operator who spent all his time listening, and who often was number one catching the DX station showing up, will often lose the advantage, as the better-equipped stations will now be informed much faster of the DX showing up on the bands. The fact is that DX Clusters and the Internet are here to stay, and we will all have to develop different skills that will help us keep this technological advantage over our "competitor" DXers in this ever faster changing world.

I am personally convinced that the DX Clusters and the reflectors and Web pages on the Internet are just a set of superb tools that have evolved from our wonderful hobby.

DX Clusters bring together so many specialists from diverse fields in Amateur Radio: the digital guys take care of the hardware, the software freaks improve on operating systems, while the UHF/SHF guys establish and maintain reliable packet radio links over large areas. And of course there are the DXers who make the best use of the wonderful system and support the DX Cluster by their financial contributions. I cannot recall any other activity in Amateur Radio where people with such different interests work together and enjoy the results of their work as much as is the case with packet radio used for DX Clusters. DX Clusters are just another step forward in our wonderful hobby, where we make use of all the newest techniques and technologies to keep moving the boundaries of the achievable.

The Internet has made it possible to achieve a super-fast exchange of information among different groups of interest. Several interest groups are potentially of great interest to the low-band DXer, eg:
• The Low-Band Reflector
• The DX Reflector
• The Contest Reflector

Reflectors are semi-open mailboxes, to which anyone can subscribe free of charge. Once subscribed, you will get copies of all the mail that is being sent to this reflector. By addressing e-mail to the reflector, you reach everyone who is currently subscribed to that particular reflector. More on these in a separate chapter.

Nowadays you can make use of all these "utilities" in a multi-tasking environment under Windows. State of the art logging and award tracking software will allow you to connect to the local DX Cluster, and simultaneously to a worldwide Internet cluster via TCP/IP if you want. DX4WIN seems to be the most popular and most advanced logging software at this time (see **http://www.erols.com/pvander/**). Other software allows you to connect, via a local packet radio node, to up to 10 DX Clusters at the same time, and will automatically reconnect whatever the path is, if the link is broken. *QW* is such a program, written by Mario Fietz, DL4MFM (see **http://www.qsl.net/dl0qw** or **http://members.aol.com/dl0qw**). *QW*, which runs under Windows 3 and Windows 95, will filter out all duplicate announcements, and besides seeing the spots on the screen of your PC, and/or announced in CW or on voice via your sound blaster. You can also feed the output via a serial port to your favorite logging program, such as *DX4WIN*, or to your contesting program.

While many years ago, DX magazines served the noble purpose of informing the DXers of upcoming activities, this role is now, evidently, taken over by the DX Clusters and the Internet.

To my knowledge, the only monthly printed publication that specializes in low-band affairs, is *The Low Band Monitor*, published by Lance Johnson (see **http://www.**

Fig 2-10—The Titanex 21-meter-tall multiband vertical, which has proven to put out a walloping signal on the low bands from different DXpeditions in the Pacific. This picture was taken in high winds during the 1998 ZL7DK Chatham Island DXpedition.

qth.com/lowband). This little magazine has monthly activity reports, stories on recent low-band DXpeditions (and logs for 160), articles on low-band antennas, etc. I look forward to reading it every month!

■ 14. THE 8 COMMANDMENTS FOR WORKING THE RARE DXPEDITION

Joerg, YB1AQS / DL8WPX (from ZL7DK, VK9CR, VK9XY, S21XX and P29XX fame), formulated the following rules:
• **Rule #1:** Listen, listen, listen! It's much harder than to transmit, but the only chance. . .
• **Rule #2:** Don't give up before the DXpedition leaves. If you're serious, you can't miss any possible opening (and your opening may come only the last day).
• **Rule #3.** Long-haul propagation is always very area selective. Don't forget to monitor closely who has been worked, and in which direction the propagation is moving to determine your skip.
• **Rule #4:** For medium range distances there are not only the gray-line openings. Don't always wait for the gray line; getting up one hour earlier has often been the winner.
• **Rule #5:** In big pileups try to avoid calling zero beat

with anybody else. One hundred Hertz up or down can readily make the difference. On 160 meters I would go even further away. If you ever have tried to work a full-blown pileup covered by two layers of tropical noise, you'll know what I mean.

- **Rule #6:** Tail-ending means Tail-ENDING. It's definitely an art and not many DXers can do it right. Don't break in with your call as long as the previous one is not 100% clear. The timing of sending your call is very critical and you have to be synchronized with the behavior of the DX station. But that means clearly you have to hear the DX station well. If not, don't try it.
- **Rule #7:** In case of turmoil on the DX frequency, stay calm and monitor. A good DX operator will soon be aware of the situation and usually try to move just a bit.
- **Rule #8:** If you call, do it with moderate speed and take into account that the DX may have much more difficulties to copy you, especially if he's in the tropics. On Top Band, sending your call only one time is often not enough if the DX operator has to interpolate your call, but more than three times in a row is also not productive.

■ 15. ACHIEVEMENT AWARDS

There are a number of low-band-only DX awards. The IARU issues 160 and 80-meter WAC (Worked All Continents) awards. These are available through IARU societies including ARRL (225 Main Street, Newington, CT 06111, USA). ARRL also issues separate DXCC awards for 160, 80 and 40 meters. More information on these awards can be found at the ARRL Web page at: **http://www.arrl.org/ awards/.**

CQ magazine issues single-band WAZ awards (for any band). Applications for the WAZ award go to: K1MEM, Jim Dionne, 31 De Marco Rd, Sudbury, MA 01776, USA.

In addition, there are 5-band awards that are very challenging: 5-Band WAS (Worked All States), 5-Band DXCC (worked 100 countries on each of 5 bands), both issued by ARRL, and 5-Band WAZ (worked all 40 CQ zones on each of the 5 bands, 10 through 80 meters), issued by *CQ* (via K1MEM, see above).

The *Low Band Monitor* (**http://www.qth.com/ lowband**) sponsors awards for the low-band DXer that begin each season on September 1 and end March 31:

- 160-Meter WAC—The first *LBM* subscriber to complete a WAC receives a beautiful plaque to commemorate the achievement. Other subscribers who qualify receive individualized, numbered 160-Meter WAC Certificates.
- 80-Meter 100—The first *LBM* subscriber to work 100 DXCC Countries on 80 meters receives a beautiful plaque. Other subscribers who qualify receive individualized, numbered 80-Meter 100 Certificates.
- 40-Meter 150—The first *LBM* subscriber to work 150 DXCC Countries on 40 meters receives a beautiful plaque. Other subscribers who qualify receive individualized, numbered 40-Meter 150 Certificates.

The achievement awards issued by the sponsors of the major DX contests that have single-band categories are also highly valued by low-band DX enthusiasts.

The major contests of specific interest to the low-band DXers are:

- The CQ World-Wide DX Contest (phone, last weekend of October)
- The CQ World-Wide DX Contest (CW, last weekend of November)
- The CQ World-Wide 160-Meter Contest (CW, usually last weekend of January)
- The CQ World-Wide 160-Meter Contest (phone, usually the last weekend of February)
- The ARRL International DX Contest (CW, third weekend of February)
- The ARRL International DX Contest (phone, first weekend of March)
- The ARRL 160-Meter Contest (first weekend of December)
- The Stew Perry Topband Distance Challenge (last weekend of December)

Continental and world records are being broken regularly, depending on sunspots and improvements in antennas, operating techniques, etc.

Collecting awards is not necessarily an essential part of low-band DXing. Collecting the QSL cards for new countries is essential, however, at least if you want to claim them. Unfortunately there are too many bootleggers on the bands, and too many unconfirmed exchanges that optimists would like to count as QSOs. These factors have made written confirmation essential unless, of course, the operator never wishes to claim country or zone totals at all. Many other achievements can be the result of a goal one has set out to reach.

The ultimate low-band DXing achievement would be to work all countries on the low bands. This goal is quite achievable on 40, possible on 80, but quite impossible on 160 meters, although we see the 160-meter scores slowly climbing steadily to the 300 mark as well. The only question now is, "Will the first one do it before the turn of the century?"

■ 16. STANDINGS

Each year, the ARRL publishes the *DXCC Yearbook*, where those who have increased their totals in the program can check their ranking in the various DXCC awards listings. If you don't see someone's call sign there, it may be because there has been no change in status during the DXCC year (October through September).

Nick, VK2ICV / VKØLX, (see **http://www.qsl.net/ 160**) publishes "*Who's Who on the Top Band*" on his very nice Web page. It lists standings for World, Europe, USA, East Coast, Midwest and West Coast.

■ 17. THE SURVEY

During the early months of 1998 a new survey was made among active low banders. This time, the Internet was used to collect the data. Nick, VK2ICV, was also helpful by incorporating the questionnaire on his popular Web page. An e-mail was sent to all top ranking (80 and 160-meter) DXCC stations as well as some of the well-known contest stations that regularly take part in major contests and score well on the low bands, and for whom I could locate an e-mail address.

Well over 200 active low-band DXers sent in the filled-out questionnaire.

The purpose in collecting the information is to show the achievements of the leading low-band DXers, as well as the equipment and antennas they are using.

The information that was gathered from the questionnaires was used to calculate the percent usage of:
• mode per band
• antennas for 40
• antennas for 80
• antennas for 160
• special receiving antennas
• equipment

For each of these items I calculated the data for the entire population of the poll (232 entries) as well as for the leading 100 entrants. The entrants are ranked based on the number of countries worked on the three low bands together (40+80+160). This differentiation allows you to see how the most successful low-band DXers differ from the average one.

17.1. Achievements

The achievement figures are listed in **Table 1** while equipment and antennas are listed in **Table 2** (seen at the end of this chapter). The listing is alphabetically by call. Columns 2, 3 and 4 gives the year that the station started chasing DX on 40, 80 and 160 meters.

The DXCC status shown is the all-time status. The WAZ status for 160 and 80 meters is shown as well. At the rate new countries are being worked on Top Band it looks like we might have our first 300 country score before the turn of the century! And who else could it be but W4ZV?

The purpose of the listing is not to give an accurate DXCC status report, but to show what some of the leading low-band DXers have achieved and what they are using to do it. A few well-known DXers are missing in the tables. They have chosen not to reply to the questionnaire, which of course is their privilege.

Rankings of Top Band early award winners can be found in K1ZM's excellent book, *DXing on the Edge* (ref 511), as well as in the *CQ Amateur Radio Almanac*, which is published every year.

The age of the participants in this poll varies from 29 to 82 years, and the average age of the low-band DXers is 50 years. This is clearly to be explained by the fact that most DXers start, when they are young, on the higher bands, and start looking for more of a real challenge in the hobby when they get on a little in age and wisdom. . .

The average low-band DXer, based on our survey sample, has been DXing on 40 meters for 22 years and on 80 meters for 20 years, but has only been on 160 for 14 years. It is undeniable that Top Band has picked up in popularity especially in the last 10 years.

Based on the data from the survey the mode split for

the three low bands is as follows:

Band	Total Group CW	SSB	Top 100 CW	SSB
160	87%	13%	94%	6%
80	64%	36%	68%	32%
40	76%	24%	81%	19%

The reasons for this CW/SSB split are obvious: CW is by far the most efficient mode when it comes to dealing with weak signals under marginal conditions. On 40 meters the split is evidently due to the fact that the US phone band is shared with broadcast services in other parts of the world.

17.2. Antennas and Equipment

Table 2 gives an overview of the antennas used by the participants in the poll.

The summary for **40 meters** is:

Antenna Type	Total Group	Top 100
2 el Yagi	33%	35%
Dipole	16%	15%
3 el Yagi	16%	18%
Other (eg, Loops)	14%	15%
Vertical Antenna	12%	10%
Phased Verticals	9%	7%

The 2-element Cushcraft and Hy-Gain antennas seem to be extremely popular on 40 meters.

For **80 meters** we find:

Antenna Type	Total Group	Top 100
Vertical Ant	24%	20%
Dipole / Inv V	20%	23%
Antenna Type	Total Group	Top 100
Vertical Array	15%	14%
1/4-Wave Sloper	11%	11%
Delta Loop / Quad	9%	12%
Yagi / Quad	8%	7%
1/2-Wave Sloper	6%	8%
Shunt-Fed Tower	4%	2%
Other	3%	3%

Verticals are kept separated from shunt-fed towers, the former being "dedicated" constructions. It is normal that not too many DXers shunt feed their tower for 80 as it would in almost all cases be way too long electrically.

The Top Band results are:

Antenna Type	Total Group	Top 100
Vertical Ant	23%	20%
Inverted L /T	22%	22%
Shunt-Fed Tower	18%	19%
Dipole/ Inv V	14%	18%
Vertical Array	9%	9%
1/4-Wave Sloper	9%	10%
Loop	3%	3%
Other	2%	0%

Verticals in all shapes and forms (dedicated verticals, inverted Ls, Ts and shunt-fed towers) are good for $^2/_3$ of the antennas!

As to the usage of special receiving antennas the situation is:

Antenna Type	Total Group	Top 100
Beverages	46%	50%
No RX Ant	20%	20%
Magnetic Loops	16%	14%
EWE	7%	8%
Other	6%	5%
K9AY Loop	2%	1%
Low Dipole	2%	2%

It was remarkable that almost everyone not using Beverages commented, "No room for Beverages," as an excuse for not using these marvelous antennas!

Note that there is not a very significant difference as to what the total group uses versus what the top 100 use. Maybe the most successful DXers are better operators or more dedicated to DXing?

17.3. The Low-Band DXer's Equipment

As far as the equipment used by the low-band DXers from the survey we came to the following results:

Manufacturer	Total Group	Top 100
Yaesu	44%	52%
Kenwood	30%	21%
ICOM	17%	20%
Ten-Tec	8%	4%
Other	2%	3%

No doubt that at the time of writing (early 1998) the FT-1000D and the FT-1000MP are by far the most popular radios with low-band DXers. In the top 100 group, the lead by Yaesu is even more outspoken than in the total group. Six years ago, Kenwood was leading this ranking, while Yaesu was then only a modest third with only 12% of users. The TS-930 is by far the most popular Kenwood transceiver, but the TS-850 with a narrow filter in the first IF is highly appreciated by its users as well.

Everybody seems to recognize what makes a good radio: good strong (close-in) signal handling capabilities (being able to copy an S1 CW signal 0.5 kHz below an S9 plus local station), low VCO noise, and excellent selectivity with many different bandwidths.

When asked what are the characteristics of their dream radio, the one thing that really excelled was, "**something that effectively eliminates both man-made as well as atmospheric QRN.**" Further it seems that many are anxiously awaiting the marvels of IF DSP. Coming as a bit of a surprise is that many want more built-in receiving antenna switching possibilities, with high-gain preamps (apparently for low gain receiving antennas). I, for one, don't see why all of this should be inside our transceiver. Another message to the designers of our new radios is that

we want better protected front ends, especially when using the separate RX-antenna inputs. A number of low-band DXers want additional CW filters with a "wider" bandwidth (and I second that), eg between 1000 and 1500 Hz (ZS6EZ called it "to give us better situational awareness").

A more advanced wish list comes from Bill, KØHA, who wrote, "It should have the ability to make readable-signals at least 20 dB *below* the noise (à la government space communications receivers). It should have RF diversity reception with inputs for at least 20 different receive antennas. It would automatically select the antenna having the best signal-to-noise ratio. A receive antenna indicator light would show which antenna was chosen. Based on this choice it would automatically show (or switch to) the best transmit antenna. Of course it should have all of these things and then be included as the free toy inside a box of Cracker Jacks! Decreasing the degree of these attributes might be an answer for the desires for near-term hopes. . ."

17.4. Propagation Software Programs

In our questionnaire we asked if any propagation prediction software was used on the low bands. Only 8% said they used a propagation program, but most of them pointed out immediately that they used it for its mapping and showing the gray line. *Miniprop Plus* appears to be the most popular program. Unfortunately this program is no longer sold. More than 50% explicitly said that propagation prediction software is of no use at all on 160, of very limited use on 80, but useable on 40. K9FD and K4ZW explicitly mentioned they use ON4UN's low-band, gray-line, and SR/SS software. (Thank you!)

17.5. Why Does the Low-Band DXer Operate the Low Bands?

"Why do you operate the low bands?" was one of the questions on the questionnaire. In general the answer was: the challenge of what's difficult, what's unpredictable, what's not proven. The satisfaction from designing your station, your antennas, and then making the "almost impossible" contact.

Here are some of the "interesting" statements:
- DXing on the high bands is like shooting fish in a barrel. (AA4MM)
- Low-band DXing is the greatest challenge in Amateur Radio. (ABØX)
- I love a good "static salad." (K1UO)
- Anyone can do it if it's easy. (K4PI)
- I experience the same thrills as 40 years ago that hooked me on radio; high bands are too easy. (K4TEA)
- Top Band is the only band that still gives me a thrill. (K6ANP)
- 160 is an addictive band; 160 is not easy to be good at. (KO1W)
- Working a new one on 160 is not so cut-and-dried. (KX4R)
- On 160, CW shines. (VO1NA)
- Fewer lids than on high bands. (WØGJ)

- See how much pain one can endure before taking the headphones off. (W7TVF)
- Fun. (WB9Z) [W7TVF and WB9Z should get together!]
- The low bands are where you can test the station and the operator's skills. (UA3AB)
- Challenge of hearing, silence the utility poles; be ready all the time. (N7RT)
- Best demonstration of operating skill, station design and knowledge of propagation (like 6 meters). (W4DR)
- Pushing the operator and the station to the limits. (N4KG)
- 160: No nets, no lists, no deliberate QRM. Moving on the edge. Alone with QRN. (IV3PRK)
- 160: This is a new mountain to climb (the tallest one). (K6RK)
- 160 requires more technical skills and operating skills—the ultimate DXing challenge. (K9RJ)
- On the low bands, success comes through knowledge (antennas), not money. Few do it well. (K1VR)
- It's not that easy but I like difficulties (easy things are for everyone). (RA3AUU)
- DX nets on high bands make many contacts phony; playing field on low bands is more level. (ZS6EZ)
- Why do you climb mountains? Because they're there. 160 is the highest mountain with no worn path. (KØHA)
- Try to get the impossible, work all countries on all bands. (HB9AMO)
- I like difficult things, and if you can't hear them you can't work them. (ON7TK)
- 160 is like the BC band, I was a BC SWL as a child. (N5SV)
- 160-meter DX requires the best of everything: antennas, equipment, QTH, operator skills. (4X4NJ)
- I think I was dropped on my head when I was a baby. (K4SB)
- On 160 you can be competitive using your hands, not your checkbook. (NW6N)
- Doing the impossible from a city. Camaraderie on West Coast. (K6SSS)
- Satisfaction of achieving the seemingly impossible. (PA3DZN)
- The intellectual challenge of dealing with all the odd variables of propagation makes it a thrilling activity. (NØAX)
- I feel more at ease with my fellow low-band DXers than some of the "stuffed shirts" that hang out on 20 meters. (WØFS)
- Ties with early pioneers who did so much with so little. (K8MN)
- 160 is the absolute end in DXing, the last frontier. (K9UWA)
- To make the impossible possible. 160 DXCC from the worst place on Earth. (YB1AQS)
- K6SE got infected at an early age. "As an 8 year old in Detroit I would stay up late at night to DX on the AM broadcast band."
- Chance to do something everybody thinks is impossible. (G4DBN)

- 160 is more a gentleman's band. Lids are too lazy to fight QRN. (W9WI)
- No pain, no gain, and no nets on 160 yet. (GW3YDX)
- Creates great friendships. (W6KW ex W6NLZ, K2RBT)
- It helps to be an insomniac. (W8RU)

17.6. QSL Cards and Contests

About 96% of the low-band DXers in the poll said they collect QSL cards, but not all seem to appreciate that others also collect QSLs. A few really shocked me by stating that they do collect QSLs but never reply to QSL cards! Some state without shame that they do not reply to bureau cards, and only bother with cards they need. Could it be that the guy who sent you the card via the bureau needs your card for a new state, zone or country? Could it be that he cannot afford sending the cards via regular mail?

Over 90% said they do take part in contests, a minority to put down a score, and a majority to look for new ones or just good DX on the low bands.

■ 18. THE SUCCESSFUL LOW-BAND DXER

If we want to analyze what's required to become a successful low-band DXer, we must first agree on what is success. Success can be very relative. If you have only a $1/8$-acre city lot and you want to work the low bands, your goals will have to be different from the guy who's got 10 acres and a well-filled bank account. But you can be successful just as well, in your own way, relative to your own goals.

There are a few essential qualities that make good low-band DXers, I think. They apply even for the modest low-band DXer.

- Knowledge of antennas: For the low bands, it just does not work like opening a catalogue and ordering an antenna. You'll have to understand the antennas, the Whys and the Why Nots. You will have to become an antenna experimenter to be successful, even more so if you'll have to do it from a tiny city lot!
- Knowledge of propagation: Don't expect to turn on the radio at any time of the day on 80 or 160 and work across the globe. You must understand that you are trying to do something that is very difficult, something that requires a lot of expertise in order to become successful. You'll have to be able to predict openings, sometimes with an accuracy of minutes. You will have to understand the basic mechanisms behind it. The successful low-band DXer will build up his propagation expertise over a long period.
- Equipment and technologies: Receivers are getting better at every vintage. Maybe not as fast as we would like, but this evolution helps us unveil the previously buried weak signals from under the mud. The successful low-band DXer uses the best equipment that is available and uses it in a professional way. He gets involved with the latest technologies in radio communication, such as packet radio and DX Clusters that will provide him with real-time information about the activity on the different low bands.

- Good QTH: If we look at non-relative success on the low bands, we see that the success stories are all written from an excellent QTH. They are not all mountaintop QTHs, but each success story, in its own way, has been written from an "above average" QTH. This does not mean that a successful low-band DXer has to be a rich landowner. I, for one, have just over half an acre, but the location is excellent. The neighbors are nice and I can use the fields in the wintertime for my Beverage antennas.
- Perseverance, persistence, dedication: If you are not prepared to get up in the middle of the night, 5 days in a row, to try to work your umpteenth country on 80 or 160, top success will not be for you. If you think it's too hard to go out at night, in the fields or through the woods, in the dark, and roll out a special one-time Beverage for the new country you have a sked with in a few hours, then you better forget about becoming successful in the game, or rather the art, of low-band DXing.
- Operating proficiency: the "know-how-to-do-it" is probably the best weapon that can make a low-band DXer with a modest station perform outstandingly.

■ 19. THE 10 LOW-BAND COMMANDMENTS

Jeff Briggs, K1ZM, published in his excellent book *DXing on the Edge* (Ref. 511) a set of rules, from the hand of Bill, W4ZV, and which had been published earlier on the Internet Top Band Reflector. It goes without saying that these rules apply equally as well to the other low bands, and in particular to 80 meters.

A chapter on operating would not be complete without these rules, which I like to call the 10 Low-Band Commandments:

Rule #1: When the DX station answers someone else, listen; do not call. Instead try to find where *he* is listening. Most good operators spread the pileup over at least 1 to 2 kHz. If you listen for the station he is working, you will maximize your probability of being heard since you will know where he is listening. You may also recognize the pattern the operator uses. That is, is he slowly moving up in frequency, down in frequency or alternating picks to either side of the pileup? You will also know when to transmit (ie, when *he* is listening). It's very hard for him to hear you calling while he is transmitting!

Rule #2: Listen carefully! He may change his QSX frequency or QSY. If you're calling continuously, you will never know it. I can't tell you all the good stuff I've worked easily because I was one of the first on a new QSX frequency. If you're transmitting continuously, you'll be one

of the last to know. For those of you with QSK, you have an advantage here. If you don't, use a foot switch so that you can listen between calls and stop sending when he starts.

Rule #3: Do not transmit on the station answering. Why? Because a good operator will stay with that station until he finishes the QSO. Repeats necessitated by your QRM just reduces the amount of time *you* will have to work him before propagation goes out. The name of the game is for the DX to work as many stations as quickly as possible. Continuously calling only slows down the whole process and reduces *your* probability of a QSO. It might also encourage some DX operators to make a mental note in their head to never "hear" you again!

Rule #4: Learn your equipment so you know how *exactly* to place your transmit signal properly on frequency. No, this does not mean exactly zero beat on the last listening frequency where all the other hams are. It's far better to offset by a few hundred Hz based on which way you think the DX is tuning (see Rule #1). Also *please* learn to use your equipment so you don't transmit on the DX frequency inadvertently. This only slows things down for everyone and wastes precious opening time on 160 meters.

Rule #5: If you have limited resources on 160, focus on your receive antenna capability. You will work far more 160 DX with good ears than with a big mouth. Being an "alligator" who cannot hear anything is not productive on Top Band.

Rule #6: Send your full call. Partial calls only slow things down on Top Band (from Rolf, SM5MX, XV7SW)

Rule #7: Use proper and consistent spacing when sending your call on CW. There are some very well known DXers who don't understand this. They will break the cadence of their calls with pregnant pauses—this can confuse the DX station trying to decipher your call through 160-meter QSB and QRN.

Rule # 8: Send the DX station's call if you are in doubt whom you are working. You will not be happy if you log a DX station while you actually worked another station! This is especially important if more than one DX station is listening QSX in the same general area of the band (from 4S7RPG)

Rule #9: Listen to the DX station's reports and match his sending speed. If he is giving 459 at 18 WPM, don't reply at 35 WPM! If the DX station is missing part of your call, or if he has incorrectly copied part of your call, repeat only that part of the call several times, at a constant pace. (from 4S7RPG).

Rule #10: Listen, . . . listen, . . . listen!

Table 2-1
Low-Band DXer's Survey

CALL	AGE	ON 40 Since	ON 80 Since	ON 160 Since	DXCC 40 WKD	DXCC 40 CFD	DXCC 80 WKD	DXCC 80 CFD	DXCC 160 WKD	DXCC 160 CFD	5B DXCC	5B WAZ	Zones 80 CFD	Zones 160 CFD
4X4NJ	55			73					236	230	Y	N		40
4Z4DX	47	68	68	68	332		282		212		Y	Y		
AA4MM	71		77	81	120	120	175	150	216	207	Y	N		
ABØX	50	76	76	80		250		241	176	173	Y	N	38	32
ACØM	62	78	78	80	275	268	172	169	38	31	Y	Y	40	
AJ1H	52		69	97			208	199	91	28	N	N	36	20
CT1EEB	30	90	90	98	225	176	182	159	9	5	Y	N	37	4
CT4NH	48	83	84	97		273		231		54	Y	Y	40	
DJ2YA	59	55	55	81	310	309	293	293	210	205	Y	Y	40	38
DJ8QP	52	88	88	89	312	289	280	267	192	174	N	N	39	35
DK6WL	39	76	87	93	320	318	268	246	194	179	Y	Y		39
DL3DXX	42	81	83	87	325	322	276	270	204	189	N	N	39	38
DL6ET	29	90	90	94	295	247	257	230	165	165	N	N	38	38
EA6ACC	32	92	92	95	111	86	141	98	115	107	N	N	25	26
EA6NB	49	85	85	85	320	317	263	255	155	146	Y	Y	40	34
EA6SX	32	83	83	84	260	238	227	211	129	95	Y	Y	40	30
F5NZO	37	95	95	97	166	112	79	60	32	4	N	N		
F6BKI	47	80	80	83	290	265	289	283	219	217	Y	Y	40	39
G3KMA	59	55	72	79	339	338	294	294	234	229	Y	Y	40	39
G3SED	63		74	63	240		208		203					
G4BWP	40	75	75	75	329	326	296	292	225	211	N	Y	39	36
G4DBN	39	94	94	92	210		185		207		N	N		
GM3PPE	53	78	80	88			210	170	138	119	N	N		32
GW3YDX	50	85	85	85	285		261		274			Y		
HAØDU	41	74	74	88	315	314	303	297	243	235	Y	Y	40	38
HB9AMO	49	75	79	83	307	301	280	279	228	227	N	N	40	40
I4EAT	48	81	82	86	328	328	310	308	222	216	Y		40	38
I8UDB	45	80	85	81	315		303		140		N	N	40	30
IC8WIC	29	88	91	95	190	120	210	133	85	52	Y	N	32	10
IV3PRK	53			86					237	233	Y	Y	40	40
JAØDAI	46	81	81	87	308	293	269	254	137	112	Y	Y	40	30
JA1EOD	54	60	71	95	326	321	267	261	102		Y	Y	40	28
JA2AAQ	59	69	69		292	285	279	279			Y	Y		
JA2VPO	45	80	80	90	337	337	311	311	120	117	Y	Y	40	30
JA3FYC	50	76	80	87		328		275		171	Y	Y	40	35
JA3ONB	50			73					208	203				39
JA4LKB	42	75	79	81		276		239	184	181	Y	N	39	38
JA4MRL	41	84	84	86	317	315	277	272	160	156	Y	Y	40	30
JA7AO	63	80	85	68	277	254	167	149	169	269	Y			
JH2TPI	48	83	85	89	235	220	225	218	125	118	N	N		
JH3VNC	39	78	78	89	316	312	275	274	163	156	Y		39	31
JH4IFF	38	80	83	93	319	316	249	242	127	120	N	N	39	28
JR3IIR	39	82	78	92	333	333	310	308	90	45	Y	Y		
JR6PGB	38	83	83	87	307	307	294	291	157	155	Y	Y	40	36
KØCKD	53								117	95				32
KØCS	50	70	78	87		226		252		276	Y	Y	40	34
KØHA	45	69	69	78										
KØLW	50	62		63										
K1FK	57	54	54	90	226	187	225	280	152	110	N	N	33	29
K1GE	64	90	90	94	211	170	176	146	146	136	Y	N	30	30
K1KY	40	97	97	97	105	45	25				Y	N	35	
K1MY	55	79	79		182	182	176	176			Y	N	35	
K1PX	55									98				22
K1UO	49	85	85	97	264	251	238	230	82	55	Y	N	40	21
K1VR	52	56	56	85							Y	Y	38	30
K1VW	53	80	83	85	243	227	199	191	180	157	Y	N	37	37
K1YT	55	87	88	93	302	250	250	225	145	130	Y	N		
K1ZM	52	55	60	73					290	290	Y	Y	40	39
K2UO	52	75	75	85	299	299	218	218	200	198	Y	N		
K3GT	44	91	92	92	240	155	167	103	44	18	N	Y		
K3LR	38	75	75	75	234	167	197	167	154	134	N	N	38	30
K3ND	50	80	80	85	278	242	233	203	124	105	Y	N	31	25
K3OSX	51	61	61	96	280	273	196	179	94	83	Y	N	32	20
K3UL/ WA3EUL	45			82					261	259	Y	Y		37
K4CIA	58	55	66	75	329	329	304	304	172	166	Y	N	38	31
K4MQG	56	58	58		330		324	324			Y	Y	40	0
K4PI	54	63	72	83	331	331	295	295	225	223	Y	Y	38	34
K4SB	60	70	70	68	273	257	219	205	121	113	Y	N	36	24
K4TEA	54			85					236	236				36
K4TO	56	57	57	68	189	133	133	104	119	109	Y	N	22	26
K4UEE	53	72	72	76	279	268	230	218	215	212	Y	N	35	
K4VX	68	47	47	73										
K4XU	52	80	70	70					158	145	Y			
K4YP	55	95	95	95	77	35	123	66			N	N		
K4ZW	36		79	80			280	280	192	192	Y	N	38	34
K5AQ	65	50	65	80	305	302	249	245	155	147	Y	Y	40	30
K5PC	58	80	80	83	300	296	264	241	123	115	Y	Y	37	29
K5QY	55		79	96	100	100	100	100	60		Y	N		

CALL	AGE	ON 40 Since	ON 80 Since	ON 160 Since	DXCC 40 WKD	DXCC 40 CFD	DXCC 80 WKD	DXCC 80 CFD	DXCC 160 WKD	DXCC 160 CFD	5B DXCC	5B WAZ	Zones 80 CFD	Zones 160 CFD
K5YF	47	93	93		132	102	18				N	N	28	
K6ANP	62	62	62	80	197	156	142	135	138	138	Y	N		
K6RK	82	50	50	85	250		110		151	138	Y	Y		
K6SE	64	55	60	66	150	149	95	80	146	135	N	N		
K6SSS	60	88	78	95	280		299	298	132	117	N	N	40	31
K7FL	48	88	88		210	196	140	138			Y	Y	36	
K7OX	60													
K7SO	55	85	88		316	325	220	215			N	N	40	
K7ZV	53	92	92	96	255	210	256	215	92	61	Y	N	40	13
K7ZV	53	92	92	96	255	210	256	215	92	61	Y	N	40	13
K8GG	58	78	75	84		200		200	204	197	Y	N	36	36
K8JJC	54	94	95	95	279	270	217	205	104	94	Y	N	35	25
K8JP	54	67	68	69	268	173	198	129	129	129	Y	Y	25	18
K8MFO	51	58	71	83	330	329	278	276	261	261	Y	N	38	37
K8NA	55	70	70	96	310	307	254	247	107	90	Y	Y	40	24
K8RF	40	74	74	93	256	250	221	215	143	142	Y	Y	40	32
K9EL	50	80	70	85	315	310	260	250	140	120	Y	Y	40	30
K9FD	53	80	80	90	320	318	303	298	195	181	Y	Y	40	35
K9KU	50	85	85	94	234	209	194	177	115	99	Y	N	30	23
K9MA	49	89	89	92										
K9MN	49	73	73	73										
K9RJ	49			80					232	232	Y		38	35
K9UWA	53	71	71	75	250	250	250	250	260	254	Y	Y	40	40
K9YY	35	85	85	95	295	274	231	207	145	116	Y		37	32
KA7T	56	78	83	84					136	126	Y	N	36	30
KB3X	33	88	89	94	295	282	252	237	97	71	Y	N	34	17
KC2TX/7	45	95	94	96	20	10	121	58	25	10	N	N		
KC7EM	34	84	84		298	256	246	221			Y	Y	40	
KC7V	43	82	85	86	310	302	245	240	154	141	Y	Y	39	32
KE9I	37	78	78	78	180	160	155	125	139	93	N	N		
KG6I	50	64	79	81	243	228	194	177	144	133	Y	N	34	33
KG7D	47	65		73										
KJ9C	51	93	94	95	198	175	120	101	65	50	N	N		
KJ9I	34			87					220	217	Y	N		
KL7RA	52	65	65	75										
KL7Y	49	78	78	79	267	246	200	187	146	134	Y	N	39	33
KN6DV	51	93	93	96	198	176	190	170	98	70	Y	Y	39	30
KO1W	50	62		63										
KX4R	44	80	80	92	260	252	220	214	210	196	Y	N	38	35
KZ5Q	61	76	76		320	275	302	263			Y	N	40	
N0AT	47	80	85	89	297	264	250	201	159	137	Y	N	34	29
N0AX	43	92	94		195		94		50		N	N		
N0FW	36	78	80	85	305	302	274	252	185	123	Y	Y	39	32
N0IJ	56	90	91	92	220	180	148	135	97	91	Y	N	28	24
N0JK	42			91						85				
N0WX	40	91	93		115	105	92	75						
N1OF	65	92	92	92	150		100		75		N	N	33	28
N2QT	43	92	92	95	299	291	263	254	162	137	Y	N	37	32
N4CC	49				328	326	296	293	177	167	Y	Y	40	33
N4JJ	48	65	64	64	336	336	324	324	263	262	Y	Y	40	36
N4KG	54	60	62	82	325	325	310	309	237	234	Y	Y	40	36
N4RJ	55	56	73	81	321		292		231		Y	Y		
N4XX	59	74	74	96	192	183	130	129	121	111	Y	N	26	28
N5FG	54	78	75	91	291	285	200	189	50	46	Y	Y	40	
N5KO	34	77	79	79										
N5OK	57	58	58	70	280	273	192	190	71	67	Y	N		
N5SV	55	93	93	95										
N5ZC	35	92	92	97	82	23	56	16	9	2	N	N	11	3
N7RK	45	65	64	68	252	198	262	255	66	48	N	N	40	
N7RT	54	80	82	84	316	313	267	260	123	108	Y	Y	40	29
N7UA	50	59	70	77										
N8JV	44	83	83	95	317	316	287	285	101	95	Y	Y	40	28
N9AU	53	61	72	87	308	302	203	197	148	142	Y	Y	33	29
NA0Y	66	83	83	85					221	218	Y	Y	40	37
NH7A	50	63	63	81										
NL7Z	35	85	90	91			134	89	100	69	N	N		
NR0X	62	60	60	95	275	225	227	202	153	132	N	N	32	32
NT5C	63	89	89		280	273	237	233			Y	Y	39	
NW6N	44	85		90					167	134			38	27
OH2BU	40	74	76	80	326	326	310	310	205	205	Y	N	40	32
OH3ES	41	85	90	93			272		207				40	40
OH3SR	50	79	79	90	327	327	309	308	200	169	Y		40	16
ON4ADZ	39	84	84	96	282	227	218	167	101	59	Y	Y	40	40
ON4UN	57	87	61	87	310	302	345	345	278	276	Y	Y	36	16
ON7NQ	32	93	93	96	292	270	232	200	92	60	Y	N	38	29
ON7TK	42	83	83	87	268	222	231	193	164	119	Y	Y	40	36
OZ1ING	32	88	88	92	160	239	277	258	177	140	Y	N	40	40
OZ1LO	55	62	62	80	333	332	292	292	244	239	Y	N	40	40
OZ5MJ	67	49	49	80	269	257	227	215	135	122	Y	N	40	30
OZ7C/														
OZ1CTK	48	77	77	93	321	316	288	283	209	193	Y	N	40	40
OZ8ABE	37	91	92	93	308	304	283	280	213	205	N	N	40	39
OZ8RO	61										Y	Y		

CALL	AGE	ON 40 Since	ON 80 Since	ON 160 Since	DXCC 40 WKD	DXCC 40 CFD	DXCC 80 WKD	DXCC 80 CFD	DXCC 160 WKD	DXCC 160 CFD	5B DXCC	5B WAZ	Zones 80 CFD	Zones 160 CFD
PA0LOU	66	56	55	65	260	250	202	186	149	148	Y	N		
PA3DZN	30	90	95	97	308	247	199	102	120	43	Y	N	25	3
RA3AUU	29	90	90	90	291	250	249	200	162	150	Y	Y		
SM2EKM	46	65	65	85	250		280		150		Y	Y	40	
SM4CAN	50	65	68	86	300	298	290	275	266	262	Y	Y	40	40
SM4CHM	37	79	79	92	293	251	302	301	250	242	N	N	40	40
SM6CVX	53	61	75	85	335	335	303	303	240	239	Y	Y	40	40
SM6DYK	52	75	80	86	287	281	246	239	129	119	Y	Y	40	33
SP5EWY	43	73	73	88	332	331	297	296	240	236	Y	Y	40	40
UA3AB	38	88	88	88	298	268	273	248	145	139	Y	Y	40	34
VA3DX	47	76	76	84	324	306	288	255	203	197	N	N	40	36
VE3IRF	59	63			184	136								
VE3OSZ	66	52	52	52	211	202	157	152	130	127	Y	N		26
VE5RA	56						202	199	120	111	Y	N		
VE6LB	57	56	56	94	171	154	123	115	82	69	Y	N	23	24
VE7BS	77			47					130	120				31
VE7FPT	54	84	84	84		227		159		121				
VE7ON	51	85	90		165	125	143	121			Y	N		
VE7SBO	60	61	61	63		332		205		152	Y	N		35
VE7SV	52	60	60	60			290	288	105	95	N	N	40	
VK3QI	49	67	67	67	260	211	220	159	95	42	Y	Y	40	25
VK6HD		69	69	70	335	335	310	308	191	188	Y	Y	40	35
VK6VZ	42	89	93	93					115	105				29
VO1NA									190					33
W0CD	80	68	68	68		150		150	245	244	Y	N	36	37
W0EJ	52	76	76	81							Y	N		
W0FS	51	90	90	93	190	189	220	214	119	115	Y	N	34	28
W0GJ	48	96	95	97	236	197	229	202	126	104	Y	N	36	26
W0HW	75										N	N		
W0RI	64	50	54	86	227	193	143	113	145	140	Y	N	30	28
W0YG	61	82	82	85	278	265	243	231	213	207	Y	N	36	37
W1FV	46	89	83	84									49	34
W1JCC	46			72					200	200	N	N		34
W1JR	61	52	52	78	340	340	310	308	232	227	Y	Y		
W1TE	44	95	95	96	160	155	130	105	98	65	Y	N	26	18
W1UK	37		91	93			388	269					39	
W1ZK	53	89	80	84	226	216	252	247	175	172	Y	N	37	35
W2UB	42	73	73	80	160	122	108	93	65	61			28	15
W2UE	71	48	72	77	260	206	245	170	141	124	Y	Y	40	32
W3UM	51	84	84	84	325	323			223	219	Y	Y	40	33
W3UR	33	80	80	83	299	284	243	220	167	144	Y	N	39	29
W4AG	50	72	72		252	160	321	318			Y	N	40	
W4CTG	65	83	83	96			305	305			Y	Y	40	
W4DR	62	50	68	70	350	350	338	337	283	283	Y	Y	40	39
W4JM	42								114					
W4PA	29	82	82	82	273	261	224	180	86	79	Y	N		
W4ZV	53		80	84			325	325	293	293	N	Y	40	39
W5AA	81	69	69		220	200	185	170			Y	N	37	
W5FI	50	83	83	95	268	232	195	164	113	83	Y	N		
W5KFT	48	90	90	93	247	201	207	173	54	29	Y	Y	36	
W5PS	36	79	79	78	205	203	263	255	213	209	Y	N	37	35
W5WP	61	83	93	89	121	119	116	114	51	49	Y	N		
W6AJJ	75		90	94			151	142	154	143	Y			36
W6BWY	64	85	85	87	180	120	140	90	94	60	N	N		
W6KW	54	81	58		150	150	321	305			N	N	40	
W6PAA	60	53	54	82			250		50		N	N		
W6RJ	60				297		306				Y	Y	40	
W6SR	57	78	78	96	266	249	189	182	44	24	Y	Y	39	17
W6YA	57	81	83		329	329	259						40	
W7AWA	64			82					160	154	N	N		34
W7EW/W7AT	48	83	91	96										
W7LR	77			95					158	149	Y			32
W7TVF	78	52	52	78		205		183		154	Y			
W8AH	81					354		328		258	Y	Y		
W8RU	39	95	95	95	155	155	42	42	57	57	N	N	16	18
W8XD	50	61	63	92					174	168	Y	N	34	35
W9AGH	59	85	85	91	250	233	190	191	113	108	Y	N	32	25
W9QA	33	77	78	78		215		145		88	Y	N	36	30
W9WI	38	74	94	94	156	123	87	62	29	19	N	N	25	12
W9XT	43	87	91	96	188	112	140	94	70	20	N	N		
W9ZR	49	70	71	79	333	333	331	331	273	273	Y	Y	40	39
WA3AFS	51	63	63	63		56		98		34	N	N		
WA5VGI	58	53	53	96	166	157	53	46	29	25	N	N	22	18
WB3AVN	51	60	60	61		200		240	193	188	Y	N		32
WB9Z	43	85	85	85			287	285	265	264	Y	Y	40	39
WD4JRA	47								142	132	N	N		32
WX0B	47	62	62	89										
YB1AQS	38	83	83	92	135		68		44		N	N		
YC0LOW	45	95	95	97	122	110	100	91	57	42			27	27
YT6A	38		82				275	215						
ZS4TX	31	84	84	95	289	217	238	202	137	113	Y	Y	40	35
ZS6EZ	33	81	82	93	280	260	215	190	108	103	Y	Y	40	27

Table 2-2
Low-Band DXer's Survey

CALL	40 ANT	80 ANT	160 ANT	RX ANTENNAS
4X4NJ	3 EL KLM	GROUND PLANE	50 M SLOPER FROM 30 M TWR	BEVERAGES, LOW DIPOLE
AA4MM	SLOPER @ 70'	DELTA LOOP	3 EL VERT (PARASITIC)	BEVERAGES
ABØX	PHSD VERTS, SLOPER	SLOPER, SH FED TWR	SH FED TWR	EWE, COAXIAL ANT
ACØM	2 EL @ 60'	1/4 WV SLOPERS	1/4 WV SLOPER	LOOPS
AJ1H		73' VERT	73' VERT	BEVERAGE
CT1EEB	KLM MONOBAND YAGI	INV V	INV L	
CT4NH	402BA @ 18 M	1/2 WV SLOPING DIPOLE (21 TO 6 M)	INV V	
DJ2YA	25 M VERT	3/4 WV INV L	3/8 WV INV L	BEVERAGES
DJ8QP	2 EL CUSH CRAFT	SH FED TWR	SH FED TWR	BEVERAGES
DK6WL	2 PHSD VERTS	1/4 WV VERT	TP LOADED VERT	BEVERAGES
DL3DXX	20 M LONG VERT	20 M LONG VERT	20 M LONG VERT	BEVERAGES
DL6ET	2 EL YAGI	2 DELTA LOOPS, VERT	TP LOADED VERT (28 M)	BEVERAGES
EA6ACC	DIPOLE @ 14 M	DIPOLE @ 14 M	25 M VERT	EWE'S
EA6NB	402CD	2 HALF SLOPERS, DELTA LOOP	72' SH FED TWR	MAG LOOP
EA6SX	402BA	INV V, DELTA, VERT, SLOPERS	DIPOLE, VERT	
F5NZO	ROT DIPOLE	DIPOLE	SLOPER	NONE
F6BKI	2 EL CUSH CRAFT @ 22 M	27 M VERT, LOADED ROT DIPOLE	27 M VERT	BEVERAGES
G3KMA	80 M DELTA LOOP (402BA DOWN)	DELTA LOOP	INV L (17 M VERT)	
G3SED	ROT DIPOLE @ 30 M	25 M VERT	28 M SH FED TWR	BEVERAGES
G4BWP	402 CD	DELTA LOOP	SH FED TWR, INV V	
G4DBN	DELTA LOOP @ 20 M	PHSD BENT WIRE ARRAY	DIPOLE @ 17 M	K9AY LOOP, BEVERAGE
GM3PPE	DELTA LOOP	1/4 WV VERT	INV L	BEVERAGE
GW3YDX	2 EL @ 110'			
HAØDU	3 EL YAGI @ 22 M	3/8 WV VERT	VERT (29 M)	
I4EAT	3 EL YAGI & GP	4 SQ	28 M VERT	
I8UDB	4 EL KLM, 3 SLOPING DIPOLES	3 EL M² YAGI	SLOPING DIPOLE, DELTA LOOP	NO
IC8WIC	2 EL YAGI	DELTA LOOP, VERT	INV V	BEVERAGE
IV3PRK			24 M SH FED TWR	4 SQ MINI ARRAY
JAØDAI	ROT SHORT DIPOLE @ 20 M	SLOPERS	SH FED TWR	SMALL LOOP
JA1EOD	2 EL YAGI	DELTA LOOP	SH FED TWR	
JA2AAQ	ROT DIPOLE @ 25 M	ROT LOADED DIPOLE @ 25 M	SH FED TWR	NONE
JA2VPO	2 EL PHSD ARRAY ROT BEAM	2 EL PHSD ARRAY BEAM @ 40M	ROT DIPOLE @ 43 M	NONE
JA3ONB			INV V, 1/4 WV SLOPER, 1/4 WV VERT	LOOPS
JA4LKB	3 EL YAGI @ 28M	LOADED ROT DIPOLE @ 33 M	33 M SH FED TWR	BEVERAGES
JA4MRL	2 EL YAGI @ 23 M	ROT DIPOLE	INV L	
JA7AO	3 EL LOADED YAGI @ 30 M	GP	30 M SH FED TWR	NONE
JH2TPI	2 EL @ 25 M	DIPOLE @ 28M	INV L (24 M VERT) , INV V @ 23 M	
JH3VNC	4 EL YAGI @ 21 M	SHORT ROT DIPOLE, 1/4 WV SLOPER	SH FED TWR	LOW WIRE
JH4IFF	2 EL PHSD HORIZ ROT ARRAY	2 EL PHSD ROT ARRAY, 1/4 WV SLOPER	1/4 WV SLOPER	NONE
JR3IIR	2 EL YAGI, GROUND PLANE @ 15 M	ROT DIPOLE, GROUND PLANE	INV V, LOADED ELEMENTS	BEVERAGE, LOOP
JR6PGB	3 EL YAGI @ 22 M	2 EL YAGI @ 28 M	1/4 WV SLOPER	BEVERAGES
KØCKD	SHORT VERT (CTSVR)	SHORT VERT (CTSVR)	SHORT VERT (CTSVR)	60' CNTR FD SWA (BEVERAGE A LA MISEK)

CALL	40 ANT	80 ANT	160 ANT	RX ANTENNAS
K0CS	402CD @ 115'	DELTA LOOP @ 110', 1/4 WV VERT	1/4 WV VERT	BEVERAGES
K0HA	8 EL VERT ARRAY	7 EL PARASITIC ARRAY	4 EL PARASITIC ARRAY	8 BEVERAGES, HI ANGLE RX ANT
K0LW	402CD		1/4 WV FULL SIZE VERT	BEVERAGES
K1FK	4 EL VERT ARRAY	4 EL VERT ARRAY		
K1GE	1/4 WV VERT	1/4 WV VERT	INV L	BEVERAGES
K1KY	LOOP, WINDOM @ 40 M	WINDOM @ 40 M	WINDOM @ 40 M	2 BEVERAGES
K1MY	HY GAIN HY TWR	HY GAIN HY TWR		
K1PX			TP LOADED FOLDED MONOPOLE	3 BEVERAGES
K1UO	2 OVER 2 CUSHCRAFT (85 & 145')	INV V @ 125'	DIPOLE @ 120', GLADIATOR 8' UP, INV L (90')	BEVERAGE, LOW DIPOLE ON 160, PLANS EWE
K1VR	402CD & 4 SQ	4 SQ LOADED	3 EL PARASITIC, SH 42' VERTS	BEVERAGES, 160 MAG LOOP FED TWR AS DRIVEN EL
K1VW	2 EL @ 60'	INV L	2 EL 1/4 WV SPACED L, INV V	BEVERAGES, PHSD BEVER-AGES
K1YT	4 SQ 3 M HIGH	PHSD INV L'S	INV L (25 M VERT)	
K1ZM		4 SQ	4 SQ , 1/4 WV VERT, INV V	BEVERAGES, LOOPS
K2UO	INV V	DIPOLE	STEALTH DIPOLE	500' BEVERAGE
K3GT	204BA @ 76'	SH FED TWR	SH FED TWR	BEVERAGES
K3LR	4 OVER 4	3 EL YAGI & 4 SQ	3 EL PARASITIC ARRAY	BEVERAGE
K3ND	HOME MADE 2 EL	60' VERT WITH LOADED YAGI LOADING	60' VERT WITH LINEAR TP HAT	EWE, LOW DIPOLE LOADING & TP HAT
K3OSX	HF2V VERT	HF2V VERT	HF2V VERT	EWE
K3UL/ WA3EUL	ROT DIPOLE @ 30 M	SH FED TWR	SH FED TWR	SHORT BEVERAGES, EWE
K4CIA	HALF SQ	42' LOADED VERT	INV L	EWE, SHORT BEVERAGE
K4MQG	402CD @ 135'	2 EL YAGI (CREATE) @ 120'		
K4PI	2 EL MOSLEY YAGI @ 100'	4 SQ (K8UT STYLE)	100' SH FED TP LOADED TWR	BEVERAGES, 10' DIA LOOP,
K4SB	ROT SHORT DIPOLE @ 80'	1/4 WV VERT & DIPOLE	T ANTENNA, 75' HIGH	SMALL LOOPS, SHORT BEVERAGES
K4TEA			1/2 SQ VERT LOOP	BEVERAGES
K4TO	PHSD 402CD'S (70 & 140')	4 SQ	1/4 WV VERT	SMALL LOOP & BEVERAGE
K4UEE	2 EL YAGI	INV V @ 90'	INV L (80' VERT)	BEVERAGES
K4XU	4 SQ	4 SQ	PHSD INV L'S	BEVERAGES, EWE'S, LOOPS
K4YP	INV V @ 65'	INV V @ 65'		NONE
K4ZW		4 SQ	INV L @ 21 M	BEVERAGES
K5AQ	VERT & HIGH DIPOLE	LOADED TWR	LOADED TWR	BEVERAGES
K5PC	ROT DIPOLE @ 88'	HY GAIN HY TWR VERT	72' SH FED TWR	EWE
K5QY	HY GAIN HY TWR	HY GAIN HY TWR	PHSD VERTS	BEVERAGES
K5YF	BUTTERNUT 6V VERT, SLOPING DIPOLE	BUTTERNUT 6V		
K6ANP	3 EL YAGI	DELTA LOOP	DELTA LOOP	BEVERAGES & EWE'S
K6RK	2 EL @ 72'	TP LOADED VERT	TP LOADED VERT	LOW HORIZ WIRES
K6SE	402BA	2 PHSD SH FED TWRS	PHSD SH FED TWRS	BEVERAGES
K6SSS	VERT	4 SQ (ELEVATED)	INV L	2 M LOOP
K7FL	DIPOLE @ 28 M,	85' VERT		
K7OX	PHSD VERTS	SLOPER	SH FED TWR	LOOPS
K7SO	4 SQ	SHORT VERT (GAP VOYAGER)		
K7ZV	ROT LOADED DIPOLE	2 EL SHORT YAGI	SH FED TWR	NONE
K7ZV	ROT LOADED DIPOLE (HI Q COILS)	2 EL SHORT YAGI (HI Q COILS)	SH FED TWR	NONE
K8GG	ROT DIPOLE	INV V	SH FED TWR	BEVERAGES
K8JJC	2 EL HY GAIN @ 72', 1/4 WV VERT	SHORT VERT, DELTA	INV L	EWE
K8JP	402BA @ 75'	INV V @ 75'	INV L	SLINKY BEVERAGE

CALL	40 ANT	80 ANT	160 ANT	RX ANTENNAS
K8MFO	2 EL MOSLEY @ 117'	INV V @ 85'	INV V @ 105'	BEVERAGE
K8NA	2 EL MOSLEY @ 40 M	SLOPING DIPOLE ARRAY AROUND TWR	INV L, INV V @ 35 M	SHORT BEVERAGES
K8RF	DELTA LOOP	DELTA LOOP	DELTA LOOP	BEVERAGES
K9EL	2 EL HY GAIN DISCOVERER @ 80'	1/4 WV VERT	1/8 WV VERT	VARIOUS WIRES
K9FD	4 SLOPERS	2 QUARTER WV SLOPERS	SH FED TWR (N4KG FEED)	BEVERAGES, LOOPS, EWE
K9KU	PHSD 1/4 WV VERTS	1/4 WV VERT	SH FED TWR	LOOPS, HOUSE GUTTERS
K9MA	ROT DIPOLE (CITY LOT)	23 M SH FED TWR	23 M SH FED TWR	TWO LOOP ARRAY
K9MN	402CD @ 70'	INV V @ 70'	INV L	
K9RJ	2 EL HY GAIN @ 100'	2 SLOPERS TP @ 75'	100' SH FED TWR	CNTR FED BEV, LOOP,
K9UWA	402CD STACKED	ROT DIPOLE @ 145'	4 SQ	BEVERAGES, MINI 4 SQ , EWE
K9YY	402 CD	4 SQ	1/4 WV SLOPERS	BEVERAGES
KA7T	2 EL @ 80'	SH FED TWR	DISCONE	LOOPS
KB3X	ROT DIPOLE @ 30 M	DELTA LOOP @ 22M	INV V	NONE
KC2TX/7	1/4 WV SLOPER	1/4 WV SLOPER	1/4 WV SLOPER	COAXIAL LOOPS FOR 80 & 160 (CITY LOT)
KC7EM	STACKED 3 EL YAGIS (DX ENG) @ 160 & 80'	4 SQ , 64' ELEMENTS		NO
KC7V	4 EL KLM @ 100'	2 EL FORCE 12 @ 100'	1/4 WV SLOPER @ 90', SH FED TWR	BEVERAGES, SMALL RX LOOP
KE9I	60' VERT	60' VERT	60' VERT	BEVERAGES
KG6I	160 M FULL WV LOOP	160 M FULL WV LOOP	160 M FULL WV LOOP	BEVERAGES, SHIELDED LOOP
KG7D	1/4 WV VERTS		1/16 WV TP LOADED VERT + PARASITIC ELEMENTS	NO
KJ9C	DIPOLE	DIPOLE	INV L	SLINKY BEVERAGE
KJ9I	3 EL FORCE 12, @ 35 M	2 EL W9JA YAGI @ 52 M	4 SLOPING DIPOLES	REVERSIBLE BEVERAGES
KL7RA	3 EL TELREX @ 190'	INV V @ 150', DIPOLE @ 170', VERT	INV V @ 180', 1/4 WV VERT	NONE
KL7Y	3 EL YAGI	ARRAY OF 1/2 WV SLOPERS	125' SH FED TWR, INV V @ 150', 1/4 WV SLOPER	NONE
KN6DV	DELTA LOOP	1/4 WV SLOPER		K9AY LOOPS & MAG LOOPS
KO1W	402CD		1/4 WV VERT	BEVERAGES
KX4R	ELEVATED GP, WIRE ANTENNAS	INV L	23 M SH FED TP LOADED TWR	BEVERAGE, SEVERAL LOOP ANTENNAS
KZ5Q	ROT DIPOLE	1/4 WV SLOPER	TWO 1/8 WV VERTS, 1/8 WV SPACING	BEVERAGES
N0AT	2 EL FORCE 12 YAGI	2 PHSD 1/4 WV VERTS		
N0AX	402CD @ 50' SLOPERS	TWO 1/4 WV	INV L 15 M VERT SHIELDED LOOP VERT	W0CM COAX ANT,
N0FW	3 EL HY GAIN	2 EL VERT ARRAY		
N0IJ	2 EL MOSLEY @ 125'	FORCE 12 ROT DIPOLE @ 116'	DXA SLOPER	NONE
N0JK	130' BALLOON VERT	130' BALLOON VERT	130' BALLOON VERT	W0CM COAXIAL RCV ANT
N0WX	VERT R7	1/4 WV SLOPER	1/4 WV SLOPER	NONE
N1OF	SH FED BOOM OF 204 BA	1/4 WV SLOPER	SH FED TWR	COAX LOOP
N2QT	402CD @ 48'	DELTA LOOPS WITH REFLECTOR	INV L TP 55'	EWE
N4CC	402CD	HALF WV SLOPER	1/4 WV SLOPER	BEVERAGES
N4JJ	2 EL CUSHCRAFT, DIPOLES	3 PHSD VERTS, INV V @ 105'	118' TP LOADED SH FED TWR, INV V @ 105'	BEVERAGES, 4 EWE'S
N4KG	2 EL CUSHCRAFT, DIPOLES	DIPOLE @ 130', INV V @ 140'	VERT (N4KG INV FEED)	BEVERAGES

CALL	40 ANT	80 ANT	160 ANT	RX ANTENNAS
N4RJ	3 EL YAGI @ 100'	DIPOLE @ 170'	SH FED TWR, 1/4 WV VERT	NONE
N4XX	HF9V VERT, TRAP	HF9V VERT, TRAP DIPOLE	INV L (VERT 25 M), TRAP DIPOLE @ 15 M	INDOOR LOOP, PHSD INV L & DIPOLE (WITH MFJ 1026)
N5FG	402CD	INV V @ 65'	INV V @ 70'	
N5KO	INV VEE IN TREE	1/4 WV GP IN TREE	1/4 WV GP IN TREE	
N5OK	2 EL LOADED YAGI @ 110'	PHSD DELTA LOOPS		
N5SV	MOSLEY YAGI, 540' DELTA LOOP	2 EL FORCE 12 YAGI, 540' LOOP	FORCE 12 ROT LOADED DIPOLE, 540' LOOP	BEVERAGE, LOOP, LOOP ARRAY
N5ZC	2 EL YAGI @ 75'	SLOPER	INV L	
N7RK	1/2 WV ELEVATED (STEALTH TYPE)	ELEVATED SLOPING VERT (BASE ON BALCONY!)	ELEVATED SLOPING VERT (BASE ON BALCONY)	SLOPER
N7RT	4 EL HY GAIN	2 EL HOME MADE YAGI	143' VERT	BEVERAGES, LOOP
N7UA	4 EL KLM	6 EL WIRE BEAM	4 SQ , INV V	BEVERAGES
N8JV	2 EL HY GAIN (DISCOVERER) @ 88'	4 SQ AROUND 160 M TWR (WIRES)		1/4 WV VERT FULL SIZE, INSULATED BASE
N9AU	2 EL FORCE 12 @ 72'	1/4 WV SLOPER	1/4 WV SLOPER	EWE, LOOPS
NAØY	INV V @ 75'	1/4 WV SLOPER	1/2 SLOPER @ 75'	SEVERAL SHORT BEVERAGES
NH7A	4 EL YAGI, 2 EL YAGI	4 SQ	SH FED TWR	NONE
NL7Z	5/8 WV VERT, INV V	72' VERT, 1/4 @ 87'	3/8 WV INV L, SLOPER @ 91'	DIPOLE @ 10' SLOPER @ 90'
NRØX	3 EL FULL SIZE	2 DIPOLES (ON4UN DESIGN)	INV V, APEX @ 36 M @ 24 M	BEVERAGES, PHSD BEVERAGES
NT5C	402CD @ 27 M	SLOPING DIPOLE (20 M APEX), HORIZ DIPOLE		
NW6N	3 EL VERT ARRAY	5/8 WV VERT	1/4 WV VERT	BEVERAGES, LOOPS, EWE
OH2BU	3 EL FULL SIZE	SLOPING DIPOLE	SLOPING DIPOLES	BEVERAGES
OH3ES	2 EL HOMEMADE YAGI	1/4 WV VERT	2 EL ARRAY (1/4 WV VERTS)	
OH3SR	2 EL FULL SIZE YAGI @ 31 M	4 HALF SLOPERS TP @ 30 M	HALF SLOPER TP @ 30 M, INV V @ 29 M	80 M SLOPERS
ON4ADZ	INV V @ 24 M	INV V @ 24 M	SH FED TWR	NO
ON4UN	3 EL FULL SIZE YAGI @ 100'	4 SQ	1/4 WV FULL SIZE VERT	BEVERAGES
ON7NQ	DELTA LOOP	1/4 WV SLOPERS	INV L	
ON7TK	1/4 WV VERT	1/4 WV VERT	INV L	BEVERAGES
OZ1ING	BOBTAIL CURTAIN, INV L	BOBTAIL CURTAIN	INV L	BEVERAGES
OZ1LO	2 EL HY GAIN @ 27 M	VERT	1/4 WV VERTICAL	BEVERAGE, EWE
OZ5MJ	INV L	INV V	INV V	
OZ7C/ OZ1CTK	MULTIBAND YAGI @ 26 M (PRO 96)	1/4 WV SLOPER	3/8 WV INV L	BEVERAGE, EWE
OZ8ABE	DIPOLE @ 12 M	1/4 WV VERT, LOW 1/2 WV DIPOLE	INV V @ 12 M	
OZ8RO				
PAØLOU	DELTA LOOP, SLOPER	DELTA LOOP, SLOPER	HALF SLOPER	
PA3DZN	DELTA LOOP	HALF SLOPER	INV L	K9AY LOOPS
RA3AUU	INV V	INV V	INV V	
SM2EKM	4 EL YAGI & 4 SQ	4 SQ	VERT	BEVERAGES, LOW DIPOLES
SM4CAN	BOBTAIL	HALF SQ	SH FED VERT (26 M, TP LOADED)	BEVERAGES
SM4CHM	3 EL KLM @ 20 M	2 EL YAGI @ 40 M 1/4 WV SLOPERS (MODIFIED CREATE)		BEVERAGES
SM6CVX	1/2 WV SLOPER	1/2 WV SLOPERS	VERT	NONE
SM6DYK	2 EL KLM @ 24 M	DISCONE	VERT DISCONE ANT	BEVERAGES
SP5EWY	2 EL DELTA LOOP	1/4 WV SLOPER	1/4 WV SLOPER	BEVERAGES
UA3AB	402CD	TWO INV V'S 15 M ABOVE ROOF	INV V 15 M ABOVE ROOF	BEVERAGES (COUNTRY QTH)
VA3DX	2 EL YAGI	INV V @ 70'	INV L	
VE3IRF	VERT HF6V			SHIELDED LOOP FOR 40

CALL	40 ANT	80 ANT	160 ANT	RX ANTENNAS
VE3OSZ	INV L	INV L IN TREES		
VE5RA	2 EL HY GAIN DISCOVER BEAM	4 SQ	2 1/4 WV SLOPERS	
VE6LB	HF2V VERT (MODIFIED)	HF2V VERT (MODIFIED)	HF2V VERT (MODIFIED)	SHIELDED LOOP
VE7BS			FULL WV HORIZ LOOP @ 30 M, HALF DIAMOND APEX 33 M	BEVERAGES
VE7FPT	DELTA LOOP	HF2V VERT	SH FED TWR	BEVERAGES
VE7ON	402CD @ 90'	1/4 WV SLOPER	1/4 WV SLOPER	
VE7SBO	STACKED 2 & 3 EL FULL SIZE YAGIS	PHSD V'S	3 EL HORIZ ARRAY @ 135'	WIRE
VE7SV	402CD @ 135', 2 EL DELTA LOOP	4 SQ (K8UR STYLE), DIPOLE	1/4 WV SLOPER	
VK3QI	GAMMA MATCHED BOOM OF 402BA	1/4 WV SLOPERS	SH FED TWR	K6STI LOOP, ETC
VK6HD	2 EL	SLOPING DIPOLES	INERTED V @ 30 M	
VK6VZ	2 EL CAPACITIVELY LOADED QUAD (A LA G3FPQ)	INV L	INV V DIPOLE, 15 20 M HIGH	BEVERAGE, 40 M QUAD
VO1NA			25 M BASE INSULATED VERT	BEVERAGES
W0CD	ROT DIPOLE @ 23 M	3 EL VERT ARRAY, K8UR STYLE	3 EL ARRAY (75' ELEMENTS), SH FED TWR	2 SETS OF PHSD LOOPS
W0EJ	INV V @ 100'	INV V @ 100'	1/4 WV SLOPER @ 100'	BEVERAGES, LOW LOOP
W0FS	DROOPING DIPOLE	2 EL DROOPING DIPOLE	190' INV L, 50' VERT	BEVERAGE
W0GJ	3 EL @ 145', 4 SQ , DIPOLE @ 90'	4 SQ , ROT DIPOLE @ 180'	1/4 WV VERT	BEVERAGES
W0HW	2 EL FORCE 12	SH FED TWR	SH FED TWR	LOOP, LOW WIRE
W0RI	INV V @ 70'	1/4 WV SLOPER @ 70'	80' SH FED TWR	LOOP
W0YG	3 EL YAGI	3 EL YAGI	3 EL PHSD ARRAY	BEVERAGES
W1FV	W8JK ARRAY	3 EL PHSD ARRAY (TRIANGLE)	3 SHORT PHSD VERTS	BEVERAGES
W1JCC	DIPOLE	VERT, DIPOLE	1/4 WV VERT	BEVERAGES
W1JR	ROT DIPOLE	DIPOLES, LOOPS	DIPOLE @ 90'	BEVERAGE
W1TE	INV V'S	INV V'S	INV V'S, BALLOON VERT, GLADIATOR VERT	BEVERAGES
W1UK	4 SQ	4 SQ	INV L	BEVERAGES, SLINKY
W1ZK	ROT DIPOLE @ 70'	1/4 WV VERT	1/4 WV SLOPERS	BEVERAGES
W2UB	4 EL INV V, VERT, DIPOLE	3 EL INV V YAGI	INV L	BEVERAGE
W2UE	GROUND PLANE	1/4 WV SLOPER	1/4 WV SLOPER	LOOP
W3UM	402CD	DELTA LOOP	INV L	BEVERAGES
W3UR	3 EL @ 200'	FOUR 2 EL QUADS	4 SQ , 1/4 WV VERT, 1/4 WV SLOPER	PHSD BEVERAGES, SINGLE BEVERAGES
W4AG	4 SQ	4 SQ (SHORT VERTS)		
W4CTG	BOBTAIL, WIRE VERTS	4 SQ WITH SLOPING DIPOLES, INV V	2 PHSD VERTS	BEVERAGES
W4DR	3 EL FULL SIZE @ 140'	4 SQ	4 SQ	BEVERAGES
W4PA	3 EL SHORT YAGI HY GAIN	PHSD HF2V VERTS	INV L	BEVERAGE
W4ZV		1/4 WV GP	0.35 WV VERT	BEVERAGES
W5AA	2 EL KLM YAGI	DELTA LOOPS WITH REFLECTOR		
W5FI	2 EL MOSLEY		HF2V VERT	BEVERAGES, LOW DIPOLE
W5KFT	402CD	DELTA LOOP	SLOPER	NONE
W5PS	2 EL YAGI @ 40 M	4 SQ	5 EL PARASITIC ARRAY	BEVERAGES
W5WP	PHSD VERTS	PHSD VERTS, DIPOLE	VERT	
W6AJJ		70' TP LOADED VERT	70' TP LOADED VERT	BEVERAGES, EWE'S, LOOPS
W6BWY	4 SQ	1/4 WV VER	INV L 75' VERT	BEVERAGE
W6KW	3 EL YAGI @ 120'	3 EL YAGI, COIL LOADED ELEMENTS 165'	NONE	NONE

CALL	40 ANT	80 ANT	160 ANT	RX ANTENNAS
W6PAA	STACKED KLM YAGIS @ 20 & 40 M	WIRES QUAD	4 SQ	BEVERAGES
W6RJ	FORCE 12 MULTIBAND 6 OVER 6`	3 EL KLM LINEAR LOADED YAGI		
W6SR	3 EL KLM	1/4 WV SLOPERS & INV V	SH FED TWR	NONE
W6YA	2 EL DELTA (HANGING FROM 20 M YAGI) (CITY LOT)	1/2 WV VERT	1/4 WV VERT	RESONATE WARC DIPOLE, 4/1 BALUN + PREAMP
W7AWA	NONE	SLANTING T ANTENNA	EWE, BUILDING BEVERAGE	
W7EW / W7AT	3 EL FULL SIZE	3 EL M² YAGI	1/4 WV VERT	BEVERAGES
W7LR	LAZY V	90' VERT	90' VERT	EWE, K6STI LOOPS, LOW DIPOLE
W7TVF	2 EL FORCE 12	SH FED TWR	PHSD SH FED TWRS	BEVERAGES, K9AY LOOPS
W8AH				
W8RU	2 EL LINEAR LOADED YAGI @ 25 M	1/4 WV SLOPER @ 15 M	INV L, 14 M VERT	COAX RECEIVING LOOP
W8XD	HF2V & R7000	HF2V VERT	INV L	LONGWIRES
W9AGH	3 EL @ 140'	DIPOLE @ 130'	DIPOLE @ 140'	BEVERAGE
W9QA	VERTICAL BEAM	DIPOLE	DIPOLE	NO
W9WI	DIPOLE	DIPOLE	3/2 WV INV L	BEVERAGE
W9XT	2 EL HY GAIN	DIPOLE & SH FED TWR	INV L & DIPOLE	BEVERAGES
W9ZR	2 EL CUSHCRAFT, DIPOLES	4 SQ	SH FED TWR	BEVERAGES
WA3AFS	PAIR OF PHSD INV L'S	PHSD INV L'S	INV L	NO
WA5VGI	VERT (GAP VOYAGER)	VERT (GAP VOYAGER)	VERT (GAP VOYAGER)	BEVERAGE
WB3AVN	DIPOLES	1/4 WV VERT IN TREES	INV L IN TREES	BEVERAGE
WB9Z	3 EL FULL SIZE @ 43 M	3 EL YAGI @ 43 M	49 M BASE FED VERT, DIPOLE @ 40 M	BEVERAGES
WD4JRA	160 M DIPOLE OPEN WIRE FED	160 M DIPOLE OPEN WIRE FED	INV L (68' VERT), DIPOLE @ 30'	
WXØB	TWO M² 2 EL YAGIS (145 & 75')	3 EL VERT ARRAY, K8UR STYLE	VERT 135'	BEVERAGES
YB1AQS	INV V	VERT	VERT	NO
YCØLOW	WINDOM @ 24 M	WINDOM @ 24 M	INV V @ 24 M	BEVERAGES
YT6A		3 EL YAGI		BEVERAGES
ZS4TX	2 EL HY GAIN @ 36 M	DIPOLE @ 24 M	SH FED 30 M LOADED TWR, 1/4 WV SLOPER	BEVERAGES
ZS6EZ	3 EL SHORT YAGI @ 36 M	2 EL LOADED @ 36 M, SH FED TWR	SH FED TWR	SHORT BEVERAGE

RECEIVING AND TRANSMITTING EQUIPMENT

■ **1. THE RECEIVER**

1.1. Receiver Specifications

1.2. Sensitivity

    1.2.1. Thermal noise

    1.2.2. Receiver noise

    1.2.3. Noise

1.3. Intermodulation distortion

1.4. Gain Compression or Receiver Blocking

1.5. Dynamic Range

1.6. Cross Modulation

1.7. Reciprocal Mixing (VCO Noise)

    1.7.1. Measuring reciprocal mixing

        1.7.1.1. Conversion to dBc/Hz

1.8. Selectivity

    1.8.1. SSB bandwidth

    1.8.2. CW bandwidth

    1.8.3. Passband tuning.

    1.8.4. Continuously variable IF bandwidth

    1.8.5. Filter shape factor

    1.8.6. Static and dynamic selectivity

    1.8.7. IF filter position

    1.8.8. DSP Filtering

        1.8.8.1. Audio DSP

        1.8.8.2. IF DSP

    1.8.9. Audio filters

1.9. Stability

1.10. Frequency Readout

1.11. Switchable Sideband on CW

1.12. Outboard Front-end Filters

1.13. Intermodulation Outside the Receiver

1.14. Noise Blanker

1.15. Receiver Evaluations

1.16. Graphical Representation

1.17. In Practice

1.18. Areas for Improvement

1.19. Adding Input Protection to your Receiver Input Terminals

1.20. Noise Canceling Devices

■ **2. TRANSMITTERS**

2.1. Power

2.2. Linear Amplifiers

2.3. Phone Operation

    2.3.1. Microphones

    2.3.2. Speech processing

2.4. CW Operation

2.5. DSP in the Transmitter

2.6. Signal Monitoring Systems

2.7. Areas of Improvement

3 RECEIVING AND TRANSMITTING EQUIPMENT

THANKS TO GEORGE, W2VJN

I would like to thank George Cutsogeorge, W2VJN, for having been my critic and proofreader for this chapter. George has been licensed with the same call sign since 1947. George retired as an electronic engineer after a 40-year career which included 17 years with RCA designing spacecraft and ground station equipment and another 17 years with Princeton University where he was involved with fusion energy research. George now owns half of International Radio and Top Ten Devices, which keeps him heavily involved with electronics and radio amateurs. He has been on the top of the DXCC Honor Roll since 1986 and is waiting for a QSO with P5.

The performance of our communication equipment has progressed by leaps and bounds over the years. However, there is no doubt that significant improvement can still be made to our present day radios. Low-band DXing and contesting are two areas of our hobby that demand the utmost from our equipment, due to the almost continuous presence of all kind of noise as well as strong nearby signals.

Before summing up which are the improvements to be made (see Section 1.18) I will review in this chapter the specifications of both the receiver and the transmitter sections of a station.

■ 1. THE RECEIVER

1.1. Receiver Specifications

Until about 20 years ago, receiver performance was most frequently and almost exclusively measured by sensitivity and selectivity. In the fifties and early sixties a triple-conversion superheterodyne receiver was a status symbol, more or less like today a transceiver with IF DSP. It was not until the mid sixties that strong-signal handling came up as an important parameter (Ref. 250). Today we consider the following topics to be most important for a communication receiver (not necessarily in order of importance):
1. Sensitivity
2. Intermodulation distortion
3. Gain compression
4. Dynamic range
5. Cross modulation
6. Reciprocal mixing (VCO noise)
7. Selectivity
8. Noise suppression
9. Stability
10. Frequency-display accuracy

I will review these items in detail while highlighting their impact on successful low-band DXing.

1.2. Sensitivity

Fig 3-1 shows the voltage and power relationships of the RF signals we generally deal with at the front end of a receiver. The chart can be used to make conversion between the many different units used to express signal strength.

Sensitivity is the ability of a receiver to detect weak signals. The most important concept related to sensitivity performance is the concept of signal-to-noise ratio. Good reception of a weak signal implies that the signal is substantially stronger than the noise. It is accepted as a standard that comfortable SSB reception requires a 10 dB signal-to-noise (S/N) ratio. CW reception may have a much lower S/N ratio, and any CW operator can deal with a 0 dB S/N ratio quite well. Experienced operators can dig CW signals out of the noise at −10 dB S/N ratio. This proves again the inherent advantage of CW over SSB for weak-signal communications.

1.2.1. Thermal noise

The noise present at the receiver audio output terminals is generated in different ways. Inherent internal receiver noise is produced by the movement of electrons in any

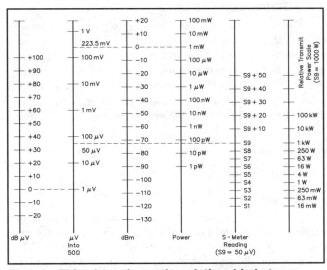

Fig 3-1—This chart shows the relationship between receiver input voltages, standard S-meter readings and transmitter output power.

substance (such as resistors, transistors and FETs, which are part of the receiver circuit that has a temperature above 0 kelvins (0 K or −273° Celsius), which is absolute zero—the temperature at which all electrons stop moving. Above 0 K, electrons move in a random fashion, colliding with relatively immobile atoms that make up the bulk of matter. The final result of this effect is that in most substances there is no net current in any particular direction on a long-term average, but rather a series of random pulses. These pulses produce what is called thermal-agitation noise, or simply thermal noise.

The Boltzmann equation expresses the noise power in a system. The equation is written as:

$$p = kTB$$

where

    p = thermal noise power in watts
    k = Boltzmann's constant (1.38×10^{-23} joules/kelvin)
    T = absolute temperature, kelvins
    B = bandwidth, Hz

Notice that the power is directly proportional to temperature, and that at 0 kelvins the thermal noise power is zero. For equivalent noise voltage, the equation is rewritten as:

$$E = \sqrt{kTBR}$$

where R is the system impedance (usually 50 ohms).

For example, at an ambient temperature of 27° C (300 K), in a 50-ohm system with a receiver bandwidth of 3 kHz, the thermal noise power is:

$$P = 1.38 \times 10^{-23} \times 300 \times 3000 = 1.24 \times 10^{-17} \text{ W}$$

This is equivalent to $10 \log (1.24 \times 10^{-17}) = -169$ dBW or −139 dBm (139 dB below 1 milliwatt), and is equivalent to 32 dB below 1 μV or −32 dBμV (Ref. 223). This is the theoretical maximum sensitivity of the receiver under the given bandwidth and temperature conditions.

1.2.2. Receiver noise

No receiver is noiseless. The internally generated noise is often evaluated by two measurements, called noise figure and noise factor. Noise factor is by definition the ratio of the total output noise power to the input noise power when the termination is at the standard temperature of 290 K (17° C). Being a ratio, it is independent of bandwidth, temperature and impedance. The noise figure is the logarithmic expression of the noise factor:

$$NF = 10 \log F$$

where F is the noise factor.

1.2.3. Noise

Besides the noise that is generated internally in the receiver, the second factor that limits the sensitivity of a receiver system is the noise supplied by the antenna. This noise is mainly atmospheric or man-made noise, and its presence limits the maximum usable receiver sensitivity.

Fig 3-2 shows the maximum usable receiver sensitivity for (A) an urban environment, (B) a quiet rural environment during the day and (C) a quiet rural environment at night. The curved lines correspond to the limits imposed by the atmospheric noise. These curves are conservative, are based on typical or average noise conditions and are for a 10 dB S/N ratio in a 3 kHz bandwidth.

Low-band DXers usually push the limits, and often copy signals quite well at the noise level (0 dB S/N ratio). Curve D shows the data for a 0 dB S/N ratio. Switching from a wide SSB passband of 3 kHz to a typical CW bandwidth of 500 Hz, lowers the curve another 7.8 dB (curve E).

The noise levels shown are typical for a receiving system consisting of an efficient half-wave dipole or quarter-wave vertical. Less efficient antennas will produce less noise and will therefore require a more sensitive receiver (Ref. 201, 202, 205, 223 and 247).

Table 3-1 shows the typical minimum required receiver sensitivity for the low bands. If you use a receiving antenna such as a Beverage, greater sensitivity is required.

A typical 1998-vintage receiver has the following sensitivity characteristics (specified for a 500-Hz bandwidth):

Band	Noise Floor	Noise Figure
1.8	−120 to −130 dBm	17 to 27 dB
3.5/7.0	−130 to −140 dBm	7 to 17 dB

The thermal noise level at 27° C and 3 kHz bandwidth is −139 dBm (see Section 1.2.1). For a bandwidth of 500 Hz it is 7.8 dB lower or approximately −147 dBm.

This means that the modern transceiver has a typical surplus sensitivity of approximately 10 to 15 dB (in a good DX location) on 160 and 80 meters, using a full-size dipole or vertical antenna. This surplus sensitivity is very welcome when using special receiving antennas, eg Beverages, which have a signal output that is typically 10 dB down from a full-size dipole or vertical.

There are of course exceptions to such "typical" circumstances. If the receiving location is located on a remote island, hundreds of miles away from any source of man-made noise, and if there are no thunderstorms within hundreds of kilometers, extremely weak signals can be heard—even on the low bands. This is also why under such circumstances DXpeditions often report not having the need for separate receiving antennas such as Beverage antennas. Under such circumstances there often is a sort of *one-way* situation where the DXpedition can hear much better than the "average" DXer, because of the total absence of any man-made noise.

When using a separate receiving antenna such as the Beverage antenna, the *surplus* sensitivity of a receiver becomes very useful, as those antennas often produce a very low-output signal, and are much less susceptible to noise (atmospheric and man-made) because of their directive characteristics. This means that the noise (and desired signal) input from these antennas are *both* lower, and the extra receiver sensitivity is required.

In my QTH, which is in a semi-rural location, using Beverage antennas, I can, most of the time, insert 6 or 10 dB of input attenuation, without audibly deteriorating the sig-

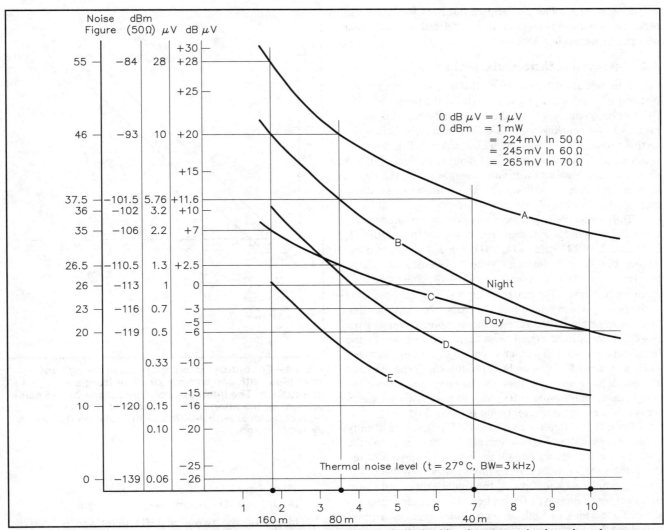

Fig 3-2—The maximum usable receiver sensitivity on the low bands is limited by the atmospheric noise plus man-made noise. This example is for a receiver with a noise floor of −129 dBm. Curve A is for a typical (noisy) city environment and for a 10 dB S/N ratio in a 3.0-kHz bandwidth. Curve B and C are for rural areas under the same S/N ratio and bandwidth conditions. On the low bands we often copy signals that have 0 dB S/N ratio. Curves D and E are for 0-dB S/N ratio for 3.0-kHz bandwidth and for 500-Hz bandwidth at night in a quiet rural area.

Table 3-1
Minimum Required Receiver Sensitivity

	Quiet Rural Day	Quiet Rural Night	Urban	Freq (MHz)
Acceptable Noise Figure	33 dB	46 dB	55 dB	1.8
Minimum Sensitivity	2.2 µV	10 µV	28 µV	
Acceptable Noise Figure	28 dB	37.5 dB	46 dB	3.5
Minimum Sensitivity	1.3 µV	3.75 µV	10 µV	
Acceptable Noise Figure	23 dB	26 dB	37.5 dB	7
Minimum Sensitivity	0.7 µV	1 µV	11.5 µV	
Noise Figure	10 dB	10 dB	10 dB	Typical
Sensitivity	0.15 µV	0.15 µV	0.15 µV	Receiver

These are typical receiver sensitivities required on the low bands. The values assume a receiving antenna such as a dipole or a vertical. Special low-gain receiving antennas require more sensitivity (up to 10 dB).

nal to noise ratio. The use of preamplifiers for low-noise receiving antennas is covered in more detail in the chapter on Special Receiving Antennas.

1.3. Intermodulation Distortion

Intermodulation distortion (IMD) is an effect caused by two (or more) strong signals that drive the front end stage (or a subsequent stage) of the receiver beyond its linear range so that spurious signals called intermodulation-distortion products are produced. Third-order IMD is the most common and annoying front-end overload effect. **Fig 3-3** shows the IMD spectrum for an example where the parent signals are spaced 1 kHz apart. (The third-order products are: $2F_1-F_2$, and $2F_2-F_1$.)

Third-order IMD products increase in amplitude three times as fast as the pair of equal parent signals (Ref. 210, 211, 213, 226, 239, 247, 255, 274, 281). **Fig 3-4** shows two examples for third-order intercept points. The vertical scale is the relative output of the receiver front end in dB, referenced to an arbitrary zero level. The horizontal axis shows the input level of the two equal-amplitude parent signals, expressed in dBm. Point A sits right on the receiver noise floor. Increasing the power of the parent signals results in an increase of the fundamental output signal at a one-to-one ratio. Between –129 dBm and –44 dBm, no IMD products are generated that are equal to or stronger than the receiver noise floor. At –44 dBm, the third-order IMD products have risen to exactly the receiver noise floor level (point B in Fig 3-4).

Point B is called the two-tone IMD point. It is usually expressed in dBm. Further increasing the power of the parent input signals will continue to raise the power of the third-order IMD products *three times faster* than that of the parent signals. At some point, the fundamental and third-order response lines will flatten because of "gain compression." Extensions of both response lines cross at a point called the third-order intercept point. The level can be read from the input scale in dBm.

The intercept point (I_P) can be calculated from the IMD point as follows.

$$I_P = \frac{2MD(noisefloor) + 3IMD_{DR}}{2}$$

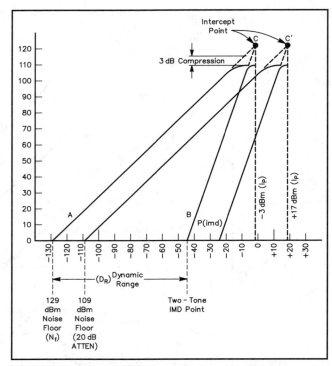

Fig 3-4—Third-order intercept point showing two examples, with and without 20 dB of front-end attenuation. The intercept point increases by the same amount as the attenuation introduced. This is for a typical receiver with 84-dB intermodulation distortion dynamic range.

where

MDS = minimum discernible signal
IMD_{DR} = IMD dynamic range

Inversely, the two-tone IMD point can be derived mathematically from the intercept point as follows.

$$P_{IMD} = \frac{2I_P + Nf}{3}$$

An example (for a receiver having a –3 dBm intercept point and a –129 dBm noise floor is

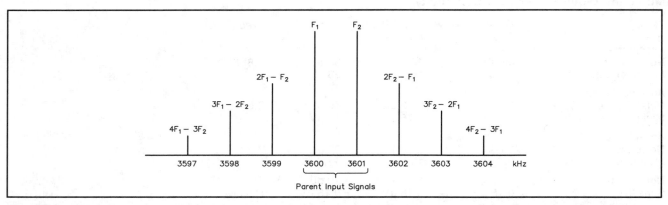

Fig 3-3—Third, fifth and seventh-order intermodulation products generated by parent input signals on 3600 and 3601 kHz.

$$P_{IMD} = \frac{2 \times (-3) + (-129)}{3} = -45 \text{ dBm}$$

What does this mean? It means that signals below –45 dBm will not create audible IMD products, while stronger signals will. This means that signals at around S9 +30 dB will start generating audible IMD products! In Europe this is an everyday situation on the 7-MHz band where 30 to 50 mV signals are common.

When evaluating third-order intercept points, we must always look at receiver noise floor levels at the same time. When we raise the noise floor from –129 dBm to –109 dBm (for example by inserting 20 dB of attenuation into the receiver input line), both response lines and the intercept point will shift 20 dB to the right as shown in Fig 3-4. This means that the intercept point has been improved by 20 dB! Any (average) receiver with a +5 dBm intercept point can be raised to +25 dBm merely by inserting 20 dB of attenuation into the input. Remember that this can frequently be done with present-day receivers, as they often have a large surplus sensitivity.

Fig 3-5 shows the better solution, however, where the improvement is obtained by using design techniques and components which can handle stronger signals before becoming nonlinear.

The frequency separation of the two parent input signals can greatly influence the intermodulation results. The worst case applies when there is no selectivity in the front end to attenuate one of the signals. This happens when both input signals are within the passband of the first-IF filter and the intermodulation products are produced in the second mixer. Most present-day receivers have a rather wide first IF so they can accommodate narrowband FM and AM signals without requiring filter changes.

This is one of the greatest problems with modern "all bells and whistles" general-coverage transceivers, using fixed-tuned input circuits. In order to obtain sufficient image rejection, a very high IF, outside the operating range of the equipment is required. Filters at IFs between 50 and 100 MHz are very inferior in shape factor to what can be obtained on much lower frequencies.

Measurements at 2-kHz spacing using a 500-Hz second-IF filter (CW filter) are often used to find the worst-case IMD performance. In this case there will be *no* selectivity from the first IF (roofing) filter. Measurements at 20 and 100-kHz spacing are also used for assessing receiver intermodulation performance, in which case the high-frequency IF filter does play its role. With this much spacing, the first-IF filter generally improves the picture considerably. The real picture, that low banders are interested in is the picture obtained with close-spacing measurements, because that is the situation we encounter when operating on crowded bands—especially during contests.

Often measurements using 100-kHz spacing are made for evaluating the strong signal handling performance of receivers where the local oscillator (LO) noise limits measurement accuracy at 20 kHz and closer spacings (see Section 1.7).

Fig 3-6 shows the third-order intercept point for a modern-day receiver versus frequency separation of the two parent signals. Two cases are shown, with and without a preamplifier.

1.4. Gain Compression or Receiver Blocking

Gain compression occurs when a strong signal drives an amplifier, or mixer, stage (for example, a receiver front end) so hard that it cannot produce any more output. The stage is driven beyond its linear operating region and is saturated. Gain compression can be recognized by a de-

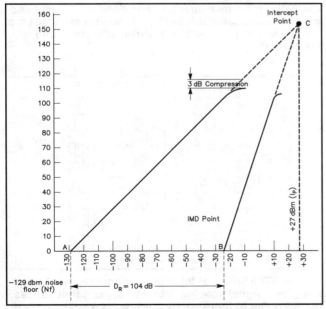

Fig 3-5—Third-order intercept point for an improved receiver with 104-dB intermodulation distortion dynamic range.

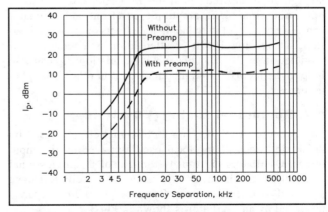

Fig 3-6—Third-order intercept point values for a typical modern general-coverage up-conversion receiver. The graph shows clearly that the first IF filter (roofing filter) has a bandwidth of approximately 10 kHz. Decreasing the bandwidth of this filter, together with the use of a lower noise local oscillator (VCO) would dramatically improve the dynamic range at closer frequency spacing.

crease in the background noise level when saturation occurs (Ref. 223, 239, 281). Gain compression can be caused by other amateur stations nearby; such as in a multioperator contest environment. Outboard front-end filters are the answer to this problem. (See Section 1.12.)

1.5. Dynamic Range

The lower limit of the dynamic range of a receiver is the power level of the minimum discernible signal (MDS) or receiver noise floor. The upper limit is the power level of the signals at which IMD becomes noticeable (intermodulation products are equal to the MDS).

Refer to Fig 3-5 for a graphical representation of dynamic range. Dynamic range can be calculated as follows:

$$DR = P_{IMD} - Nf$$

where

DR = dynamic range, dB
P_{IMD} = two-tone IMD point, dBm
Nf = receiver noise floor, dBm

If the intercept point is known instead of the two-tone IMD point we can use the following equation:

$$DR = \frac{2(I_P - NF)}{3}$$

where I_P is the intercept point in dBm.

The dynamic range of a receiver is important because it allows us to directly compare the strong-signal handling performance of receivers (Ref. 234, 239, 255), as it takes into account the sensitivity as well.

1.6. Cross Modulation

Cross modulation occurs when modulation from an undesired signal is partially transferred to a desired signal in the passband of the receiver. Cross modulation starts at the 3-dB compression point on the fundamental response curve as shown in Fig 3-5. Cross modulation is independent of the strength of the desired signal and proportional to the square of the undesired-signal amplitude, so a front-end attenuator can be very helpful in reducing the effects of cross modulation. Introducing 10 dB of attenuation will reduce cross modulation by 20 dB. This exclusive relationship can also help to distinguish cross modulation from other IMD phenomena (Ref. 223, 247).

1.7. Reciprocal Mixing (VCO Noise)

Reciprocal mixing is a large-signal effect caused by noise sidebands of the local oscillator feeding the input mixer. Oscillators are mostly thought of as single-signal sources, but this is never so in reality. All oscillators have sidebands to a certain extent. One example of the sidebands produced by an oscillator is shown in **Fig 3-7**.

The detrimental effect of these noise sidebands remained largely unnoticed until voltage-controlled oscillators (VCOs) were introduced in state-of-the art receivers. J. Grebenkemper, KI6WX, covered the effects of phase noise on amateur communications in full detail in his article in *QST* (Ref. 286 and 289). **Fig 3-8** shows the levels of the interfering signals produced versus frequency spacing, the standard method of evaluating the effects of the VCO noise in a receiver.

VCOs are much more prone to creating noise sidebands than conventional LC oscillators. The phase-locked loops in VCOs are responsible for the poor noise spectrum

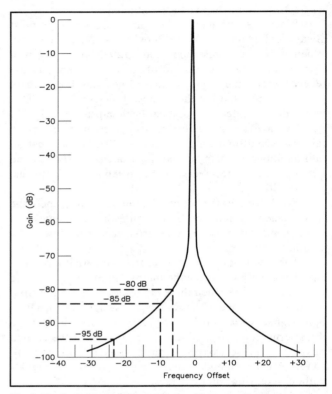

Fig 3-7—Output spectrum of a voltage-controlled oscillator. If the measurement was done at a 3-kHz bandwidth, the oscillator sideband performance referred to a 1-Hz bandwidth is 85 + 34 = 119 dBc (dB referenced to the carrier).

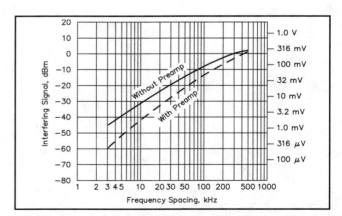

Fig 3-8—The levels of interfering signal (vertical axis) at a given signal spacing (horizontal axis) that causes the AF noise to increase by 3 dB. A 2.7-kHz IF bandwidth is assumed. This is the standard method of evaluating the effects of the VCO noise in a receiver. See text for details.

(Ref. 209). **Fig 3-9** shows the relationship in the usual superheterodyne receiver between the input signal, the IF and the local oscillator for both the ideal situation of a noiseless LO and for the case where the LO has realistic noise sidebands.

Reciprocal mixing introduces off-channel signals into the IF at levels proportional to the frequency separation between the desired signal and the unwanted signal. This effectively reduces the selectivity of the receiver. In other words, if the static response of the IF filters is specified down to –80 dB, the noise in the LO must be down at least the same amount in the same bandwidth in order not to degrade the effective selectivity of the filter.

According to the thermal-noise equation (see Section 4.2.1), the noise power is –174 dBm at room temperature for a bandwidth of 1 Hz. The noise in an SSB bandwidth of X Hz can be scaled to a 1-Hz bandwidth using the factor (10 log X). This equation yields a factor of 34.8 dB for a 3-kHz bandwidth, 34.3 dB for 2.7 kHz and 33.2 dB for 2.1 kHz. Continuing with the example where the static response of the IF filter is –80 dB and the filter has a 3-kHz bandwidth, the noise of the LO should be no more than 80 + 34 = 114 dBm referenced to a 1-Hz bandwidth. The carrier noise is usually referenced to a 1-Hz bandwidth, and is expressed in dBc/Hz.

1.7.1. Measuring reciprocal mixing

Phase noise is most easily measured with a single tone.

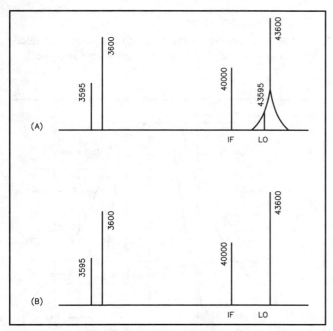

Fig 3-9—An LO with noise sidebands can produce reciprocal mixing products inside the bandwidth of the IF filter. At A, the undesired signal at 3595 kHz mixes with sideband energy from the LO at 43,595 kHz to produce an output signal at the IF of 40,000 kHz. The LO at B has no sideband energy so the unwanted signal produces a mixing product at 40,005 kHz, which is outside the IF passband.

A signal source, such as a low-noise crystal oscillator, is connected through an attenuator to the receiver input. The receiver is offset from the crystal frequency by various values and the level required to reach the receiver noise floor (noise = MDS) is recorded.

The noise floor is measured with an audio rms voltmeter on the output and a reference level is set with no input signal. The input signal level is then raised until the audio output goes up 3 dB. The generator level is then at the noise floor.

There are other measuring methods using two signals (Ref. 281, 274, 247). All equipment manufacturers are encouraged to list their phase noise specifications at 1 kHz and 10 kHz spacing.

1.7.1.1. Conversion to dBc/Hz

The dBc/Hz value can then be calculated at each offset by subtracting the level from the noise floor and adding a correction for the filter bandwidth.

Example: We have a receiver with a –130 dBm noise floor, using 500 Hz bandwidth. A –40 dBm injected signal at 10 kHz offset is just discernible (on the noise floor). The phase noise is:

dBc = –130 – (–40) = –90 dB

The correction factor is determined as follows:

Cf = 10 × log (BW)

In our example:

Cf = 10 × log(500 Hz) = 26.9 dB

If we would do the measurement with a 2.7-kHz IF filter, the correction factor would be 34.3 dB.

Adding the correction for a 500 Hz bandwidth (26.9 dB) we obtain the dBc/Hz factor:

dBc/Hz = –90 – 27 = –116.9 dBc/Hz

1.8. Selectivity

Selectivity is the ability of a receiver to separate (select) a desired signal from unwanted off-frequency signals.

1.8.1. SSB bandwidth

On a quiet band with a reasonably strong desired signal, the best sounding audio and signal-to-noise ratio can be obtained with selectivity on the order of 2.7 kHz at –6 dB. Under adverse conditions, selectivity as narrow as 1 kHz can be used for SSB, but the carrier positioning on the filter slope becomes very critical for optimum readability. The ideal selectivity for SSB reception will of course vary, depending on the degree of interference on adjacent frequencies.

1.8.2. CW bandwidth

For most DX work a 500 Hz (at –6 dB) IF filter is adequate. If the transceiver is equipped with a continuously variable bandwidth control (or shift + width control), you can also narrow the bandwidth further, but remember that the shape factor worsens as you reduce the bandwidth. If you are a serious CW operator, it is my advice to use a 250-Hz filter in addition to the 500-Hz filter for when things get rough.

Ideal would be to have 3 selectivity positions. Indeed, 800 to 1000 Hz selectivity can be very useful for checking a not so crowded band. International Radio (see Section 1.8.5) is now providing wider CW filters for some radios.

1.8.3. Passband tuning

Passband tuning (and IF shift) allows the position of the passband on the slope to be altered without requiring that the receiver be retuned. The bandwidth of the passband filter remains constant, however. In some cases interfering signals can be moved outside the passband of the receiver by adjusting the passband tuning. In better receivers, passband tuning has been replaced by a filter system with a continuously variable bandwidth

1.8.4. Continuously variable IF bandwidth

Continuously variable bandwidth is available in two configurations. In one the filter can be independently narrowed down from both sides (low pass and high pass). The other approach is to use a width plus a passband tuning control. This feature is, nowadays, available on all state-of-the-art receivers.

The mechanism in producing a continuously variable bandwidth consists in passing the signal through two separate filters, on two different IFs (eg, 9 MHz and 455 kHz). The mixing frequency is slightly altered so the two filters do not superimpose 100%, but have their passbands sliding across one another, effectively creating a continuously variable bandwidth. You must understand, however, that a variable bandwidth system as described will never have as good a shape factor as individual well-shaped crystal filters, as the shape factor always worsens when you narrow the bandwidth.

1.8.5. Filter shape factor

The filter shape factor is expressed as the ratio of the bandwidth at 60 dB to the bandwidth at 6 dB. Good filters should have a shape factor of 1.5 or better. This 1.5 figure is a typical shape factor for an 8-pole crystal filter. Too many transceivers are equipped with rather wide IF filters (typically 2.7 kHz at 6 dB) having mediocre skirt selectivity. On a quiet band these give nearly hi-fi quality, but is that what we are after? For the average operator this may be an acceptable situation, although the serious DXer and contest operator may want to go a step further.

International Radio (formerly RCI / Fox Tango), still offers modification kits for modern transceivers. The kits are very popular with Kenwood users, where the original Kenwood filters, which typically have a rather poor shape factor, can be replaced with much sharper filters. A number of years ago I used to have a matched pair of 2.1-kHz-wide filters in the 8.8-MHz and 455-kHz IF strips of my Kenwood TS-930 and TS-940, which yielded a very respectable shape factor (6/60 dB) of 1.25. The paired CW filters gave a –6 dB bandwidth of 400 Hz and a –60 dB bandwidth of less than 700 Hz. Details for all International Radio filter products (including modification notes, etc) can be obtained at **http://www.qth.com/inrad/**.

While the Yaesu FT-1000 and FT-1000MP are transceivers, which have stock filters with a much better shape factor than most other brands, there is one exception, and that is the Collins mechanical filter which they sell for the 455 kHz IF. While mechanical filters were excellent designs in the '50s, we can now achieve much better responses from crystal filters. **Fig 3-10** shows the selectivity curves of the 500 Hz stock mechanical filter sold by Yaesu and the 400 Hz replacement crystal filter, sold by International Radio.

1.8.6. Static and dynamic selectivity

Fig 3-11 shows the typical static selectivity curve of a filter system with independent slope tuning. The static selectivity curve is the transfer curve of the filter with no reciprocal mixing. The dynamic selectivity of the receiver front end is shown in **Fig 3-12**. The dynamic selectivity is the combination of the static selectivity and the effects of reciprocal mixing. Note that the static selectivity can be deteriorated by the effect of reciprocal mixing with noise from the local oscillator.

If the amplitudes of the reciprocal mixing products are greater than the stop-band attenuation of the filter, the ultimate stop-band characteristics of the filter will deteriorate. Good frequency-synthesizer designs can yield 95 dB (–129 dBc), while good crystal oscillators can achieve over 110 dB (–144 dBc) at a 10-kHz offset. This means it is pointless to use an excellent filter with a 100-dB stop-band characteristic if the reciprocal mixing figure is only 75 dB.

1.8.7. IF filter position

The filter providing the bulk of the operational selectivity can theoretically be inserted anywhere in a receiver

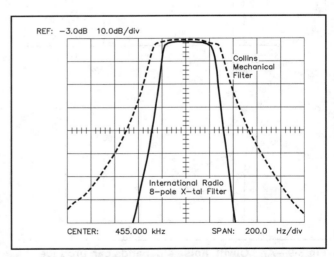

Fig 3-10— Selectivity curves for the stock 455-kHz IF 500-Hz passband Collins mechanical filter and the 400-Hz crystal filter offered by International Radio. Notice that at –60 dB the replacement crystal filter has only half the bandwidth of the stock Collins mechanical filter. The crystal filter is mounted on a small adapter board which plugs in directly to the IF board.

between the RF input and audio output. When considering parameters other than selectivity, however, it is clear that the filter should be as close as possible to the antenna terminals of the receiver. In Section 1.3 we saw that front-end selectivity will help reduce IMD products.

Most modern receivers use triple or even quadruple conversion. In order to be most effective, the selectivity (filter) should be as far ahead in the receiver as possible. The logical choice is the first IF. Most modern designs use a first IF in the 50 to 100-MHz range, for image rejection reasons. This is not the most ideal frequency for building crystal filters with the best possible shape factor. In general, we find rather simple 2-pole crystal filters with a nominal selectivity of 15 to 20 kHz (at –6 dB) in the first-IF chain. The reasons for this very wide bandwidth are twofold:

• To retain the original impulse noise shape (short rise time) in order to be able to incorporate a (more or less) useful noise blanker.
• Most of the modern (all bells and whistles transceivers) must operate on FM as well, where a selectivity of less than 15-20 kHz cannot be tolerated.

I am convinced that most successful low-band DXers live in quiet areas. No need for noise-blankers to reduce man-made noise. If we are plagued with this kind of noise, we will cure the problem at the source. Therefore I am convinced that noise blankers are of little use to most low-band DXers. I also am not at all interested in being able to receive FM on my transceiver. This means I could use better (narrower) first-IF filters. They would greatly improve the dynamic selectivity of our receivers (see Section 1.8.6), in other words the IMD behavior at close signal spacing.

This whole problem of poor roofing filters in modern transceivers is the reason why the Drake C-Line receiver, with a 600 Hz first IF filter and a 500, 250, or 125 second IF filter, is still today, 25 years after it was designed, found in a number of low-band DXers' shacks, especially those regularly venturing in contests. This should be a clear message to the designers.

In today's vintage of transceivers, the second IF is often in the 9-MHz region, the third IF usually on approx. 455 kHz. Both these frequencies are very well suited for building high-quality crystal filters with excellent shape factors. In some receivers ferrite or mechanical filters are used on 455 kHz, but they have an inferior shape factor compared to a good crystal filter (see also Section 1.8.5).

Modern transceivers often will be equipped with IF filters only in the first (roofing filter) and second IFs. Filters

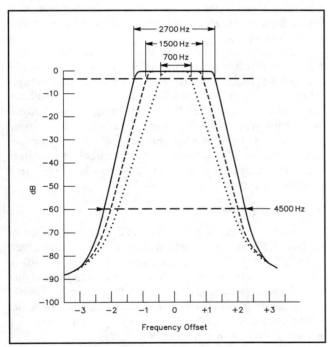

Fig 3-11—Static selectivity curve of a receiver using continuously variable bandwidth. This result is obtained by using selective filters in the first and second (or second and third) IFs, and by shifting the two superimposed filters slightly through a change in the mixing frequency. Note that the shape factor worsens as the bandwidth is reduced. It is not ideal to use this method to achieve CW bandwidth. The shape factor may deteriorate to 4 or more, while a good stand-alone CW filter can yield a shape factor of less than 2.

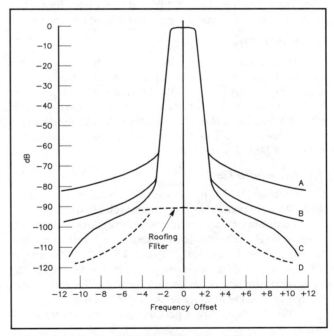

Fig 3-12—The total dynamic front-end filter response is the combination of the main selectivity filter (second IF) and the roofing filter (first IF). Curve A is the selectivity curve with an LO having 80-dB noise suppression (at 10-kHz spacing). Curve B is for 95-dB suppression, and Curve C is for an LO with 115-dB suppression. Dashed Curve D represents the noise level from the LO that yields Curve C. In Curve C, the selectivity is not influenced by the LO noise (up to ±10 kHz from the center frequency), but is made up by a combination of the main selectivity filter and the first-IF roofing filter.

in the third IF are obtainable as options. I strongly suggest not to compromise in this area. Install all the available filters in your transceiver. If you don't install the optional filters in the third IF, the shape factor of the variable bandwidth control will also be mediocre (using the slope of the standard 2.7-kHz filter).

Ideally, the last-IF filters should be placed just ahead of the product detector, in order to reduce the wideband noise generated by the IF amplifier stage beyond the last filter. Many modern receivers show an annoying wide-band noise (hiss) which is especially noticeable on narrow CW when the band is very quiet.

In recent years DSP has been introduced at the lowest IF, usually in the 10 to 50-kHz range. DSP allows unlimited flexibility as far as varying bandwidths etc, but it can never replace the very essential filters in the earliest stages of the transceivers. Transceivers that have tried to do it this way have quickly become infamous for their very bad behavior with strong signals.

Today we are all waiting for DSP systems to be moved closer to the front-end of the receivers. But as long as the concept requires a very high first IF, it is unlikely that we will see DSP in those frequency ranges (50 to 100 MHz) in the next few years.

I am convinced that somehow the designers have to move away from these very high first IF, and compensate for the loss of image rejection by going back to tuned input circuits.

1.8.8. DSP filtering

Digital signal processing consists in digitizing the analog signal (eg, an audio signal or an IF signal) so that a digital processor can handle the signal and do whatever is desired before converting it back to an analog signal (Ref. 290 and 291). In order to be able to handle the digitized signals, we use a CPU with a very high clock frequency. The heart of a DSP device is the software.

1.8.8.1. Audio DSP

Outboard DSP units for use in the receiver audio chain were the first available in the commercial market.

There are still a great number of excellent transceivers without DSP on the market that score very well on the low bands, eg the Drake R4C, Yaesu FT-1000D, Kenwood TS-830 and TS-930 (see survey results in Chapter 2). The basic performance of these receivers can be further enhanced by the use of external AF DSP systems.

DSP filters as a rule perform three different sorts of tasks:

1. Add variable, high-skirt selectivity, although obtaining selectivity this far down the receiver chain is a bad substitute for good IF filters. DSP filters can improve the reception of digital modes substantially, especially when older-technology converters are used.
2. Perform as an automatic (multi-frequency) notch filters: This is an area where DSP can excel. DSP units have been made available that can handle multiple carriers in the audio spectrum. The great advantage over the

analog (scanning-type) auto-notch filter (eg, the Datong FL3) is that the response time of the DSP unit is much shorter. This means that carriers are notched out before the user even notices a carrier came on. The great disadvantage of any notch filter at AF, is that the offending carrier is still present all through the receiver IF chain and will desensitize the receiver, through the AGC action. Ideally these notch filters should be inserted ahead of the AGC detector, and as far forward in the receiver chain as possible.

3. Noise reduction is another area where DSP can be very useful. Simplified one can say that noise reduction DSP action is generally based on the fact that information carrying signals have some pattern, while noise is totally random. We have to be very careful with algorithms based on such principles, because on the low bands we often look for extremely weak signals, that may now and then just pop out of the noise, and then disappear again, which may make some of the noise reducing systems mask these weak signals as well.

1.8.8.2. IF DSP

In the last few years, manufacturers have started putting their DSP circuits in the last IF of the receiver, mostly at 10 to 50 kHz. This low IF is still necessary as CPUs with operating frequencies high enough to allow operation at much higher IFs are either still too expensive or still under development.

The advantage of IF DSP is quite obvious: action before the detector and AC circuitry. Even at the low last-IF frequency (10 to 50 kHz) IF DSP has its definite advantages when we consider the auto-notch and the noise-reduction functions. Using the last IF for achieving the operational selectivity, without having sharp filters closer to the front end, proves to be disastrous especially when strong signals are involved on nearby frequencies. A particular transceiver using this approach turned out to be totally useless in a contest environment.

1.8.9. Audio filters

Audio filters, just like AF DSP circuits, can never replace IF filters. They can be welcome additions, however, especially on CW if your receiver lacks a good built-in filter. Introducing some AF filtering reduces any remaining wide-band IF noise, and can improve the S/N ratio. Removing some of the higher pitched hiss can also be quite advantageous, especially when long operating times are involved, such as in a contest (Ref. 237).

A wide variety of audio filters has been described in Amateur Radio literature, using either LC networks (Ref. 236, 246, 261, 265, 278, 283, 284 and 285) or op-amp systems (Ref. 200 and 264). Tong, G4GMQ, has developed a very effective auto-tune AF notch-filter system, which he incorporated in the Datong FL1 (Ref. 270) and FL3 filters.

1.9. Stability

State-of-the-art fully synthesized receivers have the

stability of the reference source. All present-day receivers have achieved a level of stability that is adequate for all types of amateur traffic.

1.10. Frequency Readout

Today's modern transceivers all have a frequency readout displaying frequency to the nearest 10 Hz, which is adequate for all modes of operation. A number of transceivers allow the user to choose exactly what the display will show. In order to avoid confusion, on CW the display should show the carrier frequency. Most older receivers and transceivers actually display the carrier frequency ± the beat note (usually 400 to 1000 Hz), and this can cause confusion when arranging a schedule on CW. In the digital modes the confusion is *Babble*-like.

1.11. Switchable Sideband on CW

Switchable CW sidebands is a very useful feature, which was first introduced in the Kenwood TS-850. The user can switch CW reception from lower sideband to upper sideband, just like in SSB. Although the terminology of lower and upper sideband is not so common on CW, CW signals are indeed received with the beat oscillator frequency either above (as an LSB signal) or below (as a USB signal). In the past, none of the commercial receivers offered the capability of switching sidebands on CW. This feature can be quite handy in our daily fight against QRM. Together with band-pass tuning, sideband switching can often move an offending signal down the skirts of your filter to a point where no harm is done. In recent years all high-class equipment has incorporated this feature.

1.12. Outboard Front-end Filters

Most of our present-day amateur receivers and receiver sections in transceivers are general coverage (100 kHz to 30 MHz). They make wide use of half-octave front-end filters, which do not provide any narrow front-end selectivity. Older amateur-band-only receivers used either tuned filters or narrow band-pass filters, which provide a much higher degree of front-end protection, especially in highly RF-polluted areas. Instead of providing automatic antenna tuners in modern transceivers, I believe that same space could more advantageously be taken up by some sharply tuned input filters that could be switched into the receiver when needed.

Excellent articles are available that describe selective front-end receiving filters (Ref. 219, 221, 251, 266 and 294). Martin (Ref. 219) and Hayward (Ref. 221) describe tunable preselector filters that are very suitable for low-band applications in highly polluted areas.

Whether or not such front-end filter will improve reception depends on the presence of very strong signals within the passband of the half-octave filters. Outboard narrow tuned filters may bring relief. Several designs have been published over the years by K4VX (Ref. 295), W3LPL (Ref. 2953), K1KP (Ref. 2954), and N6AW (Ref. 2952). Another popular and rather simple filter was designed by the members of the Bavarian Contest Club (Ref. 2951).

An excellent (but expensive) commercially made band-pass filter is available from the German Manufacturer, Braun (Karl Braun, Wiesgartenstr 21, 90559 Burgthann, Germany, tel 09183-8988, fax 09183-403434). The Braun preselector SWF1-40 covers all bands (including the 10, 18 and 24-MHz bands) from 10 through 160 meters, and includes an excellent preamp. The attenuation of the filters is 8 dB (see **Fig 3-13**). The preamp can compensate for this, and add perhaps an extra 8 dB, which may come in handy when using low-gain receiving antennas.

International Radio (formerly RCI / Fox Tango) also offers front-end crystal filters, which are the ultimate solution for multi-multi contest stations for protection against interference from a multiplier station operating at the same location on the same band. Information can be obtained at: **http://www.qth.com/inrad/**.

Many Top Banders experience problems with overload from local BC stations on 160 meters. Every situation may require a different approach to solve the problem. If the problem occurs with a special receiving antenna, such as a Beverage antenna, then a preselector as described above may bring a solution. Braun (see above) sells an excellent

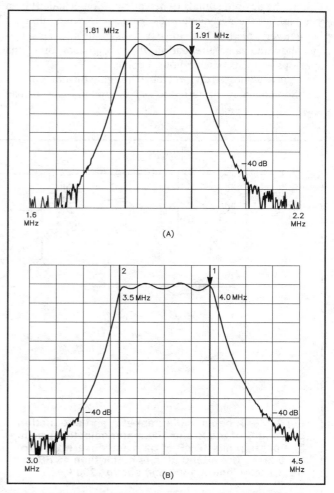

Fig 3-13—Attenuation curves for the Braun SWF 10-40 preselector at A on 160 and at B on 80 meters.

selective preamp, which uses a double-tuned input circuit plus a double-tuned output circuit, with a transistor preamp giving 8 dB of Gain, and an intercept point of +20 dBm. This preselector attenuates signals in the BC band by more than 45 dB (see **Fig 3-14** for schematic and band-pass curve). Similar circuits are available from other sources. More details on preamps in Chapter 7 (Beverage antennas).

Fig 3-15 shows the schematic diagram, the layout and the band-pass curve of a highly effective and popular BCI filter designed by W3NQN (Ref. 298, 299).

If you use no separate receiving antenna, and the problem exists when listening on your transmit antenna, you will need to install a similar filter, which you will have to bypass while transmitting.

1.13. Intermodulation Outside the Receiver

It was very interesting when John Sluymer, VE3EJ, mentioned to me the possible problem of intermodulation generated outside the receiver. If you hear what sounds like a spurious signal from a BC station in the ham bands, one way to tell if the product is occurring in your receiver is to insert an attenuator at the input of the receiver. If you observe a much greater change in the level of the spurious than in the desired signals (when attenuation is added) then you can bet the garbage comes from your own equipment being overloaded. If both the desired signals and the spurious show the same attenuation, then the generation of the spurious happens outside your receiver.

I have witnessed this problem with aging Beverage antennas. Sometimes it is referred to as bad ground loops, or bad contacts, but the effect of nonlinearity caused by corrosion can create the well-known effects of overload, cross modulation and intermodulation (in plain language—mixing). It isn't because you will not be running power into the Beverage antenna that you don't need to have good contacts in the system. If you suddenly hear all kinds of alien signals pop up in the band where they don't belong, it's time to go and check all the contacts of the receiving antenna system. Also check proper grounding of the coaxial feed line.

Such products can occur in poor electrical connections in cable TV (where aluminum cable was in contact with steel support strand), telephone wires, fences, towers, and even in your own antennas.

With broadband antennas, such as Beverages, it may be necessary to use high-pass filters or preselectors when operated in the vicinity of BC stations (see also Section 1.12).

1.14. Noise Blanker

A noise blanker, by nature of its principle of operation, is only suitable to deal with short duration ignition-type pulses. Noise blankers detect strong noise pulses, and block (gate) the receiver IF chain during the time that these pulses are present. In order to be able to detect these pulses, we use wide roofing (1st IF) filters, because narrow filters would distort these pulses and make noise blanking impossible. Noise blankers are one of the reasons why modern transceivers use very wide (much too wide) 1st IF filters, which leads to the poor close-in IMD performance as second mixers are easily overdriven by strong nearby signals. As the noise pulses are detected on our receiving frequency, rather than on any other

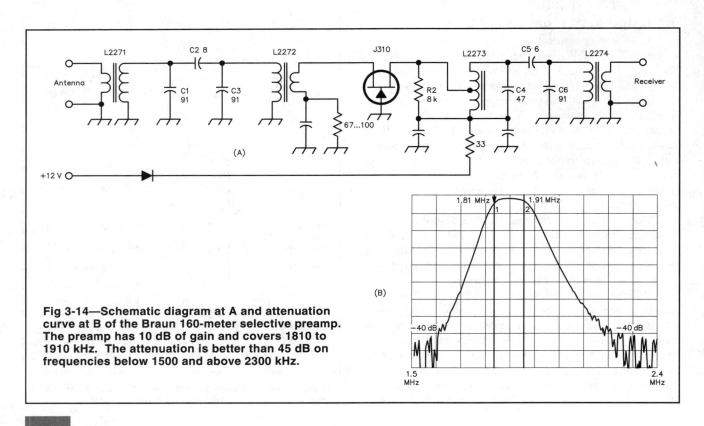

Fig 3-14—Schematic diagram at A and attenuation curve at B of the Braun 160-meter selective preamp. The preamp has 10 dB of gain and covers 1810 to 1910 kHz. The attenuation is better than 45 dB on frequencies below 1500 and above 2300 kHz.

frequency outside the busy amateur bands, noise blankers are as a rule very ineffective when the band is fully loaded, as during contests, as strong adjacent signals will gate the receiver. Use of a frequency outside the amateur bands to sense the noise, as was done by Collins in their add-on unit for the 75A-4 receiver and KWM-2 transceiver more than 30 years ago, would be a solution to that problem.

In my view, essential characteristics of a good receiver should never be compromised to achieve secondary goals. In other words, compromising on receiver close-in IMD performance to be able to have a working noise blanker, is bad engineering.

1.15. Receiver Evaluations

It is important that every avid low-band DXer understand the parameters that make a receiver good for the low bands. It is not possible for most of us to perform the tests ourselves, however.

The test methods have been very well defined in the amateur literature (Ref. 210, 211, 234 and 255), and Schwarzbeck, DL1BU, and Hart, G3SJX, have been publishing a series of excellent equipment evaluations in *CQ-DL* and *RadCom* (Ref. 400-444).

In order to minimize confusion, I have refrained from quoting test measurement data. Exhaustive test reports on the new equipment are published regularly by Schwarzbeck, DL1BU (in *CQ-DL*) and Hart, G3SJX (in *RadCom*), as well as in *QST* in recent years. Taking just a few test data out of several pages of test results does not seem fair to me. Anyone with a serious interest in these professional test reports can find all the reports listed in the Literature Review chapter. Chapter 2 (Operating) lists the results of a poll of low-band DXers, showing the equipment brands used.

1.16. Graphical Representation

Hart, G3SJX, uses an interesting graphical representation of the main receiver parameters. Two examples are shown in **Figs 3-16** and **3-17**. The following information can be found on the graphs:

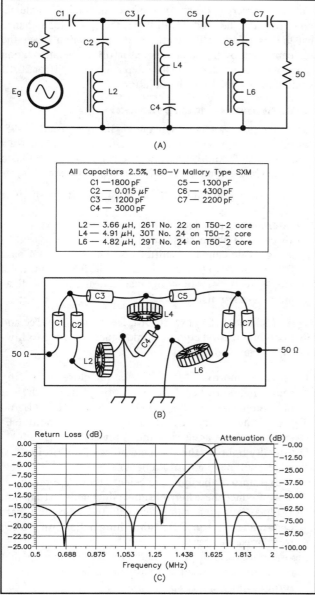

Fig 3-15—High-pass filter designed by W3NQN. Layout at A, schematic diagram at B and response curves at C.

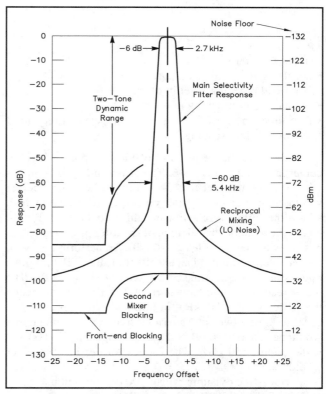

Fig 3-16—The receiver merit graph as introduced by G3SJX. This example is for an average-quality receiver with no outstanding features.

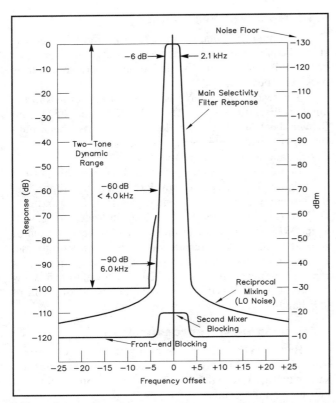

Fig 3-17—Merit graph for a "dream receiver." Note that the first-IF filter (roofing filter) has a bandwidth that is similar to the bandwidth of the main selectivity filter.

- Receiver noise floor.
- Dynamic filter response.
- Local-oscillator noise output.
- Front-end blocking level in dBm.
- Second-mixer blocking in dBm.
- Two-tone (spurious-free) dynamic range as a function of parent signal spacing.

The performance of a run-of-the-mill present-day receiver is shown in Fig 3-16. The graph in Fig 3-17 shows what a really good receiver would look like. This receiver would show a steep IF filter response (1.5 shape factor), where the ultimate rejection would be over 120 dB, and where the dynamic broadening of the filter passband would not show up before at least –100 dB. To perform this well, the receiver would need an excellent local oscillator with a noise sideband performance of greater than 134 dBc. The better receiver would have narrow first-IF filters to match the mode to be used (3 kHz for SSB and 500 Hz for CW), in order to have a two-tone spurious-free dynamic range that would be at least 100 dB both on close spacing (5 kHz) and wide spacing (50 kHz). The figures for our ideal receiver would read:

- Spurious-free dynamic range: 100 dB minimum.
- Noise floor: –130 dBm.
- Third-order intercept point: +30 dBm at full sensitivity (with preamp).

- IMD point level: –30 dBm (equivalent to over 10 mV or nearly S9 + 60 dB).
- LO sideband noise performance: Better than 135 dBc at close spacing (2 kHz).

1.17. In Practice

Now that you understand what makes a receiver good or bad for low-band DXing (and contesting) and after you study all the available equipment reviews, remember that what really counts is how the radio operates at your location, in your environment, with your antennas, how it satisfies your expectations, and how it compares to the receiver you have been using. The easiest test is to try the receiver when the band is really crowded, when signals are at their strongest. When you listen closely where it is relatively calm, you may hear weak crud that sounds like intermodulation or noise mixing products. If you insert 10 dB or 20 dB of attenuation in the antenna input line, and the crud is still there, there is a good chance that the signal is really there. As the attenuation raises the intercept point by the same amount in dB as the attenuation figure, it is likely that raising the intercept point by 10 or 20 dB would have stopped intermodulation.

1.18. Areas for Improvement

In the survey, which I sent out via Internet in early 1998, I asked: "What are the main characteristics that would make a dream receiver?" And, "What improvements would you like to see to the receiver you use now?" The top hits were:

1. Better strong signal handling capability also with close signal spacing.
2. Better and more selectivity.
3. Lower VCO phase noise.
4. Reduce wide band transmitted noise by using band-pass filters instead of low-pass filters on the PA output stage.
5. Truly effective systems against man-made as well as atmospheric noise.

Let's hope that Yaesu, ICOM, Kenwood, Ten-Tec and other communication equipment designers read the following. What they should do is:

- Drastically improve IMD behavior for close-in spacing (down to 500 Hz) eg, by using roofing filters with final operational bandwidth (2.0 kHz on SSB and 500 Hz on CW). If necessary we will be happy to trade-in full HF-spectrum coverage (use a much lower 1st IF).
- Drastically improve the sideband noise of our oscillators, which will have a twofold benefit:
  a) Improve the dynamic selectivity of the receiver.
  b) Reduce these horrible noise sidebands on transmit.
- Develop systems that can effectively reduce or eliminate man-made or static noise (something radically different from noise blankers).
- Incorporate variable make and break CW shaping that tracks the CW speed so that we can have ideal waveform shaping at any speed.
- Offer more (choices) and better (shape factor) IF filters, eg 800 Hz to 1 kHz selectivity for CW.

- Design and build quiet IF stages that don't hiss like an old steam-train.
- Develop true diversity receiving systems with automatic selection (including computer controlled auto-tune noise canceling systems).
- Move DSP forward in the IF chain, and write good and user friendly DSP software.
- Design and build quality audio stages, not the 1.5 W at 10% distortion that we have seen for years and years.
- Incorporate good front-end protection circuitry (for main as well as auxiliary antenna inputs).
- Make it possible to move the CW beat note as low as 200 Hz.

Instead of spending more money on bells and whistles like more memories, more sophisticated readouts, etc new development efforts should be channeled toward the needs of better basic performance, as specified above. If the concerned ham tells his supplier about his real demands, he will contribute to achieving this goal.

Development of modern Amateur Radio equipment is largely market driven. If the marketers keep telling the designers they want more bells and whistles, that is what the user will get. If users tell the manufacturers they want better basic performance often enough, maybe the designers will get the right message and we will see more progress toward better receiver performance.

1.19. Adding Input Protection to your Receiver Input Terminals

More and more new transceivers have a separate receiver input for use with special receiving antennas such as Beverages. If those receiving antennas are installed very close to the transmit antenna, dangerously high voltages may appear at the receiver input terminals, which may destroy the input circuitry of the receiver. As long as our equipment manufacturers do not incorporate a suitable protective circuit it may be wise to build one of your own.

Fig 3-18 shows a possible protective circuit. A small relay shorts the input of the receiver during transmit. The voltage for the relay may be obtained from any 12 V source (eg, from the transceiver itself), while the relay is switched by the amplifier control line. Two diodes make it possible to

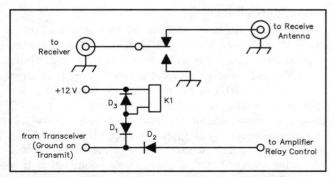

Fig 3-18—Receiver input protective circuit. Receiver input is grounded during transmit. D1, D2 and D3 are silicon diodes.

switch the amplifier and the protection circuit from the same line. It is clear that this circuit only protects your equipment from RF coming from the same transceiver. Where more than one transmitter is used (like in a multi-transmitter setup during contests), a different approach must be taken, such as using band-pass filters.

1.20. Noise-Canceling Devices

When one is plagued with a local single-source man-made noise, it often can be dramatically improved, or even eliminated all together by using a so-called noise-canceling device. In a noise-canceling unit signals from two antennas (one is the regular receiving antenna, the second one is called the noise source antenna) are combined (added) in such a way that the noise received on the noise antenna is of equal amplitude as noise received on the normal receiving antenna, and both exactly 180° out of phase with each other. Details of such noise-canceling devices can be found in Chapter 7.

■ 2. TRANSMITTERS

2.1. Power

It should be the objective of every sensible ham to build a well-balanced station. Success in DXing can only be achieved if the performance of the transmitter setup is well balanced with the performance of the receiving setup. It is true that you can only work what you can hear, but it is also true that you can only work the stations that can hear you. It is indeed frustrating when one can hear the DX very well but cannot make a QSO, and it must be frustrating for the station on the other end to hear a loud DX station calling without being able to raise it. We all know stations like that. Some characters just like to be loud. When they cannot hear the DX some even go as far as to make fictitious QSOs and "read the Callbook." Fortunately those are the rare exceptions.

A well-balanced station is the result of the combination of a good receiver, the necessary and reasonable amount of power and, most of all, the right transmitting and receiving antennas. It is, of course, handy to be able to run a lot of power for those occasions when it is necessary. In many countries in the world, amateur licenses stipulate that the minimum amount of power necessary to maintain a good contact should be used, while there is of course a limitation on the maximum power.

There are modes of communication where we have real-time feedback of the quality of the communication link, and among them are AMTOR, PACTOR as well as other similar error-correcting digital transmission systems. In CW as well as SSB, we can only go by feeling and by the reports received, and therefore we are most of the time tempted to run *power*.

There are a number of dedicated operators who have worked over 250 countries on 80 meters or 100 countries on 160 meters without running an amplifier. But it is true also, that a large majority of the active low-band DXers run some form of power amplifier, and that most of them run between 800 and about 1500-W output.

Fortunately, most low-band buffs get involved gradually in the DX game. Together with building a better receiving system (eg, with Beverage antennas), the need for a little more power will become apparent.

2.2. Linear Amplifiers

Today the newest technologies are utilized in receivers, transmitters and transceivers to a degree that makes competitive home construction of those pieces of equipment out of reach for all but a few. Most of our high-power amplifiers still use vacuum tubes, however, and circuit integration as we know it for low-power devices has not yet come to the world of high-power amplifiers. At any major flea market it is possible to buy all the parts for a linear amplifier (as in **Fig 3-19**).

Amateur Radio has come a long way in the past 30 years, from an era where the vast majority of amateur operators used all home-made equipment, to today where all but a few use state-of-the-art, high-tech (and fortunately also high-performance) equipment. Those among us who were in Amateur Radio 30 years or more ago will remember the immense degree of satisfaction we got from building our own equipment. There are at least two areas in Amateur Radio where the DXer can still get this kind of satisfaction: building his own amplifier and building his own antennas.

Those who say that homemade amplifiers are all running illegal power and that running excessive power is the only driving force behind home-brewing an amplifier have probably never built one themselves. The home builder will usually build more reserve into his design. He will have the option himself to spend a few more dollars on metal work and maybe on a larger power supply transformer in order to have a better product that runs cool all the time and never lets him down. Maybe he will use two tubes instead of one, and run those very conservatively so that the eventual cost-effectiveness of his own design will be better than for the commercial black box.

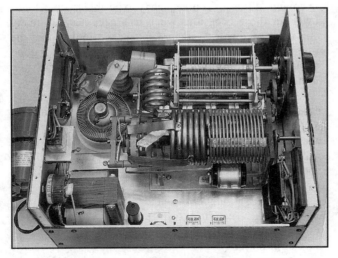

Fig 3-19—This home-built amplifier makes use of surplus parts obtained at a hamfest.

Excellent amplifier designs have been published in the Amateur Radio literature over the years. Very often the home builder will be driven by the availability of parts, especially final-amplifier tubes. Zero-bias triodes have become very popular over the past 25 years for linear service. The popular 811 was followed by the 872B, while the 3-500Z has been holding strong for over 25 years now. Triodes can usually take quite a bit of beating. The only thing really to watch is grid dissipation. Excessive grid current and early tube failure can result if the amplifier is too lightly loaded.

Whereas tetrodes such as the 4CX1000 and 4CX1500B used to be readily available at reasonable prices on the surplus market, they have become very rare and very expensive. Such tetrodes have the inherent advantage of requiring much less drive power, but on the other hand they require careful amplifier design and knowledgeable operation because of the very sensitive control grid and screen grid.

The range of modern ceramic US-made tubes have, in the last years, been supplemented by a list of Russian-made tubes that are available at much lower cost, and a large percentage of commercially made amplifiers now use these tubes as well.

Power amplifiers designers and builders now have their own reflector on the Internet, where very interesting information is exchanged between builders. To subscribe to the AMP reflector one can send a message to **amps-request@contesting.com** containing "subscribe" in the body text.

A lot of very valuable information about building your own amplifier can be obtained at AG6K's Web Page: **http://www.vcnet.com/measures/**. AG6K described the addition of 160 meters to the Heath SB-220 (Ref. 340), and in another article he covered the addition of QSK to the popular Kenwood TL-922 amplifier (Ref. 338). Information on both modifications are also available on AG6K's Web Page.

S. M. North, KG2M, described a conversion for the popular SB-200 linear amplifier to add 160-meter coverage (Ref. 341).

2.3. Phone Operation

If you choose to play the DX game on phone (SSB), there are a few points to pay great attention to.

2.3.1. Microphones

Never choose a microphone because it looks pretty. Most of the microphones that match (aesthetically) the popular transceivers have very poor audio. Most dynamic microphones have too many lows and too few highs. In some cases the response can be improved by "equalizing" the microphone output. Even if you have one of the specially designed communication microphones (eg, Shure 444), it still may be necessary to apply some tailoring to match the microphone to your voice and the transmitter you are using. The most simple form of microphone equalization is a simple RC high-pass T-filter in the microphone lead. I have successfully matched my voice to a Shure 444 microphone and a variety of transceivers by incorporating an RC filter

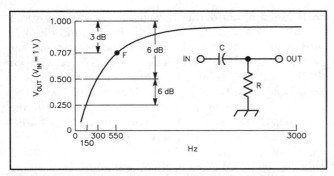

Fig 3-20—Attenuation characteristic of a single-pole high-pass RC network. The roll off for such a filter is 6 dB per octave, which means the output voltage is halved each time the frequency is halved. Identical sections can be cascaded (2 sections would produce 12-dB per octave roll off).

consisting of a 1000-pF capacitor with a 50-kΩ resistor. The 3-dB cut-off frequency is given by:

$$f = \frac{10^6}{2\pi RC} \approx 3.2 \text{ kHz}$$

where

    f = 3-dB cutoff frequency
    R = parallel arm resistor, kΩ
    C = series arm capacitor, nF

The optimum value of the capacitor can be determined by cut-and-try methods, while R should be roughly equal to the output impedance of the microphone. The curve of a single-section RC filter is given in **Fig 3-20**. Heil, K9EID, recognizes the problem of poor audio on our bands, and has tackled the problem by designing a multistage op-amp equalizer that really can do wonders for bad microphones and awful voices (Ref. 301 and 323).

Both lows and highs can be independently adjusted (enhanced or attenuated). The center adjusting frequencies for lows and highs are 500 and 2200 Hz, respectively.

In some cases the audio spectrum of a bad microphone can be drastically improved by changing the characteristics of the microphone resonant chamber. If the microphone has too many lows (which is usually the case), improvement can sometimes be obtained by filling the resonant chamber with absorbent foam material, or by closing any holes in the chamber (to dampen the membrane movement on the lower frequencies).

The most practical solution is to use a microphone designed for communications service. A typical communications microphone should have a flat peak response between 2000 and 3000 to 4000 Hz, a smooth roll-off of about 7 to 10 dB from 2000 to 500 Hz and have a much steeper roll-off below 500 Hz. **Fig 3-21** shows the typical response curve for the Heil communications microphone elements. We should caution against overkill here too, however! We know that the higher voice frequencies carry the intelligence, while the lower frequencies carry the voice power. Therefore a good balance between the lows and highs is

essential for maximum intelligibility combined with maximum power.

At this point it should also be mentioned that correct positioning of the carrier on the slope of the filter in the sideband-generating section of the transmitter is at least as important as the choice of a correct microphone. Therefore you should test your equalized microphone system into a good-quality tape recorder before doing the on-the-air tests. Incorrectly positioned carrier crystals will also produce bad-sounding receive audio in a transceiver, as the same filter is (in most transceivers) used in both the transmit and receive chain.

One way of checking to see if the USB and LSB carrier crystals have been set to a similar point on the filter slopes is to switch the rig to a dead band, turn up the audio, and switch from USB to LSB. The pitch of the noise will be a clear indication of the carrier position on the filter slope. The pitch should always be identical on both sidebands.

Modern transceivers allow tailoring of the AF bandpass curve through DSP. Some of the top range transceivers also make it possible to change the position of the carrier vs the filter curve via software programming (eg, the Yaesu FT-1000MP).

As important as the choice of the microphone is the use of the same. Communications microphones are made to be held close to the mouth when spoken into. Always keep the microphone at maximum of two inches from your lips. A very easy way to control this is to use a headset/microphone combination. Heil has various headset/microphone combinations, which can be equipped with either their HC4 or HC5 cell. The speaker phones in the original Heil Proset are notorious, however, for their poor sensitivity, "cheap" mechanical (plastic) quality and their inability to be driven to adequate volume from some Yaesu transceivers. There have been excellent reports on an RS combo (PRO-50MX) which sells for about 25% of its Heil counterpart. For years I have been using the very affordable Azden HS-03 combo, equipped with the appropriate Heil cell.

Even from a microphone point of view, nowadays what we would consider less than ideal communication cells, can be tailored with the transmitter DSP to sound like an expen-

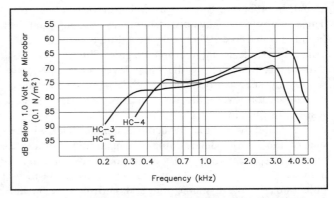

Fig 3-21—Typical response of Heil communication microphones with a 2000-ohm resistive load. Note the sharp cutoff below 300 and 500 Hz.

sive custom tailored element! Try the "cheap" mike with different DSP settings on you transmitter (if you have transmit DSP).

Using a headset/microphone combination makes DXing under the most difficult circumstances a real pleasure, even over very long periods. If you do not speak closely into the microphone, you will have to increase the microphone gain, which will bring the acoustic characteristics of the shack into the picture, and they are not always ideal. We often have high background noise levels because of the fans of our amplifiers. It is this background level, and the degree to which we practice close-talking into our microphone, that will determine the maximum level of clipping we can use in a system.

2.3.2. Speech processing

Speech processing should be applied to improve the intelligibility of the signal at the receiving station, not just to increase the ratio of average power to peak envelope power. This means that increased average power together with the introduction of lots of distortion may achieve little or nothing. Although audio clippers can achieve a high degree of average power ratio increase, the generation of in-band distortion products will raise the in-band equivalent noise power generated by harmonic and intermodulation distortion and in turn decrease the intelligibility (signal-to-noise ratio) at the receiving end.

RF clipping generates the same increase in the ratio of transmitted average power to PEP, but does not generate as many in-band distortion products. This basic difference eventually leads to a typical 8-dB improvement in intelligibility over AF clipping (Ref. 322). All commercial manufacturers of ham equipment realized this long ago, and virtually all the high end current transceivers are equipped with RF clippers.

Adjusting the speech-processor level seems to be a difficult task with some modern transceivers if you judge from what we sometimes hear on the air. Each RF clipper should have at least two controls. The first one controls the input to the clipper. Its setting will determine the amount of clipping. This control is usually called PROCESSOR or PRO-CESSOR IN. The second gain control (sometimes called PRO-CESSOR OUT or DRIVE) sets the output level of the processed signal; in other words it acts as an RF-drive level control. These controls will have different names depending on the brand of transceiver. There may be a third control on some transceivers, the actual microphone gain. This gain control will be used for setting the mike level when operating without the speech processor. All modern transceivers have a compression-level indicator, which is very handy in adjusting the clipping level.

We already said that the acoustics in the shack will be one of the factors determining the maximum allowable amount of speech clipping. By definition, a speech-clipped audio signal has a low dynamic range. In order not to be objectionable, the dynamic range should be kept on the order of 25 dB. This means that during speech pauses the transmitter output should be at least 25 dB down from the peak output power during speech. Let us assume we run 1400-W PEP output. A signal 25 dB down from 1400 W is just under 5-W PEP. Under no circumstances should our peak-reading wattmeter indicate more than 5-W peak (about 3-W average), or we will have objectionable background noise (Ref. 305). The clipping level should be increased by increasing the PROCESSOR IN control, until we come to the point where the shack ambient noise produces 5-W PEP maximum or where the transceiver compression level indicator indicates maximum 20 dB or whatever the manual recommends as a maximum clipping level.

The PROCESSOR OUT (DRIVE) control should only be used to obtain the correct amount of drive from the transceiver into the final or the correct amount of ALC. Never use the MICROPHONE GAIN or the PROCESSOR IN control to adjust the drive to the final.

2.4. CW Operation

Keying Waveform

In the old generation of transmitters, it was a simple matter of adjusting an RC network in order to adjust the leading and the tailing slope of the keying waveform. Today's transmitters are loaded with bells and whistles, useless memories, TX and TX clarifiers, memory scanners etc, but from the total lack of control possibilities of the CW waveform, it is clear that those who design our transceivers are certainly not very knowledgeable CW operators!

Most transceivers seem to have the waveform shaping adjusted for 70 wpm. It is obvious that a state-of-the art transmitter should incorporate a circuit that adjusts wave shaping to the CW speed, and this fairly simple task could easily be done automatically.

The problem of harsher keying seems to go hand in hand with the problem of oscillator phase noise. Some of the 40-year-old transmitters showed far better performance in this field than today's radios. Let this be a message to the people who review the new transceivers to put emphasis on this issue, so that the designers wake up!

QSK

QSK (full break-in) is a nice feature, but not essential, neither for the low-band DXer nor to the contester. If not properly designed and set up, QSK may be disastrous: it can generate severe key clicks, and it can ruin the antenna relay in the amplifier in no time. In a properly operating QSK setup, the amplifier relay will close as quickly as possible after keyer closure. A few ms later the transceiver will start transmitting. The opposite must be true at the end of each dot or dash: the transceiver first switches to receive, and then only the amplifier relay will switch. This prevents hot switching. The faster this whole sequence is, the better QSK will work at high speed.

In order not to send the RF too soon to the transceiver output, many designs seem to shorten the dots and dashes, which makes the CW sound very light. If a fully adjustable CW keyer is used, this may be adjusted from the keyer.

Even setups not operating in QSK often exhibit poor timing and hot switching. In every JA contest, I seem to copy a lot of OA stations calling me; these are extreme cases where the entire first dot is missing! The same problem sometimes turns a W station (USA) into an M station (England).

The better transceivers allow the user to fully adjust the delay and the hold time in QSK, which have to be adjusted when using widely varying speeds, depending on the amplifier's relay-closure timing.

Leading Edge Spikes

Another common problem with many modern transceivers is that they generate a power surge on the leading edge of the first character. This surge is in some cases twice the output as compared to the constant key-down output. This causes increased IMD distortion, sounding like key clicks, and can even trip the protective overdrive circuits from some commercial amplifiers.

In some transceivers this problem can be overcome turning the RF-out control from the transceiver down to the point where either you have enough drive or where the output power just starts dropping.

2.5. DSP in the Transmitter

The newest generations of HF transceivers make extensive use of DSP technology. Over the years we will certainly see more and more DSP-based gadgets in our transceivers. A few of the possible applications in the transmit chain are:

- DSP speech processing.
- DSP audio tailoring (nice velvet-like audio for a rag chew, and piercing sharp quality for the contest).
- CW make and break timing (hard or soft keying).
- DSP VOX control (delayed audio switching).
- Background noise elimination (kill the noisy blower).

2.6. Signal Monitoring Systems

It is essential for the station operator to have some means of monitoring the quality of his transmission. All modern transceivers have a built-in audio monitor system which allows the operator to check the transmitted signal. It should not be a mere audio output, but should be a detected SSB signal, which makes it possible to evaluate the adjustment of the speech processor. This feature allows the operator to check the audio quality and is very useful for checking for RF pickup into the audio circuits.

A monitor 'scope should be mandatory in any amateur station. With a monitor 'scope you can:

- Monitor your output waveform (envelope).
- Check and monitor linearity of your amplifier (trapezoid pattern).
- Monitor the keying shape on CW.
- Observe any trace of hot-switching on QSK.
- Check the tone of the CW signal (ripple on the power supply).

- Correctly adjust the speech processor.
- Correctly adjust the drive level of the exciter in order to optimize the make and the break waveform on CW and to avoid the leading edge overshoot.

I have been using a monitor 'scope on my stations ever since I was licensed almost 40 years ago, and without this simple tool I would feel distinctly uncomfortable when on the air. The specific monitor 'scopes (eg, Yaesu, Kenwood) are rather expensive and have one distinct disadvantage: You must route the output RF from the amplifier "through" the 'scope to tap off some RF which is fed directly to the plates of the CRT.

After having burned out three CRT tubes in such dedicated monitor 'scopes, I decided to use a good second-hand professional 'scope. Such 'scopes (eg, a Tektronix 2213) cost less than a new monitor 'scope. This approach gives you the advantage that you need to sample only a very small amount of RF to feed to the input of the 'scope. A small resistive power divider can be mounted at the output of the amplifier, from where the millivolts of sampled RF can be routed to the 'scope with a small coaxial cable.

2.7. Areas of Improvement

- Improve the third-order distortion products of the transmitter significantly (no 12-V PAs for fixed station PAs).
- Noise sidebands (VCO noise) down to at least –135 dB at 2 kHz separation.
- Easily and precisely adjustable power output.
- No leading-edge power spikes on CW.
- Fully adjustable (either by operators or speed tracking) make and break waveform on CW.
- SSB transmitter with max. 2.1-kHz bandwidth filters.
- Fully adjustable timing for QSK operation (to match the amplifier characteristics).

Equipment designers are hereby urged to aim their efforts toward improving the essential features as mentioned above, instead of adding mostly useless bells and whistles.

The bandwidth of the transmitted signal can also be significantly reduced on some transceivers without any detrimental effect. Where 20 years ago a bandwidth of 2.1 kHz (at –6 dB) was sufficient for good quality (eg, Collins mechanical filters), today most transceivers have a 2.7-kHz bandwidth. The audio quality may be more pleasing, but this can hardly be an acceptable reason for increasing the signal bandwidth on our crowded ham bands. International Radio (formerly RCI / Fox Tango), is supplying kits for popular transceivers. In most transceivers these better filters will not only dramatically improve the receiver performance but also reduce the bandwidth of the transmitted signal. Using those narrow-band filters requires more critical adjustment of the carrier position on the filter slope as a function of the operator's voice and the microphone characteristics. Check **http://www.qth.com/inrad/** for details.

ANTENNA DESIGN SOFTWARE

4 ANTENNA DESIGN SOFTWARE

■ 1. ANTENNA MODELING PROGRAMS

Until not too long ago, predicting antenna performance was more a black art than a scientific or engineering activity, especially in Amateur Radio circles. Building full-size models or scale models and testing them on wide-open test sites was out of reach to most amateurs. That was also the era when some of the old myths were born and that the rat-race for decibels was started.

Antenna modeling programs are computer programs that via mathematics calculate and predict the performance (electrical, mechanical) of an antenna, called a model. Modeling is done in all branches of science, physics and many other sciences. Modeling always has its limitations, partly because the model that we have to describe (enter into the program) can almost never be described in the same detail as the real thing (and especially its environment), and partly because of numerical limitations in the calculating code used. The final limitation is the operator, who enters the data and who interprets the results. In all cases a good deal of knowledge and experience in the field of antennas is required in order to draw the correct conclusions and make the right decisions during the process of modeling.

One of the first Yagi modeling programs that was reported in the literature was written in 1965 by J. L. Morris for his PhD dissertation at Harvard University. Others (Mailloux, Thiele, Cheng and Cheng) have elaborated on these programs to perform further analysis and optimization.

Such a program was used by Hillenbrand (N2FB) to optimize Yagis. This program was later adapted for use on the IBM PC by Michaelis (N8TR—formerly N8ATR).

1.1. *MININEC*-Based Programs

Today, every more-or-less serious amateur who has any interest in antenna building has a PC and a copy of the ever so popular antenna modeling program *MININEC* (or derivative programs such as *MN* and *ELNEC*). *MININEC* (Mini Numerical Electromagnetic Code) was developed at the NOSC (Naval Ocean Systems Center) in San Diego by J. C. Logan and J. W. Rockway. The newest version of the software (*MININEC3* at this writing) is public-domain software and can be obtained with the documentation from the NTIS, US Department of Commerce, 5285 Port Royal Rd, Springfield, VA 22161, order no. ADA 1811681.

The technical reference, describing the program (*The New MININEC, Version 3: A Mini Numerical Electromagnetic Code, NOSC TD 938*), is available from the NTIS

as well. Order document no. ADA 181682. A fee is charged for the program and its documentation.

The original *MININEC* is not a user-friendly program. Several hams have written the necessary pre and post-processing codes to make *MININEC* a user-friendly and powerful modeling tool.

The most popular version is the version known as *ELNEC*, developed by R. Lewallen, W7EL (available directly from W7EL, PO Box 6658, Beaverton, OR 97007). Another version is known as *MN* (by B. Beezley, K6STI, 3532 Linda Vista Dr, San Marcos, CA 92069). Both are regularly advertised in the major Amateur Radio magazines. Both are MS-DOS programs.

1.1.1. How it works

In *MININEC* the user splits up all the conductors (called wires) of an antenna into more or less short segments. During modeling, the HF current in each segment is considered to be constant (one current segment equals a pulse). The program calculates the self impedance and the mutual impedances for each of the pulses, as well as the field created by the contribution from each pulse with its self impedance and range of mutual impedances. I explain what mutual impedance is in Chapter 11 covering Arrays. The user can specify where he wants to excite the antenna (the source) and, if he wants, can put loads (eg, loading coils, capacitor, tuned circuits and resistors) anywhere in the antenna. Modeling can be done in free space, over perfect ground or over real ground.

Specific modeling issues, such as the required segment length, the segment length tapering technique, etc, are covered in specific antenna chapters (Verticals, Dipoles, Yagis and Quads) where relevant.

1.1.2. Limitations

The major limitation concerns calculations over real ground. The real-ground modeling capability is limited to modeling far-field patterns. In the near field (right near the antenna), a perfectly conducting ground is assumed.

Some of the consequences are:

- You cannot use *MININEC* to calculate the influence of radials on the feed-point impedance of a ground-mounted vertical. A quarter-wave vertical will yield a 36-ohm impedance over any type of ground. In reality the ground and the radials in the near field are important in collecting the return currents. This will influence the feed-point impedance and the efficiency of the antenna due to "lost return currents" in a poor ground. Radials can be specified with *MININEC*, but they will influence only the low-angle reflection-attenuation in the far field (in the Fresnel Zone). See Chapters 8 and 9 on dipole antennas and vertical antennas for details.

- The reported gain as well as the impedance of horizontally polarized antennas at low heights are incorrect. By low I mean less than 0.25 wavelength above ground for dipoles. For larger antennas the minimum height may be higher. At lower heights the reported gain will be too high and the feed-point impedance too low. The shape of the radiation patterns will remain correct, however.

This means that we have a certain handicap when using *MININEC* on the low bands, as very often we will be modeling antennas under the conditions specified above. As long as we know the limitations, and how to interpret the results, all is okay. For modeling antennas such as Yagis on the higher frequency bands, this is unlikely to be a problem. There are other modeling problems with quads. These are covered in Chapter 13 on Yagis and Quads.

Also, wires that are thicker than 0.001 wavelength may not be modeled accurately due to computational approximations in the code. While low band antennas will not be affected, this limitation may be encountered when working on antennas for 10 meters and higher frequencies.

These, and other limitations, are very well covered in good detail by R. Lewallen in "MININEC: The Other Edge of the Sword" (Ref. 678).

1.2. The *NEC* Modeling Program

NEC is the full-fledged brother of *MININEC*, which means that *NEC* also employs the method of moments to model antennas. The original versions ran on main-frame computers only, and were accessible to professionals only. While *NEC2* is public domain software, the most recent version, *NEC4* has only recently been released to the US-public. The copyright for *NEC4* is held by Lawrence Livermore National Labs (**http://www.llnl.gov**) and one must get a license from them in order to use either *NEC4* or any of the other software packages which use the *NEC4* core (eg, *EZNEC 4.0*). Non US citizens must apply for a license through their Embassy although there is a possibility that this may soon become a simpler, direct application process.

NEC2 will also model real ground in the near field. It will do away with most of the limitations I explained for *MININEC*. It can model antennas quite close to the ground, as well as radials above, and on, the ground. *NEC3* and *NEC4* also have the capability to model buried conductors such as radials and ground stakes. One notable limitation of *NEC2* is its inability to model stepped-diameter wires (such as tapered Yagi elements. This problem has been corrected in *NEC4*.

I have frequently used *NEC* to model antennas where the limitation of *MININEC* would have made the results unreliable. I will come back to specific modeling issues when discussing those antennas (eg, Beverages, low delta loops, elevated radials).

The original *NEC* programs were really intended for the professionals, and has a very unfriendly user interface. In the last years however, a number of user-friendly *NEC*-based programs have been developed. One of the most user-friendly ones is undoubtedly *EZNEC* by W7EL, Roy Lewallen (e-mail address: **w7el@teleport.com**. The looks and the user interface are identical to *ELNEC*, which means that an experienced *ELNEC* user will master *EZNEC* in no time. In both *ELNEC* as well as *EZNEC* data are entered in a spreadsheet-like environment, equipped with many tools for shortcuts. This way of entering is felt by most users as being far superior to other systems where the data have to be entered as text files, which means they are less interactive and more complicated to manipulate. **Fig 4-1** shows the "View Antenna" screen of the model representing a ^{1}/$_2$-wavelength long Beverage for 160 meters, using 2 quarter-wave in-line terminations at each end (see Chapter 7, Section 3.2).

NEC4WIN is a *NEC2*-based modeling program written in 32-bit code for Windows 95 (and later). Details and order info can be obtained at: **http://www.CAM.ORG/~mboukri**.

A most useful *NEC*-based software package is *NEC-Win Pro*, sold by Nittany Scientific. This program is based on *NEC2* and combines the accuracy of *NEC* with a user-friendly Windows interface. Additionally a program called *NEC-Win Basic* is available at lower cost. While this package does not have the full capabilities of *NEC-Win Pro*, the user interface is even simpler. Details and order information can be obtained on: **http://www.nittany-scientific.com**.

1.3. Optimizing Programs

With a regular *MININEC* or *NEC*-based program, you will have to spend quite some time if you want to optimize a design for a given parameter (could be gain, F/B or maybe impedance or SWR bandwidth). For this kind of applications so-called Optimizing Programs can be quite helpful. *YO* (Yagi Optimizer) from B. Beezley (K6STI) was the first optimizing program around, but works only for Yagi antennas. You can optimize for any of the above mentioned characteristics or any weighted combination of those parameters. *AO* (Antenna Optimizer) is a similar program for any type of antenna. Both are based on *MININEC*.

The strength of an optimizer program lies undoubtedly in the strength of the algorithm used. When using an optimizer program one must be extremely cautious to keep an eye on all performance parameters, or weigh the different performance issues very carefully. Both *YO* and *AO* use what is called an unconstrained local optimization. This means that the computer begins adjusting the antenna at user defined variables until a performance maximum is obtained. The danger here is that while the computer has reached a LOCAL maximum, there may be a better solution that has not been discovered because it is too far from the starting points. Additionally the computer might "run-away" and give false results that are not physically possible to construct.

To date, the only optimization on *NEC* has been *NEC-OPT* which was sold by Paragon Technologies. The software was very expensive and, as a result, was purchased only by

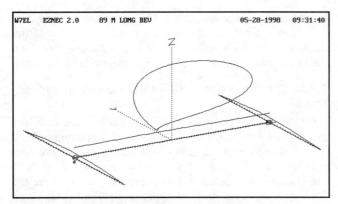

Fig 4-1—Screen dump of a View Antenna screen generated by *EZNEC*. See text for details. *NEC/Wires* is another *NEC*-based program by B. Beezley, K6STI. Both are DOS-based programs.

a few individuals and companies. Additionally their user interface is very difficult to use and can only be managed by a professional. Due to its limited sales, this software is no longer sold. There are companies, such as Nittany Scientific, that plan to introduce *NEC*-based optimization codes, but to date nothing is available.

It must be understood that an optimizing algorithm is just a mathematical code, that works purely on figures. Real optimizing must, to a large degree, come from the brain of the antenna designer. One must first have a very good idea as to what the final antenna should look like. For example, you must input a 5-element Yagi and then adjust the lengths and spacings to achieve desired goals: you cannot just tell the computer to design a good antenna for 20 meters.

The designer should understand what is happening in the modeling program, and then let the optimizing system assist him in calculating what he has conceived. Most importantly, one must retain control at all times, and realize when the program is giving answers that are not possible.

1.4. The Modeling Programs and This Book

ELNEC and *EZNEC* are used throughout this book for antenna modeling and for generating radiation pattern plots. In each chapter I will come back to the specific issues regarding modeling of the antennas covered in that particular chapter.

■ 2. THE ON4UN LOW BAND SOFTWARE

The nice thing about personal computers is that everyone can now handle the difficult mathematics that are part of doing the complicated calculations when dealing with antennas and feed lines. All you need to do is understand the question . . . and the answers. The program will do the hard mathematics for you and give you answers that you can understand. The theory of antennas and feed lines is not an easy subject. When it comes down to calculating antennas and feed lines, we are immediately confronted with complex mathematics, mathematics with real and imaginary parts, numbers that have a magnitude and an angle.

We have all been brought up to know how much is 5 times 4. But nobody can tell off the top of his head how much $5 - j3$ times $12 + j12$ is. At least I cannot. When I started studying antennas and wanted not only to "understand" the theory, but also to be able to calculate things, I was immediately confronted with the problem of complex mathematics. That's where the computer came to my aid. While studying the subject I wrote a number of small computer programs that were meant as calculating aids. They have since evolved to quite comprehensive engineering tools that should be part of the software library of every serious antenna builder. The NEW LOW BAND SOFTWARE also includes a number of low-band dedicated propagation programs.

The NEW LOW BAND SOFTWARE is based on the original "Low Band DXing Software" which I wrote in the mid 1980s, while preparing the original *Low Band DXing* book. The new software is a very much enhanced version of the original software, and it also contains numerous new programs. In addition it was written to be much more user friendly. The full-color software is now available only in MS-DOS on a single high-density 3¹/₂ inch diskette. **Fig 4-2** shows the Menu of the software package, listing the various modules that can be selected.

Each of the modules starts with a complete introduction (on screen), telling what the software is meant to do, and how to use it. All propagation-related programs are integrated into a single module. Also new are the many help screens in each of the modules. They explain what the program does, how the questions should be answered, and how the final results should be interpreted.

2.1. Propagation Software

The propagation software module is covered in detail in Chapter 1 on propagation. It contains a low-band dedicated sunrise/sunset program and a gray-line program, based on a comprehensive database containing coordinates for over 550 locations, and which can be user changed or updated. The database can contain up to 750 locations. A few screens from the propagation module are shown in Fig 32 and Fig 33 of Chapter 1 on propagation.

2.2. Mutual Impedance and Driving Impedance

From a number of impedance measurements you can calculate the mutual impedance and eventually, knowing the antenna currents (magnitude and phase), you can calculate the driving impedance of each element of an array with up to 4 elements.

2.3. Coax Transformer/Smith Chart

The original software covered only ideal (lossless) cables. Now there are two versions of the program: for lossless cables and for "real" cables (cables with losses). The real cable program will tell you everything about a feed line. You can analyze the feed line as seen from the generator (transmitter) or from the load (antenna). Impedance, voltage and currents are shown in both rectangular coordinates (as real and imaginary part) or in polar coordinates (as magnitude and phase angle). You will see the Z, I and E values at the end of the line, the SWR (at the load and at the generator), as well as the loss—divided into cable loss and SWR loss.

A number of "classic" coaxial feed lines with their transmission parameters (impedance, loss) are part of the program, but you can specify your own cable as well. Try a 200-ft RG-58 feed line on 28 MHz with a 2:1 SWR and compare it to a ³/₄-inch Hardline with the same length and SWR, and find out for yourself that a "big" coax is not necessarily there for power reasons. It makes no sense throwing away 2 or 3 dB of signal if you have to spend a lot of effort in building a top performance antenna. This program lets you juggle with facts and figures without any hassle. If you are going to design your own array, you will probably use this software module more than any other.

2.4. Impedance, Current and Voltage Along Feed Lines

Again, there are two versions of each module: loss free, and "real" cable.

2.4.1. Z, I and E Listing

A coaxial cable, when not operated as a "flat" line (SWR greater than 1:1) acts as a transformer: The impedance,

current and voltage are different in each point of the cable.

You enter the feed-line data (impedance, attenuation data), the load data (impedance and current or voltage), and the program will display Z, I and E at any point of the cable.

2.4.2. Simultaneous voltage listing along feed lines

This module was written especially as a help for designing a KB8I feed system for arrays. The program lists the voltage along feed lines. This allows the user to find points on the feed lines of individual array elements where the voltages are identical. These are the points where the feed lines can be connected in parallel (see Chapter 11 on arrays). This program is also helpful to see how high the voltage really rises on your feed line with a 4.5:1 SWR, for example.

2.5. Two and Four-Element Vertical Arrays

These two completely new modules take you step by step through the theory and the practical realization of a 2-element (cardioid) or 4-element (4-square) array, using the W7EL feed system. This tutorial and engineering program uses graphic displays to show the layout of the antenna with all the relevant electrical data. This unique module is extremely valuable if you want to understand arrays and if you are tempted to build your own array with a working feed system.

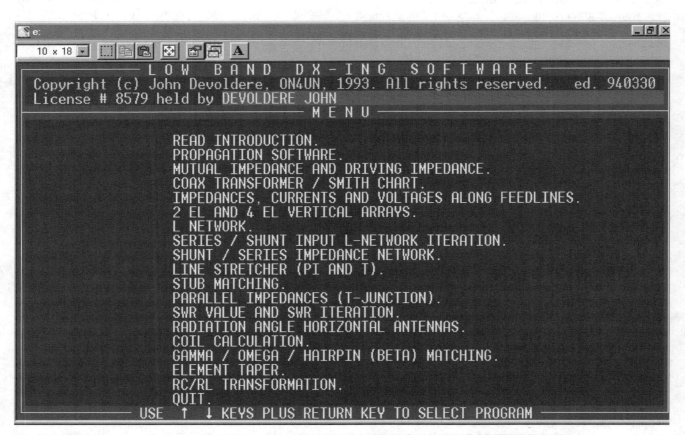

Fig 4-2—Screen dump showing the Menu page of the ON4UN NEW LOW BAND SOFTWARE package.

2.6. The L Network

The L network is the most widely used matching network for matching feed lines and antennas. The module gives you all the L-network solutions for a given matching problem. The software also displays voltage and current at the input and output of the network, which can be valuable in assessing the component ratings of the network. **Fig 4-3** shows a screen dump of the L-network module, showing all the relevant data.

2.7. Series/Shunt Input L-Network Iteration

This module was written especially for use in the K2BT array matching system, where L networks are used to provide a given voltage magnitude at the input of the network, given an output impedance and output voltage. See Chapter 11 on phased arrays for details.

2.8. Shunt/Series Impedance Network

This is a simplified form of the L network, where a perfect match can be obtained with only a series or a shunt reactive element. It is also used in the modified Lewallen phase-adjusting network with arrays that are not quadrature fed (see Chapter 11 on vertical arrays).

2.9. Line Stretcher (Pi and T)

Line stretchers are constant-impedance transformers that provide a required voltage phase shift. These networks are used in specific array feed systems (modified Lewallen method) to provide the required phase delay. See Chapter 11 on vertical arrays for details.

2.10. Stub Matching

Stub matching is a very attractive method of feed-line matching. This module describes the method of matching a feed line to a load using a single stub placed along the transmission line. The program is very handy for making a stub matching system with an open-wire line feeding a high-impedance load (2000 to 5000 ohms).

2.11. Parallel Impedances (T Junction)

This very simple module calculates the resulting impedance from connecting in parallel a number of impedances (do you really want to calculate on your calculator what $21 - j34$ and $78 + j34$ ohms are in parallel?).

2.12. SWR Value and SWR Iteration

2.12.1. SWR Value

This module calculates the SWR (eg, the SWR with a load of $34 - j12$ ohms on a 75-ohm line). The mathematics are not complicated, but it's so much faster with the program (and error free!).

2.12.2. SWR Iteration

This module was especially developed for use when designing a W1FC feed system for an array (the hybrid coupler). See Chapter 11 on arrays for details.

2.13. Radiation Angle of Horizontal Antennas

This module calculates and displays the vertical radiation pattern of single or stacked antennas (fed in phase).

2.14. Coil Calculation

With this module you can calculate single-layer coils and toroidal coils. It works in both directions (coil data from required inductance, or inductance from coil data).

2.15. Gamma-Omega and Hairpin Matching

This module is a simplified version of one of the modules of the YAGI DESIGN software (see information later in this chapter). Given the impedance of a Yagi and the diameter of the driven element (in the center), you can design and prune a gamma or omega match and see the results as if you were standing on a tower doing all the pruning and tweaking.

2.16. Element Taper

Antennas made of elements with tapering diameters show a different electrical length than if the element diameters had a constant diameter. This module calculates the electrical length of an element (quarter-wave vertical or half-wave dipole) made of sections with a tapering diameter. A modified W2PV tapering algorithm is used.

The NEW LOW BAND SOFTWARE is available in MS-DOS format on a single $3^1/2$-inch disk from: J. Devoldere, ON4UN, Poelstraat 215, B9820 Merelbeke, Belgium, or from G. Oliva, K2UO, 5 Windsor Dr, Eatontown, NJ 07724.

Price: $65 + $5 for shipping and handling worldwide. Prepayment only, US bank check or international money order. An order form can be detached from the back of this book.

■ 3. THE ON4UN YAGI DESIGN SOFTWARE

Together with Roger Vermet, ON6WU, I have written a number of software programs dealing with both the electrical and the mechanical design of monoband Yagis. These programs were used for the Yagi designs presented in Chapter 13 on Yagis and quads.

The 3-element 40-meter Yagi that I have been using since 1989, as well as all my other HF band Yagis, have been designed with the YAGI DESIGN software. The 40-meter Yagi was instrumental in setting two all-time European records in the 1992 ARRL CW and Phone Contest on 40 meters. KS9K, one of the top US midwest contest stations, has been using designs from this software program to rebuild the entire antenna farm.

YAGI DESIGN is a multifunctional software package that will take the user through all the aspects of Yagi designing (mechanical as well as electrical). It is not a modeling program, but is based on a comprehensive database containing all the dimensional and performance data for 100 different HF Yagis (2 to 6 elements). The database contains approximately 20 reference designs by W6SAI, W2PV, N2FB etc, while the majority are newly designed Yagis with a range of properties that are well described in a manual

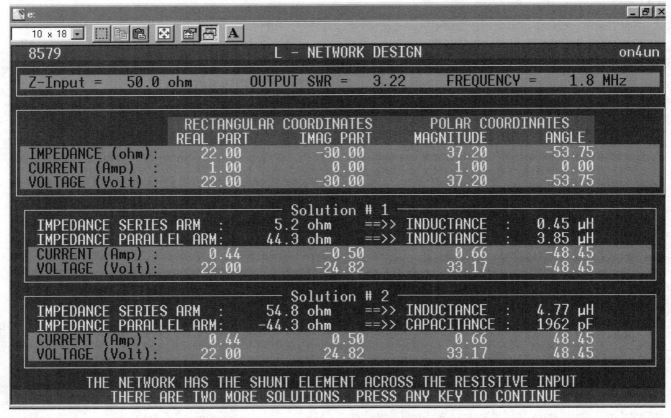

Fig 4-3—Screen dump of the L-network module of the ON4UN NEW LOW BAND SOFTWARE. This module is used very extensively in Chapter 11 on arrays. All relevant data (Impedance, Current and Voltage) are shown in both Cartesian (as a + *j*b) as well as polar coordinates (as A/*b*°).

which is available with the software. The literature standards are included so that the user has a known base of reference to compare the new designs. Most of the new designs were verified by either modeling them on a scale frequency (72 MHz) or by making full-size HF-band models.

The YAGI DESIGN database has a Yagi for every application: from low to high-Q, contest, CW only, SSB only, narrow band, wide band, gain optimized, F/B optimized, etc. One of the software modules also allows you to create text (ASCII) input files for the *MN*, *AO* and *YO* modeling programs. This allows you to further change and manipulate any of the designs from the system database.

The mechanical design modules are based on the latest issue of the EIA/TIA-222-E standard, which is a much upgraded version of the older, well known EIA RS-222-C specification. The "cross flow principle" is used to determine the effect of wind on a Yagi. Most amateur literature, as well as amateur mechanical design software, uses the principle of "variable area," which has no scientific grounds (see Chapter 13 on Yagis and quads).

The YAGI DESIGN software consists of several modules, which are briefly described. Each time you leave a module, you can save the results in a work file that you can recall from any other module. You can also view the contents of the work file at any time, using the VIEW DATA FILES module.

3.1. The Analyze Module

Unless you are very familiar with the content of the database, it would take you a long time to browse through all the performance and dimensional data to make your choice. The main-menu option print database prints out the content of the entire database, either in a tabular format (only the key characteristics) or it can generate a full-blown data sheet for all the Yagis (with two designs per printed page, which represents a little booklet of 50 pages).

In the ANALYZE module you can specify some key characteristics such as boom length (expressed in either wavelengths, feet or meters), minimum gain, minimum F/B, maximum Q factor, etc. The software will automatically select the designs that meet your criteria.

3.2. Generic Dimensions

Select the SELECT DESIGN module. After having chosen a proper design from the system database, the screen will display all the data relevant to this design—gain, F/B, impedance, etc on the design frequency and 6 other frequencies spread up to +1.5 and –1.5% of the design frequency.

You must now enter the design frequency (eg,

14.2 MHz). The screen now displays all the generic dimensions of the Yagi for the chosen design frequency. "Generic" means that the element lengths given in inches as well as centimeters are valid for an element diameter-to-wavelength ratio of 0.0010527. These are not the dimensions we use for constructing the Yagi, as the element will be made of tapered sections. The screen display also shows the amount of reactance that the driven element has on the design frequency. The element positions as listed are those that will be used in the final physical design.

3.3. Element Strength

Before we calculate the actual lengths of Yagi elements with tapering sections, we must first see which taper we will use. What are the required diameters and taper schedule that will provide the required strength at minimal cost, weight and element sag?

The ELEMENT STRENGTH module helps you build elements of the required strength at a minimum weight. Up to 9 sections of varying diameters can be specified (that's enough sections even for an 80-meter Yagi). Given the lengths (and overlap) of the different sections and the wall thickness as entered from the keyboard, the program calculates the bending moments at the critical point of every section. The module lets you specify wind speeds and ice loading as well as a vibration-suppression internal rope and several types of aluminum material.

3.4. Element Taper

It's time now to calculate the exact length of the tapered elements. We follow the taper schedule we obtained with the ELEMENT STRENGTH module. An improved version of the well-known W2PV algorithm is used to calculate the exact length. A wide range of boom-to-element clamps (flat, square, L, rectangular, etc) can be specified. These clamps influence the eventual length of the tapered elements.

3.5. Mechanical Yagi Balance

This mechanical design module performs the following tasks.

3.5.1. Boom strength

This routine calculates the required boom diameter and wall thickness. An external sleeve (or internal coupler) can be defined to strengthen the central part of the boom. If the boom is split in the center, the sleeve or the coupler will have to take the entire bending moment.

Material stresses at the boom-to-mast plate are displayed. Any of the dimensional inputs can be changed from the keyboard, resulting in an instantaneous new display of the changed stress values.

3.5.2. Weight balance

Many of the newer computer optimized Yagis have nonconstant element spacing, and hence the weight is not distributed evenly along the two boom halves. The WEIGHT BALANCE section shifts the mast plate (attachment point) on the boom until a perfect weight balance is achieved. It is nice to have a weight-balanced Yagi when laboring to mount it on the mast!

3.5.3. Yagi wind load

The program calculates the angle at which the wind area and wind load are largest. In most literature the wind area and wind load are specified for a wind angle of 45 degrees (a wind blowing at zero degree angle is a wind blowing along the boom; at 90 degrees it blows right onto the boom). This is incorrect, because the largest wind load always occurs either with the boom broadside to the wind or with the elements broadside to the wind. With large low-band antennas, it is likely that the elements broadside to the wind (wind angle equal to zero degrees) produces the largest wind thrust. With higher frequency long-boom Yagis having many elements (eg, a 5 or 6-element 10 or 15-meter Yagi), the boom is likely to produce more thrust than the elements. The wind load is calculated in increments of 5 degrees, given a user-specified wind speed.

3.5.4. Torque balancing

Torque balance ensures that the wind does not induce any undue torque on the mast. This can only be achieved by a symmetrical boom moment. When the boom-to-mast plate is not at the center of the boom, a "boom dummy" will have to be installed to compensate for the different wind area between the two boom halves. The program calculates the area and the position of the boom dummy, if required.

3.6. Yagi Wind Area

Specifying the wind area of a Yagi is often a subject of great confusion. Wind thrust is generated by the wind hitting a surface, exposed to that wind. The force is the product of the dynamic wind pressure multiplied by the exposed area, and with a so-called drag coefficient, which is related to the shape of the exposed body. The "resistance" to wind of a flat body is obviously different from the resistance of a round-shaped body. This means that if we specify or calculate the wind area of a Yagi, we must always specify if this is the equivalent wind area for a flat plate (which should be the standard) or if the area is simply the sum of the projected areas of all the elements (or the boom). In the former case we must use a drag coefficient of 2.0 according to the latest EIA/TIA-222-E standard, while for (long and slender) tubes a coefficient of 1.2 is applicable. This means that for a Yagi, which consists only of tubular elements, the flat-plate wind area will be 66.6% lower (2.0/1.2) than the round-element wind area. The WIND AREA module calculates both the flat-plate wind area and the round-element wind area of a Yagi.

3.7. Matching

The software provides three widely used matching systems: gamma, omega and hairpin. When choosing the gamma or omega system, you will be asked to enter the antenna power, as the program will calculate the voltage across and current through the capacitor(s) used in the system. If no match can be found with the given element length and diameter as well as gamma (omega) rod diameter and spacing (eg, very low radiation resistance and not enough negative reactance), then the program leaves you the choice of either changing the physical dimensions of the components (diameter of rod and rod-to-element spacing in order to change the system step-up ratio) or to shorten the element length to introduce some negative feed-point reactance (see above). In all cases a match will be found.

With a hairpin match the procedure is even simpler. The program will tell you exactly how much you will have to shorten the driven element (from the length shown in the table under "generic dimensions") and how long the hairpin should be.

The program also lists the matching data over a total frequency range of 3% (from −1.5 to +1.5% versus the design frequency, in 0.5% steps). These data include antenna impedance before matching, antenna impedance after matching, and SWR value. These data are very important for assessing the bandwidth characteristics of the antenna.

3.8. Optimize Gamma/Omega

Maybe you would like to see if other dimensions (lengths, spacings, diameters) of your gamma (omega) system would result in more favorable matching-system components?

Maybe you would like to "balance" the SWR curve? Most (not all) Yagis show an intrinsic asymmetric SWR curve, which means that the SWR rises faster above the design frequency than below. If you want to have the same SWR values on both band ends, it is obvious that the SWR will not be 1:1 at the center frequency. The OPTIMIZE GAMMA/OMEGA module allows you to change any of the matching-system variables while immediately observing the results of the output impedance and the SWR value. You can also change from gamma to omega and vice versa. Changing the variables from the keyboard simulates tuning the Yagi in practice. The module is also very well suited for "balancing" the SWR over a given frequency range.

3.9. Feed Line Analysis

When designing a Yagi, you must have a look at the feed line as well. It makes no sense to build an optimized long Yagi, where every inch of metal in the air contributes to gain (and F/B) and then to throw half of the boom length away by using a mediocre, lossy feed line.

The FEED LINE ANALYSIS module assesses the performance of the feed line when connected to the Yagi under design. The characteristics of the most current 50-ohm coaxial cables are part of the software (from RG-58 to 7/8-inch Hardline), but you may specify your own (exotic) cable as well.

3.10. Rotating Mast Calculation

A weak point in many Yagi installations is the rotating mast. The MAST module calculates the stresses in the rotating mast for a mast holding up to ten stacked antennas.

3.11. Utilities

3.11.1. Make input files for *YO*, *MN* or *AO*

The popular Yagi modeling programs *YO* (Yagi Optimizer), *MN* (MININEC) and *AO* (Antenna Optimizer) by Beezley (K6STI) require data inputs in the form of a text file (ASCII file). The YAGI DESIGN software package contains a program which automatically creates a text input file of the correct format for *YO*, *MN* or *AO*.

In the case of *MN* you can also specify a stack of two antennas that are identical (and fed in phase), or different (eg, a 15-meter and a 10-meter Yagi). In this way you can model any of the 100 designs of the database in either *YO* or *MN* without having to type those horrible text-input files where you're bound to make errors.

3.11.2. Your own database

If you'd like to add your own designs, the software package has provided an empty database that can contain up to 100 records (Yagis). The OWNDATA module is used to enter all the dimensional and performance data in the database.

The YAGI DESIGN software is available in MS-DOS format on a single 3¹/₂-inch disk from: J. Devoldere, ON4UN, Poelstraat 215, B9820 Merelbeke, Belgium, or from G. Oliva, K2UO, 5 Windsor Dr, Eatontown, NJ 07724, USA.

Price: $65 + $5 for shipping and handling worldwide. Prepayment only, US bank check or international money order. An order form can be detached from the back of this book.

ANTENNAS: GENERAL, TERMS, DEFINITIONS

5 ANTENNAS: GENERAL, TERMS, DEFINITIONS

A greeing on terms and definitions is as important as anything else. Too many technical discussions seem to take place in the tower of Babble. First make sure you speak the same language, then speak. Before we get involved in the debate on what's the best antenna for the low bands (that must be the key question for most), we define what we want the antenna to do for us and how we will measure its performance.

Antennas for the low bands are one of the areas in Amateur Radio where home building will yield results that can substantially outperform most of what can be obtained commercially.

All my antennas are homemade. Visitors often ask me, "Where do you buy the parts?" Or, "Do you have a machine shop to do all the mechanical work?" Very often I don't "buy" the parts. And no, I don't have a machine shop, just the run-of-the-mill hand tools. But my friends who are antenna builders and I keep our eyes open all the time for goodies that might be useful sometime for our next antenna project. There is a very active swap activity between us. Among friends we have access to certain facilities that make antenna building easier. It's almost like we are a team where each one of us has his own specialization.

Don't look at low-band antenna designing and building as a kit project. It requires some know-how, a good deal of imagination and inventiveness and often even more organizational talent. But unlike the area of receivers and transmitters, where as home-builders most of us do not have access to the custom-designed integrated circuits and other very specialized parts, antennas and antenna systems are built using materials that can be found locally by most of us.

Antennas are one field in Amateur Radio where the old pioneering spirit of the spark age of Amateur Radio is still alive. The most outstanding low-band antenna systems are the ones that came about through hard work and brilliant engineering done by small groups of highly engaged individuals.

A number of successful major low-band antennas are described in this book. These are not meant to be "kit-like" building projects with step-by-step instructions, but are there to stimulate thinking and put the newcomer to antenna building on the right track.

The ARRL Antenna Book (Ref. 600) contains a wealth of excellent and accurate information on antennas. The antenna chapters of this book emphasize typical aspects of low-band antennas, and explain how and why some of the popular antennas work and what we can do to get the best results given our particular constraints.

■ 1. THE PURPOSE OF AN ANTENNA

1.1. Transmitting Antennas

A transmitting antenna must radiate all the RF energy supplied to it in the desired direction with the required elevation angle (directivity).

1.1.1. Wanted direction

1.1.1.1 Horizontal directivity

We have learned in Chapter 1 (Propagation) that low-band propagation paths quite frequently deviate from the theoretical great circle direction. This is more specifically so for paths going through or very near the aurora oval (eg, West Coast or Midwest USA to Europe). This is a fact we have to take into consideration if we consider putting up a fixed-direction antenna. For paths near the antipodes, signal direction can change as much as 180° (with every direction in between) depending on the season. All this must be taken into account when designing an antenna system. Rotary systems will provide a great deal of flexibility as far as horizontal directivity is concerned.

At this time it must be emphasized that the term horizontal directivity is really meaningless without further definition. Zero wave angle directivity (perfectly parallel to the horizon) is of very little use, as practical antennas produce no signal at zero wave angle over real ground. This issue is important when designing or modeling an antenna. It often would be ideal to design an antenna which concentrates the transmitted energy at a relatively low angle, while exhibiting the highest rejection off the back at a much higher angle (to achieve maximum rejection of the stronger "local" signal, that as a rule come in at a much higher wave angle. In general, horizontal directivity should always be specified at a given elevation angle. An antenna can have quite different azimuthal directional properties at different elevation angles. We will see further that a very low dipole radiates most of its energy at 90° (zenith angle), and shows no directivity at high wave angles (60 to 90°). The same antenna, at the same height shows a pronounced directivity (hardly any signal off the ends of the dipole) at very low wave angles, but hardly radiates at all at these low elevation angles. These issues must be very clear in our mind if we want to understand radiation patterns of antennas.

1.1.1.2. Vertical directivity

In the past few years a lot of modeling has been done using various propagation software. D. Straw, N6BV, used *IONCAP* (Ionospheric Communications Analysis and Prediction Program) to calculate the wave angle for various paths on the different amateur bands. *IONCAP* is based on a mass of data collected over more than 35 years. The summary of these wave angles is listed in **Table 5-1**. **Fig 5-1** shows the distribution of the wave angles for a few common paths and are based on average figures for varying types of solar activity and for varying times of year and day.

Very little has been published in amateur literature on optimum wave angles for given paths on the lower bands. Results of tests between England and North America have long been used to determine the wave angles for 80 and 40 meters. These predict wave angles that are higher than what the more recent studies using *IONCAP* show. Neither of those studies give us any information about 160 meters.

Most of the professional literature deals with research near the MUF. For commercial links it is most advantageous to operate near the MUF. In our hobby, when working DX on the low bands we are certainly not operating near the MUF under most circumstances. We operate far below the MUF and FOT, and in many cases near the LUF (lowest usable frequency).

There is a range of optimum radiation angles for different directions and different paths, and they are not the same for the three lower bands. Looking at Table 5-1 and keeping in mind what is possible we can resume the wave angle situation for the low bands as follows:

Table 5-1
Range of Radiation Angles for 40 and 80 Meters for Various Paths
The values are averages across the sunspot cycle and across the seasons. The value between brackets is the most common radiation angle (peak value in the distribution).

From	Path to	40 meters	80 meters
W. Europe	Southern Africa	6-13 (11)	10-19 (15)
(Belgium)	Japan	3-15 (9)	11-17 (12)
"	Oceania	5-10 (10)	no data
"	South Asia	9-19 (15)	13-18 (15)
"	USA (W1 - W6)	3-18 (14)	15-25 (18)
"	South America	10-18 (18)	10-17 (15)
USA	Southern Africa	8-18 (12)	10-18 (10 and 18)
East	Japan	10-15 (12)	10-18 (12)
Coast	Oceania	4-10 (10)	6-7 (7)
"	South Asia	10-15 (10)	10-18 (10 and 18)
"	South America	6-14 (10)	10-19 (15)
"	All of Europe	6-20 (15)	16-26 (20)
USA	Southern Africa	10-14 (10)	10
Midwest	Japan	3-17 (13)	10-27 (10)
"	Oceania	5-10 (10)	5-10 (10)
"	South Asia	8-14 (10)	no data
"	South America	7-13 (8)	8-15 (12)
"	Europe	3-19 (16)	13-32 (13)
USA	Southern Africa	5-10 (10)	no data
West	Japan	5-19 (11)	11-27 (14)
Coast	Oceania	8-14 (10)	10-14 (10)
"	South Asia	7-14 (10)	10-11 (10)
"	South America	5-15 (10)	10-12 (11)
"	Europe	4-16 (11)	8-18 (9)

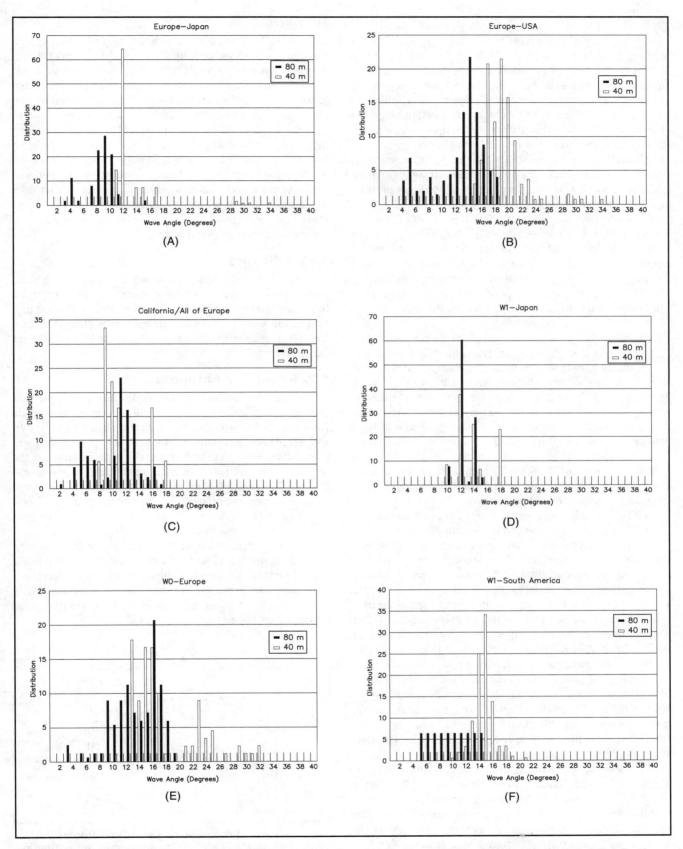

Fig 5-1—Distribution of the wave angle for a few common paths on 80 and 40 meters. Notice that the distribution is not a Gaussian one. This is because many mechanisms are involved that are totally unrelated.

40 meters

For most paths an average wave angle between 10 and 15° seems to be optimum. A dipole or Yagi (or quad) at a height of 1 λ (40 m) produces a main radiation angle of approx. 15°, and appears to be the optimum height for 40 meters. However, if you can put this antenna "only" 24 m (0.6 λ) up, the loss at 15° wave angle (vs 40 m height) is only approx. 1.5 dB. This difference in signal strength (1.5 dB) remains the same also at very low radiation angles (eg, 5°). Similar low radiation angles can be obtained with vertical antennas over very good ground, using many long radials, or by ½ λ or ⅝ λ verticals over excellent ground, again using many long radials (see Chapter 10 and 11).

Conclusion: for horizontally polarized antennas optimum height is 40 m (1 λ), minimum height for decent low-angle radiation is 20 m (0.5 λ).Verticals and vertical arrays are excellent provided they are put up over good to excellent ground, which helps to lower the radiation angle (see Chapter 9). If the reflecting ground has poor conductivity, it is better to stay away from verticals and try a horizontal antenna.

80 meters

According to *IONCAP* analysis, most DX paths use wave angles between 15 and 20° for Europe and US East Coast and 10 - 15° for US Midwest and West Coast. With horizontally polarized antennas you can't really get high enough: 40 m (½ λ) seems to be a bare minimum (radiation angle approx. 25°), but we know of 80-meter Yagis at 30 m that perform very well indeed. Quarter-wave verticals over good and excellent ground also achieve nearly the ideal radiation angle for DXing on 80 meters. If your ground is poor, then you will obtain better results by using a high, horizontally polarized antenna.

It is clear that any energy radiated at more than 35° is a waste, at least if you want to work long distance.

Many DX operators have experienced, however, that under certain, very specific circumstances, a relatively low dipole (eg, at 25-m height, which produces a wave angle of approx. 60°) produces better signals on DX than a low-angle (eg, 20°) radiator. The phenomenon causing these specific circumstances is described in Chapter 1 (Propagation), and is called ducting. This phenomenon is even more pronounced on Top Band. An even lower dipole (eg, at 12 m), which has its main radiation angle at 90° (zenith angle) will do the trick as well. **Fig 5-2** shows the vertical radiation angle of dipoles at heights of 12, 14, 20, 25 and 30 m. For radiation at angles between 60 and 70° all these dipoles produce similar signals, with the exception of the 30-m high dipole, which is clearly inferior at these high angles. A 12-m high dipole is all that is required.

Conclusion: the old saying, "You can't have enough antennas," is true. Besides an efficient low-angle radiator (15 - 25°) a high-angle antenna is certainly a welcome addition for these specific circumstances. Many DXers have concluded however, that—as an average—the low-angle antenna will outperform the high-angle radiator in 95 to 98% of the cases.

160 meters

On 160 meters, most of us have the choice between an antenna that shoots straight up a horizontal dipole or inverted-V dipole even at 30 m (100 ft) height will produce a 90° radiation angle, and a vertical (it may be shortened) that produces a good low radiation angle (20 to 40° depending on the ground quality). This means we have little chance to experience the differences in signal strength between different radiation angles. Here too, the antenna with the low wave angle will be the best in maybe 99% of the cases. Again, there are (even more than on 80 meters) the exceptional cases where a high radiation angle is required to set up a ducting mechanism, which around sunset or sunrise produces much stronger signals than can be achieved by using a low wave-angle antenna at these times.

1.1.2. Efficiency

The efficiency of a transmitting antenna is simply the ratio of power radiated from an antenna to the power applied to it. Any energy that is not radiated will be converted into heat in the lossy parts of the antenna. It is clear that for a transmitting antenna, radiation efficiency is a paramount parameter.

1.2. Receiving Antennas

For a receiving antenna, the requirements are somewhat different. Here we expect the antenna to receive only signals from a given direction and at a given wave angle (directivity), and we expect the antenna to produce signals that are substantially stronger than the internally generated noise of the receiver, taking into account losses in matching networks and feeders. This means that the efficiency of a receiving antenna is not the main requirement. An important asset of a good receiving antenna system is its directivity, its ability to null out certain directions and to change directions very rapidly.

In most amateur applications, the transmitting antenna is used as the receiving antenna, and the transmitting requirements of the antenna outweigh the typical receiving requirements. On the low bands, however, generous use is made of special receiving antennas, as we will see in Chapter 7 on Special Receiving Antennas.

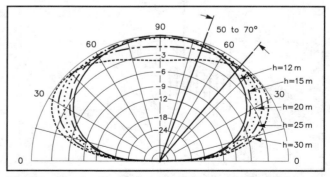

Fig 5-2—Vertical radiation angle of "low dipoles." A 12-m high dipole is all that is required to get optimum reception of signals arriving at a wave angle of 60-70°.

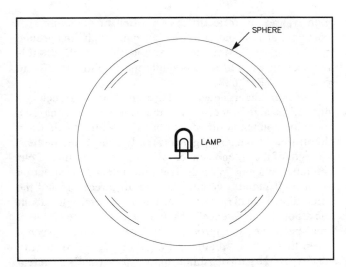

Fig 5-3—In this drawing the isotropic antenna is simulated by a small lamp in the center of a large sphere. The lamp illuminates the sphere equally well at all points.

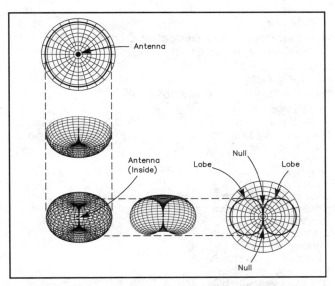

Fig 5-4—Vertical (left) and horizontal (right) radiation patterns as developed from the three-dimensional pattern of a horizontal dipole.

■ 2. DEFINITIONS

2.1. The Isotropic Antenna

An isotropic antenna is a theoretical antenna of infinitely small dimensions that radiates an equal signal in all directions. This concept can be illustrated by a tiny light bulb placed in the center of a large sphere (see **Fig 5-3**). The lamp illuminates the interior of the sphere equally at all points. The isotropic antenna is often used as a reference antenna for gain comparison, expressed in decibels over isotropic (dBi). The radiation pattern of an isotropic antenna is a sphere, by definition. A dBi is no more and no less than a convenient abbreviation for power per unit area over the volume of a sphere.

2.2. Antennas in Free Space

Free space is a condition where no ground or any other conductor interacts with the radiation from the antenna. In practice, such conditions are approached only in VHF and UHF, where very high antennas (in wavelengths) are common. Every real-life antenna has some degree of directivity and, if placed in the center of a large sphere, illuminates certain portions better than others. In antenna terms: The antenna radiates energy better in certain directions. A dipole has maximum radiation at right angle to the wire, and minimum off the ends. Such a dipole, in free space, has a gain of 2.15 dB over isotropic (2.15 dBi).

Radiation patterns are collections of all points in a given plane, having equal field strength. **Fig 5-4** shows the radiation pattern of a dipole in free space as seen three dimensionally and in two planes, the plane through the wire and the plane perpendicular to the wire.

2.3. Antennas over Ground

In real life, antennas will be near the ground. We can best visualize this situation by cutting the sphere in half with a metal plate going through the center of the sphere. This plate represents the ground, a perfect electrical mirror. **Fig 5-5** shows what happens with an antenna near the ground: Direct and reflected waves will combine and illuminate the sphere unequally in different points at different angles. For certain angles the direct and reflected waves will be in phase and will reinforce one another. The field is doubled, which means a gain of 6 dB.

Over ground the radiation patterns are often identified as vertical (cutting plane perpendicular to the ground) or horizontal (cutting plane parallel to the ground). The latter is of very little use, as practical antennas over real ground produce no signal at zero wave angle. The so-called horizontal

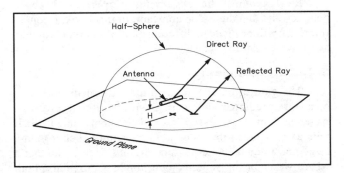

Fig 5-5—The effect of ground is simulated in a sphere by putting a plate (the reflecting ground plane) through the center of the sphere. As the power in the antenna is now radiated in half the sphere volume, the total radiated field in the half sphere is doubled. This means that the ground reflection can add up to 6 dB of signal increase as compared to free space. A smaller total gain is caused in practice, as part of the RF energy is absorbed in the (poorly reflecting) ground.

Table 5-2
Conductivities and Dielectric Constants for Common Types of Earth

Surface Type	Dielectric Constant	Conductivity (mS/m)	Quality
Fresh Water	80	1.0	
Salt Water	81	5000.0	Sea Water
Pastoral, low hills, rich soil (typ Dallas, TX to Lincoln, NE areas)	20	30.3	Very good
Pastoral, low hills, rich soil (typ OH, IL)	14	10.0	Good
Flat country, marshy, densely wooded (typ LA nr MS river)	12	7.5	
Pastoral, medium hills and reforestation (typ MD, PA and NY, exclusive of mtns and coastline)	13	6.0	
Pastoral, medium hills and reforestation, heavy soil (type central VA)	13	5.0	Average
Rocky soil, steep hills (typ mountainous area)	12-14	2.0	Poor
Sandy, dry flat, coastal	10	2.0	
Cities, industrial areas	5	1.0	Very poor
Cities, heavy industrial	3	0.1	Extremely poor

These definitions are used throughout this book.

directivity should in all practical cases be specified as directivity in a plane making a given angle with the horizon (usually the main wave angle).

Low-band antennas always involve real ground. With real ground, the above-mentioned gain of 6 dB will be lowered, as part of the RF will be dissipated in the lossy ground. For evaluation purposes, we often use perfect ground, a ground consisting of an infinitely large, perfect reflector.

Real grounds have varying properties, in both conductivity and dielectric constant. In this book, frequent reference will be made to different qualities of real grounds, as shown in **Table 5-2**.

2.4. Antenna Gain

The gain of an antenna is a measure of its ability to concentrate radiated energy in a given direction (less losses). Antenna gain is expressed in decibels, abbreviated dB. It tells us how much the antenna in question is better than a reference antenna, under certain circumstances. And that's where we enter the antenna gain jungle. Commonly, both the isotropic as well as the "real" dipole are used as reference antennas. In the former case the gain is expressed as dBi, in the latter as dBd. But that's only part of the story. We can do a comparison in free space, or over perfect or over real ground. The only situation that makes generic comparison possible is to compare in free space. The dBi in free space is what can always be compared; there is no inflation of

gain figures by reflection. Very often manufacturers of commercial antennas will calculate gains including ground reflections, and often not mention that at all. In this publication we will always quote gain figures in dBi, in free space. (Ref. 688).

You might argue why not use a real antenna, such as a dipole, as a reference, since the isotropic antenna is a theoretical antenna that does not exist, while a half-wave dipole does. Comparing gains is really comparing the field strength of the antenna under investigation with that of our reference antenna. With the isotropic antenna the situation is clear: It radiates equally well in all directions and the three-dimensional radiation pattern is a sphere. What about the dipole as a reference? The gain of a half-wave dipole in free space over an isotropic is 2.15 dBi. But that does not mean that a "real" dipole has a gain of 2.15 dBi. It only means that the gain of a dipole in free space (that's an unreal condition as well because nothing is really in free space) is 2.15 dB over an isotropic radiator. If we put the dipole over a perfect ground, it suddenly shows a gain of 8.15 dBi! You pick up 6 dB by radiating the power in half a hemisphere instead of a whole hemisphere as in the theoretical case of free space. With less than perfect ground, part of the power will be absorbed in the ground and the ground reflection gain will be less than 6 dB. It is clear that the only generic way of comparing antenna gains is in dBi, the isotropic antenna (in free space) being the only generic reference antenna that is not influenced by height or ground conditions.

2.5. Front-to-back Ratio

Being a ratio (just like gain), we would expect front-to-back ratio (F/B) to be expressed in decibels, which it is. The front-to-back ratio is a measure expressing an antenna's ability to radiate a minimum of energy in the back direction of the antenna.

Free-space F/B is always measured at a zero-degree wave angle. Over ground the F/B depends on the vertical radiation angle being considered. In most cases a horizontal radiation pattern over real ground is not really the pattern in the horizontal plane, but in a plane which corresponds to the main wave angle. If we look at the back lobe at that angle, it may be okay, but at the same time there may be a significant back lobe at a much different angle.

With the advent and the widespread use of modeling programs, especially some of the optimizer programs, started the rat race for the most ludicrous F/B figure. Let's not forget: mathematics is one thing, antennas are another thing. It is totally possible to calculate an antenna exhibiting a F/B of eg, 70 dB in a given direction, at a given wave angle. But that's all there is to it. One degree away, the rejection may be down 40 or 50 dB. When one understands the physics behind all of this, it will be clear that F/B above a certain level (maybe 35 dB) are rather meaningless.

Geometric front-to-back

In the past, front-to-backs were usually defined in the sense of a geometric front-to-back: the radiation 180° off the front (lobe) of the antenna. At the same time we compare the "forward" power at the main (forward) radiation angle, with the power radiated at the same radiation angle in the backward direction.

Front-to-back is a property we use in the real world to discriminate against unwanted signals coming from "other" directions. It is very unlikely that unwanted signals will be generated exactly 180° off the beam direction or at a radiation angle that is the same as the main forward-lobe radiation angle. Therefore, the geometric F/B can be ruled out immediately as a meaningful way of defining the antenna's ability to discriminate against unwanted signals.

Average front-to-back (integrated front-to-back) or Front to Rear (F/R)

The average front-to-back can be defined as the average value of the front-to-back as measured (or computed) over a given back angle (both in the horizontal as well as the vertical plane). In Chapter 7 (Special Receiving Antennas), I use this concept for evaluating different antennas.

Worst lobe front-to-back

Another meaningful way to quantify the F/B ratio of an antenna is to measure the ratio of the forward power to the power in the "worst" lobe in the entire back (to be specified) of the antenna.

Front-to-back and gain

Is there a link between gain and the front-to-back ratio of an antenna? Let's visualize a three-dimensional radiation pattern of a (simple) Yagi. The front lobe resembles a long stretched pear, the back lobe (let's assume for the time we have a single back lobe) a (much) smaller pear. The antenna sits where the stems of the two pears touch. The volume of the two pears (the total volume of the three-dimensional radiation pattern) is determined only by the power fed to the antenna. If you increase the power, the volume of the large, as well as the small, pear will increase in the same proportion. Let's take for definition of front-to-back the ratio of the power radiated in the back versus the power radiated in the front. This means that the F/B ratio is proportional to the ratio of the volume of the two pears.

By changing the design of the Yagi (by changing element lengths or element positions), we will change the size and the shape of the two pears, but as long as we feed the same power to it the sum of the volumes of the two pears will forever remain unchanged. It's as if the two pear-shaped bodies are connected with a tube, and are filled with a liquid. By changing the design of the antenna, we merely push liquid from one pear into the other. If the antenna were isotropic, the radiation body would be a sphere having the volume of the sum of the two pears.

Assume we have 100 W of power with 10% of this power applied to the antenna in the back lobe. The F/B ratio will be $10 \times \log (10 / 1) = 10$ dB. Ninety percent of the applied power is available to produce the forward lobe.

Let's take a second case, where only 0.1% of the applied power is in the back lobe. The F/B ratio will be $10 \times \log (100 / 0.1) = 30$ dB. Now we have 99.9% of the power available in the front lobe.

The antenna gain realized by having 99.9 W instead of 90 W in the forward lobe is $10 \log (99.9 / 90) = 0.45$ dB. Pruning an antenna with a modest F/B pattern (10 dB) to a supreme 30 dB value, "can" give us 0.45 dB more forward gain, provided that the extra liquid is used to lengthen the cone of the big pear.

The mechanism of obtaining gain and F/B is much more complicated than that described above. I am only trying to explain that optimizing an antenna for F/B does not necessarily mean that it will be optimized for gain. What is always true is that a high-gain antenna will have a narrow forward lobe. You cannot concentrate energy in one direction without taking it away from other directions! We will see later that maximum-gain Yagis show a narrow forward lobe, but often a poor front-to-back. This is the case with very high-Q gain-optimized 3-element Yagis.

Conclusion: There is no simple relationship between front-to-back ratio and gain of an antenna.

2.6. Radiation Resistance

Radiation resistance (referred to a certain point in an antenna system) is the resistance which, inserted at that point, would dissipate the same energy as is actually radiated from the antenna. In other words, radiation resistance is the total power radiated divided by the square of the current at some point in the system. This definition does not state where the antenna is being fed, however. There are two common ways of specifying radiation resistance:

• The antenna being fed at the current maximum $R_{rad(I)}$.

• The antenna being fed at the base between the antenna lower end and ground $R_{rad(B)}$.

$R_{rad(I)} = R_{rad(B)}$ for verticals of $^1/_2$ wavelength or shorter. $R_{rad(B)}$ is the radiation resistance used in all efficiency calculations for vertical antennas. Fig 9-10 shows the radiation resistance according to both definitions for four types of vertical antennas:

• a short vertical (< 90°)
• a quarter-wave vertical
• a $^3/_8$-wave vertical (135°)
• a $^1/_2$-wave vertical

2.7 Antenna Efficiency

The efficiency of an antenna is expressed as follows:

$$Eff = R_{rad} / (R_{rad(B)} + R_{loss})$$

where $R_{rad(B)}$ is the radiation resistance of the antenna as defined in Section 2.6, and R_{loss} is the total equivalent loss resistance of all elements of the antenna (resistance losses, dielectric losses, ground losses, etc) normalized to the same point where R_{rad} was defined.

2.8. Standing-wave Ratio

Standing-wave ratio (SWR) is a measure of how well the feed-point impedance of the antenna is matched to the characteristic impedance of the feed line. If a 50-ohm feed line is terminated in a 50-ohm load, then the impedance at any point of the cable, thus also the impedance at the end of a cable (of any length), is 50 ohms.

If the same feed line is terminated in an impedance different from 50 ohms, the impedance will vary along the line. The SWR is a measure of the match between the line and the load. Changing the length of a feed line does not change anything regarding the SWR on the line (apart from minute changes due to the feed-line loss). The only thing that changes is the impedance at the input end of the line.

If changing the line length (slightly) changes the SWR reading on your SWR meter, then your SWR meter is not measuring correctly (many SWR meters fall into this category). It is a good test for an SWR meter to insert short cable lengths between the end of the antenna feed line and the SWR meter (a few feet at a time). If the SWR reading changes, throw away the meter, or just use it as a relative output and SWR indicator, but don't use it expecting to obtain correct SWR values.

Only when we have currents on the outside of the coaxial cable shield can a change in position on the line change the SWR reading (see Chapter 6). That's why we need a balun when feeding balanced feed points with a coaxial cable. In addition current baluns are a good idea to install on any coaxial feed line. One can insert a current balun (eg, a short length of coax equipped with a stack of eg, 50 or 100 appropriate ferrite cores) at the SWR meter. If this balun changes the SWR value as measured with the SWR meter, this indicates that RF currents were flowing on the outside of the coaxial cable.

Changing the feed-line length never changes the performance of the antenna. A feed line is an element which is supposed *not* to radiate. Changing the line length can have an impact on the feed-system transformation, though. SWR on a feed line has no relation whatsoever to the radiation characteristics of an antenna. SWR of an antenna is something that does *not* exist. What exists is the SWR on a feed line, feeding an antenna.

A perfect match results in a 1:1 SWR. What are the reasons we like a 1:1 SWR or the lowest possible SWR value?

- Showing a convenient impedance: It is clear that we would like to live in a world where, unless we want to use the line as an impedance transformer, we would like all feed lines to show a 1:1 SWR. This would be perfect as far as presenting the ideal load impedance (50 ohms) for transistor-final transceivers.
- Minimizing losses: All feed lines have inherent losses. This loss is minimal when the feed line is operated as a flat line (SWR = 1:1) and increases with the SWR value. On the low bands this will seldom be a criterion for working with a very low SWR, because the nominal losses on the low frequencies are quite negligible, unless very long lengths are used.

The SWR value describes the relationship between the antenna (the load) and the feed line. It does not describe an intrinsic (radiation) property of the antenna.

SWR is, for many hams, the only property they can measure. Measuring gain and F/B with any degree of accuracy is beyond the capability of most. That is why most hams pay attention only to SWR properties.

The amount of SWR that can be tolerated on a line depends on:

- Additional attenuation caused by SWR; in other words the quality of the feed line. A good-quality feed line can tolerate more SWR from an additional loss point of view than a mediocre-quality line.
- How much SWR the transceiver or linear amplifier can live with.
- How much power we will run into a line of given physical dimensions (for a given power, a larger coax will withstand a higher SWR without damage than a smaller one).

It must be said that a poor quality line (a "small" cable with high intrinsic losses), when terminated with a load different than its characteristic impedance, will show at the input end a lower SWR value than if a good (low loss, big) cable is used. Remember that a very long, poor (having high losses) coaxial cable, whether terminated, open or shorted at the end, will exhibit a 1:1 SWR at the input (a perfect dummy load).

From a practical point of view an SWR limit of 2:1 is usually employed. It is clear that, from a loss point of view, higher values can easily be tolerated on low frequencies. Coaxial feed lines used in the feed systems of multi-element arrays sometimes work with an SWR of 10:1!

An antenna tuner can always be used if near the band edges the SWR value is such that the transceiver or the amplifier would rather see a lower value (usually above 2:1). Remember that the antenna tuner will not change the SWR on the line; it will merely transform the impedance existing at the line input and present the transceiver (or linear amplifier) with a "reasonable and more convenient" SWR value. While this approach is valid on the low bands, I strongly suggest not using it on the higher frequencies, as the additional line losses caused by the SWR can become quite significant.

2.9. Bandwidth

The bandwidth of an antenna is the difference between the highest and the lowest frequency on which a given property exceeds or meets a given performance mark. This can be gain, front-to-back ratio or SWR. In this book, "bandwidth" refers to SWR bandwidth unless otherwise specified. In most cases the SWR bandwidth is determined by the 2:1 points on the SWR curve. In this text the SWR limits will be specified when dealing with antenna bandwidths. Many amateurs only think of SWR bandwidth when the term bandwidth is being used. In actual practice, the bandwidth as referred to other properties is at least as important if not more important. Consider a dummy load that has a very "good" SWR bandwidth but a very poor gain (does not radiate at all!).

Bandwidth is an important performance criterion on the low bands. The relative bandwidth of the low bands is large compared to the higher HF bands. Special attention must be given to all bandwidth aspects, not only SWR bandwidth.

2.10. Q factor

2.10.1. The tuned-circuit equivalent

An antenna can be compared to a tuned LCR circuit. The Q factor of antenna is a measure of the SWR bandwidth of an antenna. The Q factor is directly proportional to the difference in reactance on two frequencies around the frequency of analysis, and inversely proportional with the radiation resistance and relative frequency change.

$$Q = \frac{F_0 \times (X1 - X2)}{2 \times R \times \Delta F}$$

where

F_0 = geometric mean frequency between higher and lower frequency of analysis

X1 = reactance at the lower frequency

X2 = reactance at the higher frequency

R = average value of resistive part of feed-point impedance at frequencies of analysis (R_{rad} + R_{losses})

ΔF = relative frequency change between the higher and the lower frequency of analysis

Example:

F_{low} = 3.5 MHz

F_{high} = 3.6 MHz

F_0 = 3.55

$\Delta F = (3.6 - 3.5) = 0.1$

R_{feed} (aver) = 50 ohms

X1 = –20 ohms

X2 = +20 ohms

$$Q = \frac{3.55 \times (20 - (-20))}{2 \times 50 \times 0.1} = 14.2$$

It is clear that a low Q can be obtained through:
- A high value of radiation resistance.
- A high loss resistance.
- A flat reactance curve.

An antenna with a low Q will have a large SWR bandwidth, and an antenna with a high Q will have a narrow SWR bandwidth. Antenna Q factors are used mainly to compare the (SWR) bandwidth characteristics of antennas.

2.10.2. The transmission-line equivalent

A single-conductor antenna (vertical or dipole) with sinusoidal current distribution can be considered as a single-wire transmission line on which a number of calculations can be done, just as on a transmission line.

Surge Impedance

The characteristic impedance of the antenna seen as a transmission line is called the surge impedance of the antenna.

The surge impedance of a vertical is given by:

$$Z_{surge} = 60 \times \ln\left[\frac{4h}{d} - 1\right]$$

where

h = antenna height (length of equivalent transmission line)

d = antenna diameter (same units)

The surge impedance of a dipole is:

$$Z_{surge} = 276 \times \log\left[\frac{S}{d \times \sqrt{1 + \frac{S}{4h}}}\right]$$

where

S = length of antenna

d = diameter of antenna

h = height of antenna above ground

Q-factor

The Q-factor of the transmission-line equivalent of the antenna is given by:

$$Q = \frac{Z_{surge}}{R_{rad} + R_{loss}}$$

Example 1:

A 20-m (66-ft) vertical with OD = 5 cm (1.6 inches), and R_{rad} + R_{loss} = 45 ohms.

$$Z_{surge} = 60 \times \ln\left[\frac{4 \times 2000}{5} - 1\right] = 443 \text{ ohms}$$

Q = 443/45 = 9.8

Example 2:

A 40-m (131 ft) long dipole, at 20 m (66 ft) height is made of 2 mm OD wire (AWG 12). The feed-point impedance is 75 ohms.

$$Z_{surge} = 276 \times \log\left[\frac{4000}{0.2 \times \sqrt{1 + \frac{4000}{4 \times 2000}}}\right] = 1211 \text{ ohms}$$

Q = 1211/75 = 16

THE FEED LINE AND THE ANTENNA

6 THE FEED LINE AND THE ANTENNA

The feed line is the inevitable link between the antenna and the transmitter/receiver. It may look strange that I cover feed lines and antenna matching before discussing any type of antenna. The reason is that I want to make clear that antenna matching and feeding has no influence on the characteristics or the performance of the antenna itself (not unless the matching system and/or feed lines also radiate). Antenna matching is something generic, which means that any matching system can in theory be used with any antenna. Antenna matching must therefore be treated as a separate subject.

The following topics are covered:
• Coaxial lines, open-wire lines.
• Loss mechanism.
• Real need for low SWR.
• Quarter-wave transformers.
• L networks.
• Stub matching.
• Wide-band transformers.
• 75-Ω feed lines in 50-Ω systems.
• Baluns.
• Connectors.

Before we discuss antennas from a more or less theoretical point of view and describe practical antenna installations, let us analyze what matching the antenna to the feed line really means and how we can do it.

■ 1. PURPOSE OF THE FEED LINE

The feed line "transports" RF energy from a source to a load (eg, from a transmitter to an antenna). A feed line, when terminated in a resistor having the same value as its own characteristic impedance, will operate under ideal circumstances: The line will be "flat;" there will be no standing waves on the line. The value of the impedance will be the same in each point of the line. If the feed line were lossless, the magnitude of the voltage and the current would also be the same along the line. The only thing that would change is the phase angle of these values, and the phase angle would be directly proportional to the line length. All practical feed lines have losses, however, and the values of current and voltage decrease along the line in an exponential way.

In our real world the feed line will rarely if ever be terminated in a load ensuring a 1:1 SWR. Since the line is most frequently terminated in a load with a complex impedance, in addition to acting as a transport vehicle for RF

energy, the feed line will also act as a transformer, whereby the impedance (also the voltage and current) will be different at each point along the line. A feed line working under these circumstances is not "flat," but has standing waves.

Besides transporting energy from the source to the load, feed lines can also be used to supply feed current to the elements of an antenna array, whereby the characteristics of the feed lines (with SWR) will be used to supply current at each element with the required relative magnitude and phase angle. This application is covered in detail in Chapter 11, Vertical Arrays.

■ 2. FEED LINES WITH SWR

The typical characteristics of a line with SWR are:
• The impedance at every point of the line is different; the line acts as an impedance transformer. (While the impedances in a lossless line repeat themselves every half wavelength, the impedances in a real-earth lossy line do not repeat.)
• The voltage and the current in every point of the feed line are different.
• The losses of the line are higher than for a flat line.

Most transmitters, amplifiers and transceivers are designed to work into a nominal impedance of 50 Ω. Although they will provide a match to a range of impedances which are not too far from the 50-Ω value (eg, within the 2:1 SWR circle on the Smith Chart), it is generally a proof of good engineering and workmanship that an antenna, on its design frequency, shows a 1:1 SWR on the feed line. This means that the feed-point impedance of the antenna must be "matched" to the characteristic impedance of the line at the design frequency. The SWR bandwidth of the antenna will be determined in the first place by the Q factor of the antenna, but the bandwidth will be largest if the antenna has been matched to the feed line (1:1 SWR) at some (the design) frequency within that passband, unless special broadband matching techniques are employed. This means we want a low SWR for convenience reasons: We don't want to be forced to use an antenna tuner between the transmitter and the feed line in order to obtain a conjugate match.

• Conjugate match

A conjugate match is a situation where all power is effectively coupled from the transmitter into the line, and where the wave, reflected from the load (antenna) back to the transmitter due to SWR is reflected back toward the load again. A conjugate match is automatically achieved when we match the transmitter for maximum power transfer into the line. In transmitters or amplifiers using vacuum tubes, this is done by properly adjusting the common pi or pi-L network. Modern transceivers with fixed-impedance solid-state amplifiers do not have this flexibility, and an external antenna tuner will be required in most cases if the SWR is higher than 1.5:1 or 2:1. Many of the present-day transceivers have built-in antenna tuners that automatically take care of this situation.

But this is not the main reason for low SWR. The above reason is a "reason of convenience." The real reason is one of losses or attenuation. A feed line is usually made of two conductors with an insulating material in between. Open-wire feeders and coaxial feed lines are the two most commonly used types of feed lines.

2.1. The Coaxial Cable Case

Coaxial feed lines are by far the most popular type of feed lines in amateur use, for one specific reason: Due to their coaxial (unbalanced) structure, all magnetic fields caused by RF current in the feed line are kept inside the coaxial structure. This means that a coaxial feed line is totally "inert" from the outside, when terminated in an unbalanced load (regardless of SWR). An unbalanced load is a load where one of the terminals is grounded. This means you can bury the coax, affix it on the wall, under the carpet, tape it to a steel post or to the tower without in any way upsetting the electrical properties of the feed line. Sharp bending of coax should be avoided, however, to prevent impedance irregularities and permanent displacement of the center conductor caused by cable dielectric heating and induced stresses. A minimum bending radius of five times the cable outside diameter is a good rule of thumb for coaxial cables with a braided shield.

Like anything exposed to the elements, coaxial cables deteriorate with age. Under the influence of heat and ultraviolet light, some of the components of the outer sheath of the coaxial cable can decompose and migrate through the copper braid into the dielectric material, causing degradation of the cable. Ordinary PVC jackets used on older coaxial cables (RG-8, RG-11) showed migration of the plasticizer into the polyethylene dielectric. Newer types of cable (RG-8A, RG-11A, RG-213 and so on) use non-contaminating sheaths that greatly extend the life of the cable.

Also, coaxial cables love to drink water! Make sure the end connections and the connectors are well sealed. Because of the structure of the braided shield, the interstices between the inner conductor insulation and the outer sheath will literally suck up liters (quarts) of water, even if only a pin hole is present. Once water has penetrated the cable, it is ruined. Here is one of the big advantages of the larger coaxial cables using expanded polyethylene and a corrugated solid copper outer conductor: As the PE sticks (bonds) to the copper, water penetration is impossible even if the outer jacket is damaged.

It is always a good idea to check the attenuation of feed lines at regular intervals. This can easily be done by opening the feed line at the far end. Then feed some power into the line through an accurate SWR meter (such as a Bird wattmeter), and measure the SWR at the input end of the line. A lossless line will show infinite SWR (Ref. 1321). From the measured value the attenuation of the line can be deduced using the graph in **Fig 6-1**. It will often be difficult to do this test at low frequencies because the attenuation on the low bands is such that accurate measurements are difficult. For best measurement accuracy the loss of the cable to be measured should be in the order of 2 to 4 dB (SWR between 2:1 and 4:1). The test frequency can be chosen accordingly. Use a professional type SWR meter such as a Bird wattmeter. Many of the cheaper SWR meters are inadequate.

2.2. The Open-wire Case

Even when properly terminated in a balanced load, an open-wire feeder will exhibit a strong RF field in the immediate vicinity of the feed line (try a neon bulb close to an open-wire feeder with RF on it!). This means you cannot "fool around" with open-wire feeders as you can with coax. During installation the necessary precautions should be taken to preserve the balance of the line: The line is to be kept away from conductive materials, etc. In one word, generally it's a nuisance to work with open-wire feeders.

But apart from this mechanical problem, open-wire feeders outperform coaxial feed lines in all respects on HF (VHF/UHF can be another matter).

2.3. The Loss Mechanism

The intrinsic losses of a feed line are caused by two mechanisms:
- Conductor losses (losses in the copper conductors).
- Dielectric losses (losses in the dielectric material).

An excellent insulator is (dry) air. From that point of view the open-wire line is unbeatable. Coaxial feed lines

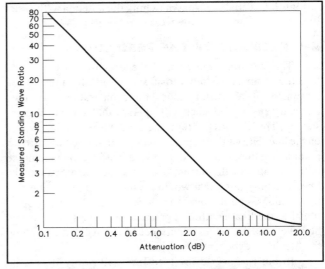

Fig 6-1—Cable loss as a function of SWR measured at the input end of an open or short-circuited feed line. For best accuracy, the SWR should be in the 1:1 to 4:1 range.

generally use polyethylene (PE) as a dielectric, or polyethylene mixed with air (cellular PE or foam PE). Cables with foam or cellular PE have lower losses than cables with solid PE. They have the disadvantage of potentially having less mechanical (impact and pressure) resistance. Cell-flex cables using a solid copper or aluminum outer conductor are the top-of-the-line coaxial feed lines used in amateur applications. Sometimes Teflon is used as dielectric material. This material is mechanically very stable and electrically very superior, but very expensive. Teflon-insulated coaxial cables are often used in baluns. (See Section 7.)

Coaxial cables generally come in two impedances: 50 and 75 Ω. For a given cable (outer) diameter, 75-Ω cable will show the lowest losses. That's why 75 Ω is always used in systems where losses are of primary importance, such as CATV. If power handling is the major concern, a much lower impedance is the optimum (35 Ω). The standard of 50 Ω has been created as a good compromise between power handling and attenuation.

Fig 6-2 shows the typical flat-line attenuation characteristics for many commonly used transmission lines. Note how the open-wire line outperforms even the biggest coaxial brother by a large margin. But these attenuation figures are only the "nominal" attenuation figures for lines operating with a 1:1 SWR.

Frank Donovan, W3LPL, put together a listing of the most commonly used coaxial cable types in the US (see **Table 6-1**). The table was made in two versions, one giving the classic attenuation/100 ft. The second list gives the cable length for 1 dB attenuation.

When there are standing waves on a feed line, the voltage and the current will be different at every point on the line. Current and voltage will change periodically along the line and can reach very high values at certain points (antinodes). The feed line uses dielectric (insulating) and conductor (mostly copper) materials with certain physical properties and limitations. The very high currents in the antinodes along the line will be responsible for extra conductivity-related losses. The voltages associated with the voltage antinodes will be responsible for increased dielectric losses. This is the mechanism that makes a line with a high SWR have more losses than the same line when matched. **Fig 6-3** shows the additional losses caused by SWR. By the way, the losses of the line are the reason why the SWR we measure at the input end of the feed line (in the shack) is always lower than the SWR at the load.

The extreme example is that of a very long cable, having a loss of at least 20 dB, where you can either short or open the end and in both cases measure a 1:1 SWR at the input. Such a cable is a perfect dummy load!

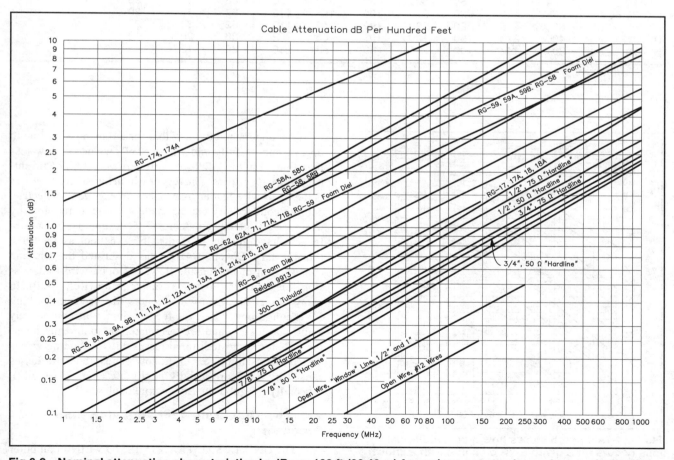

Fig 6-2—Nominal attenuation characteristics in dB per 100 ft (30.48 m) for various commonly used transmission lines.

Table 6-1
Attenuation for commonly used coaxial cables

Cable Attenuation (dB per 100 ft)

	1.8	3.5	7.0	14.0	21.0	28.0	50.0	144	440	1296
LDF7-50A	0.03	0.04	0.06	0.08	0.10	0.12	0.16	0.27	0.5	0.9
FHJ-7	0.03	0.05	0.07	0.10	0.12	0.15	0.20	0.37	0.8	1.7
LDF5-50A	0.04	0.06	0.09	0.14	0.17	0.19	0.26	0.45	0.8	1.5
FXA78-50J	0.06	0.08	0.13	0.17	0.23	0.27	0.39	0.77	1.4	2.8
¾" CATV	0.06	0.08	0.13	0.17	0.23	0.26	0.38	0.62	1.7	3.0
LDF4-50A	0.09	0.13	0.17	0.25	0.31	0.36	0.48	0.84	1.4	2.5
RG-17 0.10	0.13	0.18	0.27	0.34	0.40	0.50	1.3	2.5	5.0	
SLA12-50J	0.11	0.15	0.20	0.28	0.35	0.42	0.56	1.0	1.9	3.0
FXA12-50J	0.12	0.16	0.22	0.33	0.40	0.47	0.65	1.2	2.1	4.0
FXA38-50J	0.16	0.23	0.31	0.45	0.53	0.64	0.85	1.5	2.7	4.9
9913	0.16	0.23	0.31	0.45	0.53	0.64	0.92	1.6	2.7	5.0
RG-213	0.25	0.37	0.55	0.75	1.0	1.2	1.6	2.8	5.1	10.0
RG-8X	0.49	0.68	1.0	1.4	1.7	1.9	2.5	4.5	8.4	

Cable Attenuation (ft per dB)

	1.8	3.5	7.0	14.0	21.0	28.0	50.0	144	440	1296
LDF7-50A	3333	2500	1666	1250	1000	833	625	370	200	110
FHJ-7	2775	2080	1390	1040	833	667	520	310	165	92
LDF5-50A	2500	1666	1111	714	588	526	385	222	125	67
FXA78-50J	1666	1250	769	588	435	370	256	130	71	36
¾" CATV	1666	1250	769	588	435	385	275	161	59	33
LDF4-50A	1111	769	588	400	323	266	208	119	71	40
RG-17	1000	769	556	370	294	250	200	77	40	20
SLA12-50J	909	667	500	355	285	235	175	100	53	34
FXA12-50J	834	625	455	300	250	210	150	83	48	25
FXA38-50J	625	435	320	220	190	155	115	67	37	20
9913	625	435	320	220	190	155	110	62	37	20
RG-213	400	270	180	130	100	83	62	36	20	10
RG-8X	204	147	100	71	59	53	40	22	12	

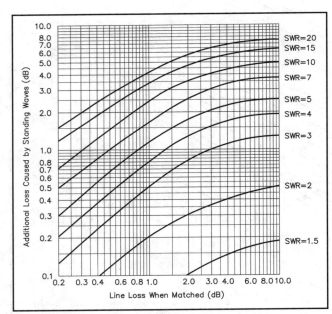

Fig 6-3—This graph shows how much additional loss occurs for a given SWR on a line with a known (nominal) flat-line attenuation.

We understand by now that for a transmission line to operate successfully under high SWR, we need a low-loss feed line with good dielectric properties and high current-handling capabilities. The feeder which has such properties is the open-wire feeder. Air makes an excellent dielectric, and the conductivity can be made as good as required by using heavy gauge conductors. Good-quality open-wire feeders have always proved to be excellent as feed-line transformers. Elwell, N4UH, has described the use and construction of homemade, low-loss open-wire transmission lines for long-distance transmission (Ref. 1320). In many cases, the open-wire feeders are used under high SWR conditions (where the feeders do not introduce many additional losses) and are terminated in an antenna tuner. Fig 6-2 shows the additional losses due to standing waves on a transmission line. On the low bands, the extra losses caused by SWR are usually negligible (Ref. 1319, 322), especially for good-quality coaxial cables.

2.4. The Universal Transmission Line Program

The UNIVERSAL TRANSMISSION LINE computer program, which is part of the NEW LOW-BAND SOFT-

WARE, is an ideal tool for evaluating the behavior of feed lines under any circumstance.

Let us analyze the case of a 50-m (164-ft) long RG-213 coax, feeding an impedance of 36.6 Ω (without matching network). The frequency is 3.5 MHz.

Fig 6-4 shows the screen print obtained from the UNIVERSAL SMITH CHART module which is part of the NEW LOW BAND SOFTWARE. All the operating parameters are listed on the screen: impedance, voltage and current at both ends of the line, as well as the attenuation data split in nominal coax losses (0.61 dB) and losses due to SWR (0.03 dB). We also see the real powers involved. In our case we need to "pump" 1734 W into the 100-m long RG-213 cable to obtain 1500 W at the load, which represents a total efficiency of 86%. Note also the difference in SWR at the load (1.4:1) and at the feed line end (1.3:1). For higher frequencies, longer cables or higher SWR values, this software module is a real eye-opener.

2.5. Conclusions

Coaxial lines are generally used when the SWR is less than 3:1. Higher SWR values can result in excessive losses when long runs are involved, and also in reduced power-handling capability. Many of the popular low-band antennas have feed-point impedances which are reasonably low, and can result in an acceptable match to either a 50 or a 75-Ω coaxial cable.

In some cases we will intentionally use feed lines with high SWR as part of a matching system (eg, stub matching) or as a part of a feed-system for a multi-element phased array.

Let us conclude that it is good engineering practice to use a feed line with the lowest possible attenuation (in a concept of money versus performance) and that we want it to operate at a unity SWR at the design frequency of our antenna system (a valid exception is described in Section 5).

■ 3. THE ANTENNA AS A LOAD

It has been proven that very small antennas are able to radiate the supplied power nearly as efficiently as much larger ones (see Chapter 9 on vertical antennas). Small antennas have two disadvantages, however. On one hand, since their radiation resistance is very low, the antenna efficiency will be lower than it would be if the radiation resistance were much higher. If the short antennas are to be loaded along the elements, the losses of the loading devices will have to be taken into account when calculating the antenna efficiency. On the other hand, if the short antenna (dipole or monopole) is not loaded, the feed-point impedance will have a large amount of capacitive reactance in addition to the resistive component.

One solution is to install a transformer at the antenna feed point to match the complex antenna impedance to the feed-line impedance. In this case, the feed line will no longer act as a transformer. Conversion will be done in the transformer at a given efficiency, and transforming extreme impedance ratios inevitably results in poor transformation efficiencies. Transforming the impedance of a very short vertical with an imped-

```
  7921                 UNIVERSAL  SMITH   CHART  PROGRAM                      on4un
    CALCULATING  THE  IMPEDANCE  AT  THE  END  OF  COAX,  KNOWING  THE  LOAD  (ANTENNA)  DATA.

    Z-CABLE: 50.0 ohm          VELOCITY FACTOR: 0.66              FREQ:  3.500 MHz
    WAVELENGTH = 85.66 METERS.                     WAVELENGTH IN CABLE = 56.53 METERS
    CABLE LENGTH =   50.00 m.  or     164.0 feet  or     318.4 deg  or     5.56 Rads
    TOTAL FLAT CABLE ATTENUATION =  0.49 dB      or        0.057 NEPERS

                      RECTANGULAR COORDINATES        POLAR COORDINATES
                      REAL PART      IMAG PART      MAGNITUDE      ANGLE
    IMPEDANCE (ohm) =    36.60          0.00          36.60          0.00       A
    CURRENT  (Amp) =     6.40          0.00           6.40          0.00       N
    VOLTAGE  (Volt) =  234.31          0.00         234.31          0.00       T

    IMPEDANCE (ohm) =    46.63        -13.04          48.42        -15.63       E
    CURRENT  (Amp) =     4.99         -3.36           6.02        -33.91       N
    VOLTAGE  (Volt) =  189.07       -221.67         291.35        -49.54       D

    POWER INTO COAX    =  1688.37 W.
    POWER INTO LOAD    =  1500.07 W.                    EFFICIENCY  =   88.85 %
    TOTAL SYSTEM LOSS  =   188.29 W.  (- 0.51 dB)      INPUT SWR   =   1.3/1
    FLAT COAX LOSS     =   180.88 W.  (- 0.49 dB)    SWR AT LOAD   =   1.4/1
    SWR COAX LOSS      =     7.42 W.  (- 0.02 dB)

    H:HELP   X:EXIT   R=RUN   Z=Z-ant  I=I-ant  F=Feedl  V=V.fact  A=Att/Fq  L=Lgth
```

Fig 6-4—Example of screen display of the UNIVERSAL SMITH CHART, a module of the NEW LOW BAND SOFTWARE that covers all technical aspects of a transmission line. See text for details.

ance of $0.5 - j3000 \, \Omega$ to $50 + j0 \, \Omega$ is a very difficult task, and it cannot be done without a great deal of loss. In military applications where very short antennas are often required, one technique used to reduce circuit losses is the cooling of network components to near absolute zero (to achieve super conductivity of the metals involved).

One can also supply power to this feed point without inserting a transformer. In this case the feed line itself will act as a transformer. In the case of the above example, an extremely high SWR would be present on the feed line. The transmission-line transformer is not a lossless component, and the losses will be determined by the quality of the materials used to make the feed line. In the pre-war days, when coaxial cables were still unknown, everybody used 600-Ω open-wire lines, and nobody knew what SWR was.

If we are not particularly interested in the transformation aspect of such a feed line, the line can be terminated in a low-loss antenna tuner. What is a quality antenna tuner? The same qualifications for feed lines apply here: one that can transform the impedances involved, at the required power levels, with minimal losses.

Many of the modern antenna tuners, which are essentially unbalanced to unbalanced tuners, use a toroidal transformer/balun to achieve a high-impedance balanced output. This principle is cost effective, but has its limitations where extreme transformations are required. The "old" tuners, eg, the Johnson Matchboxes, are ideally suited to match a very wide range of impedances. Unfortunately these Matchboxes are no longer available commercially, and are not designed to cover 160 meters.

■ 4. THE MATCHING NETWORK AT THE ANTENNA

Let's analyze a few of the most commonly used matching systems.

4.1. Quarter-wave Matching Sections

For a given design frequency you can transform impedance A to impedance B by inserting a quarter-wave long coaxial cable between A and B having a characteristic impedance equal to the square root of the product $A \times B$.

$$Z_{1/4\lambda} = \sqrt{A \times B} \qquad \text{(Eq 6-1)}$$

Example:

Assume we have a short vertical antenna that we wish to feed with 75-Ω coax. We have determined that the radiation resistance of the vertical is 23 Ω, and the resistance from earth losses is 10 Ω (the feed-point resistance is 33 Ω). We can use a $1/4$-wave section of line to provide a match, as shown in **Fig 6-5**. The impedance of this line is determined as $\sqrt{33 \times 75} = 50 \, \Omega$.

Coaxial cables can also be paralleled to obtain half the nominal impedance: A coaxial feed line of 35 Ω can be made by using two parallel 70-Ω cables.

One way to adjust $1/4$- or $1/2$-wavelength cables exactly for a given frequency is shown in **Fig 6-6**. Connect the transmitter through a good SWR meter (the author uses a Bird model 43)

to a 50-Ω dummy load. Insert a coaxial T connector at the output of the SWR bridge. Connect the length of coax to be adjusted at this point and use the reading of the SWR bridge to indicate where the length is resonant. Quarter-wave lines should be short-circuited at the far end, and half-wave lines left open. On the resonant frequency, a cable of the proper length represents an infinite impedance (assuming lossless cable) to the T junction. At the resonant frequency, the SWR will not change when the quarter-wave shorted line (or half-wave open line) is connected in parallel with the dummy load. At slightly different frequencies, the line will present small values of inductance or capacitance across the dummy load, and these will influence the SWR reading accordingly. I have found this method very accurate, and the lengths can be trimmed precisely, to within a few kHz.

Odd lengths, other than $1/4$ or $1/2$ wavelength, can also be trimmed this way. First calculate the required length

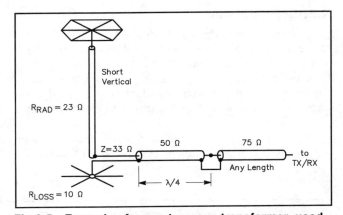

Fig 6-5—Example of a quarter-wave transformer, used to match a short vertical antenna (R_{rad} = 23 Ω, R_{ground} = 10 Ω, Z_{feed} = 33 Ω) to a 75-Ω feed line. In this case a perfect match can be obtained with a 50-Ω quarter-wave section.

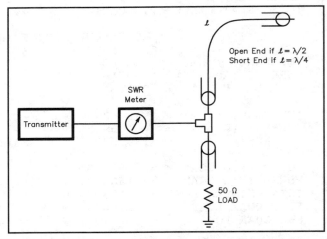

Fig 6-6—Very precise trimming of $1/4$ λ and $1/2$ λ lines can be done by connecting the line under test in parallel with a 50-Ω dummy load and watching the SWR meter while the feed line length or the transmit frequency is changed. See text for details.

difference between a quarter (or half) wavelength on the desired frequency and the actual length of the line on the desired frequency. For example, if you need a 73-degree length of feed line on 3.8 MHz, that cable would be 90 degrees long on (3.8 × 90 / 73) = 4.685 MHz. The cable can now be cut to a quarter wavelength on 4.685 MHz using the method described above.

The dip oscillator method isn't the most accurate way to cut a 90-degree length of feed line, and it often accounts for length variations of 2 or 3 degrees (due to the inductance of the link). One can also use a noise bridge and use the line under test to effectively short-circuit the output of the noise bridge to the receiver.

4.2. The L Network

The L network is probably the most commonly used network for matching antennas to coaxial transmission line. In special cases the L network is reduced to a single-element network, being a series or a parallel impedance network (just an L or C in series or in parallel with the load).

The L network is treated in great detail by W. N. Caron in his excellent book *Antenna Impedance Matching* (ARRL publication). W. Caron exclusively used the graphical Smith Chart technique to design antenna matching networks. The book also contains an excellent general treatment of the Smith Chart and other basics of feed lines, SWR and matching techniques.

Graphic solutions of impedance-matching networks have been treated by I. L. McNally, W1NCK (Ref 1446), R. E. Leo, W7LR (Ref 1404) and B. Baird, W7CSD (Ref 1402).

Designing an L network is something you want to do using a computer program. I have written a computer program (L-NETWORK DESIGN) that will just do that for you. The program is part of the NEW LOW BAND SOFTWARE. Similar computer programs have been described in amateur literature (Ref 1441).

So-called shunt-input L networks are used when the resistive part of the output impedance is lower than the required input impedance of the network. The series-input L network is used when the opposite condition exists. In some cases, a series-input L network can also be used when the output resistance is smaller than the input resistance (in this case we have four solutions). All possible alternatives (at least two, but four at the most) will be given by the program.

Fig 6-7 shows the eight possible L-network configurations. **Fig 6-8** shows the four different regions of the Smith

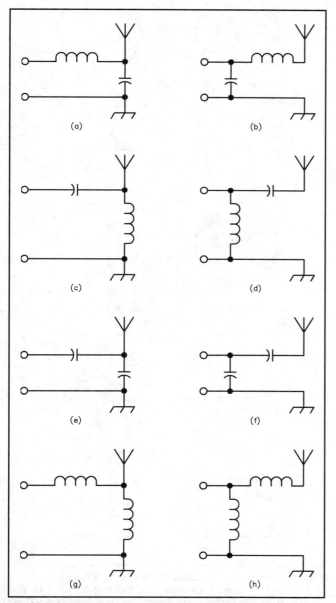

Fig 6-7—Eight possible L-network configurations. (After W. N. Caron, *Antenna Impedance Matching*.)

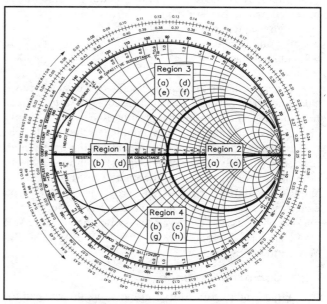

Fig 6-8—The Smith Chart subdivided in four regions, in each of which two or four L-network solutions are possible. The graphic solution methods are illustrated in Fig 6-9. (After W. N. Caron, *Antenna Impedance Matching*.)

Chart and which of the solutions are available in each of the areas. **Fig 6-9** shows the way to design each of the solutions. For more details on the Smith Chart and how to use it for the graphic design of L networks, *Antenna Impedance Matching* by W. N. Caron is a must (ARRL publication). If you have an IBM or compatible PC, an even easier way to design L networks with an on-screen Smith Chart is with the program ARRL MICROSMITH by W. Hayward, W7ZOI (also an ARRL publication). A detailed knowledge of the Smith Chart is not required to use MICROSMITH.

The choice of the exact type of L network to be used (low pass, high pass) will be up to the user, but in many cases, component values will determine which choice is more practical. In other instances, performance may be the most important consideration: Low-pass networks will give some additional harmonic suppression of the radiated signal, while a high-pass filter may help to reduce the strength of strong medium-wave broadcast signals from local stations.

Some solutions provide a direct dc ground path for the antenna through the coil. If dc grounding is required, such as in areas with frequent thunderstorms, this can be achieved by placing an appropriate RF choke at the base of the antenna (between the driven element and ground).

The L-NETWORK software module from the NEW LOW BAND SOFTWARE also calculates the input and output voltages and currents of the network. These can be used to determine the required component ratings. Capacitor current ratings are especially important when the capacitor is the series element in a network. The voltage rating is most important when the capacitor is the shunt element in the network. Consideration regarding component ratings and the construction of toroidal coils are covered in Section 4.3.

The L-NETWORK software module inputs to be provided by the user are:
• Design frequency.
• Cable impedance.
• Load resistance.
• Load reactance.

Fig 6-10 shows the screen display of a case where we calculate an L network to match $36.6 - j0\ \Omega$ to a 50-Ω transmission line. From the prompt line we can easily change any of the inputs. If the outcome of the transformation is a network with one component having a very high reactance (low C value or high L value), then we can try to eliminate this component all together. The SERIES NETWORK or SHUNT NETWORK programs will tell you exactly what value to use, and if the match is not perfect you may want to assess the SWR by switching to the SWR CALCULATION module of the NEW LOW BAND SOFTWARE to do just that.

4.2.1. Component ratings

What kind of capacitors and inductors do we need for building the L networks?

• Capacitors

The transmitter power as well as the position of the

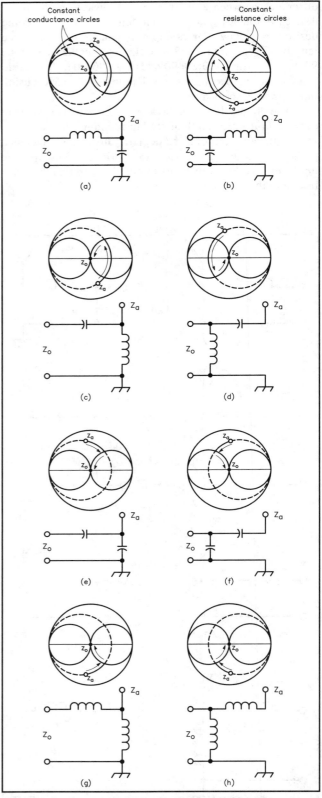

Fig 6-9—Design procedures on the Smith Chart for solutions *a* through *h* as explained in Fig 6-8. (After W. N. Caron, *Antenna Impedance Matching*.) If you have a PC you can use the program *ARRL MICROSMITH* to quickly and easily calculate the matching values graphically on screen.

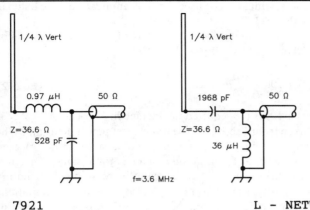

Fig 6-10—Design of an L-network to match a resonant quarter-wave vertical with a feed-point impedance of 36.6 Ω to a 50-Ω line. Note that in practice we must add the ground resistance to the radiation resistance to obtain the feed-point impedance. Therefore, in most cases the impedance of a quarter-wave vertical will be fairly close to 50 Ω.

```
7921                    L - NETWORK DESIGN                        on4un

Z-Input =   50.0 ohm       OUTPUT SWR = 1.37       FREQUENCY =   3.6 MHz

                   RECTANGULAR COORDINATES        POLAR COORDINATES
                   REAL PART      IMAG PART     MAGNITUDE        ANGLE
IMPEDANCE (ohm) =    36.60          0.00         36.60          0.00
CURRENT  (Amp)  =     6.40          0.00          0.00          0.00
VOLTAGE (Volt)  =   234.31          0.00        234.31          0.00

                       ──────── Solution # 1 ────────
   IMPEDANCE SERIES ARM      = -22.1 ohm   ==>> CAPACITANCE =   1996 pF
   IMPEDANCE PARALLEL ARM =  82.6 ohm   ==>> INDUCTANCE   =   3.65 µH
   CURRENT (Amp)    =   4.69     -2.84          5.48        -31.18
   VOLTAGE (Volt)  = 234.31   -141.78        273.87        -31.18


                       ──────── Solution # 2 ────────
   IMPEDANCE SERIES ARM    =  22.1 ohm   ==>> INDUCTANCE  =   0.98 µH
   IMPEDANCE PARALLEL ARM = -82.6 ohm   ==>> CAPACITANCE =   535 pF
   CURRENT (Amp)   =   4.69      2.84          5.48         31.18
   VOLTAGE (Volt)  = 234.31    141.78        273.87         31.18

       THE NETWORK HAS THE SHUNT ELEMENT ACROSS THE RESISTIVE INPUT
   X:EXIT   N:NEW RUN   R:Z-out   Z:Z-load   E:load volt   I:load curr   F:Freq
```

component in the L network will determine the voltage and current ratings that are required for the capacitor.

- If the capacitor is connected in parallel with the 50-Ω transmission line (assuming we have a 1:1 SWR), then the voltage across the capacitor is given by $E = \sqrt{P \times R}$. Assume 1500 W and a 50-Ω feed line.

$$E = \sqrt{1500 \times 50} = 274 \text{ V RMS}$$

The peak voltage is $274 \times \sqrt{2} = 387$ V

- If the capacitor is connected between the antenna base and ground, we can follow a similar reasoning. But this time we need to know the absolute value of the antenna impedance. Assume the feed-point impedance is $90 + j110\,\Omega$ ($R_r = 90$). The magnitude of the antenna impedance is

$$Z_{ant} = \sqrt{90^2 + 110^2} = 142.1 \ \Omega$$

The voltage across the antenna feed point is given by:

$$E = I \times Z_{ant} = \sqrt{\frac{P}{R_r}} \times Z_{ant} = \sqrt{\frac{1500}{900}} \times 142.1$$

$$= 580 \text{ V RMS} = 820 \text{ V peak}$$

- If the capacitor is the series element in the network, and if the parallel element is connected between the feed line and ground (transmitter side of the network), then the current through the capacitor equals the antenna feed current. Assume a feed-point impedance of $120 + j190\,\Omega$. The magnitude of the antenna feed-point impedance is

$$Z = \sqrt{120^2 + 190^2} = 225 \ \Omega$$

Again assume 1500 W. The magnitude of the feed current is

$$I = \sqrt{\frac{P}{Z}} = \sqrt{\frac{1500}{225}} = 2.58 \text{ A}$$

Assume the capacitor has a value of 200 pF and the operating frequency is 3.65 MHz. The impedance of the capacitor is

$$X_C = \frac{10^6}{2\pi f C} = \frac{10^6}{2\pi \times 3.65 \times 200} = 218\ \Omega$$

where f is in MHz and C is in pF. The voltage across the capacitor is

$$E = I \times Z = 2.58 \times 2.18 = 562\ V\ RMS\ or\ 795\ V\ peak$$

- If the capacitor is the series element in the L network and if the parallel element is connected between the feed point of the antenna and ground, then the current through the capacitor is the current going in the 50-Ω feed line. Assuming we have a 1:1 SWR in a 50-Ω feed line and a power level of 1500 W, the current is given by

$$I = \sqrt{\frac{P}{Z}} = \sqrt{\frac{1500}{50}} = 5.48\ A$$

Assume the same 200-pF capacitor as above, whose impedance at 3.65 MHz was calculated to be 218 Ω. The voltage across the capacitor now is

$$E = I \times Z = 5.48 \times 218 = 1194\ RMS\ or\ 1689\ V\ peak$$

In practice we should always use at least a 100% safety factor on these components. For the capacitors across low-impedance points, transmitting type mica capacitors can be used, as well as BC-type variables such as normally used as the loading capacitor in the pi network of a linear amplifier.

For the series capacitors, only transmitting type ceramic capacitors (eg, doorknob capacitors) can be used. For fine tuning, high-voltage variables or preferably vacuum variables can be used. I normally use parallel-connected transmitting-type ceramics across a low-value vacuum variable (these can usually be obtained at real bargain prices at flea markets).

• Coils

Up to inductor values of approximately 5 µH, air-wound coils are usually the best choice. A roller inductor comes in very handy when trying out a new network. Once the computed values have been verified by experimentation, the variable inductor can be replaced with a fixed inductor. Large-diameter, heavy-gauge Air Dux coils are well suited for the application.

Above approximately 5 µH, powdered-iron toroidal cores can be used. Ferrite cores are not suitable for this application, as these cores are much less stable and are easily saturated. The larger size powdered-iron toroidal cores, which can be used for such applications, are listed in **Table 6-2**.

The required number of turns for a certain coil can be determined as follows:

$$N = 100 \times \sqrt{\frac{L}{A_L}} \qquad \text{(Eq 6-2)}$$

where L is the required inductance in µH. The A_L value is taken from Table 6-2. The transmitter power determines the required core size. It is a good idea to choose a core somewhat on the large side for a margin of safety. One may also stack two identical cores to increase power-handling capability, as well as the A_L factor. The power limitations of powdered-iron cores are usually determined by the temperature increase of the core. Use large-gauge enameled copper wire for the minimum resistive loss, and wrap the core with glass-cloth electrical tape before winding the inductor. This will prevent arcing at high power levels.

Consider this example: A 14.4-µH coil requires 20 turns on a T400A2 core. AWG 4 or AWG 6 wire can be used with equally spaced turns around the core. This core will easily handle well over 1500 W.

In all cases you must measure the inductance. A_L values can easily vary 10%. It appears that several distributors (eg, Amidon) sells cores under the name type number coming from various manufacturers, which accounts for the spread in characteristics.

When measuring the inductance of a toroidal core, it is important to do this on the operating frequency, especially when dealing with ferrite material. The impedance versus frequency ratio is far from linear for this type of material. Be careful when using a digital L-C meter, which usually uses one fixed frequency for all measurements (eg, 1 MHz).

Accurate methods of measuring impedances on specific frequencies are covered in Chapter 11 (Arrays).

• The smoke test

Two things can go wrong with the matching network:
- Capacitors will flash over (short circuit, explode, vaporize, catch fire, burn up, etc) if their voltage rating is too low.
- Capacitors or coils will heat up (and eventually be destroyed after a certain time), if the current through the components is too high or the component current capabilities are too low.

In the second case the excessive current will heat up either the conductor (coil) or the dielectric (capacitor).

One way to find out if there are any losses in the

Table 6-2
Toroid Cores Suitable for Matching Networks

Supplier	Code	Permeablity	OD (in)	ID (in)	Height (in)	A_L
Amidon	T-400-A2	10	4.00	2.25	1.30	360
Amidon	T-400-2	10	4.00	2.25	0.65	185
Amidon	T-300-2	10	3.05	1.92	0.50	115
Amidon	T-225-A2	10	2.25	1.41	1.00	215

capacitor, resulting from large RF currents, is to measure or feel the temperature of the components in question (not with power applied!) after having stressed them with a solid carrier for a few minutes. This is a valid test for both coils and capacitors in a network. If excessive heating is apparent, consider using heavier duty components. This procedure also applies to toroidal cores.

4.3. Stub Matching

Stub matching can be used to match resistive or complex impedances to a given line impedance. The STUB MATCHING software module, a part of the NEW LOW BAND SOFTWARE, allows you to calculate the position of the stub on the line and the length of the stub, and whether the stub must be open or shorted at the end. This method of matching a (complex) impedance to a line can replace an L network. This approach saves the two L-network components, but necessitates extra cable to make the stub. Also, the stub may be located at a point along the feed line which is difficult to reach. **Fig 6-11** shows the screen of the computer program where we

are matching an impedance of 36.6 Ω to a 50-Ω feed line. Note that between the load and the stub the line is not flat, but once beyond the stub the line is now matched. The computer program gives line position and line length in electrical degrees. To convert this to cable length you must take into account the velocity factor of the feed line being used!

4.3.1. Replacing the stub with a discrete component.

Stub matching is often unattractive on the lower bands because of the lengths of cable required to make the stub. The module STUB MATCHING also displays the equivalent component value of the stub (in either µH or pF). Nothing prevents one from replacing the stub with an equivalent capacitor or inductor, which is then connected in parallel with the feed line at the point where the stub would have been placed. The same program shows the voltage where the stub or discrete element is placed. In order to know the voltage requirement for a parallel capacitor, one must know the voltage at the load.

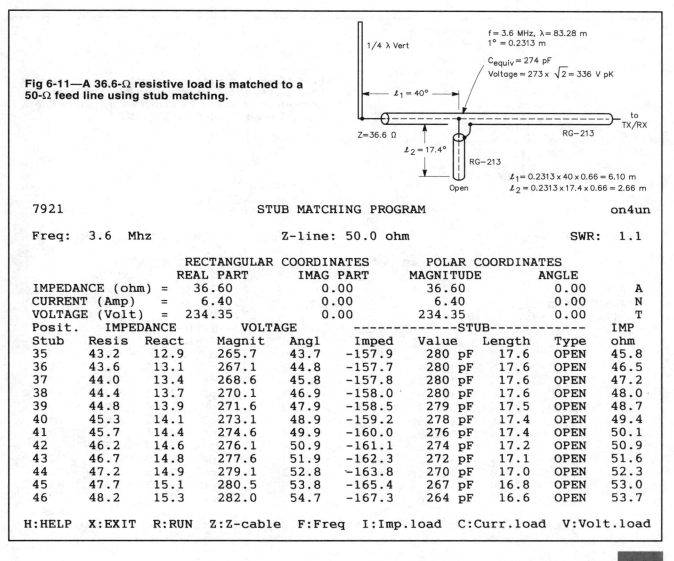

Fig 6-11—A 36.6-Ω resistive load is matched to a 50-Ω feed line using stub matching.

```
7921                         STUB MATCHING PROGRAM                      on4un

Freq:  3.6  Mhz             Z-line: 50.0 ohm                     SWR:  1.1

                 RECTANGULAR COORDINATES          POLAR COORDINATES
                 REAL PART      IMAG PART       MAGNITUDE      ANGLE
IMPEDANCE (ohm) =   36.60          0.00           36.60         0.00      A
CURRENT (Amp)   =    6.40          0.00            6.40         0.00      N
VOLTAGE (Volt)  =  234.35          0.00          234.35         0.00      T
Posit.    IMPEDANCE         VOLTAGE        -------------STUB-----------   IMP
Stub   Resis   React    Magnit   Angl    Imped   Value   Length   Type    ohm
35     43.2    12.9     265.7    43.7   -157.9   280 pF   17.6    OPEN    45.8
36     43.6    13.1     267.1    44.8   -157.7   280 pF   17.6    OPEN    46.5
37     44.0    13.4     268.6    45.8   -157.8   280 pF   17.6    OPEN    47.2
38     44.4    13.7     270.1    46.9   -158.0   280 pF   17.6    OPEN    48.0
39     44.8    13.9     271.6    47.9   -158.5   279 pF   17.5    OPEN    48.7
40     45.3    14.1     273.1    48.9   -159.2   278 pF   17.4    OPEN    49.4
41     45.7    14.4     274.6    49.9   -160.0   276 pF   17.4    OPEN    50.1
42     46.2    14.6     276.1    50.9   -161.1   274 pF   17.2    OPEN    50.9
43     46.7    14.8     277.6    51.9   -162.3   272 pF   17.1    OPEN    51.6
44     47.2    14.9     279.1    52.8   -163.8   270 pF   17.0    OPEN    52.3
45     47.7    15.1     280.5    53.8   -165.4   267 pF   16.8    OPEN    53.0
46     48.2    15.3     282.0    54.7   -167.3   264 pF   16.6    OPEN    53.7

H:HELP   X:EXIT   R:RUN   Z:Z-cable   F:Freq   I:Imp.load   C:Curr.load   V:Volt.load
```

Consider the following example: The load is 50 Ω (resistive), the line impedance is 75 Ω, and the power at the antenna is 1500 W. Therefore, the RMS voltage at the antenna is

$$V = \sqrt{P \times R} = \sqrt{1500 \times 50} = 274 \text{ V}$$

Running the STUB MATCHING software module, we find that a 75-Ω impedance point is located at a distance of 39 degrees from the load. See **Fig 6-12** for details of this example. The required 75-Ω stub length, open-circuited at the far end, to achieve this resistive impedance is 22.2 degrees (equivalent to 230 pF for a design frequency of 3.6 MHz). The voltage at that point on the line is 334 V RMS (472 V peak).

Note that the length of a stub will never be longer than ¼ wavelength (either open-circuited or short-circuited).

4.3.2. Matching with series-connected discrete components

In stub matching with a 50-Ω system, we look on a line with SWR for a point where the impedance on the line, together with the impedance of the stub (in parallel) will produce a 50-Ω impedance.

A variant consists in looking on the line for a point where the insertion of a series impedance will yield 50 Ω. At that point the impedance will look like $(50 + jX)$ Ω or $(50 - jY)$ Ω. All we need to do is to put a capacitor or inductor in series with the cable at that point. The capacitor will have a reactance of X Ω, or the inductor of Y Ω.

Example: Match a 50-Ω load to a 75-Ω line (same example as above).

The software module IMPEDANCE ITERATION from the NEW LOW BAND SOFTWARE lists the impedance along the line in 1-degree increments, starting at 1 degree from the load. Somewhere along the line we will find an impedance where the real part is 75 Ω (see details in **Fig 6-13**). Note the distance from the load. In our example this is 51 degrees from the 50-Ω load. The impedance at that point is $75.2 + j30.7$ Ω.

If we want to assess the current through the series

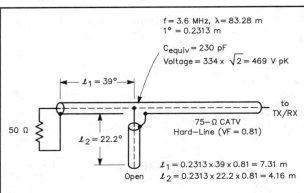

f = 3.6 MHz, λ = 83.28 m
1° = 0.2313 m

C_{equiv} = 230 pF
Voltage = 334 × $\sqrt{2}$ = 469 V pK

ℓ_1 = 39°

50 Ω

ℓ_2 = 22.2°

to TX/RX

75−Ω CATV Hard−Line (VF = 0.81)

Open ℓ_1 = 0.2313 × 39 × 0.81 = 7.31 m
ℓ_2 = 0.2313 × 22.2 × 0.81 = 4.16 m

Fig 6-12—Example of how a simple stub can match a 50-Ω load to a 75-Ω transmission line. Note that between the load and the stub the SWR on the line is 1.5:1. Beyond the stub the SWR is 1:1.

```
7921                       STUB MATCHING PROGRAM                      on4un

Freq:  3.6  Mhz                 Z-line: 75.0 ohm                 SWR:  1.1
```

Posit.	IMPEDANCE		VOLTAGE		----------STUB----------				IMP
	RECTANGULAR COORDINATES				POLAR COORDINATES				
	REAL PART	IMAG PART			MAGNITUDE	ANGLE			
IMPEDANCE (ohm) =	50.00	0.00			50.00	0.00			A
CURRENT (Amp) =	5.47	0.00			5.47	0.00			N
VOLTAGE (Volt) =	273.50	0.00			273.50	0.00			T
Stub	Resis	React	Magnit	Angl	Imped	Value	Length	Type	ohm
34	60.5	23.4	322.6	45.3	-180.0	246 pF	22.6	OPEN	67.8
35	61.2	24.0	324.9	46.4	-180.2	245 pF	22.6	OPEN	69.2
36	61.9	24.5	327.3	47.5	-180.7	245 pF	22.5	OPEN	70.6
37	62.6	25.1	329.6	48.5	-181.4	244 pF	22.5	OPEN	71.9
38	63.3	25.6	332.0	49.5	-182.3	243 pF	22.4	OPEN	73.3
39	64.1	26.1	334.4	50.5	-183.4	241 pF	22.2	OPEN	74.7
40	64.9	26.6	336.8	51.5	-184.8	239 pF	22.1	OPEN	76.0
41	65.7	27.1	339.2	52.5	-186.4	237 pF	21.9	OPEN	77.4
42	66.6	27.6	341.6	53.5	-188.2	235 pF	21.7	OPEN	78.7
43	67.4	28.0	343.9	54.4	-190.2	232 pF	21.5	OPEN	80.0
44	68.3	28.4	346.3	55.4	-192.5	230 pF	21.3	OPEN	81.3
45	69.2	28.8	348.6	56.3	-195.0	227 pF	21.0	OPEN	82.5

```
H:HELP  X:EXIT  R:RUN  Z:Z-cable  F:Freq  I:Imp.load  C:Curr.load  V:Volt.load
```

element (which is especially important if the series element is a capacitor), we must enter actual values for either current or voltage at the load when running the program. Assuming an antenna power of 1500 W, the current at the antenna is

$$I = \sqrt{\frac{P}{R}} = \sqrt{\frac{1500}{50}} = 5.47 \text{ A}$$

All we need to do now is connect an impedance of –30.7 Ω (capacitive reactance) in series with the line at that point. Also note that at this point the current is

$$I = \sqrt{\frac{1500}{75.2}} = 4.46 \text{ A}$$

The software module SERIES IMPEDANCE NETWORK can be used to calculate the required component value. In this example, the required capacitor has a value of 1442 pF for a frequency of 3.6 MHz (**Fig 6-14**). The required voltage rating (RMS) is calculated by multiplying the current through the capacitor times the capacitive reactance, which yields a value of

$E = I \times Z = 4.46 \times 30.7 = 136.9 \text{ V RMS} = 193.6 \text{ V peak}$

In the case of a complex load impedance, the procedure is identical, but instead of entering the resistive load impedance (50 Ω in the above example), we must enter the complex impedance.

4.4. High-impedance Matching System

Unbalanced high-impedance feed points, eg, a half-wave vertical fed against ground, a voltage-fed T-antenna, the Bobtail antenna, etc) can best be fed using a parallel-tuned circuit on which the 50-Ω cable is tapped for the lowest SWR value. See **Fig 6-15**, drawings at A, B and C. Symmetrical high-impedance feed points, such as for two half-wave (collinear) dipoles in phase, the bisquare, etc, can be fed directly with a 600-Ω open-wire feeder into a quality antenna tuner, Fig 6-15D.

Another attractive solution is to use a 600-Ω line and stub matching, Fig 6-15E. Assume the feed-point impedance is 5000 Ω. Running the STUB MATCHING software

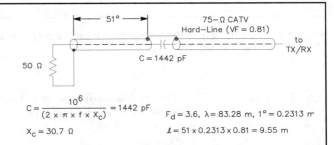

Fig 6-13—Example of how a series element can match a 50-Ω load to a 75-Ω transmission line. See text for details.

$$C = \frac{10^6}{(2 \times \pi \times f \times X_c)} = 1442 \text{ pF}$$

$X_c = 30.7 \text{ Ω}$

$F_d = 3.6, \lambda = 83.28 \text{ m}, 1° = 0.2313 \text{ m}$

$\ell = 51 \times 0.2313 \times 0.81 = 9.55 \text{ m}$

```
7921                    STUB MATCHING PROGRAM                    on4un

Freq:  3.6  Mhz              Z-line: 75.0 ohm              SWR:  1.1

              RECTANGULAR COORDINATES        POLAR COORDINATES
              REAL PART     IMAG PART      MAGNITUDE     ANGLE
IMPEDANCE (ohm) =  50.00       0.00          50.00        0.00     A
CURRENT (Amp)   =   5.47       0.00           5.47        0.00     N
VOLTAGE (Volt)  = 273.50       0.00         273.50        0.00     T
Posit.   IMPEDANCE      VOLTAGE      -------------STUB------------  IMP
Stub  Resis  React   Magnit  Angl   Imped  Value  Length  Type   ohm
45    69.2   28.8    348.6   56.3  -195.0  227 pF  21.0   OPEN   82.5
46    70.2   29.2    351.0   57.2  -197.7  224 pF  20.8   OPEN   83.7
47    71.1   29.6    353.3   58.1  -200.7  220 pF  20.5   OPEN   84.9
48    72.1   29.9    355.6   59.0  -204.0  217 pF  20.2   OPEN   86.0
49    73.1   30.2    357.9   59.9  -207.5  213 pF  19.9   OPEN   87.2
50    74.2   30.4    360.1   60.8  -211.2  209 pF  19.5   OPEN   88.2
51    75.2   30.7    362.3   61.6  -215.3  205 pF  19.2   OPEN   89.3
52    76.3   30.9    364.5   62.5  -219.7  201 pF  18.9   OPEN   90.3
53    77.4   31.0    366.7   63.3  -224.4  197 pF  18.5   OPEN   91.3
54    78.6   31.1    368.8   64.2  -229.4  193 pF  18.1   OPEN   92.2
55    79.7   31.2    370.9   65.0  -234.8  188 pF  17.7   OPEN   93.2
56    80.9   31.2    372.9   65.8  -240.6  184 pF  17.3   OPEN   94.1

H:HELP   X:EXIT   R:RUN   Z:Z-cable   F:Freq   I:Imp.load   C:Curr.load   V:Volt.load
```

```
   7921                    SERIES IMPEDANCE NETWORK (L OR C)              on4un


                      RECTANGULAR COORDINATES        POLAR COORDINATES
                      REAL PART     IMAG PART      MAGNITUDE       ANGLE
   IMPEDANCE (ohm) =    75.25        30.67           81.26         22.17     O
   CURRENT  (Amp)  =     3.44         2.83            4.46         39.46     L
   VOLTAGE  (Volt) =   172.12       318.82          362.32         61.64     D

   IMPEDANCE (ohm) =    75.25         0.00           75.25          0.00     N
   CURRENT  (Amp)  =     3.44         2.83            4.46         39.46     E
   VOLTAGE  (Volt) =   259.04       213.25          335.52         39.46     W


      ┌────────────────────────────────────────────────────────────────┐
      │  CAPACITANCE =  1442 pF          FREQUENCY =    3.60 MHz         │
      │                                                                  │
      └────────────────────────────────────────────────────────────────┘




   X = EXIT    R = RUN   Z = Z-load    E = E-load    I = I-load    F = Freq
```

Fig 6-14—Calculation of the value of the series element required to tune out the reactance of the load 75.248 + j30.668 Ω. See text for details.

module, we find that a 200-Ω impedance point is located at a distance of 81 degrees from the load. The required 600-Ω stub to be connected in parallel at that point is 14 degrees long (X = 154 Ω). The impedance is now a balanced 200 Ω. Using a 4:1 balun, this point can now be connected to a 50-Ω feed line.

Let me sum up some of the advantages and disadvantages of both feed systems.

• Tuned open-wire feeders

- Fewest components, which means the least chance of something going wrong.
- Likely least losses.
- Very flexible (can be tuned from the shack).
- Open wire lines are mechanically less attractive.

• Stub matching plus balun and coax line

- Coaxial cables are much easier to handle.

4.5. Wide-band Transformers

4.5.1. Low-impedance wide-band transformers

Broadband transformers exist in two varieties: the classic autotransfomer and the transmission-line transformer. The first is a variant of the Variac, a genuine autotransformer. The second is making use of transmission-line principles. What they have in common is that they are often wound on toroidal cores. It is not the scope of this book

to go into details on this subject. More details can be found in Chapter 7 where such broadband transformers are commonly used to feed special receiving antennas. *Transmission Line Transformers* by J. Sevick, W2FMI, is an excellent textbook on the subject of transmission-line transformers. It covers all you might need in the field of wideband RF transformers.

4.5.2. High-impedance wide-band transformer

If the antenna load impedance is both high and almost perfectly resistive (such as for a half wavelength vertical fed at the bottom), one may also use a broadband transformer such as is used in transistor power amplifier output stages. **Fig 6-16** shows the transformer design used by F. Collins, W1FC. Two turns of AWG 12 Teflon-insulated wire are fed through two stacks of 15 half-inch (OD) powdered-iron toroidal cores (Amidon T50-2) as the primary low impedance winding. The secondary consists of 8 turns. The turns ratio is 4:1, the impedance ratio 16:1.

The efficiency of the transformer can be checked by terminating it with a high-power 800-Ω dummy load (or with the antenna, if no suitable load is available), and running full power to the transformer for a couple of minutes. Start with low power. Better safe than sorry. If there are signs of heating of the cores, add more cores to the stack. Such a transformer has the advantage of introducing no phase shift between input and output, and therefore can easily be incorporated into phased arrays.

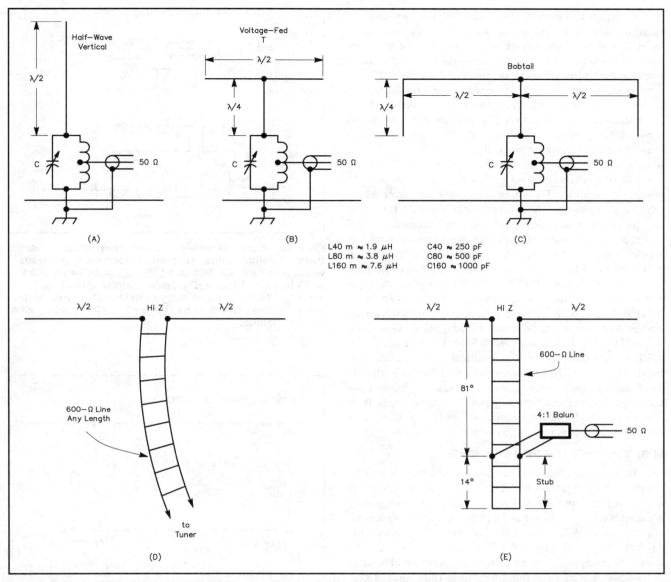

Fig 6-15—Recommended feed methods for high-impedance (2000 to 5000-Ω) feed points. Asymmetrical feed points can be fed via a tuned circuit. The symmetrical feed points can be fed via an open-wire line to a tuner, or via a stub-matching arrangement to a 4:1 (200 to 50-Ω) balun and a 50-Ω feed line.

■ 5. 75-Ω CABLES IN 50-Ω SYSTEMS

Lengths of 75-Ω Hardline coaxial cable can often be obtained from local TV cable companies. If very long runs to low-band antennas are involved, the low attenuation of Hardline is an attractive asset. If one is concerned with providing a 50-Ω impedance, a transformer system must be used.

Transformers using toroidal cores (so-called ununs) have been described in various places (Ref 1307, 1517, 1518, 1521, 1522, 1523, 1524, 1525, 1526, 1527, 1528, 1829, 1830). Ununs have been designed for a very wide range of impedance ratios. One application is as a matching system for a short loaded vertical. If the short loaded vertical is used over a good ground radial system, its impedance will be lower than 50 Ω. Ununs have been described that will match, eg, 25 to 50 or 37.5 to 50 Ω.

Transformer systems can also be made using only coaxial cable, without any discrete components. If 60-Ω coaxial cable is available (as in many European countries), a quarter-wave transformer will readily transform the 75 Ω to 50 Ω at the end of the Hardline.

Carroll, K1XX, described the non-synchronous matching transformer and compared it to a stub matching system (Ref 1318). While the toroidal transformer is broadbanded, the stub and non-synchronous transformers are single-band devices.

Compared to quarter-wave transformers, which need coaxial cable having an impedance equal to the geometric mean of the two impedances to be matched, the non-synchronous transformer requires only cables of the same impedances as the values to be matched (see **Fig 6-17**).

In reality, on the low bands (and even up to 30 MHz), the losses caused by using 75-Ω Hardline in a 50-Ω system (50-Ω antenna and 50-Ω transceiver/amplifier) are generally negligible. A real problem is that a 75-Ω feed line works as a transformer, and even when terminated with a perfect 50-Ω load, will show 100 Ω at the end of the line if the line is an odd multiple of quarter-waves long. This may cause problems to your linear amplifier. There is an easy solution to that problem, which is using $^1/_2$-wave lines (or multiples thereof). If you use a multiband antenna, make sure that the line is a number of half waves on all the frequencies used. For an antenna that works on 80 and 160, make the coaxial line a multiple of half waves on 160. Assume a 75-Ω Hardline with a velocity factor of 0.8, then the line should be 0.8 (300/1.83) / 2 = 65.6 m or any multiple thereof. You can trim the length by terminating the line with a 50-Ω load, and adjusting the length for minimum SWR on the highest frequency (in the above case: 3.66 MHz). Don't fool yourself though, in this case the SWR on the 75-Ω line is still 1.5:1, but the consequences are minimal as far as additional losses are concerned (because we use a feed line with intrinsic low losses) and are compensated for as far as the transformation effect is concerned, by using $^1/_2$-wave lengths. To be fully correct the transformation is not a perfect 1:1 transformation with a real line, but close (1:1 is only with a lossless line).

■ 6. THE NEED FOR LOW SWR

In the past, SWR was not understood by many radio amateurs. Unfortunately it still is not. Reasons for low SWR are often false, and SWR is often used as the outstanding parameter telling us all about the performance of an antenna.

Maxwell, W2DU, published a series of articles on the subject of transmission lines. They are excellent reading material for anyone who has more than just a casual interest in antennas and transmission lines (Refs 1308-1311, 1325-1330 and 1332). These articles were combined and, with information added, published by ARRL as a book, *Reflections: Transmission Lines and Antennas* [out of print].

J. Battle, N4OE, wrote a very instructive article "What is your Real Standing Wave Ratio" (Ref 1319), treating in detail the influence of line loss on the SWR (difference between apparent SWR and real SWR).

Everyone has heard comments like, "My antenna really gets out because the SWR does not rise above 1.5:1 at the band edges." Low SWR is no indication at all of good antenna performance. It is often the contrary. The "antenna" with the best SWR is a quality dummy load. Antennas using dummy resistors as part of loading devices come next (Ref. 663). It may be easily concluded from this that low SWR is no guarantee of radiation efficiency. The reason that SWR has been wrongly used as an important evaluation criterion for antennas is that it can be easily measured, while the important parameters such as efficiency and radiation characteristics are more difficult to measure.

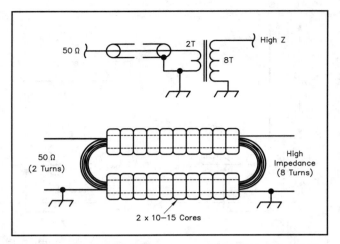

Fig 6-16—A wideband high-power transformer for large transformation ratios, such as for feeding a half-wave vertical at its base (600 to 10,000 Ω), uses two stacks of 10 to 15 half-inch-OD powdered-iron cores (eg, Amidon T502-2). The primary consists of 2 turns and the secondary has 8 turns (for a 50 to 800-Ω ratio). See text for details.

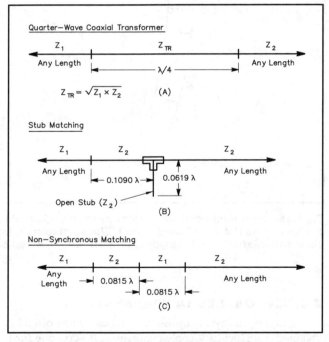

Fig 6-17—Methods of matching 75-Ω cables in 50-Ω systems. The quarter-wave transformer at A requires a cable having an impedance that is the geometric mean of the values being matched. The stub matching system at B and the non-synchronous matching system at C require only cables of the impedances being matched. The stub can be replaced with a capacitor or an inductor. All these matching systems are frequency sensitive.
Z_{TR}—60-Ω line.
Z_1—50-Ω line (or load).
Z_0—75-Ω line.

Antennas with lossy loading devices, poor earth systems, high-resistance conductors and the like, will show flat SWR curves. Electrically short antennas should always have narrow bandwidths. If they do not, it means that they are inefficient.

In Chapter 5, Section 2.8, I explained the valid reasons for a low SWR.

■ 7. THE BALUN

Balun is a term coming from the words balanced to unbalanced. It is a device we must insert between a symmetrical feed line (eg, an open-wire feeder) and an asymmetric load (eg, a ground-mounted vertical monopole) or an asymmetric feed line (eg, coax) and a symmetric load (eg, a center-fed half-wave dipole). If we feed a balanced feed point with a coaxial feed line, currents will flow on both the outside of the coaxial braid as on the inside (that's where we want to have them). Currents on the outside will cause radiation from the line.

Unbalanced loads can be recognized by the fact that one of the terminals is at ground potential. Examples: the base of a monopole vertical (the feed point of any antenna fed against real ground), the feed point of an antenna fed against radials (that's an artificial ground), the terminals of a gamma match or omega match, etc.

Balanced loads are presented by dipoles, sloping dipoles, delta loops, quad loops, collinear antennas, bisquare, cubical quad antennas, split-element Yagis, the feed points of a T match, a delta match, etc.

Many years ago I had an inverted-V dipole on my 25-m tower, and the feed line was just hanging unsupported alongside the tower, swinging nicely in the wind. When I took down the antenna some time later, I noticed that in several places, where the coax had touched the tower in the breeze, holes were burned through the outer jacket of the RG-213, and water had penetrated the coax, rendering it worthless. The phenomenon of burning holes illustrates that currents (thus also voltages) are present on the coax if no balun is used. Currents create fields, and fields from the feed line upset the field from the antenna.

How much radiation there is from such a feed line depends on several factors, the main one being its length. In most cases the feed line outer conductor will be (RF) grounded at the station. Assume the feed line is an odd number of quarter-waves long: in that case the impedance of the long wire (which is the outer shield of the feed line) will be very high at the antenna feed point, and hence the currents will be minimal, as well as parasitic radiation from it. If, however, the feed line is a number of half-waves long (and the outer shield grounded at the end), then we have a low-impedance point at the antenna end. Consequently, a large current can flow. In actual practice, unless the feed lines are a multiple of half-waves long, the impedance of the "long-wire" will be reactive, which in parallel with the resistive and low impedance of the real antenna (at resonance) will result in a relatively small currents flowing on the outer shield of the coaxial feed line. The best answer is "take no chances," and use a current balun, especially if you use (multiples of) half-wave long feed lines (see also Section 5).

Baluns have been described in abundance in the amateur literature (Refs 1504, 1505, 1502, 1503, 1515, 1519 and 1520 through 1530).

- In the simplest form a balun can consist of a number of turns of coaxial cable wound into a close coil. In order to present enough reactance at the low-band frequencies, a fairly large coil is required.

- The newest approach, introduced by Maxwell, W2DU, several years ago, is to slip a stack of high-permeability cores over the outer shield of the coaxial cable at the load terminals. In order to reduce the required ID of the toroids or beads, one can use a short piece of Teflon insulated coaxial cable, eg, RG-141, RG-142 or RG-303, which has an OD of approximately 5 mm. A balun covering 1.8 to 30 MHz uses 50 no. 73 beads (Amidon no. FB-73-2401 or Fair-Rite no. 2673002401-0) to cover a length of approximately 30 cm (12 inches) of coaxial cable. The stack of beads on the outer shield of the coax creates an impedance of one to several kΩ, effectively suppressing any current from flowing down on the feed line. Amidon beads type 43-1024 can be used on RG-213 cable. Ten to thirty will be required, depending on the lowest operating frequency. In general we can state that a choking effect of at least 1 kΩ is required for the common-mode current arrestor to be effective. This type of balun transformer is a true transmission line just like the beaded balun. but it can gain a much higher choking action from the transformation of the N turns power.

The two above approaches are the so-called current baluns. They are called current-type baluns because even when the balun is terminated in unequal resistances, it will still force equal, opposite-in-phase currents into each resistance.

Current baluns made according to this principle are commercially available from Antennas Etc., PO Box 4215, Andover, MA 01810; The Radio Works Inc, Box 6159, Portsmouth, VA 23703; and from The Wireman, Inc, 261 Pittman Road, Landrum, SC 29356. This last supplier also sells a kit at a very attractive price consisting of a length of Teflon coax (RG-141 or RG-303) plus 50 ferrite beads to be slipped over the Teflon coax.

The traditional balun (eg, the well known W6TC balun) is a *voltage* balun, which produces equal, opposite-in-phase voltages into the two resistances. With the two resistances we mean the two "halves" of the load, which are "symmetrical" with respect to ground (not necessarily in value!). If the load is in perfect common-mode balance and of a controlled impedance, a voltage-type balun is as good as a choke-type balun. But the choke-type balun is almost always much better in the real world. The toroidal-core type baluns as covered in *Transmission Line Transformers* by J. Sevick, W2FMI, are also voltage-type baluns.

Fig 6-18 shows the construction details for the W6TC voltage-type balun designed for best performance on 160, 80 and 40 meters, as well as a current-type balun as described in the text above.

We have stated on several occasions that if the reading of

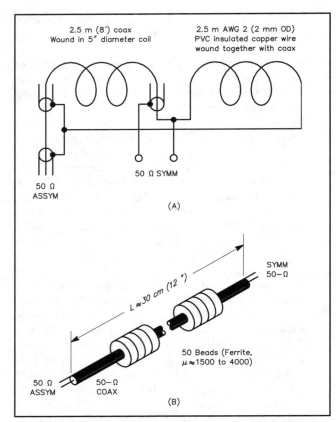

Fig 6-18—At A, details of a W6TC voltage-type balun for 160-40 meters, and at B, a current transformer for 160-10 meters. See text for details.

an SWR meter changes with its position on the line (small changes in position, not affected by attenuation) this means the SWR meter is not functioning properly. The only other possible reason for a different SWR reading with position on the line is the presence of RF currents on the outside of the coax. For that reason it is common practice in professional SWR measuring setups to put a number of ferrite cores on the coaxial cable on both sides of the measuring equipment.

We've touched upon three good reasons for using a balun with a symmetrical feed point:

• We don't want to distort the radiation pattern of the antenna.
• We don't want to burn holes in our coax.
• We want our SWR readings to be correct.

Are there good reasons to put a so-called current balun on a feed line attached to a asymmetrical feed point? Yes, there are: assume a vertical antenna using two elevated-radials. The feed point is an asymmetric one, but the end of the two radials are *not* the real ground which usually is some distance below it. If we do not connect a current balun at the antenna feed-point, antenna-return currents will flow on the outside of the coaxial feed line in addition to flowing in the elevated radials, which is not what we want with elevated radials (see also Chapter 9).

Does it harm to put a current balun on all the coaxial

antenna feed lines of all your antennas? Not at all. If the feed point is asymmetric, there will be no current flowing (exception: see above), and the beads will do no harm. As a matter of fact they may help reduce unwanted coupling from antennas into feed lines of other nearby antennas. A good thing is to use an RF current meter (see Chapter 11) and check currents on the outside of any feed line while transmitting on any nearby (within $1/2$ wavelength) antenna. These currents should be zero, if not they act as parasitically excited elements, which will influence the radiation pattern of your antenna.

How many ferrite beads (toroidal cores) are required on a coaxial cable to make a good current balun? From a choking impedance point of view you need at least 1 kΩ on the lowest operating frequency. The ferrite cores are not lossless, and depending on the mix used, they can be quite lossy. Where no *power* is involved (such as for solving EMC problems) this is never a problem. The total loss of the RF choke is then made up by the impedance of the inductance in series with the loss resistance. In other words, you have a low-Q coil. Where we use such ferrite cores to choke off potentially high RF currents (this is mostly the case with current baluns on transmitter feed lines), the resistive losses of the ferrites may actually heat those up to the point where they either become totally ineffective (permanently destroyed!) or actually crack or explode! This problem can be avoided by using ferrite material that is not very lossy on the transmit frequency. In actual practice one can successfully combine two sorts of ferrite cores in a current balun: low-resistive (high-Q) cores at the "hot side" of the balun, and lower Q beads at the "cold side". In practice, the touch and feel method is a very adequate test method: first run reduced power. If some of the cores get warm with 100 W, chances are you will destroy them with a kW.

■ 8. CONNECTORS

A good coaxial-cable connector, even a PL-259 connector, has a loss of less than 0.01 dB, even at 30 MHz, and typically 0.005 dB or less on the low bands). This means that for 1 kW of power, you will have a heat loss of typically 1 W per connector. Given the mass of a connector, and the heat-dissipating capacity of the cable, this will produce a barely noticeable temperature increase. If you feel a connector getting hot (with "reasonable" power) on the low bands, then there is something wrong with that connector. Connectors are *not* to be avoided for their high intrinsic losses as claimed by some. But when using connectors, make sure they are well installed and properly waterproofed. Despite what some may claim, N connectors will easily take 5 kW on the low bands, and over 2 kW on 30 MHz. N connectors are intrinsically waterproof and the newer models are extremely easy to assemble (much faster than a PL-259). A PL-259 connector is not a constant-impedance connector, but that is not relevant on the low bands. It is, however, a connector that is difficult to waterproof without external means. I always use a generous amount of medical-grade petroleum jelly (Vaseline) inside the connector to keep moisture out.

Some cheaper coax, as well as semi air-insulated coax may see the inner conductor retract or protrude after time. Such coaxial cables are best used with PL-259 connectors, where you can mechanically anchor the inner conductor in the connector by soldering. In an N connector, the retracting inner conductor sometimes will retract the connector pin to the point of breaking the contact.

■ 9. BROADBAND MATCHING

A steep SWR curve is due to the rapid change in reactance in the antenna feed-point impedance as the frequency is moved away from the resonant frequency. There are a few ways to try to broadband an antenna:

• Implement elements in the antenna that will counteract the effect of the rapid change in reactance. The so-called "Double Bazooka" dipole is a well known example. This solution is dealt with in more detail in the chapter on dipoles.

• Instead of using the simple L network, use a multiple-pole matching network which has the property of flattening the SWR curve.

The second solution is covered in great detail in *Antenna Impedance Matching*, by W. N. Caron, published by the ARRL.

ANTMAT is a computer program described in technical Document 1148 (Sep 1987) of the NOSC (Naval Ocean Systems Center). The document describing the matching methodology as well as the software is called "The Design of Impedance Matching Networks for Broadband Antennas." The computer program assists in designing broadband matching networks. These programs are very useful for designing broadband networks to match, eg, small whip antennas over a very wide frequency spectrum.

SPECIAL RECEIVING ANTENNAS

7 SPECIAL RECEIVING ANTENNAS

THANKS W3LPL

Frank Donovan, W3LPL, will not need much introduction, unless maybe to the newcomers at our wonderful hobby. Frank, who is a professional engineer, has been one of the most successful station and antenna builders on the East Coast. The W3LPL contest station stands as an example for all hams. What has always carried away my admiration, is the fact that Frank has the urge to share his knowledge with his fellow amateurs. Frank has published a lot of his knowledge on several ham-radio Internet reflectors. This has been helpful and informative for many. Frank's yearly open-house is a symbol of his willingness to share with others. When asked, Frank immediately volunteered to be the Godfather for this chapter. Who could have been a better critic, supporter and guide, than Mister W3LPL? Thank you Frank.

K9RJ wrote in an e-mail to me, *"The challenge of 160-meter (low-band) DXing is receiving. It should be no surprise that the highest DXCC totals on this band are achieved only by those who have the space for good receiving antennas or live in a location where much of the DX is close by (there is no such spot, the scores on 160 prove it: From the US East Coast much of Asia is the 'black hole,' and for Europe, most of the Pacific is, and look—the top scores from both areas are much the same). I'm not aware of any exceptions to this. Thus, the greatest need is for creative development of low-noise directional receiving antennas or techniques such as active noise canceling that can be used to improve receiving capability."*

I explained earlier what the differences are between the main requirements for a receiving and for a transmitting antenna. Efficiency and radiation pattern are the main issues with transmitting antennas, while signal-to-noise ratio is the only issue with receiving antennas on the low bands. Only highly directive transmitting antennas such as Yagis, quads or phased verticals will perform adequately as receiving antennas. Verticals are known to be excellent low-angle radiators, and therefore they pick up a lot of man-made noise when used as receiving antennas. They are very prone to rain (and snow) static as well, and they hear equally well (poorly?) in all directions. That's why a good low-band setup, using a single vertical as a transmitting antenna, needs to be complemented with specialized receiving antennas.

He who says "receiving antenna" usually thinks "Beverage antenna." As a rule Beverage antennas are not used by amateurs as transmit antennas, but they have been extensively used by the military either for their stealth properties or as elements forming large arrays. Their good directivity and ease of installation (low to the ground, no tower construction) makes them an ideal candidate when low probability of intercept is desired. For example, US Special Forces routinely use the Beverage for HF transmission and reception by man-portable HF communications equipment operating over single-hop distances, typically field unit to a fixed net-control station. HF-Beverage arrays are also used for transmitting applications requiring extremely narrow beamwidths and ease of maintenance are required. The efficiency of a multiple Beverage array (typically from 16 Beverages to as many as 128 Beverages in an array) is much better than the efficiency of a single Beverage antenna. The individual Beverage elements of such a Beverage array are typically spaced by a distance of about twice their height above ground (*source: W3LPL*).

■ 1. THE BEVERAGE ANTENNA: SOME HISTORY

The Beverage antenna, named after Harold Beverage, W2BML, made history in 1921. In fact, a Beverage antenna was used in the first transatlantic tests on approximately 1200 kHz. For many decades, the Beverage antenna wasn't used very much by hams, but in the last 25 years it has gained tremendous popularity with low-band DXers. The early articles on the Beverage antenna (Ref 1200-1204) are excellent reading material for those who want to familiarize themselves with this unique antenna.

■ 2. BEVERAGE PRINCIPLES

Fig 7-1 shows the basic configuration of the Beverage antenna (also called a "wave antenna"). It consists of a long wire (typically 1 to 4 λ) erected at a low height above the ground. The Beverage antenna has very interesting directional properties for an antenna so close to the ground, but it is relatively inefficient. This is why the antenna is primarily used for reception only on the amateur low bands.

The Beverage antenna can be thought of as an open-wire transmission line with the ground as one conductor and the antenna wire as the other. In order to have a unidirectional pattern, the antenna must be terminated at the far end

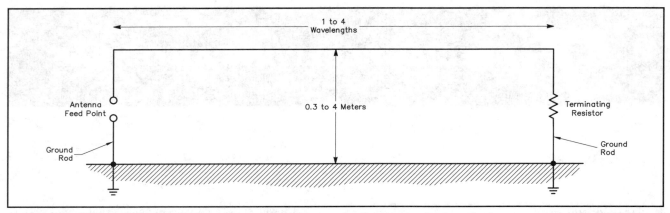

Fig 7-1—The basic Beverage antenna is a straight wire, typically 1 to 4 wavelengths long, constructed parallel to the ground at a height of 0.01 to 0.03 wavelength.

in a resistor equal to the characteristic impedance of the antenna.

If the Beverage antenna is to be used on VLF (where it was originally used), the velocity of propagation in the "two wires" (one is the antenna conductor, the other one is its image in the earth) has to be different, so that the arriving wave front (at zero wave angle for VLF signals) inclines onto the wire and induces an EMF in the wire. Therefore, the ground under the antenna must have rather poor conductivity for best performance.

On the amateur low bands, the situation is different because the wave angle is not zero. It is typically 10 to 50° for DX signals on 160, 80 and 40 meters. In this case the wave angle of the arriving signal itself is responsible for inducing voltage in the antenna wire, so it is not essential that the antenna be installed over a poorly conducting ground. We will see, however, that the poorer the ground quality, the higher the output of the Beverage antenna.

On our low HF bands, the tilted wave (tilted by the difference of velocity at VLF, and tilted due to the arriving wave angle on the HF bands) will induce signals in the wire. It may seem that the longer the Beverage antenna, the greater the induced signal. This is not the case, however. The gain increases with length, but beyond a certain length, the gain actually begins to drop off. This length limitation varies with the velocity factor of the antenna, but is also dependent on the angle of the incoming signal. The drop in gain is caused by the current in the wire increasingly lagging the tilted wave in space as the length of the wire is increased (due to the different velocity of propagation in space and in the wire). A point is eventually reached where the current in the wire is more than 90° out of phase with the space wave, and the wave begins to subtract from the signal on the wire, causing a reduction in gain.

The theoretical maximum length for a zero wave angle is:

$$L_{max} = \frac{L \times V_f}{4 \times (100 - V_f)} \qquad \text{(Eq 7-1)}$$

where

L = wavelength of operation (in meters).

V_f = velocity factor of the antenna in percent (eg, 95).

For non-zero wave angles, the above equation includes the wave angle factor as follows:

$$L_{max} = \frac{L \times V_f}{4 \times [100 - (V_f \times \cos\alpha)]} \qquad \text{(Eq 7-2)}$$

where α = wave angle

The maximum length can be determined this way because it is a function of the maximum gain of the antenna. Making the antenna longer will result in reduced output, but the horizontal main lobe can be further narrowed (with substantial side-lobes) and the vertical angle further lowered with greater lengths (at the expense of signal strength). The velocity factor is the ratio of velocity of propagation of the electromagnetic wave in the antenna wire to the velocity of propagation of electromagnetic energy in air.

The velocity factor of a Beverage will vary typically from about 90% on 160 meters to 95% on 40 meters. These figures are for a height of 3.0 to 3.5 m. At 1-m height, the velocity factor can be much lower (typically 85%). This is a major drawback of very low Beverage antennas.

The velocity of propagation of your Beverage antenna can be determined experimentally as follows: Measure the physical length of the antenna, then calculate the theoretical wavelength (L_{qw}) and frequency (F_{qw}) on which the antenna is 0.25 λ (assuming 100% velocity factor). Fill the figures in on the worksheet as shown in **Table 7-1**. Open one end of the antenna and feed the antenna via a two-turn link from a dip oscillator. Tune through the spectrum starting at approximately 70% of the calculated quarter-wave frequency (approximately 350 kHz in the example case) and note the exact frequency of all the dips up to the maximum frequency of interest. Repeat the same procedure with the antenna end short circuited to ground. Again note the dip frequencies. Note that the velocity factor changes with frequency of operation.

The antenna from Table 7-1 shows a velocity factor of approximately 95% on 160 meters and 95% on 80 meters. Applying the maximum-length formula, we find a maximum length of $L_{max} = 480$ m for 80-meter operation and

Table 7-1
Beverage Antenna Velocity of Propagation Worksheet

Physical antenna length: 200 meters
Calculated $\frac{1}{4} \lambda$ L_{qw}: $4 \times (300/L) = 600$ meters
$\frac{1}{4} \lambda$ frequency (F_{qw}): $300/600 = 0.5$ MHz

Length	Open-End Freq	Shorted-End Freq	Velocity Factory (Dip freq/N × F_{qw})
$1 \times \frac{1}{4} \lambda$	0.46	–	$0.40 / 1 \times 0.5 = 92$
$2 \times \frac{1}{4} \lambda$	–	0.94	$0.94 / 2 \times 0.5 = 94$
$3 \times \frac{1}{4} \lambda$	1.42	–	$1.42 / 3 \times 0.5 = 95$
$4 \times \frac{1}{4} \lambda$	–	1.90	$1.90 / 4 \times 0.5 = 95$
$5 \times \frac{1}{4} \lambda$	2.38	–	$2.38 / 5 \times 0.5 = 95$
$6 \times \frac{1}{4} \lambda$	–	2.87	$2.87 / 6 \times 0.5 = 96$
$7 \times \frac{1}{4} \lambda$	3.36	–	$3.36 / 7 \times 0.5 = 96$
$8 \times \frac{1}{4} \lambda$	–	3.85	$3.85 / 8 \times 0.5 = 96$
$9 \times \frac{1}{4} \lambda$	4.34	–	$4.34 / 9 \times 0.5 = 96$
$10 \times \frac{1}{4} \lambda$	–	4.84	$4.84 / 10 \times 0.5 = 97$
$11 \times \frac{1}{4} \lambda$	5.34	–	$5.34 / 11 \times 0.5 = 97$
$12 \times \frac{1}{4} \lambda$	–	5.83	$5.83 / 12 \times 0.5 = 97$

Example of a worksheet for determining the velocity of propagation of a Beverage antenna. The example is for 200-m long Beverages. A number of resonant frequencies are measured for both open-ended as well as short-circuited far-end conditions.

maximum length of 768 m for 160 meters (for a zero wave angle). If we consider wave angles lower than 20° on 160 meters, the maximum length becomes 214 m (Eq 7-2). For 80 meters (assuming a wave angle of 20°), the maximum length is 196 m. The antenna under evaluation was 200-m long, which is certainly a good compromise for use on both 160 and 80 meters.

I will cover the performance as a function of length in detail in Section 4.1.

■ 3. MODELING BEVERAGE ANTENNAS

3.1. Modeling with *MININEC*

You can do some Beverage modeling with *MININEC*, but you have to be aware of the shortcoming of *MININEC*:

While modeling, *MININEC* assumes a perfect ground right under the antenna. This means that data obtained from modeling with *MININEC* must be interpreted with caution.

Modeling Beverages with *MININEC* has the following consequences:

- *MININEC* works with a 100% velocity factor (because of the perfect ground under the antenna): the gain will increase with length, without reaching a maximum, which is not correct.
- *MININEC* will show an antenna impedance that is typically 10 to 20% lower than the actual impedance over real ground.
- *MININEC* patterns will show deep nulls in between the different lobes. This is not correct. The various lobes merge into one another due to the real ground conditions.
- Gains reported with *MININEC* are too high.

If you want to use *MININEC* and include the vertical

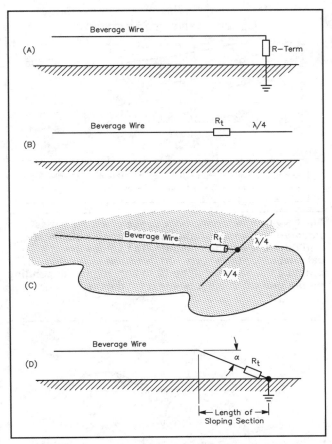

Fig 7-2—Different terminating systems for Beverage antennas. Version A suffers from stray pickup because of the vertical down lead. This configuration is only recommended when using very low to the ground antennas (eg 1 m). Method B uses a single quarter-wave terminating wire (like a radial with ground-plane antennas) in-line with the antenna. Method C uses two in-line quarter-wave lines, by which the radiation from these lines is effectively canceled. This is not a very practical solution because of its area requirements. This configuration is used throughout the book for modeling Beverage antennas. Method D is most widely used, and reduces the omnidirectional pickup, by using a long sloping section. The same techniques can be used for feeding Beverages, by substituting a matching transformer for a terminating resistor in a mirror reflection of the antenna.

down leads, you will be confronted with the problem of correctly dimensioning the pulses near the right-angle connection. This can be avoided by modeling the Beverage with 0.25-λ terminations as shown in **Fig 7-2** at C and D.

In order to determine how many pulses you need, first model a relatively short (eg, 1 λ) Beverage, and note the impedance. The resistive part should conform to the values given in **Fig 7-3**. If the value is very different, then you must increase the number of pulses.

The characteristic impedance of a single-wire Beverage over a perfect ground is given by:

$$Z_k = 138 \times \log\left(\frac{4 \times h}{D}\right)$$

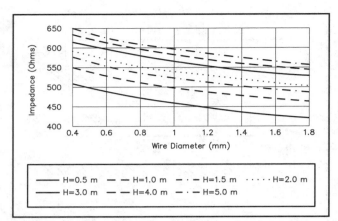

Fig 7-3—Characteristic impedance of a single-wire Beverage antenna for different conductor diameters and different antenna heights. The values are calculated for the single-wire feed line equivalent. In practice the values can be 10 to 30% higher, depending on the ground quality. See text for details.

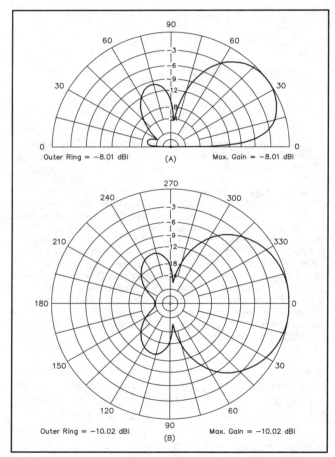

Fig 7-4—Vertical (A) and Horizontal (B) radiation patterns modeled with *EZNEC*, using two in-line single ¹/₄-wave termination wires at each end as described in the text (See also Fig 7-2C). The Beverage is 176-m long, and is modeled for a height of 2 m over good ground. The horizontal pattern is calculated for a wave angle of 30°. Note the excellent front to back of 35 dB. The front to rear worst lobe in the rear 180°, is only 14 dB, however.

where
D = wire diameter
h = height above ground (same units)

3.2. Modeling with *NEC*

We can do much more accurate modeling using *NEC*-based modeling programs. But even with *NEC*-based programs (eg, *EZNEC*, *NEC4WIN*, *NEC WIN PRO* or *NEC WIRES 2.0*) one must be careful.

Using a *NEC*-based program, the antenna velocity factor is taken into consideration. Modeling various Beverage lengths will show the gain going through a maximum at approx. 5 to 8 λ, depending on height and ground quality.

- An easy way to terminate the Beverage *in our computer model,* is to use 0.25-λ wires as ground terminations. There are two options:

1. Use two 0.25-λ wires (at each end of the Beverage) in line with each other, and at right angles to the Beverage wire (see Fig 7-2C). The radiation from these two wires is canceled in our model. This is the correct way of modeling a real short to ground. **Fig 7-4** shows a 1-λ Beverage modeled with this termination.

2. Use a single 0.25-λ termination at each end, the 0.25-λ wire being in line with the antenna (see Fig 7-2 B). **Fig 7-5** shows the pattern obtained with a 1-λ Beverage using this termination. It is clear that the 0.25-λ terminations, extending in line with the Beverage add to the creation of the radiation pattern. This modeling technique is to be avoided if accurate results are to be obtained.

In the real world, however, 0.25-λ terminations are rarely used, and if used, it will mostly be a single wire, in line with the antenna wire.

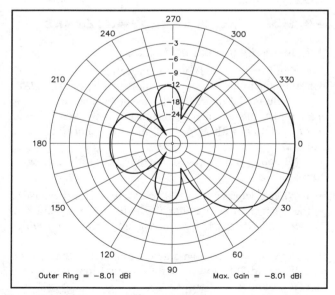

Fig 7-5—Radiation pattern modeled for the same antenna as modeled in Fig 7-4, but this time terminated using a single ¹/₄-wave termination wire at each end (as shown in Fig 7-2B). Note the much poorer F/B and the slightly higher gain, which indicates that the quarter-wave section contributes to the overall Beverage length.

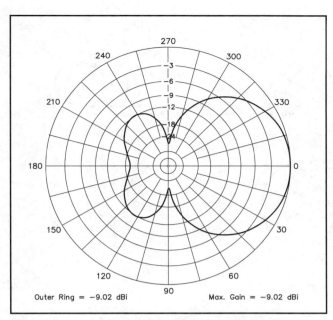

270

240 · 300

210 · 330

−3
−6
−9
−12
−18
−24

180 · 0

150 · 30

120 · 60

90

Outer Ring = −9.02 dBi   Max. Gain = −9.02 dBi

Fig 7-6—Radiation pattern modeled for the same antenna as modeled in Fig 7-4, but using sloping terminations (as shown in Fig 7-2D). Note that the F/B is slightly inferior as compared to the model shown in Fig 7-4.

- When modeling Beverages with sloping terminations (see Fig 7-2C), make sure you have lots of "pulses" on the sloping wires. For a 175-m long Beverage with two 20-m sloping end sections, accurate results are obtained using a total of 200 pulses, 20 on each sloping section, and 160 on the central section. Compare the patterns from **Fig 7-6** using the sloping terminations, with the pattern from the model in Fig 7-4. The small loss in F/B may be due to the horizontal component in the radiation pattern, caused by the sloping wire.
- Using a vertical ground connection (see Fig 7-2A) for our Beverage *model* can easily give erroneous results, Even with a great number of pulses on the down leads, and a tapered pulse length approach on the horizontal wire near the feed point and the load point, erroneous gains and impedances are reported.

All Beverages, and Beverage patterns, described in this chapter were modeled using *EZNEC* (by W7EL) using the real, high-accuracy ground option, which employs the NEC Sommerfeld method. All models were terminated using two in-line 0.25-λ terminations (at right angles to the Beverage wire as described earlier) at each end (see also Fig 7-4).

■ 4. DIRECTIONAL CHARACTERISTICS AND GAIN

4.1. Influence of Length

I calculated the gain of a series of Beverage antennas for 160 and 80, for lengths varying from 89 meters to 890 m, all over good ground. The choice of lengths was indicated by the fact that these happen to be the ideal target lengths

for obtaining best F/B (see Section 4.2). The influence of ground quality is discussed in Section 4.4.

Fig 7-7 shows the horizontal and the vertical radiation pattern for the different Beverage antenna lengths. The patterns are calculated for a "common" antenna height of 2.0 m and over good ground. The horizontal radiation patterns are calculated for the maximum-radiation angle.

The radiation angle only changes marginally (a few degrees) between very poor ground and very good ground. The radiation angles listed in **Fig 7-8** are for average ground. Note the large difference in best radiation angle between long and short Beverages: 17° for a 3-λ antenna and approx. 40° for a 1-λ antenna!

4.1.1. 160-meter analysis (see Fig 7-8A)

The major observation is that the gain tops off at a given length. This length is determined by the velocity factor of the antenna, which, in turn is determined by the height and especially the quality of the ground (see Section 2).

Over very poor ground (VPG)

- Maximum gain is obtained at 3 λ (535 m).
- The very popular length of 2 λ (176 m) has approx. 4 dB less gain than the maximum-gain antenna of 3 λ.

Over average ground (AVG)

- Maximum gain occurs at approx. 4 λ (710 m).
- The gain difference between 3 and 4 λ is less than 2 dB.

Over very good ground (VGG)

- The gain peaks beyond 4 λ.
- For a given "practical" length (176 to 353 m) a Beverage over very poor ground has a gain of 3 to 6 dB over the same Beverage over very good ground.

Conclusion: to provide substantial low-angle coverage, a length of at least 1.5 λ (268 m) is recommended.

4.1.2. 80-meter analysis (see chart in Fig 7-8B)

The curves are similar to the 160-meter curves.

Over very poor ground (VPG)

- Maximum gain is obtained at 4 λ (353 m). Refer to the radiation patterns in Fig 7.7, and note that the 3-dB forward angle becomes very small (less than 30°), which may make this too sharp an antenna for most amateur purposes.
- The 268-m long Beverage (3 λ on 80 meters) has only approx. 0.5 dB less gain than the maximum gain antenna of 4 λ length.

Over average ground (AVG)

- Maximum gain occurs at approx. 6 λ (535 m).
- The gain difference between 3 and 6 λ is only about 1.3 dB.

Over very good ground (VGG)

- The gain peaks at approx. 10 λ.
- Such long antennas will be of little practical use as

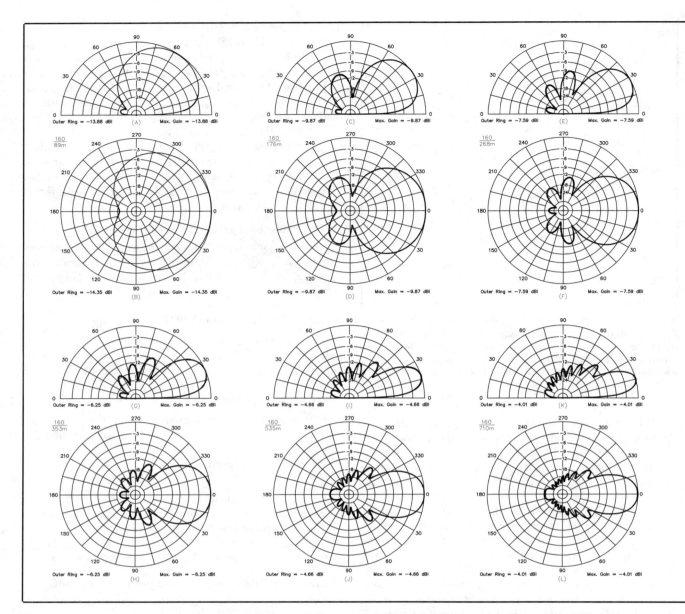

Fig 7-7—Vertical and horizontal radiation pattern of a 2-m high Beverage antenna over good ground, for different Cone-of-Silence lengths for 160 and 80 meters.

receiving antennas for 80 meters as the forward lobe becomes very sharp, in both the horizontal as well as the vertical plane (see Fig 7-7).

For a given "practical" length (176 to 353 m) a Beverage over very poor ground has a gain of 3 to 5 dB over the same Beverage over very good ground.

Conclusion: to provide substantial low-angle coverage, a length of at least 2.0 λ seems to be the best choice.

4.1.3. Calculating the velocity factor

The velocity factor can easily be calculated from the charts in Fig 7-8. Let's take the example of the 160-meter Beverage (Fig 7-8A):

- The maximum gain is at 3.5 λ over poor ground (from chart).

- This means that at 3.5 λ the physical length of the antenna is 90° longer than the electrical length of the antenna.
- The physical length is 3.5 λ or 3.5 × 360 = 1260°.
- The electrical length is 1260 – 90 = 1170°.
- The velocity factor = 1170/1260 = 92.8%.

Second example (80 meters), see Fig 7-8B:
- The maximum gain is at 6.5 λ over average ground (from chart).
- The physical length is 6.5 λ or 6.5 × 360 = 2340°.
- The electrical length is 2340 – 90 = 2250°.
- The velocity factor = 1170/1260 = 96%.

4.1.4. 3-dB beamwidth

A Beverage of 1 λ has a 3-dB beamwidth of almost 90°, which is "a lot." In order to obtain a more selective forward

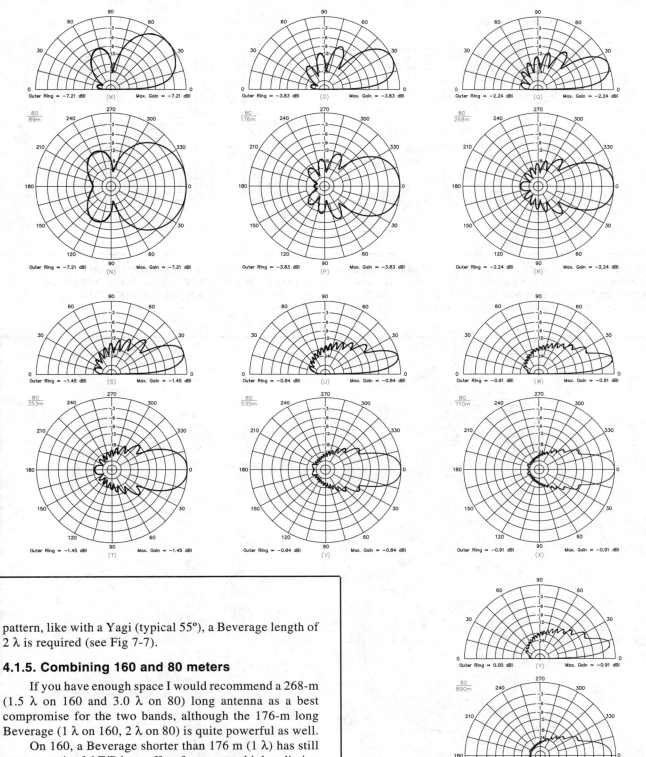

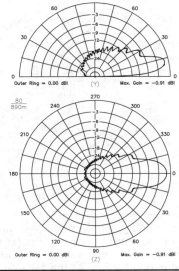

pattern, like with a Yagi (typical 55°), a Beverage length of 2 λ is required (see Fig 7-7).

4.1.5. Combining 160 and 80 meters

If you have enough space I would recommend a 268-m (1.5 λ on 160 and 3.0 λ on 80) long antenna as a best compromise for the two bands, although the 176-m long Beverage (1 λ on 160, 2 λ on 80) is quite powerful as well.

On 160, a Beverage shorter than 176 m (1 λ) has still some meaningful F/B but suffers from a very high radiation angle (typical >50°) as well as fairly low gain (−14 dBi over average ground).

4.2. The "Cone-of-Silence" Length

The Beverage antenna, being an aperiodic antenna (also called traveling-wave or non-resonant antenna), has

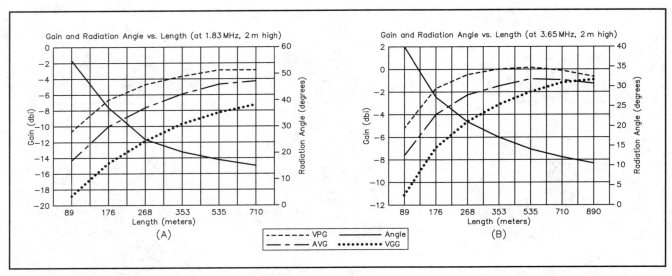

Fig 7-8—Gain and radiation angle for a 2-m high Beverage antenna for 160 meters (A) and 80 meters (B), as a function of the antenna length. Three curves are shown: over very poor ground (VPG), over average ground (AVG), and over very good ground (VGG). The radiation angle is computed for average ground. This angle only changes marginally between very poor ground and very good ground.

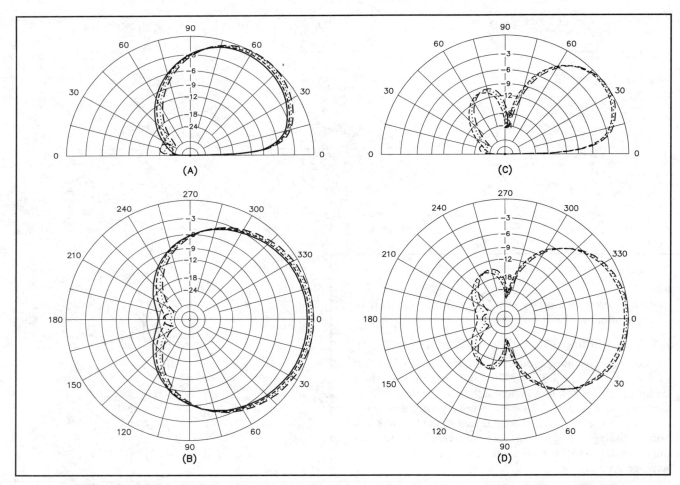

Fig 7-9—The 0.5-λ long Beverage was modeled for lengths of 82, 85, 89 and 92 m (at A and B), and the 1-λ Beverage for 165, 172 and 178 m (at C and D). Notice that a range of Cone-of-Silence lengths can be defined, depending on the wave angle considered. The horizontal radiation patterns were computed for 30° for the 0.5-λ Beverage and for 25° for the 1-λ long antenna.

the tremendous advantage of being very broadbanded, and does not—in principle—require exact lengths to perform well.

In Section 4.1, I covered gain and radiation angle. But there is more, of course. The Beverage antenna, being almost exclusively used for receiving purposes with radio amateurs, its directivity, and more importantly its front-to-back or front-to-rear ratio (see Chapter 5 for definitions) are the paramount performance parameters of this antenna.

Front-to-back ratio (F/B) is usually defined as the geometrical ratio of signal at zero and at 180° azimuth. For an antenna which is used to null out an interfering signal (usually a rotary antenna) the F/B is very important, it determines how deep you can null out the offending signal (see also Section 26). When fighting noise and general QRM "off the back," the average F/B, or the integrated F/B is more important. In this case we have to consider all the lobes in the back area of the antenna. We usually refer to this as the F/R (front-to-rear ratio).

It appears that the F/B goes through maximum values for lengths that are a multiple of electrical 0.5 λ. These "secret" lengths are often referred to as the *Cone-of-Silence* lengths.

The mechanism for the Cone-of-Silence length is that the F/B is not only determined by the absorption of the induced voltage from the back of the antenna in the load resistor, but in addition to the canceling effect of out-of-phase signals on the antenna wire. This is also why we can "tune" a non-Cone-of-Silence length antenna by using a complex load (R + L) which establishes the required phase shift to enable signal cancellation on the antenna wire (see Section 6.3).

To obtain the electrical wavelength we have to take into account two things:
1. The velocity factor of the antenna (depends on ground quality and antenna height).
2. The wave angle we are considering.

Fig 7-9 shows the horizontal and vertical radiation angles for a 0.5-λ and a 1.0-λ 160-meter Beverage for different lengths, all in the Cone-of-Silence area.

This means we cannot define *one* length to be *the* Cone-of-Silence length. The range of C.O.S. lengths is given in **Table 7-2**.

As the Beverage gets longer (in wavelengths), the C.O.S. effect is less outspoken. Above 3 λ the effect is negligible. It is very outspoken between 0.45 and 2.5 λ. If

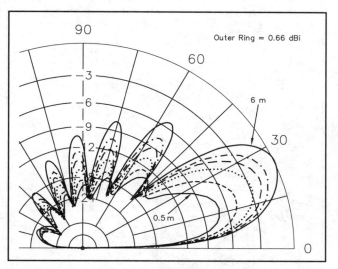

Fig 7-10—Vertical radiation pattern for a 268-m long (3-λ) Béverage on 3.65 MHz, for heights ranging from 0.5 to 6.0 m.

you have a Beverage measuring that long, it is best to cut it to the lengths given in Table 7-2.

4.3. Influence of Antenna Height

The general rules are as follows:
• higher antennas produce higher output.
• higher antennas have larger side lobes.
• higher Beverages have higher wave angles.

Fig 7-10 shows the vertical radiation patterns for a 268-m long Beverage operating on 3.65 MHz (3 λ) for heights ranging from 0.5 to 6 m. The horizontal pattern was calculated at the forward lobe peak angle, which ranges from 17° for 0.5-m height to 25° at 6 m as shown in **Fig 7-11**. All calculations were done over average ground.

It is obvious that an 80-meter Beverage at a height of 6 m is way too high, as the secondary lobe at about 55° wave angle (and at approx. 50° of the main azimuth direction) is only 10 dB down from the main lobe. The very low Beverages (0.5 m) shows the best directivity pattern but are far less sensitive.

The high-angle lobes that appear at higher heights are due to the increasing horizontal radiation component. The horizontal and vertical components as well as the total radiation pattern of a 1-m high and a 6-m high Beverage (for 80 meters) are shown in **Fig 7-12**. Note the important horizontal component at 6-m height. Still higher the Beverage will start behaving like a terminated long wire, and not as a Beverage antenna.

This high-angle response of a *high* Beverage is often used by those who don't believe in important path skewing, to explain "apparent" path skewing. Though I do not deny that reception of these high-angle side-lobes may cause some confusion at times, the existence of an important degree in direction skewing has been confirmed repeatedly through the use of other directive antennas such as phased arrays.

If one suspects receiving signals from such high-angle

Table 7-2
Cone-of-Silence Ranges for 160, 80 and 40 Meters
The lengths are computed for Beverages 2-m high over good ground. The influence of height and ground quality is rather limited.

160 meters	80 meters	40 meters
82 - 92 m (0.5 λ)	43 - 47 m (0.5 λ)	45 - 48 m (1.0 λ)
165 - 178 m (1.0 λ)	85 - 92 m (1.0 λ)	67 - 72 m (1.5 λ)
250 - 270 m (1.5 λ)	126 - 136 m (1.5 λ)	87 - 92 m (2.0 λ)
332 - 352 m (2.0 λ)	169 - 179 m (2.0 λ)	135 - 140 m (3.0 λ)
410 - 430 m (2.5 λ)	212 – 222 m (2.5 λ)	
	258 – 268 m (3.0 λ)	

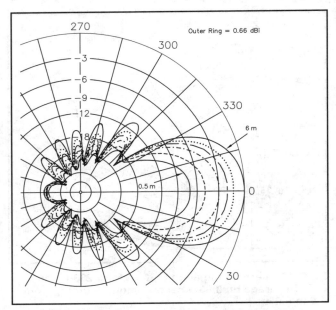

Fig 7-11—Horizontal radiation pattern at the main radiation angle for the same antenna as Fig 7-10.

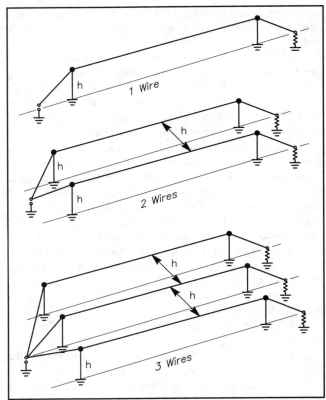

Fig 7-13—Layout for the single-wire, 2-wire and 3-wire Beverage with sloping termination. In this example the multi-wire Beverages are spaced their height above ground. The multiple-wire Beverages in this example use individual terminating resistors but are fed from a common point.

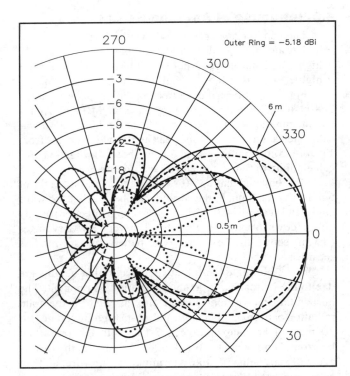

Fig 7-12—Radiation patterns for a high (6 m) and a low (0.5 m) Beverage of identical length (268 m). The patterns show the total fields (solid curves) as well as the horizontal and vertical components (dotted and dashed lines). Most of the secondary lobes are made of horizontal radiation components, which become more pronounced as the antenna height is increased. Note that the secondary lobe for the 6-m high Beverage (at approx. 50° off the main direction) is almost as strong as the main lobe of the 0.5-m high Beverage!

side lobes, this can usually be verified by switching to a high-angle antenna such as a low dipole. As explained in the chapter on propagation this may typically occur at sunrise or sunset (gray-line propagation).

Conclusion: A height of 2 m seems to be a reasonable compromise for a Beverage to be used on 80 as well as 160 meters. For 160 meters only, I would recommend 3 m height. If directivity is the main target, and not so much sensitivity, then one can consider using Beverages at 0.5 or 1 m height.

4.3.1. The multi-wire Beverage

If very low Beverages are practical at your QTH, and if you want to reduce side lobes to a minimum, efficiency of low Beverages can be significantly improved by using a multi-wire Beverage rather than a single-wire Beverage (this is of course true for Beverages at any height). In this configuration several close-spaced Beverages (typically three) are used. One can simply space two or three Beverages side-by side, spaced from each other by a distance equal to their height.

Fig 7-13 shows a possible practical configuration, where the receiving ends slope to a common feed point. At the termination end, individual resistors are used. In such a configuration the feed-point impedance for a two-wire parallel Beverage is half of that of a single-wire Beverage (200 to 250 Ω), while the three-wire version exhibits a feed impedance of 130 to 170 Ω.

Fig 7-14 shows the radiation patterns of a single-wire, a two-wire and a three-wire Beverage, 1-m high and 176-m long, for a frequency of 1830 kHz. Going from 1 to 2 wires adds 3 dB to the signal, while going from 1 to 3 wires increases the gain by 4.5 dB.

The disadvantage is that the F/R (front-to-rear ratio) is slightly deteriorated, due to the feed points joining in one point. One could also use a common termination resistor, but this would even further deteriorate the F/R.

In fact this configuration is the same as an array of Beverages fed in phase. Each of the individual Beverages could also be fed individually and in phase (see Section 15.2). In this case the F/B deterioration will not occur.

4.4. Influence of Ground Quality

The general mechanism is:
- The better the ground, the lower the output from the antenna.
- The better the ground, the better the low-angle performance. The peak of the main lobe hardly changes between VPG, AVG and VGG (28 to 30°).
- The poorer the ground quality, the less pronounced the nulls will be between the different lobes. This is similar to what we notice with horizontally polarized antennas over real ground.
- The front-to-back ratio as well as the front-to-side ratio

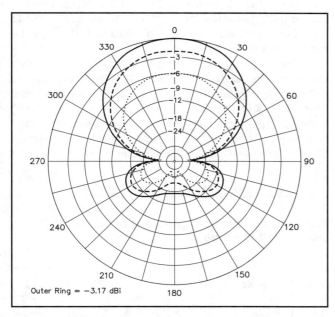

Fig 7-14—Radiation patterns calculated for the single-wire, 2-wire and 3-wire Beverages as shown in Fig 7-13. Note that the 3-dB forward beamwidth remains unchanged. Gain is obtained merely by reducing the losses of the system by using multiple wires for the radiator.

of the Beverage remains almost constant for grounds ranging from very good to very poor.
- The vertical radiation angle does not change very much between very good ground and very poor ground
- The Beverage does not work well over sea water. Its output is down 15 dB compared to the same antenna over poor ground and the main radiation angle is at 45°. This confirms the observations made by Ben Moeller, OZ8BV, that his Beverages near the sea never worked well at all. The Beverages at VKØIR, put up over a salt water marsh never worked either—as I told them!

■ 5. BEVERAGE IMPEDANCE (SURGE IMPEDANCE)

Over perfect ground the single-wire Beverage impedance can be calculated using the formula of the single-wire transmission line over ground:

$$Z = 138 \log\left(\frac{4 \times h}{d}\right)$$

where
    h = height of wire.
    d = wire diameter (in the same units).

The theoretical impedance values listed in Fig 7-3 are generated over perfect ground. They are useful for estimating the terminating resistor for a single-wire Beverage and for designing matching transformers and networks. Note that the impedance does not change drastically with height or wire size. Very low Beverages do not have a very low impedance as is sometimes said. Belrose (Ref 1236) reports impedances varying between approximately 420 Ω and 550 Ω for a 110-m (360-ft) long Beverage at frequencies ranging from 2 to 10 MHz (height above ground = 1.1 m).

Over real ground the impedance appears to be higher than over perfect ground. The following correction figures as compared to the impedance over perfect ground (Fig 7-3) can be used:
- Good ground: +12%
- Average ground: +20%
- Very poor ground: +30%

The Beverage surge impedance can be determined experimentally in several ways:
- Couple a dip meter to one end of the antenna via a two-turn link. Terminate the other end with a 300-Ω resistor and tune the dip meter from 1 to 7 MHz. Measure the depths of the resonant points as you scan through the frequency range. Repeat the same test with a 400 Ω and then a 500-Ω terminating resistor. This process will allow you to find a resistor value for which almost no dips will be found between 1 and 7 MHz. That will be the resistor value for which the antenna is fully aperiodic (non-resonant). The exact impedance can vary greatly with ground conditions (season, humidity and so on).
- Sweep the antenna over a wide frequency range, and adjust the termination input impedance for minimum input impedance change at the ANTENNA terminals (but not down any length of feed line). Virtually any SWR

meter at the Beverage feed point could be used for this; a battery-powered antenna analyzer is ideal.

- Excite the antenna with a small signal, and measure the current along the antenna with a clamp on RF-current meter or RF-voltage meter. Adjust the termination resistance until the voltage or current has a uniformly smooth taper toward the far end (typically 25% to 50% depending on ground quality and antenna length).
- Measure the feed-point impedance across a spectrum (eg, 1.5 to 3 MHz) and note the lowest (Z_{min}) and the highest impedance value (Z_{max}) you measure. The Beverage surge impedance is then given by:

$$Z_{Bev} = \sqrt{Z_{max} \times Z_{min}}$$

Misek, W1WCR, described the use of wires under the Beverage to stabilize ground conditions (Ref 1206) and to obtain more stable impedances with varying weather conditions.

It is my own experience that the exact value of the terminating resistor is not so critical. I use a 470 or 560-Ω resistor on all my Beverage antennas.

■ 6. TERMINATING THE SINGLE-WIRE BEVERAGE

6.1. Vertical Down-Lead, Sloping Beverage or Quarter-Wave Terminations

A common way of terminating the single-wire Beverage is to connect the proper terminating resistor between the end of the Beverage and the earth. This, however, is not necessarily the best solution, especially on frequencies higher than the very low ones, where the vertical down lead that connects the Beverage to the terminating resistor becomes a significant part of a wavelength and thus picks up a lot of signal. This effect can significantly impair the front-to-back ratio of the antenna. For Beverages at relatively high heights (1 meter or higher) this is not the most suitable way of terminating the antenna.

Fig 7-2 shows the classic resistor termination system using the vertical down lead as well as alternative systems that eliminate vertical down-lead pickup.

The systems using 0.25-λ terminations are not very practical as they use a lot of real estate. Using single 0.25-λ terminations in-line with the Beverage (Fig 7-2B) greatly impairs the F/R of the antenna through pick up by the termination wires. This extra length does not add any gain (although adding an extra $1/2$ λ of wire), and hence appears to be very impractical.

Using two in-line 0.25-λ wires at each end (Fig 7-2C) overcomes the poor F/B problem, but makes it a very complex system to build and requires even more wire and real estate, again not adding any signal to the antenna! This configuration is however, the easiest and most accurate model for evaluating Beverages using a *NEC*-based computer program.

Quarter-wave terminations are obviously single-band terminations. Traps can be included for multiband opera-tion. Wires shorter than 0.25 λ can be used together with a coil to add the proper loading (as done with elevated radials, see chapter on radials).

I have used the sloping termination (Fig 7-2D) extensively and found it to work very well. The slope angle should be lower than the minimum expected arrival angle of the signal. **Table 7-3** gives the length of the sloping wire as a function of the minimum wave angle and the Beverage height. A 20-m long sloping section is a good ball park figure under all circumstances. The sloping wire section is part of the antenna proper, which means that sloping termination is fully broadbanded. The terminating resistor (approximately 500 Ω) is now connected between the end of the sloping wire and the ground system.

It appears that the military extensively uses sloping terminations in their Beverage arrays, used in HF-radar systems. Knowing that directivity is the main issue with such arrays (a clean pattern), this confirms that sloping terminations are the way to go.

It is obvious that the discussion of the terminating methods (vertical, sloping or using 0.25-λ wires) *applies to both ends* of the Beverage.

6.2. The Termination Resistance

If the termination resistance is not exact, the F/B will not be the maximum obtainable. In many instances you will not notice this very much, unless the noise or the signal you want to reject is located exactly in the null-direction of the Beverage. Assuming a lossless antenna, the F/B ratios for different terminations are given in **Table 7-4**.

If the theoretical termination resistance is 500 Ω and we use 450 Ω, the SWR is 500 / 450 = 1.12, and the maximum obtainable F/B is 25 dB. If we use a 250-Ω terminating resistor, the max. obtainable F/B would be reduced to approx. 10 dB.

The purpose of the terminating resistor is to absorb energy propagating down the wire from the rear (for good F/B ratio). The termination should *not* be adjusted to achieve the best SWR on the feed line to the receiver. If everything is perfect, a proper termination resistance value and a proper transformer (at the receiving end) will achieve a perfect SWR, but it is advisable to do things in the right sequence:

Table 7-3
Wave Angle Versus Slope Length (m) for Beverage Antennas

	Wave Angle					
Beverage Height	*10°*	*15°*	*20°*	*25°*	*30°*	*35°*
0.3 m (1 ft)	1.7	1.2	0.9	0.7	0.6	0.5
1.0 m (3.3 ft)	5.8	3.9	2.9	2.4	2.0	1.7
2.0 m (6.6 ft)	11.5	7.7	5.8	4.7	4.0	3.5
3.0 m (10 ft)	17.3	11.6	8.8	7.1	6.0	5.2

The length of the sloping termination is shown as a function of the antenna height and the lowest wave angle that must be received without loss of gain.

- Experimentally determine the antenna surge impedance by one of the methods explained in Section 5. This value may be lower than what you expect, especially over poor or very poor ground as the ground resistance will be in series with the termination resistance. K6SE has reported

Table 7-4
F/B Ratios for Different Terminations

Termination SWR	F/B
3.6	5 dB
1.9	10 dB
1.4	15 dB
1.21	20 dB
1.1	25 dB

Table 7-4A
Variation of Termination Resistance for a 1.2 m High Beverage Antenna as a Function of Ground

Type of Ground	Optimum Termination Resistance
very good ground	580 Ω
good ground	560 Ω
poor ground	500 Ω
very poor ground	320 Ω

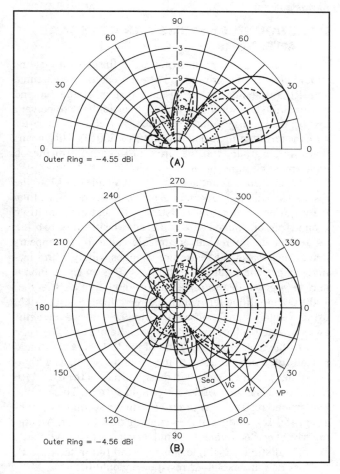

Outer Ring = −4.55 dBi

(A)

Outer Ring = −4.56 dBi

(B)

Fig 7-15—Vertical (A) and horizontal (B) radiation patterns of a 268-m long, 2-m high Beverage operating on 160 meters, for different grounds (very poor ground, average ground, very good ground and sea water). Note the poor performance over salt water, as well as a solid 8 dB difference in gain between very poor ground and very good ground! The horizontal pattern was calculated for a wave angle of 30° in all cases.

finding a 300-Ω terminating resistor to be the proper value over very poor ground at his QTH.

- Attach a 9:1 transformer at the receiving end. If you measure the SWR (using an antenna analyzer) it will likely be between 1 and 1.5:1, which is very adequate for receiving. A very low SWR value (<1.2:1) is only required if you want to use lengths of coaxial feed line to achieve given phase shifts in arrays of Beverages (see Section 15).

The typical variation of termination resistance for a 1.2-m high Beverage antenna as a function of ground is shown in Table 7-4. The termination resistor must be a non-inductive (or low-induction) type resistor, which means that it cannot be a wire-wound resistor. In principle any wattage will do, but if you have the Beverage close to transmitting antennas (which is bad, see Section 13), you may want to use a few 1 or 2-W resistors in parallel. I use a single 2-W resistor and have never seen one discoloring or burning up.

If you live in an area with a high thunder storm occurrence, the use of genuine carbon resistors (Allen Bradley still makes them) is highly recommended. Further protection can be added by using small gas-discharge tubes connected across the resistor, or even by putting a pair of home made small air-gap electrodes across the resistor. The air gap should be no more than the thickness of a sheet of paper, and should be well protected in a small box.

6.3. Inductive Load Terminations

In Section 4.2, we learned about the Cone-of-Silence length. A 120-m long (approx. 0.75-λ) Beverage, terminated in a termination resistance equal to its characteristic impedance (500 Ω), yields approx. 13-dB F/B. This is not very good, and Cone-of-Silence length Beverages can yield 25 dB of F/B.

If we now terminate the same Beverage in a complex load, a resistor in series with a coil, we can obtain the same F/B as with a Beverage measuring the Cone-of-Silence length. **Fig 7-16** shows a 120-m long Beverage (on 160 meters) which, when terminated in its "classic" load resistance of 500 Ω (12-dB F/B), and the same antenna terminated in a complex impedance of 500 + j250 Ω (an inductance of 22 μH on 160 meters), the F/B jumps to 25 dB, while the gain goes down only a fraction of 1 dB. At the same time the feed impedance becomes complex (500 − j220 Ω). We can take care of this capacitive reactance by inserting a coil with an impedance of +220 Ω between the 9:1 transformer and the Beverage antenna (approx. 38 turns on a T106-2 powdered-iron core). This last coil is only there to provide a match for the feed line.

Another example shown in Fig 7-16 is for a 200-m long Beverage (again on 160 meters). In this case a coil with a reactance of 100 Ω was required to obtain maximum F/B.

In practice it is best to use a variable inductor and to adjust the value for maximum F/B, using a small signal generator, placed off the back of the antenna, a few wavelengths away. I use a small roller inductor that can be adjusted to max. 34 μH. Once the exact value is determined, the variable inductor can be replaced with a powdered-iron

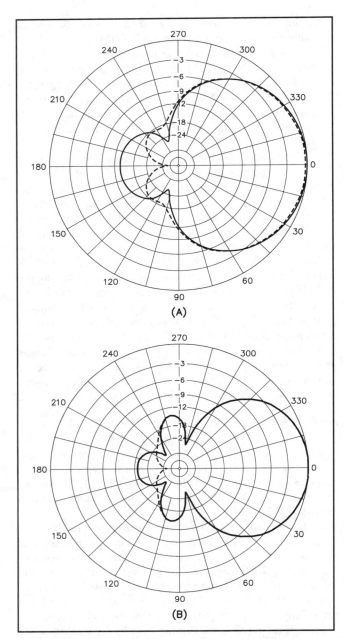

Fig 7-16—Radiation patterns (at 25° wave angle) of a 120 m (A) and a 200 m long (B) Beverage antenna on 160 meters. In each case the pattern for the antenna terminated in a 550-λ resistor exhibits a larger lobe to the back. The patterns with improved F/B are for a termination consisting of a 550-λ resistor in series with an inductor. The inductor at A has a reactance of 250 Ω, and the inductor at B 150 Ω at 1.83 MHz. Note that the F/B can be drastically improved.

toroidal coil of the same inductance. Do not use ferrite core for this application. The value of the series inductor at the feed point can be adjusted to obtain minimum SWR on the primary of the 9:1 transformer.

Remember that when you use such a complex termination, that the Beverage becomes a single-band antenna, and that you will have to switch loads for different bands.

6.4. Adjusting the Termination for Best F/B

In order to obtain a good F/B, make sure your Beverage length corresponds to a Cone-of-Silence length (Section 4.2). In this case a simple resistor will do. If not, you will need an R + L to obtain the best F/B (Section 6.3). This R + L termination is a single-band termination.

Adjustment of the terminating impedance can easily be done using a small signal generator with a small whip antenna, which should be installed at least 5 λ from the antenna.

Unless you are interested in maximum F/B at zero degree elevation angle (nobody does!), you should *not* put the signal generator in-line (off the back, of course) with the antenna. The signal generator must be set up at an azimuth (ie, angle with respect to the line of the Beverage wire) equal to the elevation angle at which we want the antenna to produce the best F/B. In general an angle of 25 to 30° can be used.

This principle does not only apply to Beverage antennas, but to most of the special receiving antennas described in this chapter.

■ 7. GROUND SYSTEMS FOR BEVERAGE ANTENNAS

If you have real ground (not rock), the ground systems should consist of a ground rod at each end of the antenna. The RF resistance of the ground system at the far end (termination end) of the Beverage does not have to be very low, as the (high) ground resistance is effectively in series with the terminating resistance (approx. 500 Ω). This means that a simple ground rod of approx. 1 m length will do the job under most circumstances.

Where ground conductivity is really bad (Ref 1260) the resistance of a single ground rod may actually be higher than the required terminating resistance. In that case multiple ground rods or chicken wire will need to be installed to bring the resistance down to an acceptable value. It also appears that in such cases, the resistance changes widely with varying weather conditions. If optimum F/B is an issue, then it may be a good idea to install a non-inductive (AB J-type) 1000-Ω potentiometer, which can be adjusted in the field using a small signal source installed several wavelengths away off the back of the antenna (see Section 6.4).

At the receiving (feed) end of the Beverage we should pay a little more attention to the ground system, as a high-resistance ground system will dissipate part of the useful RF in the loss resistance of a poor ground system. The equivalent ground resistance is in series with the secondary (high-Z, eg, 450-Ω) winding of the matching transformer. Assuming we have an equivalent ground resistance of 300 Ω (that's not unrealistic for a single short ground rod in the desert), the loss due to this 300-Ω resistance would be:

$$20 \times \log\left(\frac{550}{550 + 300}\right) = -3.8 \text{ dB}$$

This means that whenever possible we should provide a decent RF ground at this point. If you have determined the termination resistance of the Beverage by one of the experimental methods outlined in Section 5, you can use the value of the termination resistance to evaluate the quality of the ground. If a typical 2-m high Beverage requires a 300-Ω termination resistor in order to be aperiodic (no standing waves on the antenna), then the above calculation applies, unless you take measures to establish a lower resistance ground at the receiving end.

Multiple ground rods, spaced at least the length of the ground rods, are one approach. Where we cannot use a ground rod (rocky ground), we can use a large ground mat made of large strips of chicken wire (eg, the shape of a cross measuring approx. 6 × 6 m), or a large number of short (interconnected) radials laid on the ground. Laying 0.25-λ wires on the ground as radials makes little sense. On the ground, these wires are far from 0.25-λ electrically, and it is much better to use the same amount of wire to make 4 shorter radials.

Long radials should also be avoided as they inevitably will alter the F/B of the Beverage and mess up the outstanding F/B ratios if Cone-of-Silence lengths are used (Section 4.2).

Where a sloping termination is used, the ground rod can extend slightly above ground to act as a mechanical termination post for the sloping wire section (Fig 7-2D).

■ 8. THE BEVERAGE FEED-POINT TRANSFORMER

The easiest way to match the Beverage impedance (typically 450 to 600 Ω) to the commonly used coaxial cable (50 or 75 Ω) is to use a wideband toroidal transformer. Such transformers are usually wound on magnetic-material cores.

The magnetic material can in principle be either ferrite or powdered iron. Ferrites attain much higher permeability (up to 10,000) than powdered-iron materials (only up to 100), but are less stable at higher frequencies and saturate more easily. For wideband transformers which are not typically confronted with high power, such as in the Beverage antenna situation, ferrites are the most logical approach.

Such transformers are commercially available from various sources, but can easily be wound for a fraction of the price of commercial units.

8.1. Core Material and Size

Several core sizes and core materials can be used for this job. As far as the size is concerned, the 0.5-inch cores (or even smaller ones) will do the job if the builder is not tempted to transmit on the Beverage. I have been using the high-permeability ferrite material very successfully. To note only one advantage, it is much easier to wind the cores when there are only a few turns!

Don't use just any core you find in your scrap box. Transformation ratio or SWR is not the only issue. I have seen transformers that showed a perfect SWR with the secondary terminated with 500 Ω, but they also exhibited a prohibitive loss of not less than 5 dB!

Many people report having successfully used toroidal cores using 75, 43 and even the low permeability 61 ferrite material. I have used for decades the Ceramic Magnetics MN8CX high permeability cores without a single failure.

I cannot recommend the 61 mix as this material requires too many turns and has too much interwinding

capacitance. Powdered-iron toroids all have too low permeability to be used for this application.

Measuring the inductance for a given number of turns, and then deducing the required turns for a transformer works well with powdered-iron cores, but *not* with ferrite cores. The best approach to design a transformer using a ferrite material core is to wind a transformer and test its performance while gradually reducing the number of turns (keeping the primary-to-secondary ratio constant). This is how I determined the turns required to make a 9:1 transformer. A good transformer has an insertion loss of typically less than 0.5 dB from 1.8 though 7 MHz and a return loss of better than 25 dB (SWR 1.1:1).

Several transformers were made using three types of cores. The required number of turns are shown in **Table 7-5**. Alternative part numbers for toroid cores can be found in **Table 7-6**.

8.2. How to Wind Your Own Beverage Transformer

A 9:1 impedance transformer (3:1 turns ratio) will give a more than acceptable match for both 50 and 75-Ω lines. A transmission-line transformer using a trifilar winding is well suited to this purpose. The enameled wires can be twisted together.

8.2.1. Common ground transformers

The classic configuration as well as the winding information for a 9:1 wideband transformer is shown in **Fig 7-17**. Note that the ground for the coaxial feed line as well as the ground for the antenna are connected together *in* the transformer. This means that the antenna and the feed line use a common ground.

Table 7-5
Number of Turns for a 50 or 75-Ω Winding
This means that a 9:1 transformer made on a MN8CX core requires 4 trifilar turns on the core.

Core type	Turns
MN8CX (*)	4
FT50-75, FT82-75, FT114-75(*)	8
FT50-43(*), FT82-43(*), FT114-43(*)	8

Number of turns for a 50 or 75 Ω winding. This means that a 9:0 transformer made on an MN8CX core requires 4 trifilar turns on the core. The cores indicated with (*) give the lowest attenuation and are recommended.

Table 7-6
Alternative Manufacturer's Part Numbers for Toroid Cores

Type	Alternative Supplier and Code
Amidon: FT-50-75	Fair Rite 5975000301
	Ferroxcube 768T188/3E2A
Amidon: FT-82-75	Fair Rite 5975000601
	Ferroxcube 846T250/3E2A
Amidon: FT-114-75	Fair Rite 59750001001
	Ferroxcube 502T300/3E2A
Amidon: FT-50-43	Fair Rite 5943000301
	Ferroxcube 768T188/3D2
Amidon: FT-82-43	Fair Rite 5943000601
Amidon: FT-114-43	Fair Rite 59430001001

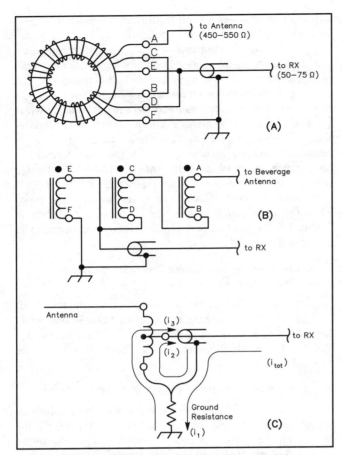

Fig 7-17—The transmission-line transformer (A) consists of three parallel-wound wires, equally spaced across a toroid core. The proper connecting scheme (B) is shown for a 9:1 impedance transformation ratio. The enameled copper can be securely fixed to the core using tape or Q-dope. Antenna currents (C) on the outside of the feed line (acting as a sort of "on the ground Beverage") are routed into the inside of the coax via the common ground resistance and via the low-Z link of the transformer.

Refer to Table 7-5 for the winding information for some of the more common core materials and toroids ranging from 13 to 25-mm outside diameter. The same number of turns can be used for a 50-Ω and a 75-Ω system impedance. Data are given for the frequency range 1.8 to 7 MHz.

Winding such a transformer is very easy. Take three lengths of 0.4-mm enameled wire (AWG 26) and twist them together with a pitch of approx. 2 turns per cm. Use a hand drill to get an even twist. Fix one end of the wires in a vise, and the other end in the head of your hand drill. Make sure the tension in the three wires is identical and that the three wires run perfectly parallel (no wire crossings!). If you have access to enameled wire in different colors, then the task of identifying the wires is easy. Otherwise you will have to scrape off (or burn off) some insulation material and identify the three wires with an ohmmeter. Connect the wires exactly as shown in Fig 7-16. After soldering, cover the soldered wires with some shrink sleeve.

I have used both single MN8CX cores as well as two stacked MN8CX cores for these transformers. **Table 7-7** shows the insertion loss as well as the SWR for transmission-line transformers made with different types of cores. The measurements were done on an R&S Network Analyzer. Insertion loss was measured using two transformers back-to-back. In both cases 4 trifilar turns were wound on the cores.

8.2.2. Isolated grounds

The outside of the coaxial feed line acts as a very low (on the ground) lossy long-wire or unterminated Beverage, receiving signals from directions other than those we are interested in. These signals cause currents to flow on the outside of the feed line (i_{tot}), which can be fed into the *inside* of the feed line through coupling via the common Earth resistance (i_1 i_3) as well as via the low-Z winding of the transformer (i_2)! See Fig 7-17C.

This is a common cause of spurious-signal reception, especially if long feed lines are involved. In order to prevent

---

Table 7-7

Insertion loss and SWR for a transmission-line type autotransformer wound on various cores.

(A) = single MN8-CX core (3 wires × 4 turns) (B) = stack of 2 MN8-CX cores (3 wires × 4 turns)
(C) = single FT-114-43 core (3 wires × 8 turns) (D) = single FT-114-75 core (3 wires × 8 turns)

Type	1.8 MHz	3.65 MHz	7.1 MHz	Type	1.8 MHz	3.65 MHz	7.1 MHz
[A] Insertion Loss (dB)	0.50 dB	0.45 dB	0.45 dB	[C] Insertion Loss (dB)	0.28 dB	0.20 dB	0.18 dB
SWR	<1.1	<1.1	<1.1	SWR	<1.1	<1.1	<1.1
[B] Insertion Loss (dB)	0.25 dB	0.25 dB	0.25 dB	[D Insertion Loss (dB)	0.20 dB	0.21 dB	0.28 dB
SWR	<1.1	<1.1	<1.1	SWR	<1.1	<1.1	<1.1

Table 7-8

Insertion Loss and SWR for the Beverage Transformer With Isolated Primary, Wound on MN8CX Toroidal Cores

(A) = single core (B) = stack of 2 cores.

Type		1.8 MHz	3.65 MHz	7.1 MHz
(A)	Insertion Loss (dB)	0.55 dB	0.50 dB	0.47 dB
	SWR	<1.1	<1.1	<1.1
(B)	Insertion Loss (dB)	0.30 dB	0.30 dB	0.28 dB
	SWR	<1.1	<1.1	<1.1

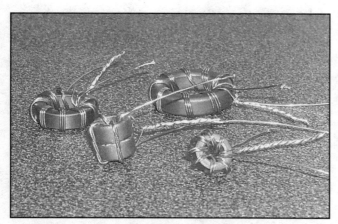

A series of 9:1 transmission-line type Beverage transformers. From left to right: FT-82 size core, two stacked MN8CX cores, FT-114 core and FT-50 core. Twisted wires were used on the smaller cores, while three parallel conductors were used on the larger cores.

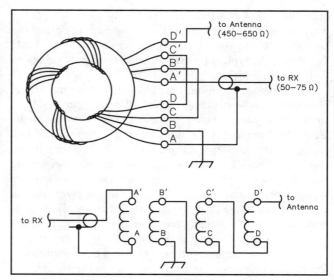

Fig 7-18—Modified transmission-line transformer, where the primary is galvanically separated from the secondary. See Fig 7-19 for the correct connection of ground and feed line.

this from happening the Beverage antenna and the coaxial feed line can be connected to separate grounds. This requires a transformer where the primary and the secondary grounds are galvanically insulated. **Fig 7-18** shows a modified transmission-line transformer, with a trifilar secondary winding plus a separate primary winding, each one having its own ground terminal. The correct wiring of the transformer, feed line and ground system is explained in detail in Section 9.3.

Table 7-8 lists measured insertion loss and SWR for a transmission-line type transformer with a separate primary winding, using MN8CX cores, using 4 turns for the primary and 3 × 4 turns for the secondary.

8.2.3. A Faraday shield?

Some authors advocate the use of a Faraday shield in the transformer. Tom, W8JI, covered this issue on the Internet: "*A Faraday shield is really just an extra plate of a capacitor placed between the primary and secondary. It makes the original small capacitance between the windings look like two larger values (more than double the original value). The junction of these new capacitors can be connected to ground. This allows us to "shunt" capacitively coupled currents to ground at the common point. A Faraday shield is most effective in multiple-turn or resonant transformers that have very high primary and secondary impedances. That's because the very small amount of primary to secondary capacitance can only couple a noticeable amount of signal if both windings have very high impedances. A Faraday shield, in the case of a Beverage, will do nothing to eliminate common-mode excitation of the antenna by the feed line. It makes the problem worse since it more than doubles unwanted capacitance to the common ground. For a Faraday shield to be effective in this application, it needs to connect to its own separate ground system that is not common to either the feed line or antenna.*"

8.3. Checking SWR and Insertion Loss

After winding the transformer, it's a good idea to check its performance. Connect a 450-Ω non-inductive resistor (a small metal-film unit will do) across the secondary and check the impedance or SWR using a noise bridge or an antenna analyzer (or still better, a network analyzer). With a well-made transformer the SWR should be less than 1.2:1 from 1.5 to 10 MHz.

You can easily evaluate the insertion loss of a home-made transformer—build two transformers and connect them back to back. Insert the back-to-back configuration in the feed line to your transceiver. You should not be able to detect any change in signal strength, as two well-built transformers should not exhibit more than 0.5 dB of insertion loss, which is just about non-measurable without special test equipment. If you have access to professional test equipment you can do an actual insertion loss measurement in a 50-Ω system. A network analyzer will give you accurate results.

■ 9. THE FEED LINE TO THE BEVERAGE

It is very important that as much attention is being paid to the feed line as to the Beverage antenna itself. Bad feed-line practice can completely annihilate the directive properties of the antenna.

The principles explained in this paragraph apply equally as well to all other weak-signal receiving antennas described in this chapter.

9.1. Which Coax Cable?

Various feed-line issues are covered in this section.

9.1.1. Attenuation

As the Beverage antennas will most likely be operated on relatively low frequencies, we need not use a feed line with the lowest possible loss, especially where all but the very longest feed-line runs are being considered. For runs up to 100 m, RG-58 (or 59) sized coax will be just fine. Table 6-1 shows the typical losses for common coaxial cables on 1.8, 3.5 and 7 MHz.

A strange reasoning was published some time ago on the Internet, where someone wrote, "Unless your feed lines are extremely long, there is no reason to use low-loss cable. Since the Beverages have loss, not gain, a few more dB loss makes no difference, just use a preamp in the shack if you really need it." The issue of "absolute" gain is *not* an issue with Beverage antennas. The real issue (see the chapter on equipment) is to have enough signal at the receiver antenna terminal to surpass the internal noise of the receiver "significantly." If the antenna feed line is very long, or if the attenuation of the feed line is too high, then the drop in signal strength can be overcome by using a preamplifier at the receiving antenna terminals, and *not* in the shack!

9.1.2. Mechanical properties

Running standard coax on the ground can be a problem for some. Many low banders report having lost small or medium-sized RG-type feed lines, as well as CATV-drop type feed lines due to animal bites. Small bites will usually not open or short the feed line, but do cause enough damage to allow moisture migration and corrosion.

There are three ways to prevent this from happening:
• Use CATV Hardline ($^1/_2$ or $^5/_8$-inch stuff is very sturdy—especially the *figure-8*, which has mechanical support cable–type cable).
• Use quad shielded and flooded RG-6 type coax. It appears that the critters don't like the flooding compound.
• Bury the cables in a closed cable duct underground.

9.1.3. Availability

For short runs any coax you can buy at the local flea market will do, provided the shield or the inner conductor is not corroded (green and black) from moisture ingress.

Often, 75-Ω CATV-type coax leftovers (often lengths up to 100 m!) can be bought at reasonable prices from the local cable TV company. The flexible coax used for drop lines is good for anything but very long runs. Hardline is the ultimate choice for long runs, as it offers the lowest attenuation and best shielding characteristics.

George, K8GG, uses $^1/_2$-inch 75-Ω CATV from his shack to the Beverage antenna park "head-end" which is 1200 m (yes, 4000 feet) from the shack. From the head-end, he uses RG-59 flooded cable for all the connection runs to the remote antenna selector.

9.1.4. Coax impedance

This is definitely the least important issue: it is totally irrelevant whether you use 75 or 50-Ω coaxial cable. The purist may want to design the matching transformer according to the impedance of the coax used, but even that is a bit far fetched.

9.1.5. Shielding effectiveness

It is important to use well shielded coax in order to have a quiet feed system under all circumstances and on all frequencies. Some of the very cheap (non mil-standard) coax has a very poor shield coverage factor (50% or so). This cable should *not* be used. Hardline is intrinsically the best choice, provided the solid shield is not broken. CATV companies use so-called *figure-8* Hardline, with an incorporated support cable. It is a well known fact that such cables, from swinging in the wind (between the support poles) often develop cracks and eventually breaks (opens) in the solid shield. In a CATV-trunking net these breaks are responsible for radiating cables, which many of us have witnessed. Watch out when buying *used* surplus CATV cables, and always check the cable for visual shield damage as well as for electrical shield continuity.

9.2. Is Your Coaxial Cable a "Snake" Antenna?

In Section 8.2.2, I explained how the outside of the feed line can act as a lossy, on-the-ground antenna (sort of Beverage), and how signals from this "Snake" antenna (see Section 19.2) can sneak into our Beverage feed system and introduce a lot of garbage. What we now, often jokingly, call a Snake antenna is in fact the shape of the very first Beverage antenna in history. The very first antenna Harold Beverage used was simply several miles of wire lying on a sandy path on Long Island. Only later was the Beverage antenna refined to be an elevated wire with a resistive termination!

In areas where the quality of the ground is poor, and where it is difficult to obtain a low-impedance ground, this coupling, via the common ground path can be quite significant. In fact, what we do is effectively add the signals of the Beverage with the signals of the Snake antenna, which is our feed line.

There are actually several ways of reducing the possible ill effects of your Snake antenna / feed line:
• Make sure your receiver is well grounded (not through the shield of one or more coaxial feed lines, but with an independent low-Z ground.
• Make sure the coax is well grounded where it enters the shack. Just relying on grounding at the receiver (chassis) is bad practice. It is quite common that the power mains feeding your receiver are carrying conducted RFI. This RFI is bypassed to the receiver chassis (mains decoupling capacitors) and the shield of the cable becomes the new path for this unwanted noise to leak, via the outside of the cable all the way to the feed point of the Beverage. Therefore it is always good to use good-quality low-pass mains filters with a good RF ground system in the shack, and to ground the feed lines to another good-quality ground system where they leave the house or the shack.
• Make the impedance of the ground system as low as possible. Use several ground rods, spaced at least their length, and interconnected in the ground. If ground rods cannot be installed (rock) use chicken-wire mats on the

largest possible area. These mats form a capacitor to ground. The larger their area, the lower the effective ground resistance. A very low ground impedance is *not* required at the far end, where the ground resistance can be part of the termination resistance.

- Use separate ground rods (separated at least the length of the rods) for the antenna return (high-Z ground end) and the coaxial feed-line ground (low Z). Keep the connections to these separate grounds as far away from one another as possible.
- Use a stack of ferrite cores (eg, 100 of the Wireman's cores) at the end of the feed line as shown in **Fig 7-19**. This common-mode filter can also be made by winding, eg, 50 turns of miniature 50 or 75-Ω coax on a stack of ferrite high-μ toroids. How many cores, how many turns? I use a simple digital LC meter and use as many as to achieve an impedance of approx. 1500 Ω on 160 meters (L = 130 μH). Make sure the LC meter actually operates near the frequency of interest (mine operates at 1 MHz). Ferrite core material type 73 is most suitable. A number of small-diameter coax turns through a large high-μ ferrite core does the same job. Do the wiring exactly as shown in Fig 7-19.
- Install a stack of beads on the feed line at regular distances (make it a Snake antenna with even more losses). This is not practical if you use a large-diameter feed line.
- In the most obnoxious cases of ground currents, it may be necessary to ground the shield of the feed line approx.

every 50 m with independent ground rods (ground rods without any other connection).
- Do not run feed lines parallel in close proximity to elevated radials. In this case there will be field coupling, from induction or radiation fields.
- Bury the feed line deep, which is not always practical.

Fig 7-19 shows how the feed line should be connected to achieve the greatest possible attenuation of signals picked up by the outside of the feed line. The stack of beads (B_1) will form a high impedance (Z_1 = typical 1500 Ω on 1.8 MHz for 100 stacked Wireman beads) for the common mode currents. In the equivalent schematic Z_1 and RG_1 (the ground resistance of the coaxial cable grounding rod) form a voltage divider. Assume this ground resistance (RG_1) is 10 Ω. This means that we have a voltage divider of 10/1510 or 1/151 which results in a signal attenuation of $20 \times \log (1/151) = 43$ dB. The spurious signals left across RG_1 will now be fed *into* the feed line via another voltage divider made up by the impedance of the second bead stack (B_2) in series with the impedance of the low-Z winding of the transformer (Z_3) and in series with the impedance of the coaxial cable ($Z_{coax} = 75$ Ω, assuming the cable is terminated in its own characteristic impedance at the receiver end). This time we have a voltage divider made up by $Z_2 = 1500$ Ω (bead stack) + $Z_3 = 250$ Ω (transformer low-Z winding) and $Z_{coax} = 75$ Ω. The attenuation of the voltage divider is $20 \times \log (75/(1500+250+75)) = 28$ dB. The total attenuation of this setup is approx. 70 dB for Snake antenna signals on the

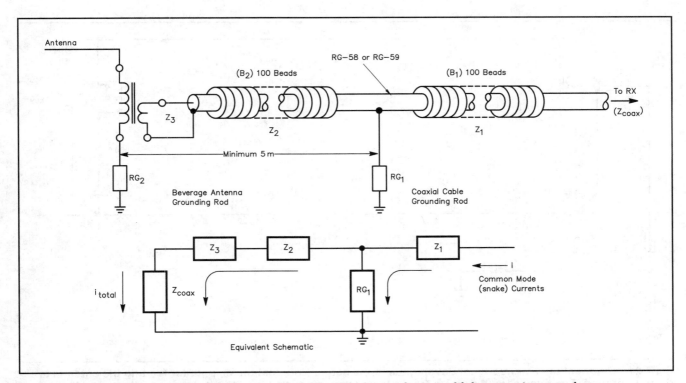

Fig 7-19—Method of connecting the feed line to the Beverage transformer which guarantees maximum suppression of common-mode signals induced on the outside of the feed line. See text for details.

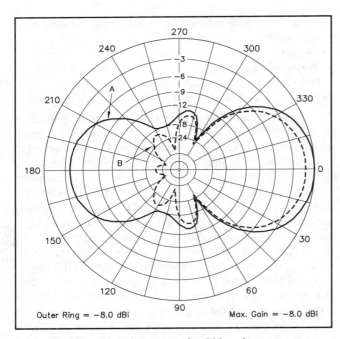

Fig 7-20—Directivity pattern of a 268-m long non-terminated Beverage antenna at a height of 2 m over good ground (A), as compared to the same antenna properly terminated (B). Note the difference in gain from the back, resulting from the extra losses encountered by the reflected wave.

outside of the feed line. This should be sufficient even for the most stubborn cases.

Note that the stacks of ferrite cores across the coaxial feed line can be replaced by coiled-up lengths of coax. In order to achieve a choking impedance of approx. 1500 Ω on 160 meters, the inductance must be approx. 125 μH. This requires a coil of coaxial cable measuring 30 cm in diameter and having approx. 20 close-wound turns. Such a coil requires not less than 20 meters of coax, and two such coils are required in the system! If you use this approach, make sure the two coils are at right angles in order to minimize coupling between them. Another alternative is to wind miniature type coaxial cable on a large high-permeability core. Five turns of miniature coax through a FT-150A-F core ($\mu = 3000$ and $A_L = 5000$) achieves an impedance of 1500 Ω (7 turns = 3 kΩ on 160 meters).

9.3. Evaluating the Feed System

Make sure your receiver has a good RF ground, and make sure the Beverage feed line is grounded to a separate RF ground where entering the shack (house). Then connect the feed line to the receiver. For the time being, do *not* connect the far end of the feed line to the ground, but connect the shield to the inner conductor (in other words, short the line). By doing this we simply connect the common-mode signals (the signals from the Snake antenna that is your feed line) directly into the feed line. If the receiver does not hear

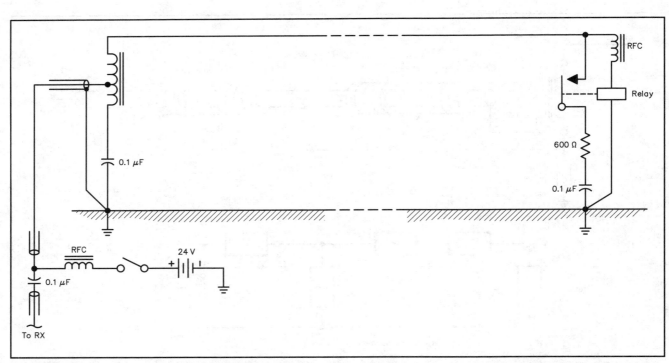

Fig 7-21—Unidirectional/bidirectional Beverage antenna setup. Note that the dc path for feeding the antenna is through ground. The relay should operate from a low voltage (max 24 V) so that dangerous potentials will not be present on the antenna wire.

any signals (try this on the BC band), then your feed line is not picking up much at all. This is, however, quite unlikely, unless the feed line is either very short or maybe buried.

Next, connect the end of the feed line to a ground rod as shown in Fig 7-19, and install the common-mode chokes at the end of the feed line. Now the receiver should be completely dead. Once you have obtained a perfectly quiet feed line, connect the termination box with the transformer, and connect the ground leads. Without the Beverage wire(s) connected, the receiver should still be "dead" on all frequencies. This procedure will prove the performance of a good feed system with no reduction in directivity from stray pickup.

9.4. Routing the Feed Line

If the feed lines are properly decoupled for common-mode signals, as explained above, there are very few limitations as to how to route the feed line. If they are not properly decoupled they can upset the effective directivity pattern of the Beverage and turn a fantastic receiving antenna into a mediocre one.

■ 10. BIDIRECTIONAL BEVERAGE ANTENNA

When the Beverage antenna is not terminated, the directivity will be essentially bidirectional. **Fig 7-20** shows the horizontal directivity pattern for a 1-λ unterminated Beverage antenna. Notice the slight attenuation from the back direction because of the extra loss of the reflected wave in the wire. A method of switching the Beverage from unidirectional to bidirectional is described in **Fig 7-21**. A relay at the far end of the antenna wire is fed through the antenna wire and the earth return via an RF choke and blocking capacitors (TV preamplifiers are fed in a similar manner via the coaxial cable).

■ 11. TWO DIRECTIONS FROM ONE WIRE

Two directions can also be obtained from a single-wire Beverage by feeding coax to both ends of the antenna, with the appropriate matching network (**Fig 7-22A**). In this case, one end is terminated in the shack with an appropriate impedance, while the other is in use for reception. (See Ref 1210.) The "appropriate" terminating impedance can be found as follows. First, determine the characteristic impedance of the Beverage by the method explained earlier in this chapter. Next, connect an impedance bridge or noise bridge across the high-impedance secondary of the matching transformer. Adjust the terminating impedance at the end of the feed line (inside the shack) until a value is found which is the same as the Beverage impedance. In many cases, the impedance will be close to real, and a simple resistor with a value in the range of 50 to 75 Ω will provide the proper termination. An alternative method is to use a small signal source in the back of the antenna (remember the offset angle! see Section 6.4) and simply adjust a 200-Ω potentiometer, at the end of the second feed line for maximum F/B. Switching directions is done by merely interchanging the receiver and the terminating resistor.

■ 12. TWO-WIRE SWITCHABLE-DIRECTION BEVERAGE ANTENNAS

Two-wire switchable Beverage antennas are covered in great detail by V. Misek, W1WCR (Ref 1206). I strongly recommend that you read this excellent book if you want to get serious with open-wire-type Beverage antennas.

12.1. A Coaxial Cable as Feed Line and as an Antenna Wire

We have seen in Section 9.2 that a coaxial feed line can act as a feed line (in that case everything happens inside the cable, between the center conductor and the inside of the

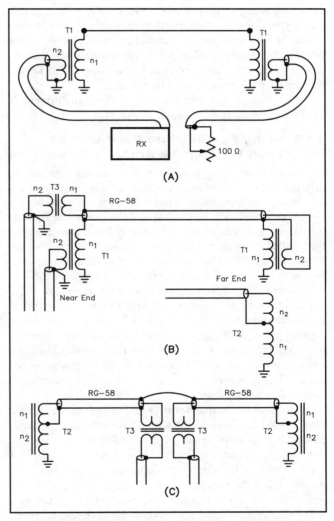

Fig 7-22—The Beverage with individual coaxial feed lines brought to both ends of the antenna (A). The array where the antenna wire is a coaxial cable (B), which is also used to transport the signal from the "back direction" of the array to the front. In a variant (C), the coaxial Beverage is fed somewhere along its length. See text for details.

shield), but it can also act as a single (fat) conductor, when you look at the coaxial cable "from the outside."

In Fig 7-22B, I have used a small coax cable as the antenna wire of a Beverage. This small coax cable serves as a Beverage antenna wire, and as feed line for the far-end of the array. Transformer T1 is the classic 9:1 transformer. The reflection transformer at the far-end provides the right impedance transformation to feed the RF via a good match (proper turns ratio) to the inside of the coax. This transformer is identical to the transformer T1 used at the near-end. At the near-end a 1:1 isolating transformer (T3) must be used (1:1 turns ratio).

The far-end transformer can also be made in the form of an auto-transformer as shown (T2).

The same antenna can also be fed in the center, or anywhere along its length, as sketched in Fig 7-22C. In this case a transformer, similar to the one used at T2 in case B, is used at both ends, and the coaxial antenna wire / feed line is opened at the appropriate point, where the isolating transformers (1:1 ratio) are used in both directions (T3). This is the same transformer as used in case B.

The design and construction of the transformers is covered in detail in Sections 11.3 and 11.4.

12.2. Using Parallel-Wire Feed Line as Feed Line and as Antenna Wire

Instead of using a (small) coaxial cable (which is relatively heavy and expensive), one can also use a parallel-conductor transmission line to achieve the same purpose.

This situation is somewhat more delicate using parallel-wire transmission lines, as the only thing that stops the line (when used as a transmission line) from picking up unwanted signals is the balance of the system. If one conductor is closer to ground over the entire length of the line it unbalances the system. This is also the reason why the parallel transmission line must be parallel to the ground. It cannot be constructed in a "vertical fashion," using single support poles, with one wire under the other. The closer the wire spacing of the line (the lower the impedance of the line), the less critical installation-related unbalance becomes. The higher the line, the less unbalance the unequal coupling to earth creates.

In its most simple form (**Fig 7-23A**) we can use a homemade open-wire feeder with a wire spacing of approx. 30 cm. Signals arriving off the near end of the antenna will induce equal in-phase voltages in both wires. Because of the close spacing of the wires, there is no space diversity effect. At the end of the antenna, one wire is left open, while the other wire is short-circuited to ground. This provokes a 100% reflection of the wave, but with a 180° phase reversal. The signals received off the back of the antenna are now fed in a push-pull mode along the open-wire feeders toward the other end of the antenna.

Earlier I stated that the ground resistance of the far-end ground system was not very critical. This is only true, of course, where the terminating resistor is connected to this ground. Here the quality of the ground is as important as at

the near-end, as otherwise part of the signals received from the near-end would be dissipated in the ground.

At the near-end a properly designed push-pull transformer (T4) transforms the RF from the push-pull mode (open-wire feeder) to the unbalanced coaxial cable impedance. The well-known 9:1 transformer (T1) is now connected to the center tap of T4, where the push-push signals from signals received off the far-end of the Beverage are available. These signals received off the far-end will not produce any output at the low-Z side of T4 provided a good balance is achieved in the transformer. Outputs from both directions are simultaneously available from outputs J1 and J2 of this system.

When only one of the feed lines is connected to a receiver, make sure the other feed line is terminated. Using a potentiometer in the shack can be helpful to optimize the F/B under all circumstances.

The far-end configuration of this antenna (one conductor open and one conductor shorted to ground) is a good solution, however; only in case the line impedance and the antenna impedance are the same. **Table 7-9** shows the characteristic impedance of the open-wire transmission line, and **Table 7-10** the characteristic impedance of the parallel-wire Beverage, using different types of parallel transmission lines. Only when using a 450-Ω transmission line approx. 3-m high, can we achieve this goal.

Another, more common way to overcome this problem

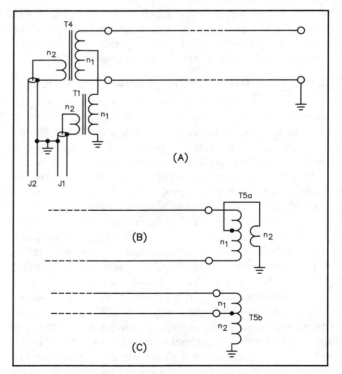

Fig 7-23—The most simple variant of the open-wire two-direction Beverage antenna (A). Reflecting transformers used to obtain reflection and perfect impedance matching (B and C—see text for details).

Table 7-9

The Impedance of a Transmission Line Made of Parallel Conductors

Wire Spacing	For 1.3-mm wire (16 AWG)	For 1.6-mm wire (14 AWG)	For 2.0-mm wire (12 AWG)
25 cm (10 in)	713 Ω	688 Ω	661 Ω
30 cm (12 in)	735 Ω	710 Ω	683 Ω

This is the impedance the two wires exhibit when used to transport the RF energy back to the feed point from the far end.

is by using a transformer as shown in Fig 7-22B. Reflection transformer, T3, is built to have the correct transformation ratio from the transmission-line impedance (710 Ω balanced for the open-wire line) to the antenna impedance (approx. 332 Ω if the open-wire Beverage is 2-m high).

Using twin-lead transmission line is obviously the better choice, as it intrinsically guarantees better balance. When using twin-lead, it is appropriate to twist the line 2 or 3 turns per m. The reflection transformer can also be made as an autotransformer as shown in Fig 7-22C.

The same precautions as explained in Section 9.2 regarding the intrusion of common-mode currents from the feed lines into our antenna system must be applied. Therefore, use separate ground rods for the antenna and for the feed lines, and install common-mode chokes on the feed lines as shown in Fig 7-19.

In order to avoid stray pickup from the vertical down leads of the earth connections, sloping terminations can be used as explained for the single-wire Beverage antenna. At both ends the box containing the transformers is directly attached to the ground rod at ground level.

The design and construction of the transformers is covered in detail in Section 12.3.

The push-pull impedance of the open-wire line is given by:

$$Z = 276 \times \log \frac{2S}{d}$$
(Eq 7-3)

where

S = spacing between conductors

d = diameter of wires (in the same units)

The impedance of the parallel wires over ground is given by:

$$Z = 69 \times \log \left[\frac{4h}{d} \sqrt{1 + \frac{(2h)^2}{S}} \right]$$
(Eq 7-4)

where

S = spacing between wires

d = diameter of wires

h = height of wires above ground (all in the same units)

12.3. Designing the Transformers

The exact number of turns that you will need to wind your transformers depends on the permeability of the material used. Using cores with a high permeability has the advantage of requiring few turns, which means little capacitive coupling and wide frequency coverage (see Section 8.1 and 8.2).

Table 7-11 lists the turns ratios for the transformers used with the two-wire Beverage antennas shown in Fig 7-22. For case (A) 3 Beverage antenna impedances are listed: 450 Ω for a low Beverage (0.5-m high), 525 Ω for a 2.0-m high Beverage and 600 Ω for a 4-m high Beverage.

The procedure to determine the exact number of required turns is as follows:

1. Find the core to be used in Table 7-5.
2. Note the numbers of turns required for the 50-Ω winding (min. 3 times the coax impedance, or min. 150 Ω).
3. Multiply the turns ratios from Table 7-8 or 7-9 with the number of turns taken from Table 7-5.

Example: using Ceramic Magnetics MN8CX cores, the turn ratios must be multiplied by a factor of 4.

Table 7-12 lists the turns ratios for the open-wire Beverages from Fig 7-23. Case (A) is for a two-wire Beverage made of 30-cm spaced open-wire feeders (Z = 710 Ω). Case (B) is for 450-Ω twin-lead, case (C) for 300-Ω twin-lead, and case (D) for 75-Ω twin-lead. The calculations were made for the antenna wire 2 m above ground.

Whereas some think that the high-μ cores such as the MN8CX cores (which I have used for many years) are no good for this application, the truth is that the insertion loss of transformers made with these cores is far less than 1 dB.

If you are confronted with extremely strong BC signals,

Table 7-10

Impedance for a Beverage Antenna Made of Two Parallel Wires (Open-Wire, Spaced 30 cm and 300 Ω Twin-Lead)

Antenna Height	Impedance 30 cm Spaced Open Wire (Z=710 Ω)	Impedance for 450 Ω Twin Lead	Impedance for 300 Ω Twin Lead	Impedance for 75 Ω Twin Lead
0.5 m	252	341	365	393
1.0 m	295	383	407	435
2.0 m	332	424	449	476
3.0 m	357	449	473	500
4.0 m	375	466	490	518

it may be better to use a larger core. Under extreme conditions the MN8CX cores have been reported to saturate.

12.4. Constructing the Transformers

I will briefly discuss the construction method of each of the transformers listed in Table 7-11 and Table 7-12, and list typical performance data. The transformers were all built using MN8CX cores, but transformers made on #75 and #43 material should yield comparable results (see Section 8.1).

Transformer T1

This transformer can be wound as explained in Fig 7-18. An alternative way of constructing the transformer is shown in **Fig 7-24A**.

Transformers using a single core were wound and measured for insertion loss and SWR:

T1	1.8 MHz	3.65 MHz	7.1 MHz
Attenuation(dB)	0.8 dB	0.6 dB	0.6 dB
SWR	<1.2:1	<1.2:1	<1.2:1

Transformer T2

The layout of this reflection transformer is shown in Fig 7-24B. The measured performance data for a transformer with n2 = 3, n1 = 9 are:

T2	1.8 MHz	3.65 MHz	7.1 MHz
Attenuation(dB)	0.6 dB	0.5 dB	0.5 dB
SWR	<1.2:1	<1.2:1	<1.2:1

Transformer T3

Fig 7-24C shows the layout of this 1:1 isolation transformer. The performance data (n1 = 6, n2 = 6) are:

T3	1.8 MHz	3.65 MHz	7.1 MHz
Attenuation(dB)	0.6 dB	0.9 dB	1.8 dB
SWR	1:1	1:1	1:1

Transformer T4

Fig 7-24D shows the winding layout transformer for the balanced line two-wire Beverage antenna. The measured performance data are:

T4	1.8 MHz	3.65 MHz	7.1 MHz
Attenuation(dB)	0.8 dB	0.6 dB	0.6 dB
SWR	<1.2:1	<1.2:1	<1.2:1

When properly adjusted, the suppression of the common-mode signal was measured as >45 dB.

Transformer T5 (a and b)

The open-wire reflection transformer with separate primary and secondary windings is shown in Fig 7-24E. Fig 7-24F shows the equivalent transformer wound as an autotransformer.

12.5. Housing the Transformers

To house the transformers I have used both aluminum and PVC-type die cast boxes of appropriate size. The PVC boxes are inexpensive and do not need feedthrough stand-

---

Table 7-11
Winding Information for the Different Transformers Used in Fig 7-22
The numbers are turns ratios, whereby the reference (1) is the number of turns required for 50 Ω. See text for details.

Case Fig 7-22	Z_{ant}	Z_{coax}	T1 n1	T1 n2	T2 n2	T2 n1	T3 n2	T3 n1
(A)	450	50	1.0	3.0				
	450	75	1.2	3.0				
	525	50	1.0	3.2				
	525	75	1.2	3.2				
	600	50	1.0	3.5				
	600	75	1.2	3.5	-	-	-	-
(B)	450	50	1.0	3.0	3.0	1.0	2.0	2.0
	450	75	1.2	3.6	3.6	1.2	2.4	2.4
(C)	450	50	-	-	3.0	1.0	2.0	2.0
	450	75	-	-	3.6	1.2	2.2	2.4

Table 7-12
Winding Information for the Different Transformers used in Fig 7-23
The numbers are turn ratios, whereby the reference (1) is the number of turns required for 50 Ω. See text for details.

Case Fig 7-23	$Z_{feed\ line}$	Z_{Ant}	Z_{coax}	T4 n2	T4 n1	T1 n2	T1 n1	T5a n2	T5a n1	T5b n2	T5b n1
(A)	50	330	50	1.0	3.8	1.0	2.6	2.6	3.8	0.7	3.8
(A)	75	330	75	1.2	3.8	1.2	2.6	2.6	3.8	0.7	3.8
(B)	50	425	50	1.0	3.0	1.0	2.9	2.9	3.0	1.4	3.0
(B)	75	425	75	1.2	3.0	1.2	2.9	2.9	3.0	1.4	3.0
(C)	50	450	50	1.0	2.4	1.0	3.0	3.0	2.4	1.8	2.4
(C)	75	450	75	1.2	2.4	1.2	3.0	3.0	2.4	1.8	2.4
(D)	50	475	50	1.0	1.2	1.0	3.1	3.1	1.2	2.5	1.2
(D)	75	475	75	1.2	1.2	1.2	3.1	3.1	1.2	2.5	1.2

offs for the output to the Beverage antenna. A stainless steel screw with nuts and bolts mounted right through the plastic wall is all that is needed. Do not economize on coax connectors. One excellent way of waterproofing coaxial connections (and any others) is to squeeze plenty of medical-grade petroleum jelly into both connector parts before mating. Some petroleum jelly can be applied to the box seams to waterproof the whole assembly once the cover has been put on. Petroleum jelly is an excellent electrical insulator (it has a very low dielectric constant similar to Teflon and polyethylene). I have used the same material for protecting non-stainless steel antenna hardware. After many years of use in a humid and aggressive climate, the hardware still comes apart with no problems.

Fig 7-25 shows possible layouts for transformer units for single-wire and two-wire Beverage antennas.

12.6. Testing the Transformers

The transformers T1, T2, T3 and T4 can easily be tested for SWR by terminating the secondary with a terminating resistor and checking the SWR with an antenna analyzer or a noise bridge.

The balance of the push-pull transformer T4 can be tested by connecting the two secondary outputs together and feeding a small amount of RF from any low-power source to this connection. A perfectly balanced transformer should yield no output at the low-impedance secondary from this configuration. The physical symmetry of the transformer may be adjusted (slightly adjusting turn spacings on the toroidal core) while performing this test, until the lowest

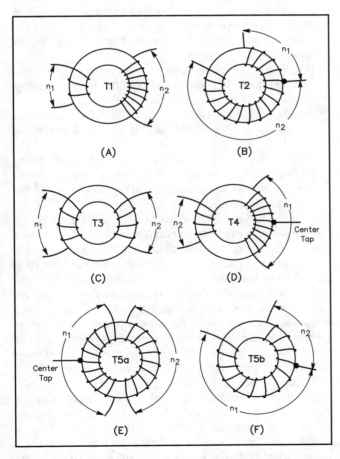

Fig 7-24—Winding layouts for the various transformers described in the text.

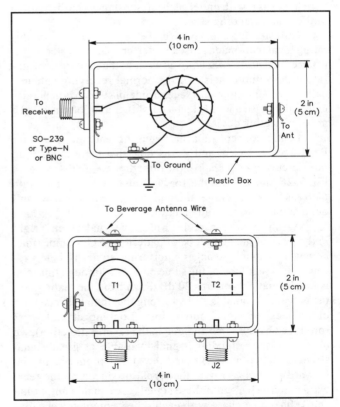

Fig 7-25—Suggested layouts for single and two-wire receiving-end terminations. Plastic boxes are inexpensive and allow simple feedthrough systems to be used (screw and nut), while the toroidal transformers can be secured directly to the plastic material with silicone sealant.

possible signal output is achieved. The balance can be assessed by temporarily disconnecting one of the secondary leads and measuring the signal-strength difference on the receiver. Better than 40-dB difference should be easily obtainable.

The insertion loss can be checked as explained in Section 8.3. It should be well under 1 dB for all transformers.

12.7. Electrical Null Steering

Electrical null steering is a technique used to obtain virtually infinite rejection of an unwanted signal. This technique is most commonly used with small-loop antennas to obtain a unidirectional radiation pattern. Signals from the sense antenna and the loop are combined in the correct phase to achieve a cardioid pattern. The article by Webb, W1ETC, on electrical antenna null steering (Ref 1239) is an excellent reference work for understanding the principles of this technique and its limitations.

Null steering consists in combining signals from two different antennas in such a way that an offending signal is reduced to (almost) zero by adding the two signals 180° out of phase (and of equal amplitude). This implies that the signals received by the two antennas must be of *constant phase and amplitude*. As a rule, signals that arrive via ionospheric propagation have a continuously varying phase, which means it will, under such circumstances be difficult to null out an offending signal.

Null steering can, however, be very efficient in annihilating local man-made noise or interference. Null steering is often called *noise canceling* as well. In the generic configuration the setup consists of the normal receiving antenna, called the main antenna, a second antenna which is generally called the noise antenna and a combiner unit, often called a noise-canceling device.

In the concept of a null-steering antenna, two-wire Beverages are used, and we assume that there is a strong interference *off the back* of the direction we are listening to. **Fig 7-26** shows how the forward and rearward radiation patterns of a Beverage antenna can be combined to null out a signal source at any angle.

Assume we are receiving a desired signal which is right in the peak of the main lobe of pattern A. At the same time, however, a much stronger interfering signal is being received in a rear lobe of the same pattern. Although this lobe is much smaller (perhaps 20 dB down from the main lobe), the unwanted signal is still so strong that it makes reception of the desired signal impossible. Looking at pattern B (produced by the same antenna in the reverse direction), we now have the offending signal available at great signal strength. If we now combine the signal produced by the forward pattern (A) with the signal produced by the rearward pattern (B), and if these signals are equal in amplitude and 180° out of phase with each other, complete cancellation of the offending signal will result. It is clear that the offending signal must be received on the "noise antenna" at least as strong as on the main antenna.

After many hours of experimentation on 80 and 160

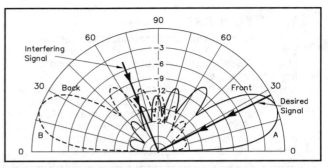

Fig 7-26—An interfering signal off the back of the Beverage, coming in at a wave angle of approx. 70°, can be completely suppressed by taking the signal off the back antenna and adding it, 180° out of phase to the signal received off the front antenna. (See text for details.)

meters as well as the top end of the medium-wave broadcast band, the following observations were made:

Misek, W1WCR, (Ref 1206) has described in detail the design and the construction of a steerable wave antenna. I have experimented quite extensively with null steering on Beverage antennas, using the original null-steering circuit as described by Misek. The findings were as follows:

- Extremely deep nulls (up to 50 dB or better) can be obtained on local (ground wave) broadcast stations in the medium-wave band. Because of the nature of the signals (AM), critical adjustment of both the amplitudes and phasing is possible.

- It becomes much more tricky to null out even a ground-wave SSB or CW signal on 80 or 160 meters because of the non-constant amplitude nature of the signals.

- Nulling out sky-wave signals is generally impossible, because they often arrive via multiple paths of varying lengths.

- Nulling out sources of man-made noise can be accomplished quite easily, as these sources are usually vertically polarized and are generated in the vicinity of the receiving antenna.

In recent years various so-called noise-canceling devices have been made commercially available. These units can be used to perform the so-called null steering.

I have had extremely good results with the MFJ-1026 unit. This unit has a wide-range phasing control, and adjustable gain on both channels. You can use it with a two-wire Beverage, but it can also be used with single-direction Beverages, using a separate antenna as the noise antenna. I normally use my transmit antenna as the noise antenna. As the real gain of this antenna is always superior to the gain of a Beverage antenna, and pick up locally generates noises well (being vertically polarized), the offending signal will always be available with sufficient amplitude to be able to do a canceling job without having to reduce the signal on the main antenna.

I have successfully used the MFJ unit to null out a broadband S7 noise on 160, generated by PWM controllers for electrical motors at a chemical plant some 10 km away.

Another very interesting application is to use the unit to

eliminate the noise pickup from a close-by transmitting antenna by your Beverage antenna. In this case you *must* use the offending transmitting antenna as a noise antenna. It is amazing how a noisy Beverage can be turned into a super quiet one.

Fig 7-27 shows how the MFJ unit is connected to a transceiver like the FT-1000MP. The signal for the noise antenna is picked off from the RX-Out jack on the back of the transceiver, while the output from the noise-canceling device is injected into the RX-Ant jack.

Eliminating man-made noise this way requires precise phase and gain adjustment. When you switch Beverage antennas, you will have to readjust the controls as the phase of the offending signal will likely be different on all antennas.

■ 13. LOCATION OF THE BEVERAGE

How far should I keep the Beverage from my vertical transmit antenna? How close can I run two Beverages in parallel (one shooting in the opposite direction)? Can Beverages cross one another? How close can they cross? These are all very valid questions, and failing to understand what and why may turn a potentially wonderful antenna into a really lousy performer.

Beverages that are close to large transmit antennas, which are resonant (or close to resonance), pick up noise and signals retransmitted by the transmit antennas. The real key is in the coupling between the systems. If the conductors (antennas) have a high level of mutual coupling, they will affect each other in some way. Resonance, or more properly even being near resonance, increases mutual coupling dramatically. The Beverage is a nonresonant antenna, but the transmit antenna can be resonant during reception.

If you have a number of Beverages, and one or two are all the time noisier than the others, then there is a good chance those are the ones closest to your resonant transmit antenna. Another indication may be that your (only) Beverage antenna is not really much quieter than your vertical.

Is that noise really coming from your transmit antenna? To check this proceed as follows:

- If you have a series-fed ground-mounted vertical, disconnect the coax, so that the vertical part of the antenna is now "floating." If it is a full-size 0.25-λ element, this will now be resonant at twice the frequency, and mutual coupling to the Beverage will be minimal. You probably will see the noise drop, and the noisy Beverage become as quiet as the other Beverages.
- If you have a shunt-fed vertical (eg, a loaded tower), just disconnecting the feed line may not help! Once you shunt feed a nonresonant structure, it becomes a resonant structure at the frequency it's tuned to. To detune it you will need to intervene in the matching section.

If you have a vertical with elevated radials, then you are really in trouble, unless you use only 1 radial. Systems with 2 or 4 verticals are very nasty. Two radials (more or less in line) are a perfect 0.5-λ element, reradiating like hell! Two more at right angles makes things even worse. **Fig 7-28** shows the radiation pattern of a 268-m long Beverage without and with a 0.25-λ vertical with 4 elevated (4-m high) radials. Note how an F/B of 30 dB went down to a mere 12 dB!

With a 160-meter vertical using only 1 elevated 0.25-λ radial, decoupling the feed line leaves only two $^1/_8$-λ wires (one vertical, one horizontal), which will do no harm, except when on 80 meters, if the Beverage is used for that band as well!

Even if you don't notice an observable "noise increase," a pass near another conductor (antenna, radials, etc) that is coupled to the receive antenna can still hurt S/N ratio by changing the directional pattern of the receive antenna. This may not always be easily detectable by just listening to an antenna. A Beverage may still "out hear" a vertical, even if it is severely compromised by the presence of another antenna

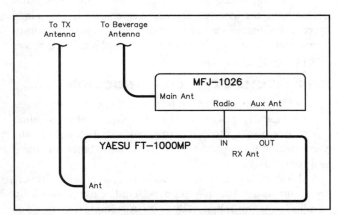

Fig 7-27—The FT-1000MP can easily be interfaced with the MFJ-1026 noise canceling device. The RX ANT OUT on the FT-1000MP supplies the transmit antenna signal, which is used as "noise" antenna. The output from the MFJ-1026 can be fed directly in the RX ANT IN jack on the transceiver.

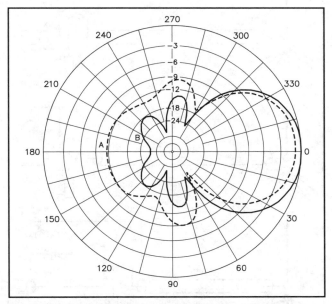

Fig 7-28—At A, influence of the ¼-wave 160-meter vertical with 4 elevated radials on the radiation pattern of our 268-m long Beverage (B). In this example the Beverage came within 10 m of the vertical.

in its neighborhood. But you can do some tests to find out.

This can be checked as follows:

- Set up a signal source off the back of a Beverage antenna, right in the null. This signal source can be another local ham or a small signal generator, whatever it takes to get a good S-meter reading on your receiver.
- Detune the transmit antenna using one of the methods described above. It is very likely that doing this you will be able to improve the null off the back by more than 10 dB!

Now that you have detected how the proximity of your transmit antenna ruins the F/B of your Beverage and renders it very noisy, what is the permanent solution to this problem?

There are two ways to cope with the problem:

- Increase the distance between the antennas.
- Shift the resonant frequency as far away as possible from the operating frequency.

The simplest rule of thumb is to stay away at least $\frac{1}{4} \lambda$ from a large resonant transmit antenna or resonant elevated radials. If this is impossible provisions must be taken to detune

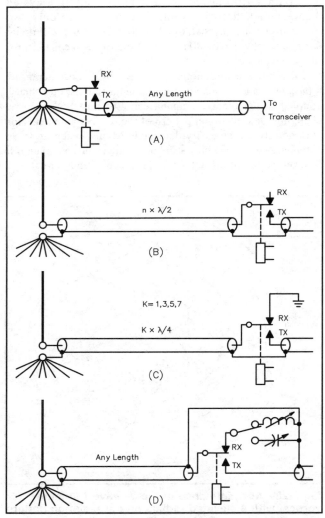

Fig 7-29—Various methods for decoupling a series-fed vertical. See text for details.

the transmit antenna during reception. If the transmit antenna is a series fed 0.25-λ vertical, then this antenna can be made anti-resonant by floating the 0.25-λ vertical element above ground. This can be done at the base of the antenna (see **Fig 7-29A**) with appropriate hardware (relays), or one can do it at the end of the feed line. If the feed line has an electrical length equal to an odd number of quarter waves, then the feed line should be short circuited (in the shack) during reception (see Fig 7-29B). If the feed-line length is a multiple of half-waves long, then the end of the feed line should be left open in the shack during transmit (see Fig 7-29C).

If you don't know the feed-line length, just terminate the feed line in a variable capacitor (eg, a 4-gang BC variable) or a variable inductor and tune these to obtain minimum noise level in the Beverage. This capacitor or inductor can then be switched across the end of the feed line on reception, by a relay as shown in Fig 7-29D.

If your Beverage is close to the transmit antenna, and if you have an apparent noise problem, this can be for one of the following reasons:

1. Your transmit antenna couples into the Beverage.
2. You have an extremely quiet location (Heard Island?). If there is no noise to speak of being picked up on the vertical, then it would be very hard to discern any change in the Beverage performance.

Under any circumstance it is better to keep antennas separated as far as possible. This is also true for transmit antennas, of course. A situation where a wire (or more wires) are stuck randomly in the near field of an antenna ends up in results that are a matter of pure blind luck.

Parallel Beverages

Beverages are nonresonant and exhibit very little mutual coupling. The rule of thumb is to keep two parallel Beverages separated by a distance equal to their height above ground. This separation guarantees a decoupling of at least 30 dB.

Crossing Beverages

Beverage antennas for different directions may cross each other if the wires are separated by at least 10 cm (4 in.). Beverage antennas should also not be run in close proximity to parallel conductors such as fences, telephone lines, power lines (even if quiet) and the like.

■ 14. MECHANICAL CONSTRUCTION

Almost any type of support will work for a Beverage antenna: metal pipes, wooden poles, plastic tubes, bamboo sticks, tree trunks, etc. For permanent installations polyethylene insulators, commonly used for electric fences are ideal and inexpensive.

Most of my Beverages are winter-type-only Beverages, put up on friendly farmers' land. I use 2.4-m long bamboo sticks and support the wire in a simple loop made of electric tape. Do not wrap the Beverage wire around the stick, or around branches. This creates an inductance which has to be avoided. Bamboo supports are cheap ($25 per 100) but do the job wonderfully well (see **Fig 7-30** and **Fig 7-31**). Their life-time is about 4 years before they rot at the bottom.

Fig 7-30—Bamboo sticks 2.4-m long are used to support the winter-time Beverages at ON4UN. This picture shows heavy ice-loading on the antenna, which makes the antenna visible.

For permanent Beverages I use single-strand bronze wire, 1.6-mm OD (14 AWG), which is pulled to approximately 45-kg tension. Copper-clad steel wire is also a good performer, as well as hard drawn (non-annealed) copper. If ice loading is to be expected, the wire should only be pulled to 15 kg (30 lb.). With such strong wire and 45-kg tension on the wire, spans of up to 100 m can be covered without much sag.

Soft-drawn copper wire cannot be tensioned to any great degree without causing much stretching, and it is not suitable for very long unsupported spans of wire. If one does not mind using many supports, a soft wire may be used for the antenna. I use soft-drawn single-strand copper wire of 0.8-mm OD (AWG 20) for my winter-type Beverages, supported every 30 m by 2.4-m long bamboo sticks, which are inserted about 15 cm in the ground.

Insulated wire is better than bare conductor as this will render the antenna less sensitive to rain static and will also ensure fully quiet operation even when the antenna wire touches branches or leaves in the wind.

Long Beverages made of thin soft-drawn copper wire have the disadvantage of stretching even in wind. A solution to that problem is to end the Beverage in a well-tensioned bungee cord.

Some people have been using surplus telephone paired wire. This wire appears to be often available at hamfests. Don't try to separate the wires, just connect them in parallel at both ends. Using a pair as the Beverage antenna conductor has the advantage that you can check continuity from one end, using an ohmmeter. All connections must be soldered, preferably with silver solder. Common Sn/Pb solder rots away after years of exposure to the weather.

I would like to warn every Beverage builder about a potential danger. People who are not familiar with the local situation may stumble across low Beverages or may be hurt in another way from higher Beverage wires. If you use permanent Beverages, I would recommend that you put them at least 2-m high (lowest point) especially if you use strong and heavily tensioned wire (eg, copper-clad steel).

Fig 7-31—Method of affixing the Beverage wire on the bamboo stick, using a loop made of electrical tape.

Make sure nobody can ride a horse on the terrain where you have your Beverage antennas. If necessary post signs, warning trespassers of the possible danger.

■ 15. ARRAYS OF BEVERAGES

What are the shortcomings of single-wire Beverages? Let me sum up a few:
• They require a lot of real estate (especially on 160 meters).
• They have a lot of side-lobes.
• Long Beverages have a good output, but a narrow 3-dB beamwidth (a 3-λ Beverage has a beamwidth of less than 40°, which means you will require many Beverage anten-

Fig 7-32—View of a rotatable (or should I say "mobile") Beverage antenna. Must be somewhere in Texas? I wonder how they terminated this monster.

nas to cover all directions).

Beverage antennas can be combined in antenna groups, just like any other antenna, with the aim of either increasing their gain or their spatial selectivity (directivity).

15.1. Beverages and Mutual Coupling

Beverage antennas are very lossy and tightly coupled into the nearby earth. As a result there is virtually no mutual coupling among even very close-spaced Beverages. Being nonresonant antennas emphasizes this characteristic. Elements on or near resonance exhibit the highest degree of mutual coupling (that's why Yagi antennas work!).

This can be confirmed experimentally by putting some RF into one Beverage and measure the loss into an adjacent Beverage. Parallel Beverages spaced from each other by the same distance as their height above ground have about 30 dB of isolation.

This is confirmed by modeling as well. Two Beverages, fed in phase were modeled and the feed impedance was monitored while changing the separation between the antennas.

Fig 7-33 shows the radiation pattern for a 168-m long (and 2-m high) Beverage, with 20-m long sloping terminations on both ends, as well as for a group of two side-by-side identical Beverages (fed in phase), spaced 10, 5, 2, 1 and 0.5 m. Going from 1 to 2 Beverages increases the gain by (the expected) 3 dB for a spacing of 10 and 5 m. For closer spacing the gain drops somewhat but not significantly. At 50-cm spacing the gain loss is only 0.9 dB versus 5-m spacing (see **Table 7-13**).

Note that the shape of the radiation pattern has barely changed. The extra gain is *not* obtained by more directivity, but by reducing the losses of the antenna. Conclusion: we

Table 7-13
Gain and 3-dB Forward Angle of Two Broadside Beverages (fed in phase) as a Function of Spacing.

Configuration	Gain (dBi)	3-dB Angle
single Beverage	−9.65	80°
pair in phase 10 m spaced	−6.68	80°
pair in phase 5 m spaced	−6.68	80°
pair in phase 2 m spaced	−7.00	80°
pair in phase 1 m spaced	−7.25	80°
pair in phase 0.5 m spaced	−7.51	79°

Table 7-14
3-dB Forward Angle as a Function of (wide) Spacing of Two Beverages Fed in Phase (Broadside)

Configuration	3 dB Angle
single antenna	80°
5 m (close spacing)	80°
40 m ($^1/_4 \lambda$)	71°
80 m ($^1/_2 \lambda$)	53°
120 m ($^3/_4 \lambda$)	39°

can space Beverages quite closely together (side by side) without trading in much performance.

15.2. In-phase Arrays of Beverages (Broadside Array)

If you want to reduce the response off the side of the array, you would expect that spacing the Beverages ½ λ

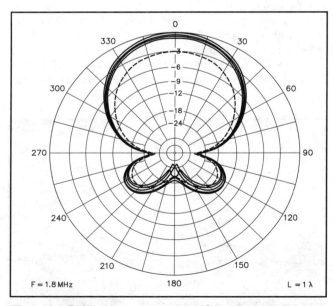

Fig 7-33—Radiation patterns at 25° wave angle for two close-spaced Beverages (side by side) fed in phase as spacing is changed. The smaller pattern is for the single (reference) antenna. Note that the pattern barely changes as the side-by-side separation is changed from 10 m to 0.5 m. (See text.)

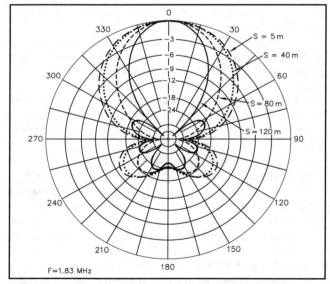

Fig 7-34—Radiation pattern for two wide-spaced Beverages (side by side) fed in phase. The smaller pattern is for the single (reference) antenna. Note that the pattern gets narrower as the spacing is increased. The gain remains the same in all cases, and is the same as with narrow spacing—3 dB more than a single (identical) Beverage. (See text.)

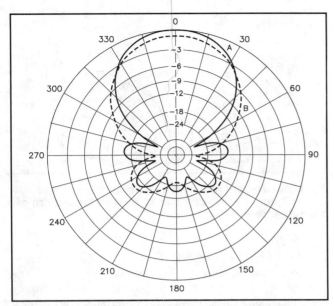

Fig 7-35—Radiation pattern for two ¹/₂-λ spaced, 1-λ long Beverages (side by side) fed in phase (A), compared to the pattern of a single 1.5-λ long Beverage (B). Both patterns at 25° wave angle.

would be ideal, as in that case signals coming off the side are now 180° out of phase.

Table 7-14 shows how the 3-dB angle changes as the spacing is increased. The gain remains exactly the same as for the close-spaced pair but the gain now comes from a narrower beamwidth rather than from reduced losses.

Fig 7-34 shows the radiation patterns (at 25° wave angle) for the same 168-m long (2-m high) Beverage pair (fed in phase) for various degrees of "wide" spacing. Spacings over ¹/₂ λ introduce secondary lobes and yield a forward 3-dB angle that gets very narrow. If reduced forward beamwidth is the goal, then a spacing of ¹/₂ λ is the answer.

Fig 7-35 shows the comparison between two 0.5-λ spaced, 1-λ long Beverages and a 1.5-λ long single Beverage. Gains are within 0.5 dB of one another, but the phased

pair has a somewhat narrower forward 3-dB beamwidth (53° vs 66°). Conclusion: even with close-spaced pairs, the gain of a 50% longer antenna can be approached. The radiation angle, though, cannot be lowered, and the pair of short Beverages will still exhibit a higher radiation angle than the single long one.

Feeding the in-phase array

The easiest way to feed these antennas (with 50-Ω feed points) is via two 75-Ω, 0.25-λ transformers to a common T-junction (**Fig 7-36**). The 75-Ω cables transform the 50-Ω impedance to 100 Ω which, in parallel, again gives 50 Ω. You can of course also feed via two equal lengths of 50 or 75-Ω cable (exact length is irrelevant) to a T-junction, where the impedance will then be 25 or 37.5 Ω. From that point we can use an L network to match either a 50 or a 75-Ω feed line to the receiver. Other matching systems can of course be used (see Chapter 6, The Feed Line and the Antenna).

15.3. Two-Element End-Fire Beverage Array

You can put Beverages in arrays like you put verticals in arrays. The Beverages that are elements of the array must be identical in length, height, etc. The easiest setup uses two Beverages 90° out of phase, spaced 0.25 λ (we call the spacing "stagger" distance similar to cardioid configuration with verticals). Actually if you want to optimize the F/B at a 30° wave angle, in that case the physical stagger distance must be 90° × cos (30°) = 78°.

Fig 7-37 shows the layout of the array and the pattern obtained by phasing two identical 168-m long Beverages, staggered 0.25 λ (40 m) and fed 90° out of phase (equal current amplitude), for varying lateral spacings (2 m, 10 m, 20 m and 40 m). The single (identical) Beverage is the reference. Note that as we increase the lateral spacing, the rejection off the side where we have the Beverage with the leading current increases. At a spacing of 40 m (¹/₄ λ) the forward lobe has become clearly asymmetrical, being wider to the right (in our figure) than to the left. The biggest difference, however, is in the back lobes. At a spacing of

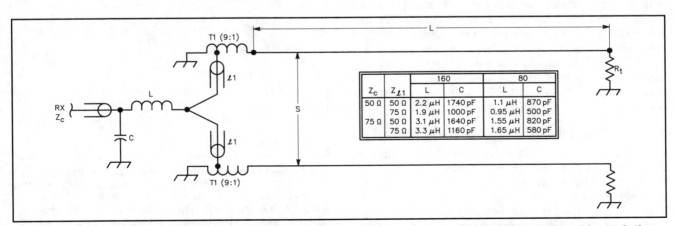

Z_C	$Z_{\ell 1}$	160 L	160 C	80 L	80 C
50 Ω	50 Ω	2.2 μH	1740 pF	1.1 μH	870 pF
	75 Ω	1.9 μH	1000 pF	0.95 μH	500 pF
75 Ω	50 Ω	3.1 μH	1640 pF	1.55 μH	820 pF
	75 Ω	3.3 μH	1160 pF	1.65 μH	580 pF

Fig 7-36—A pair of in-phase Beverages are very simple to feed: just connect equal coax lengths and match the junction to the feed line with an L network (or other matching device). If 75-Ω feed lines are used at ℓ1, they will convert the impedance to 100 Ω at the conjunction of the lines. The two lines in parallel result in an impedance of 50 Ω (100/2), and a 50-Ω feed line may be connected directly—no need for the L network

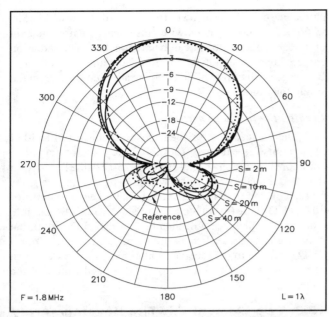

Fig 7-37—Radiation patterns at 25° wave angle for two 1-λ long Beverages in end-fire configuration (¼-λ stagger as shown in Fig 7-40, 90° phase shift) for various lateral spacings. The pattern for the single-element identical-length Beverage is added as reference. See text for details.

¼ λ the back lobe has actually increased in amplitude, because for waves coming off the right, the summing of the signals is fully in phase. At 2-m spacing the F/B is not optimized, but starting with 5-m spacing (and up) the F/B is increased from 20 dB for a single Beverage to 30 dB for the pair. Unless you really want an asymmetrical pattern, the recommended spacing of the 2-element end-fire array for 160 meters is 5 to 10 m. The vertical pattern of this array is shown in **Fig 7-38**.

So far most of the Beverages I have shown have the so-called Cone-of-Silence length (Section 4.2), which means that their length is optimized for best F/B. **Fig 7-39** shows a single Beverage and an end-fire pair (spaced 5 m) with a random length (in this case 200 m). The single Beverage has a rather poor F/B of only 17 dB, while the pair achieves not less than 30-dB F/B. It is clear that due to the subtractive nature of the phased array, the improvement in F/B will be most outspoken with individual elements that have a rather poor F/B. This also means that it is *not* necessary to use Cone-of-Silence lengths with end-fire arrays in order to obtain a good F/B.

15.3.1. Feeding the End-Fire Beverage Array

When spaced at 2 m, the effects of mutual coupling are clearly reflected in the feed-point impedances of the two Beverages, which are quite different. At 10-m spacing, the coupling has all but gone and the impedances are the same. This is an important consideration for designing a proper feed system. If we use close spacing, it will be necessary to use the so-called current-forcing method, such as using ¼-λ feed lines into a hybrid coupler. This is explained in

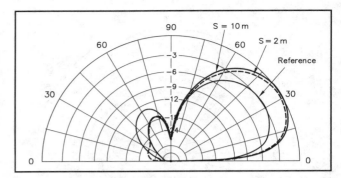

Fig 7-38—Vertical radiation pattern for the array of Fig 7-37, for spacing of 2 and 10 m, with the pattern of a single element as reference.

detail in Chapter 11 (Arrays). The hybrid coupler is easy to construct and requires no high-power components. The reason why we use current forcing, is that we cannot simply equal phase shift (in degrees) to feed-line length (in degrees) unless the SWR is 1:1 on that line. In our 2-element array the impedances will always be different if the elements are close together. When we use wide spacing (>0.05 λ) mutual coupling is so small that the feed-point impedances are identical and can be matched to obtain a very low SWR on the feed lines. Under such circumstances, we can simply insert a 90° long "extra" cable length to the element that needs the current to lag by 90°.

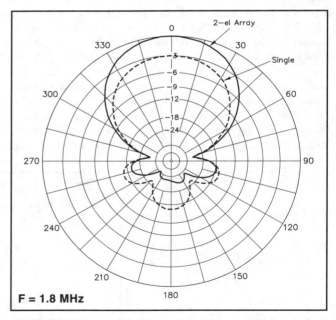

Fig 7-39—The single-element Beverage of 200-m length is not a Cone-of-Silence Beverage and exhibits a rather poor F/B. In the 2-element end-fire array the F/B goes from 17 to over 30 dB! This illustrates the fact that in order to obtain spectacular F/B in end-fire arrays it is *not* necessary to use Cone-of-Silence lengths for the individual elements (patterns calculated for 25° wave angle).

Fig 7-40 shows three alternative ways of feeding the array. If you want to use the system using an extra 90° long cable length, it is very important that the individual Beverages are properly terminated and that the transformer at the feed point achieves a 1:1 SWR to your coaxial feed line. Adjust the value of the termination resistance for best F/B. Then, if necessary, adjust the tap on the feed-point transformer for a 1:1 SWR on the feed line. Only when each of the individual feed lines show an SWR better than 1.2:1 can we

assume that we will achieve the desired phase shift by inserting a length of coax with the same electrical length (in degrees) as the desired shift. As the physical length-wise spacing is $^1/_4 \lambda$, a simple $^1/_4$-λ long delay line will not reach between the transformers of the two Beverages. Therefore we have to insert two equal lengths of coax (ℓ_1). Where the coax cables are connected in parallel, the impedance is 25 Ω. In order to obtain a good match to a 50 or 75-Ω line to the shack we can use an L network, which can easily be

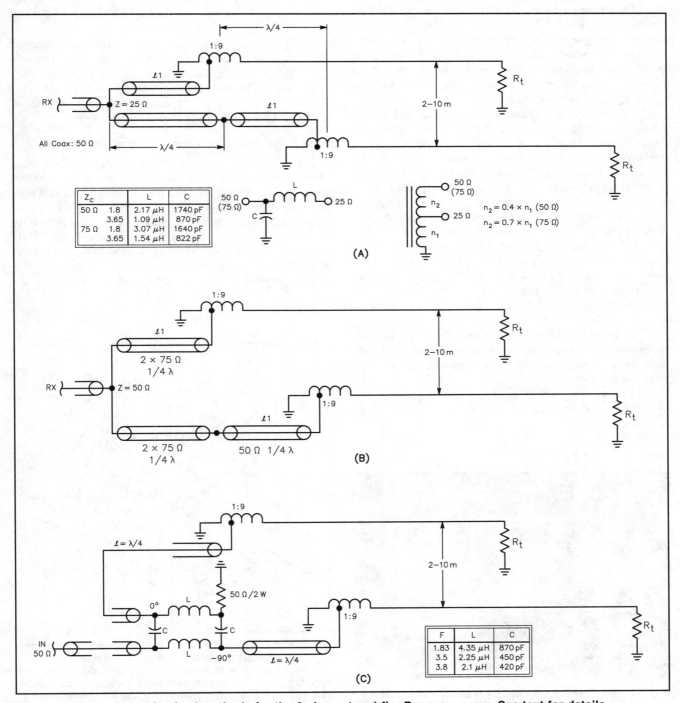

Fig 7-40—Three alternative feed methods for the 2-element end-fire Beverage array. See text for details.

calculated with the L-network module from the ON4UN LOW-BAND SOFTWARE.

Fig 7-40B shows an attractive alternative. Transformer T1, with a primary feed impedance of 50 Ω, is fed via a 0.25-λ, 75-Ω coax. At the end of the line the impedance is 100 Ω (¼-λ transformer, see Chapter 6). Transformer T2 is fed via the 50-Ω ½-λ delay line, and then via another 75-Ω ¼-λ line to the common point. The two 100-Ω impedances in parallel result in 50 Ω, so at the junction we can continue with a 50-Ω line to the receiver.

If you have no means of making sure that the SWR on the feed lines is <1.2:1, then I recommend you use the current-forcing method with the hybrid coupler. The hybrid-coupler feed method is shown in Fig 7-40C. The feed lines to the coupler should each be ¼-λ long. The hybrid coupler can be wound on a relatively small core as this is only used for receiving. A bifilar winding on a small powdered-iron core may be used. For a 160-meter hybrid coupler, I used a T106-2 core (1-inch OD), wound with 0.6-mm (AWG 22) enameled wire. This is quite a large core for the application but makes it easier to make a nicely symmetric winding. This type of core has an A_L factor of 135. The turns are calculated as explained in Section 8.2.1. The formula indicates 18 turns for 3.45 µH. After winding the core, I had to reduce the number of windings to 14 to obtain the correct value. A small digital LC meter is an invaluable tool for such jobs.

An 80-meter model uses a T5-2 core (0.5-inch diameter), wound with 0.3-mm (AWG 28) wire. I twisted the parallel wires using a hand drill (approx. 2 to 3 twists per cm). Especially with small wire it is much easier to wind twisted wire than parallel wire. Exactly 18 turns were required to obtain 2.2 µH.

Ferrite material should *not* be used because it is much too unstable for applications where a precise and stable inductance value is required. The capacitors should be mica capacitors.

Fig 7-41 shows the 80 and the 160-meter hybrid coupler. The 50-Ω terminating resistor is mounted inside the little box.

Fig 7-41—Home-made receiving-type hybrid couplers for 160 and for 80 meters. See text for details.

15.3.2. Other spacings and phase shifts

If you don't use the hybrid coupler where you are forced to use quadrature configuration (increments of 90°), other configurations can be tried, as you will obtain the required phase shift by cutting an extra feed-line length (length in degrees equal to phase shift in degrees, on condition that the feed-line SWR is 1:1). Using an antenna-modeling program based on *NEC2*, you can easily play around and tune your own array. Slight deviations from the 0.25-λ stagger spacing as well as from the 90° phase increments can result in better directivity, but usually at a slight sacrifice of gain.

Fig 7-42 shows the same array on 160 with a stagger distance of only 32 m and for phase shifts of 95, 105 and 155°. We can conclude that the exact phase shift is not very important and has only little influence on the F/B and gain. These are arrays that really want to work, and are not at all critical for obtaining good performance.

15.3.3. An 80-meter 2-element Beverage array

Fig 7-43A shows the horizontal pattern for a 2-element 166-m long end-fire array (quadrature fed) with the reference single Beverage for comparison. As expected, the gain is 3 dB. The 166-m long single Beverage has a Cone-of-Silence length, and as such exhibits a very good F/B on its own. The phased array adds a little bit to that. In Fig 7-43B we have the same situation with a length of 188 m, which is *not* a Cone-of-Silence length. As expected, in this case, the improvement in F/B is much more spectacular.

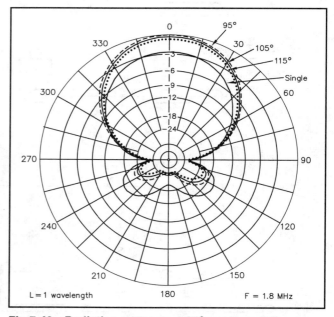

Fig 7-42—Radiation patterns at 25° wave angle for 2-element end-fire 160-meter array with reduced stagger spacing (32 m instead of 40 m) and various phase shifts (95, 105 and 115°). The pattern for the single-element identical-length Beverage is added as reference.

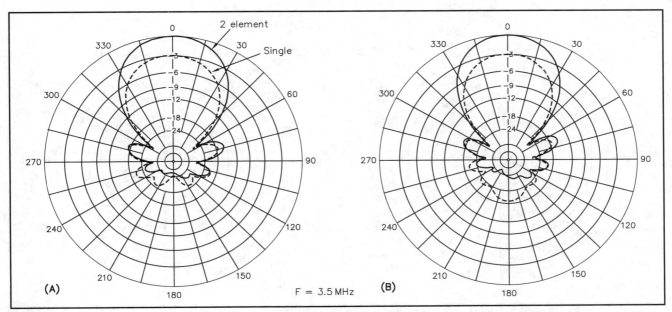

Fig 7-43—At A, the radiation pattern for a single 166-m long Beverage (Cone-of-Silence length) and for the 2-element array (same length) on 80 meters. At B, we have the same comparison for a Beverage that is 20 m longer (no longer Cone of Silence). Note that the performance of the array does not suffer from the fact that the elements are not Cone-of-Silence length. All patterns are calculated for 25° elevation angle.

15.3.4. A 160/80-meter 2-element Beverage array

If we want to make a Beverage array that operates on 80 as well as160, we will need to compromise just slightly. Let's work out an example for two 168-m long phased Beverages, spaced 5 m (not critical).

Optimized for 80 meters

Our starting point is the physical layout of the 80-meter

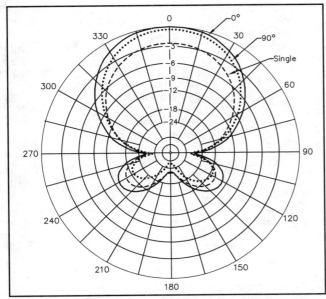

Fig 7-44—Radiation pattern at 25° wave angle for the 2-element array with 20-m stagger distance, on 160 meters (see text for details).

array, which means a stagger distance of 20 m ($^1/_4$ λ on 80 meters). The result of this array is shown in Fig 7-43 (90° phase shift). The problem is that on 160 meters, with this layout we only have $^1/_8$-λ stagger distance. I tried different phasing angles: with 135° phase shift we obtain a great back pattern, but the gain is 1 dB *less* than for a single element! Feeding both Beverages in phase yields 2.5-dB gain versus a single Beverage, but with no improvement off the back (F/B is already 30 dB!). If we introduce 90° phase shift we improve the attenuation in the back by approx. 10 dB, at less than 1-dB gain sacrifice. **Fig 7-44** shows the situation.

Optimized for 160 meters

In this case the length-wise stagger distance is 40 m ($^1/_4$ λ on 160). However, we can reduce the stagger distance from 40 to 30 m without sacrificing anything. **Fig 7-45** shows the pattern for the array with 90° phase shift and 30-m stagger distance. The gain is almost 3 dB over the single reference Beverage, while the F/B is increased by about 10 dB. With a stagger distance of 30 m, it appears that on 80 meters a phase shift of 90 to 105° yields almost 3-dB gain over a single element as well as 5 to 10-dB improvement in the F/R situation. If we choose 90° phase shift we can use a hybrid coupler as feed system (see Fig 7-45).

The feed system

Both above designs have an interesting common characteristic: the elements being fed in quadrature, which means we can use two hybrid couplers and 0.25-λ (current-forcing) feed lines. Phased arrays, current forcing and hybrid couplers are explained in greater detail in Chapter 11 (Arrays).

Fig 7-46 shows the feed system. For the system opti-

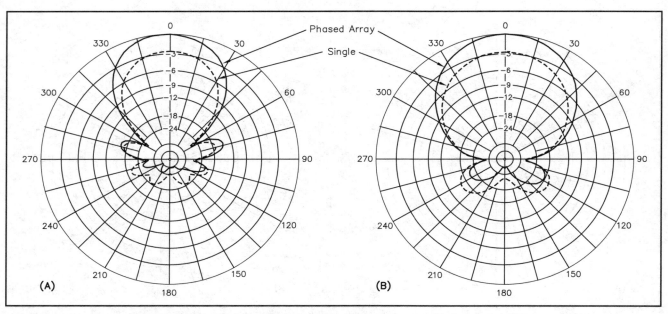

Fig 7-45—At A, radiation pattern at 25° wave angle, for the 2-element Beverage array for 80 meters. At B, the same for 160 meters. On both bands approx. 2.5 dB of gain is realized versus a single Beverage, while the F/R is increased by 5 to 10 dB.

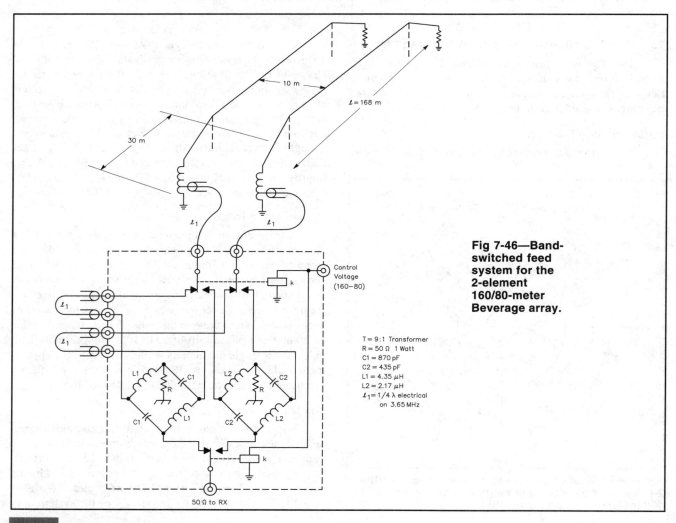

Fig 7-46—Band-switched feed system for the 2-element 160/80-meter Beverage array.

T = 9:1 Transformer
R = 50 Ω   1 Watt
C1 = 870 pF
C2 = 435 pF
L1 = 4.35 μH
L2 = 2.17 μH
ℓ_1 = 1/4 λ electrical
        on 3.65 MHz

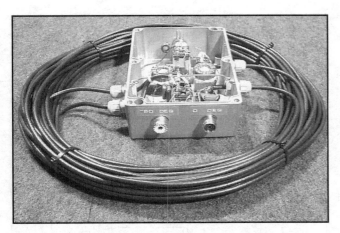

Fig 7-47—Two-band hybrid coupler and band switching system built by the author. Relays for band switching can be controlled via the coax cable or via a separate small cable. The two lengths of RG-58 cable are $^1/_4$ wave on 3.85 MHz, and serve to lengthen the 80 meter $^1/_4$-λ lines to become $^1/_4$-λ lines on 160 meters.

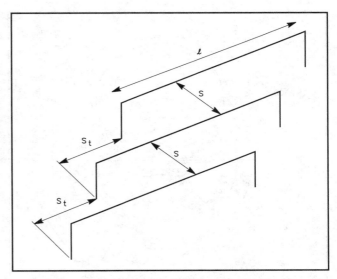

Fig 7-48—Layout of the 3-element end-fire Beverage array. In the most classic configuration, the stagger distance (S_t) is $^1/_4$ λ. The elements are fed in 90° increments while the center element is fed with twice as much current as the outer ones. For simplicity the Beverage antennas are shown without the sloping terminations.

mized for 80 meters described above, we can use regular solid PE-dielectric 0.25-λ coax lines to reach the hybrid coupler/switching box, located halfway between the feed points of the 2 elements (linear distance between the two elements is 22.3 m in case of 10-m lateral spacing (two $^1/_4$-λ feed lines of solid-PE coax on 80 are 2 × 13.55 = 27.1 m).

When optimized for 160 meters, (30-m stagger distance) we must use foam coaxial cable (eg, RG-8 foam). Assuming a V_f of 0.8 for this cable, a quarter wavelength in the cables is 0.8 × 300 / (3.65 × 4) = 16.44 m. With a lateral spacing of 10 m, and a longitudinal stagger distance of 30 m, the distance between the feed points of the two antennas is 31.62 m, which means we can just reach the hybrid coupler in the middle. When using foam coax, do not merely go by the published velocity factor figures. Measure a quarter-wavelength (see Chapter 4) by electrical means!

The two hybrid couplers can be built into one box (see **Fig 7-47**), together with two relays that will do the band switching. On 160, extra lengths of coax are added to the feed lines to arrive at 0.25-λ (electrical) lines on 160 meters.

15.4. Three-Element End-Fire Beverage Array

The 3-element Beverage is similar to the 3-element vertical end-fire array using binomial current distribution (see Chapter 11). Binomial means that the center element is fed with twice the current amplitude as compared to the outer elements. The phase shift is in quadrature, which means in steps of 90°. The general layout of a 3-element end-fire array is shown in **Fig 7-48**.

160-meter array

Fig 7-49 shows the 160-meter pattern for the array, made up from 168-m long elements, 2-m high, with 5-m spacing between the elements. The array has just over 1-dB gain over the 2-element array, and adds another 5-dB F/R.

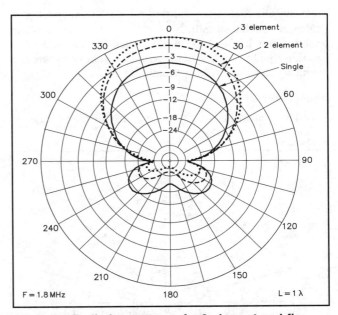

Fig 7-49—Radiation pattern of a 3-element end-fire Beverage array, with elements measuring 168-m long. The 2-element and the single Beverage patterns are included for comparison. All patterns are calculated for a 25° wave angle.

It appears that this array, even with fairly large lateral spacing exhibits far more mutual coupling than the 2-element array. In fact, it is best to feed this array with a hybrid coupler rather than to rely on coaxial cable length to provide the proper phase shift.

I also analyzed the influence of lateral spacing on gain and F/B. **Fig 7-50** shows the patterns for various spacings.

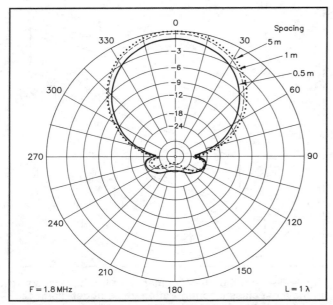

Fig 7-50—Radiation patterns (at 25° wave angle) for the 2-element 160-meter Beverage array for various lateral spacings.

Note that the 3-element end-fire, with a lateral spacing of 0.5 m is equaled in gain by a 2-element array with 2-m spacing, but has approx. 6 dB better F/R. A minimum spacing equal to the height of the antenna above the ground seems a reasonable rule of thumb in this case.

80-meter array

The 3-element array (also with 168-m long elements)

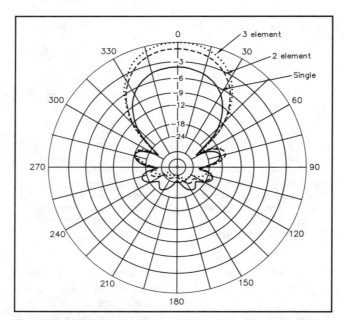

Fig 7-51—Radiation pattern on 80 meters for the 168-m long 3-element end-fire Beverage array with 5-m lateral spacing, and ¼-λ stagger distance. The 2-element array and the single Beverage patterns are added as reference.

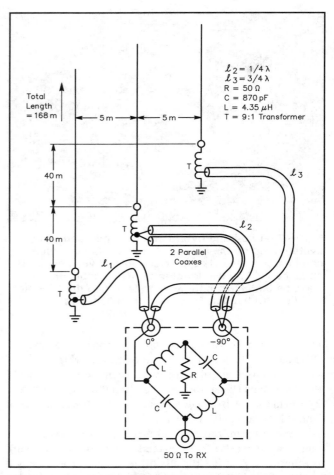

Fig 7-52—Feed system for the 3-element end-fire 160-meter array. The coax length ℓ_3 consists of the ¼-λ current-forcing section followed by a 180° phasing line, making it ¾-λ long. ℓ_2 is made up by 2 parallel ¼-λ lines (making 25-Ω impedance), in order to "force" twice the current into the center element. Using regular solid PE coax (eg, RG-58) the line lengths are: $\ell_1 = \ell_2 = 27.03$ m and $\ell_3 = 81.09$ m. For 80 meters, all dimensions can be divided by 2.

modeled for 80 meters, has a pattern that looks very much like a 3-element Yagi (see **Fig 7-51**). The increase in gain and F/B as compared to the 2-element array is rather marginal.

It is obvious that you can build 3-element arrays with other lengths. The length is not at all critical to obtain good F/B, as it is with the single Beverage (Cone-of-Silence).

Feeding the array

These 3-element arrays must be fed via a hybrid coupler. The principle is the same as for the 2-element arrays. See **Fig 7-52** for dimensions for 160 meters. For 80 meters all dimensions should be halved. An 80 and 160-meter 3-element array could be made by inserting a fourth Beverage between element 1 and 2 of the 160-meter array.

15.5. Summing up on Beverage Arrays

Important things to remember are:

In-phase arrays (broadside)

- Are very easy to feed.
- If you are not after F/B you can feed 2 or 3 close-spaced Beverages from a common feed point (see Fig 7-13).
- Give 3-dB gain for a pair, 4.5-dB gain for 3 Beverages in phase.
- Does not improve F/B ratio (so you need to use elements that have a Cone-of-Silence length).
- 3-dB forward lobe remains essentially the same with close spacing.
- 3-dB forward lobe gets narrower as spacing gets wider (see table in Section 15.2).
- When tuning your Beverage for best F/B, set up the signal generator at an angle with respect to the back angle which is equal to the wave angle where you want max. F/B (eg, 25°).

End-fire arrays

- F/B on a non-Cone-of-Silence length Beverage can be dramatically improved (10 to 20 dB) by phasing two Beverages.
- Arrays of Beverage increase the gain by 3 dB (2 el) or 4 dB (3 el).
- Forward lobe gets narrower (more forward directivity).
- The high elevation angle of short Beverages is *not* improved by phasing Beverages.
- Cone-of-Silence lengths are *not* important with Beverage arrays.
- The difference between a 2 and a 3-element end-fire array is minimal.
- A 2-band, 2-element array is easy to construct and to feed using a duo-band hybrid coupler.
- The performance difference between a 2 and a 3-element end-fire array is limited and not really worth all the effort.

■ 16. VARIABLE PHASING

In Section 12.7, I reviewed the steerable Beverage concept, and how the MFJ noise-canceling device could be used for the job. You could use the device quite well to combine the outputs of a 2-element Beverage array. In that case the outputs from the two elements should be fed in separate cables (preferably of equal length) to the main antenna and the noise-antenna input of the noise-canceling device. The unit would allow some kind of control to further reduce interfering signals from a given direction, as compared to a fixed configuration with a more classic feed system as described above.

■ 17. PREAMPLIFIERS FOR BEVERAGE ANTENNAS

17.1. A Preamp, Yes or No?

Beverages are low-gain antennas, producing a signal output which is generally 5 to 15 dB down from a single vertical transmit antenna. Other receiving antennas such as described in Sections 20, 21, 22 and 23 may have an output up to 30 dB down!

If you want to see your S-meter kick as high when using a Beverage antenna as it does while receiving on your transmit antenna, then you obviously will need a 5 to 15-dB gain preamplifier. Fortunately, we don't (usually) listen with our eyes.

The easy rule to follow is: if when receiving on a low-output (receiving) antenna you can hear the band (external) noise over the internal noise of the receiver, then you are in fine shape and do not need a preamp. The entire issue is one of signal-to-noise ratio.

As a rule we should keep the signal level as low as possible to prevent chances of intermod and overload. This is also why our receivers have a front-end attenuator as well as a switchable preamp (on the top-range rigs).

A preamplifier will amplify everything that's available at its input: signals *and* noise as well! As the readability issue is an issue of signal-to-noise ratio, a preamp should only be used for one of the following reasons:

- **The receiver has a high noise floor:** In this case the preamp can be situated right at the receiver. It appears however, that most receivers have a noise figure of 5 to 10 dB on the low bands, which means that there is ample sensitivity (see Chapter 3, Fig 3-2), under almost all circumstances.
- **The feed line to the Beverage is long and lossy:** Under such circumstances, it is not uncommon that during gray-line times the noise produced by the antenna drops to such a low level that the receiver noise becomes stronger than the antenna noise. Under such conditions a 10-dB gain preamp can be a welcome help. In such case the preamp should be located at the antenna, and not at the receiver, in order to compensate for the losses of the long feed line. A 300-m long RG-58 feed line would give approx. 5-dB loss on 160 meters and 7.5 dB on 80. That would be unacceptable, and a 10-dB preamp at the receiving antenna would be a good choice. If you use low-loss feed line such as $^5/_8$-inch CATV coax, then 300 m of it will only make you lose approx. 0.5 dB on 160 and 0.75 dB on 80. Here, a remote preamp is a wasted effort!
- **Compensate for the insertion loss of a good band-pass or BCI filter:** Such filters often have between 5 and 10-dB insertion loss that you have to compensate for.

In my Beverage antenna setup, I included at one time 10-dB selective preamps, but I have removed them, as they never gained anything. All my Beverages are fed via $^5/_8$-inch Hardline (some feed lines are more than 300-m long), so the line losses are totally negligible.

In the shack I can use the Braun SWF 1-40 (multiband) preselector preamp, which can give up to 8-dB gain, and a lot of selectivity (see Fig 3-14 in Chapter 3), if required.

For antennas with outputs of −15 to −30 dB a preamplifier is a must.

17.2. Homebuilt Preamps

Excellent preamplifiers for this purpose have been described in the literature (Ref 257, 267, 1232, 1251, 1254

Fig 7-53—The KD9SV preamp, a popular 160 or 80-meter preamp, which is available as a kit.

and 1256). The KD9SV preamp (Ref 1251) has an excellent reputation, and is available as a kit (both 160 and 80-meter units). For details check the Web page: **http://home1.gte.net/ crlewis/160MPREAMP.HTM** or contact C&S Engineering, 9229 Goldenrod Dr., Ft. Wayne, IN 46835 (219) 485-1458. **Fig 7-53** shows the unit, assembled from the kit.

Untuned preamps using MMICs are not the right choice. MMICs have poor IMD performance and are very poor with even-order harmonic distortion. They have a fixed gain, which is usually way too high anyhow. Last but not least, they withstand little abuse.

17.3. Commercial Sources

Commercial units are available from Ameco, ZJ Electronics (K2ZJ), Ramsey, Ten-Tec, Palomar, ICE, Mini Circuits, Advanced Receiver Research and Braun (Germany).

■ 18. BEVERAGE PERFORMANCE

I have been using Beverages since 1968. For me, Beverage antennas have undoubtedly been the key to working my last 50 countries on 80 meters. As far as 160 meters is concerned, I would not like to think what it would be without them.

I use the Beverages all the time on 80 and 160. Unfortunately, I have to take down most of my Beverages during summer time, which for me was "off-time" on the low bands, time to go bike riding, until I discovered the EWE, the K9AY loop and especially the rotatable 2-element end-fire array.

In winter I have 12 Beverages up, ranging from 180 to 300 m in length. In all cases, the longer ones are better than the shorter ones. Some people have said that 300-m long Beverages are too long. I strongly state that this is not true, not on 80 meters, and not on 160 meters. Of course, if you would have only 300-m long Beverages, you would need 10 or 12 Beverages to cover all directions equally well.

Some people have argued that their 4-square hears

better than their Beverages. Beverages that are at least 1-λ long have a narrower 3-dB forward lobe than a 4-square. If well built and of Cone-of-Silence length, their F/B can be as good as for a perfect 4-square. The F/R for a well designed Beverage is always better than the F/R of a 4-square (see also Section 26.2).

The only time that a 4-square can outhear a good Beverage is when signals are extremely weak or when the noise is very low. Then it is a question of overall gain, rather than signal-to-noise ratio. This, however, is a fairly rare situation on 160 and 80 meters. Long Beverages (2 or 3-λ long) will always outhear a 4-square when there is a lot of noise to be rejected.

Hams who have experienced situations where their 4-square hears as well or better are very fortunate to live in very quiet areas, or perhaps their Beverages are not properly working, or maybe too short. Indeed, a 100-m "short" Beverage is not really a Beverage on 160 meters!

Have a look at the results of the poll amongst low-band DXers, published in Chapter 2. See for yourself how many of the 4-square users use Beverages as well. These results speak for themselves.

K4ISV's statement *"Can you imagine a fisherman going out with only one bait?"* of course makes a lot of sense. But I would immediately like to add that if you choose to have a bunch of antennas, make sure they do not couple with one another, or you will have a totally uncontrolled condition where anything can happen.

In Chapter 1, I mentioned a fairly typical phenomenon of high-angle propagation before sunset or after sunrise. Under such circumstances a low dipole will outperform a (long) Beverage because of its angle of radiation. A low dipole is definitely a useful arm in the gallery of your weapons. Like Frank, W3LPL, said "You never can have too many antennas."

A special word about Beverage arrays: The only way to test a 2 or 3-element array is to have a single-element antenna of equal length alongside. I have seen some most spectacular improvements, like signals generally off the back dropping from S7 to S1 or less! Simply amazing.

■ 19. ANTENNAS DERIVED FROM BEVERAGE ANTENNAS

19.1. The Slinky Beverage

Carl, K1MH, had to explain to me what a Slinky is. Apparently it's a child's toy, but I still don't know how children play with it. . . Anyhow, it looks like Top Banders now also play with it. *"It is a continuously wound coil about 7 cm in diameter and about 90 turns for each unit. I look at it as a helically wound Beverage that exhibits very low-noise pickup and is still very efficient as compared to a regular wire Beverage."*

Although Carl does not claim to be the "inventor" of the Slinky he says its initial use goes back to 1985 and started more as a joke and dare as the result of an "after midnight" 160-meter SSB roundtable. At the same time he claims that the Slinky equals the performance of a Beverage of 1.5 to 2 times the length of the Slinky.

He uses the 7-cm diameter version (available from K1MH) and has connected 5 of them in series stretched over a total length of approx. 53 m. This makes 8 turns per meter, which according to Carl gives a velocity of propagation (V_P) of 0.58. A V_P of 0.5 means the wave travels half the speed of light, so the antenna looks twice as long as it is physically. A V_P of 0.58 means a 55-m long Slinky is 95-m long electrically (which is actually too long to be a Cone-of-Silence $1/2$-λ length).

Obviously, what we are after is directivity, or more specifically F/B. We know that the Cone-of-Silence lengths (multiples of electrical 0.5 λ) give the best F/B. The minimum length of a Beverage to show a "null" in the back is $1/2$-λ (electrical). This means that the Slinky, being shorter than $1/2$ λ will be loaded (electrically lengthened) to become $1/2$-λ long. A half wave on 1.83 MHz (free space) measures 82 m. This means that, if we like to achieve a $1/2$-λ electrical (Cone-of-Silence length) we shall have a V_P of 53/82 = 0.65 (at zero wave angle).

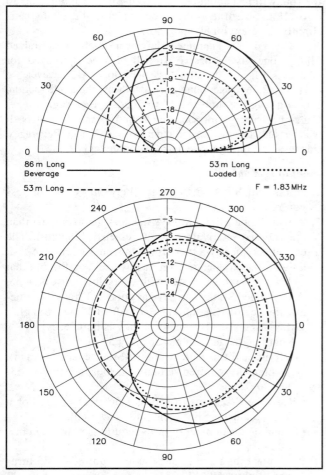

Fig 7-54—Vertical and horizontal radiation pattern (at 30° wave angle) of a loaded short Beverage (Slinky) as compared to the $1/2$-λ long Beverage and the unloaded short Beverage.

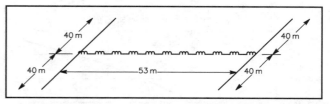

Fig 7-55—View of the model used to evaluate the effect of a loaded Beverage. Continuous loading was simulated by using 10 coils equally spaced along the length of the short Beverage. The 40-m long wires at both ends of the antenna are the quarter-wave terminations.

Let me first analyze and compare two "real" Beverages:
• $1/2$-λ long Beverage.
• 53-m long Beverage (no loading).

Fig 7-54 shows the radiation patterns (vertical and horizontal at 30° wave angle) of these short Beverage antennas. As I said before, this is really too short to enjoy the intrinsic possibilities of a Beverage. A minimum length of 1 λ is recommended. But if you have limited space, a short one will be better than nothing. The $1/2$-λ long Beverage has a gain of approx. −14 dBi over good ground (see also Fig 7-7) and 25 dB F/B. The non-loaded 53-m long Beverage has 4-dB less gain, and has only 6-dB F/B.

In principle it is possible to get good directivity in short antennas, but phase delay and current distribution along the wire needs to be correct. If we now properly load the 53 m short Beverage, so that it becomes 180° electrically, we again achieve a good F/B. This can be done by making the Beverage one long coil (a Slinky) or alternatively by inserting coils at regular (short distances). The third radiation pattern shown in Fig 7-54 was obtained by inserting 10 coils equally spaced along the length of the 53 m short Beverage. These coils were dimensioned to obtain the highest possible F/B. **Fig 7-55** shows the layout (from *EZNEC*). Note that, as explained in Section 3.2 (and Fig 7-2), I always use two 0.25-λ wires (in-line) as terminations. The reactance of each of these coils was 165 Ω (equals 14.3 µH on 160 meters). The impedance of the antenna was 500 Ω, and it was terminated with a 500-Ω load resistor. All models were at a height of 2 m.

Loading the wire has achieved an F/B which is as good as for the full-size $1/2$ λ but we notice a loss of approx. 5 dB. Instead of loading the Beverages with discrete coils every 5 m, we can "distribute" the inductance in the form of a Slinky. The question of course is: "How much Slinky, how much stretching?" As we are shooting for an electrical length that gives us the best F/B, the easiest approach to solve this problem is to install a small signal source in line with the antenna (on an angle off the back), and adjust the stretching of the Slinky until the best F/B (approx. 25 dB) is obtained.

If we want to use these Slinkies instead of discrete coils, how many do we need? Tom, W8JI, measured a Slinky setup on 160 meters and computed the relationship between the Velocity of Propagation and the ratio coil-pitch to coil

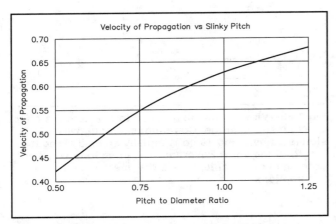

Fig 7-56—Velocity of Propagation for 160 meters as a function of the pitch to diameter ratio of a Slinky.

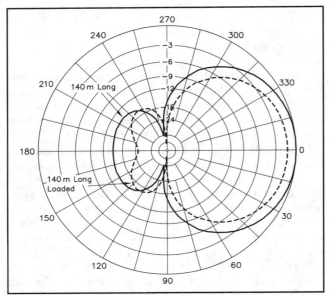

Fig 7-57—A non-Cone-of-Silence length Beverage (140-m long) is lengthened by loading to achieve 10 dB better F/B ratio, at a sacrifice of 1 dB in forward gain.

diameter (see **Fig 7-56**) for 160 meters.

In our above example we had a Vp of 0.65, which, according to the table, gives us a pitch to diameter ratio of 1:1. For a Slinky diameter of 7.7 cm this means 1 coil turn every 7.7 cm. For a total antenna length of 53 m this means (5300 / 7.7 =) 688 turns. Each Slinky has 85 turns, so we require 688/85 = 8 Slinkies.

Note that this principle can be used to electrically stretch a "longer" Beverage in order to achieve a Cone-of-Silence length. **Fig 7-57** shows the case of a 140-m long Beverage (far from a Cone-of-Silence length), which exhibits a poor F/B of only 15 dB. By loading the Beverage with 10 coils, equally spaced along its length, the F/B is increased to better than 25 dB, at a sacrifice of about 1 dB in gain. In this particular case, each coil has a reactance of 80 Ω, which represents a coil of 7 µH on 160 meters.

By now it should be clear that by loading the "short" Beverage (with individual coils at regular intervals, or by one long stretched coil—a Slinky) all we do is adjust its electrical length so that it becomes a multiple of 0.5 λ, performing as a Cone-of-Silence Beverage. However, we do this at a slight sacrifice of gain. Loading the Beverage does *not* improve the gain, to the contrary! Here is why: In Section 2, I explained the principle of operation of a Beverage, and how a wave induces signals into the antenna wire. Both having a different Vp, it is clear that the greater the difference, the faster the phase difference will grow between the wave in the air and the wave in the wire. For any given length of Beverage, a lower Vp thus means lower gain. This is also why lower Beverages have lower gain—they have a much lower Vp.

The same Slinky on 80 meters?

If we have electrically lengthened a short Beverage, and adjusted it for best F/B on 160 meters, how will it perform on 80? With a full-size Beverage, and given the harmonic relationship of 160 and 80 meters, an antenna that's been dimensioned to be a Cone-of-Silence antenna for 160, will just about be a Cone-of-Silence antenna on 80 also. (Not 100% true as the velocity factor of the antenna is slightly different on 80 and 160.)

There is no guarantee, however, that an optimized Slinky (one giving best F/B) will also be optimized for 80 meters. Chances are slim this can be done. As, however, we can put up 2 Beverages quite close together without inter-action, I would suggest not to try to load a Beverage to get Cone-of-Silence performance on both 160 and 80 meters. It will certainly be much easier to put two such loaded Beverages, side by side, and adjust each one for optimum F/B.

What to remember

1. A random Slinky will most likely *not* work—unless you're *very* lucky!
2. A well tuned Slinky always has less gain (output) than the full-size equivalent (same electrical length). The output is typically 3 to 5 dB down from its full-size $\frac{1}{2}$-λ long brother (depending on its physical length), but that can mostly be compensated with a good low-noise preamp.
3. A well tuned Slinky can have as good an F/B as its Cone-of-Silence full-size equivalent (25 dB at 25° wave angle).
4. Calculating your 160-meter Slinky is quite easy:
   • What length can you put up? Say 53 m.
   • Determine if this is the Cone-of-Silence length (86 m).
   • Calculate the required velocity of propagation: Vp = 53/86 = 0.62.
   • Read the S/D ratio from Fig 7-56: S/D = 1.1. For a 7-cm diameter Slinky the pitch is 7.7 cm.
   • Calculate the number of turns required: 53 m / 7.7 cm = 688 turns.
   • Calculate the number of Slinkies required (at 85 turns per Slinky): 688/85 = 8.
5. Tune the Slinky for best F/B. This is a *must*. Use a signal source exactly in line with the geometric back of the antenna (several wavelengths away). Adjust the pitch to

obtain maximum F/B.

6. You *can* put two in phase even at relative close spacing and 3-dB gain!

7. A Slinky must be terminated in a 500-Ω resistor, and fed via a 9:1 transformer as described in Section 8.

8. As it is an even lower output antenna than a Beverage, decoupling and proper grounding of the feed line is very important (see Section 9.2).

Last but not least, you can achieve exactly the same effect (an excellent F/B), with less loss, by simply terminating the straight-wire antenna (of the same length) in a complex load (R + L in series). See Section 6.3. Bye-bye Slinky?

19.2. The Snake Antenna, or Beverage On Ground (BOG) Antenna

The Snake antenna gets its name from what it looks like: it is usually a coaxial cable just stretched out on the ground like a giant reptile. The antenna is said to have properties similar to those of a genuine Beverage antenna; it receives off the far end of the Snake, at least if we may believe all that's been published. DeMaw, W1FB (Ref 1254) described different versions, and gives us some rather interesting to read material. .

One type DeMaw describes is short circuited at its end. He states the cable will represent a 50-Ω load at the receiver, provided the line is lossy enough. It happens that at least 20 dB is required for a short circuited line to appear like a matched line at the other end. Well, that's going to have to be a very long Snake, if you know that even a mediocre RG-8 cable exhibits 0.5-dB per 100 ft on 1.8 MHz. Yes, to get the 20-dB attenuation you would need a 4000-ft long Snake. DeMaw also says that the velocity factor of the coaxial line should be taken into account when constructing the antenna.

Let's leave the tales for what they are worth and look at the facts. Seen from the outside (by the radio-wave we want to capture) the shield of the coax acts as a single (fat) wire antenna. Obviously all goes on the *outside* of the coax. The velocity factor of the coax has nothing to see with the coax acting as a fat-wire antenna, contrary to what DeMaw states. The velocity factor of the coax relates to the coax acting as a transmission line, and then everything happens *inside* the coax.

Let us look at this very low, fat Beverage. Its impedance is in the 200-Ω range. It is that high because of the losses of the ground. Over a perfect conductor it would be as low as 50 Ω. The Snake has a very low velocity factor and of course, very low output, but it still retains essentially the same directivity characteristics as a Beverage at more usual heights. A historic note: in the very first test H. Beverage did, he laid the antenna flat on the ground. Only later did he elevate the wire!

Fig 7-58A shows how we can feed a Snake antenna and use the coax as antenna and as a transmission line. In Section 12.1, I explained how a coax could be used to feed a Beverage from the far end. The Snake is such an antenna, laying on the ground. At the end, we need a reflection transformer as described in Section 12.1. (200 Ω to 50 Ω) At the near end of the Snake we need to put the terminating resistance, which will dissipate the induced voltage from the waves received from the opposite direction (back). This shall be a 200-Ω resistor connected to a good RF ground. As we are dealing with an antenna having very low output we must take all precautions for removing common-mode current on the outside of the coaxial line between the termination point (T) and the receiver. A large number of ferrite beads on both sides of another good RF ground will do the job (see also Section 9.2 and Fig 7-19).

If we want to receive off the other end, we need to

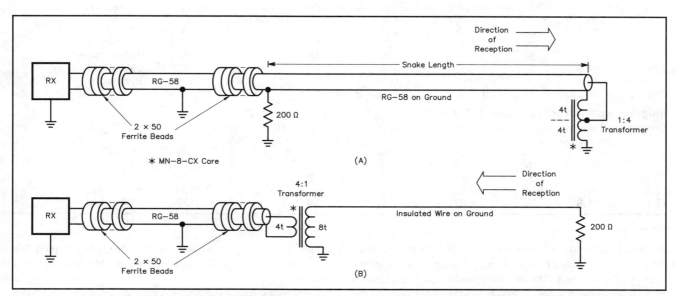

Fig 7-58—At A, the so-called Snake antenna where the coax on the ground actually serves as feed line bringing the received signals off the back of the antenna to the receiver. Careful decoupling of the feed line is required (see text). At B is the simplified Snake using a simple insulated wire on the ground. In this case reception is off the far end of the antenna.

Table 7-15
Data for 160-Meter EWEs

The EWE antenna typically exhibits 1 deep null in the back and a back lobe. The position of the null and the back lobe and their rejection are listed in the tables. See also Fig 7-59. (Best lengths are shown in **bold**.)

Height (length vertical members): 3.0 m, Frequency: 1.83 MHz, Ground: Good

Length (m)	Gain (dBi)	Wave Angle (°)	Null Angle (°)	Null Reject (dB)	Back Lobe Angle (°)	Back Lobe Reject. (dB)	R_{term} Ω	Z_{ant} Ω
5 m	−30.4	30	57	26	19	12	1300	704 −j 492
7.5 m	−27.6	31	47	33	18	16	1300	580 −j 520
10 m	−25.4	32	38	>35	21	15	1300	486 −j 477
12.5 m	**−23.7**	**33**	**29**	**31**	**14**	**27**	**1300**	**414 −j 426**
15 m	−22.1	33	25	25	14	26	1300	350 −j 383

Height (length vertical members): 5.0 m, Frequency: 1.83 MHz, Ground: Good

Length (m)	Gain (dBi)	Wave Angle (°)	Null Angle (°)	Null Reject (dB)	Back Lobe Angle (°)	Back Lobe Reject (dB)	R_{term} Ω	Z_{ant} Ω
5 m	−26.3	29	62	33	20	10	1000	597 −j 357
7.5 m	−23.8	30	53	>35	19	13	1000	538 −j 338
10 m	−21.9	31	46	>35	17	17	1000	479 −j 311
12.5 m	**−20.4**	**32**	**38**	**33**	**15**	**20**	**1000**	**435 −j 278**
15 m	−19.1	33	30	30	15	25	1000	401 −j 244

Height (length vertical members): 3.0 m, Frequency: 1.83 MHz, Ground: Very Good

Length (m)	Gain (dBi)	Wave Angle (°)	Null Angle (°)	Null Reject (dB)	Back Lobe Angle (°)	Back Lobe Reject (dB)	R_{term} Ω	Z_{ant} Ω
5 m	−30.9	22	58	>35	16	11	900	678 −j 295
7.5 m	−28.2	23	51	>35	15	14	900	610 −j 303
10 m	−26.1	24	46	34	14	16	900	550 −j 265
12.5 m	**−24.6**	**24**	**39**	**30**	**12**	**18**	**900**	**497 −j 281**
15 m	−23.3	25	34	26	11	21	900	453 −j 258

Height (length vertical members): 5.0 m, Frequency: 1.83 MHz, Ground: Very Good

Length (m)	Gain (dBi)	Wave Angle (°)	Null Angle (°)	Null Reject (dB)	Back Lobe Angle (°)	Back Lobe Reject (dB)	R_{term} Ω	Z_{ant} Ω
5 m	−26.2	22	60	>35	15	9	750	594 −j186
7.5 m	−23.8	23	55	>35	14	12	750	561 −j183
10 m	−22.0	23	49	>35	13	14	750	530 −j175
12.5 m	**−20.6**	**24**	**45**	**33**	**13**	**17**	**750**	**504 −j163**
15 m	−19.4	24	38	31	12	19	750	481 −j147

Height (length vertical members): 3.0 m, Frequency: 1.83 MHz, Ground: Poor

Length (m)	Gain (dBi)	Wave Angle (°)	Null Angle (°)	Null Reject (dB)	Back Lobe Angle (°)	Back Lobe Reject (dB)	R_{term} Ω	Z_{ant} Ω
5 m	−30.1	33	57	24	21	12	1800	649 −j 763
7.5 m	−27.1	35	47	29	20	16	1800	501 −j 678
10 m	−24.8	36	37	>35	16	22	1800	400 −j 592
12.5 m	**−22.9**	**37**	**22**	**>35**	**11**	**32**	**1800**	**329 −j 511**
15 m	−21.4	38	20 (*)	20 (*)	10	24	1800	278 −j 437

(*) no back lobe, values measured at 20° and 10° wave angle

Height (length vertical members): 5.0 m, Frequency: 1.83 MHz, Ground: Poor

Length (m)	Gain (dBi)	Wave Angle (°)	Null angle (°)	Null Reject (dB)	Back Lobe Angle (°)	Back Lobe Reject (dB)	R_{term} Ω	Z_{ant} Ω
5 m	−26.3	33	62	28	23	9	1300	560 −j 504
7.5 m	−23.8	34	54	34	20	13	1300	477 −j 458
10 m	−21.8	35	47	>35	19	16	1300	413 −j 407
12.5 m	−20.2	36	39	>35	17	21	1300	364 −j 355
15 m	**−18.9**	**37**	**27**	**32**	**14**	**27**	**1300**	**327 −j 304**

construct our Snake as shown in Fig 7-58B. In this case we can save some money and use any *insulated* wire (not a coax) as the antenna. The on-the-ground antenna wire will now be terminated at its far end by a 200-Ω resistor. At the near end the antenna wire can be connected via a 4:1 transformer (bifilar wound) to the inner conductor of the coaxial feed line. Here too, we will need to ground the end of the feed line, and a series of ferrite beads on the feed line will discourage any RF currents from "messing up" our Snake.

We can make the Snake exactly as the antenna from Fig 7-22B, in which case we will have 2 directions available at the same time. The differences between this and the "high" Beverage antenna as described in Section 12.1 is that

Table 7-16
EWE Antenna Data for 80 Meters
Best lengths are shown in bold. Lengths shown in italics should be avoided.

Height (length vertical members): 3.0 m, Frequency: 3.65 MHz, Ground: Good

Length (m)	Gain (dBi)	Wave Angle (°)	Null Angle (°)	Null Reject (dB)	Back Lobe Angle (°)	Back Lobe Reject (°)	R_{term} Ω	Z_{ant} Ω
5 m	−20.4	34	4	>35	18	16	975	413 −j 275
7.5 m	**−17.7**	**36**	**31**	**>35**	**15**	**26**	**975**	**351 −j 208**
10 m	−15.7	38	20 (*)	27	10	>35	975	315 −j 137
12.5 m	*−14.2*	*42*	*20 (*)*	*18*	*10 (*)*	*22*	*975*	*297 −j 67*
15 m	−13.1	46	20 (*)	18	10 (*)	14	975	295 −j 3

(*) no back lobe, values measured at 20° and 10° wave angle

Height (length vertical members): 5.0 m, Frequency: 3.65 MHz, Ground: Good

Length (m)	Gain (dBi)	Wave Angle (°)	Null Angle (°)	Null Reject(dB)	Back Lobe Angle (°)	Back Lobe Reject (°)	R_{term} Ω	Z_{ant} Ω
5 m	−16.0	33	55	>35	20	12	800	396 −j 140
7.5 m	**−14.0**	**35**	**43**	**>35**	**17**	**14**	**800**	**376 −j 82**
10 m	−12.4	37	30	>35	13	12	800	371 −j 25
12.5 m	*−11.2*	*41*	*30*	*20 (*)*	*10 (*)*	*35*	*800*	*380 −j 31*
15 m	−10.3	45	21	20 (*)	10 (*)	26	800	404 −j 84

(*) no back lobe, values measured at 20° and 10° wave angle

Height (length vertical members): 3.0 m, Frequency: 3.65 MHz, Ground: Very Good

Length (m)	Gain (dBi)	Wave Angle (°)	Null Angle (°)	Null Reject (dB)	Back Lobe Angle (°)	Back Lobe Reject (°)	R_{term} Ω	Z_{ant} Ω
5 m	−20.2	26	49	>35	15	14	700	476 −j 156
7.5 m	**−17.7**	**27**	**38**	**>35**	**14**	**19**	**700**	**437 −j 124**
10 m	−15.9	29	27	>35	11	26	700	410 −j 85
12.5 m	*−14.6*	*31*	*20 (*)*	*30*	*10 (*)*	*35*	*700*	*398 −j 42*
15 m	−13.8	35	20 (*)	22	10 (*)	27	700	398 +j 1

(*) no back lobe, values measured at 20° and 10° wave angle

Height (length vertical members): 5.0 m, Frequency: 3.65 MHz, Ground: Very Good

Length (m)	Gain (dBi)	Wave Angle (°)	Null Angle (°)	Null Reject (dB)	Back Lobe Angle (°)	Back Lobe Reject (°)	R_{term} Ω	Z_{ant} Ω
5 m	−15.5	25	56	> 35	15	16	600	486 −j 41
7.5 m	**−13.3**	**27**	**46**	**35**	**14**	**14**	**600**	**484 −j 28**
10 m	−11.8	29	39	33	13	13	600	489 −j 7
12.5 m	*−10.8*	*31*	*32*	*30*	*12*	*12*	*600*	*501 +j 11*
15 m	**−10.0**	**34**	**27**	**26**	**11**	**11**	**600**	**517 +j 25**

Height (length vertical members): 3.0 m, Frequency: 3.65 MHz, Ground: Poor

Length (m)	Gain (dBi)	Wave Angle (°)	Null Angle (°)	Null Reject (dB)	Back Lobe Angle (°)	Back Lobe Reject (°)	R_{term} Ω	Z_{ant} Ω
5 m	−19.9	41	42	>35	19	19	1600	284 −j 401
7.5 m	**−16.7**	**44**	**49**	**>35**	**7**	**>35**	**1600**	**225 −j 283**
10 m	−14.4	47	20 (*)	18	10 (*)	23	1600	194 −j 181
12.5 m	*−12.5*	*52*	*20 (*)*	*14*	*10 (*)*	*18*	*1600*	*179 −j 86*
15 m	−11.0	58	20 (*)	11	10 (*)	15	1600	177 +j 4

(*) no back lobe, values measured at 20° and 10° wave angle

Height (length vertical members): 5.0 m Frequency: 3.65 MHz, Ground: Poor

Length (m)	Gain (dBi)	Wave Angle (°)	Null Angle (°)	Null Reject (dB)	Back Lobe Angle (°)	Back Lobe Reject (°)	R_{term} Ω	Z_{ant} Ω
5 m	−16.7	39	52	>35	22	14	1150	293 −j 207
7.5 m	**−14.2**	**42**	**35**	**>35**	**16**	**23**	**1150**	**269 −j 121**
10 m	−12.3	46	20 (*)	29	10 (*)	>35	1150	260 −j 37
12.5 m	*−10.9*	*50*	*20 (*)*	*19*	*10 (*)*	*24*	*1150*	*167 +j 43*
15 m	−9.8	55	20 (*)	14	10 (*)	19	1150	291 +j 125

the antenna impedance is typically around 200 Ω instead of 500 Ω, the velocity factor is much lower, typically 0.7, and its output is down about 10 to 15 dB.

It is obvious that in both cases the coaxial feed lines will be properly connected (as a feed line) to the receiver, which needs to be properly grounded (see Section 9.2).

This Snake antenna, like every Beverage antenna, will have an optimum Cone-of-Silence length. Because of its low height (low Vp), we can expect 0.5-λ Cone-of-Silence operation on 160 meters with lengths as short as 57 m. Paul, N5FY, reported a length of about 67 m being optimal at his QTH with his ground conductivity.

What to remember about the Snake antenna

- A piece of coax thrown on the ground, in any of the fashions described by DeMaw will not work like a Beverage antenna.
- A proper Snake requires good grounds at both ends.
- To provide directivity, a Snake needs to be tuned to its Cone-of-Silence length.
- Feeding a Snake is even more delicate than feeding a regular Beverage as signals are much weaker, and common-mode currents can more easily do harm.
- The properly dimensioned Snake has the same pattern as the Beverage.
- The Snake will more than likely require a preamp.

■ 20. THE EWE RECEIVING ANTENNA

Floyd Koontz, WA2WVL, is considered the "father" of the EWE. Reading his publications (Ref 1263 and 1264) are musts for anyone who wants to get his feet wet with this novel receiving antenna.

An EWE can be designed to cover 80 and 160, is highly directional and is small. Moreover, the EWE is low profile and can be built for little money. In appearance, the EWE resembles a very short Beverage, though in fact it is an array of two short vertical antennas, where the horizontal wire is part of the (radiating) feed system. The horizontal wire is only about 7.5 to 15-m long and is about 3 to 5 m above ground. In other words, an EWE can fit in many tiny yards. If there is a lot of space available, several can be phased for additional directivity.

But before getting deeper into EWEs, don't forget that, just like with Beverage antennas, the directivity patterns of EWEs can be totally destroyed by mutual coupling with transmit antennas (see Section 13).

20.1. What is a EWE?

The EWE has been described in the few publications that have covered this antenna, as a simple pair of verticals with a horizontal feed line, one being base fed, the other fed via the top. That is considered to provide crossfire phasing, which, when the element spacing is less than $1/4 \lambda$, always fires toward the feed-point end of the array. Some observers have noticed that the current in the reflector is only approx. 70% of the current in the driven vertical, and wonder how, under such circumstances, such a good F/B can be obtained. They even invoked the radiation of the top wire as being responsible for that. But this is not the full story.

Looking at the current data of the antenna (obtained through modeling), we note that for a typical EWE the current amplitude in the reflector is indeed only about 70% of the current amplitude in the fed element. With a base-fed vertical this would mean that you could not get more than approx. 8-dB F/B.

With the EWE though, the terminating resistor makes the Vp (velocity of propagation) in the reflector leg much lower than in the driven vertical. In the case (over good ground) of an EWE measuring 5-m high by 15-m (separation), the 5-meter long driven element has an electrical

length of 9.3° ($V_f = 0.83$), the horizontal wire has an electrical length of approx. 40° ($V_f = 0.83$) while the reflector, also measuring 5 m in length has an electrical length of not less than 18° ($V_f = 0.63$). This is accomplished by the terminating resistor. The higher the value of the resistor, the slower the wave in the reflector.

The net effect of the slower wave is that, despite the lower current amplitude in the reflector, a very good F/B can be achieved. Over poor ground the velocity factor in the fed vertical and the horizontal wire will be lower to start with, and consequently the terminating resistor will need to be of a higher value to achieve conditions where a good F/B can be obtained.

It is impossible to make a simple comparison with two phased bottom-fed verticals. Due to the terminating resistor the antenna works much more as a traveling-wave antenna than as a standing-wave antenna. The antenna current is nearly uniform or slowly tapers (due to loss from radiation and resistance) in the EWE antenna system and contains no, or greatly suppressed, "end reflection." Therefore it has no standing waves and can be called a traveling-wave system. This is also why the antenna is very broadbanded, and exhibits a good F/B on both 160 and 80 meters.

20.2. Modeling EWEs

EWEs are easy to model, and any *MININEC*-based program (eg, *ELNEC*) will do the trick. EWEs have been modeled with *NEC-4* and the results confirm what we obtain with *MININEC*. When you model grounded verticals, with the feed point at the ground, you normally must insert a resistance at that point, which must be equivalent to the ground losses (*MININEC* assumes the ground in the near field, used for computing impedances) are zero. With a EWE model this is not really necessary as you already have a 700 to 2000-Ω load anyhow. Adding a 30 or 50-Ω (loss) resistance in series with the feed point will hardly change anything in signal output of the antenna.

There is nothing magic about the dimensions of the EWE. You can easily brew your own if you have a *MININEC*-based modeling program. Just enter the dimensions, and change the value of the terminating resistance until the best F/B is obtained.

The verticals are usually 3 to 5-m high, and the spacing (length of the horizontal wire) between 5 and 15 m.

I modeled a range of configurations over 3 types of ground (Good, Very Good and Poor). The relevant data are shown in **Table 7-15** (160 meters) and **Table 7-16** (80 meters).

Discussion of 160-meter EWEs

- Going from 3 m to 5-m vertical elements increases the gain by 3 to 4 dB.
- Taller EWEs require a lower R-term (resistance termination).
- Going from good ground to very good ground lowers the wave angle by approx. 8° (+ lower R-term).
- Going from good to poor ground raises the wave angle by approx. 4° (+ higher R-term).

- Close spacing (short horizontal wire): better high-angle rejection.
- Wide spacing (long horizontal wire): better low-angle rejection.

Discussion of 80-meter EWEs

- Going from 3 m to 5-m vertical elements increases the gain by 2 to 4 dB.
- Taller EWEs require a lower R-term.
- Going from good ground to very good ground lowers the wave angle by approx. 8° (+ lower R-term).
- Going from good to poor ground raises the wave angle by approx. 7° (+ higher R-term).
- Close spacing (short horizontal wire): better high-angle rejection.
- Wide spacing (long horizontal wire): better low-angle rejection.
- Spacings over 10 m are to be avoided for 80 meters.

For a single-band 160-meter EWE, I would recommend the configuration with 5-m long vertical elements and 12.5-m spacing. For 80 meters, the 3-m tall, 7.5-m long EWE seems to be the best choice. If you go for a duo-band EWE receiving antenna, a spacing of 10 m with a height of either 3 m or 5 m is the logical compromise.

Fig 7-59 shows the vertical and horizontal radiation pattern for a 12.5 × 5 m EWE and a 10 × 3 m EWE as compared to a 0.5-λ Beverage. Note that the shapes of the radiation patterns are very similar, but the Beverage has substantially more output. The peak radiation angle is actually somewhat higher in the short Beverage (50°) than it is with the 2 EWEs (30°).

20.3. Feeding the EWE

The feed-point impedance of all EWEs varies from approx. 300 Ω to approx. 700 Ω, and in many cases the feed point shows a substantial reactive component. In most cases a 9:1 transformer will achieve an acceptable match to the feed line, provided you take care of canceling the reactive component. In most cases, the capacitive reactance can be "tuned-out" by connecting a coil between the feed point of the EWE and the 9:1 transformer.

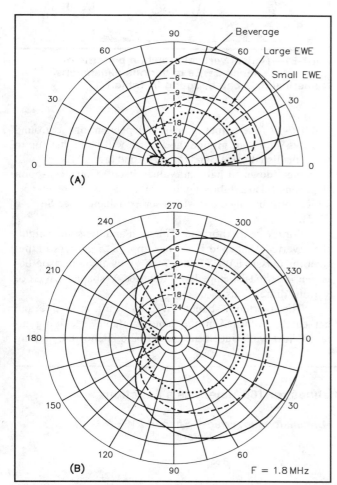

Fig 7-59—Typical radiation patterns for a "small" (3 × 11 m) and a large (5 × 15 m) EWE compared to the pattern of a half-wave Beverage. The patterns are quite similar, but the Beverage has substantially more output.

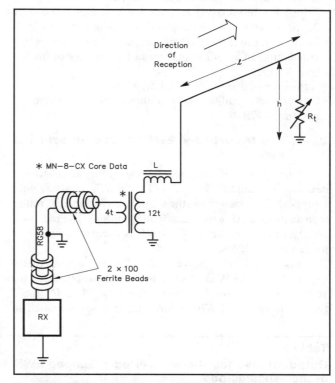

Fig 7-60—Typical setup of a EWE antenna. Note that the reception is off the end where the antenna is fed. In order to obtain a reasonably low SWR on the feed line one can tune out (in most cases) the capacitive reactance in the feed-point impedance by adding a series coil as shown. The classic 9:1 transformer, as used with Beverage antennas will provide an adequate impedance match to the feed line. The necessary precautions must be taken to prevent common-mode (Snake) currents on the outside of the coaxial feed line to "sneak" into the feed system.

The value of the coil is:

$$L = \frac{X}{2\pi F}$$

where:

X = the reactance in Ω (the value after the "*j*" in the complex impedance)

F = operating frequency in MHz

L = coil inductance in μH

Example: Z = 437 $-j$124 Ω. The coil value will be:

$$\frac{124}{2\pi \times 1.83} = 10.8\ \mu H$$

As the output of a EWE is even lower than that from a Beverage, the precautions described in Section 9.2 apply even more. **Fig 7-60** shows a typical layout. To adjust the EWE proceed as follows:

• Connect the 9:1 transformer as shown, but leave the series inductor out for the moment.

• Put a small signal source in the direction of the null (right off the back, not at a wave angle) at several wavelengths. Contrary to what's the case with a Beverage antenna, the back is where the feed point is!

• Adjust the terminating resistor for best F/B (should be 35 dB or better). Do *not* adjust this resistor for best SWR!

• Insert a variable coil at the antenna feed point, while connecting an antenna analyzer at the primary of the 9:1 transformer.

• Adjust the coil value for minimum SWR.

• If necessary, readjust the terminating resistor value for maximum F/B.

20.4. The Inverted-V EWE vs the Inverted-U EWE

During one of his modeling sessions, Earl, K6SE, developed a simplified version of the EWE, requiring only 1 support. The antenna has the shape of an inverted V, rather than an inverted U in its classic configuration. The radiation pattern of this version of the EWE is exactly the same as for the classic EWE.

Fig 7-61 shows the vertical radiation patterns of inverted-V shaped EWEs of various dimensions (modeled for 160 meters over good ground). The other data for these typical inverted-V EWEs (over good ground) are listed in

Table 7-17

Data for the Inverted-V Shaped EWE Antenna of Various Dimensions (Data for 160 Meters Over Good Ground)

For 80 meters, divide dimensions by 2 for the same results

Dimensions (H × L)	Gain	R_{term} (Ω)	Z_{ant} (Ω)
3 m × 9 m	−30.1 dBi	2100 Ω	556 −j823
5 m × 12 m	−25.0 dBi	1400 Ω	471 −j506
6 m × 15 m	−22.2 dBi	1300 Ω	404 −j405
8 m × 16 m	−19.8 dBi	1100 Ω	399 −j297

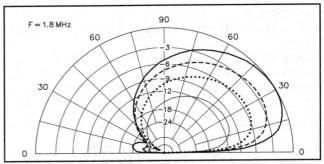

Fig 7-61—Typical vertical radiation patterns of inverted-V shaped EWEs of various dimensions. Numerical data can be found in Table 7-17.

Table 7-17. The output is quasi-independent of the ground quality. With poor ground the R-term will be higher (up to twice the listed value), and for very good ground R-term will be lower (down to half the values listed). The feed-point impedance also exhibits a good deal of capacitive reactance, which must be tuned out with a series inductor (see Section 20.3).

For the same (horizontal) length and the same height, the inverted-V shaped EWE has approx. 5 dB less output than the classic EWE. **Table 7-18** gives a comparison between inverted-U shaped and inverted-V shaped EWEs of various dimensions.

The inverted-V shaped EWEs look quite attractive, and can easily be turned into a "hand-rotatable" receiving antenna: just walk out in the garden and anchor the sloping

Table 7-18

Output of Inverted-U and Inverted-V Shaped EWE Antennas for 160 Meters as a Function of their Dimensions

Gain	Dimension Inverted-U-Shaped EWE (H × L) In Meters	Dimensions Inverted-V-Shaped EWE (H × L) In Meters
−20 dBi	5 × 12.5	8 × 16
−22 dBi	3 × 15 and 5 × 10	6 × 15
−24 dBi	3 × 12.5 and 5 × 7.5	
−25 dBi	3 × 10	5 × 12
−26 dBi	5 × 5	
−27 dBi		6 × 24
−29 dBi		6 × 20
−30 dBi	3 × 5	3 × 9

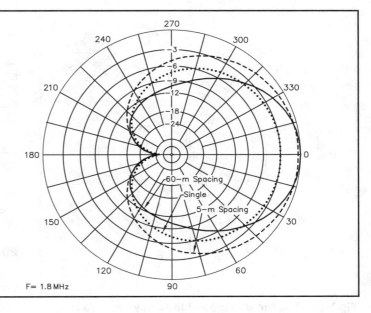

Fig 7-62 Horizontal radiation pattern (at 25° wave angle) for a broadside (fed in phase) pair of EWEs. Close spacing gets you 3-dB signal gain but no lobe shaping (like with Beverages). Wide spacing gets you the gain and narrower forward lobe, at the expense of a lot of real estate.

F = 1.8 MHz

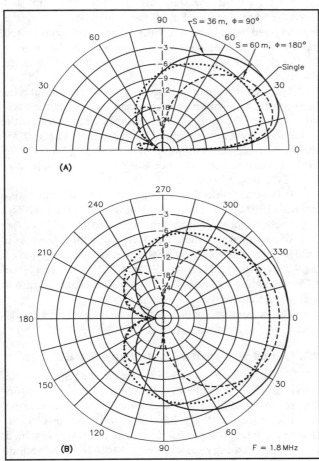

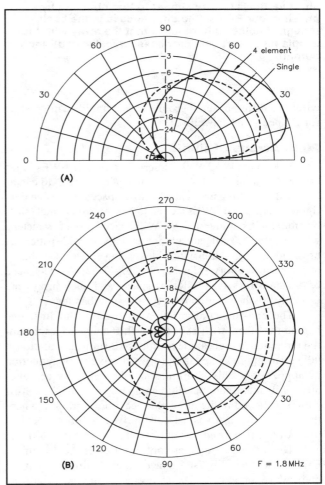

Fig 7-63—Two examples of end-fire EWE arrays. The classical quarter-wave spacing, 90° phase shift configuration gets you 3-dB gain and a little narrower lobe. Increasing the spacing to 0.375 λ with a 180° phase shift reduces the high-angle sensitivity, and creates a dep null at 90° in the horizontal pattern. In this case the gain is limited to approx. 1.5 dB over a single EWE.

Fig 7-64—With a 4-element EWE you can obtain some staggering horizontal directivity patterns, but at the expense of a lot of real estate. The vertical pattern, however, keeps suffering from the high angle radiation caused by the horizontal wire, and is far from ideal.

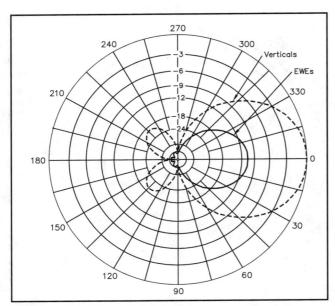

Fig 7-65—The forward lobe of the array using four short verticals is very similar to that obtained by phasing four EWEs. The difference is in the back, though. Another difference is that the array with four simple lossy verticals produces typically 10 dB more output!

wires to a different set of ground rods. Should be interesting for DXpeditions as well.

20.5. Arrays of EWEs

F. Koontz, WA2WVL, described arrays of EWEs in his *QST* publication (Ref 1264). A signal gain of 2 to 3 dB can be obtained by putting up a pair. In close-spaced broadside configuration (the two EWEs are fed in phase) two parallel (side-by-side) EWEs or inverted-V EWEs spaced just a few meters, will give almost 3-dB gain, with essentially the wide pattern. A broadside configuration with wide spacing is helpful in obtaining a narrower forward lobe (see **Fig 7-62**). Wide-spaced broadside arrays take a lot of space, and such arrays are certainly not the solution for a typical suburban lot.

The end-fire configuration (two EWEs in line, one behind the other) with 180° phase shift achieves a substantial reduction in high-angle reception (infinite attenuation at 90°) resulting in an apparent forward wave angle reduction of approx. 5°. This configuration though, only achieves 50% of its maximum possible gain (1.5 vs 3 dB). The 3-dB gain is achieved in the more classical configuration, where the two arrays are spaced (in-line) by a little less than 0.25 λ (36 m) and fed 90° out-of-phase (see **Fig 7-63**).

The ultimate on-paper configuration is the 4-element EWE as shown in **Fig 7-64**. The horizontal directivity at the main wave angle of 25° is awesome. If anyone would be tempted to build this array, let me point out that its 3-dB forward opening angle is only 53°. This means that, if you want to cover the entire azimuth, you will require some 10 to 12 such arrays! Even if you feel tempted to set up a single such array to your main target area (eg, from W6 to

Europe), where will you point it? Don't forget signals often skew as much as 45° on these paths (see Chapter 1).

If we go through the trouble of putting up 4 EWEs, and have the space to do it, then just 4 short verticals are almost as good, and certainly will produce much more signal. **Fig 7-65** shows the radiation pattern obtained by phasing four 6-meter short verticals (eg, Gladiator verticals). Over a normal ground system (a few radials) these verticals have an efficiency of a few percent, which is ideal for a receiving antenna (not for a transmit antenna though!), as the effect of mutual coupling is very low (most of the feed-point impedance is losses). This simple array uses quadrature-fed elements (in increments of 90°), and can be fed via 0.75-λ feed lines to a hybrid coupler as described in Section 15.3.1. (See Fig 7-40-C.)

20.6. A Loop-Shaped EWE?

EA3VY was the first to come up with the idea of adding a ground wire between the bottom of the two verticals as an effective way to minimize the effect of different soil conductivities. I went one step further, added the wire, and lifted the whole antenna 1 m off the ground. It now was like an elongated loop. I was not really surprised when it turned out that the radiation pattern remained exactly the same as for the classic EWE. I immediately considered a rotatable EWE.

The radiation pattern looks exactly like the classic (grounded) EWE, but its performance, as well as the optimum value of the termination resistance, seems to be little influenced by varying ground conditions. **Table 7-19** shows the main data compiled from modeling loop EWEs for 3 different dimensions. See **Fig 7-66** for the configuration of the loop EWEs.

It is remarkable that the optimum termination resistance is 1100 Ω for all situations. Note that the antenna is fed in the bottom corner, but termination is in the *middle* of the opposite vertical member. The feed-point impedance is a little high for a simple 9:1 transformer, so a 16:1 version of the 9:1 transformer is required. The SWR on either a 50 or a 75-Ω feed line will be below 1.3:1 on both bands, without tuning out the reactive component.

The bottom wire which closes the loop, makes it possible to install a rotatable EWE!

20.7. And the Delta-Loop Type EWE

The delta-loop type EWE has as clean a pattern as the classic EWE. I think this antenna is really attractive for DXpeditions and for semi-rotatable setups. Just moving the base line around is all you have to do to change directions. During modeling it appeared that the best F/B was obtained with the terminating resistor mounted 10% from the bottom corner of the delta loop. The antenna is fed in the opposite bottom corner. **Table 7-20** shows some typical data for this novel antenna. The same remarks apply for feedings as in the case of the loop-shaped EWE (Section 20.6).

More shapes are available. Earl Cunningham, K6SE, and Jose Matta Garriga, EA3VY, developed what they call the pennant-shaped loops, which is really like half a delta

Table 7-19
Typical Characteristics of Loop-Type EWEs for 160 and 80 Meters

Dimension (H × L) in m.	Gain (dBi)	1.83 MHz R_{term}	Z (Ω)	Gain (dBi)	3.65 MHz R_{term}	Z (Ω)
5 × 10	−30.2	1100	$1050 - j121$	−19.6	1100	$931 - j107$
6 × 12	−26.8	1100	$1039 - j111$	−16.4.	1100	$943 - j64$
8 × 15	−22.2	1100	$1038 - j86$	−12.2	1100	$1012 - j21$

Table 7-20
Typical Characteristics of Delta Loop-Shaped EWEs for 160 and 80 Meters

Dimension (H × L) in m.	Gain (dBi)	1.83 MHz R_{term}	Z (Ω)	Gain (dBi)	3.65 MHz R_{term}	Z (Ω)
5 × 12	−32.3	900	$967 + j12$	−21.6	900	$1008 + j9$
7 × 15	−27.6	1100	$1019 - j54$	−17.0	1000	$1006 - j34$
9 × 20	−23.1	1100	$1006 - j34$	−13.1	1000	$1098 - j50$

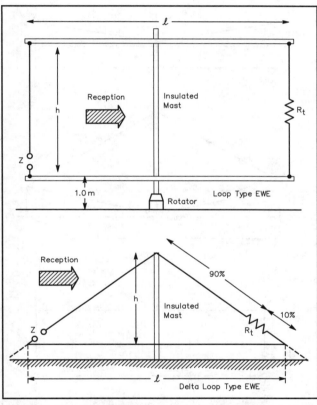

Fig 7-66—Configuration of the rectangular and delta-shaped EWE loop. See text for details.

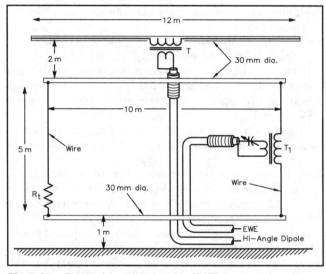

Fig 7-67—Rotatable rectangular EWE loop with compensating dipole mounted above the loop. Both rotate together. The output from the two antennas are fed to a combiner where signals are added with variable phase and amplitude. An MFJ-1026 Signal Enhancer can be used for this application.

loop. The loop is fed in the center of the vertical section, while the load is situated where the horizontal and the sloping wire meet (bottom corner). It looks like the only limit as to shapes and dimensions of these kind of loops is one's imagination.

20.8. Suppressing High Wave Angle Response

The major drawback of any EWE is its high-angle response, caused by the radiation from the horizontal wire(s), or from the horizontal component in case of slant wires. We can compensate this by putting a small dipole above the EWE, and by feeding the output of this low dipole 180° out of phase with the output of the EWE.

Fig 7-67 shows a possible setup. The rotatable EWE measures 5 m (high) by 10 m (wide), and has the bottom wire 1 meter above ground. A 10-m long dipole is mounted 1 m above the EWE, oriented parallel to the horizontal wires of the EWE. This antenna should pick up enough high-angle horizontally polarized signal to allow canceling with a unit such as the MFJ-1026 device. The dipole must be parallel to the loop and rotate together with the loop.

The dipole is resonated to the frequency of operation. The 10-m long dipole, with a diameter of 3.0 cm requires an inductor of approx. 260 µH to be resonated to 1.83 MHz. The easiest way to do this is to wind the inductor on a high-µ ferrite core or rod and wind a link to achieve a good match to the feed line. The feed line should run inside the aluminum rotating mast to the bottom of the array.

The loop must be fed in the center of the vertical wire. The coaxial feed line (equipped with common-mode chokes—eg, ferrite beads) runs perfectly horizontal right to the rotating mast, in order to reduce unwanted coupling and unbalance.

According to the modeling results, the impedance of the loop is approx. $750 + j500 \ \Omega$, and a terminating resistance of 600 Ω is required for best F/B. This means that we will require a series capacitor to tune out the inductive reactance and achieve a low SWR on the feed line.

Note that, without the dipole, the optimum terminating resistor is approx. 1000 Ω, with a feed-point impedance of 900 Ω (no reactive component).

Tuning the rotatable array with compensating dipole:
- Construct the loop, connecting the 9:1 transformer and the terminating resistor (use a 1-kΩ potentiometer). Set the value to 525 Ω.
- Install the 10-m long dipole 2 m above the loop, and resonate it with the center-loading coil. Feed the dipole via a link. Use an antenna analyzer to obtain the proper turns for the link. If necessary (if the Q of the loading coil is "too" good), add a series resistor to achieve a good SWR (<1.5:1).
- With the dipole tuned, adjust the value of the series capacitor (approx. 200 pF) for a good SWR (<1.5:1) on the feed line to the loop.
- Put a signal generator at least 5 λ from the antenna (1 km!) at an angle of 30° with respect to the geometric back of the array (150 or 210°), and change the terminating resistor for max. rejection.
- Adjust the value of the series capacitor (loop feed point) for minimum SWR.
- Run two feed lines to a phasor/combiner; eg, the MFJ 1026.
- Adjust the phasor unit for best high-angle rejection. This adjustment is critical, as you can completely ruin the directional characteristics of the antenna when the phasing is not done correctly.

■ 21. THE K9AY LOOP

Gary Breed, K9AY, covered the loop very well in his *QST* article (Ref. 1265). In Section 20, I sketch different variants of the classic EWE: the inverted-V shaped EWE, the square-loop EWE and the delta-loop EWE.

K9AY designed another variant of the EWE, where the bottom wire of the loop is grounded in the center. He points out in his article that the loop can really be any shape. The diamond shape he uses is probably dictated by practical construction considerations rather than anything else. All of the K9AY loops can be used both on 160 and 80 meters, although optimal performance may require slight adjustment of the terminating resistance, as pointed out in his article.

21.1. The Evolution

Fig 7-68 shows how we can evolve from the inverted-V shaped EWE (see Section 20.4) to the well-known K9AY loop.

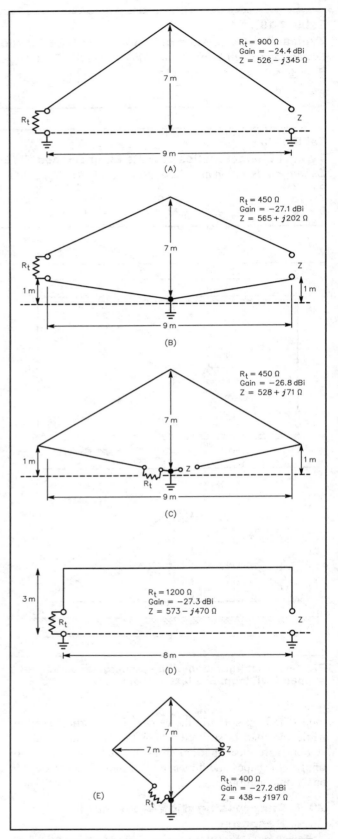

Fig 7-68—Variations on an identical theme: the EWE (D) and some of its derivatives, leading to the classic K9AY loop (at C). See text for details.

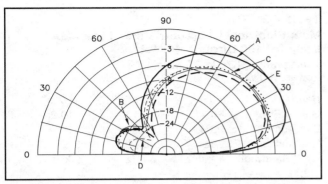

Fig 7-69—Vertical radiation patterns of various forms of the EWE. At A: inverted-V EWE, at B: corner-fed closed delta-loop EWE, at C: the classic K9AY configuration, at D: classic EWE configuration and at E: diamond-shaped equilateral K9AY loop. See also Fig 7-68 for the configurations.

The vertical radiation patterns for the different antennas (modeled on 1.83 MHz over good ground) are shown in **Fig 7-69**. Note that pattern and gain for all are very similar. At A is the inverted-V EWE, which, for the dimensions shown requires a termination resistance of 900 Ω for best directivity. In B the ends of the inverted-V are lifted 1 m off the ground and connected with two sloping wires to a ground post in the center of the loop. The termination and feed point are still at the end of the sloping wires. It is remarkable that we lose almost 3 dB in signal! In C, which is the K9AY configuration, we moved the termination point and the feed point to the "bottom" of the loop, near the ground stake. The loop impedance is only slightly reactive, and a simple 9:1 transformer will achieve an acceptable SWR (typical <1.3:1). Performance-wise, the K9AY measuring 7-m high and 9-m wide (the original K9AY dimensions) is comparable in performance to a classic EWE (over good ground) measuring 3-m high and 8-m long as shown at D. The diamond-shaped loop is an interesting configuration for making a rotatable loop, it would look much like a 20-meter diamond-shaped quad loop.

21.2. The Rotatable K9AY Loop

K9AY, in his publication, uses 2 loops set up at right angles, and switches the array in 4 quadrants to cover all directions. The strength of this type of array is obtained from the null (rejection) off the back, rather than from a narrow forward lobe. From a forward lobe point of view (typical 150° at –3 dB points) 4 quadrant direction switching is just fine, but it may be advantageous to be able to point the null more precisely in the direction of an interfering source. I developed a variant, looking much like a single-element 20-meter diamond-shaped quad antenna (Fig 7-68E and Fig 7-69E). Notice the position of the feed point and the load resistor. This antenna can easily be made rotatable. The feed line can be guided along a bamboo spreader to the 9:1 matching transformer. A series capacitance can be inserted to cancel the inductive reactance at the feed point. Do not forget to put the usual common-mode suppressers on the feed line

(lots of beads will do). This antenna shows only 12 dB of F/B at low angle, but has a very respectable rejection at higher angles (which is an advantage for rejecting "local" signals).

21.3. Adjusting the K9AY Loop

The K9AY loop can be adjusted just like any EWE loop, by varying the termination resistance for best F/B ratio. As the highest rejection is at a relatively high angle, it is not possible to find a sharp null when testing on ground wave (almost-zero wave angle). It may therefore take some cut and try on sky-wave signals to find the best value.

A 9:1 transformer will match the K9AY loop very well to a 50 or 75-Ω coax cable.

■ 22. MAGNETIC LOOP RECEIVING ANTENNAS

In the early days of radio, small loop antennas were used extensively as receiving antennas. A small loop antenna is a magnetic antenna, which means that the antenna is excited by the magnetic rather than the electric component of a radio wave. Most other antennas such as dipoles, ground planes, rhombics, Yagis, large loop antennas, and so on, are called electric antennas, responding to the electric component of the wave. Full-size loops (quads and delta loops) are not loop antennas in the strictest sense, but rather an array of close-spaced stacked dipoles. With large loops, the directivity is broadside to the loop plane. With small loops, the directivity is 90° from this, in the plane of the loop (end-fire). **Fig 7-70** shows directivity patterns of both large and small loop antennas.

The windings of a small loop antenna can best be compared to the windings of a transformer. The antenna is tuned to resonance with a tuning capacitor. The energy can then be coupled from the "transformer" by a link, or by an inductive or capacitive tap. Most receive-only designs use several turns for the loop. The radiation resistance and the loop efficiency are directly proportional to the loop diameter and the number of turns. Receive-only loops have been described with dimensions ranging from very small to quite large, using a variety of feed systems, and often employ built-in preamps (Ref 1219, 1226, and 1229). Such receive-only loops have proved to be especially valuable for reception on 160 meters, where they are often extremely helpful in eliminating local sources of interference.

Loop antennas can show a great deal of horizontal directivity, but when operated vertically (with the plane

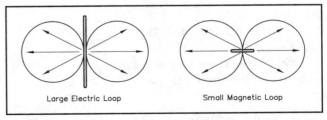

Fig 7-70—Horizontal radiation patterns of a large electric loop (1-λ circumference) and a small magnetic loop.

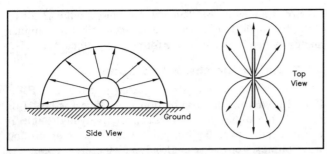

Fig 7-71—Vertical and horizontal radiation pattern of a small magnetic loop erected in a plane perpendicular to the ground (classic configuration).

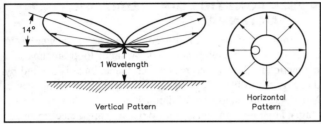

Fig 7-72—Vertical and horizontal radiation patterns of a small magnetic loop erected in a plane parallel to the ground. The radiation angle now depends on the height of the antenna above ground (as with a dipole).

of the loop perpendicular to the ground) the loop has no vertical directivity at all (see **Figs 7-71** and **7-72**). This means that the loop receives high as well as low-angle signals. Its performance is not influenced by the quality of the ground (there is no pseudo-Brewster angle involved!).

McCoy, W1ICP, described a small, single-turn loop in amateur literature (Ref 1227). The antenna is a loop for 80 meters, with a diameter of 3.5 m (12 ft), and is resonated by a series connection of three capacitors with the feed line connected across the middle one (forming a capacitive voltage divider). Other designs of magnetic loops have been published more recently (Ref 1255 and 1250). Ohmic losses appear to be the main problem in obtaining reasonable efficiency, in view of the very low radiation resistance. The radiation resistance of a single-turn loop is given by:

$$R_{rad} = 197 \times \left(\frac{C}{\lambda}\right)^4$$

where

C = circumference of the loop
λ = wavelength

For a (quite large) loop with a diameter of 6 m (20 ft) operating on 160 meters, R_{rad} is:

$$197 \times \left(\frac{18.8}{164}\right)^4 = 0.034 \ \Omega$$

The design parameters of a single turn loop are:

Table 7-21

Main Electrical Characteristics of the AMA 7 Loop for 160 through 40 Meters

Tuning range: 1.75 to 8.0 MHz
Power: 100 W

Band	40 m	80 m	160 m
Gain (dBd)	−0.71	−3.13	−10.2
Efficiency	93%	53%	11%
R_{rad} (milohms)	762	48	3.5
SWR Bandwidth (2:1)	14 kHz	5 kHz	2 kHz

$$L_{\mu H} = 2S \times \frac{\ln(S/d) - 1.07}{10}$$

where

$L_{\mu H}$ = inductance of the single-turn coil (loop)
S = circumference of the loop in meters
d = diameter of the loop conductor in meters

Example:

Take a loop of 6-m diameter, using a 30-mm tube as a conductor:

$$L = 2\pi \times 6 \times \frac{\ln(6/0.03) - 1.07}{10} = 15.9 \ \mu H$$

The required capacitor to resonate the loop at a given frequency is:

$$C = \frac{25,300}{f^2 \times L}$$

where

C = tuning capacitance, pF
f = design frequency, kHz
L = inductance of the loop, μH

If we want to resonate the 6-m (diameter) loop (15.9-μH self inductance) on 1.835 kHz, the required capacitor is:

$$\frac{25,300}{1835^2 \times 15.9} = 473 \ pF$$

In recent years, Wuertz, DL2FA; Kaeferlein, DK5CZ; and Schwarzbeck, DL1BU (Ref 1218), have experimented extensively with single-turn transmission-type loops (Refs 1215-1217).

Commercial versions of these loops are manufactured by C. Kaeferlein, DK5CZ, and sold worldwide. The largest model (AMA 7) has a diameter of 3.4 meters and is tunable from 1.7 to 8.0 MHz. **Table 7-21** shows the main characteristics of this loop antenna, which tunes 160, 80 and 40 meters. The loop has been built for transmitting as well (max. 100-W power), but on 160 meters it would make a rather poor transmitting antenna in view of the intrinsic loss of 10 dB. For receiving, however, this is not important at all.

Note the very low radiation resistance of the loop on

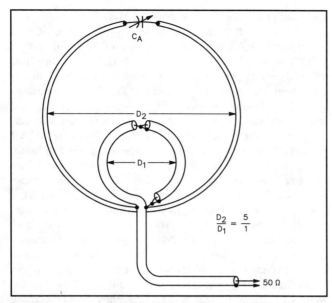

$$\frac{D_2}{D_1} = \frac{5}{1}$$

$\rightarrow 50\ \Omega$

Fig 7-73—Layout of the coupling loop as used by DK5CZ on his AMA magnetic loops. Changing the shape of the small inner loop makes it possible to obtain a nearly 1:1 SWR over the entire spectrum covered.

160 meters: 3.5 milohms! This corresponds with a calculated 3-dB bandwidth of just over 2 kHz. Here we come to another advantage of a small loop: The narrow bandwidth adds a high degree of front-end selectivity, which can be an advantage in reducing intermodulation distortion in the receiver, for instance when strong nearby BC stations are a problem. Although one cannot expect such a small loop antenna to be a competitive transmitting antenna for DX work on the low bands, it can be a worthwhile aid in obtaining better reception. For man-made noise, the electric component of the radiated wave is most often predominant in the near-field. Because a loop responds to the magnetic component of the signal only, loops are much quieter receiving antennas than dipoles or monopoles. Man-made noise sources are almost always of local nature (ground-wave signals), and as such, the signal polarization and the phase are constant (assuming a stationary noise source).

These are the necessary prerequisites for achieving a stable null on an interfering signal by orienting the loop in-line with the noise source. A practical rejection of 20 dB or better is easily achievable.

When mounted vertically (the classic configuration), the height of the loop above ground does not influence the radiation pattern or the efficiency to any great degree. Poorly conducting ground will not influence the efficiency of a loop, as the magnetic field lines are parallel to the ground. When mounted horizontally (plane of the loop parallel to ground), the horizontal directivity pattern becomes omnidirectional, and the vertical pattern shows a radiation angle which depends on the height of the antenna above ground (the radiation angles given for horizontal dipoles in Chapter 8 can be used).

Excellent articles have appeared in literature on home

building magnetic loops (Ref 1215, 1216, 1219, 1220, 1221, 1229, 1252, 1253, 1254, and 1255). It must be said however, that the problems in constructing an efficient loop for 160 meters are not easy to overcome. Multiband loops must be small enough to cover the highest frequency with the tuning capacitor set at minimum. For the lowest frequency, the loop may then require a sizeable capacitor. Also at the lowest frequency the R_{rad} will be lowest (perhaps 5 to 50 milliohms), which means the loss resistance in the loop must be kept very low.

Currents as well as voltages involved are very high, even with 100 W of transmitter power. On the other hand the high voltages are not lethal: As soon as you come close to the loop it will be detuned by the proximity effect so that voltages will drop to a low and safe level. Another problem is you want to have a remotely driven tuning capacitor (which really must be a split-stator capacitor if avoiding losses is important) that tunes slowly enough to be practical. Don't forget, we are talking of 3-dB bandwidths of 2 kHz on 160 meters! The AMA 7 magnetic loop antenna, available from Kaeferlein-Electronic, DK5CZ, Germany, combines all these requirements in a very well engineered unit. The AMA antennas are also extensively used in commercial service, such as in embassies.

Coupling the RF into the loop can be done in different ways. In his commercial brochure, DK5CZ, describes an inductively coupled loop, made of coaxial cable. This loop has a diameter of approximately one-fifth of the main loop diameter. **Fig 7-73** shows the layout of the coupling loop. Adjusting for lowest SWR is done simply by reshaping the smaller loop inside the larger loop. A shape can be found that gives an acceptable SWR over the entire operating bandwidth (eg, 40 through 160 meters for the model AMA 7 loop).

Mozzochi, W1LYQ, has been using a loop with a circumference of 24 m (80 ft). He reports excellent results on 160 meters even for transmitting, where he has worked numerous European stations with the loop.

Petry, HB9AMO, reports using the shielded loop described in *The ARRL Antenna Book* (18th edition, pages 5-19 to 5-20) in conjunction with his 27-m top-loaded vertical to bring him the first ever 160-meter WAZ. He uses a 15-dB preamp to boost the signal. **Fig 7-74** shows the dimensions of the loop. For details see the full description in *The ARRL Antenna Book*.

S. Ritchie, KC2TX, has made these loops available at very reasonable prices. He has a single-band 80 meter, single-band 160 meter and a duoband (80 meter inside the 160-meter) loop. The output of these loops is approx. –15 dBi, and the feed-point impedance at resonance is close to 50 Ω. The 80-meter loop is approx. 1.2 meters in diameter, the 160-meter loop is twice as large. A preamp is required in most cases, and if the coax run to the shack is fairly long, the preamp must be installed at the loop (**Fig 7-75**). Contact KC2TX at **http://www.qsl.net/kc2tx/** for prices and other details.

It is obvious that two or more such loops can be phased together for added directivity and gain. One can run the outputs from two loops to a signal combiner with variable

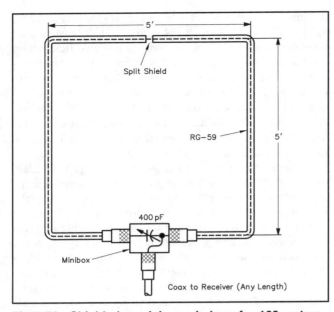

Fig 7-74—Shielded receiving-only loop for 160 meters as described in *The ARRL Antenna Book*. This is the receiving antenna used by HB9AMO together with a 15-dB preamplifier.

gain and phase controls, such as the MFJ-1026.

Fenwick, K5RR, described an array made of two small loops (8.5-m circumference) spaced 18 m and fed via a Wilkinson hybrid coupler and a coaxial delay line (Ref 1248). Fenwick reports a front-to-back ratio of 35 dB on and near the design frequency and nearly 20 dB in a wide range around the design frequency. The output of the array is said to be down approximately 29 dB from a reference vertical.

During the winter of 1992-1993, I installed the AMA 7 loop (1.8 through 7 MHz) in my front garden (see **Fig 7-76**). The base of the loop is about 1.5 m (5 ft) above ground. The antenna is fed with 30 m (100 ft) of RG-213. The purpose was to evaluate the loop, especially in comparison with the 200 to 300-m long Beverages that I normally use. It immediately became clear that on 160 meters, under all circumstances, the loop is quieter than the quarter-wave transmit antenna, while signals are down 10 to 15 dB (which is really totally irrelevant). On 80 the same is true, but the "loss" vs the vertical is only approximately 1 S unit. On 40 meters, the loop is down about 10 dB vs the 3-element Yagi. On all three lower bands I have found the loop to be a worthwhile asset in receiving. In no case however, have I found the loop to be

Fig 7-75—The KC2TX magnetic loop for 80 and 160 meters.

Fig 7-76—The AMA 7 magnetic loop as erected in the front garden of ON4UN. It is located approximately 60 m from the 160-meter transmit vertical, the large spacing reduces coupling to the vertical.

as good a receiving antenna as any of my Beverages. The loop is sensitive to rain static, unless it is covered by an insulating material.

Magnetic loops can easily be modeled using *MININEC*, provided precautions are taken to have enough pulses per side (if a square loop).

■ 23. COMPACT RECEIVING END-FIRE ARRAYS

Two-element end-fire arrays have some interesting characteristics that make them attractive for the home builder. They can almost match Beverage antennas!

23.1. Using Grounded Vertical Elements

23.1.1. Trigonometry: the ideal paper case

Let's work out an example of a 2-element array using vertical ground-mounted elements. Take two very short verticals, say 6-m long (2.5-cm diameter). You don't have much space in the garden and can space them 8 m, not more. You start playing with the modeling program, feeding the first element with a current of 1 A and a phase angle of zero. After playing around a while you will find by pure cut and try (fortunately you don't have to do real "cutting" on the computer) that if you feed the second element with the same current amplitude, but phase difference of 164°, you get a perfect cardioid pattern with 45-dB F/B. If you get a kick out of even higher F/B numbers, you can push this ridiculous game and get your computer to say 60 or 70 dB. These are of course paper (or screen?) dBs. I never understood how anyone could get excited about these high figures. It just proves some people don't seem to be able to distinguish mathematics from physics. Anything above 35 dB is better taken with a "large" grain of salt in real life!

Anyhow, our computer said 164° and this is logical because, taking into account a wave angle of say 25°, the spacing of 8 m equals 16°, and 164 = 180 – 16.

How do we calculate this?

1 wavelength on 1.83 MHz = 299.8/1.83 = 163.8 m = 360°

1° = 0.455 m

Assume we want maximum F/B for a wave angle of 25° (cos 25° = 0.91).

Taking into account the wave angle of 25° we can calculate the required phase shift to obtain an infinite F/B at 25° wave angle:

$$0.91 \times \frac{8}{0.455} = 16°$$

Important remark

If we design an array for maximum F/B at a wave angle of 25°, then when tuning the array with a remote signal generator (should in principle be at least 5 λ away), we should put the generator at an angle of 25° with respect to the geometric back of the antenna, which means at an angle of 155° or 205°. This is simply because we (usually) cannot do the measurement at a 25° wave angle (we would need to have our signal source at a height of 466 m if the signal generator is 1 km

away...). An antenna which is optimized for best F/B (180°) for a 25° wave angle, will also have the best rejection at (almost) zero degree wave angle at 155° (180 – 25) and at 205° (180 + 25). The mathematics above explain why. The wave angle for a signal generator at 1.5-m height and 1000 m from the antenna produces a wave angle of approx. 0.1°.

23.1.2. The drive impedances

So far we've only done some trigonometry. Let's look at impedances. The drive impedances at the bottom of the 2 elements are 4.0 –j1349 Ω and –3.9 –j1349 Ω. Negative drive impedances are normal; see Chapter 11 (Arrays). These figures can be obtained by using any *MININEC*-based program.

Any idea how you will get those currents (1A /0° and 1A /164°) in the element feed points showing such "unreal" impedances? That's where the problem lies. If we could do that with no losses this 2-element array would work as well on transmit as the 2-element array with elements measuring 40-m high and spaced 40 m.

What we need to achieve is to feed the elements with a matched feed line (SWR <1.2:1) over as wide a spectrum as

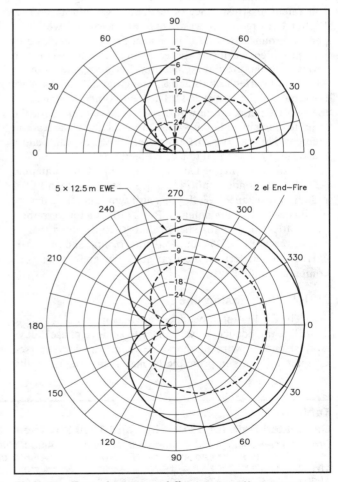

Fig 7-77—Two-element end-fire array patterns as compared to a large (5 × 12.5 m) EWE. Notice the better high-angle rejection for the 2-element end-fire array.

possible. This means we need to have a feed-point impedance equal to the feed-line impedance over a given frequency span.

There are several ways to achieve this. One can load the short vertical with an inductor, but not resonate it, and then do a parallel feed with a transformer, plus tune out the remaining reactance in the secondary of the transformer in order to present a conjugate match to the feed line (see Section 33.1.9).

The only real issue is to obtain over a given frequency span, by whatever matching system, a (fairly) constant feed-point impedance for all elements in an array.

For receiving antennas this is normally done by adding a lot of resistance in the system. We can resonate the elements (get rid of the capacitive reactance) and then add series resistance, if necessary, to obtain, eg, 75-Ω resistive at resonance. The feed-point impedance will be made up by 99%-Ω resistance and (typically) 1% radiation resistance. That's why these antennas are 20-dB down in sensitivity!

23.1.3. Loading the elements

Let's "first load" these short elements, and tune them to resonance. We put a husky coil (117 µH), to tune out the negative reactance of −1349 Ω. These are real coils, with a real Q. Assume for a minute we achieve a Q of 30. The equivalent loss resistance of each coil is 1349/30 = 45 Ω. Earlier I (in reality it was *MININEC*) assumed we had a perfect ground. *MININEC* is perfect for doing this kind of job, provided we insert the equivalent loss resistance in series with the ground connection. Let's assume we have some short buried radials and a good ground, and I estimate the ground resistance at 26 Ω. The feed-point impedances of the two elements are now 4.0 + 45 + 26 = 75.0 Ω for the lagging element and −3.9 + 45 + 26 = 67.1 Ω for the leading element. No reactive part anymore, 100% resonant, and a reasonably good match to a 75-Ω feed line.

Look what happened to the output of the antenna: −27 dBi. No wonder with 74 Ω of losses and less than 1 Ω of radiation resistance! Well, you can't win 'em all, and after all, this is a receiving antenna. But 27 dB is a lot (or rather, very little) of signal. That's almost 20-dB less than a 1-λ Beverage! **Fig 7-77** shows the horizontal and the vertical radiation pattern for the array at the design frequency, as compared to the EWE.

23.1.4. The phasing line

The feed-point impedances are reasonably close to the 75-Ω line impedance, so we can try to achieve phase delays by just cutting wire of the same length in degrees (see Chapter 11). This in principle is okay if the line has unity

SWR and if there are no losses.

As the line is not perfectly flat, and not without losses, in reality the transformation will not be perfect. Assume we make a delay line, using 75-Ω cable. It must be 164° long (our target to obtain infinite rejection at a wave angle of approx. 25°). How do you cut a delay line 164° long? Cut it for 180° (see Chapter 6, Section 4.1), and once you have an exact 0.5-λ line, use the rule of 3 to cut off 16°.

23.1.5. Bandwidth considerations

In order to maintain the 164° phase shift away from the design frequency (eg, on 1810 and on 1860 kHz):

• The impedance of the elements should stay constant.
• The feed line should be loss free.
• The feed line should remain of constant electrical length (164°) on all frequencies.

Assuming no line losses and unity SWR, the fact of using the 164° long phasing line (cut for the middle of the band: 1830 kHz) will be 162.2° on 1.850 kHz and 165.8° on 1810 kHz (simple rule of three). We will have to take these phase angles into account when calculating the radiation patterns on 1810 and 1850 kHz.

To investigate the effect of this slight change in phase delay, as well as the effect of SWR and cable losses, I calculated the impedances of the lagging element (the element fed through the phasing line) using *ELNEC*. The real current phase shift (as well as amplitude) was then calculated using the module "Real Cable Z/I/E Listing Program," which is part of the ON4UN LOW-BAND SOFTWARE (see Chapter 4). Finally the radiation pattern was calculated using these "real" values for feed current of the second element in order to see the impact of all these variables. **Table 7-22** shows the results.

The maximum F/B (in the geometric back of the array) is listed with the wave angle at which it occurs. Again, the max. value is limited to 35 dB.

The average rejection is the mathematical average of 30 values, taken in 10° increments of elevation angle (from 10 to 60°), and in 20° increments from 140 through 220° (azimuth). All values greater than 35 dB were taken as 35 dB. The integration of the rejection values over an area measuring 80° wide and 60° wide gives a good idea of the average rejection by the antenna of signals "in the back."

How good are these figures? In order to be able to assess them, I also calculated the average rejection (0 to 60° elevation, 140 to 220° azimuth) for an ideal antenna (an antenna using a lossless phasing line with unity SWR, yielding a phase angle 164° and amplitude 1.00) the average rejection is 22.2 dB. This means that our practical model,

Table 7-22
Calculated Data for 1810, 1830 and 1850 kHz (See Text)

Freq.	Theoretical Phase	Impedance Lag Antenna	Actual Current Phase	Actual Current Amplitude	SWR on Phasing Line	Avg. Rejection 60° × 80°	Max F/B at ° Elev.
1810	165.8°	74.6 −j 15	164.9°	1.02	1.22	21.7 dB	29 dB/30°
1830	164.0°	75.0 +j 0	164.0°	0.97	1.00	20.0 dB	25 dB/27°
1850	162.2°	75.4 +j 16	163.5°	1.03	1.25	19.4 dB	26 dB/10°

Table 7-23

In Wet Weather (R$_{ground}$ is 10 Ω Lower). Compare with Table 7-22

Freq	Theoretical Phase	Impedance Lag Antenna	Actual Current Phase	Actual Current Amplitude	SWR on Phasing Line	Avg Rejection 60° × 80°	Max F/B at ° Elev
1810	165.8°	64.4 −j15	166.7°	1.03	1.30	20.8 dB	25 dB/40°
1830	164.0°	65 +j0	165.9°	0.98	1.16	22.4 dB	29 dB/40°
1850	162.2°	65.4 +j16	165.5°	0.94	1.32	17.8 dB	21 dB/10°

over a span of 40 kHz is within about 3 dB as far as average rejection from our purely mathematical model, which is quite good.

23.1.6. Specifying the performance of an array

You can specify this array as having an F/B of well over 35 dB on its design frequency, at a wave angle of 25°. What are the chances that the interfering signal you want to suppress will be right at that wave angle and exactly off the back of the array? The wave angle is something we cannot easily control as we would have to use a delay line of variable length. The azimuth angle (direction) we could control by making the array rotatable (see Section 22.3).

It is much more realistic, however, to have a good look at the average rejection of a receiving antenna off its back! Consider yourself lucky if the interfering signal is somewhere in the cone measuring 60° high and 80° wide. Once it's in there, it's a question of luck that your signal targets a valley or a peak of the cone. You know, however, what the average rejection is. In Section 26 I will compare Beverages with EWEs and end-fire arrays from this point of view.

23.1.7. The effect of variable ground losses

Ground loss is another variable: so far I have assumed the ground loss would be 25 Ω. But it rains, and losses go down to 15 Ω, or during a dry summer to 35 Ω. What then?

I have calculated the theoretical performance for a system using different ground-loss resistances. **Table 7-23** shows the results for a case where the ground resistance was 10 Ω lower than anticipated (in the middle of a wet winter?).

The same calculations were done for a loss resistance 10 Ω higher than estimated (dry summer), and the results are listed in **Table 7-24**.

From these tables we can conclude that the pattern does change with varying ground (loss) conditions (quite significantly when the losses are higher).

23.1.8. Conclusions

There are two mechanisms that deteriorate F/B in the models I investigated:

- The varying ground losses raise or lower the real part of the element impedances. Although the radiation pattern is not really falling apart under these varying conditions, it is important to control this parameter if best performance is expected all year around. We can diminish the influence of varying ground conductivity by burying many long radials under each element or by installing nets of chicken wire. This is not always very practical. Another solution is to get rid of the ground: instead of using ground-mounted electrical quarter-wavelength antennas, use electrical half-wave antennas, elevated slightly above ground (see Section 23.3).

- The varying feed-point reactance changes with frequency. We can make the elements thicker, and we can cancel the reactance with a coil that we switch in series with the feed point as we go lower in frequency. One could imagine using some small relays and retune the elements to full resonance every 10 kHz.

I changed the element diameter from 2.5 cm to a wire cage measuring 15 cm in diameter. The results are shown in **Table 7-25**.

Using such a "fat radiator" reduces the reactive part from 16 Ω to approx. 10 Ω at 1810 and 1850 kHz. Surpris-

Table 7-24

In Dry Weather (R$_{ground}$ is 10 Ω Higher). Compare with Table 7-22

Freq	Theoretical Phase	Impedance Lag Antenna	Actual Current Phase	Actual Current Amplitude	SWR on Phasing Line	Avg Rejection 60° × 80°	Max F/B at ° Elev
1810	165.8°	84.4 −j15	163.1°	1.01	1.25	18.2 dB	20 dB/10°
1830	164.0°	85 +j0	162.1°	0.96	1.13	20.7 dB	25 dB/10°
1850	162.2°	85.4 +j17	161.1°	0.9	1.28	13.7 dB	17 dB/10°

Table 7-25

Using 15 cm Diameter Wire Cage Elements. Compare with Table 7-22

Freq	Theoretical Phase	Impedance Lag Antenna	Actual Current Phase	Actual Current Amplitude	SWR on Phasing Line	Avg Rejection 60° × 80°	Max F/B at ° Elev
1810	165.8°	75 −j10	165.1°	1.00	1.14	23.4 dB	>35 dB/32°
1830	164.0°	75 −j0	164.0°	0.97	1.00	20.0 dB	27 dB/27°
1850	162.2°	74 +j10	163.3°	0.95	1.16	17.6 dB	23 dB/10°

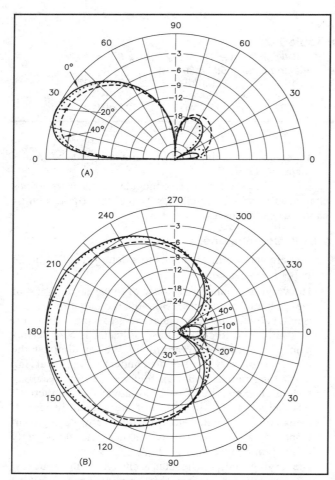

Fig 7-78—Vertical radiation patterns for 0, 20 and 40° azimuth, and horizontal patterns for 10, 20, 30 and 40° elevation angle, for the 2-element vertical ground-mounted end-fire array (pattern for 1.81 MHz).

ingly enough this did not seem to improve the overall pattern situation. The array designed with 2.5-cm diameter elements holds its pattern just as well as its brother made with wire-cage elements measuring 15 cm in diameter! (compare Table 7-25 with Table 7-22). Using "fat" elements will also drastically reduce the reactance of the element for a given physical length. For a receiving antenna this is of little importance as we are introducing additional resistance anyhow. **Fig 7-78** shows the horizontal and the vertical radiation patterns at various angles.

23.1.9. Feeding the array

I have already explained how important it is to remove any common-mode currents from the outside of the coaxial feed line, mainly because we are dealing with low signal levels. If currents that inevitably exist on the outside of a coaxial cable are not completely removed at the point where the feed line is connected to the short elements, then these currents will be somehow added to the currents from our

short elements and will completely upset all directivity. **Fig 7-79** shows a possible approach for minimizing common-mode currents on the feed line and the phasing section.

If the ground conductivity must be improved or stabilized it is probably best to use strips of chicken wire, as a few radials on the ground may be more susceptible to acting like antennas and to develop an RF voltage across an intrinsically poor ground rod.

The antenna could also be built around a 50-Ω impedance. In this case the output of the antenna will be slightly higher (approx. 2 dB) but the bandwidth will be narrower.

23.1.10. Alternative feed method

Some advocate using nonresonant elements to make it easier to obtain good directivity over a wide bandwidth. Unless there is a compensating element in the matching network, the change in reactance (imaginary component) at the feed point is only determined by the L/D ratio (in other words, the diameter) of the element. A further complication of using such systems and networks is that the extra components create uncontrollable additional phase shifts.

The relative change in reactance can be lowered by introducing resistance (losses) in the system. This is what we do in most receiving arrays.

23.1.11. Modeling feed lines with *EZNEC*

In the previous sections, I have analyzed in detail the behavior of real feed lines, which have to provide the correct phasing to achieve maximum directivity. When using *EZNEC* as a modeling program you can specify the feed or phasing lines as an integral part of the mode, but you have to be aware of the fact that *EZNEC* uses lossless lines. It means we can never analyze the situation as accurately as I have done above.

23.2. The K9UWA, W7EL, KD9SV Mini 4-Square for 160

The receiving 4-square described in *The ARRL Antenna Compendium Volume 3* (Ref 1266, page 33) is the first small-area receiving array I have seen. There are a few successful builders of this array. The article is a must for anyone who wants to venture into this kind of receiving antenna.

The array used 4 elements, but the basics are the same as for a simple 2-element design. To avoid the problems of feeding the elements in the center, and having to deal with common-mode currents on the feed lines, the authors chose to feed the element through the lower half right and bury the cable right in the ground and get them out of the way of the elements. This caused other problems, which are dealt with in detail in the article (decoupling the feed line from the very high impedance ends of the loaded dipoles).

IV3PRK has built a copy of the array and swears by it (see **Fig 7-80**). He said it was fairly difficult to get the feed impedances of the 4 elements equal, and he had to use coil values which were different for each element. A positive point is that very satisfactory results were obtained using a small Autec RF-1 Antenna Analyzer as the only measuring equipment.

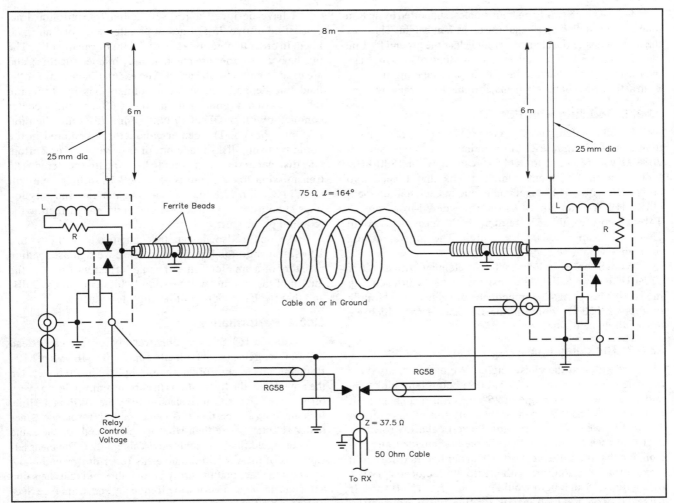

Fig 7-79—Suggested feed system for the ground-mounted 2-el end-fire array. The impedance at the junction of the feed lines is 37.5 Ω. Purists can add an L network or you may content yourself by feeding it with a 50-Ω feed line and take a 1.4 SWR as acceptable.

Fig 7-80—Photo of the 4-element receiving array as constructed by IV3PRK (*photo by IV3PRK*).

23.3. A Rotary 2-Element End-Fire Receiving Array for 80 or 160 Meters

Two-element end-fire arrays always have a very broad forward lobe but if well designed can have a very deep null off the back. This means you get most of the advantage from turning the back toward a noise or QRM source, rather than from getting relative gain from a given (narrow) forward direction. This means that ideally these arrays should be made rotatable. There is no reason why we could not do that!

23.3.1. A design example

The idea was to simply transform the array described in Section 22.1 to an array using half-wave elements, in order to eliminate the variables caused by ground radials, and to make it a rotating array.

I started with two 12-m long dipoles, supported by an 8-m long boom installed on a small tower, at 7-m height. This way the bottom tips of the elements would clear the ground by 1 m.

I used fairly fat elements, consisting of a wire cage measuring 30-cm diameter at the tip, and tapering down to 3 cm at the center of the elements, forming a conical wire cage.

23.3.2. Modeling with *EZNEC*

For the calculations I assumed a constant element diameter of 150 mm. *EZNEC* calculated that a reactance of 2688 Ω was required to achieve resonance on 1830 kHz. I chose a split element, which means that I need two coils each with 1344 Ω reactance, one in each half element (117 μH). We can wind this on a powdered-iron core. A T-184-2 core would require approx. 70 turns, and would yield a Q of approx. 200, quite high! The equivalent loss resistance would be 1344 / 200 = 7 Ω.

We will not only load each half element with a coil of 117 μH, but we will add a resistor of 68 Ω in each leg, in series. This brings the impedance of each vertical dipole to approx. 150 Ω, which is the impedance for which we designed this antenna.

23.3.3. The delay line

Let's look at the phase angles. We have already done the calculation in Section 23.1. For a boom length of 8 m, and assuming a wave angle of 25°, we require a phase shift of 180 − 16 = 164°. We can do this with lots of coax (about 50 m!) like we did in the case of the vertical array (Section 23.1.4), or we can do it with a simple homemade transmission line that is a little bit shorter than our boom. If we could make a transmission line with a velocity factor (V_f) of 1, then a line of 8-m length would give us 17.6° (8 / 0.455). If we make an open-wire transmission line, with air dielectric, we will have a V_f of approx. 0.98 (due to the plastic spreaders used), which means that an 8-m long transmission line will be 17.6° / 0.98 = 18°. If we twist the transmission line, the phase shift will be 180 − 18 = 162°. In fact for exactly 164° shift we require a transmission line of 7.12 m length (8 × 16 / 18).

I have designed a 150-Ω parallel transmission line, consisting of two 75-Ω lines in parallel, each one air insulated, in order to obtain a velocity factor of (nearly) 1.0. The two heavy feed lines are mechanically bonded together, and are used as boom for the array. The exact dimensions for the feed line depends on available dimensions in aluminum tubing. I used a tubing with an ID of 60 mm and a center conductor with an OD of 17 mm. (For a 75-Ω line the ratio D/d must be 3.5. The center conductor is centered in the boom by using disks made out of polyethylene or Teflon. One disk per m is used in order to keep the line centered. The transmission-line / boom is exactly 7.12 m long. We will learn in Section 23.8 that we can lengthen the transmission lines at both ends by an equal amount without affecting the required phase shift.

We could also design this antenna to work with a 100-Ω transmission line, in which case we will use loading resistors of a smaller value. The net effect will be that the output of the antenna will be slightly higher (approx. 2 dB) and that the bandwidth will be slightly reduced.

23.3.4. Performance

Use of a 16° long line of extremely low losses, instead of a 164° long line (with typically approx. 0.5-dB loss per 100 ft) enables us to control the required phase much better. On the other hand, the fluctuations in current amplitude are every bit as much there, and are caused only by the SWR on the line. It is amazing to see that this calculated performance is actually slightly worse than what we calculated for the same array using vertical elements (see Table 7-22). The clear advantages of the vertical dipole array (over the ground plane array) is the fact that the array is rotatable and that the same antenna can cover a very wide frequency spectrum (see Section 23.3.5). **Table 7-26** shows the relevant data for the array.

23.3.5. And 80 meters?

The nice feature of using a twisted transmission line of near unity velocity factor is that this line always achieves the right delay, for whatever frequency we use this antenna, as long as the spacing between the elements is less than

Table 7-26
160-Meter Data for The Vertical Dipole Array with a Conductor Diameter of 15 cm

Freq	Theoretical Phase	Impedance Lag Antenna	Actual Current Phase	Actual Current Amplitude	SWR on Phasing Line	Avg Rejection 60° × 80°	Max F/B at ° Elev
1810	164.2°	150 −j 29	163.3°	1.05	1.16	17.5 dB	>23 dB/10°
1830	164.0°	150 −j 0	164.4°	1.00	1.00	22.1 dB	>35 dB/25°
1850	163.8°	150 +j 30	164.8°	1.03	1.15	18.4 dB	21 dB/30°

Table 7-27
80-Meter Data for the Vertical Dipole Array with a Conductor Diameter of 15 cm

Freq	Theoretical Phase	Impedance Lag Antenna	Actual Current Phase	Actual Current Amplitude	SWR on Phasing Line	Avg Rejection 60° × 80°	Max F/B at ° Elev
3500	149.0°	150 −j 8	148.1°	1.02	1.05	20.4 dB	>35 dB/20°
3520	149.2°	150 +j 0	149.2°	1.00	1.01	22.0 dB	>35 dB/28°
3540	149.4°	150 +j 0	150.2°	0.98	1.03	22.1 dB	>35 dB/29°

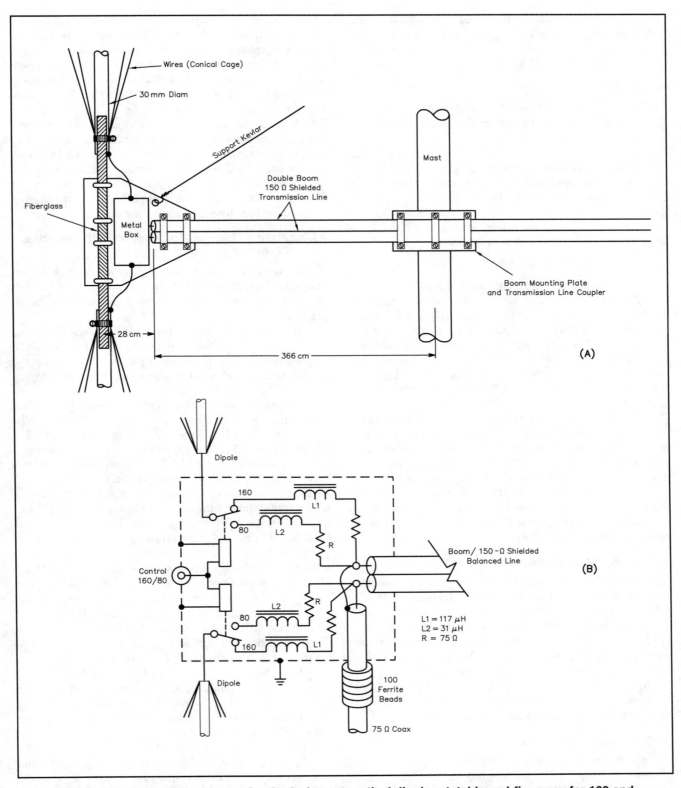

Fig 7-81—Suggested mechanical layout for the 2-element vertical dipole rotatable end-fire array for 160 and 80 meters is shown at A and B. The boom is exactly 7.12-m long, while the two elements are spaced exactly 8.00 m. The element to boom plate provides ample room for mounting a metal box containing the loading coils, the resistors and two relays. The 75-Ω feed line is attached to the end of the phasing line via an efficient current balun (stack of 100 beads) representing at least 1500-Ω impedance on 160 meters.

¼ λ! All we have to do to use the same antenna on 80 meters, is to insert another set of loading coils with matched loading resistors.

Calculation for 3.52 MHz:

$$\lambda = 299.8 / 3.52 = 85.17 \text{ m}$$

$1°$ in free space ($V_f = 1.0$) = 0.2366 m

Assume we want maximum F/B for a wave angle of $25°$

$$\cos(25°) = 0.91$$

Taking into account the wave angle of $25°$ we can calculate the required phase shift to obtain an infinite F/B at $25°$ wave angle: $180 - [0.91 \times (8 / 0.2366)] = 180 - 30.8° = 149.2°$.

Let us now check the electrical length of our 7.12 m long transmission line (with a V_f of 0.98):

$$L = (7.12 / 0.2366) / 0.98 = 30.7°$$

That's close enough to our target of $30.8°$.

EZNEC tells us that for 3.520 MHz we require 4 loading coils with Z = 683 Ω (L = 31 μH) and 4 loading resistors each of 75 Ω. On 80 meters this 2-element array has a gain of −21 dBi over good ground (−32 dBi on 160 meters).

Table 7-27 shows the relevant data for the end-fire array on 80-meter CW. We can calculate a set of values for 3.8 MHz as well, and for any other frequency between 1.5 and 4 MHz.

The directional characteristics are quite impressive on 80 meters, and better than 20 dB average rejection (over 60° × 80°) can be obtained over more than 50 kHz, using 12-m long elements, with an effective element diameter of 150 mm. This makes this a "killer" receiving antenna!

There is no reason why the same antenna cannot be used on 40 meters as well. For each band the correct set of loading coils must be inserted. This can be done by a set of relays, as shown in **Fig 7-81**.

23.3.6. Mechanical issues

The 7.12-m long shielded, parallel transmission line acts as a (double) boom for the array, holding the 2 elements 8.0 m apart. Fig 7-81 shows how this problem can be solved. Two aluminum die cast boxes house the loading coils (117 μH) and the 68-Ω resistors, with room for relays in case we want to do frequency switching. Make sure the loading coils do *not* radiate (put them in a metal enclosure), as this might upset the array. Short pieces of equal-length 75-Ω feed line can be used at both sides of the array without upsetting the exact phase shift (See Section 23.3.8).

23.3.7. Tuning the array

Put up the array in its final position (bottom tips of the elements at least 1 m off the ground). While element 2 is open-circuited, adjust element 1. If necessary, change the turns on the loading coils symmetrically to achieve resonance on 1.83 MHz. An antenna analyzer will do the job, but a network analyzer will probably get you more precise

results. Make sure the two loading coils have exactly the same number of turns and that they are kept away from one another as much as possible. Next connect the series resistors. Both should be of equal value, and they should result in an antenna impedance of 150 Ω at resonance. The resistors should be carbon or metal film. Do not use wire-wound resistors.

Next, do the same thing with element 2, while element 1 is open-circuited (nonresonant). Mutual coupling will be minimal because of the large amount of resistance involved, but, for beauty's sake, do it the correct way. If you will use the antenna for 80 as well, repeat the same procedure for 80 meters. Once the elements are tuned, connect the phasing line and the feed line, and you are all set!

23.3.8. Using Regular Coax vs. the Air-Dielectric Line

I explained in Section 23.3 how we can use an air-spaced feed line ($V_f \approx 1$) and achieve the required phase shift between the two elements of the end-fire array almost perfectly.

If we use cable with a lower velocity factor, then we cannot reach the elements with the feed line. Let us analyze what happens if we add two short pieces of feed-line at the feed point of each element, so that we can reach our phasing line (see **Fig 7-82**).

I worked out a case using semi-air-spaced coaxial cable with a Velocity Factor of 0.85 (eg, for 95-Ω RG-62).

Using a V_f of 0.85 the length of the line is: (299.8 × 16 × 0.85) / (1.83 × 360) = 6.18 m. This is for achieving a null at a wave angle of $25°$. For a wave angle of $0°$ the length is 6.8 m. In any case, we are short by 1.2 to 1.82 m to reach the elements. What if we add equal lengths of coax to the feed points of the array as shown in Fig 7-83? Let us consider the $25°$ wave angle case. If we add 0.91 m of feed line at both ends, representing delay lines of $2°$ each (0.91 / 0.455):

• The phase at element 1 is now $-2°$ (minus means lagging).
• The phase to element 2 is $-16° - 2° + 180°) = -162°$. The $-16°$ is the 6.18 m length of line, $-2°$ is the length of the extra 0.91 m, and $+180°$ is for the phase reversal.
• The phase difference is $-2° - 162° = -164°$.

Conclusion: by adding two pieces of coax of equal length we make no error in the desired phase shift.

This also means that, even with the added pieces of coax, the feed system remains frequency independent. To cover a wide range of frequencies, it is only necessary to tune the antennas for resonance at those frequencies, so that a flat SWR can be obtained, and so that line length equals phase delay as closely as possible.

In actual life the antenna elements with really short elements show a lot of reactance once you depart from the resonant frequency. In order to reduce the ill effects of SWR, it is necessary to use a transmission line with the lowest possible losses. Therefore, using a transmission line with an air dielectric is highly recommended (see Section 23.3.3).

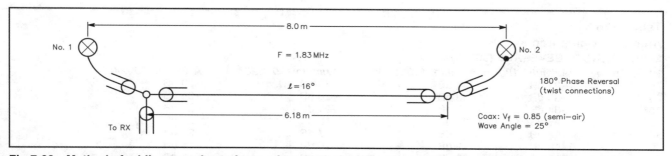

Fig 7-82—Method of adding two short pieces of transmission line to reach the elements. See text for details.

23.4. The Titanex Rotary End-Fire Array for 80 and 40 Meters

Titanex, a German company that has made its reputation in the last years with low-band vertical antennas has recently introduced an 80 meter and a 160-meter rotatable end-fire array (model SES 80 and SES 160). For details consult **http://www.qth.com/titanex/** on the Internet (see **Fig 7-83**).

These antennas follow exactly the principles outlined in Section 23.3. Two 5-m long vertical dipoles are spaced 6 m. The height of the rotating boom is approx. 5 m above ground. The two dipoles (actually two back-to-back helically loaded mobile whips) are tuned to resonance with a coil and additional resistance is added to match the dipoles to the transmission lines used to feed the dipoles and to achieve the correct phase delay. The Titanex antenna uses a custom made (tube in tube) 50-Ω RG-58 feed line. Using a system impedance of 50 Ω, the gain of the array is approximately –19 dBi on 80 meters, which is about 8 dB less output than a well matched 1-λ long Beverage antenna. This should not cause a problem, but a 10-dB preamp seems to be indicated. The Model 80-SES exists in two versions, one for the CW band, centered on 3.520 MHz, and one for SSB, centered on 3.780 MHz.

The 160-meter array, with its 10-m long boom, and also using 5-m long vertical dipoles has a calculated gain of approx. –30 dBi. The 40-meter model has a 4-m long boom, and also uses a 5-m long dipole. The outputs of this array is calculated

as –18 dBi. The 40, 80 and 160-meter antennas can all be mounted on a single boom, as they do not couple at all because of the very high (intentional) losses in the system.

The antenna seems to be a good choice for DXpeditions, as it is very light (7.5 kg) and easily transportable. The deep null off the back, which is the major advantage of such an array, can be turned toward offending noise sources such as thunderstorms.

I had hoped to be able to test a SES-160 array before going to print, as both the 80 and especially the 160-meter antenna appears to be a very attractive alternative to much bigger receiving antennas for those that lack the space. Unfortunately some unforeseen delays have made this imposible.

23.5. Achievable bandwidth with 2 short-element end-fire arrays

In bandwidth considerations, it is most important to use dipole elements of a fat diameter. Dipoles made of wire just won't give you much bandwidth. I modeled the Titanex 80-meter antenna with a dipole having an effective equivalent diameter of 20 mm. This seems to be a bare minimum to obtain acceptable results.

Using a lossless phasing line optimized for 3.52 MHz, the bandwidth for 20 dB F/B is approx. 15 kHz, as shown in **Fig 7-84A**. Fig 7-84B shows the case where we tuned the element when moving frequency. In practice this can be accomplished by adding a small inductor as one moves lower in frequency. This inductor could be switched with small relays.

The models were calculated for an array system impedance of 95 Ω, which of course yields a perfect feed impedance of approx. 50 Ω where the two feed lines are joined.

For the 160-meter array, the usable bandwidth for an acceptable F/B should be approx. half of what is achieved on 80 meters.

Conclusion: Bandwidth is without a doubt the main issue with these 2-element end-fire antennas. Fortunately because of the intrinsically very high losses of the antenna, the resonant frequency is not much affected by the ground or other nearby objects. This makes it possible in principle to fine tune the antenna off site (eg, in a production environment in the case of commercial antennas), and not worry about customer problems in the field.

Fig 7-83—Photo of Titanex SES 80 rotatable array.

Table 7-28
Antennas Compared
2-WAVELENGTH BEVERAGE (352 m):

Wave Angle	Azimuth: 180°	Azimuth: 160°/200°	Azimuth:140°/220°	Azimuth:120°/240°
10°	–27 dB	–31 dB	–33 dB	–29 dB
20°	–33 dB	–33 dB	–26 dB	–26 dB
30°	–34 dB	–35 dB	–21 dB	–24 dB
40°	–26 dB	–22 dB	–18 dB	–28 dB
50°	–18 dB	–17 dB	–17 dB	–25 dB
60°	–18 dB	–19 dB	–25 dB	–18 dB
70°	–28 dB	–26 dB	–18 dB	–14 dB
80°	–20 dB	–14 dB	–13 dB	–12 dB
Average	**–24.3 dB**	**–24.6 dB**	**–21.4 dB**	**–21.9 dB**

Total cone average rejection: 23 dB
Typical output: –6 dBi

1.5-WAVELENGTH BEVERAGE (268 m):

Wave Angle	Azimuth: 180°	Azimuth: 160°/200°	Azimuth: 140°/220°	Azimuth: 120°/240°
10°	–35 dB	–35 dB	–33 dB	–26 dB
20°	–29 dB	–35 dB	–26 dB	–20 dB
30°	–35 dB	–32 dB	–21 dB	–17 dB
40°	–25 dB	–22 dB	–17 dB	–16 dB
50°	–17 dB	–16 dB	–15 dB	–17 dB
60°	–14 dB	–14 dB	–15 dB	–22 dB
70°	–16 dB	–17 dB	–20 dB	–26 dB
80°	–26 dB	–25 dB	–21 dB	–17 dB
Average	**–24.6 dB**	**–24.5 dB**	**–21.0 dB**	**–20.1 dB**

Total cone average rejection: 22.3 dB
Typical output: –8 dBi

1-WAVELENGTH BEVERAGE (178 m):

Wave Angle	Azimuth: 180°	Azimuth: 160°/200°	Azimuth: 140°/220°	Azimuth: 120°/240°
10°	–29 dB	<–35 dB	–23 dB	–14 dB
20°	–34 dB	<–35 dB	–20 dB	–13 dB
30°	–33 dB	–26 dB	–17 dB	–12 dB
40°	–23 dB	–20 dB	–16 dB	–12 dB
50°	–19 dB	–17 dB	–15 dB	–12 dB
60°	–17 dB	–16 dB	–15 dB	–13 dB
70°	–17 dB	–17 dB	–16 dB	–15 dB
80°	–21 dB	–21 dB	–20 dB	–20 dB
Average	**–24.1 dB**	**–24.8 dB**	**–17.8 dB**	**–13.9 dB**

Total cone average rejection: 19 dB
Typical output: –10 dBi

2-ELEMENT END-FIRE MINI ARRAY spacing 8 m, element length 6 m, ¹/₂ wave elements

Wave Angle	Azimuth: 180°	Azimuth: 160°/200°	Azimuth: 140°/220°	Azimuth: 120°/240°
10°	–27 dB	–32 dB	–34 dB	–25 dB
20°	–27 dB	–33 dB	–25 dB	–19 dB
30°	–34 dB	–37 dB	–21 dB	–15 dB
40°	–28 dB	–23 dB	–16 dB	–13 dB
50°	–18 dB	–16 dB	–13 dB	–11 dB
60°	–13 dB	–12 dB	–11 dB	–10 dB
70°	–10 dB	–10 dB	–10 dB	–10 dB
80°	–10 dB	–10 dB	–9 dB	–12 dB
Average	**–20.9 dB**	**–20.4 dB**	**–17.3 dB**	**–14.4 dB**

Total cone average rejection: 19.5 dB
Typical output: –32 dBi

■ 24. LOW HORIZONTAL ANTENNAS

Over the last few years, Top-Band DXers have found out that high-angle propagation often happens during gray line. Therefore it is not a bad idea to have a low dipole, or loop that covers such high angles. On 160, a 20-m high inverted V is such a "low" dipole. Watch out though, make sure it does not couple with your other transmit (or receiving) antennas! Test it! You can do this by alternatively opening and shorting the end of the feed line while listening or transmitting on the other antenna. If the background noise

EWE (15 × 5 m)

Wave Angle	Azimuth: 180°	Azimuth: 160°/200°	Azimuth: 140°/220°	Azimuth: 120°/240°
10°	−27 dB	−22 dB	−16 dB	−12 dB
20°	−29 dB	−20 dB	−14 dB	−9 dB
30°	−37 dB	−19 dB	−12 dB	−8 dB
40°	−23 dB	−17 dB	−11 dB	−8 dB
50°	−16 dB	−14 dB	−10 dB	−11 dB
60°	−12 dB	−11 dB	−9 dB	−7 dB
70°	−9 dB	−8 dB	−10 dB	−7 dB
80°	−7 d B	−6 dB	−6 dB	−5 dB
Average	**−19.5 dB**	**−14.6 dB**	**−14.7 dB**	**−7.75 dB**

Total cone average rejection: 13.5 dB
Typical output: −21 dB

EWE (10 × 5 m) + Compensation through phased dipole

Wave Angle	Azimuth: 180°	Azimuth: 160°/200°	Azimuth: 140°/220°	Azimuth: 120°/240°
10°	−27 dB	−28 dB	−21 dB	−14 dB
20°	−27 dB	−26 dB	−18 dB	−12 dB
30°	−28 dB	−26 dB	−18 dB	−12 dB
40°	−29 dB	−24 dB	−17 dB	−12 dB
50°	−29 dB	−23 dB	−17 dB	−12 dB
60°	−32 dB	−24 dB	−18 dB	−13 dB
70°	−35 dB	−25 dB	−19 dB	−15 dB
80°	−24 d B	−21 dB	−19 dB	−17 dB
Average	**−28.9 dB**	**−24.1 dB**	**−18.4 dB**	**−13.7 dB**

Total cone average rejection: 20.2 dB
Typical output: −30 dBi

K9YA loop (9 × 7.5 m)

Wave Angle	Azimuth: 180°	Azimuth: 160°/200°	Azimuth: 140°/220°	Azimuth: 120°/240°
10°	−13 dB	−12 dB	−11 dB	−9 dB
20°	−12 dB	−11 dB	−9 dB	−7 dB
30°	−13 dB	−12 dB	−9 dB	−7 dB
40°	−16 dB	−13 dB	−10 dB	−7 dB
50°	−21 dB	−16 dB	−11 dB	−8 dB
60°	−21 dB	−16 dB	−12 dB	−8 dB
70°	−14 dB	−13 dB	−11 dB	−9 dB
80°	−10 d B	−10 dB	−9 dB	−8 dB
Average	**−10.2 dB**	**−12.8 dB**	**−10.3 dB**	**−7.8 dB**

Total cone average rejection: 11.0 dB
Typical output: −25 dBi

Full-size 4-square (0.22 λ spacing, quadrature fed)

Wave Angle	Azimuth: 180°	Azimuth: 160°/200°	Azimuth: 140°/220°	Azimuth: 120°/240°
10°	−26 dB	−35 dB	−16 dB	−8 dB
20°	−33 dB	−30 dB	−13 dB	−6 dB
30°	−28 dB	−20 dB	−11 dB	−5 dB
40°	−19 dB	−15 dB	−9 dB	−5 dB
50°	−15 dB	−12 dB	−9 dB	−6 dB
60°	−13 dB	−12 dB	−9 dB	−7 dB
70°	−14 dB	−13 dB	−11 dB	−10 dB
80°	−18 dB	−17 dB	−14 dB	−16 dB
Average	**−20.8 dB**	**−19.2 dB**	**−11.5 dB**	**−7.4 dB**

Total cone average rejection: 14 dB
Typical output: +5 dBi

does change on a receiving antenna, or if the SWR changes on a transmit antenna, then you have mutual coupling, and you should either move the antennas further apart or take measures to decouple them.

If you are not going to use the antenna for transmitting, you may be better off with a loaded (physically short) dipole, as it will produce weaker overall signals and have less chances of coupling to other nearby antennas.

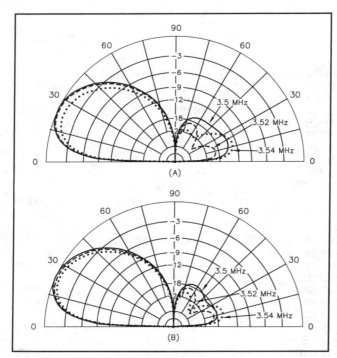

Fig 7-84—At A, the patterns for a fixed phase delay cable length, and a fixed dipole loading inductance. At B, we have improved the situation by tuning the dipoles when changing frequency. This at least doubles the 20 dB F/B bandwidth. See text for details.

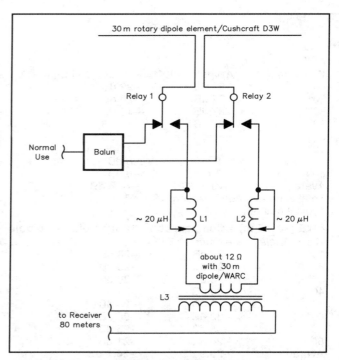

Fig 7-85—Diagram that shows how W6YA uses a WARC-band dipole for receiving on 80 meters. (See text for details.)

■ 25. OTHER RECEIVING ANTENNAS

A small (10.37-m long) split dipole

Jim McCook, W6YA, uses his Cushcraft D3W WARC-band dipole as a center-loaded receiving antenna on 80 meters. (See **Fig 7-85**). He uses two coils (approx. 20 μH) to resonate the dipole, and matches the low impedance to the feed line with a 4:1 transformer (2:1 turns ratio), wound on a type 77 ferrite toroid. The 20-μH coils are made from Air Dux coil stock (20 turns #18, 2″ diameter, 1.25″ long). Jim reports several S units S/N improvement over the transmit antenna, and he can null a single noise source 25 dB by rotating the antenna. He uses a preamp and band-pass filter to help avoid intermod from a local BC station.

■ 26. WHICH RECEIVING ANTENNA?

Receiving antennas work well only for one reason: their *directional selectivity*, in other words they don't receive equally well in all directions and all elevations.

26.1. High Lateral Directivity

Going by the shape of the receiving antennas I've reviewed, we can first distinguish those that get a lot of forward directivity, a small forward lobe, usually in addition to good F/B (front to back, the geometrical back at −180°) and or F/R (front to rear, the averaged picture) as well. Long Beverages (1.5 λ and longer), and wide-spaced broadside arrays of short Beverages (0.5 to 1.5 λ), EWEs, K9AY loops or subgroups of end-fire antennas, can have a 3-dB forward lobe angular width which can be compared with that of a 3-element Yagi. If you want to take full benefit of these antennas, you must be beaming accurately in the direction the signal comes from (which is not always the great circle direction, see Chapter 1 on Propagation). In other words, it is not a luxury to have 12 such antennas to cover the entire range of azimuths.

26.2. High F/B

The second group of special receiving antennas are those that rely on a deep zero off the back. Short Beverages (tuned for maximum F/B), well adjusted EWEs, K9AY loops and end-fire groups of 2 short verticals (eg, the Titanex SES 80 and SES 160 rotary 2-el array). With these antennas we rely in the first place on the (deep) null off the back to eliminate interference and achieve a better S/N ratio. This means that we really ought to be able to rotate those antennas, not for reasons of their narrow forward angle, but in order to be able to put their null right into the direction of the offending signal or noise source.

The EWE and the K9AY loop can, in general, be dimensioned to exhibit a very deep null at a fairly high wave angle (40 to 60°), which may be more practical than getting a very deep null at a lower wave angle (eg, 20°). Interference is usually from fairly local stations, arriving at higher wave angles.

For all receiving antennas we can find a direction (azimuth or elevation) where the rejection is very high

Table 7-29
Restricted Area Receiving Antennas Compared

EWE	K9AY Loop	2-Element Close-Spaced End-Fire Array (dipoles)
Positive:	**Positive:**	**Positive:**
• Simple to make	• Simple to make	• Can be made rotatable (also on 160)
• Small area	• Smallest area (by far)	• Good high angle rejection
• Not critical, easy to adjust	• Not critical, easy to adjust	• Commercially available
• Reasonably good output	• Rejection mostly at high angle (where you want it!)	
	• Easy to make rotatable	
Negative:	**Negative:**	**Negative:**
• Much high angle response (unless compensated with dipole)	• Less average rejection than EWE	• Fairly complex to build
• Fixed null direction		• Critical to adjust (unless commercial)
		• Very low output

(infinite). Quoting only this one figure is of course meaningless. What is the chance that your interference will be a single point source, exactly in the direction of this null? Nil!

I think it is much more relevant to look at the average F/B (rejection) in a cone (see also Section 22), which I defined as 80° wide (from 120° to 240° azimuth) and 60° high (from 10° to 80° elevation). Note that the definition of the cone is slightly different from what I used in Section 22.

I compared 4 antennas on 160 meters, and the results are found in **Table 7-28**.

- 2-λ long Beverage
- 1.5-λ long Beverage
- 1-λ long Beverage
- EWE (15 × 5 m)
- K9AY loop (9 × 7.5 m)
- 2-element close-spaced end-fire array
- 4-square (0.22-λ spacing, quadrature-fed)

The conclusions are simple:

- Nothing can beat a 1.5 or 2-λ (Cone-of-Silence length) Beverage antenna, considering its simplicity. It has a reasonably narrow forward lobe (forward and lateral discrimination, a spectacular back area rejection and a high output. It is also unbeatable as far as simplicity. All you need is space, and wire. . .
- The 1-λ Beverage and the 2-element end-fire array have fairly similar patterns, although the lateral discrimination of the Beverage is obviously better. The Beverage has 22 dB more output than the end-fire array!

- The 2-element end-fire array has very good back rejection, over a fairly wide angle, but suffers from very low output, and is frequency sensitive. It can be made rotatable, and to cover 80 as well as 160 meters. (This is my summer-time replacement for the Beverage antennas.)
- The EWE and K9AY loop have very poor lateral and wide-angle back-side rejection. It is clearly the compromise antenna (compromise between utmost simplicity and performance).
- The Compensated EWE has a spectacular high-angle rejection, and is second only to a Beverage. This array can be made rotatable, but has low output (typically −30 dBi) and is critical to adjust.
- Those who say that their 4-square is better than a Beverage have a Beverage that is not working properly. A 2-λ Beverage is largely superior to the 4-square!

26.3. Comparing Restricted-Area Receiving Antennas

EWEs, K9AY loops and close-spaced 2-element end-fire arrays are probably the best candidates for a relatively small lot. These are compared in **Table 7-29**. Again, don't forget to avoid mutual coupling with your transmit antenna (see Section 13). The lower the output of a receiving-type antenna, the easier its pattern is upset by coupling into other antennas or conductors (downspouts, electrical wiring, etc). The first thing to suffer is the deep null off the back, and for the family of smaller receiving antennas this is what we rely on to realize an advantage!

THE DIPOLE

The first antenna most amateurs are confronted with is a dipole. I remember how, as a young boy, I put up my first 20-meter dipole between a second floor window of our house and a nearby structure. It was fed with 75-ohm TV coax, and it worked, whatever that meant. For a while, my whole antenna world was limited to a dipole. But there is more to dipoles.

Although we often think of dipoles as $1/2$-λ, center-fed antennas, this is not always the case. The definition used here is that of a center-fed radiator with a symmetrical sinusoidal standing-wave current distribution.

■ 1. HORIZONTAL HALF-WAVE DIPOLE

1.1. Radiation Pattern of the Half-Wave Dipole in Free Space

The pattern in the plane of the wire has the shape of a figure 8. The pattern in a plane perpendicular to the wire is a circle (see **Fig 8-1**). The three-dimensional representation of the radiation pattern is shown in the same figure and is a ring (torus). The gain of this dipole over an isotropic radiator is 2.14 dB. This means that the dipole, at the tip of the ring where radiation is maximum, has a gain of 2.14 dB vs the theoretical isotropic dipole, which radiates equally well in all directions (its radiation pattern is a sphere).

1.2. The Half-Wave Dipole Over Ground

In any antenna system, the ground acts more or less as an imperfect or lossy mirror that reflects energy. Simplifying, and assuming a perfect ground, we can apply the Fresnel reflection laws, which state that the angles of incident and reflected rays are identical.

1.2.1. Vertical radiation pattern of the horizontal dipole

The vertical radiation pattern determines the wave angle of the antenna; the wave angle is the angle at which the radiation is maximum. Since obtaining a low angle of radiation is one of the main considerations when building low-band antennas, we will usually consider only the lowest lobe in case the antenna produces more than one vertical lobe. In free space, the radiation pattern of the isotropic antenna is a sphere. As a consequence, any plane pattern of the isotropic antenna in free space is a circle. In free space, the pattern of a dipole in a plane perpendicular to the antenna wire is also a circle. Therefore, if we analyze the vertical radiation pattern of the horizontal dipole over ground, its behavior is similar to an isotropic radiator over ground.

1.2.1.1 Ray analysis

Refer to **Fig 8-2**. In the vertical plane (perpendicular to the ground), an isotropic radiator radiates equal energy in all directions (by definition). Let us now examine a few typical

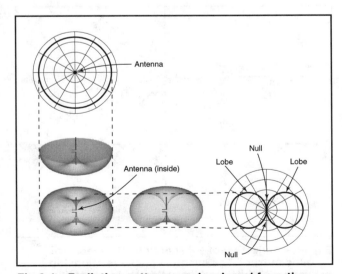

Fig 8-1—Radiation patterns as developed from the three-dimensional pattern of a half-wave dipole in free space. Upper left, vertical-plane pattern, and right, horizontal plane.

rays. A and A′ radiate in opposite directions. A′ is reflected by the ground (A″) in the same direction as A. B″, the reflected ray of B′, is reflected in the same direction as B.

The important issue is now the phase difference between A and A″, B and B″, C and C″, etc. Phase difference is made up by path length difference (length is directly proportional to time, as the speed of propagation is constant) plus possible phase shift at the reflection point. It is known that horizontally polarized rays undergo a 180° phase shift when reflected from perfect ground. This can be simulated by "feeding" the image antenna with an equal-amplitude current 180° out of phase (I = –I′) with the current feeding the (real) antenna (I = I′).

If at a very distant point (in terms of wavelengths) the rays at points A and A″ are in phase, then their combined field strength will be at a maximum and will be equal to the sum of the magnitudes of the two rays. If they are out of phase, the resulting field strength will be less than the sum of the individual rays. If A and A″ are identical in magnitude and 180° out of phase, total cancellation will occur.

If the dipole antenna is at a very low height (less than ¼ λ), A and A″ will reinforce each other. Low-angle rays will be almost completely out of phase, resulting in cancellation, and thus there will be very little radiation at low angles. At increased heights, A and A″ may be 180° out of phase (no radiation at zenith angle), and lower angles may reinforce each other. In other words, the vertical radiation pattern of a dipole depends on the height of the antenna above the ground.

1.2.1.2 Vertical radiation pattern equations

The radiation pattern can be calculated with the following equation.

$$F_\alpha = \sin h \left(h \sin \alpha \right) \qquad \text{(Eq 8-1)}$$

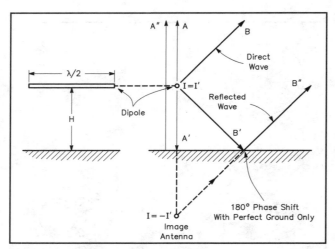

Fig 8-2—Reflection of RF energy by the electrical "ground mirror." The eventual phase relationship between the direct and the reflected horizontally polarized wave will depend primarily on the height of the dipole over the reflection ground (and to a small degree on the quality of the reflecting ground).

where
    F_α = normalized field intensity at vertical angle α
    h = height of antenna in degrees
    α = vertical angle of radiation

One wavelength equals 360°. The above equation is valid only for perfectly reflecting grounds. For real ground the ideal reflected wave must be multiplied by the complex reflection coefficient as given in **Fig 8-3**; its total phase difference is then ≥ 180°, its magnitude ≤ 1

Equation 8-1 can also be rewritten as follows

$$H_1 = \frac{74.95}{f \sin \alpha} \qquad \text{(Eq 8-2)}$$

where
    H_1 = height antenna in meters
    f = frequency, MHz
    α = vertical angle for which the antenna height is sought.

When more lobes are of interest, replace 90 with 270 for the second lobe, with 450 for third lobe, etc. If the nulls are sought, replace 90 with 180 for the first null, with 360 for the second null, etc.

Table 8-1 gives the major lobe angles as well as reflection-point distances for heights ranging from 18 m

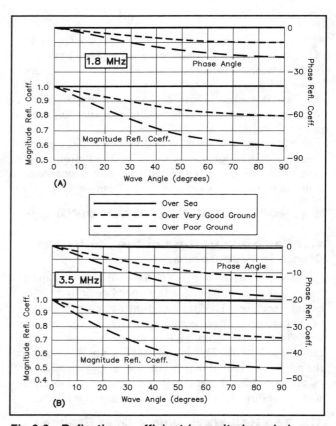

Fig 8-3—Reflection coefficient (magnitude and phase angle) of horizontally polarized waves over three types of ground: sea, average and very poor. See text for details.

Table 8-1

Major Lobe Angles and Reflection Point for Various Dipole Antenna Heights

Antenna Height (ft)	(m)	40 Meters Angle (deg)	Distance (ft)	(m)	80 Meters Angle (deg)	Distance (ft)	(m)	160 Meters Angle (deg)	Distance (ft)	(m)
60	18	36	83	25	90	0	0	90	0	0
80	24	26	163	50	54	58	18	90	0	0
100	30	20	266	81	40	118	36	90	0	0
120	36	17	391	119	33	187	57	90	0	0
140	42	15	540	148	28	268	82	77	31	9
160	48	13	710	217	24	362	110	59	97	30
180	54	–	–	–	21	467	142	49	154	47
200	60	–	–	–	18	584	178	43	213	66

(60 ft) to 60 m (200 ft) for 40, 80 and 160 meters.

1.2.1.3. Sloping ground locations

In many cases, an antenna cannot be erected above perfectly flat ground. A ground slope (Ref 630) can greatly influence the wave angle of the antenna. The RADIATION ANGLE HORIZONTAL ANTENNAS module of the NEW LOW BAND SOFTWARE calculates the radiation pattern of dipoles (or Yagis) as a function of the terrain slope.

Table 8-2 shows the influence of the slope angle on the required antenna height for a given wave angle on 80 meters. The table lists the required antenna height and the distance to the reflection point for a horizontally polarized antenna. A positive slope angle is an uphill slope. The results from this table can easily be extrapolated to 40 or 160 meters by simply dividing or multiplying all of the results by 2. For 40 meters, all dimensions should be halved; for 160 meters, all dimensions should be doubled.

1.2.1.4. Antennas over real ground

Up to this point, for most of the presented results, a perfect ground has been assumed. Perfect ground does not exist in practical installations, however. Perfect ground conditions are approached only when an antenna is erected over salt water.

Radiation efficiency and reflection efficiency

Contrary to the case with vertical antennas, a horizontal antenna does not rely on the ground for providing a return path for antenna currents. The physical "other half" takes care of that. In practice, this means that the ground plays no important role in determining the radiation efficiency of a horizontal antenna. The radiation efficiency is related only to the losses in the antenna itself (conductor, insulator, loading coils, etc), although of course some of the total radiated energy can be dissipated in the ground losses.

Both horizontally as well as vertically polarized antennas rely on the ground for reflection of the RF in the so-called Fresnel zone to "build up" the radiation pattern in combination with the direct wave, as shown in Fig 8-2. The efficiency of the reflection depends on the quality of the ground, and is called the "reflection efficiency."

Reflection coefficient

The reflection from real ground is not like on a perfect mirror. The reflection coefficient is a complex number that describes the reflection from real ground:

- With a perfect mirror, all energy is reflected. There are no losses; the reflection coefficient magnitude is 1.
- With a perfect mirror, the phase of the reflected horizontal wave is shifted exactly 180° vs the incoming wave.
- With real ground, part of the RF is absorbed, and the reflection coefficient magnitude is less than 1.
- With real ground, the phase angle of the reflection coefficient is greater than 180°. Except when the antenna wave angle is quite high, the departure from 180° is very small. This departure typically varies between 0 and 25° for reflection angles (equal to wave angles) between 0 and 90°.

The magnitude of the reflection coefficient, which becomes smaller as the ground quality becomes poorer, is the reason that the dipole over real ground shows less gain than over perfect ground.

The reflection coefficient is a function of the wave angle. The smaller the wave angle, the closer the reflection coefficient magnitude will be to 1. This explains why the loss with a dipole (poor ground vs perfect ground) is higher at high angles (eg, zenith) than at low angles. See **Fig 8-4**.

The fact that the dipole over poor ground seems to have a lower radiation angle than over perfect ground is because at lower angles there is less loss. In other words, over poor ground it just has less loss at low angles than at high angles.

The filling in of the deep notch at a 90° wave angle for the dipole at $\frac{1}{2}$ (and 1) λ (Fig 8-4B and D) is because the reflected wave is considerably attenuated and additionally phase shifted and can no longer cancel the direct wave. Note that the effect of the additional phase shift could be compensated for by changing the height of the antenna.

Table 8-2

Slope Angle Versus Antenna Height at 3.5 MHz

Slope Angle (deg)	20° Wave Angle Height (ft)	Distance (ft)	30° Wave Angle Height (ft)	Distance (ft)	40° Wave Angle Height (ft)	Distance (ft)
35	–	–	–	–	906	10,364
30	–	–	–	–	430	2441
25	–	–	819	9367	275	1029
20	–	–	396	2249	201	553
15	768	8789	258	966	158	340
10	378	2146	192	528	131	227
5	251	937	153	329	113	161
0	189	520	129	224	100	120
−5	153	329	113	161	91	91
−10	131	227	102	121	85	72
−15	116	166	94	94	81	57
−20	107	127	89	75	79	45
−25	101	101	87	61	78	36
−30	97	91	86	49	78	28

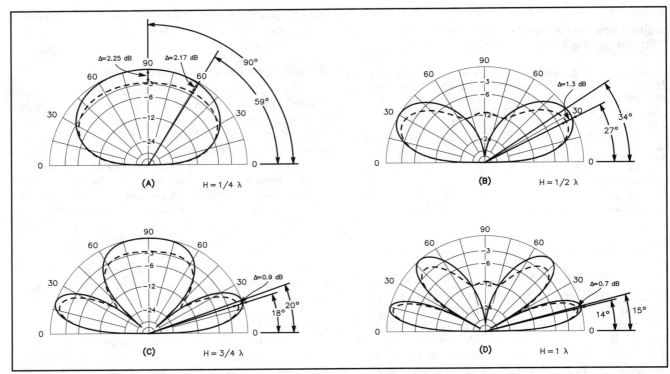

Fig 8-4—Vertical radiation patterns over two types of earth: sea water (solid line in each set of plots) and very poor ground (broken line in each set of plots). The wave angle as well as the gain difference between sea water and poor ground are given for four antenna heights.

Again refer to Fig 8-3, showing the reflection coefficient (magnitude and phase) for a horizontally polarized wave. Information is given for a horizontally polarized antenna over sea, average ground and very poor ground, for both 160 and 80 meters. Eq 8-1 multiplied by this reflection coefficient gives the vertical radiation pattern over real ground.

Radiation patterns

Fig 8-4 shows vertical patterns of a horizontal dipole over both near-perfect ground (salt water) and desert, the two extremes. **Table 8-3** lists the wave angle and the relative loss for a half-wave dipole over 5 different types of ground, and for two antenna heights. Note that for a dipole at $1/2\,\lambda$, the peak wave angle drops from 30° over sea water to 26° over desert. At the same time there is a radiation loss of 1.21 dB.

For an antenna at $1/4$-λ height, maximum radiation occurs at 90° over a perfect conductor (see Fig 8-4). Over very poor ground (desert), the maximum radiation is at 59°. This is not because more RF is concentrated at this lower angle, but only because more RF is being dissipated in the poor ground at the 90° angle than at 59° (the reflection coefficient is much lower at 90° than at 59°). The difference, however, between the radiation at 90° and at 59° is very small (0.08 dB). The difference in radiated power at 90° between salt water and a desert type of reflecting ground is 2.25 dB. As 90° is a radiation angle of little practical use, the relatively high loss at the zenith angle should not bother us.

With a vertical antenna, poor ground results in loss at the low angles in the first place. With horizontal dipoles the loss due to poor ground is in the first place at high angles!

Notice that for a height of $1/2\,\lambda$ (Fig 8-4B), the sharp null at a 90° elevation angle has been degraded to a mere 12-dB attenuation over desert-type ground.

Conclusion

We can conclude that the effects of absorption over poor ground are quite pronounced with low antennas and become less pronounced as the antenna height is increased. Artificial improvement of the ground conditions by the installation of ground wires is only practical if one wants maximum gain at a 90° wave angle (zenith) from a low

---

Table 8-3

Relative gain (vs dipole over perfect ground) and wave angle (max vertical radiation angle) for $1/2$-λ dipoles at heights of $1/4$ and $1/2\,\lambda$

	Height = $1/4$ Wave		Height = $1/2$ Wave	
	Rel. Loss (dB)	Wave Angle (deg)	Rel. Loss (dB)	Wave Angle (deg)
Perfect Ground	0	90	0	30
Sea	−0.05	90	−0.01	30
Very Gd Grnd	−0.57	71	−0.16	29
Avg Grnd	−1.23	62	−0.52	28
Very Poor Grnd	−2.17	53	−1.21	26

dipole ($\frac{1}{8}$ to $\frac{1}{4}$-λ height). This can be done by burying a number of wires ($\frac{1}{2}$ to 1-λ long) underneath the dipole, spaced about 60 cm apart, or by installing a parasitic reflector wire ($\frac{1}{2}$-λ long plus 5%) just above ground (2 m), about $\frac{1}{8}$ to $\frac{1}{4}$ λ under the dipole.

Improving the efficiency of the reflecting ground for low-angle signals produced by high dipoles is impractical and yields very little benefit. The active reflection area can be as far as 10 or more wavelengths away from the antenna!

Dipoles, unlike verticals, do not suffer to a great extent from poor ground conditions. The reason is that for horizontally polarized signals, when reflected by the ground, the phase shift remains almost constant at 180° (within 25°), whatever the incident angle of reflection (equal to the wave angle) may be. For verticals, the phase angle varies between 0° and 180°. For vertical antennas, the pseudo Brewster angle is defined as the angle at which the phase shift at reflection is 90°. This means that there is no pseudo Brewster angle with horizontally polarized antennas, such as a dipole, because there never will be a 90° phase shift at the reflection point.

The effects of this mechanism are proved daily by the fact that on the low bands, big signals from areas with poor ground conditions (mountainous, desert, etc) are always generated by horizontal antennas, while from areas with fertile, good RF ground, we often hear big signals from verticals and arrays made of verticals.

1.2.2. Horizontal pattern of horizontal half-wave dipole

The horizontal radiation pattern of a dipole in free space has the shape of a figure 8. The horizontal directivity of a dipole over real ground depends on two factors:
1. Antenna height.
2. The wave angle at which we measure the directivity.

Fig 8-5 shows the horizontal directivity of half-wave horizontal dipoles at heights of $\frac{1}{4}$, $\frac{1}{2}$, $\frac{3}{4}$ and 1 λ over average ground. Directivity patterns are included for wave angles of 15° through 60° in increments of 15°. At high angles, a low dipole shows practically no horizontal directivity. At low angles, where it has more directivity, the low dipole hardly radiates at all. Therefore, it is quite useless to put two dipoles at right angles for better overall coverage if those dipoles are at low heights.

At heights of $\frac{1}{2}$ λ and more, there is discernible directivity, especially at low angles. **Fig 8-6** gives a visual representation of the three-dimensional radiation pattern of a half-wave dipole at $\frac{1}{2}$ λ above average ground.

1.3. Half-Wave Dipole Efficiency

The radiation efficiency of an antenna is given by the equation

$$\text{Eff} = \frac{R_{rad}}{R_{rad} + R_{loss}}$$

where

R_{rad} = radiation resistance, ohms

R_{loss} = loss resistance, ohms

1.3.1. Radiation resistance

The radiation resistance is a fictive resistance in which the same current flows as in the center of the dipole, and whereby this resistor dissipates all the RF applied to it. For a half-wave dipole at or near resonance, the radiation resistance is equal to the real (resistive) part of the feed-point impedance (assuming a perfectly lossless antenna system).

The relationship of the radiation resistance and reactance of a half-wave dipole to its height above ground is shown in **Fig 8-7**. The radiation resistance varies between 60 and 90 Ω for all practical heights on the low bands. For determining the reactance, the dipole was dimensioned to be resonant in free space (72 Ω). We can conclude that the resonant frequency changes with half-wave-dipole height above ground. Where the reactance is positive, the dipole appears to be too long, and too short where the reactance is negative.

1.3.2. Losses

The losses in a half-wave dipole are caused by:
- RF resistance of antenna conductor (wire)
- Dielectric losses of insulators
- Ground losses

Table 8-4 gives the effective RF resistance for common conductor materials, taking skin effect into account. The resistances are given in ohms per kilometer. The RF resistance values in the table are valid at 3.8 MHz. For 1.8 MHz the values must be divided by 1.4, while for 7.1 MHz the values must be multiplied by the same factor. The RF resistance of copper-clad steel is the same as for solid copper, as the steel core does not conduct any RF at HF. The dc resistance is higher by 3 to 4 times, depending on the copper/steel diameter ratio. The RF resistance at 3.8 MHz is 18 times higher than for dc (25 times for 7 MHz, and 13 times for 1.8 MHz). Steel wire is not shown in the table; it has a much higher RF resistance. Never use steel wire if you want good antenna performance.

1.3.2.1. Dielectric losses in insulators

Dielectric losses are difficult to assess quantitatively. Care should be taken to use good quality insulators, especially at the high-impedance ends of the dipole. Several insulators can be connected in series to improve the quality.

1.3.2.2. Ground losses

Reflection of RF at ground level coincides with absorption in the case of non-ideal ground. With a perfect reflector, the gain of a dipole above ground is 6 dB over a dipole in free space (the field intensity doubles due to a perfect ground reflection).

The ground is never a perfect reflector in real life. Therefore part of the RF will be dissipated in the ground. The effects of power absorption in the real ground have been covered in paragraph 1.2.1.4. and illustrated in Fig 8-4 and Table 8-3.

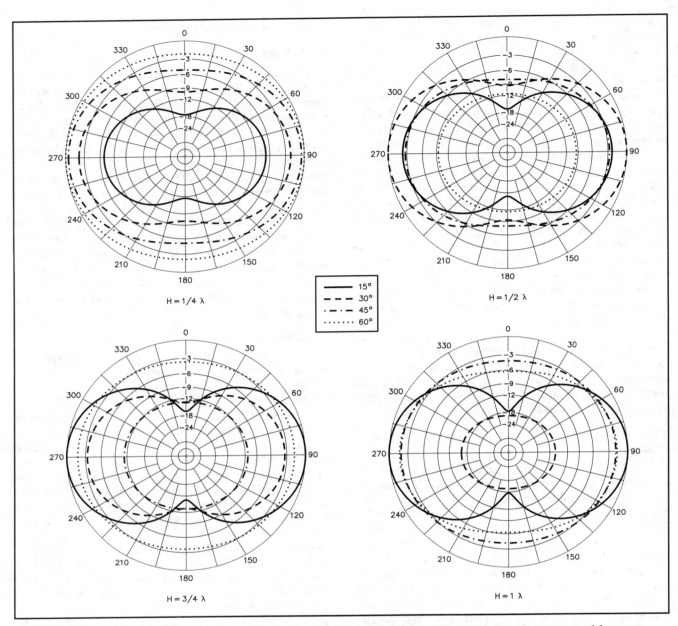

Fig 8-5—Horizontal radiation patterns for ¹/₂-wave horizontal dipoles at various heights above ground for wave angles of 15, 30, 45 and 60° (modeled over good ground).

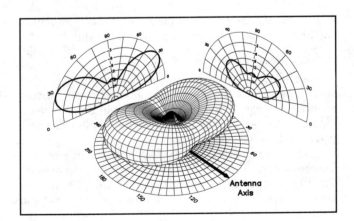

Fig 8-6—Three-dimensional representation of the radiation pattern of a half-wave dipole, ¹/₂ λ above ground.

Attempting to improve ground conductivity for improved performance is a common practice with grounded vertical antennas. One can also improve the ground conductivity with dipoles, although it is not quite as easy, especially if one is interested in low-angle radiation and if the antenna is physically high. From Table 8-1 we can find the distance from the antenna to the ground reflection point. For the major low-angle lobe this is 36 m (118 ft) for an 80-meter dipole at 30 m (100 ft). Consequently, this is the

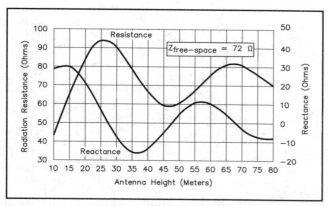

Fig 8-7—Radiation resistance and feed-point reactance of a dipole at various antenna heights. Calculations were done at 3.65 MHz using a 2-mm OD conductor (AWG 12 wire) over good ground.

place where the ground conductivity must be improved. Because of the horizontal polarization of the dipole, any wires that are laid on the ground (or buried in the ground) should be laid out parallel to the dipole. They should preferably be at least 1-λ long. However, in view of the small gain that can be realized, especially with high antennas and for low wave angles, it is very doubtful that such improvement of the ground is worth all the effort. The only really worthwhile improvement will be obtained by moving to the sea coast or to a very small island surrounded by salt water. It cannot be emphasized enough, however, that the quality of the reflecting ground with horizontal antennas is of far less importance than with vertical antennas.

The efficiency of low dipoles (¹/₄-λ high and less), which essentially radiate at the zenith angle (90°), can be improved by placing wires under the antenna running in the same direction as the antenna.

1.4. Feeding the Half-Wave Dipole

In general, half-wave dipoles are fed in the center. This, however, is not a must. The Windom antenna is a half-wave dipole fed at approximately ¹/₆th from the end of the half-wave antenna, with a single-wire feed line. It has been proved (and can be confirmed by modeling) that careful placing of the feed point results in a perfect symmetrical and sinusoidal current distribution in the antenna (Ref 688).

The disadvantage of the single-wire fed antenna (Windom) is that the feed line does radiate, and as such distorts the radiation pattern of the dipole. Belrose described a multiband "double Windom" antenna using a 6:1 balun and coaxial feed line in the above-mentioned publication.

1.4.1. The center-fed dipole

The feed point of a center-fed dipole is symmetrical. The antenna can be fed via an open-wire transmission line, if it is to be used on different frequencies (eg, as two half-waves in phase on the first harmonic frequency), or with a coaxial feed line via a balun. The balun is mandatory in order not to upset the radiation pattern of the antenna. Baluns are covered in detail in the chapter on transmission lines. A current-type balun, consisting of a stack of high-permeability ferrite beads, slipped over the coaxial cable at the load, is recommended. The exact feed-point impedance can be found from Fig 8-7.

Bandwidth

The SWR bandwidth of a full-size half-wave dipole is determined by the diameter of the conductor. **Fig 8-8A** shows the SWR curves for dipoles of different diameters. Large-conductor diameters can be obtained by making a so-called wire-cage (Fig 8-8B). I used the wire-cage approach on my 80-meter vertical, 6 wires forming a 30-cm (12-inch) diameter cage.

Fig 8-9 shows the effective equivalent diameter of such a cage conductor as a function of the number of wires making up the cage. Instead of using a wire cage, one can also use configuration existing of a number of equally spaced wires in a plane. **Fig 8-10** shows the effective equivalent diameters of such a flat multi-wire configuration. (source: Kurze Antennen, by Gerd Janzen, ISBN 3-440-05469-1). Example: three parallel wires, each measuring 2 mm OD, and equally spaced 5 cm have an effective equivalent diameter of a solid conductor of $50 \times 0.65 = 32.5$ mm.

A folded dipole shows a much higher SWR bandwidth than a single-wire dipole. A folded dipole for 80 meters, made of AWG no. 12 wire, with a 15 cm (6 in.) spacing between the wires, will cover the entire 80-meter band (3.5-3.8 MHz) with an SWR of approximately 1.75:1, as compared to 2.5:1 or more for a straight dipole.

1.4.2. Broadband dipoles

Instead of decreasing the Q factor of the antenna, one can also devise a system whereby the inductive part of the impedance is compensated for as one moves away from the resonant frequency of the antenna. The "double Bazooka dipole" is probably the best known example of such an antenna. In this antenna, part of the radiator is made of coaxial cable, connected in such a way as to present shunt impedances across the dipole feed point when moving away from the resonant frequency. F. Witt, AI1H, covered a similar broadband dipole antenna in detail (Ref 1012). **Fig 8-11** shows the dimensions of Witt's 80-meter DX-special antenna, which has been

Table 8-4
Resistance of Various Types of Wire Commonly Used for Constructing Antennas

Wire Diameter	Copper dc (Ω/km)	Copper 3.8 MHz (Ω/km)	Copper-clad dc (Ω/km)	Copper-clad 3.8 MHz (Ω/km)	Bronze dc (Ω/km)	Bronze 3.8 MHz (Ω/km)
2.5 mm (AWG 10)	3.4	61	8.7	61	4.5	81
2.0 mm (AWG 12)	5.4	97	13.8	97	7.2	130
1.6 mm (AWG 14)	8.6	154	22.0	154	11.4	206
1.3 mm (AWG 16)	13.6	246	35.0	246	18.2	326
1.0 mm (AWG 18)	21.7	391	55.6	391	29.0	521

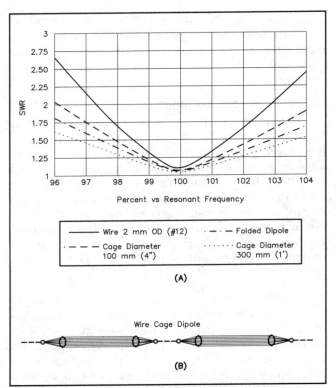

Fig 8-8—At A, SWR plots for 3.75-MHz half-wave dipoles (in free space) of various conductor diameters. The total bandwidth of the 80-meter band (3.5-3.8 MHz) is 8%. The 100-mm (4-inch) and 300-mm (12-inch) diameter conductors can be made as a cage of wires, as shown at B. Note that the SWR bandwidth of a folded dipole is substantially better than for a straight dipole. The spacing between the wires of the folded dipole does not influence the bandwidth to a large extent.

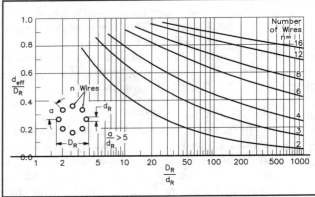

Fig 8-9—Normalized effective diameter d_{eff}/DR for a wire cage conductor, made out of n conductors (diameter d_R) spaced uniformly on a ring with a diameter of D_R. Example: a wire cage made out of 6 wires of 2-mm diameter, spaced equally on a circle measuring 20 cm in diameter, had an equivalent diameter of a single solid conductor of 0.62 × 20 = 125-mm diameter. (source: *Kurze Antennen*, by Gerd Janzen, ISBN 3-440-05469-1)

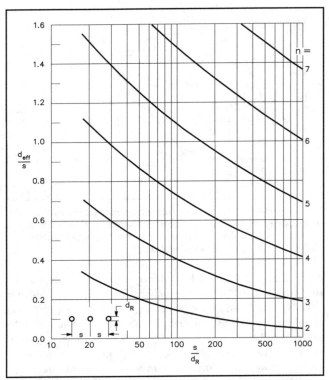

Fig 8-10—Normalized effective diameter d_{eff}/DR for a flat multi-wire conductor, made out of n conductors (diameter d_R) spaced uniformly with a spacing S. (Source: *Kurze Antennen*, by Gerd Janzen, ISBN 3-440-05469-1)

dimensioned for minimum SWR at both the CW as well as SSB end of the band.

Another innovative broadbanding technique was described by M. C. Hatley, GM3HAT (Ref 682).

R. Severns, N6LF, described a broadband 80-meter folded dipole using the principle of the open-sleeve antenna (Ref 1014). **Fig 8-12** shows the layout of the folded dipole, where Severns inserted another nearly half-wave long wire between the legs of the folded dipole. The resulting SWR curve is also shown in Fig 8-12B.

Of course there is no reason why one could not apply switched inductive or capacitive loading devices, such as described in detail in the chapters on verticals and large loop antennas, although this is seldom done.

1.4.3. Does a resonant dipole radiate better than a dipole off-resonance?

No, an infinitely short dipole would radiate as well as a full-size $1/2$-λ dipole, provided you can get the same power in the dipole, and provided the (normalized) losses are the same. Such an infinitely short dipole is called a Hertz dipole. It has a constant current distribution and therefore a slightly different radiation pattern than a half-wave dipole. But that should not concern us as it is a theoretical antenna, anyhow.

If you have a really short dipole, the radiation resistance will be very low, maybe a few ohms, and the feed-point impedance (in the center) will be extremely capacitive (several thousand ohms). This makes it quite difficult to

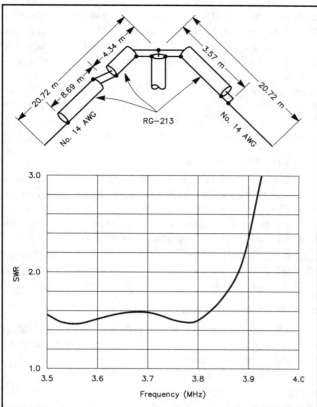

Fig 8-11—Dimensions and SWR curve of the 80-Meter DX Special, a design by F. Witt, AI1H.

feed this very short antenna with a good efficiency (see Section 2). However, if the antenna is only slightly shorter (or longer) than a (resonant) half-wave, its feed-point impedance will vary only slightly (a few percent max.) from what it is at resonance, and the reactive components will still be manageable as far as feeding this nonresonant dipole.

Take the example of a typical center-fed open dipole (a single wire) for 80 meters tuned for 3.65 MHz at a height where its impedance at resonance is exactly 50 Ω (see Fig 8-7). The antenna impedance at 3.5 and at 3.8 MHz will be such that the SWR will be approx. 2:1 (still referred to our 50-Ω system impedance), and this is mainly caused by the reactive component. Whether or not this nonresonant antenna will radiate as much power as its resonant counterpart, depends on how much loss there will be in the feed system, which has to match a complex impedance (eg, 40 + $j70$ Ω on 3.5 MHz and 60 + $j70$ Ω on 3.8 MHz). One can safely say that on the low bands, feed systems can be made

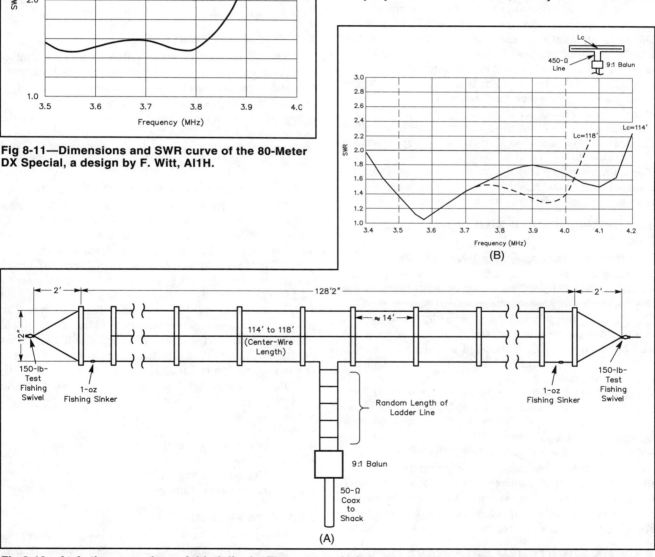

Fig 8-12—At A, the open-sleeve folded dipole. The antenna is fed with a random length of 450-Ω open-wire transmission line through a 9:1 balun. At B, SWR curve of the open-sleeve dipole, showing curves for different lengths of the center wire (L C).

and used that will show negligible losses when operated with SWRs below 2 or even 3:1.

Summarizing, the off-resonance dipole (and the same is true for a monopole) will radiate just as well as the resonant dipole. The only issue is the "ease" of feeding this dipole, when it is way off resonance.

1.4.4. Does frequency of lowest SWR equal resonant frequency?

Is it true that the resonant frequency of a dipole is the frequency where the SWR is lowest? No, it is not, but is it important to know the exact resonant frequency of a dipole? No, it is neither. What is generally important is to know the frequency where the dipole will cause the lowest SWR on the feed line.

Let me explain with an example: A dipole which, on its resonant frequency, has an impedance of 70 Ω, has an SWR of 1.4:1 on its resonant frequency. Somewhat lower in frequency, a combination of a lower resistive part with some capacitive reactance can result is a lower SWR than on the antenna's resonant frequency! It really depends on how fast the reactance changes compared to the resistance. If the resistance change is slow, and the reactance change is fast, that statement can be true. But if the resistance changes rapidly around resonance that statement can be false. In other words, lowest SWR is achieved where the feed impedance is at the point closest to the center of the Smith Chart, and this isn't necessarily on the line of zero reactance.

But all of this should not bother us; we should cut our dipole for lowest SWR in the center of the (portion of the) band we want to cover. Whether or not this is the dipole's resonant frequency is here totally irrelevant.

1.5. Getting the Full-Size Dipole in Your Backyard

The ends of the half-wave dipole can be bent (vertically or horizontally) without much effect on the radiation pattern or efficiency. The tips of the dipoles carry little current; hence, they contribute very little to the radiation of the antenna.

Bending the tips of a dipole is the same as "end loading" the dipole (equal to top-loading with verticals). The folded tips can be considered as capacitive loading devices. For more details, see Section 2 of Chapter 9 (Vertical Antennas).

N. Mullani, KØNM, calculated the gain and the impedance of a half-wave dipole with its end hanging down vertically (Ref 691). He concluded that with horizontal lengths as short as 40% of full size, the trade-offs are rather insignificant, being only about 0.6 dB in gain, and some reduction of SWR bandwidth. Bending the end may actually somewhat improve the match to a 50-ohm feed line, depending on antenna height. KØNM concludes: "Don't be afraid to bend your dipole antennas if you are cramped for space." (See also Chapter 14, Antennas for the Small Lot.)

■ 2. THE SHORTENED HALF-WAVE DIPOLE

On the low bands, it is sometimes impossible to use full-size radiators. This section describes the characteristics of short dipoles, and how they can be successfully deployed. Short dipoles are often used as elements in reduced-size Yagis (see the chapter on Yagis) or to achieve manageable dimensions whereby the antenna can be fit into a city lot.

Short antennas are the subject of an excellent book (in German) by Gerd Janzen, DF6SJ/VK2BJZ (Ref: 7818). This book is highly recommended for anyone who does not fear a formula and a graph, and who really wants to dig a little deeper into the subject.

2.1 The Principles

You can always look at a dipole as two back-to-back connected verticals, whereby the "vertical" elements are no longer vertical. Instead of having the ground make the mirror image of the antenna (this is always the case with verticals), we supply the mirror half ourselves.

All principles about radiation resistance and loading of short verticals, as explained in the chapter on vertical antennas, can be directly applied to dipoles as well.

2.2. Radiation Resistance

The radiation resistance of a dipole (made of an infinitely small diameter conductor) in free space will be twice the value of the equivalent monopole. For instance, the R_{rad} for the half-wave dipole made of an infinitely thin conductor is approximately 73.2 Ω, which is twice the value of the quarter-wave vertical (36.6 Ω). Over ground, the radiation resistance will also vary in a similar way as the full-size half-wave dipole (see Fig 8-7).

2.3. Tuning or Loading the Short Dipole

Loading a short dipole consists of bringing the antenna to resonance. This means eliminating the capacitive reactance component in the feed-point impedance. Different loading methods yield different values of radiation resistance.

It is not necessary, however, to load a shortened antenna to resonance in order to operate it. You can also connect a feed line to it, directly or via a matching network, without tuning out the capacitive reactance. Therefore you can consider the dipole together with its feed line as the dipole system, and analyze the system of a short dipole to see what the alternatives are. Sometimes this situation is referred to as a dipole with "tuned feeders."

There are different ways to operate the short dipole system:
• Tuned feeders
• Matching at the dipole feed point
• Coil loading to tune out the capacitive reactance
• Linear loading
• Capacitive end loading
• Combined loading methods

2.3.1. Tuned feeders

Tuned feeders were common in the days before the arrival of coaxial feed lines. Very low-loss open-wire feeders can be made. The Levy antenna is an example of a short dipole fed with open-wire line; its overall length is $^1/_4 \lambda$. This antenna (R_{rad} = approximately 13 Ω and X_C = approximately $-j1100$ Ω), can be fed with open-wire feeders (450-600 Ω) of any length into the shack, where we can match it to 50 Ω via an antenna tuner. An outstanding feature of this approach is that the system can be "tuned" from the shack via the antenna tuner, and is not narrow banded, as is the case with loaded elements.

Let us calculate the losses in such a system. The losses of the antenna can be assumed to be zero (provided wire elements of the proper size and composition are used). The loss in a flat open-wire feeder is typically 0.01 dB per 100 feet at 3.5 MHz. The SWR on the line will be an unreal value of 280:1 (this value was calculated using the program SWR RATIO which is part of the NEW LOW BAND SOFTWARE. The additional line loss due to SWR for a 30-m (100 ft) long line will be in the order of 1.5 dB (Ref 600, 602). On a line with standing waves, the impedance is different at every point. Slightly changing the feeder length can produce more manageable impedances which the tuner can cope with more easily. This can be done with the software module COAX TRANSFORMER/SMITH CHART or IMPEDANCE, CURRENT AND VOLTAGE ALONG FEED LINES. A good antenna tuner should be able to handle this matching task with a loss of less than 0.2 dB. The total system loss depends essentially on the efficiency with which the tuner can handle the impedance transformation. The typical total loss in the system should be under 2.0 dB.

2.3.2. Matching at the dipole feed point

You can of course install the matching network at the dipole feed point, although this will be highly impractical in most cases. In the case of a vertical antenna this solution is practical, as the feed point is at ground level.

2.3.3. Coil loading

Loading coils can be installed anywhere in the short dipole halves, from the center to way out near the end. Loading near the end will result in a higher radiation

A short 80-meter rotary dipole sits atop the mast of the near tower at JH3VNC.

resistance, but will also require a much larger coil, and hence introduce more coil losses.

2.3.3.1. Center loading

The inductive reactance required to resonate the Levy dipole from the previous example is approximately 1100 ohms. To achieve this, two 550-ohm (reactance) coils need to be installed in series at the feed point. We should be able to realize a coil Q (quality factor) of 300.

$$R_{loss} = \frac{550}{300} = 1.83 \ \Omega$$

The total equivalent loss resistance of the two coils is 3.66 Ω.

The antenna efficiency will be:

$$Eff = \frac{13}{13 + 3.66} = 78\%$$

The equivalent power loss is $-10 \log (0.78) = 1.08$ dB.

The feed-point resistance of the antenna is $13 + 3.66 \approx 16.7 \ \Omega$ at resonance. This assumes negligible losses from the antenna conductor (heavy copper wire). If the use of coaxial feed lines is desired, an additional matching system will be needed to adapt the 15-ohm balanced feed-point impedance to the 50 or 75-ohm unbalanced coaxial cable impedance. The above example was calculated assuming free-space impedances. Over real ground the impedances can be different, and will vary as a function of the antenna height.

Another way to determine the necessary inductance is to model the antenna using a *MININEC* program (eg, *ELNEC* or *MN*). Let us work out the example of the $\frac{1}{4}$-λ dipole using *ELNEC*:

Input data:
f = 1.83 MHz
h = 25 m
ℓ_{ant} = $\frac{1}{4}$ λ

Procedure

1. Find a dipole length that is resonant at 1.83 MHz. (This turns out to be 80.1 m.) Do not model the dipole at any lower height, as the results obtained with *MININEC* will be erroneous. It makes no difference what type of earth you model, as *MININEC* always reports the impedance over a perfect reflector. However, do not model the dipole in free space as this will give incorrect results.
2. The $\frac{1}{4}$-λ dipole is half the above length: 40.05 m. Model the dipole again: The impedance is $12.5 - j1094$ Ω.

The required center loading coil has a reactance of 1094 ohms. Assuming a loading-coil Q of 300, the total equivalent loss resistance is 3.64 ohms. The feed-point resistance becomes $12.5 + 3.64 = 16.14$ ohms.

Matching to the feed-line impedance

One way of matching this impedance to a 50-ohm feed line is to use a quarter-wave transformer. The required impedance of the transformer is

$$Z_0 = \sqrt{16.04 \times 50} = 28.4$$

We can "construct" a feed line of 25 Ω (that's close) by paralleling two 50-Ω feed lines. Don't forget you need a 1:1 balun between the antenna terminals and the feed line.

Another attractive way that has been used by a number of commercial manufacturers of short 40-meter Yagis is to use a single central loading coil, on which we install a link in the center. The link turns are adjusted to give a perfect match to the feed line.

Comparing losses

A good current-type balun should account for no more than 0.1 dB of loss. The loading coils (Q factor = 300) give a loss of 1.3 dB. Including 30 m (100 ft) of RG-213 (with 0.23 dB loss), the total system loss can be estimated at 1.63 dB. The resulting efficiency is very close to the result obtained with open-wire feeders.

There are certain advantages and disadvantages to this concept, however. An advantage is that coaxial cable is easier to handle than open-wire line, especially when dealing with rotatable antenna systems.

The high Q of the coils will make the antenna narrow-banded as far as the SWR is concerned. In the case of the open-wire feeders, retuning the tuner will solve the problem. With the coaxial feed line you may still need a tuner at the input end if you want to cover a large bandwidth, in which case the extra losses due to SWR in the coaxial feed line may be objectionable.

Another disadvantage is that the loading-coil solution requires two more elements in the system—the coils. Each element in itself is an extra reliability risk, and even the best loading coils will age and require maintenance.

Instead of modeling the antenna with *MININEC*, we can calculate the required loading coils as in the following example:

Length of the dipole = 22.5 m
f = 3.8 MHz
Wire diameter = 2 mm OD (AWG no. 12)
Antenna height = 20 m

The full-size dipole length (2.5% shorting factor) is 38.5 m. We first calculate the surge impedance of the transmission-line equivalent of the short dipole using the equation

$$Z_s = 276 \log \left[\frac{S}{d \times \sqrt{1 + \frac{S}{4h}}} \right] \qquad \text{(Eq 8-3)}$$

where
S = dipole length = 2250 cm
d = conductor diameter = 0.2 cm
h = dipole height = 2000 cm
Z_s = 1103 ohms

The electrical length of the 22.5-m long dipole is

$$\ell = 180 \times \frac{22.5}{38.5} = 105.2° \qquad \text{(Eq 8-4)}$$

The reactance of the dipole is given by

$$X_L = Z_S \times \cot \frac{\ell}{2} = 1103 \times \cot 52.6° = 843 \ \Omega$$

Separate *MININEC* calculations show a reactance of 785 Ω, which is within 7% of the value calculated above. The required inductance is

$$L = \frac{X_L}{2\pi \times f}$$

where L is in μH and f is in MHz.
For 3.8 MHz,

$$L = \frac{843}{2\pi \times 3.8} = 32.6 \ \mu H$$

There are two ways of loading and feeding the shortened dipole with a centrally located loading coil:
1. Use a single 33.5-μH loading coil and link couple the feed line to the coil. This method is used by Cushcraft for their shortened 40-meter antennas.
2. The 33.5-μH loading coil can be "opened" in the center where it can be fed via a 1:1 balun.

2.3.3.2. Loading coils away from the center of the dipole

The location of the loading devices has a distinct influence on the radiation resistance of the antenna. This phenomenon is explained in detail in the chapter on short verticals.

Clearly, it is advantageous to put loading coils away from the center, provided the benefit of higher radiation resistance is not counteracted by higher losses in the loading device.

As loading coils are placed farther out on the elements, the required coil inductance increases. With increasing values of inductance, the Q factor is likely to decrease, and the equivalent series losses will increase.

I have calculated the case where the 22.5-meter long dipole (for 3.8 MHz) from Section 2.3.3.1 was loaded with coils at different (symmetrical) positions along the half-dipole elements. In all cases I assumed a Q factor of 300.

The results of the case are shown in **Fig 8-13**. The chart includes the reactance value of the required loading coils, the radiation resistance (R_{rad}), and the feed-point impedance at resonance (Z). The radiation efficiency is given by R_{rad}/Z. Note that the efficiency remains practically constant at 88% over the entire experiment range. This means that the advantage we gain from obtaining an increased radiation resistance by moving the coils out on the dipole halves is balanced out by the increased ohmic losses of the higher coil values. In the experiment I assumed a constant Q of 300, which may not be realistic, as it is likely that the Q of the lower inductance coils will be higher than for the higher inductance ones.

The experiment was done in free space. Over real

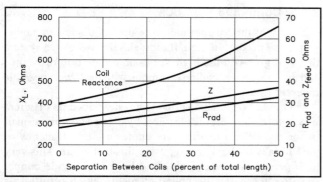

Fig 8-13—Design data for loading a short dipole. The radiation resistance, feed-point impedance at resonance, and the required reactance for the loading coils are given as a function of the separation between the coils (percentage of total dipole length). Calculations are for a coil Q of 100 and a frequency of 3.8 MHz. See text for further details.

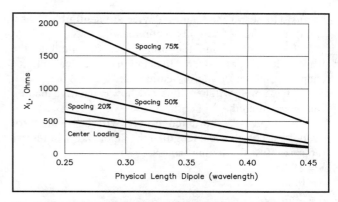

Fig 8-14—Required inductance for loading coils as a function of dipole length and position of the coils. Calculations were made using a 2-mm OD conductor diameter for a design frequency of 3.6 MHz.

ground the radiation resistance (and Z) will vary to a rather large extent as a function of the height (see Fig 8-7).

We may also conclude that if we can achieve a Q factor which is higher than 300, there will be an advantage in moving the coils out on the elements. We should not expect, however, to see several decibels of difference. The difference will be minute, and theoretical rather than noticeable in practice.

Calculating the loading coil value

The method for calculating the loading coil value is described in detail in Section 2.1.3 and 2.6.8 of the chapter on vertical antennas. In short the procedure is as follows:

- Calculate the surge impedance of the wire between the loading coil and the center of the antenna (Z_{S1})
- Calculate the surge impedance of the wire between the loading coil and the tip of the antenna (Z_{S2})
- Calculate the electrical length of the inner length (coil to center) = ℓ_1
- Calculate the electrical length of the tip (coil to tip) = ℓ_2
- Calculate the reactance of the ℓ_1 part using: $X = Z_{S1} \tan(\ell_1)$
- Calculate the reactance of the inner part of the half dipole using: $X_1 = +j Z_{S1} \times \tan(\ell_1)$
- Calculate the reactance of the tip using: $X_2 = -j Z_{S2} / \tan(\ell_2)$
- Add the reactances (the sum will be a negative value, eg, $-1000 \, \Omega$)
- The loading coil will have a reactance with the same absolute value.

It is much faster to use a *MININEC*-based modeling program, such as *MN* or *ELNEC*, to calculate the elements of a short dipole. Results obtained with *MININEC* match the results obtained by the above procedure.

Fig 8-14 shows the values of the required coils for a dipole (2 mm OD, AWG no. 12 wire, design frequency = 3.6 MHz) as a function of total antenna length (varying from 0.25 to 0.45 λ) and loading coil location.

2.3.4 Linear loading

In the commercial world, we have seen linear loading used on shortened dipoles and Yagis for 40 and 80 meters. Linear loading devices are usually installed at or near the center of the dipole. The required length of the loading device (in each dipole half) will be somewhat longer than the difference between the quarter-wave length and the physical length of the half dipole. The farther away from the center that the loading device will be inserted, the longer the "stub" will have to be. The "stub" must run in parallel with the antenna wire if we want to take advantage of the radiation off the stub (see the chapter on vertical antennas).

Example: A short dipole for 3.8 MHz is physically 28 meters (91.9 feet) long. The full half-wavelength is 39 meters. The missing electrical length is $39 - 28 = 11$ meters (36.1 feet). It is recommended that the linear loading device be constructed approximately 30% longer than half of this length:

$$L = \frac{11}{2} + 30\% = 5.5 \times 1.3 \approx 7 \, \text{m}$$

Trim the length of the loading device until resonance on the desired frequency is reached. When constructing an antenna with linear-loading devices, make sure the separation between the element and the folded linear-loading device is large enough, and that you use high-quality insulators to prevent arc-over and insulator damage.

Modeling the linear loaded dipole: Modeling antennas which use very close-spaced conductors (eg, the linear loading device which looks like a stub made of an open-wire transmission line) is very tricky. I would not recommend trying this with *MININEC*.

2.3.5. Capacitive (end) loading

Capacitive loading has the advantage of physically shortening the element length at the end of the dipole where the current is lowest (least radiation), and without introducing noticeable losses (as inductors do). End-loaded short dipoles have the highest radiation resistance, and the intrinsic losses of the loading device are negligible. Thus, end or top loading is highly recommended.

Top loading, and the procedures to calculate the loading devices, are covered in detail in the vertical antenna chapter (Chapter 9) and in Sections 2.1.2 and 2.6.3.

An example will best illustrate how capacitive loading can be calculated.

Example: A shortened dipole will be loaded for 80 meters. The physical length of the dipole is 18.75 meters (approx. 40% shortening factor).

S = 18.75 m

d = 0.2 cm

h = 20 m

We calculate the surge impedance from Eq 8-3: $Z_S = 1084 \, \Omega$. The antenna length to be replaced by a disk is

$$t = 90° \times 40\% = 36°$$

This means we must replace the outside 36° of each side of the dipole with a capacitive hat (a half-wave dipole is 180 degrees).

The inductive reactance of the shorted transmission-line equivalent) is given by

$$X_L = + j \frac{Z_S}{\tan t} = + j \frac{1084}{\tan 36°} = + j1492 \, \Omega$$

A capacitive reactance of the same value (but opposite sign) will resonate the "equivalent" transmission line. The required capacitive reactance is $X_C = -j1492$ ohms. The capacitance (f = 3.8 MHz) is

$$C = \frac{10^6}{2\pi \times f \times X_C} = 28.1 \text{ pF}$$

The required diameter of the hat disk is given by

$$D = 2.85 \times C$$

where

D = hat diameter in cm

C = the required capacitance in pF

In our example, D = 2.85 × 28.1 = 80.1 cm.

The above formula to calculate the disk diameter is for a solid disk. A practical capacitive hat can be made in the shape of a wheel with at least eight large-diameter spokes. This design will approach the performance of a solid disk. For ease in construction, the spokes can be made of four radial wires, joined at the rim by another wire in the shape of a circle.

In its simplest form, capacitive end loading will consist of bending the tips of the dipole (usually downward), in order to make the antenna shorter. By doing so we create extra capacitance between those two tips, which will load the antenna (make it electrically longer). These wire tips have an approximate capacitance of 6 pF/m. For the above example where a loading capacitance of 28.1 pF is required at each dipole end, vertically drooping wires of 28.1/6 = 4.7 m would be required.

2.3.6. Combined methods

Any of the loading methods already discussed can be employed in combination. It is essential to develop a system that will give you the highest possible radiation resistance and that employs a loading technique with the lowest possible inherent losses.

Gorski, W9KYZ (Ref 641) has described an efficient way to load short dipole elements by using a combination of linear and helical loading. He quotes a total efficiency of 98% for a two-element Yagi using this technique. This very high percentage can be obtained by using a wide copper strap for the helically wound element, which results in a very low RF resistance. Years ago, Kirk Electronics (W8FYR, SK) built Yagis for the HF bands, including 40 meters, using this approach (fiberglass elements wound with copper tape).

2.4. Bandwidth

The bandwidth of a dipole is determined by the Q factor of the antenna. The antenna Q factor is defined by

$$Q = \frac{Z_S}{R_{rad} + R_{loss}}$$

where

Z_S = surge impedance of the antenna

R_{rad} = radiation resistance

R_{loss} = total loss resistance.

The 3 dB bandwidth can be calculated from

$$BW = \frac{f_{MHz}}{Q}$$

The Q factor (and consequently the bandwidth) will depend on

- The conductor-to-wavelength ratio (influences Z_S)
- The physical length of the antenna (influences R_{rad})
- The type, quality, and placement of the loading devices (influences R_{rad})
- The Q factor of the loading device(s) (influences R_{loss})
- The height of the dipole above ground (influences R_{rad})

For a given conductor length-to-diameter ratio and a given antenna height, the loaded antenna with the narrowest bandwidth will be the antenna with the highest efficiency. Indeed, large bandwidths can easily be achieved by incorporating pure resistors in the loading devices, such as in the Maxcom dipole (Ref 663). The worst-radiating antenna one can imagine is a dummy load, where the resistor can be seen as the ohmic loading device while the radiating component does not exist. Judging by SWR bandwidth, this "antenna" is a wonderful performer, as a good dummy load can have an almost flat SWR curve over thousands of megahertz!

2.5. The Efficiency of the Shortened Dipole

Besides the radiation resistance, the RF resistance of the shortened-dipole conductor is an important factor in the antenna efficiency. Refer to **Table 8-4** for the RF resistances of common wire conductors used for antennas. For self-supporting elements, aluminum tubing is usually used. Both the dc and RF resistances are quite low, but special care should be taken to ensure that the best possible electrical RF contact between parts of the antenna is made. Some makers of military-specification antennas go so far as to gold plate the contact surfaces for low RF resistance! As a rule, loading coils are the most lossy elements, and capacitive

end loading should always be employed if at all possible. Linear loading is also a better choice than inductive loading. It is very important to minimize the contact losses at any point in the antenna, especially where high currents are present. Corroded contacts can turn a good antenna into a radiating dummy load. All these aspects are covered in more detail in the chapter on vertical antennas.

■ 3. LONG DIPOLES

Provided the correct current distribution is maintained, long dipoles can give more gain and increased horizontal directivity as compared to the half-wave dipole. The "long" antennas discussed in this paragraph are not strictly dipoles, but arrays of dipoles. They are the double-sized equivalents of the "long-verticals," as covered in the chapter on verticals.

The following antennas are covered:
• Two half-waves in phase
• Extended double Zepp

3.1. Radiation Patterns

Center-fed dipoles can be lengthened to approximately $1.25\ \lambda$ in order to achieve increased directivity and gain without introducing objectionable side lobes. **Fig 8-15** shows the horizontal radiation patterns for three antennas in free space: the half-wave dipole, two half-waves in phase (also called "collinear dipoles"), and the extended double Zepp, which is a $1.25\ \lambda$ long. Further lengthening of the dipole will introduce major secondary lobes in the horizontal pattern unless phasing stubs are inserted to achieve the correct phasing between the half-wave elements.

As we know, the dipole has 2.14-dB gain over the isotropic antenna (in free space). It is interesting to overlay the patterns of the two "long dipoles" on the same diagram, using the same dB scale; the extended double Zepp beats the dipole with almost 3 dB of gain. Note, however, how much more narrow the forward lobe on the pattern has become. This may altogether be a disadvantage in view of the varying propagation paths. The two-half-waves antenna sits right between the dipole and the extended double Zepp, with 1.5-dB gain over the half-wave dipole.

Fig 8-16 and **Fig 8-17** show the horizontal radiation patterns for the half-waves-in-phase dipole and for the extended double Zepp at various heights and wave angles. As with the half-wave dipole, the vertical radiation pattern depends on the height of the antenna above ground.

3.2. Feed-Point Impedance

The charts from Figs 9-8, 9-9, 9-11 and 9-12 can be used for estimating the feed-point impedances of "long" dipoles. The values from the charts that are made for monopoles must be doubled for dipole antennas.

The antennas can also be modeled with *MININEC*, but care should be taken with the results of long dipole antennas at less than $0.35\ \lambda$ above ground. Remember, too, that (except for free space) *MININEC* always reports the impedance for the antenna above a perfect ground conductor, but if the height is specified as mentioned above, the results should be reasonably close to the actual impedance over real ground.

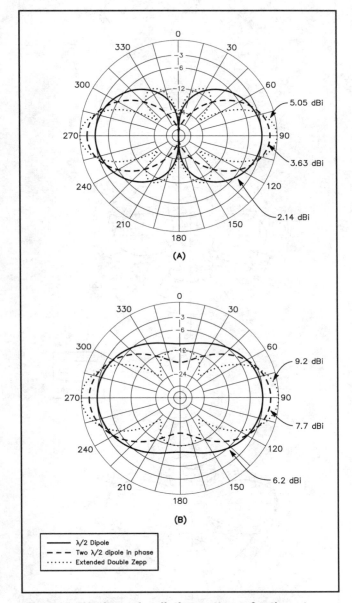

Fig 8-15—Horizontal radiation patterns for three types of "dipoles": the half-wave dipole, the collinear dipole (2 half waves in phase) and the extended double Zepp. At A, the radiation patterns at a 0° wave angle with the antennas in free space. At B, the patterns at a 37° wave angle with the antennas ³/₈ λ above good-quality ground. Notice the sidelobes apparent with the extended double Zepp antenna.

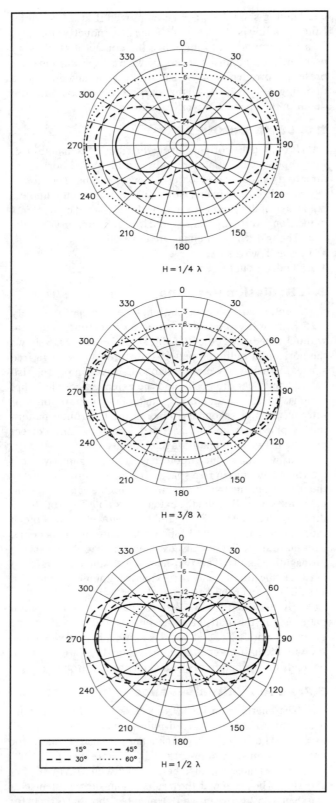

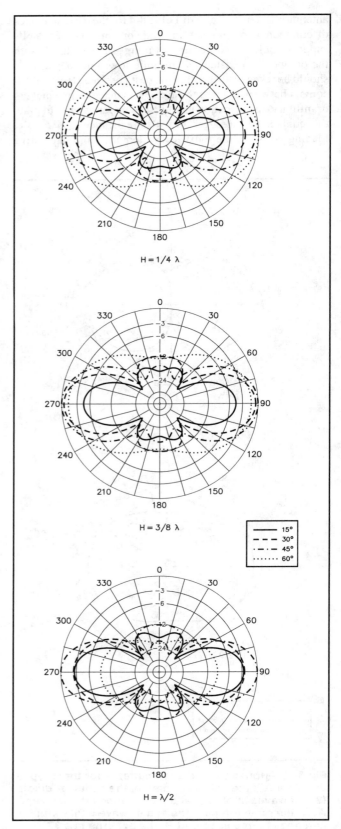

Fig 8-16—Horizontal radiation patterns for collinear dipoles (two half waves in phase) for wave angles of 15, 30, 45 and 60°. As with a half-wave dipole, directivity is not very pronounced at low heights and at high wave angles.

Fig 8-17—Horizontal radiation patterns for the extended double Zepp for wave angles of 15, 30, 45 and 60°.

Since the center-fed long antennas will not be loaded with lossy elements that would reduce their efficiency, long dipoles can have efficiencies very close to 100% if care is taken to use the best material for the antenna conductor.

3.3. Feeding Long Dipoles

The software module "COAX TRANSFORMER/ SMITH CHART" from the NEW LOW BAND SOFTWARE is an ideal tool for analyzing the impedances, currents, voltages and losses on transmission lines. The "STUB MATCHING" module can assist you in calculating a stub-matching system in seconds. In any case, we need to know the feed-point impedance of the antenna. Measuring the feed-point impedance is quite difficult, as you cannot use a noise bridge unless it is specially configured for measuring balanced loads.

3.3.1. Collinear dipoles (two half-wave dipoles in phase)

The impedance at resonance for two half-waves in phase is several thousand ohms. With a 2-mm OD conductor (AWG 12), the impedance is approximately 6000 ohms on 3.5 MHz. The shortening factor (in free space) for that antenna is 0.952. The SWR bandwidth of the two half-waves in phase is given in **Fig 8-18**. The antenna covers a frequency range from 3.5 to 3.8 MHz with an SWR of less than 2:1.

The antenna can be fed with open-wire feeders into a tuner, or via a stub matching system and balun as shown in Fig 6-15. Using "tuned feeders" with a tuner can of course ensure a 1:1 SWR to the transmitter (50 Ω) at all times.

3.3.2. Extended double Zepp

The intrinsic SWR bandwidth of the extended double Zepp is much narrower than for the collinear dipoles. For an antenna made out of 2 mm OD wire (AWG 12) and with a total length of 1.24 λ, the feed-point impedance is approximately $200 - j1100$ Ω. The SWR curve (normalized to R_{rad} at the design frequency) is given in Fig 8-15. For lengths varying from 1.24 to 1.29 λ, the radiation resistance will vary from 200 to 130 ohms (decreasing resistance with increasing length).

The exact length of the antenna is not critical, but as we increase the length, the amplitude of the sidelobes increases. The magnitude of the reactance will depend on the length/diameter ratio of the antenna: An antenna made of a thin conductor will show a large reactance value, while the same antenna made of a large diameter conductor will show much less reactance.

The impedance of the extended double Zepp also changes with antenna height, as with a regular half-wave dipole. For the 1.24-λ long extended double Zepp, the resistive part changes between 150 and 260 Ω, and settles

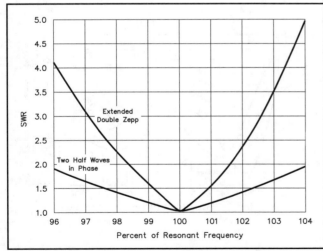

Fig 8-18—SWR curves for an extended double Zepp and for two half waves in phase (collinear array). The calculation was centered on 3.65 MHz using a conductor of 2-mm OD (AWG no. 12), and the results normalized to the radiation resistances. The SWR bandwidth of the collinear array is much higher than for the extended double Zepp.

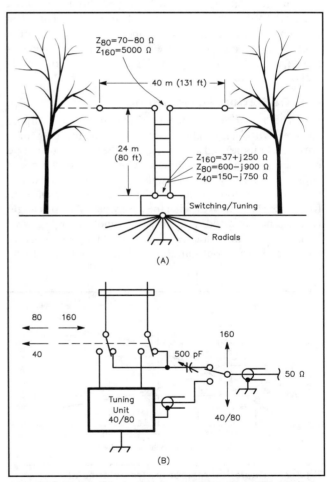

Fig 8-19—Three-band antenna configuration (40, 80 and 160 meters). On 40 meters the antenna is a collinear (two half waves in phase), on 80 meters a half-wave dipole, and on 160 meters a top-loaded vertical (T antenna). The bandswitching arrangement is shown at B.

at 200 Ω at very high elevations.

In principle we can feed this antenna in exactly the same way as the collinear, but as the intrinsic bandwidth is much more limited, it is better to feed the antenna with open-wire lines running all the way into the shack and to the open-wire antenna tuner.

3.4. Three-Band Antenna (40, 80, 160 Meters)

Refer to the three-band antenna of **Fig 8-19**. On 40 meters the antenna is a collinear array (two half-waves in phase) at 24 m (80 ft). On 80 meters, it is a half-wave dipole. For 160, we connect the two conductors of the open-wire feeders together, and the antenna is now a flat-top loaded vertical (T-antenna). The disadvantage is that we must install the switchable tuning network at the base, right under the antenna. Some slope can of course be allowed. As the antenna is a vertical on 160, its performance will largely depend on the quality of the ground and the radial system.

■ 4. INVERTED-V DIPOLE

In the past, the inverted-V shaped dipole has often been credited with almost magical properties. The most frequently claimed "special" property is a low radiation angle. Some have more correctly called it a poor man's dipole, as it requires only one high support. Here are the facts.

4.1. Radiation Resistance

The radiation resistance of the inverted-V dipole changes with height above ground (as in the case of a horizontal dipole) and as a function of the apex angle (angle between the legs of the dipole). Consider the two apex-angle extremes. When the angle is 180°, the inverted V becomes a flat-top dipole, and the radiation resistance (in free space) is 73 Ω. Now take the case where the apex angle is 0°. The inverted-V dipole becomes an open-wire transmission line, a quarter-wavelength long and open at the far end. This configuration will not radiate at all (the current distribution will completely cancel all radiation, as it should in a well-balanced feed line), and the input impedance of the line is 0 Ω (a quarter-wave stub open at the end reflects a dead short at the input). This zero-angle inverted V will have a radiation resistance of 0 Ω and consequently will not radiate at all.

I modeled a range of inverted-V dipoles with different apex angles at different apex heights. This was done using *NEC-2*. **Fig 8-20** shows the radiation resistance of the inverted V as a function of the apex angle for a range of angles between 90° and 180° (straight dipole). The curve also shows the physical length which produces resonance (feed point purely resistive). Decreasing the apex angle raises the resonant frequency of the inverted V.

Fig 8-21 shows the feed-point resistance and reactance for inverted-V dipoles with apex angles of 120° and 90°. The antennas were first resonated in free space (zero reactance). Then the reactances were calculated over ground at various heights with the antenna lengths that produced resonance in free space. Notice that the shape of both curves is similar to the shape of the straight dipole

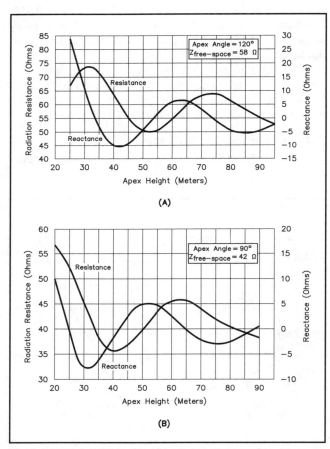

(A)

(B)

Fig 8-21—Impedance (feed-point resistance and reactance) of inverted-V dipoles as a function of height above ground. Analysis frequency is 3.75 MHz, with 2 mm OD wire (AWG no. 12). Resistances at resonance are: 120° apex angle, 58 Ω; 90° apex angle, 42 Ω. *NEC-2* was used for these calculations, as *MININEC* is unreliable for doing impedances at low heights.

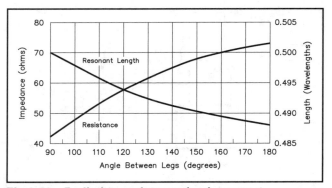

Fig 8-20—Radiation resistance (resistance at resonance) of the inverted-V dipole antenna in free space as a function of the angle between the legs of the dipole (apex angle). Also shown is the physical length (based on the free-space wavelength) for which resonance occurs.

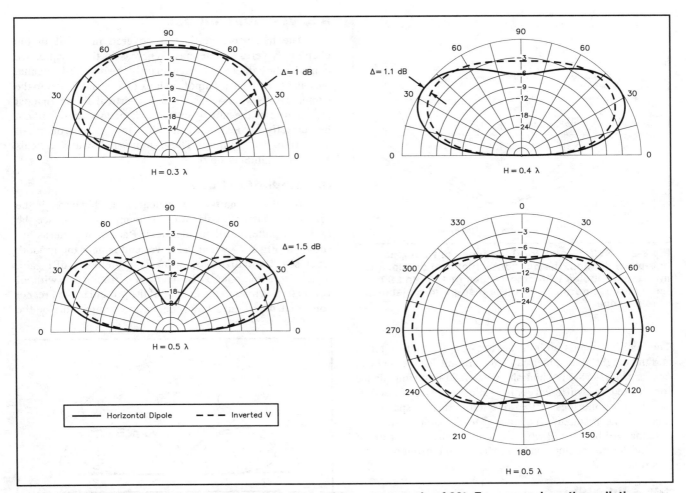

Fig 8-22—Radiation patterns for an inverted-V dipole with an apex angle of 90°. For comparison the radiation pattern for a horizontal dipole is included in each plot, on the same dB scale. The horizontal pattern is shown for the main wave angles (28° for the straight dipole and 32° for the inverted V). The height of the inverted V is referred to as the height of its apex.

curve, Fig 8-7. Bringing the inverted V closer to ground lowers its resonant frequency. This is a fairly linear function between 0.25 λ and 0.5 λ apex height.

4.2. Radiation Patterns and Gain

Previous paragraphs compare the inverted V to a straight dipole at the same apex height. It is clear that the inverted V is a compromise antenna as compared to the straight horizontal dipole. At low heights (0.25 to 0.35 λ), the gain difference is minimal, but at heights that produce low-angle radiation the dipole performs substantially better.

The 90° apex angle inverted-V dipole

Fig 8-22 shows the vertical and horizontal radiation patterns for inverted Vs with a 90° apex angle at different apex heights. Modeling was done over good ground. For comparison, I have included the radiation pattern for a straight dipole at the same (apex) height. In the broadside direction, the inverted-V dipole shows 1 to 1.5 dB less gain than the flat-top dipole, and also a slightly higher wave angle.

The 120° apex angle inverted-V dipole

The flat-top dipole is still 0.6 dB better than the inverted V at a height of 0.4 λ, 0.7 dB at 0.45 λ and 0.8 dB at 0.5 λ. In addition, the wave angle for the horizontal dipole is slightly lower than for the inverted V (approx. 3° for heights from 0.35 λ to 0.5 λ). The difference is not spectacular, but it is clear that the inverted-V dipole has no magical properties.

4.3. Antenna Height

In many situations it will be possible to erect an inverted-V dipole antenna much higher than a flat top dipole, in most cases because there is only one high support structure available. In this respect the high inverted V is certainly superior to a low horizontal dipole. The V loses from the dipole at the same apex height, but who's got two such high supports? And are they in the right direction?

4.4. Length of the Inverted-V Dipole

The usual formulas for calculating the length of the straight dipole cannot be applied to the inverted-V dipole.

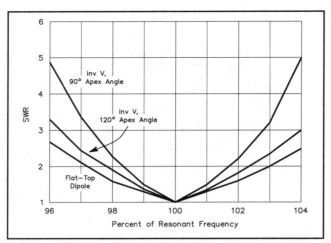

Fig 8-23—SWR curves for three types of free-space half-wave dipoles: the horizontal (flat top) dipole, and inverted-V dipoles with apex angles of 120° and 90°. Each curve is normalized to the feed-point resistance at resonance.

The length depends on both the apex angle of the antenna and the height of the antenna above ground. This effect can be seen from Fig 8-21, which shows the feed-point impedance for inverted Vs of different configurations at different heights.

Closing the legs of the inverted V in free space will increase the resonant frequency. On the other hand, the antenna will become electrically longer when closer to the ground due to the end-loading effect of the ground on the inverted V ends.

4.5. Bandwidth

Fig 8-23 shows the SWR curves for three inverted-V dipoles with different apex angles: 90°, 120° and 180° (flat-top dipole), for a conductor diameter of 2 mm (AWG no. 12) and a frequency of 3.65 MHz. As expected, the SWR bandwidth decreases with decreasing apex angle. The computed figures are for free space. The SWR values from Fig 8-23 are normalized figures. This means that the SWR at resonance is assumed to be 1:1, whatever the impedance (resistance) at resonance is. In practice the SWR will almost never be 1:1 at resonance because the line impedance will be different from the feed-point impedance (see the impedance charts in Figs 8-7 and 8-21).

Over ground the reactive part of the impedance remains almost the same value as in free space (after having re-resonated the inverted V for no reactance at the center frequency). This means that the SWR bandwidth will be largest for heights where the radiation resistance is highest. For the inverted-V dipole this is at an apex height of approximately 0.35 to 0.4 λ. Practically speaking, it means that for an apex height of 0.3 λ to 0.5 λ, the SWR curve will be somewhat flatter over ground than in free space.

The SWR bandwidth of the inverted V can be increased significantly by making a folded-wire version of the antenna. The feed-point impedance of the folded-wire version is four times the impedance shown in Figs 8-20 and 8-21.

■ 5. VERTICAL DIPOLE

The half-wave vertical is covered in detail in the chapter on vertical antennas. Whereas in that chapter we consider the half-wave vertical mainly as a base-fed antenna, we can of course use a dipole made of wire, and feed it in the center. This is what we usually call a vertical dipole. In most practical cases the half-wave vertical, made of wire, will not be perfectly vertical, but generally slope away from a tall support (tower, building). These sloping half-wave verticals are covered in Section 6.

5.1. Radiation Pattern

Whether the half-wave vertical is base fed or fed in the center, the current distribution is identical, and hence the radiation pattern will be identical. Radiation patterns are shown in **Fig 8-24** when the lower end is near the ground. Over sea, the half-wave vertical can yield 6.1-dBi gain, which drops to about 0 dBi over good ground. As with all verticals, it is mainly the quality of the ground in the Fresnel zone that determines how good a low-angle radiator the

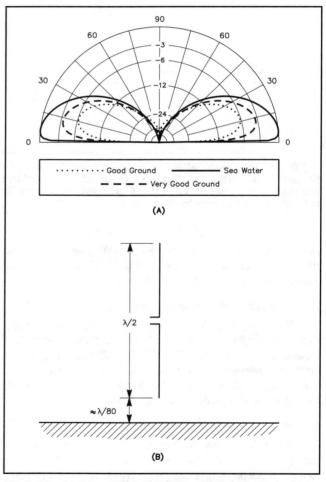

Fig 8-24—At A, vertical radiation patterns over various grounds for a vertical half-wave center-fed dipole with the bottom tip just clearing the ground, as shown at B. The gain is as high as 6.1-dBi over good ground. The feed-point impedance is 100 ohms.

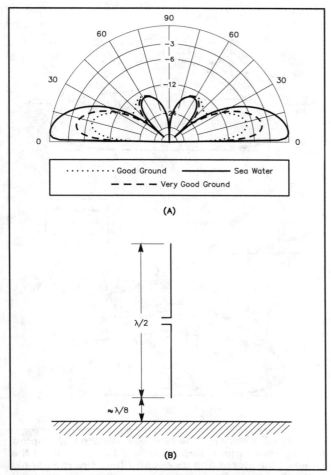

Fig 8-25—At A, vertical radiation patterns of the half-wave vertical dipole with the bottom tip $^1/_8$ λ off the gournd, as shown at B. This is the vertical equivalent of the extended double Zepp antenna.

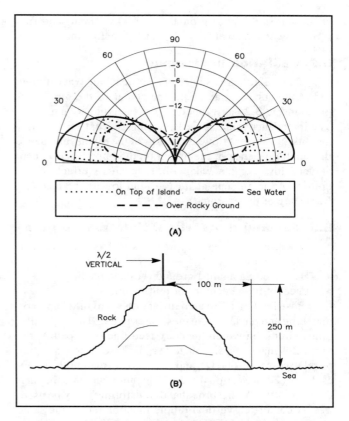

Fig 8-26—At A, the serrated radiation pattern of a half-wave vertical overlooking a slope of very poor ground (eg, an island with volcanic earth) next to the sea, as shown at B. Because of the antenna height above the sea, multiple lobes show up in the pattern. The radiation patterns of the half-wave vertical at sea level and the pattern over very poor ground are superimposed for comparison.

vertical dipole will be (see Section 3 of the chapter on verticals). Half-wave verticals produce excellent (very) low-angle radiation when erected in close proximity to salt water. As a general-purpose DX antenna, the vertical dipole may, however, produce too low an angle of radiation for the run-of-the-mill DX path.

Raising the half-wave vertical higher above the ground introduces multiple lobes. **Fig 8-25** shows the patterns for a half-wave center-fed vertical with the bottom $^1/_8$ λ above ground. Note the secondary lobe, which is identical to the lobe we encountered with the extended double Zepp. As a matter of fact, this slightly elevated half-wave vertical is the half-size equivalent of the extended double Zepp antenna.

I also modeled a half-wave vertical on top of a rocky island with very poor ground, 250 m (820 ft) above sea level, and some 100 m (330 ft) from the sea. **Fig 8-26** shows the layout and the radiation pattern. Superimposed on the pattern are the patterns for the same antenna at sea level, as well as over very poor ground. Note that whereas the extra height does not give any gain advantage over sea level, in this case the extra height does help the low angle rays to

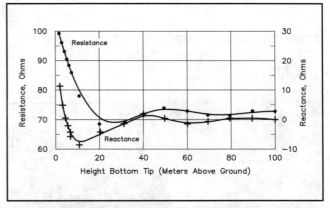

Fig 8-27—Radiation resistance and reactance of the half-wave vertical as a function of height above ground. The height is taken as the height of the bottom tip. Calculations are for a design frequency of 3.5 MHz.

shoot across the poor ground (the rocky island) and find reflection at sea level some 250 meters below the antenna.

5.2. Radiation Resistance

The radiation resistance of a vertical half-wave dipole, fed at the current maximum (the center of the dipole), is given in **Fig 8-27** as a function of its height above ground. The impedance remains fairly constant except for very low heights. No current flows at the tips of the dipole, and hence the small influence of the height on the impedance, except at very low heights where the capacitive effect of the bottom of the antenna against ground lowers the resonant frequency of the antenna.

5.3. Feeding the Vertical Half-Wave Dipole

There are two main approaches to feeding a vertical half-wave dipole:
- Base feeding against ground (voltage feeding)
- Feeding in the center (current feeding)

Base feeding is covered in Section 4.4 of the chapter on matching and feed lines. In most cases one will use a parallel tuned circuit on which the coax feed line is tapped. If the vertical is made of a sizable tower, the base impedance may be relatively low (600 Ω), and a broadband matching system as described in Section 4.5.2 of Chapter 6 on matching and feed lines (the W1FC broadband transformer) may be used.

A center-fed vertical dipole must be fed in the same way as a horizontal dipole. It represents a balanced feed point, and can be fed via an open-wire line to a tuner, or via a balun to a coaxial feed line (see Section 1.4.)

■ 6. SLOPING DIPOLE

Sloping half-wave dipoles are used very successfully by a number of stations, especially near the sea. FK8CP is using a half-wave sloper on 160, with the end of the antenna connected about 15 m above sea level, less than 50 m from the salt water. I8UDB is using a sloper on 160 from his mountaintop QTH near Naples, where the electrical ground is nonexistent, but where the sea is only 100 m away and a

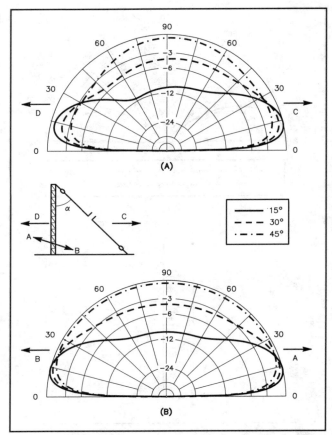

Fig 8-28—Elevation-plane radiation patterns of sloping dipoles with various slope angles. At A, patterns in the plane of the sloper and its support (end-fire radiation), and at B, perpendicular to that plane (broadside radiation). End-fire radiation is 100% vertically polarized, while the broadside radiation contains a horizontal as well as a vertical component. The horizontal pattern shows a very small amount of directivity.

few hundred meters below the antenna.

The half-wave sloper radiates a signal with both horizontal and vertical polarization components. Unless it is very high above the ground (eg, I8UDB), you need not bother with the horizontal component. The low angle will be produced only by the vertical component. All modeling in this section was done on 80 meters, over a very good ground.

6.1. The Sloping Straight Dipole

Due to the weight of the feed line, a sloping dipole will seldom have two halves in a straight line. Let us nevertheless analyze the antenna as if it does.

Radiation patterns

Fig 8-28 shows the radiation patterns of sloping half-wave dipoles for apex angles of 15, 30 and 45° over three types of ground (poor, good and sea). For the dipole with a 45° slope angle I include the pattern showing the vertical and the horizontal radiation separately, **Fig 8-29**. It is obvious that the steeper the slope, the less the horizontal radiation

View from the top of I8UDB's tower. Such an awesome view needs no comment.

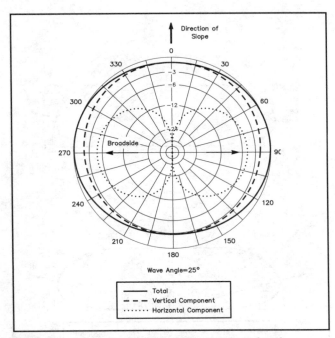

Fig 8-29—Azimuth-plane radiation pattern for the sloping dipole with a 45° slope angle, taken at a 25° wave angle. Patterns for the vertical and horizontal components of the total are also shown. The directivity is very limited. Actually, the sloping dipole radiates best about 70° either side of the slope direction!

component will be. The high-angle radiation is only due to the horizontal radiation component. For the vertical component the same rules apply as for the half-wave vertical: In order to exploit the intrinsic very low-angle capabilities, you must have an excellent ground around the antenna. Don't forget, the Fresnel zone (the area where the reflection at ground level takes place) can stretch all the way out to 10 wavelengths or more from the antenna!

Fig 8-29 shows the horizontal pattern for a sloping dipole with a 45° slope angle. The sloper is almost omnidirectional, but radiates best broadside (perpendicular to the plane going through the sloper and the support). In the end-fire direction (in the plane of the sloper and its support), it has less than 1 dB F/B at a wave angle of 25°. The antenna radiates a little better in the direction of the slope. The fact that it radiates best in the broadside direction is due to the horizontal component, which only radiates in the broadside direction.

Impedance

The radiation resistance of the sloping dipole with the bottom wire 1/80 λ above ground (1 m for an 80-m antenna) varies from 96 Ω for a 15° slope angle to 81 Ω for a 45° slope angle.

6.2. The Bent-Wire Sloping Dipole

Most real-life sloping half-wave dipoles have a bent-wire shape, because of the weight of the feed line. **Fig 8-30** and **Fig 8-31** analyze a sloping vertical with a slope angle of 20° for the top half of the antenna, and slope angles of 40° and 60°

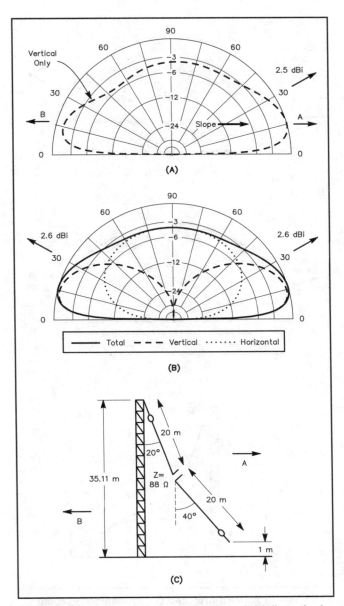

Fig 8-30—At A, "end-fire" and B, "broadside" vertical radiation patterns of a bent-wire half-wave sloper for 3.6 MHz. The horizontal and vertical components of the total pattern are also shown at B. the bottom 0.25-λ section slopes at an angle of 40°, as shown at C. Modeling is done over very good ground.

respectively for the bottom half of the dipole. Using a 60° slope angle reduces the height requirement for the support.

The sloping dipole with a relatively horizontal bottom ¹/₄-λ wire yields almost the same signal as the straight sloping dipole. It is important to keep the top half of the sloping dipole as vertical as possible. Analysis shows the angle of the bottom half of the antenna is relatively unimportant.

Feed point

Is the feed point of such a bent sloping dipole a

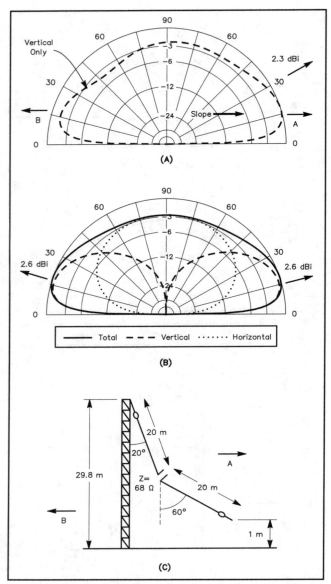

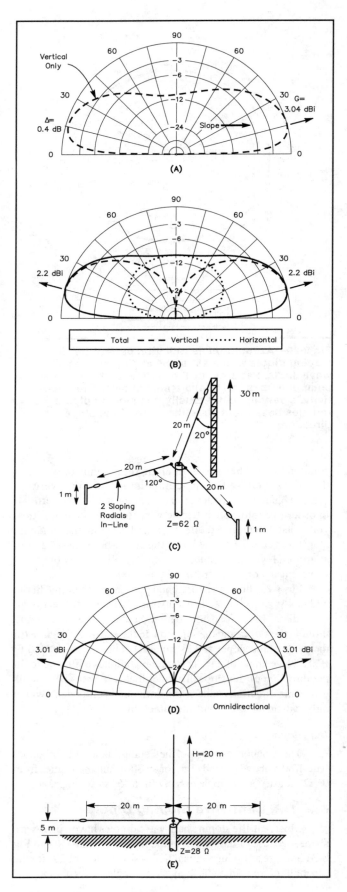

Fig 8-31—At A, "end-fire" and B, "broadside" vertical radiation patterns of a bent-wire half-wave sloper for 3.6 MHz. The horizontal and vertical components of the total pattern are also shown at B. the bottom 0.25-λ section slopes at an angle of 60°, as shown at C. This configuration and that of Fig 8-30 are just as valid as the configuration using a straight sloper; the loss in gain is negligible This arrangement requires less support height than that of a straight sloper or that of Fig 8-30.

Fig 8-32—Transition from a sloping dipole to a 0.25-λ vertical with two radials. At C the bottom half of the dipole is replaced by two 0.25-λ wires sloping to the ground; the resulting patterns are shown at A and B. At E the radials are lifted to the horizontal, with the resulting pattern at D. This change eliminates all the horizontal radiation component that was originated by the sloping wires. Analysis frequency: 3.65 MHz.

Table 8-5
MININEC Pulses Versus Calculated Impedance for a Straight Dipole Antenna

Pulses	Impedance
5	$71 - j14$
10	$67 - j26$
20	$68 - j28$
30	$68.5 - j28$
50	$68.6 - j27.3$
80	$68.7 - j27.1$
100	$68.7 - j27.0$

Table 8-6
MININEC Pulses Vs Calculated Impedance for an Inverted-V Dipole Antenna

Pulses	Impedance
5	$43.6 - j23.7$
10	$44.3 + j10.3$
20	$44.6 + j28.4$
30	$44.6 + j34.3$
50	$44.7 + j38.1$
80	$44.7 + j39.8$
100	$44.8 + j42.0$
20 tapered, min 0.4 m, max 3.0 m	$44.2 + j36.1$
26 tapered, min 0.3 m, max 2.0 m	$44.4 + j37.6$
26 tapered, min 0.4 m, max 2.0 m	$44.4 + j36.3$
28 tapered, min 0.4 m, max 2.0 m	$44.4 + j38.9$
46 tapered, min 0.2 m, max 1.0 m	$44.2 + j40$

symmetrical feed point? Not strictly speaking. If you use such an antenna, don't take any chances. It does not hurt to put a current balun at a load even when the load is asymmetric. Use a current type balun to remove any current from the outside of the coaxial cable. A coiled coax or a stack of ferrite beads is the way to go (see the chapter on feed lines and antenna matching).

6.3. Evolution into the Quarter-Wave Vertical

We can go one step further and bring the bottom $^1/_4 \lambda$ all the way horizontal. If the top half were fully vertical, we now would have a quarter-wave vertical with a single elevated radial. This configuration is described in detail in the chapter on vertical antennas (see Fig 9-18).

To transform the half-wave sloper into a quarter-wave vertical, we first replace the sloping bottom half of the antenna with two wires, now called radials. Both radials are "in line" and slope toward the ground, **Fig 8-32C**. A and B of Fig 8-32 show the radiation patterns for this configuration. Note that the high-angle radiation has been attenuated some 10 dB, and we picked up 0.5 to 0.8 dB of gain. The little horizontally polarized radiation left over is, of course, caused by the sloping radials. The configuration shows a gain in the direction of the sloping wire of approximately 0.4 dB.

Next we move the radials up, so they are horizontal,

and move the antenna down so the base is now 5 m (16 ft) above ground (Fig 8-32E). All the horizontal radiation is gone, and the gain has settled halfway between the forward and the backward gain of the previous model, which was to be expected. We now have a quarter-wave vertical with two radials, which is how the original ground plane was developed (see Section1.3.3 of the chapter on verticals).

The quarter-wave vertical with two radials definitely has an asymmetrical feed point. However, as the feed line will be exposed to the strong fields of the antenna, and as the feed line will likely be installed on the ground under the two radials, it is strongly recommended that you fully decouple the feed line from the feed point by using a current-type balun (coiled coax or stack of ferrite beads).

6.4. Conclusion

The 6.1-dBi gain can be obtained with a half-wave vertical only over nearly perfect ground (sea). Even over very good ground, the half-wave vertical will not be any better than a quarter-wave vertical (3-dBi gain). This means that unless you are near the sea, you may as well stick with a quarter-wave vertical. The sloping vertical (make the sloping wire as vertical as possible) with two radials (5-m high for 3.6 MHz) will produce as good a signal as a half-wave vertical or sloping half-wave vertical over very good ground. It will, however, only require a 25 m (82 ft) support instead of a 35 (114 ft) or 40 m (131 ft) support for the half-wave vertical.

■ 7. MODELING DIPOLES

MININEC-based modeling programs are well suited for modeling dipoles. Straight dipoles can be accurately modeled with a total of 10 to 20 pulses.

Inverted-V dipoles require more pulses, depending on the apex angle, in order to obtain accurate impedance data. **Table 8-5** shows the impedance data for a straight dipole, and **Table 8-6** for an inverted-V dipole as a function of the pulses, wires and segments. An inverted V with a 90° apex angle requires at least 50 equal-length segments for accurate impedance data. By using the TAPERING technique (see the chapter on Yagi and quad antennas), accurate results can be obtained with a total of only 26 segments. ELNEC (MININEC-based modeling program by W7EL) provides an automatic feature for generating tapered segment lengths, which is a great asset when you model antennas with bent conductors.

Knowing the exact impedance is important only if you want to calculate the exact resonant length (or frequency) of a dipole, or if the dipole is part of an array.

In order to obtain reliable results, the dipoles should not be modeled too close to ground. For half-wave horizontal dipoles, the antenna should be at least 0.2-λ high. For longer dipoles, the minimum height ensuring reliable results is somewhat higher. Vertical dipoles and sloping dipoles (with a steep slope angle) can be modeled quite close to the ground, as there is very little radiation in the near-field toward the ground (a dipole does not radiate off its tips).

The NEC modeling program is required if accurate gain and impedance data are required for dipoles very close to ground.

VERTICAL ANTENNAS

■ **1. QUARTER-WAVE VERTICAL**

1.1. Radiation Patterns

    1.1.1. Vertical pattern of monopoles over ideal ground

    1.1.2. Vertical radiation pattern of monopoles over real ground

        1.1.2.1. The Reflection Coefficient

        1.1.2.2. The Pseudo-Brewster Angle

        1.1.2.3. Ground-Quality Characterization

        1.1.2.4. Brewster Angle Formula

        1.1.2.5. Brewster Angle and Radials

        1.1.2.6. Vertical Radiation Patterns

    1.1.3. Horizontal pattern of a vertical antenna

1.2. Radiation Resistance of Monopoles

1.3. Radiation Efficiency of the Monopole Antenna

    1.3.1. Conductor RF resistance

    1.3.2. Parallel losses in insulators

    1.3.3. Ground losses

■ **2. GROUND AND RADIAL SYSTEM FOR VERTICAL ANTENNAS: THE BASICS**

2.1. Buried Radials

    2.1.1. Near-field radiation efficiency

    2.1.2. Modeling buried radials

    2.1.3. How many buried radials now, how long?

    2.1.4. Two-wavelength-long radials and the far field

    2.1.5. Ground rods

    2.1.6. Burying depth of buried radials

    2.1.7. Some practical hints

2.2. Elevated Radials

    2.2.1. Modeling vs measuring? Elevated vs ground radials?

    2.2.2. Modeling vertical antennas with elevated radials

    2.2.3. How many elevated radials?

    2.2.4. Only one radial

    2.2.5. How high the radials

    2.2.6. Why quarter-wave radials?

    2.2.7. Making quarter-wave radials of equal length

    2.2.8. The K5IU solution to unequal radial currents

    2.2.9. How do I cut radials to equal (nonquarter wave) lengths?

    2.2.10. Should the vertical be a quarter-wave?

    2.2.11. Elevated radials on grounded tower

    2.2.12. Elevated radials combined with radials (screen) on the ground

    2.2.13. Avoiding return currents via the ground

    2.2.14. Elevated Radials in Vertical Arrays

    2.2.15. "Wonder" radials for 160 Meters (?)

2.3. Buried or Elevated?

2.4. Evaluating the Radial System

■ **3. SHORT VERTICALS**

3.1. Radiation Resistance

    3.1.1. Base loading

    3.1.2. Top loading

    3.1.3. Center loading

    3.1.4. Combined top and base loading

    3.1.5. Linear loading

3.2. Keeping the Radiation Resistance High

3.3. Keeping Losses Associated with Loading Devices Low

3.4. Short-Vertical Design Guidelines

    3.4.1. Verticals with folded elements

3.5. SWR Bandwidth of Short Verticals

    3.5.1. Calculating the 3-dB bandwidth

    3.5.2. The 2:1 SWR bandwidth

9 VERTICAL ANTENNAS

The effects of the ground and the artificial ground system (if used) on the radiation pattern and the efficiency of vertically polarized antennas is often not understood, and has, until recently, not been covered extensively in the amateur literature.

The effects of the ground and the ground system are twofold. Near the antenna (in the near field), there is a need for a good ground system to collect the antenna return currents without losses. This will determine the radiation efficiency of the antenna.

At distances farther away (the far field, or the Fresnel zone), where the wave is reflected from the earth to make up the low-angle radiation, some of the energy will be absorbed by the ground. The absorption is a function of the ground quality and the incident angle. This mechanism will determine the reflection efficiency of the antenna.

Vertical monopole antennas are often called ground-mounted verticals, or simply verticals. They are, by definition, mounted perpendicularly to the earth, and they produce a vertically polarized signal. Verticals are very popular antennas for the low bands, as they can produce very good low-angle radiation without requiring the very high supports needed for horizontal antennas to produce the same low-angle radiation.

■ 1. QUARTER-WAVE VERTICAL

1.1. Radiation Patterns

1.1.1. Vertical pattern of monopoles over ideal ground

The radiation pattern produced by a ground-mounted quarter-wave vertical antenna is basically one-half of the radiation pattern of a half-wave dipole antenna in free space (with twice the physical size of the vertical and with symmetrical current distribution). As such, the radiation pattern of a quarter-wave vertical over perfect ground is half of the figure-8 shown for the half-wave dipole in free space. The representation is shown in **Fig 9-1**.

The relative field strength of a vertical antenna with sinusoidal current distribution and a current node at the top is given by:

$$E_f = k \times I \left[\frac{\cos\left(L \sin \alpha\right) - \cos L}{\cos \alpha} \right] \qquad \text{(Eq 9-1)}$$

where

k = constant related to impedance
E_f = relative field strength
α = angle above the horizon
L = electrical length (height) of the antenna
I = antenna current $\left(\sqrt{P/R}\right)$

This formula does not take imperfect ground conditions into account, and is valid for antenna heights between 0° and 180° (0 to $\frac{1}{2}\lambda$). The "form factor" containing the trigonometric functions is often published by itself for use in calculating the field strength of a vertical antenna. If used in this way, however, it appears that short verticals are vastly inferior to tall ones, as the antenna length appears only in the numerator of the fraction.

Replacing I in the equation with the term

$$\sqrt{\frac{P}{R_{rad} + R_{loss}}}$$

gives a better picture of the actual situation. For short

verticals, the value of the radiation resistance is small, and this term largely compensates for the decrease in the form factor. This means that for a constant power input, the current into a small vertical will be greater than for a larger monopole.

In practice, however, the current is not determined by just the radiation resistance (R_{rad}), but rather by the sum of the radiation resistance and the loss resistance(s). This is why, with a less-than-perfect ground system and less-than-perfect loading elements (lossy coils in the case of lumped-constant-loaded verticals), the total radiation can be significantly less than in the case of a larger vertical (where R_{rad} is large in comparison to the ground loss and where there are no lossy loading devices).

Interestingly, short verticals are almost as efficient radiators as are longer verticals, provided the ground system is good and there are no lossy loading devices. When the losses of the ground system and the loading devices are brought into the picture, however, the sum $R_{rad} + R_{loss}$ will get larger, and as a result part of the supplied power will be lost in the form of heat in these elements. For instance, if $R_{rad} = R_{loss}$, half of the power will be lost. Note that with very short verticals, these losses can be much higher.

1.1.2. Vertical radiation pattern of monopoles over real ground

The final three-dimensional radiation pattern from an antenna is made up of the combination of the direct radiation and the radiation via reflection from the earth. The following explanation is valid only for reflection of vertically polarized waves. See Chapter 8 on dipole antennas for an explanation of the reflection mechanism for horizontally polarized waves.

In case of a perfect earth, there is no phase shift of the vertically polarized wave at the reflection point. The two waves add with a certain phase difference, due only to the different path lengths. This is the mechanism that creates the radiation pattern. Consider a distant point at a very low angle with the horizon. As the path lengths are almost the same, reinforcement of the direct and reflected wave will be maximum. In case of a perfect ground, the radiation will be maximum at an angle just above zero degrees.

1.1.2.1. The reflection coefficient

Over real earth, reflection causes both an amplitude and a phase change. It is the reflection coefficient that describes how the incident (vertically polarized) wave is being reflected. The reflection coefficient of real earth is a complex number, and varies with frequency. In the polar-coordinate system the reflection coefficient consists of:

• The magnitude of the reflection coefficient: It determines how much power is being reflected, and what percentage is being absorbed in the lossy ground. A figure of 0.6 means that 60% will be reflected, and 40% absorbed.
• The phase angle: This is the phase shift that the reflected wave will undergo as compared to the incident wave. Over real earth the phase is always lagging (minus sign). At zero wave angle, the phase is always −180°. This causes the total radiation to be zero (no absorption of reflected wave, and the sum of the incident and reflected waves that are 180° out of phase). At high wave angles, the reflection phase angle will be close to zero (typically −5° to −15°, depending on the ground quality).

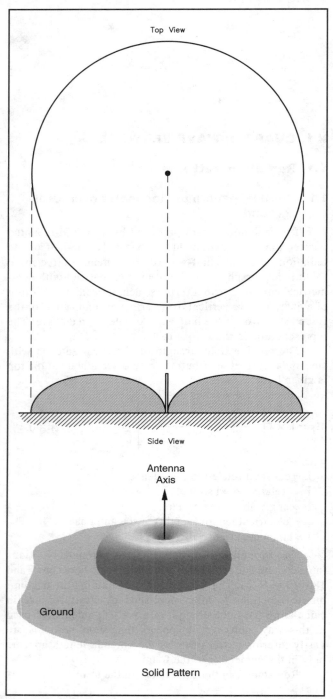

Fig 9-1—The radiation patterns produced by a vertical monopole. The top view is the horizontal pattern, and the side view is the vertical (elevation plane) pattern.

1.1.2.2. The pseudo-Brewster angle

The magnitude of the reflection coefficient is minimum at 90° phase angle. This is the reflection-coefficient phase angle at which the so-called pseudo-Brewster wave angle occurs. It is called the pseudo-Brewster angle because the RF effect is similar to the optical effect from which the term gets its name. At this pseudo-Brewster angle the reflected wave changes sign. Below the pseudo-Brewster angle the reflected wave will subtract from the direct wave. Above the pseudo-Brewster angle it adds to the direct wave. At the pseudo-Brewster angle the radiation is 6 dB down from the perfect ground pattern (see **Fig 9-2**).

All this should make it clear that knowing the pseudo-Brewster angle is important for each band at a given QTH. Most of us go to a vertical to achieve good low-angle radiation.

Fig 9-3 shows the reflection coefficient (magnitude and phase) for 3.6 MHz and 1.8 MHz for three types of ground.

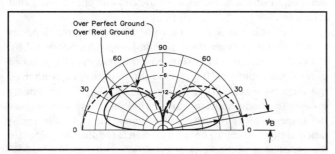

Fig 9-2—Vertical radiation patterns of a ¼-λ monopole over perfect and imperfect earth. The pseudo-Brewster angle is the radiation angle at which the real-ground pattern is 6 dB down from the perfect-ground pattern.

Over sea water the reflection-coefficient phase angle changes from –180° at a 0° wave angle to –0.1° at less than 0.5° wave angle! The pseudo-Brewster angle is at approximately a 0.2° wave angle.

1.1.2.3. Ground-quality characterization

Ground quality is defined by two parameters: the dielectric constant and the conductivity, expressed in milliSiemens per meter (mS/m). Table 5-2 (in Chapter 5) shows the characterization of various real-ground types. The table also shows five distinct types of ground, labeled as very good, average, poor, very poor and extremely poor. These come from Terman's classic *Radio Engineers' Handbook*, and are also used by Lewallen in his *ELNEC* and *EZNEC* modeling programs. The denominations and values as listed in Table 5-2 are the standard ground types used throughout this book for modeling radiation patterns.

1.1.2.4. Brewster angle formula

Terman (*Radio Engineers' Handbook*) publishes a formula which gives the pseudo-Brewster angle as a function of the ground permeability, the conductivity, and the frequency. The chart in **Fig 9-4** has been calculated using the Terman formula. Note especially how salt water has a dramatic influence on the low-angle radiation performance of verticals, where a pseudo-Brewster angle of less than 1° (actually 0.2°) exists on the low bands. In contrast, a sandy, dry ground will yield a pseudo-Brewster angle of 13° to 15° on the low bands, and a city (heavy industrial) ground type will yield a pseudo-Brewster angle of no less than nearly 30° on all frequencies! This means that under such circumstances the radiation efficiency for angles under 30° will be severely degraded.

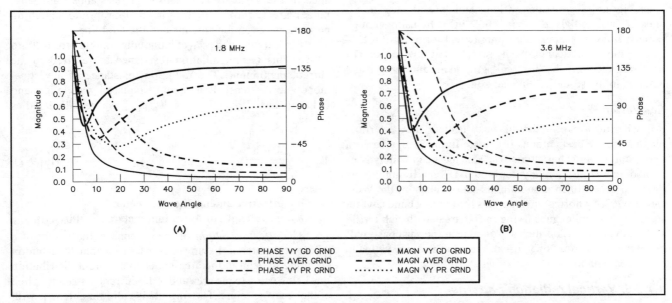

Fig 9-3—Reflection coefficient (magnitude and phase) for vertically polarized waves over three different types of ground (very good, average and very poor).

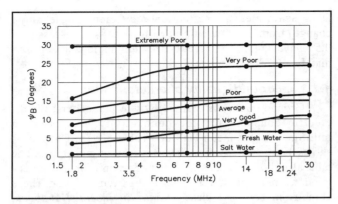

Fig 9-4—Pseudo-Brewster angle for different qualities of reflecting ground. Note that over salt water the pseudo-Brewster angle is constant for all frequencies, at less than 0.1°! That's why stations on the seacoast get out so well with vertical antennas.

1.1.2.5. Brewster angle and radials

Is there anything one can do about the pseudo-Brewster angle? Very little. Ground-radial systems are commonly used to reduce the losses in the near field of a vertical antenna. These ground-radial systems are usually quite short (0.1 to 0.5 λ), and are too short to improve the earth conditions in the area where reflection near the pseudo-Brewster angle takes place.

For quarter-wave verticals the Fresnel zone (the zone where the reflection takes place) is 1 to 2 λ away from the antenna. For longer verticals (eg, a half-wave vertical) the Fresnel zone will extend up to 100 wavelengths away from the antenna (for the radiation at a wave angle of about 0.25°).

This means that a good radial system (mostly ¹/₄ λ to max. ¹/₂ λ long) will improve the efficiency of the vertical in collecting return currents, but will not influence the radiation by improving the reflection mechanism in the Fresnel zone. Not unless you add 5 λ long radials, and keep the far ends of these radials less than 0.05 λ apart by using enough radials. But that seems rather impractical for most of us.

In most practical cases the radiation at low angles will be determined only by the "real" ground around the vertical antenna in the so-called Fresnel zone.

Conclusion

This information should make it clear that a vertical may not be the best antenna if you are living in an area with very poor ground characteristics. This has been widely confirmed in real life; many top-notch DXers living in the Sonoran desert or in mountainous rocky areas on the West Coast swear by horizontal antennas for the low bands, while some of their colleagues living in flat areas with rich fertile soil, or even better, on such a ground near the sea coast, will be living advocates for vertical antennas and arrays made of vertical antennas.

1.1.2.6. Vertical radiation patterns

It is important to understand that gain and directivity are two different things. A vertical antenna over poor ground may show a good wave angle for DX, but its gain may be very poor. The difference in gain at a 10° wave angle for a quarter-wave vertical over very poor ground, as compared to the same vertical over sea-water, is an impressive 6 dB. **Fig 9-5** shows the vertical-plane radiation pattern of a quarter-wave vertical over four types of "real" ground:

- Sea
- Excellent ground
- Average ground
- Extremely poor ground

The patterns in Fig 9-5 are all plotted on the same scale.

1.1.3. Horizontal pattern of a vertical antenna

The horizontal radiation pattern of both the ground-mounted monopole and the vertical dipole is a circle.

1.2. Radiation Resistance of Monopoles

The IRE definition of radiation resistance says that radiation resistance is the total power radiated as electromagnetic radiation, divided by the *net* current causing that radiation.

The radiation resistance value of any antenna depends on where it is fed (see definition in Chapter 6, Section 2.6). I'll call the radiation resistance of the antenna at a point of current maximum as $R_{rad(I)}$ and the radiation resistance of the antenna when fed at its base as $R_{rad(B)}$. For verticals greater than one quarter-wave in height, these two are not the same. Why is it important to know the radiation resistance of our vertical? The information is required to calculate the efficiency of the vertical:

$$Eff = \frac{R_{rad}}{R_{rad} + R_{loss}}$$

The radiation resistance of the antenna plus the loss resistance R_{loss} is the resistive part of the feed-point impedance of the vertical. The feed-point resistance (and reactance) are required to design an appropriate matching network between the antenna and the feed line.

Fig 9-6 shows $R_{rad(I)}$ of monopoles ranging from 20° to 540°. (This is the radiation resistance as referred to the current maximum.) The radiation resistance of a vertical monopole shorter than or equal to a quarter wavelength and fed at its base [thus $R_{rad(I)} = R_{rad(B)}$] can be calculated as follows:

$$R_{rad} = \frac{1450 \, h^2}{\lambda^2} \qquad \text{(Eq 9-2)}$$

where

h = effective antenna height, meters
λ = wavelength of operation, meters (= 300 / f_{MHz})

The *effective height* of the antenna is the height of a theoretical antenna having a constant current distribution all along its length, the area under this current distribution line being equal to the area under the current distribution line of the "real" antenna. The formula is valid for antennas with a ratio of antenna length to conductor diameter of greater than 500:1 (typical for wire antennas).

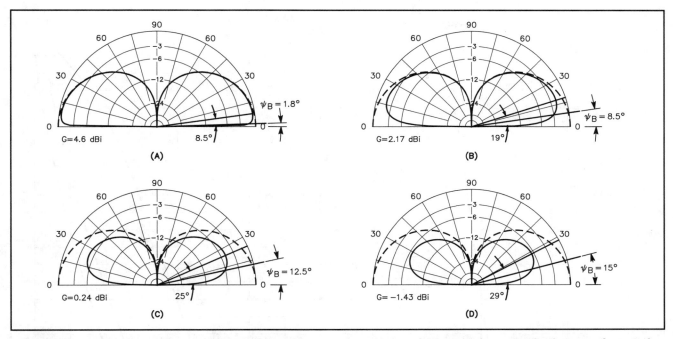

Fig 9-5—Vertical-plane radiation patterns of 80-meter quarter-wave verticals over four standard types of ground, in each case using 64 radials each 20 meters long. The perfect ground pattern is shown in each pattern as a reference (broken line, with a gain of 5.0 dBi). This reference pattern also allows us to calculate the pseudo-Brewster angle. The patterns and figures were obtained using the *NEC-4* modeling program. Modeling was done by R. Dean Straw, N6BV. A—Over salt water. B—Over very good ground. C—Over average ground. D—Over very poor ground.

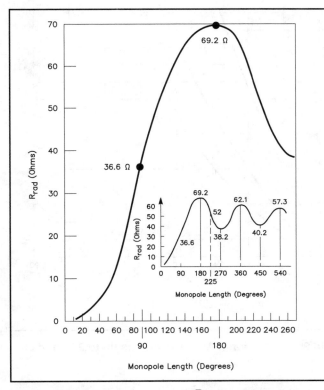

Fig 9-6—Radiation resistances ($R_{rad(l)}$, at the current maximum) of monopoles with sinusoidal current distribution. The chart can also be used for dipoles, but all values must be doubled.

For a full-size, quarter-wave antenna the radiation resistance is determined by:

Current at the base of the antenna = 1 A (given)
Area under sinusoidal current-distribution curve =
  1 A × 1 radian = 1 A × 180/π = 57.3 A-degrees
Equivalent length = 57.3° (= 1 radian)
Full electrical wavelength = 300/3.8 = 78.95 m
Effective height = (78.95 × 57.3)/360 = 12.56°

$$R_{rad} = \frac{1450 \times 12.56^2}{78.95^2} = 36.6 \ \Omega$$

The same procedure can be used for calculating the radiation resistance of various types of short verticals.

Fig 9-7 shows the radiation resistance for a short vertical (valid for antennas with diameters ranging from 0.1 to 1°). For antennas made of thicker elements, **Fig 9-8** and **Fig 9-9** can be used. These charts are for antennas with a constant diameter.

For verticals with a tapering diameter, large deviations have been observed. W. J. Schultz describes a method for calculating the input impedance of a tapered vertical (Ref 795). It has also been reported that verticals with a large diameter exhibit a much lower radiation resistance than the standard 36.6-Ω value. A. Doty, K8CFU, reports finding values as low as 21 Ω during his extensive experiments on elevated radial systems (Ref 793). I have measured a similar low value on my quarter-wave 160-meter vertical (see Section 5.6.) Section 2.1 shows how to calculate the radiation resistance of various types of short verticals.

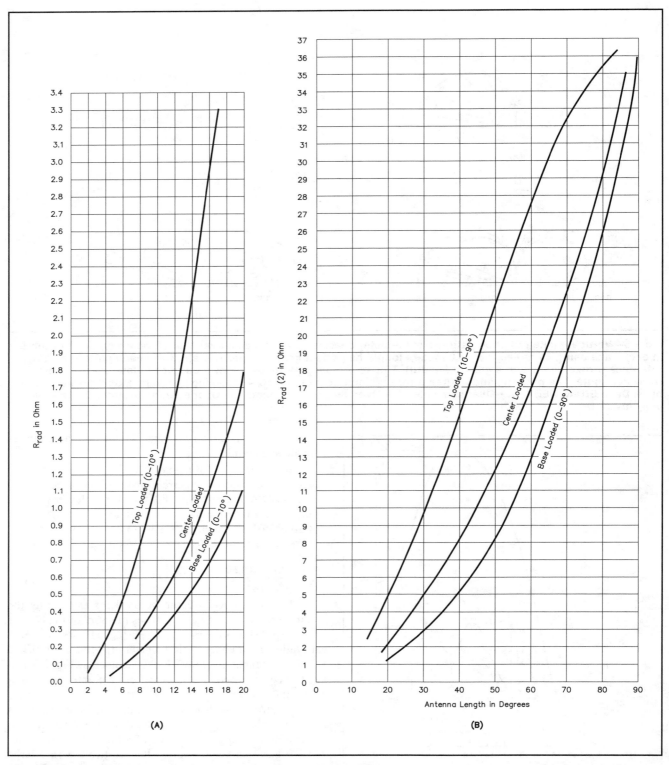

Fig 9-7—Radiation resistance charts (R_rad) for verticals up to 90° or ¼-λ long. At A, for lengths up to 20°, and at B, for greater lengths.

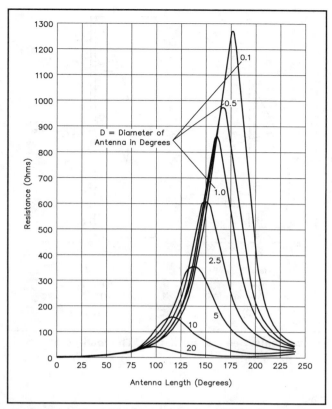

Fig 9-8—Radiation resistances for monopoles fed at the base. Curves are given for various conductor (tower) diameters. The values are valid for perfect ground only.

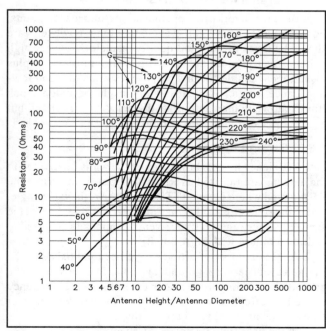

Fig 9-9—Radiation resistances for monopoles fed at the base. Curves are given for various height/diameter ratios over perfect ground.

Longer monopoles are usually not fed at the current maximum, but rather at the antenna base, so that $R_{rad(I)}$ is no longer the same as $R_{rad(B)}$. $R_{rad(B)}$ for long verticals is given in Figs 9-8 and 9-9. (Source: Henney, *Radio Engineering Handbook*, McGraw-Hill, NY, 1959, used by permission.)

$R_{rad(I)}$ is illustrated in **Fig 9-10**. The value can be calculated from the following formula (Ref 722):

$$R_{rad(I)} = \varepsilon - 0.7L + 0.1[20 \sin(12.56637L - 4.08407)] + 45$$

where

ε = the base for natural logarithms, 2.71828 . . .
L = antenna length in radians (radians = degrees times $\pi/180° $ = degrees divided by 57.296).
The length must be greater than $\pi/2$ radians (90°).

Fig 9-10C shows the case of a 135° (³/₈-wavelength) antenna. Disregarding losses, $R_{rad(B)} = R_{feed} \approx 300\ \Omega$, but the value of 2R, the theoretical resistance at the maximum current point, will be lower (57 Ω). If P1 (radiated power) = P2 (power dissipated in 2R), then $R_{rad(I)} = 2R$.

These values of $R_{rad(I)}$ are given in Fig 9-6, while $R_{rad(B)}$ can be found in Figs 9-8 and 9-9.

Figs 9-11 and **9-12** show the reactance of monopoles (at the base feed point) for varying antenna lengths and antenna diameters (Source: E. A. Laport, *Radio Antenna Engineering*, McGraw-Hill, NY, 1952, used by permission.)

1.3. Radiation Efficiency of the Monopole Antenna

The radiation efficiency for short verticals has been defined as

$$Eff = \frac{R_{rad}}{R_{rad(B)} + R_{loss}}$$

For the case of any vertical, short or long, when fed at its base this equation becomes

$$Eff = \frac{R_{rad(B)}}{R_{rad(B)} + R_{loss}} \qquad \text{(Eq 9-3)}$$

The loss resistance of a vertical is composed of:
• Conductor RF resistance
• Parallel losses from insulators
• Equivalent series losses of the loading element(s)
• Ground losses part of the antenna current return circuit
• Ground absorption in the near field

1.3.1. Conductor RF resistance

When multisection towers are used for a vertical antenna, care should be taken to ensure proper electrical contact between the sections. If necessary, a copper braid strap should interconnect the sections. Rohrbacher, DJ2NN, provided a formula to calculate the effective RF resistance of conductors of copper, aluminum and bronze:

$$R_{loss} = \left(1 + 0.1L\right)\left(f^{0.125}\right)\left(0.5 + \frac{1.5}{D}\right) \times M \qquad \text{(Eq 9-4)}$$

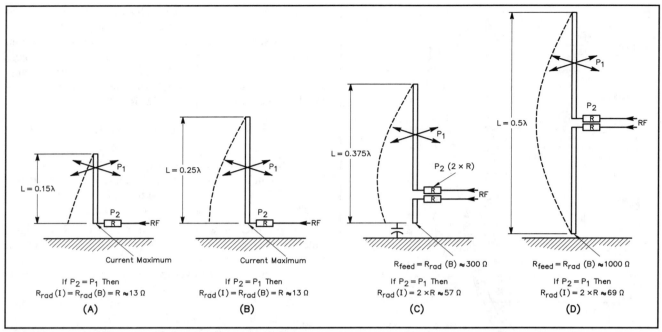

Fig 9-10—Radiation resistance terminology for long and short verticals. See text for details. The feed-point resistances indicated assume no losses.

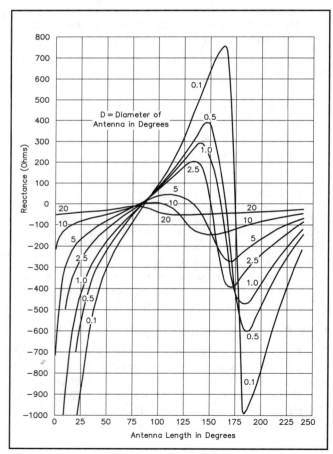

Fig 9-11—Feed-point reactances (over perfect ground) for monopoles with varying diameters.

where
L = length of the vertical in meters
f = frequency of operation in MHz
D = conductor diameter in mm
M = material constant (M = 0.945 for copper, 1.0 for bronze, and 1.16 for aluminum)

1.3.2. Parallel losses in insulators

Base insulators often operate at low-impedance points. For monopoles near a half-wavelength long, however, care should be taken to use good-quality insulators, as very high voltages can be present. There are many military surplus insulators available for this purpose. For medium and low-impedance applications, insulators made of nylon stock (turned down to the appropriate diameter) are excellent, but a good old Coke bottle may do just as well!

1.3.3. Ground losses

Efficiency means: how many of the watts I delivered to the antenna are radiated as RF. Effectiveness means: is the RF radiated where I want it? That is, at the right wave angle, and in the right direction. You can be very efficient but at the same time very ineffective. Even the opposite is possible (killing a mouse with an A-bomb).

A large number of articles have been published in the literature concerning ground systems for verticals. The ground plays an important role in determining the efficiency as well as effectiveness of a vertical in two very distinct "areas": the near field and the far field.

In the near field (efficiency issues):
• I^2R losses: Antenna return currents travel through the ground, and back to the feed point, right at the base of the

antenna (see Fig 9-36). The resistivity of the ground will play an important role if these antenna RF return currents travel through the (lossy) ground. Unless the vertical antenna uses elevated radials, the antenna return current will flow through the ground. These currents will cause I^2R losses.

• Absorption losses: the conductivity and the dielectric properties of the ground will play an important role in absorption losses, caused by an electromagnetic wave penetrating the ground. These losses are due to the interaction of the near-field energy-storage fields of the antenna (or radials) with nearby lossy media, such as ground. These type of losses are present whether elevated radials are used or not.

Losses in the near field are losses causing the radiation

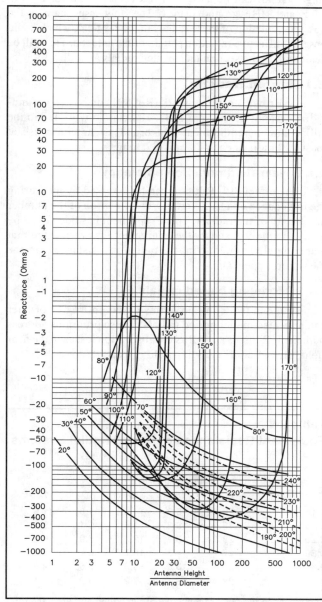

Fig 9-12—Feed-point reactances (over perfect ground) for monopoles with different height/diameter ratios.

efficiency to be less than 100%.

In the far field (efficiency and effectiveness issues):

• Up to many wavelengths away, the waves from the antenna are reflected by the ground and will combine with the direct waves to form the radiation at low angles, the angle we are concerned with for DXing. The reflection mechanism, which is similar to the light reflection mechanism (in a mirror) is described in Section 1.1.2. The real part of the reflection coefficient determines what part of the reflected wave is being absorbed. The absorbed part is responsible for the Fresnel-zone reflection losses (efficiency).

• The ground characteristics in the Fresnel zone will also determine the low-angle performance of the vertical, and this is an effectiveness issue.

The effect of ground in these two different zones has been well covered by P. H. Lee, N6PL (Silent Key), in his excellent book, *Vertical Antenna Handbook*, p 81 (Ref 701).

The next section will cover these and various other aspects of the subject.

■ 2. GROUND AND RADIAL SYSTEM FOR VERTICAL ANTENNAS: THE BASICS

Ground-plane antennas

We all know that a VHF vertical antenna usually employs four radials as a "ground-plane," hence its popular name. But in fact, two radials would do the same job. All you need with a $^1/_4$-λ vertical radiator is a $^1/_4$-λ wire connected to the feed-line outer conductor in order to have an RF ground at that point, sometimes called "another terminal to push against." This is how the antenna return currents are collected. But, if you had only one radial, this radial would radiate a horizontal wave component. Two $^1/_4$-λ radials in a straight line have their current distributed in such a way that radiation from the radials is essentially canceled (in the far field). This is similar to what happens with top-wire loading (T antennas). Using three wires (at 120° intervals) or four radials at right angles does just the same.

It was George Brown, himself, Mr. 120-buried-radials, who invented elevated resonant radials. He invented the ground-plane antenna. The story goes that when Brown first tried to introduce his ground-plane antenna that had only two radials, he had to add two extra radials because few of his customers believed that with only two radials the antenna would radiate equally well in all directions! In the case of a VHF ground plane, there is no "poor ground" involved and all return currents are collected in the form of displacement currents going through the two, three or four radials.

The above (VHF) case is the case where the detrimental effects of the real ground are eliminated by raising the antenna "very high above ground." There are no I^2R losses, because the return currents are entirely routed through the low-loss radials. There also are no near-field absorption losses, as the real ground is several wavelengths away from the antenna. Thirdly, as on VHF/UHF we are not counting on reflection from the real earth to form our vertical radiation pattern; we are not confronted by losses of Fresnel reflection (in the far field) either. In other words, we have totally eliminated "poor" earth.

Verticals with an on (or in) the ground radial system

The other approach in dealing with the "poor" earth is going to the other extreme—bring the antenna right down to ground level, and, by some witchcraft, turn the ground into a perfect conductor. This is what we try to do in the case of grounded verticals.

We put down radials, or strips of "chicken wire" in order to improve the conductivity of the ground, and to reduce the I^2R losses as much as possible. This mechanism is well known to us. We can also measure its effect: we know that as we gradually increase the number and the length of radials, the feed-point impedance is lowered, and with a fairly large number of long radials (eg, 120 radials, 0.5-λ long) we will reach the radiation resistance of the vertical. In the worst case, when no measures are taken to improve ground conductivity, losses can be incurred that range from 5 to well over 10 dB with $1/4$-λ long radiators, and much higher with shorter verticals.

The other mechanism—absorption by the lossy earth is less well known in amateur circles. This is partly because we cannot directly measure its effects (see also Section 2.4), as in the case of I^2R losses. But the effect is nevertheless there, and some have quoted that it can typically result in 3 to 6 dB of signal loss, if not properly handled. In the case of the very high ground-plane antenna we handled the situation by moving the near field of the antenna way above ground. In the case of a vertical with its base near (< approx. $3/8$ λ above) ground, or on the ground, we can only handle this situation by screening (literally hiding) the lossy ground from the near field of the antenna. This means that in the case of buried (or on the ground) radials, their number and length (both together equal density) must be so that the ground underneath is effectively made invisible to the antenna. It has been experimentally established that, for a $1/4$-λ vertical, this means that we must use at least $1/4$-λ radials, and in sufficient number so that the tips of the radials are separated no more than 0.015 λ (1.2 m on 80 meters and 2.4 m on 160 meters), which means we require approximately 100 radials to achieve this goal. With half the number, we will lose approximately 0.5 dB due to near-field absorptive losses, from RF "seeping" through our imperfect screen.

In real life, taking good care of the I^2R losses also means taking good care of the near field absorption losses when using buried radials.

Verticals with a close-to-earth elevated radial system

In some cases it is difficult or impossible to build an on-the-ground radial system that meets this requirement, in most cases because of local terrain constraints. In this case a vertical with a radial system *barely* above ground may be an alternative. The question is: how good is this alternative, and how should we handle this alternative? With radials at low height (typically less than 0.1 λ) we also have to address the two near-field issues: effectively collecting return currents, and dealing with absorption losses in the real ground.

It is clear that if we raise the almost perfect on-ground radial system (50 to 100 $1/4$-λ long radials) above ground, we will have an almost perfect elevated radial system. The

screening effect that was determined to be good for radials laying directly *on* the lossy ground, will evidently be more than good enough if the system is raised somewhat above ground. That the screening of such a dense radial system is close to 100% effective, was witnessed by Phil Clements, K5PC, who reported on the Internet that, while walking below the elevated radial system (120 elevated radials) of a BC transmitter in Spokane, Washington, he could hardly hear the transmitted signal on a small portable receiver. The question, of course, is: Do we really need so many elevated radials, or can we live with many less? This question is one of the topics that I deal with in detail in the section on elevated radial systems.

As far as dealing with the antenna return currents, it is clear that simple radial systems (in the most simple form a single radial) can be used, this has been proven for ages in VHF and UHF ground planes. The only issue here is the possible radiation of these radials in the far field, which could upset the effective radiation pattern of the antenna. This will be dealt with in the section on elevated radials as well.

2.1. Buried Radials

Dr Brown's original work (Ref 801) on buried ground-radial systems dates from 1937. His classic work led to the still common requirement that broadcast antennas use at least 120 radials, each at least 1 λ long.

2.1.1. Near-field radiation efficiency

The effect of I^2R losses can be assessed by measuring the impedance of a $1/4$-λ vertical, as a function of the number and length of the radials. This experiment has been done by many. **Table 9-1** shows the equivalent loss resistance that was computed by deducting the radiation resistance from the measured impedance.

2.1.2. Modeling buried radials

Antenna modeling programs based on *NEC-3* or later can now also model buried radials. These programs address both the I^2R losses as well as the absorption losses in the near field, plus of course any far-field effects, if any.

Table 9-1
Equivalent Resistances of Buried Radial Systems

| Radial Length (λ) | Number of Radials | | | | |
	2	15	30	60	120
0.15	28.6	15.3	14.8	11.6	11.6
0.20	28.4	15.3	13.4	9.1	9.1
0.25	28.1	15.1	12.2	7.9	6.9
0.30	27.7	14.5	10.7	6.6	5.2
0.35	27.5	13.9	9.8	5.6	2.8
0.40	27.0	13.1	7.2	5.2	0.1

The values in the body of this table are in ohms, and are valid for "good" ground.

R. Dean Straw, N6BV, ran a large number of models using *NEC-4* for me (*NEC-4* is not available to non US-citizens). Separate computations were done for 80 and 160 meters. The radiators were ¼ λ long and the radials were buried 5 cm in the ground. The variables used were:
- Ground: very poor, average, very good
- Radial length: 10, 20 and 40 m (for 80 meters), and 10, 40 and 80 m (for 160 meters)
- Number of radials: 4, 8, 16, 32, 64 and 120

We computed the gain, the wave angle and the pseudo-Brewster angle.

Although we ordinarily talk about ¼-λ buried radials, buried radials by no means need to be of a resonant (¼ λ) length. A wire which is resonant (¼ λ) above ground, is no longer resonant in the ground—not even on or near the ground. Typically for a wire on the ground, the physical length for ¼-λ resonance will be approx. 0.14 λ (the exact length depending on ground quality and height over ground). Quarter-wave radials, in the context of buried radials, are wires measuring ¼ λ over ground (typically 20 m long on 80 meters and 40 m on 160 meters).

The gains of the modeling are shown in **Figs 9-13** through **9-18**. The wave angle as well as the Brewster angle are almost totally independent of the radial system in the near field. The values are listed in **Table 9-2**.

When modeling the antenna over poor ground using only four buried radials, it was apparent that the gain was slightly higher using 15-m long radials rather than 20 m or even 40-m long radials (the gain difference being 0.7 dB, quite substantial). It happens that the resonant length of a ¼-λ radial in such lossy ground is approximately 15 m (and not ≈ 20 m as it would be in air). In case of a small number of radials, there is hardly any screening effect, and antenna return currents flow back through lossy, high-resistance earth to the antenna, as well as through the few radials (there are two parallel return circuits, a low-resistance one (the radials) and a high-resistance one (the lossy ground). If the radials are made resonant, their impedance at the antenna

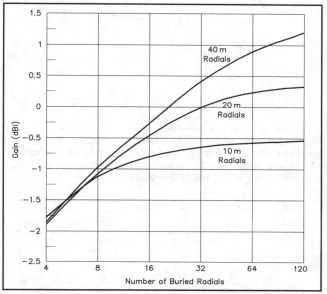

Fig 9-14—Gain of 0.25-λ 80-meter vertical over average ground as a function of radial length and number of radials.

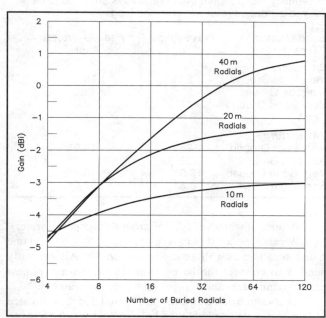

Fig 9-13—Gain of 0.25-λ 80-meter vertical over very poor ground as a function of radial length and number of radials. For short (10-m long) radials there is not much point in going above 16 radials. With 20-m radials you are within 0.5 dB of maximum gain with 32 radials. If you want maximum benefit from 0.5-λ radials (40 m), 120 radials are for you.

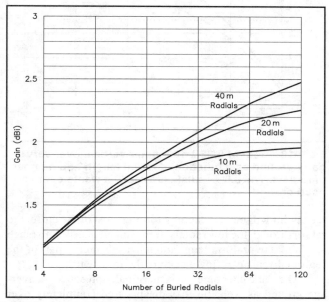

Fig 9-15—Gain of 0.25-λ 80-meter vertical over very good ground as a function of radial length and number of radials.

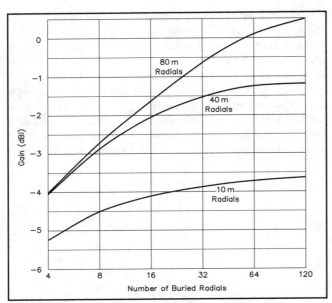

Fig 9-16—Gain of 0.25-λ 160-meter vertical over very poor ground as a function of radial length and number of radials. Note that 10-m radials, no matter how many, are really too short for 160 meters.

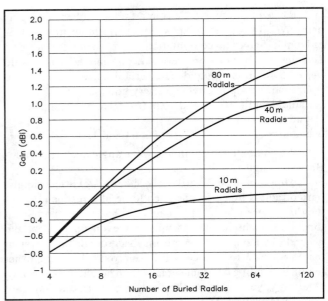

Fig 9-17—Gain of 0.25-λ 160-meter vertical over average ground as a function of radial length and number of radials.

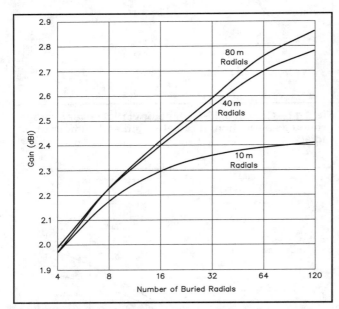

Fig 9-18—Gain of 0.25-λ 160-meter vertical over very good ground as a function of radial length and number of radials. The 0.5-λ radials are really a waste over very good ground.

Table 9-2

Wave Angle and Pseudo-Brewster Angle for Ground-Mounted Vertical Antennas over Different Grounds

The wave angle and the pseudo-Brewster angle are essentially independent of the radial system used, unless the radials are several wavelengths long.

Band/Ground Type	Wave Angle	Pseudo-Brewster Angle
80 meters		
Very Poor Ground	29°	15.5°
Average Ground	25°	12.5°
Very Good Ground	17°	7.0°
Sea Water	8.5°	1.8°
160 meters		
Very Poor Ground	28°	14.5°
Average Ground	23°	11°
Very Good Ground	19.5°	8.5°
Sea Water	8.5°	7.0°

feed point will be low, thereby forcing most of the current to return through the few radials. If the impedance is high (such as with 20 m or 40-m long radials), a substantial part of the return currents can flow back via the lossy earth.

The same phenomenon is marginally present over average ground as well, but has disappeared completely over good ground. These observations tend to confirm the mechanism that originates this apparent anomaly. All of this is of no real practical consequence, as four radials are largely insufficient, in whatever type of ground (except sea water).

We also modeled radials in sea water. As expected one radial does just as well as any other number. All we really need is to connect the base of the vertical to the almost perfect conductor (and screen) that's the sea water.

Years ago Brian Edward, N2MF, modeled the influence of buried radials (Ref 816), and discovered that for a given number of radial wires, there is a corresponding length beyond which there is no appreciable efficiency improvement. This corresponds very well with what we find in Figs 9-13 through 9-18. Brian found that this length is (maybe surprisingly at first sight) nearly independent of earth conditions. This indicates that it is the screening effect which is

Table 9-3
Optimum Length Versus Number of Radials

Number of Radials	Optimum Length (λ)
4	0.10
12	0.15
24	0.25
48	0.35
96	0.45
120	0.50

This table considers only the effect of providing a low-loss return path for the antenna current (near field). It does not consider ground losses in the far field, which determine the very low-angle radiation properties of the antenna.

more important than the return-current I^2R loss effect. Indeed, the effectiveness of a screen only depends on its geometry and not on the quality of the ground underneath. **Table 9-3** shows the optimum radial length as a function of the number of radials.

Conclusion

To me, the results obtained when modeling verticals using buried radials with *NEC-4* seem to be rather optimistic, but the trends are clearly correct.

Take the example of an 80-meter vertical over average ground: going from a lousy eight 20-m long radials to 120 radials would only buy you 1.4 dB of gain, which is less than what I think it is in reality. In very good ground that difference would be only 0.7 dB!

There has been some documented proof that *NEC-4* does not handle very low antennas correctly, and that the problem is a problem associated with near-field losses (see Section 2.2.2). Maybe this same limitation of *NEC-4* causes the gain figures, as calculated with buried radials to be optimistic as well. Future will tell. No doubt further enhancements will be added to future *NEC* releases, which may well give us gain (loss) figures that I would feel more comfortable with.

2.1.3. How many buried radials now, how long?

The following rule was experimentally derived by N7CL and seems to be a very sound and easy one to follow: Put radials down in such a way that the distance between their tips is not more than 0.015 λ (≈ 1.3 m for 80 meters and ≈ 2.5 m on 160 meters).

Using this rule with ¼-λ long radials, will make you use approx. 104 radials. The circumference of a circle with a radius of 0.25 λ is 2 × π × 0.25 = 1.57 λ. At a spacing of 0.015 λ this circumference can accommodate 1.57/0.015 = 104 radials. With this configuration you are within 0.1 dB of maximum gain. If you space the tips 0.03 λ you will lose about 0.5 dB (all this over average to good ground).

For radials that are only ⅛-λ long, 0.03-λ tip spacing yields 52 radials. Here too, if you use only half that number,

you will give up 0.5 dB of gain.

Let us apply this simple rule to some real-world cases:

Example 1

If your lot is 20 m by 20 m, and you want to install a radial system for 80 and 160 m, proceed as follows: draw a circle in which you fit your lot. In the case of a 20 m by 20 m lot, the circle has a radius of $\sqrt{20^2 / 2} \approx 14$ m. On each 20 m side of your lot you will space the ends 1.3 m, this means you will have 16 radials. The longest will be 14 m, the shortest will be 10 m. The average radial length is 12 m. You will install a total of 16 (radials) × 4 (sides) × 14 m (average length) = 896 m of radial conductor, used in a total of 64 radials. A radial system using 32 evenly spread radials, and using "only" 450 m of wire, will compromise you about 0.5 dB.

In actual practice, when laying radials on an irregular lot where the limits are the boundaries of the lot, the practical way to make best use of the wire you have is just walk the perimeter of the lot and start a radial from the perimeter (inward toward the base of the antenna) every 0.015 λ (1.3 m for 80 meters or 2.5 m for 160) as you walk along the perimeter.

Example 2

You have only 500 m of wire, but space is not a problem. How many radials and how long should they be (to be used on both 80 and 160 meters)?

The formula to be used is:

$$N = \frac{\sqrt{2 \times \pi \times L}}{A}$$

where
  N = number of radials
  L = total wire length available
  A = distance between wire tips (1.3 m for 80, 2.5 m for 160, or twice that if 0.5 dB loss is tolerated)

The dream QTH of VE1JF. View of the Atlantic Ocean in the direction of Europe. The 80-meter 4-square with elevated radials is located right on the edge of a bluff and has the sea as its Fresnel zone toward Europe!

For this example use L = 500 m, A = 1.3 m, and you calculate:

$$N = \frac{\sqrt{2 \times \pi \times 500}}{1.3} = 43 \text{ radials}$$

Each radial will have a length of 500/43 = 11.6 m.

You could also use A = 2.6 m, in which case you wind up with 22 radials each 18 m long. However, the first solution will give you slightly less loss.

In general one can say that, for a given length of wire, it is better to use a larger number of short radials than a smaller number of long radials, the limit being that the tips should not be closer than 0.015 λ.

Example 3

How much radial wire (number and length) is required to build a radial system (for a ¼-λ vertical) that will be within 0.1 dB of maximum gain. How much to be within 0.5 dB?

The answer to the first question is 104 radials, each 0.25-λ long, the total wire length for 80 meters is: 2080 m (4000 m for 160). With 52 radials each 0.25-λ long, you are within 0.5 dB of maximum, this means that 1000 m will be required for 80 meters and 2000 m for 160 meters.

Example 4

I can put down 15-m long radials in all directions. How many should I put down, and how much radial wire is required?

The circumference of a circle with a radius of 15 m is: $2 \times \pi \times 15 = 94.2$ m. With the tips of the radials separated by 1.3 meters, we have 94.2 / 1.3 = 72 radials. In total I will use $72 \times 15 = 1080$ m of radial wire. There is no point in using more than 72 radials.

2.1.4. Two-wavelength-long radials and the far field

Everything that happens in the near field determines the total radiated field strength. Radials, screens, I^2R losses have very little influence on the radiation pattern of the vertical (except maybe at very high angles, which we are not interested in, anyhow). Basically any method of improving ground conductivity in the near field (up to ¼ λ from the base of the vertical in case of a ¼-λ vertical) improves all of the radiation pattern, not selectively (not more at certain radiation angles than at others).

In the far field the ground characteristics greatly influence the low-angle characteristics of a vertical antenna. For ¼-λ verticals the area where Fresnel reflection occurs starts about 1 λ from the antenna and extends to a number of wavelengths.

For current collecting and near-field screening there is really little or no point in installing radials longer than ¼ λ. With 104 such radials you are within 0.1 dB of what is theoretically possible. The Brown rule (120 radials, 0.4-λ long) shoots for less than 0.1 dB and has some extra reserve built in.

If you want to influence the far field, and pull down the radiation angle somewhat, or reduce the reflection loss, then we are talking about radials that are about 2 λ long. However the same rule still applies, for a ground screen top to be

The Battle Creek Special that made Heard Island available on 160 for over 1000 different stations. Ghis, ON5NT, is not holding up the antenna, it is very capable of standing up by itself. The antenna was located near the ocean's edge, on salt-water-soaked lava ash.

effective, the wires should not be separated more than approx. 0.015 λ, even 2 λ away. This would mean that we would require (2 λ × 2 × π) / 0.15 λ = 837 radials, each 2 λ long, which for 160 m totals more than 27 km of radial wire, and you need an area of 660 × 660 m (43 hectares!) to install this radial system! Hardly practical, of course.

The only practical way of influencing the far-field reflection efficiency and effectiveness is to move to the coast or to a salt-water lake and install your vertical in the middle of salt water. In that case you will have a peak radiation angle of between 5 and 10° and a pseudo-Brewster angle of less than 2°! The elevation pattern becomes very "flat" showing a 3-dB opening angle ranging from 1 to 40°. All this is due to the wonderful conductivity properties of salt water. No wonder such a QTH does wonders (ask Ben Moeller, OZ8BV why he's so loud on 80 meters).

Tom Bevenham, DU7CC (also SM6CNS), testified: "At my beach QTH on Cebu Island, I use all vertical antennas standing out in salt water. Also, at high tide, water comes all the way underneath the shack. On Top Band, I use a folded monopole attached alongside a 105-ft bamboo pole. This antenna is a real winner. I use not much of a ground system, only a few hundred feet of junk wire at sea bottom. At the other QTH, less than half a mile from the beach, the same antennas with ground radials don't work at all."

Of course, we have all heard how well the over-salt-water vertical antennas perform, remember the operation from Heard Island (VK0IR) for one. The Battle Creek Special (see Section 6.7) was standing with its base right in the salt water.

2.1.5. Ground rods

Ground rods are important for achieving a good dc ground which is necessary for adequate lightning protection, but ground rods contribute very little to an RF ground. A ground rod will seldom constitute an acceptable minimum RF ground. It is always a good idea though to properly ground your towers and verticals. If a series-fed (insulated-base) vertical is used, a lightning arrestor (spark gap) with a good dc ground is a good idea as well. In addition one can install a 10 to 100-kΩ resistor or an RF choke between the base of the antenna and the dc ground to drain static charges.

2.1.6. Burying depth of buried radials

C. J. Michaels, W7XC (silent key), calculated the depth of penetration of RF current in ground of different properties. He defined the depth of penetration as the depth at which the current density is 37% of what it is at the surface. Under those conditions, for 80 meters, a depth of penetration of an amazing 1.5 m has been calculated for very good ground. For very poor ground the depth reaches 12 m!

This would mean that from an I^2R loss point of view, we can bury the radials "deep" without any ill effects. However, from near-field screening effect point of view, we need to have the radial system above the lossy material!

Leo, W7LR (Ref 808) reports that burying the radials a few inches below the surface does not detract from their performance. Al Christman, K3LC (ex KB8I), confirmed this when modeling his elevated radial systems. He found only hundredths of a dB difference between burying radials at 5 cm or 15 cm. I would not bury them much deeper though. The sound rule here is "the closer to the surface, the better."

2.1.7. Some practical hints

Local ground characteristics

It is impossible to make a direct measurement of ground characteristics. The most reliable source of information about local ground characteristics may be the engineer of your local AM broadcast station. The so-called "full proof-of-performance" record will document the soil conductivity for each azimuth out to about 30 km (20 mi).

Radial bus bar / low loss connections

It has been my experience, and has been confirmed by many others, that using a solid metal plate of reasonable size right under the antenna can result in a notable decrease in ground resistance, and consequently in ground loss. Above all it makes low-loss connections between the radials and the bus-bar (plate) possible. It is also a good idea to make all connections accessible so that individual radial-current measurements can be made.

Chicken-wire strips

Sherwood, WBØJGP, has described and compared ground systems consisting of wide strips of ground screens (Ref 809). Anyone tempted to try the "screen" approach should be warned of one thing: never use steel wire for a buried ground system, whether it be a single wire or chicken

wire. Steel is a very poor conductor at RF. The steel wire will also corrode in a very short time, although a thick layer of galvanization may improve the resistance to corrosion. I have found, in our Belgian (wet!) climate, chicken wire rusts away after just a few years!

Soldering / welding radial wires

Tin-lead (Sn-Pb), which is often used to solder copper wires, will deteriorate in the ground and may be the source of bad contacts. Therefore it is recommended to solder all your copper radials using silver, or even better to weld the radials (info about CADWELD welding products from The RF Connection in Maryland is available on their Web page: **http://www.therfc.com**).

Sectorized radial systems

Very long radials (several wavelengths long) in a given direction have been evaluated and found to be effective for lowering the wave angle in that direction, but seem to be rather impractical for just about all amateur installations. A similar effect occurs when verticals are mounted right at the salt-water line (Ref the pseudo-Brewster angle). Similar in result to a sectorized radial system is the situation where an elevated radial system is used with only one radial (see Section 2.2.3).

Radial wire gauge

When only a few (less than six) radials are used, the gauge of the wires is important for maximum efficiency. The heavier the better, #16 wires are certainly no luxury when only a few buried radials are used. With many radials, the wire size becomes unimportant since the return current is divided over a large number of conductors. DXpeditions using temporary antennas just have to take a small spool of #24 or 26 (0.5 or 0.4-mm diameter) enameled magnet wire. This is inexpensive, and can be used to establish a very efficient RF ground system.

A radial plow

Installing radials can be quite a chore. Hyder, W7IV, (Ref 815) and Mosser, K3ZAP (Ref 812) have described systems and tools for easy installation of radials. **Fig 9-19** shows such a radial plow as made by Ghis Penny, ON5NT, to bury the radials in his lawn. A small carriage, made of wood, supports a sharp knife which cuts a slot in the ground. A small aluminum feed tube deposits the radial wire at the bottom of the slot, about 5-cm deep. A person sitting on the carriage takes care of the required weight to drive the knife into the ground.

An alternative to plowing

Radials can also be laid on the ground in areas that are suitable. Another neat way of installing radials in a lawn-covered area is to cut the grass really short at the end of the season (October), and lay the radials flat on the ground, anchored here and there with metal hooks. By the next spring, the grass will have covered up most of the wires, and by the end of the following year the wires will be completely covered by the grass. This will also guarantee that your radials are "as close as possible" to the surface of the ground, which is ideal from a near-field screening point of view!

Fig 9-19—Small home-made radial plow as used by Ghis Penny, ON5NT, for burying the radials in the lawn. That's Heidi, ON5NT's youngest daughter, acting as driver for the wire plow, while the OM himself takes care of the required "horsepower" to cut the slot in the lawn.

2.2 Elevated Radials

With real, high-up VHF and UHF ground-plane antennas the three or four radials are more an electrical counterpoise (a zero-Ω connection point high above ground), than a ground plane. The ground is so far away that any term including the word "ground" is really not in place. The radials of such antennas (and all of them radiate in the near field) do not, however, cause near-field absorption loses in the ground, because of their relative height above ground.

Such HF and VHF/UHF ground planes have been in use for many years. Studies that were undertaken in the past several years, however, are concerned with vertical antennas using radials at much lower heights, typically 0.01 to 0.04 λ above ground. That there is still quite a bit of controversy on this subject is no secret to the insiders. It appears that a number of real life results do confirm the current modeling results, while others don't. The jury may still be out. I will try to represent both views in this book.

A. Doty, K8CFU, concluded from his experimental work (Ref 807 and 820) that a $^1/_4$-λ vertical using an elevated counterpoise system can produce the same field strength as a $^1/_4$-λ vertical using buried bare radials. The reasoning is that in the case of an elevated radial or counterpoise system, the return currents do not have to travel for a considerable diustance through high-resistance earth, as is the case when buried radials are used. His article in April 1984 *CQ* also contains a very complete reference list of just about every publication on the subject of radials (72 references!).

Frey, W3ESU, used the same counterpoise system with Minipoise short low-band vertical (ref 824). He reports that connecting the elevated and insulated radial wires together at the periphery definitely yields improved performance. If a counterpoise system cannot be used, Doty recommends using insulated radials lying right on the ground, or buried as close as possible to the surface.

Quite a few years after these publications, A. Christman, R. Redcliff, D. Adler, J. Breakall and A. Resnick used computer modeling to come to conclusions which are very similar to the findings brought forward after extensive field work by A. Doty.

The publication in 1988 by A. Christman, K3LC (ex KB8I), has since become the standard reference work on elevated radial systems (Ref 825), work which has stirred up quite a bit of interest and further investigation.

The results from Christman's study were obtained by computer modeling using *NEC-GSD*. It is interesting to understand the different steps he followed in his analysis (all modeling was done using average ground):

1. Modeling of the $^1/_4$-λ vertical with 120 buried radials (5-cm deep). This is the 1937 Brown reference.

2. The $^1/_4$-λ vertical was modeled using only four radials at different radial elevations. For a modeling frequency of 3.8 MHz, Christman found that 4.5 m was the height at which the four-radial systems equaled the 120-buried-radial systems as far as low-angle radiation performance is concerned.

Christman's studies also revealed that, as the quality of the soil becomes worse, the elevated radial system must be raised progressively higher above the earth to reach performance on par with that of the reference 120-buried-radial vertical monopole. If the soil is highly conductive, the reverse is true.

The elevated-radial approach has become increasingly

popular with low-band DXers since the publication of the above work, and it appears that elevated radials represent a viable alternative to digging and plowing, especially where the ground is unfriendly for such activities.

It is important to critically analyze the elevated-radial concept and therefore to understand the mechanism that governs the near-field absorptive losses (see Section 1.3.3) connected with elevated radials. In the case of an elevated-radial system these near-field losses can be minimized in only three ways:

1. By raising the elevated radials as high as possible (move the near field of the antenna away from the real lossy ground).
2. By installing many radials, so that these radials screen the near fields from "seeing" the underlying lossy earth.
3. By improving ground conductivity of the real ground.

Although the experts all agree on the mechanisms, there appears to be a good deal of controversy about the exact quantification of the losses involved (see Section 2.2.1).

Incidentally, an elevated radial system does not imply that the base of the vertical must be elevated from the ground. The radials can, from ground level, slope up at a 45° angle to a support a few meters away, and from there run horizontally all the way to the end. It is a good idea to keep the radials high enough so no passersby can touch them. This is also true when radials are quite high. In an IEEE publication (Ref 7834) it was reported that significantly better field strengths were obtained with elevated radials at 10-m height than at 5-m height. In both cases the radials were sloping upward at a 45° angle from the insulated base of the vertical at ground level.

2.2.1. Modeling vs measuring? Elevated vs ground radials?

The performance of an elevated radial system can be assessed by either computer modeling or by real-life testing and field-strength measuring. It is of course ideal if the results from modeling and field-strength (FS) measurements match.

Al Christman, K3LC (ex KB8I), used NEC-4 to study the influence of the number of elevated radials and their height on antenna gain and antenna wave angle (Ref 7825) and came to the conclusion that if the height of the radials is at least 0.0375 λ (3 m on 80, 6 m on 160) there is very little gain difference between using 4 or up to 36 radials. He also concluded that the gain of antennas with an elevated radial system compared in gain to the same antenna with about 16 buried radials. Incidentally, the modeling also showed that for buried 1/4-λ radials the difference in gain between 16 radials and 120 radials is only about 0.74 dB. When raising the elevated radials to a height of 0.125 λ (20 m on 160), the gain actually approached the gain of a vertical with 120 buried radials.

The publication of these results (1988) gave a tremendous impetus in the use of elevated-radial systems.

In another study, Jack Belrose, VE2CV (Ref 7821 and 7824) also concluded that there was a good correlation between measured and computed results. In this study Belrose used a 1/4-λ vertical, as well as 1/4-λ (resonant) radials.

A good correlation between the modeled results and FS measurements was established in several study cases. One of them was an extremely well documented case with thousands of FS measurements, which matched very well the figures obtained with modeling (NEC-4). Belrose's studies revealed that radials should be at least 0.03-λ high (2.5 m on 80 meters, 5 m on 160 meters) to avoid excessive near-field absorption ground losses, especially so if fewer than eight radials are used. With a large number of radials (>16) the radials can be much lower.

Another well documented case was reported in a technical paper delivered by Clarence Beverage (nephew of Harold Beverage) at the 49th NAB Broadcast Engineering Conference entitled: "New AM Broadcast Antenna Designs Having Field Validated Performance." The paper covered antenna tests done in Newburgh, New York under special FCC authority. The antenna system consisted of a tower 120 feet in height with an insulator at the 15-foot level and six elevated radials a quarter wavelength in length spaced evenly around the tower and elevated 15 feet above the ground. The system operated on 1580 kHz at a power of 750 W. The efficiency of the antenna was determined by radial field-intensity measurements (in 12 directions) extending out to distances up to 85 kilometers. The measured RMS efficiency was 287 mV/m (normalized) to 1 kW at 1 kilometer, which is the same measured value as would be expected for the tower above with 120 buried radials.

In a number of other cases however, it was reported that field-strength measurements indicated a discrepancy of 3 to 6 dB with the NEC-4 computed results. Tom Rauch, W8JI, published the following results, which he measured:

Number of Radials	On the ground	Elevated 0.03 λ
4	−5.5 dB	−4.3 dB
8	−2.7 dB	−2.4 dB
16	−1.3 dB	−0.8 dB
32	−0.8 dB	−0.7 dB
60	Reference (0 dB)	−0.2 dB

Calculations with NEC-4 show a difference of only about 2 dB going from 4 to 60 buried radials, which is 3.5 dB less that Rauch's experiment showed. The 5 dB he found inspired the following comment: "Consider that going from a single vertical to a four square only gained me 5 dB! I got almost that just by going from four radials to 60 radials."

Eric Gustafson, N7CL, reported (on the Top-Band Internet Reflector) that several experiments comparing signal levels of a ground mounted 1/4-λ vertical with 120 radials with those produced by the same radiator with an elevated radial system (using a few radials) have been done a number of times by various researchers for various organizations ranging from the broadcast industry and universities to the military. He reported that the results of these studies always have returned the same results: the correctly sized, sufficiently dense screen is superior to four resonant radials in close proximity to earth. The quantification of the difference has varied. The largest difference Eric personally measured during research for the military was 5.8 dB, the smallest difference 3 dB. The latter one was measured over really good ground, being a dry salt-lake bed (measured conductivity approx. 20 mS during the test). It is clear that the quality of the ground plays a very important role in the exact amount of loss.

For those who would like to duplicate these tests, understand that you cannot do these tests on one and the same vertical, switching between elevated radials to ground-mounted radials, *unless* you remove (physically) the ground-mounted radials when you use the elevated ones. If not, you have an elevated radial system *plus* a screen, effectively screening the near fields from the underlying real ground.

It seems to me that elevated-radial systems are indeed a valid alternative for buried ones, especially if buried ones are not possible or very difficult to install for whatever reason. Fact is that even the broadcast industry now also uses elevated-radial systems quite extensively and successfully where local soil conditions make it impossible to use the classic 120 buried $1/2$-λ radials. It must be said though that most of these systems use more than just a few radials. I also know of many amateur antenna systems successfully using elevated radial systems. Whether they get optimum performance or lose maybe 2 to 5 dB because of near-field absorption losses, is hard to tell. As a matter of fact, there is still the possibility to improve the ground conductivity under the elevated radial system. More on that in Section 2.2.11.

The discrepancy between measured and modeled gain figures has been recognized by a number of expert *NEC* users. All of the current modeling programs have flaws, but most are known and can be compensated for by experienced users. It seems to be that modeling of very low wires even with current versions of *NEC-4* may be affected by such a flaw.

We should also recognize that the total losses due to mechanisms in the near field can amount to much more than 5 dB. Antenna return-current losses (sometimes also called "connection" losses) can amount easily from 10 to even 40 dB over poor ground. These losses can, however, easily be mastered with elevated radials and reduced to zero. The remaining 4 or 5 dB, accountable to near-field absorption losses are indeed somewhat more difficult to deal with using elevated radials.

2.2.2. Modeling vertical antennas with elevated radials

As mentioned before only *NEC*-based programs can model antennas with elevated radials close to ground. Roy Lewallen improved the general public version of *EZNEC* (using the *NEC-2* engine) by incorporating a "high-accuracy" (NEC Sommerfeld) ground which he claims should be accurate for low horizontal wires down to 0.005-λ high (about 2.7 feet on 160 meters).

Despite this patch many cases have been reported indicating a difference of up to 6 dB in gain for antennas very close to ground. A similar flaw was already present in *NEC-2* and has been documented by John Belrose, VE2CV,

who compared the experimentally obtained results, published by Hagn and Barker in 1970 (Gain Measurements of a Low Dipole Antenna Over Known Soil) with the *NEC-2* predictions. At 0.01 λ above ground, *NEC-2* showed 5 dB more gain than the actual measured values.

All of this goes to say that modeling software is a tool, in most cases experimentally developed. Most modeling programs have well known, but also sometimes little known or barely documented limitations. Field-strength measurements are the real thing (eating is the proof of the pudding). But we should be thankful for having access to antenna-modeling programs. They have undoubtedly helped the non-professionals to gain an enormous amount of insight which they would miss without these tools. It is the role of the professionals and the experts to show the nonexpert users how to use them correctly, and make corrections if necessary.

2.2.3. How many elevated radials?

Through antenna modeling, K3LC (ex KB8I), calculated (for 80 meters), the $1/4$-λ antenna gains for elevated radial heights of 5, 10, 15, 20, 25 and 30 m, while varying the number of ($1/4$-λ) radials between 4 and 36 (Ref 7825). According to these calculations, at a height of 4.5 m (which is roughly what I have) it made less than 0.1 dB of difference between 4 and 32 radials, and this was within 0.3 dB of a buried radial system using 120 quarter-wave radials! These results were confirmed by Jack Belrose, VE2CV (Ref 7821) also through antenna modeling.

Eric Gustafson, N7CL, in a well documented e-mail addressed to the Top-Band Reflector, explained that (for a $1/4$-λ vertical radiator) a radial system, with 104 radials $1/4$-λ long (which means wire ends separated not more than 0.015 λ) achieves a 100% shielding effectiveness. His experimental work (radials about 5-m high) further indicates that the screening effectiveness of a $1/8$-λ-long radial system does not improve above 52 radials. This means that the shielding effectiveness of the $1/8$-λ radial system with 52 radials by itself is 100%, but that some loss will be caused by near fields "spilling over" the screen at its perimeter. (In other words, the screen is dense enough, but not large enough.) Using just 26 radials $1/4$-λ long you will typically lose about 0.5 dB due to near-field absorption losses in the ground.

N7CL goes on to say that a $1/4$-λ vertical with merely four elevated radials can indeed produce the same signal as a ground-mounted vertical with 120 radials $1/4$-λ long provided that:
1. The base of the vertical is at least $3/8$-λ high.
2. The quality of the ground under the elevated radials has been improved so that it acts as an efficient screen, preventing the nearby field to interact with the underlying lossy ground.

Unless such measures are effectively taken, N7CL calculated that the extra ground absorption losses can be as high as 5 or 6 dB. Loss figures of this order have been measured in a number of cases (eg, by Tom Rauch, W8JI) reported on the Internet Top-Band Reflector (see Section 2.2.1).

Conclusion

The jury is still out. According to the *NEC*-based modeling results, there is really no point to use more than four radials. With four radials over good ground the gain of a $^{1}/_{4}$-λ monopole is –0.1 dBi, two radials give an average of –0.15 dBi (+0.14 and –0.47 dBi due to slight pattern squeezing) and one radial gives a gain of +1.04 dBi in the direction of the radial, and –2.3 dBi off its back, resulting in an integrated gain of 0.65 dBi. If you have the room, use four elevated radials, but don't expect to beat the pileup because you use four radials instead of one or two (if you did not already). Four radials seems to have become the standard for elevated radials, as the additional gain from going to more is said to be minimal. But if you are restricted in space, two or three will be almost as good.

In my opinion, it is likely that the current *NEC*-based modeling programs are somewhat optimistic when it comes to dealing with near-field absorption losses. Three or four elevated radials over a poor ground, in my humble opinion, can never be as good as 120 ground-mounted (or elevated for that matter) radials. There is simply no free lunch! If you need to use an elevated radial system, maybe it's not a bad idea after all to use 26 radials, which according to N7CL would put you within 0.5 dB of the Brown standard.

Radial layout

If you use a limited number of elevated radials (two, three or four), a symmetrical layout is necessary for the radiation from radials to cancel (in the far field). One radial is not symmetrical, two and more are symmetrical, provided the radials are spread out evenly over the 360°. When using more than four radials the exact layout as well as the exact radial length becomes of little importance as to creating high-angle radiation.

2.2.4. Only one radial

In his original article on elevated radials (Ref 825) Christman showed the model of a $^{1}/_{4}$-λ vertical using a single elevated radial. This pattern shown in **Fig 9-20** is for a radial height of 0.05 λ over three different types of ground. He showed this vertical, with a single elevated radial, as having (within a minor fraction of a dB) the same gain (in its favored direction) as a ground-mounted vertical with 120 buried radials!

Note however that the pattern is non-symmetrical. The radiation favors the direction of the radial (typically 3-4 dB F/B over average ground). Modeling the same vertical over very good ground results in much less directivity, and over salt water the antenna becomes perfectly omnidirectional.

I expect that it is sufficient to install radials on the ground, under the antenna to improve the properties of the ground in the near field of the antenna to a point where the directivity, due to the single radial, is reduced to less than 1 dB.

The slight directivity can be used to advantage in a setup where one would have a vertical with eg, four radials, which are then connected one at a time to the vertical antenna. Another application (Ref 7824) is where the verti-

cal is part of a (fixed) array, and where one makes use of the initial directivity of each element to provide some added directivity (see **Fig 9-21**).

The single radial does not only create some horizontal directivity, it also introduces some high-angle radiation, caused by the radiation from the single radial. If two or more radials are used, they can be set up in such a way that the horizontal radiation of these radials is effectively canceled by the interaction with one another. Notice from Fig 9-20 that most of the high-angle pattern energy is at or near 90°.

It is obvious that, if one is looking for a maximum of low-angle radiation (which is normally the case for DXing), using only one radial is not the best choice, especially if the antenna is going to be used for reception as well. In a contest station environment, however, creating some high angle radiation, in order to have some "presence" on the band with locals as well,

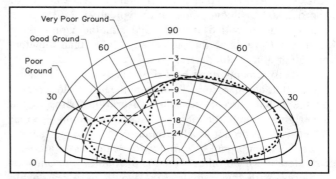

Fig 9-20—Vertical radiation pattern of a quarter-wave vertical with one horizontal $^{1}/_{4}$-λ radial at a height of 0.05 λ over different types of ground.

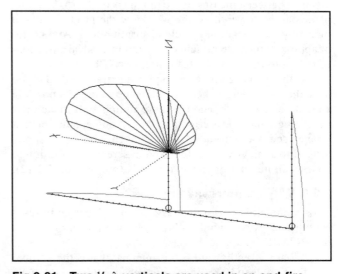

Fig 9-21—Two $^{1}/_{4}$-λ verticals are used in an end-fire configuration (see Chapter 10), producing a cardioid pattern. By placing the single radial in the forward direction of the array, some additional gain can be achieved. This technique makes it impossible to switch directions.

is often desirable. If separate directive low-angle receiving antennas (eg, Beverages) are used, using a single radial on a vertical may well be a logical choice of preference. I am using a single 5-m high elevated radial on my 80-meter 4-square (radials pointing out of the square). At the same time I have a decent shielding effect on the real ground by more than 200 radials of the 160-meter vertical which supports the 80-meter 4-square (see Chapter 11).

A vertical with a single radial can also be a logical choice for a DXpedition antenna, and this for two reasons:
1. Ease of adjusting resonance from the CW to the phone end of the band, by just lengthening the radial.
2. Extra gain by putting the radial in the wanted direction (toward areas of the world with high amateur population density).

2.2.5. How high the radials

The *NEC*-modeling results, published by Christman, K3LC (ex KB8I), indicate that radials above a height of approx. 0.03 λ achieve gains within typically 0.2 dB of what can be achieved with 64 buried radials. In other words, there is no point in raising the radials any higher than 6 meters on 160 or 3 m on 80 meters.

Measurements done by Eric Gustafson, N7CL, however, tell us another story. To prevent the near fields created by the radial currents to cause absorption losses in the underlying ground, the radials must be high enough so that the near fields do not touch ground. With up to six radials, this is between ¼ and ⅜ λ. Below ⅛ λ, the losses are very considerable (if no other screen is available). For amateur purposes, with four radials, ¼ λ would be a reasonable minimum limit to use. This minimum height decreases as the density of the radial screen is increased. With a density of approx. 100 quarter-wave long radials (in which case the distance between the tips of the radials is 0.015 λ) the radial plane can be lowered all the way *onto* the ground without incurring significant near-field absorption loss. At a height of approx. 0.03 λ, 26 radials will result in an absorption loss of not more than 0.5 dB, according to N7CL.

Conclusion: If you want to play it extra safe, and if you have the tower height, get the radials up as high as possible and add a few more. Having more radials will make their exact length much less critical as well. Another solution that I have used is to put radials and chicken-wire strips on the ground to achieve an "on-the-ground screen" in addition to your small number of elevated radials (see Section 2.2.12).

2.2.6. Why quarter-wave radials?

In modeling it is quite easy to "make" perfectly resonant quarter-wave radials. Why do we really want resonant radials? Let's examine this issue.

What we really want is the antenna, that is the vertical plus the radials to be resonant, not because this would make the antenna radiate better, but only because that makes it easier to feed the antenna.

Dick Weber, K5IU, found through a lot of measuring and testing of real-life verticals with elevated radials that using ¼-λ radials has a big disadvantage. We use four radials

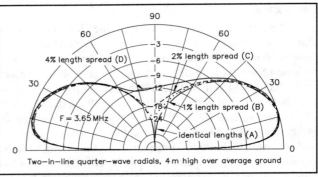

Fig 9-22—Vertical radiation patterns (over good ground) for a ¼-λ long 80-meter vertical, with two in-line radials 4-m high, for various radial lengths around ¼-λ. See text for details.

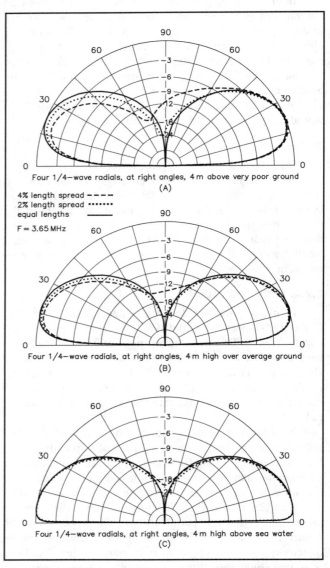

Fig 9-23—Vertical radiation patterns of an 80-meter ¼-λ vertical with four elevated radials (4-m high) over various types of ground. Patterns are for: (A) average ground, (B) very good ground and (C) sea water. See text for details.

(one per 90°) because we want the radiation from these radials to be completely canceled: no pattern distortion and no high-angle (horizontally polarized) radiation. Well, he found out that this is very difficult, if not impossible to achieve in the real world. Of course this works well on a computer model: you can define four radials that have exactly the same electrical length. But not in the real world: one radial will always be, perhaps a minute amount, electrically longer or shorter than the other one. And there lies the problem. We want these four radials all to carry exactly the same current, in order for their radiation to balance out.

The question of course is how important is it to have equal currents in the radials. I modeled several cases of intentional radial current imbalance. **Fig 9-22** shows the vertical radiation pattern of a $^1/_4$-λ vertical (F = 3.65 MHz), with two elevated radials, 4-m high. Pattern A is for two radials showing no reactance (both perfectly 90°, which can never be achieved in real life). For pattern B, I have intentionally shortened one radial about 20 cm (approx. 1% of the radial length), which introduced a reactance of $-j\,8\,\Omega$ for this radial. One radial now carried 62% of the antenna current, the other the remaining 38%. Over good ground this imbalance causes the horizontal pattern to be skewed about 0.6 dB (inconsequential), but we see a fill-in of the high-angle rejection (around 90°) that we have when the currents are equal. Pattern C is for a case where one radial is 20 cm too short, and the other one 20 cm too long (reactance $-j\,8\,\Omega$ and $+j8\,\Omega$). In this case the relative current distribution was very similar as in the first case (63% and 36%). The horizontal pattern skewing was the same as well. Pattern D is for a rather extreme case where radials differ 80 cm in length ($+j16\,\Omega$ and $-j16\,\Omega$). Current imbalance has now increased to 76% versus 24%.

A similar computer analysis was done for a vertical using four elevated radials. In this case, I did the analysis over three different types of ground: good ground, very good ground and sea water (ideal case).

Fig 9-23 shows the results of these models. Case A is for equal currents in the four radials (theoretical case), case B is for radials showing reactances of $+j8\,\Omega$, 0Ω, $-j8\,\Omega$, and $+j10\,\Omega$. The relative current distribution in the four radials was: 51%, 39%, 5% and 5%, which are values very similar to what has been measured experimentally (see **Fig 9-24**). Pattern C shows rather extreme imbalance with radial reactances of $-j16\,\Omega$, 0Ω, $+j16\,\Omega$ and $+j8\,\Omega$ (a total length spread of 4% of the nominal radial length). In this case the relative currents in the radials are 54%, 28%, 8% and 10%. Plot 1 is for the antenna over good ground, plot 2 over very good ground, and plot 3 over sea water.

It is interesting to note that the pattern deformation depends to a very high degree on the quality of the ground under the antenna! Over sea water the current imbalances practically cause no pattern deformation at all. The horizontal pattern squeeze is max. 0.6 dB over good ground, and max. 0.6 dB over very good ground (computed at main wave angle).

From this it appears that in addition to using a few (one, two, four) elevated radials, it is a good idea to improve the ground conductivity right under the radials by installing a

ground screen (eg, radials) there as well. This confirms the experimental results related by N7CL (see also Section 2.2.11).

I have done some modeling analysis myself, and have noticed that radials that are nominally 90° long, any variation in the exact electrical length will *not* result in high-angle radiation or pattern squeezing provided we use a large

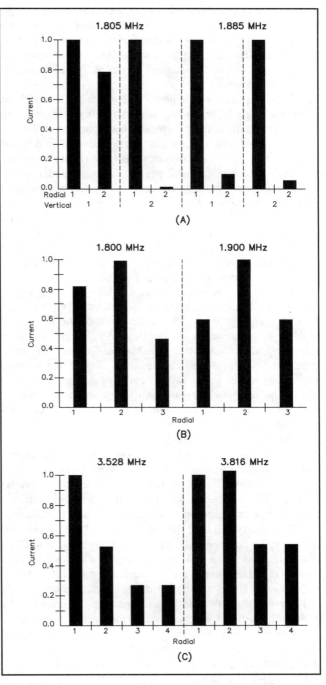

Fig 9-24—Normalized values of current distribution as measured on 80 and 160 meter full-size $^1/_4$-λ verticals using resonant $^1/_4$-λ radials. At A, two in-line radials are used by WXØB, at B, three radials (one per 120°) by KE7BT and at C, four radials by K5IU (data supplied by K5IU).

enough number of radials. With 16 radials, length variations of ±1.5%, and angular variations of ±5° (not evenly spread), the effect was of no consequence (horizontal radiation component down > 40 dB).

Dick Weber, K5IU, has measured many real-life installations with either two, three or four elevated radials, and it was not uncommon to find one radial taking 80% of the antenna current, one radial 20% and the other two almost zero! A few of the recorded current distributions are shown in Fig 9-24. The current values shown correspond very well with the computed values, from which we can deduct that the patterns presented in Fig 9-23 are valid representations of what these current imbalances cause as far as radiation pattern is concerned.

The question now is whether or not you can live with the extra high-angle fill in, (mostly around the 90° elevation angle) and slight pattern-squeeze (typically not more than 1 dB).

If you want maximum low-angle radiation, and if you don't want to lose a fraction of a dB (you realize what this is?), and if you don't want to put up a few more radials, then equal radial currents may be for you. Or maybe you want some high angle radiation as well. Maybe you are not using your vertical or vertical array for reception, and you want some high angle radiation. If you are a contest operator, this is a good idea (you want some local presence as well). In that case, don't bother with equal radial currents, maybe just one radial is the answer for you (like I did).

However, even radials that are laid perfectly (symmetrical configuration), and that carry identical currents, are no full guarantee for 100% cancellation of the horizontal high-angle radiation in the far field! Slight differences in ground quality under the radial wires (or environment, trees, bushes, buildings) can result in different near-field absorption losses under radials carrying identical RF currents. The result will be incomplete cancellation of their radiated fields in the far field. Measuring radial currents does not, indeed, tell you the full story!

It is interesting though to understand why slight differences in radial lengths can cause such spectacular differences in radial current. Connecting a $^1/_4$-λ radial (which means it is a dead short on its resonant frequency) in parallel with another $^1/_4$-λ radial, can be compared to connecting a short circuit across another short circuit, and then expect that both shorts will take the same current. We have similar situations in electronics when we parallel devices such as power transistors in power supplies, or when we parallel stubs to reject harmonics on the output of a transmitter. If one stub gives us 30 dB of attenuation, connecting a second one right across the first one will increase the attenuation by 3 dB at the most. If we take special measures ($^1/_4$-λ lines at the offending frequency) between the two stubs, then we get much greater attenuation (almost double that of the single stub). **Fig 9-25** shows the equivalent schematic of the situation using $^1/_4$-λ radials.

Conclusions

1. In case of elevated radial systems using two, three or four (resonant) $^1/_4$-λ radials, slight differences in electri-

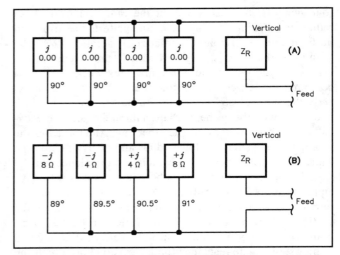

Fig 9-25—At A the ideal (not of this world) case where all four radials are exactly 90° long. They are all a perfect short and exhibit zero reactance. At B the real life situation, where it now is clear that in this circuit, where the current divides into four branches, these currents are now very unequal.

cal length cause radial current imbalances, resulting in some high-angle radiation as well as some pattern squeezing, especially over less than very good ground. However, even perfectly balanced currents are not a 100% guarantee for zero high-angle radiation (due to unequal near-field losses under different radials).

2. Starting with eight radials (or more) the influence of unequal radial current on the generation of high-angle radiation is almost nonexistent. Maybe if you are (overly) concerned about the little high-angle radiation, you should simply increase the number of radials to eight.

3. Adding a good ground screen under the antenna totally annihilates the effects of unequal radial currents, and in addition it will raise the gain of the antenna by up to 5 dB!

2.2.7. Making quarter-wave radials of equal length

Despite all of that, it's nice to know how you can make $^1/_4$-λ radials of identical electrical length! In the past, one of the standard methods of making resonant (90°) radials, was to connect them as a (low) dipole and prune them to resonance. It is evident that resonance does not mean that both halves of the dipole have the same electrical length not even if both halves are of the same physical length. One half could exhibit +j20-Ω reactance (longer than $^1/_4$ λ) while the other half could exhibit a so-called conjugate reactance, −j20 Ω. At the same time the dipole would be perfectly resonant.

Nevertheless, there is a more valid method of constructing radials that have the same electrical length. Whether these are perfect $^1/_4$-λ radials is not so important, we can always tune out a small remaining reactance by a small series coil or a capacitor (if too long).

This method is as follows:

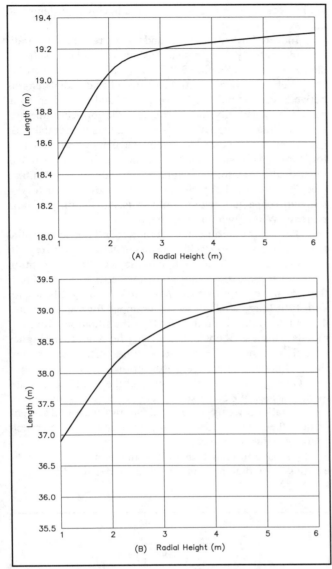

Fig 9-26—Length of a ¼-λ radial as a function of the height above ground. For 80 meters at A, for 160 meters at B.

- Model the length of the vertical to be ¼ λ on the design frequency.
- Put up the elevated vertical of the computed length.
- Use one of the charts in **Fig 9-26** to determine the theoretical radial length. Note that the length is very dependent on radial height.
- Connect one radial.
- Trim the radial to bring the vertical to resonance.
- Disconnect the radial.
- Put up the second radial in line with number one.
- Trim this second radial for resonance.
- If you use four radials, do the same with the remaining two radials.

Then connect all radials to the vertical and check its resonant frequency. It is likely that the vertical will no longer

be resonant at the design frequency. Is it necessary to have the vertical at exactly ¼-λ? No, but if you want, here are two procedures to make the antenna plus radials perfectly resonant on your design frequency:

First method

The first method requires changing the length of the vertical to bring the system to resonance. Do not change any radial length, but change the length of the vertical to get resonance on the design frequency.

Second method

Change all radials in length by exactly the same amount (all together, not one at a time) until you establish resonance.

Neither of the two methods mentioned above guarantees that both the radial system and the vertical are exactly a quarter wavelength, they only guarantee that both connected together are resonant. Again, it is totally irrelevant whether both are 90° long or not. It is not unusual that radials of different physical length result in identical electrical lengths. This is mainly due to the variation of ground conductivity which can vary to a wide degree over small distances. Other causes are coupling to nearby conductors.

On the other hand, radials of exactly the same electrical length are still no guarantee for identical radial current, though, because of near-field losses being possibly different under different radials (see Section 2.2.6).

2.2.8. The K5IU solution to unequal radial currents

D. Weber, K5IU, inspired by Moxon (Ref 693 pages 154-157 in 1st edition, pages 182-185 in 2nd edition, and Ref 7833) installed radials shorter than ¼ λ and tuned the radial assembly to resonance with a coil. It appears that slight changes in electrical length of these "short" radials have little influence on the current in the various radials (Ref 7822 and 7823).

Fig 9-27 shows the equivalent schematic of the vertical with four elevated radials measuring 45° long. It is clear that the same 1° difference in electrical length now will cause

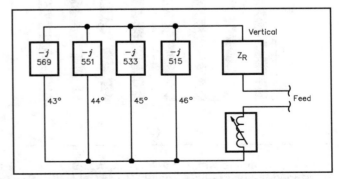

Fig 9-27—Equivalent schematic of a vertical using four 45° radials. Note that the relative rate of change of the reactance is much less here than it is near resonance (Fig 9-25).

much less difference in current flowing in each of the four branches.

With 45° radials, the gain difference between using four and eight radials is less than 0.1 dB, so there is no need to use more than four radials, according to the modeling results.

Fig 9-28 shows the measured radial currents at the same station as shown in Fig 9-24 after modifying their length to stay away from 90°.

Indeed, further modeling and testing revealed that radials substantially longer than 90° give similar results as shorter ones (see Section 2.2.7).

Weber's modeling studies showed that radial lengths between 45° and 60° and between 115° and 135° resulted in minimum sensitivity to high angle radiation from unequal electrical radial lengths. When using radials longer than 90° the system will be tuned to resonance using a series capacitor, which is easier to adjust than a coil, and also has intrinsically less losses, which is an additional advantage (see **Fig 9-29**). The purist may even use a motor-driven (vacuum) capacitor, which could be used to obtain an almost perfect SWR anywhere in the band.

The gain of the antenna is somewhat related to the radial length as shown in **Fig 9-30**. The loss in gain beyond a certain radial length must be attributed to the near-field absorption losses of these extra-long radials over lossy ground. From this graph we learn that for 1/4-λ verticals we get best gain with 120° long radials. If we are cramped for space, and use 45° radials we lose about 0.35 dB vs the situation with 120° radials. This loss (if not higher in real life) is due to the higher near-field intensities associated with shorter radials, and their influence on near-field absorption losses.

Over good or very good ground these differences are evidently less. We can also expect that these differences are much smaller if we use additional radials (or ground screen) on or in the ground under the antenna to improve the ground characteristics in the near field.

We have to realize that all of this is more pure mathematics than practical engineering, and gain differences

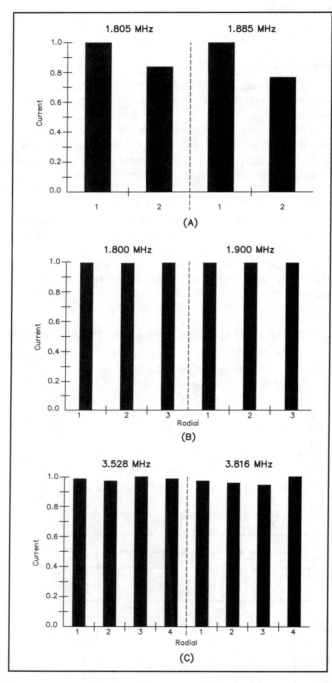

(A)

(B)

(C)

Fig 9-28—Compare with Fig 9-24. At A, WXØB shortened the two radials from 90° to 45°, at B, KE7BT lengthened his radials from 130 to 154 ft (107°), while at C, K5IU shortened his four radials to approx. 45°. Note that the current distribution has been equalized to a rather spectacular degree (data from K5IU).

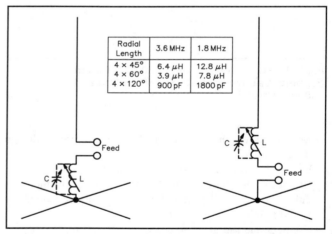

Radial Length	3.6 MHz	1.8 MHz
4 × 45°	6.4 μH	12.8 μH
4 × 60°	3.9 μH	7.8 μH
4 × 120°	900 pF	1800 pF

Fig 9-29—When radials shorter than 90° are used, the system must be tuned to resonance using a coil. With radials longer than 90° the tuning element is a capacitor. Typical values for the tuning elements are also shown. The feed line can be connected in two different ways: between the tuning element and the radiator or between the tuning element and the radials. The result is exactly the same. In both cases, a coaxial feed line connected to the feed point *must* be equipped with a current balun.

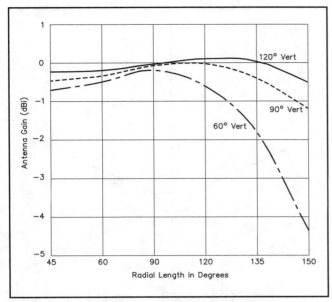

Fig 9-30—Gain as a function of radial length for verticals measuring 60°, 90° and 120° over average ground (all using four elevated radials at about 0.012-λ height) as calculated by K5IU.

of less than 1 dB are, by themselves, of no practical consequence. It becomes a different issue if we add 0.3 dB here, 0.5 dB there, etc until we come to a system where the difference is significant.

I did some modeling myself (using *EZNEC*) and found that:

- The fewer the radials, the greater the current imbalance due to length var2iations.
- The closer to 90° radial length, the more sensitive.
- If radials are ≤ 70° or ≥ 110° there is practically no sensitivity (< 0.1 dB pattern squeeze) even with as few as two radials.
- Starting with 16 radials, the effect of current imbalance is totally gone even with 90° radials.

Conclusion

- If you are really concerned about the equal current and the horizontal-component high-angle radiation, lengthen the radials to about 110° or 120°.
- If you are space limited, the 45° long radials may be for you. Don't shorten the radials to less than approx. 60°-70° if not really necessary. It is clear that we cannot indefinitely shorten radials, and expect to get the same results. If that were true we should all use two in-line loaded mobile whips on our 160-meter tower as a radial (current collecting) system. T. Rauch, W8JI, put it very clearly in the Internet Top-Band Reflector: "The *last* thing in the world I'd want to do is concentrate the current and voltage in smaller areas. Resonant radials, or especially shortened resonant radials, concentrate the electric and magnetic fields in a small area. This increases loss greatly. The ideal case is where the ground system carries current

that evenly, and slowly, disperses over a large physical area, and has *no* large concentrated electric fields from high voltage." This is clearly another plea for the classic, multi-radial ground system.

- Don't forget that equal radial currents, especially with just two, three or four radials, does *not* fully guarantee total far-field cancellation of radiation (because of local differences in near-field absorption losses).

2.2.9. How do I cut radials to equal (non-quarter-wave) lengths?

Although not strictly necessary to have these non-quarter-wave radials as closely matched in length as it would be for $1/4$-λ radials, it nevertheless is "nice," and a proof of good engineering practice to match them as closely as possible.

The procedure is simple. Let's work out the example for approx. 135° radials on 160 meters:

- Cut four (or two or eight, whatever number you have in mind) radials to approx. 135° on 1.83 MHz. Use data from Fig 9-26 to determine the physical length of a $1/4$-λ radial and convert to 135° by simple division/multiplication. (Divide by 90 and multiply by 135.)
- Connect a radial to the antenna.
- Tune the series C for resonance on 1.83 MHz as measured with a noise bridge, antenna analyzer, etc. (The series element would be a coil if the radials were shorter than 90°.)
- Disconnect the radial.
- Connect a second radial and, without changing the C, adjust the length to obtain resonance on the same frequency.
- Do the same with the other radials.
- Connect all radials in parallel.
- Adjust antenna back to resonance by retuning the capacitor.
- You are all done!

Note that the electrical length of the radials is very dependent on height (Fig 9-26), ground quality and nearby objects (antennas, feed lines, towers, large metal objects).

2.2.10. Should the vertical be a quarter-wave?

Over a buried ground-radial system, which inherently is a nonresonant low-Z system, the vertical is usually made resonant (90° long electrically), in order to present a nonreactive feed-point impedance. In many cases however, we have seen verticals slightly longer than 90° often called $3/8$-λ vertical, whereby the antenna, over a buried ground-radial system, must be tuned to resonance by using a series capacitor.

Over an elevated radial system with typically four radials, which should preferably *not* use $1/4$-λ long radials (see Section 2.2.6), the vertical can really be way off 90° length, without any effects as to the feed-point impedance, as its reactance can easily be cancelled by the compensating reactance (coil or capacitor) used to tune the nonresonant radials, together with the vertical to resonance. This is the half-size equivalent of the off-center-fed dipole.

The $3/8$-λ vertical, mentioned above, can be used in conjunction with eg, 45° long radials. An 80-meter vertical

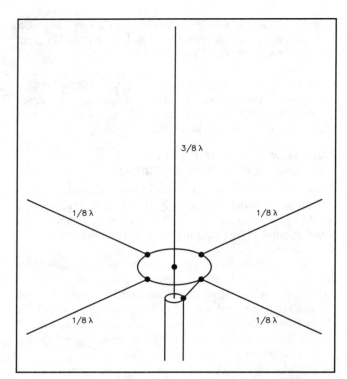

Fig 9-31—A ³/₈-λ vertical used in conjunction with 45° long radials does not require any series coil to tune the antenna; hence, losses are minimized.

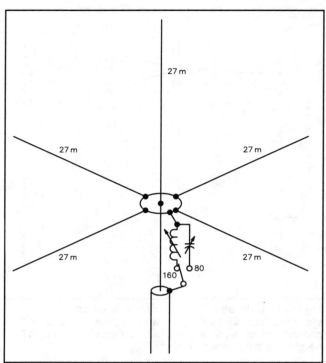

Fig 9-32—A 27-m long vertical with 27-m long radial makes an excellent antenna for both 80 and 160. Band switching only requires the switching of the loading element from a coil (160 meters) to a capacitor (80 meters).

according to these principles (for a design frequency of 3.75 MHz) is shown in **Fig 9-31**. The combination of a ³/₈-λ long radiator and ¹/₈-λ long radials does not require a coil to "tune" the antenna. The radiator length shown for a (wire) element diameter of 2 mm is 26.9 m. With four 10-m long radials, the feed impedance is exactly 52 Ω, a perfect match for our classic 50-Ω feed line. With four radials of exactly 10 m, the currents are of course equal, being 25%, 25%, 25% and 25%. With radials way off the intended 45° length (eg, 9.6-m, 10.4-m, 9.8-m and 10.2-m long), the currents are still very well balanced: 24%, 26%. 24% and 25%! This also proves that it is *not* the presence of the series coil or capacitor which helps equalize the currents in the radials. It is clear that with such relative current distribution between the radials there is no trace of high angle radiation nor any pattern squeezing.

An 80/160-meter vertical could use 27-m long radials (60° on 160 meters and 120° on 80 meters) as shown in **Fig 9-32**. The total system length on 160 meters is 60° + 60° = 120°, which is less than 180°, hence a coil is required to resonate the antenna. On 80 meters, the total length is 120° + 120° = 240°, which is longer than ¹/₂ λ; hence, a capacitor is required.

2.2.11. Elevated radials on grounded towers

The N4KG antenna

T. Russell, N4KG, eminent low-band DXer, has described a method of shunt feeding grounded towers in con-

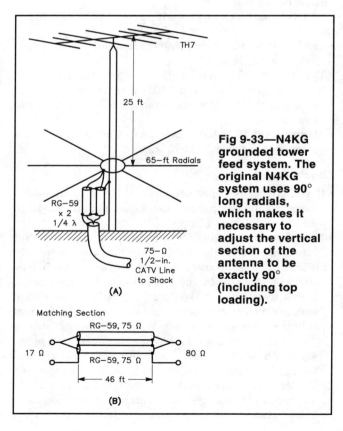

Fig 9-33—N4KG grounded tower feed system. The original N4KG system uses 90° long radials, which makes it necessary to adjust the vertical section of the antenna to be exactly 90° (including top loading).

junction with elevated radials (Ref 7813 and 7832). The tower, using a TH7 as top loading is approx. 90° long with respect to the feed point (see **Fig 9-33**). The 10 ¼-λ long radials are 4.5 m above ground. It is of course important to find the attachment point of the radials on the tower whereby the part of the tower above the feed point becomes resonant in conjunction with the radials. Russell installed the radials, and moved the ring to which the radials were attached, up and down the tower until he found the system in resonance. He found this point 7.5 m (25 ft) below the TH7 capacitance hat.

John Belrose, VE2CV, analyzed N4KG's setup using *NEC-4* (Ref 7821). He simulated the connection to earth of the tower (at the base) by using a 5-m long ground rod (a decent dc ground). It is obvious that RF current is flowing through the tower section below the feed point. This current causes the gain of the antenna to be somewhat lower than that of a ¼-λ base-fed tower. Belrose calculated the difference as approx. 0.8 dB.

A typical configuration like the one described by N4KG will yield a 2:1 SWR bandwidth of approx. 100 to 150 kHz. There are several approaches to broadband the design. Sam Leslie, W4PK, designed a system whereby he uses two sets of two radials, installed at right angles. One set is cut to resonate the system at the low end of 80 meters (CW band) and the other at the phone end. The SWR curve has two dips now, one on 3.5 and the other on 3.8 MHz.

Another approach is to design the antenna for resonance on 80 meter CW, and tune it to resonance in the SSB portion by inserting a capacitor between the feed line and the radials or the vertical conductor (tuning out the inductive reactance on 3.8 MHz).

Using non-90° radials

We have learned that 90° long radials are not the ideal choice if we want minimum radiation from the radials. The approach of using either shorter (45° to 60° long) or longer (110° to 130° long) radials is of course applicable here as well. In this case you would resonate the antenna with either a series coil or capacitor as explained in Section 2.2.8. This approach would also make it much less critical to find a point for resonance on the tower: resonance can be achieved by tuning the series element. or even by pruning the length of the radials (in a symmetrical way!).

Decoupling the tower base from the real ground

It is possible to minimize the loss by decoupling the base of the vertical from ground. Methods of doing so were described by Moxon (Ref 693 and 7833). **Fig 9-34** shows the layout of a so-called linear trap which turns the tower section between the feed point and ground into a high impedance, effectively isolating the antenna feed point from the dc-ground (rod). The trap is constructed as follows:
- Connect a shunt arm of approx. 50 cm to the tower, just below the antenna feed point.
- Connect a drop wire, parallel with the tower, from the end of the arm to ground level and connect it back to the base of the tower, to form a loop.

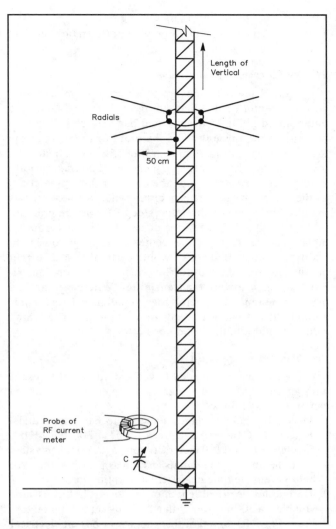

Fig 9-34—The grounded tower section below the antenna feed point can be made a resonant linear trap, which inserts a high impedance between the antenna feed point and the bottom of the tower. Tune C for minimum current in the loop.

- Insert a variable capacitor in the drop wire (wherever convenient).
- Excite the vertical antenna (above the linear stub) with some RF.
- Use an RF current probe (eg, Palomar type PCM1) and tune the capacitor for minimum current in the drop wire.
- You're done!

The loop tower + drop wire + C now form a parallel circuit resonant on the antenna operating frequency, which ensures that no RF currents can flow through the bottom tower section to the lossy ground.

Summing up

Using grounded towers with an elevated radial system can readily be done. The principles are simple:
- The vertical (top loaded or not) together with the radial system must be resonant.

- It is best not to use 90° radials.
- Provisions must be taken for minimum RF (return) current to flow in the ground.

The N4KG reverse feed system

Russell feeds his design in a very uncommon way, with the center of the coax going to the radials, and the outer shield going to the vertical part. He claims it is to prevent arcing through from the braid of the coax to the tower. Tom is coiling up his parallel 75-Ω coax inside the tower leg, and that forms an RF choke, of course. I would strongly suggest *not* to tape the coax (or the coiled coax) to the leg of the tower, especially when a linear trap is installed there may be a rather steep RF voltage gradient on that leg. I would keep the coax a few inches from all metal, and route it in the center *inside* the tower. In addition to the coiled coax I would certainly use a current balun made of a stack of ferrites (installed beyond the 1/4-λ transformer, toward the transmitter. Whether or not the braid or the inner conductor goes to radials is irrelevant if a good current balun is used. It is unimportant to which side of the dipole the braid goes.

Practical design guidelines

If you have a grounded tower and you want to use it with an elevated radial system with eg, four radials, you can proceed as follows:

1. Define the height where you want to have the radials. Take 6 m. Convert to degrees (360° = 300 / F MHz). 6 meters = 13° on 160 meters. If you have enough physical tower height, put the radials as high as possible, this will help in reducing the near-field absorption losses.
2. Define the electrical length of the tower. Let us assume you have a 30-m tower with a 5-el 20-meter Yagi on top. From Fig 9-83 we learn that this tower has an electrical length of approx. 123°.
3. The electrical length of the tower above the radial attaching point is 123° – 13° = 110°
4. Cut 4 radials to identical electrical length as explained in Section 2.2.7.
5. Whether or not you will require a coil or a capacitor to tune the system to resonance depends on the *total* length of the antenna vertical part *plus* radials. If the length is greater than 180°, a capacitor will be required. An inductor will be required if the total length is less than 180°. Assuming we use 120° long radials, in which case the total antenna length is 110° + 120° = 230°, hence a series capacitor is required to tune the system to resonance.
6. Measure the impedance at resonance using an antenna analyzer. If necessary use an unun or a quarter-wave transformer (or other suitable impedance matching system) to get a perfect (acceptable) match to your feed line impedance.
7. Install the linear trap on the tower section under the feed point and tune the loop to resonance by adjusting the loop variable capacitor (see procedure above)
8. You are all done!

Fig 9-35 shows the final configuration of the antenna

we designed above. It is obvious that the tower must use dielectric guy wires, or if steel guy wires are used, they must be broken up in short lengths so that they do not interfere with the vertical antenna.

Finally, maybe it's not such a good idea after all to have elevated radials on your grounded tower. It is much more complicated: You need the linear trap to decouple the bottom of the tower from the real ground, and you need to have radials above ground. Maybe 10 or 20 radials *on* the ground will do the job as well.

2.2.12. Elevated radials combined with radials (screen) on the ground

All publications I have seen so far on the subject of elevated radials use either one of the modeling standard grounds (average, good, etc—see Table 5-2) or have been done in real life over whatever ground there was in the site the tests were done.

The modeling I have done has suggested that improving the ground right under the vertical and its elevated radials can increase the gain, especially if only one to four elevated radials are used (see Section 2.2.3 and Fig 9-20). In the case of a single radial, or when using ≈90° long radials, improving the ground quality right under the antenna will

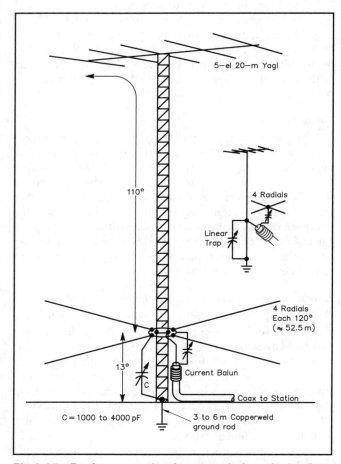

Fig 9-35—Design example of a grounded vertical using an elevated radial system (see text for details).

0.3 dB) down from a $^1/_4$-λ radius ground screen).

On 80 meters, I have, for over 5 years now, very successfully used $^1/_4$-λ verticals (in a 4-square array configuration), using a single $^1/_4$-λ radial at about 5-m height. Judging an antenna's performance by the DX worked with it certainly makes no sense. But judging the same antenna's performance by the repetitive results obtained in world-class DX contests, may be a good indication indeed, if the antenna works well. With this 80-meter 4-square (single elevated radial) array over a ground which is literally swamped with copper wire, I have never scored less than a first place Europe in the ARRL International DX Contest (single-band 80 meters), both CW and SSB and that in nine contests since 1994. In addition a new European record was set with the antenna as well. Taking into account that my QTH is certainly not the best for working Ws (Normandy or Wales should be much better), this, in my humble opinion, means that verticals, even with a single elevated radial, can be top performers.

2.2.13. Avoiding return currents via the ground

Fig 9-36 shows the vertical antenna return paths for different radial configurations. (A) shows the case where a simple ground rod is used, whereby the antenna return currents have to travel entirely through the lossy ground. This reduces the radiation efficiency of the vertical to a very high degree, because of the (I^2 R ground) losses. Burying radials in the ground can greatly reduce the losses as the return currents can now travel, to a great extent (depending on the number and the length of the radials) through the low loss radial conductors in the ground (B).

In (C) we have two radials elevated above ground. There are now two current return paths: the lossless path through the two radials, and a lossy path through the lossy ground, and the capacitance between the radials and the ground, into the radials.

We can minimize the currents in this parasitic path by:
- Raising the radials high above ground. It must be said that once the radials are a few m above ground, the capacity to ground is rather small.
- Using fewer radials: more radials means more capacitance, thus more current in the ground and hence more ground losses.
- Using more radials: more radials means a better screen: 100 radials $^1/_4$ λ long will perfectly screen the earth underneath the vertical. (This seems to contradict the previous item, but it doesn't—see Section 2.3.)
- Improving ground conductivity under the elevated radials by installing buried radials or a ground screen (not galvanically connected to the elevated radials, though!)

Another important issue is currents on the outside of the coaxial feed line. Fig 9-36D shows how unwanted currents can flow on the outside of the coaxial cable. In this situation, the coaxial feed line, as seen from the outside world, is just another conductor, and acts as just another (random length radial). Return currents will flow in that conductor, unless we disconnect it at the antenna feed point. The question is now how can we disconnect the coaxial

The Titanex V160E antenna on the "beach" at 3B7RF (St Brandon Island). Note the two elevated radials about 2 m above salt water. The combination of one or two elevated radials with a perfect reflector (screen) underneath is hard to beat.

greatly reduce the horizontally polarized high-angle radiation and increase the antenna gain. This can be done by putting down radials or ground screens on the lossy ground.

It is important to understand that these on-the-ground radials (or screen in whatever shape) should *not* be galvanically connected in any way to the elevated radials in any way. They should be connected to nothing (we don't want any antenna return currents to flow in the ground).

If you have the space, and 4 to 5 dB is worth the expense and effort to you, by all means provide a ground screen. In the case you do *not* want to use the screen for antenna current collecting, the screen does not have to have the shape of radial wires. A net of copper wires, with a mesh density measuring less than approx. 0.015 λ (1 m on 80, 2 m on 160), or even 0.03 λ if you are willing to lose 0.5 dB max., is all that is needed to provide an effective near-field screen. Make sure that the crossing copper wires make good and permanent electrical connections at their joints (see Section 2.1.8). If you use but one elevated radial you may want to increase the ground net density in the area under the single elevated radial. In principle the screen should have a radius of $^1/_4$ λ (in case of a $^1/_4$-λ vertical), but a screen measuring only $^1/_8$ λ in radius will only be a fraction of a dB (typically

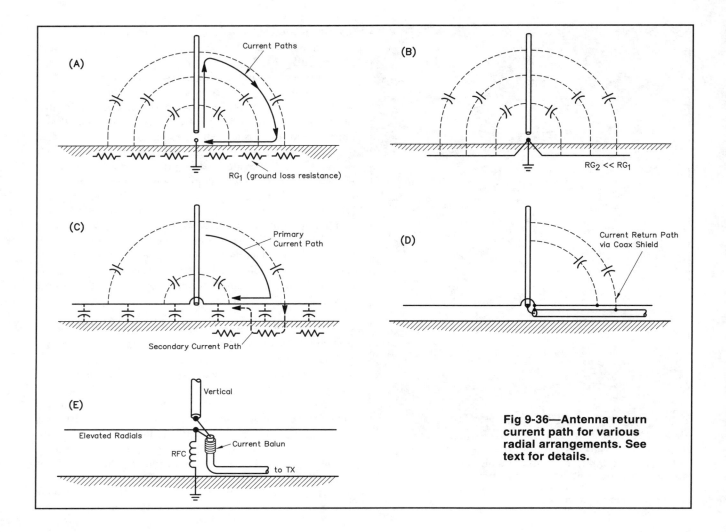

Fig 9-36—Antenna return current path for various radial arrangements. See text for details.

"radial" wire, and not the coaxial feed line?

The insertion of a current balun at the antenna feed point is a must (see Fig 9-36E). The high impedance that the current balun represents to any currents on the outside of the coax shield effectively suppresses currents on the cable. Several types of current baluns are described in Chapter 6, Section 7. If you are forced to use (for layout reasons) $^3/_4$-λ feed lines in a 4-square array, you will wind up with a lot of surplus coax length. Wind it all up in a coil and mount it as close as possible to the antenna feed point. This should make an excellent current balun. It is always better to run the coax on or preferably in the ground, rather than supported on poles at a certain height, in order to prevent coupling and parasitic currents on the outer shield.

It also makes common sense to provide a dc ground for the common radial points. This can be done by connecting an RF choke (100 µH or more) between the radial common point and a safety ground (ground rod) below the antenna feed point, as shown in (E).

If only a few radials are used, each of them can radiate considerable near-field energy and can induce currents in the feed line beyond the point where the current balun has been inserted (at the feed point). Buried feed lines improve the situation. Feed lines supported off the ground are very sensitive to this kind of coupling. Also if you run only two radials, run the feed line at right angles to the two in-line radials. In other words, keep the feed line away from the near fields of the radials.

When using a large number of elevated radials (eg, >20), it is unnecessary to use a current balun as the screening effect of the radials will be sufficient to prevent parasitic antenna return currents of any significant magnitude to flow on the coax outer shield.

2.2.14. Elevated radials in vertical arrays

When a vertical is to be used as an element in an array, an additional parameter comes into line when choosing the ideal radial length, at least if we are concerned about reducing the horizontally polarized high-angle radiation of the array to a strict minimum.

Whereas we have seen that long radials (eg, 120° to 130°) may give an extra gain edge (as compared to the 45° radials), in arrays it is always best to use the shorter radials, as this will help reduce the coupling between radials from different array elements.

The careful layout of the radials is also very important. Never run radials belonging to two different array elements in parallel. Design your layout such that coupling is minimized.

Zero coupling is of course achieved by using buried radials, terminated in bus bars where radials of adjacent elements meet one another.

Fact is that, if you use four 90° long radials on each element of an array, and have them laid out in such a way that coupling does exist between radials of adjacent elements, it may be just as good to use a single radial. Zero coupling is, of course, achieved by using buried radials, terminated in bus bars where radials of adjacent elements meet one another (see Chapter 13, Section 9.10).

2.2.15. "Wonder" radials for 160 meters (?)

CQ magazine (June '92, p 57) published a mysterious "wonder" solution to the space problem of full-size $1/4$-λ radials for 160 meters. The author makes his radials of old (hopefully not lossy) coax cable, fashioning each one into a $1/4$-λ resonant transmission line, making the velocity factor of the line work for him. For ordinary (solid PE insulated) cable the velocity factor is 0.66, so a radial for 1.85 MHz is only about 88 ft long. He leaves the end of each radial unshorted, while the shields of the radials are connected to a ground rod at the antenna base.

What happens in reality? The shield of the coax will be exposed to the field of the antenna, and will take care of the return currents. The electrical length of the shield of the coax is not related to the velocity factor of the feed line when used as a transmission line. If the radial is raised a few meters above ground, it will certainly not be resonant on 160 meters, although this is not really a necessity (see Section 2.2.4). If the radial lies on the ground, however, it may well be that its resonant frequency was almost re-established. Indeed, the resonant length for a radial barely over the ground (a cm or so) is about 0.15 λ, and not about 0.25 λ as it is at greater heights. It is *not* the velocity factor of the cable ($0.25 \times 0.66 = 0.15$) that causes the $1/4$-λ resonance of the radial but its proximity to the ground that makes it resonant at about 0.15 λ.

Furthermore, the open-circuited $1/4$-λ line represents a short circuit at the antenna base (only at the exact $1/4$-λ frequency though), which means that the radial is connected to both the ground rod and to the shield of the coaxial feed line (see **Fig 9-37**). We have learned that if we use elevated radials we want no connection to the real ground, which is certainly the case here. Maybe the radial is intended to be laid on the ground though, in which case it may be nearly resonant as described above.

The use of the coaxial cable as a radial, in the way it is connected, reduces the bandwidth of the antenna: at the exact resonant frequency of the radial, it represents a short (not really true because in a real world there is always some R, some losses in the cable) but above the resonant frequency the impedance at the end of the radial becomes inductive; below its resonant frequency it is capacitive. This is the same

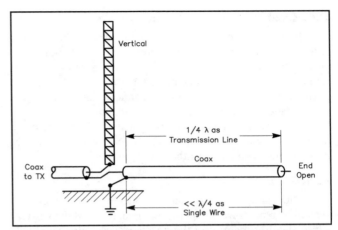

Fig 9-37—Representation of the short radials described by Artigo, KN6J. It is clear that the radial is not $1/4$-λ long, but only a $1/4$-λ times the velocity factor of the coaxial cable ($0.25 \times 0.66 = 0.165$ λ). The velocity factor of the cable (shield) used as a single wire is not the same as the velocity factor of the cable used as a transmission line. The velocity factor of the "wire" is between 95 and 98% at reasonable heights (a few meters), but can be as low as 70% for an insulated wire lying on the ground.

as what happens to the impedance of the vertical element itself. These reactances (both having the same sign) add up in the antenna feed impedance, thereby reducing the effective bandwidth, and increasing the losses of the antenna.

2.3. Buried or Elevated?

It is clear, and it has been proven over and over in the real world, that the elevated radial system (at low height) is a valid alternative for a system of buried radials. If only a very small number is used (typically 1 to 8), their task will be (almost exclusively) to efficiently collect the return currents of the vertical. With a larger number the screening effect becomes important, and near-field ground losses can be reduced by making use of the screening effect of a large number of radials.

Elevated radials can have advantages such as:
- Raising the base of the vertical above nearby obstacles.
- Providing the possibility of installing a performing ground system under very unfriendly circumstances eg, over rocky ground.
- More flexibility in matching (the real ground is not resonant, an elevated radial system using only a few radials—max. four—can be made inductive or capacitive, which may be an asset in designing a matching system.

For using elevated radials I would propose the following guidelines:
- Put up the radials as high as possible.
- Use as many radials as possible.
- If a small number (<20) is used, install a ground screen.

If you have the place and if the ground is not too unfriendly, I would suggest you use buried radials however.

2.4. Evaluating the Radial System

Evaluating means measuring antenna field strength (FS), or measuring certain parameters for which we know the correlation with radiation FS. You cannot evaluate an antenna just by modeling. You can develop, design and predict performance by modeling, but you cannot evaluate the actual performance of the antenna on the computer.

The real proof is field-strength measuring, there are no two ways about it. However, there are some indirect measurements and checks that can be done:

Buried radial system

The classic way to evaluate the losses of a ground system is to measure the feed-point resistance of the vertical while steadily increasing the number of radials. The feed-point resistance will drop consistently and will approach a lower limit when a very good ground system has been installed. Be aware, however, that the intrinsic ground conductivity can vary greatly with time and weather, so it is recommended that you do such a test in a very short time frame in order to minimize the effects of varying environmental factors on your tests (Ref 818, 819).

Peter Bobeck, DJ8WL, performed such a test on his 23-m long top-loaded (T) antenna. He added 50-m long radials (on the ground) while measuring the feed-point impedance and found the following:

No. of radials	2	5	8	14	20	30	50
Impedance	122 Ω	66 Ω	48 Ω	39 Ω	35 Ω	32 Ω	29 Ω

Incidentally, nine radials look like a perfect match to 50-Ω coax, but the system efficiency in that case is way down below 50%!

Don't be surprised if the impedance gets lower than 36 Ω with a full-size $1/4$-λ vertical. It first surprised me when I measured about 20 Ω for my 160-meter full-size $1/4$-λ vertical, but that was because of its very large effective diameter.

For calculating antenna efficiency, you can use the values from Table 9-1, which lists the equivalent resistance of buried radial systems in good-quality ground. For poor ground, higher resistances can be expected, especially with only a few radials.

It is clear that measuring the impedance of the verticals and watching it decrease as you add radials tells us nothing about the near-field absorption ground losses. It only gives us an indication of the $I^2 R$ losses (return current collecting efficiency).

Periodic visual inspections of the radial system for broken wires and loose or corroded connections, etc will assure continued efficient operation. **Fig 9-38** shows DJ6QT examining the radials of the ON4UN 160-meter vertical. If the radials are buried, it is a good idea to make them accessible anyhow just where they connect to the bus bar. This way you can periodically check with a snap-on current meter if the radial still carries any current on transmit. If it doesn't, maybe the radial is broken at a short distance from the connection point.

Elevated radial system:

Whether you have 1, 2 or 16 elevated radials, if these radials are the only antenna current return paths (the elevated radials are *not* connected to the lossy ground), the measured real part of the antenna impedance will not change. There is no gradual decrease of feed-point impedance as you increase the number of radials.

Measuring the antenna impedance does *not* give you any indication of near-field absorption ground losses.

The only test one can perform on an elevated radial system is to measure the radial current, although this has little (if any) correlation with produced low-angle field strength. Nevertheless, when using only a few radials (2 to 8) it is a good idea to check the radial currents, and to make sure they are similar (± a few percent of one another).

Do regular inspections of your current balun. I would recommend to periodically measure its effectiveness by checking its inductance. This should be measured on the

Fig 9-38—Walter Skudlarek, DJ6QT, inspecting some of the radials used on the 160-meter vertical at ON4UN. Half of the radials are buried (where the garden is), and half are just lying on the ground in the back of the garden behind the hedge (where the XYL can't see the mess from the house!). In total, some 250 radials are used, ranging from 15 m to 75 m.

operating frequency. The new AEA-CIA Analyzer is an ideal instrument for this purpose.

■ 3. SHORT VERTICALS

Short verticals can be "loaded" to be resonant at the desired operating frequency. Different loading methods are covered in this section, and the radiation resistance for each type is calculated. Design rules are given, and practical designs are worked out for each type of loaded vertical. The design of loading coils is covered in detail. The different methods are compared as to their efficiency.

Short verticals have been described in abundance in amateur literature (Ref 771, 794, 746, 7793 and 1314). Gerd Janzen published an excellent book on this subject, *Kurze Antennen* (in the German language) (Ref 7818).

The radiation pattern of a short vertical is essentially the same as for a full-size ¼-λ vertical. **Fig 9-39** shows the vertical radiation patterns of a range of short verticals over perfect ground, as calculated using *ELNEC*. Notice that the gain is essentially the same in all cases (the theoretical difference is less than 0.5 dB).

A short (shorter than ¼-λ) monopole exhibits an impedance with a real part that is smaller than 36.6 Ω, and in addition a reactive part that is capacitive. Loading a short vertical means canceling the reactive part of the impedance to bring the antenna to resonance. The simplest way is to add a coil at the base of the antenna, a coil with an inductive reactance equal to the capacitive reactance shown by the short vertical. This is the so-called base-loading method. **Fig 9-40** shows a number of classic loading schemes for short verticals, along with the current distribution along the antenna. Remember from Section 1.2 that the radiation resistance is a measure of the area under the current distribution curve. Also remember from Section 1.3 that the radiation efficiency is given by:

$$\text{Eff} = \frac{R_{rad}}{R_{rad} + R_{loss}}$$

It is clear now that the real issues with short verticals are *efficiency* and *bandwidth*. Let us examine these issues in detail.

With short verticals the numerator of the efficiency formula decreases in value (smaller R_{rad}), and the term R_{loss} in the denominator is likely to increase (losses of the loading devices such as coils). This means we have two terms which tend to decrease the efficiency of loaded verticals. Therefore maximum attention must be paid to these terms by
- Keeping the radiation resistance as high as possible (which is *not* the same as keeping the feed-point impedance as high as possible).
- Keeping the losses of the loading devices as low as possible.

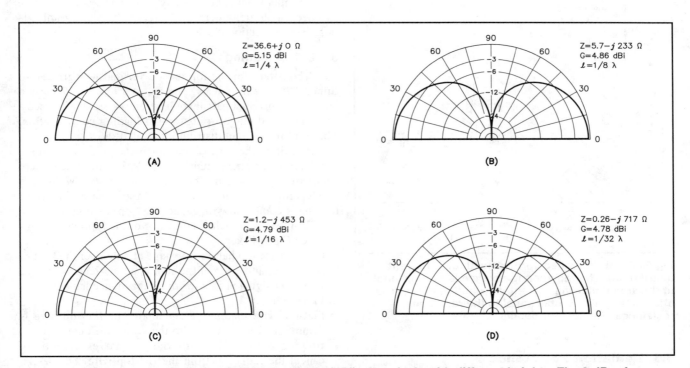

Fig 9-39—Elevation-plane radiation patterns and gain (dBi) of verticals with different heights. The 0-dB reference for all patterns is 5.2 dBi. Note that the gain as well as the shape of the radiation patterns remain practically unchanged with height differences. The patterns were calculated with *ELNEC* over perfect ground, using a modeling frequency of 3.5 MHz and a conductor diameter of 2 mm. A—Height=¼ λ. B—Height=⅛ λ. C—Height= ¹/₁₆ λ. D—Height=¹/₃₂ λ.

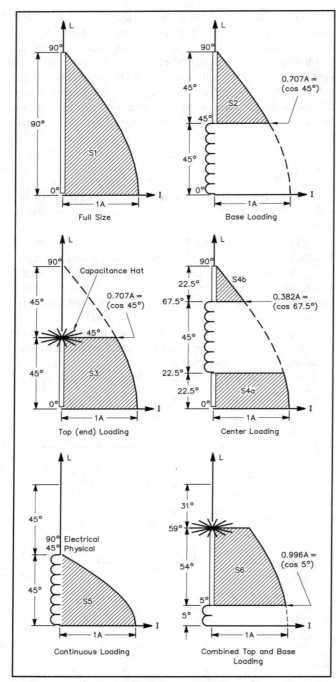

Fig 9-40—Short loaded verticals with their current distribution. The radiation resistance is proportional to the area under the current-distribution curve. The efficiency is also proportional to the radiation resistance. See text for details.

3.1. Radiation Resistance

The procedure for calculating the radiation resistance was explained in Section 1.2, where we found that for a $^1/_4$-λ vertical (made of a very small size conductor) is 36.6 Ω. We will now analyze the following types of short verticals:

1. Base loaded.
2. Top loaded.
3. Center loaded.
4. Base plus top loaded.
5. Linear loaded.

3.1.1. Base loading

Base loading consists of adding a loading coil at the base of the monopole. The base-loaded monopole shown in Fig 9-23C is physically 50% shorter than the full-size, $^1/_4$-λ monopole. Base loading makes it electrically the same size.

The radiation resistance can be calculated as defined in Section 1.2. A trigonometric expression, which gives the same results, is given below (Ref 742).

$$R_{rad} = 36.6 \times \frac{\left(1 - \cos L\right)^2}{\sin^2 L} \qquad \text{(Eq 9-5)}$$

where L = the length of the monopole in degrees (1 λ = 360°).

According to this formula, the radiation resistance of the base-loaded vertical (50% size reduction) is 6.28 Ω.

Hall, K1TD, derived another equation (Ref 1008):

$$R_{rad} = \frac{L^{2.736}}{6096} \qquad \text{(Eq 9-6)}$$

where L = electrical length of the monopole in degrees.

This simple formula yields accurate results for monopole antenna lengths between 70° and 100°, but should be avoided for shorter antennas. A practical design example is described in Section 3.6.1.

3.1.2. Top loading

The patent for the top-loaded vertical was granted to Simon Eisenstein of Kiev, Russia, in 1909. **Fig 9-42** is a copy of the original patent application, where one can see a combined loading coil plus top-hat loading configuration. The resulting current distribution is also shown.

The tip of the vertical antenna is the place where there is no current, and maximum voltage. This is the place where *capacitive* loading is most effective, and *inductive* loading (eg, loading coils) is least effective. In some cases, inductive loading is combined with capacitive (top) loading.

Top loading is achieved by one of the following methods (see **Fig 9-43**):

- Capacitance top hat: in the shape of a disk, or the spokes of a wheel at the top of the shortened vertical. Details of how to design a vertical with a capacitance hat are given in Section 3.6.3.
- Flat-top wire loading (T antenna): The flat-top wire is symmetrical with respect to the vertical. Equal currents flowing outward in both flat-top halves essentially cancel the radiation from the flat-top wire. For design details see Section 3.6.5.
- Coil with capacitance hat: In many instances a loading coil is used in combination with a capacitance hat to load a short monopole. This may be necessary, as otherwise an unusually large capacitance hat may be required to estab-

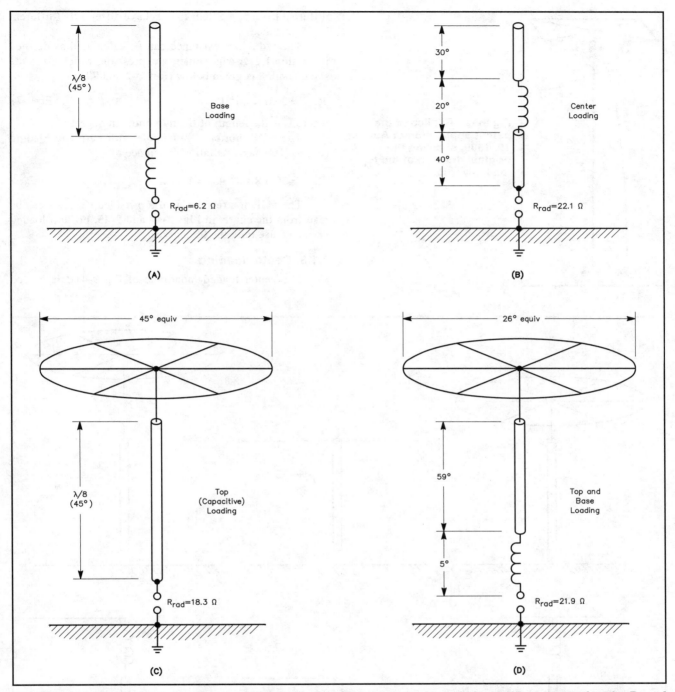

Fig 9-41—Four of the most common types of loading used to resonate short verticals. When comparing the R_rad of the four examples, be sure to also consider the differences in the physical lengths of the radiators; they are not all equal.

lish resonance at the desired frequency.
- Coil with flat-top wire: This loading method is similar to the coil with capacitance hat (see Section 3.6.7 for design example).
- Inverted L: This configuration is not really a top-loaded vertical, as the horizontal loading wire radiates along with the vertical mast to produce both vertical and horizontal polarization. Inverted-L antennas are covered sepa-

rately in Section 9.
- Coil with wire: This too is not really a loaded short vertical, but a form of a loaded inverted L.

For calculating the radiation resistance of the top-loaded vertical, it is irrelevant which of the above loading methods is used. For a given vertical height, all achieve the same radiation resistance. However, when we deal with efficiency (where both R_{rad} and R_{loss} are involved) the

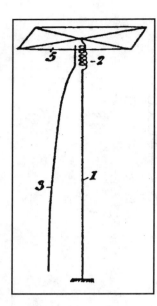

Fig 9-42—Replica of the patent application of August 10, 1909, showing the original drawing of the top-loaded vertical.

different loading methods may behave differently (different loss resistances).

The radiation resistance can be calculated as defined in Section 1.2. A trigonometric expression, which gives the same results, is given below (Ref 742 and 794):

$$R_{rad} = 36.6 \times \sin^2 L \qquad \text{(Eq 9-7)}$$

where L is the length of the monopole in degrees.

The 50% shortened monopole with pure end loading (Fig 9-41C) has a radiation resistance of

$$R_{rad} = 36.6 \times \sin^2 45° = 18.3\ \Omega$$

The radiation resistance of top-loaded verticals can be read from the charts in **Figs 9-44** and **9-45**. For top-loaded verticals, use only the 0% curves.

3.1.3. Center loading

The center-loaded monopole of Fig 9-41B is loaded

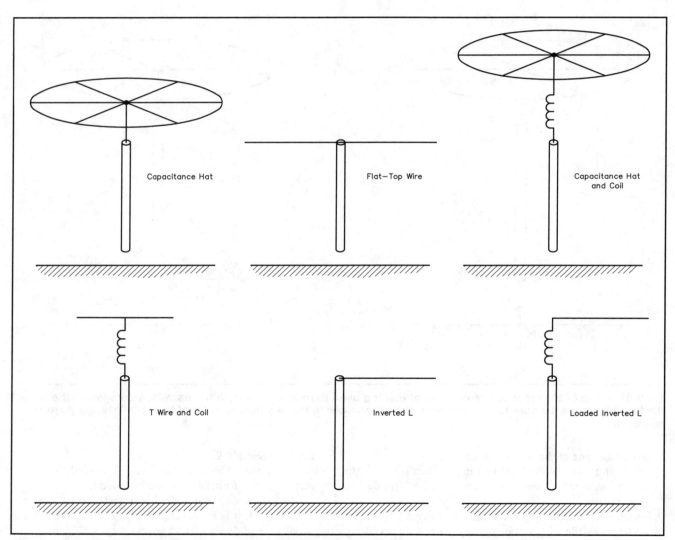

Capacitance Hat

Flat—Top Wire

Capacitance Hat and Coil

T Wire and Coil

Inverted L

Loaded Inverted L

Fig 9-43—Common types of top loading for short verticals. The inverted L and loaded inverted L are not true verticals, as their radiation contains a horizontal component.

with a coil positioned along the mast. The antenna section above the coil is often called the whip.

Inputs (from Fig 9-41B):
- Length of mast below the coil = 40°
- Length of mast (whip) above the coil = 30°

The radiation resistance can be calculated as defined in Section 1.2. A trigonometric expression which gives the same results is given below (Ref 742 and 7993):

$$R_{rad} = 36.6 \times \left(1 - \sin t2 + \sin t1\right)^2 \qquad \text{(Eq 9-8)}$$

where

t1 = length of vertical below loading coil (40°)
t2 = 90° – length of vertical above loading coil
(= whip, 30°) = 60°

Using this formula, R_{rad} is calculated as = 22.1 Ω.

3.1.4. Combined top and base loading

Top and base loading are quite commonly used together, as shown in Fig 9-41D. Top loading is often done with capacitance-hat loading, or even more frequently in the shape of two or more flat-top or sloping wires. If a wide frequency excursion

is required (eg, 3.5 to 3.8 MHz), one can load the vertical to resonate at 3.8 MHz using the top-loading technique. When operating on 3.5 MHz, a little base loading is added to establish resonance at the lower frequency.

The radiation resistance can be calculated as defined in Section 1.2. A trigonometric expression, which gives the same results, is given below (Ref 742 and 7993):

$$R_{rad} = 36.6 \times \frac{\left(\sin t1 - \sin t2\right)^2}{\cos^2 t2} \qquad \text{(Eq 9-9)}$$

where

t1 = electrical height of vertical mast
t2 = electrical length of base-loading coil
In our example shown in Fig 9-41D, t1= 59° and t2 = 5°

$$R_{rad} = 36.6 \times \frac{\left(\sin 59° - \sin 5°\right)^2}{\cos^2 5°} = 21.9 \ \Omega$$

Fig 9-44 shows the radiation resistance for monopoles with combined top and base loading. The physical length of the antenna (L) plus top loading (T) plus base loading (B)

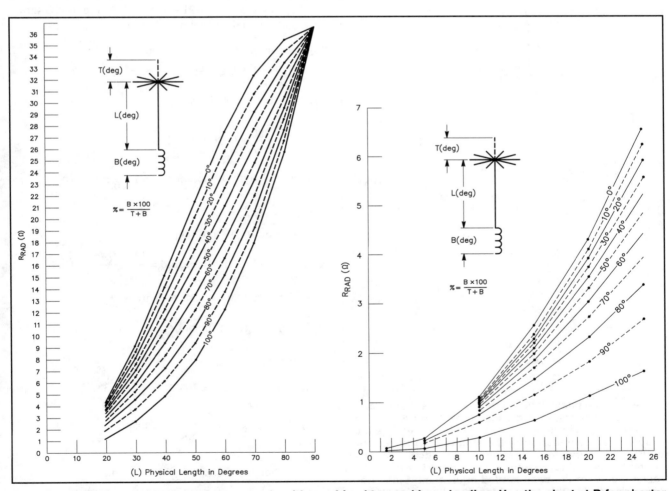

Fig 9-44—Radiation resistances of a monopole with combined top and base loading. Use the chart at B for shorter monopoles to obtain better accuracy.

must total 90°. The calculation of the required capacitance and the dimensions of the capacitance hat are explained further in Section 4.6.3.

3.1.5. Linear loading

Designing linear-loaded elements can best be done graphically as shown in Fig 9-45. Let us use as an example the vertical monopole of Fig 9-45A, 28 meters long, and linearly loaded for 1.8 MHz.

A quarter-wave vertical (90° electrical, using a 96% shortening factor) for 160 meters measures:

$$\frac{300 \times 0.96}{1.8 \times 4} = 40 \text{ m}$$

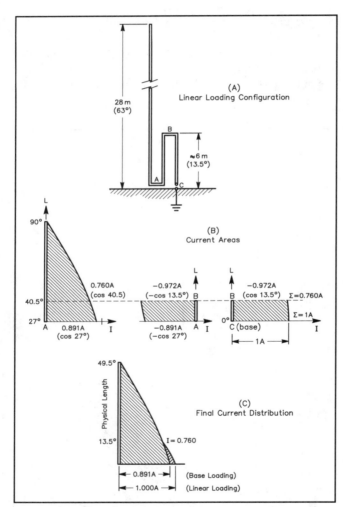

Fig 9-45—The linear-loading device at A is merely a part of the antenna folded back on itself in order to shorten the physical length of the antenna. At B is the graphical representation of the area under the current distribution curve for a linear-loaded vertical. The shaded area in C represents the increase in radiation resistance over the base-loaded equivalent. Since no coils are used in this form of loading, linear loading can be used with almost no loss.

and 28 meters represents $90 \times 28/40 = 63°$.

The remaining $90° - 63° = 27°$ are made up by a linear-loading device, which is simply a folded length of radiator.

B and C of Fig 9-45 show the current distribution for the antenna. Assuming an antenna current of 1 A at the feed point, the current can be calculated at key points:

At 13.5°: $I(a) = \cos 13.5 = 0.972$ A
At 27.0°: $I(b) = \cos 27 = 0.891$ A
At 40.5°: $I(c) = \cos 40.5 = 0.760$ A

The folded linear-loading "stub" may look like an open-wire transmission line, but, whereas in an open transmission line with a fully balanced load the currents in the two wires are equal in magnitude but 180° out of phase and hence produce no radiation, in the case of a linear-loading stub the radiation is not canceled out. To assess the effect of radiation from the linear-loading stub we analyze the area under the current distribution (Fig 9-45). Summing the areas under the current distribution along the two wires of the loading stub yield the effective area that is responsible for radiation from the loading stub, and hence for an increase in radiation resistance.

Summing the currents at these points: I_{total} at 40.5° from the top = $0.760 - 0.972 + 0.972 = 0.760$ A. I_{total} at base of antenna = $0.891 - 0.891 + 1.0 = 1.0$ A. Fig 9-45C shows the resulting total current distribution.

Top section

Area of the top section:

$$\int_{40.5}^{90} \cos L \, dL = \sin 90 - \sin 40.5$$

$$= 0.35 \times \frac{180°}{\text{radians}} = 20.1 \text{ A} - \text{degrees}$$

Equivalent length of the top section = $\frac{20.1}{0.760} = 26.44°$

Bottom section

Area of the bottom section:

Average width of the trapezoid = $\frac{1.0 + 0.76}{2} = 0.880$

$S2 = 13.5° \times 0.880 = 11.88$ A-degrees

Equivalent length of the bottom section = 11.88°

Total length

Total effective length = $26.44 + 11.88 = 38.32°$

Full electrical length = $\frac{300}{3.8} = 78.95$ m

Effective height = $78.95 \times \frac{38.32}{360} = 8.40$ m

From Eq 9-2, $R_{rad} = 16.4 \ \Omega$

Let us compare linear loading with base loading for the same antenna structure. Using Eq 9-5:

$$R_{rad} = 36.6 \times \frac{(1 - \cos 63)^2}{\sin^2 63} = 13.7 \ \Omega$$

Thus it is clear that linear loading has two advantages over base loading:

1. The radiation resistance is slightly higher with linear loading.
2. Linear loading can be done with lower losses than those of a (large) base-loading coil.

The linear-loading technique described above is used with great success on the Hy-Gain 402BA shortened 40-meter beam, where linear loading is used at the center of the dipoles. It is also used successfully on the KLM 40 and 80-meter shortened Yagis and dipoles, where linear loading is applied at a certain distance from the center of the elements.

3.2. Keeping the Radiation Resistance High

As stated before, this is *not* the same as keeping the feed-point impedance high! Using any kind of transformers, such as folded elements or any other type of matching systems do *not* change the radiation resistance as defined in Section 3.6 of Chapter 5.

The rule for keeping the radiation resistance high is simple:

1. Use as long a vertical as possible (up to 90°).
2. Use top (capacitive) loading rather than center or bottom loading.

Fig 9-44 gives the radiation resistance for monopoles with combined base and top loading. The graphs clearly show the tremendous advantage of top loading.

The values of R_{rad} given in these figures can be used for antennas with diameters ranging from 0.1° to 1° (360° = 1 λ). Sevick, W2FMI (Ref 818), obtained very similar results experimentally, while the values in the figures mentioned above were derived mathematically.

3.3. Keeping Losses Associated with Loading Devices Low

- Capacitance hat: The losses associated with a capacitance hat are negligible. When applying top-capacitance loading, especially on 160 meters, the practical limitation is likely to be the size (diameter) of the top hat. Therefore, when designing a short vertical, it is wise to start by dimensioning the top hat.
- T-wire top loading: This method is lossless, as with the capacitance hat. It may not always be possible, however, to have a perfectly horizontal top wire. Slightly drooping top-loading wires are just as effective, and when used in pairs (each wire of a pair being in-line with the second wire) the radiation from these loading wires is nil.
- Linear loading: Linear-loading systems can be made virtually loss free, just as with a capacitance hat. Linear loading does not have to be implemented at the bottom end of the vertical. Applying it higher up on the vertical

does, however, make it necessary to electrically open up the vertical at the point of loading.
- Loading coil: Loading coils are intrinsically lossy devices. The equivalent series loss resistance is given by:

$$R_{loss} = \frac{X_L}{Q} \tag{Eq 9-10}$$

where

X_L = inductive reactance of the coil
Q = Q (quality) factor of the coil

Q factors of 200 to 300 are easy to obtain without special measures. Well designed and carefully built loading coils can yield Q factors of up to 800 (see Section 3.6.6).

Base loading requires a relatively small coil, so the Q losses will be relatively low, but the R_{rad} will be low as well. See Section 3.6.2 for practical design examples with real-life values.

Top loading requires a large-inductance coil, with correspondingly larger losses, while in this case the R_{rad} is much higher (see Sections 3.6.6 to 3.6.9 for practical design values).

3.4. Short-Vertical Design Guidelines

From the above considerations we can conclude the following:
- Make the "short" vertical physically as long as possible.
- Make use of top loading (capacitance hat or T wires, horizontal or sloping) to achieve the highest radiation resistance possible.
- Use linear loading instead of a base-loading coil whenever possible to reduce losses.
- Use the best possible ground system (radials or a counterpoise).
- Try to avoid loading coils, or design and build them with great care (high Q).
- Take extremely good care of electrical contacts, contacts between antenna sections, between the antenna and the loading elements. This becomes increasingly important as the radiation resistance is lower.

Though we may be able to build small verticals with low intrinsic losses, it may not always be possible to improve the losses in the ground-return circuit (radials and ground) to a point where the loaded verticals achieve a good efficiency. Small loaded verticals will often be imposed by area restrictions, which may also mean that an extensive and efficient ground (radial) system may well be excluded. Keep in mind that with short, loaded verticals, the ground system is even more important than with a full-size ¹/₄-λ vertical.

It is a widespread misconception that vertical antennas don't require much space. Nothing is farther from the truth. Verticals take a lot of space! A good ground system for a short vertical takes much more space than a dipole, unless you live right at the coast, over salt water, where you might get away with a simple ground system. By the way, it is the salt water that makes the short, loaded verticals produce such excellent signals on many DXpeditions. Remember VKØIR (Heard Island) and ZL7DK (Chatham Island) just to name a couple of them.

3.4.1. Verticals with folded elements

Another common misconception is that folded elements increase the radiation resistance of an antenna, and thus increase the system efficiency. However, the radiation resistance of a folded element is not the same as its feed-point resistance.

A folded monopole with two equal-diameter legs will show a feed-point impedance with the resistive part equal to $4 \times R_{rad}$. The higher feed-point impedance does not help to reduce the losses due to low radiation resistance, however, since with the folded element the lower feed current now flows in one more conductor, totaling the same loss. In a folded monopole, the same current ends up flowing through the lossy ground system, resulting in the same loss whether a folded element is used or not.

This is illustrated in **Fig 9-46**. In the non-folded situation (A) it is clear that the total 1 A current flows through the 10 Ω equivalent ground-loss resistance. The ground loss is $I^2 \times R = 10$ W.

Figure 9-46B shows the folded-element situation. In this example we used equal diameter conductors, hence the feed impedance is 4 times the impedance of the single-conductor-equivalent vertical and the current half the value of the same antenna with a single conductor: 0.5 A flows in the folded-element wire and from the feed point down to the 10 Ω resistor. But there is another 0.5 A coming down the folded wire and also going to the top of the 10 Ω resistor. In the ground system, through the 10 Ω ground loss resistor, we have a total current of 1 A flowing, the same as with the unfolded vertical! The loss is again $I^2 \times R = 10$ W.

The widespread misconception is likely to arise because one assumes (incorrectly) that the current in the feed line equals the current into the lossy ground system. But the current from the folded radiator is also flowing into the lossy ground system, making the total ground current exactly the same as that before.

In other words: The impedance transformation of the folded monopole also transforms the ground loss part of the equation in the same way and there is no net improvement. It is just a transformer and has no different effect than adding a toroidal step-up transformer at the base of a regular monopole.

Although the folded monopole does not gain anything in efficiency due to the impedance transformation it does have some advantages. The impedance transformation will result in a higher impedance that might be more easily matched by a more efficient network than would be required by a plain monopole. The folded monopole has some advantages in lightning protection due to the possibility of dc grounding the structure. And the folded monopole may have a wider bandwidth due to the larger effective diameter of the two conductors (see also Chapter 8, Section 1.4.1).

Fig 9-47 shows the effective normalized diameter of two parallel conductors, as a function of the conductor diameters and spacing (source: *Kurze Antennen*, by Gerd Janzen, ISBN 3-440-05469-1). A folded element consisting of a 5-cm OD tube and a 2-mm OD wire (d1/d2 = 25), spaced 25 cm has an effective round conductor diameter of $0.6 \times 25 = 15$ cm!

One way to improve the radiation efficiency of a vertical over poor ground is to replace the single radiating element with several close-spaced verticals, all fed in phase. This is very similar to what we do with Beverage antennas, when we put two or more in parallel (see Chapter 7, Section 5.3.1).

3.5. SWR Bandwidth of Short Verticals

3.5.1. Calculating the 3-dB bandwidth

One way of defining the Q factor of a vertical is (see Section 3.10.2):

$$Q = \frac{Z_{surge}}{R_{rad} + R_{loss}} \qquad \text{(Eq 9-11)}$$

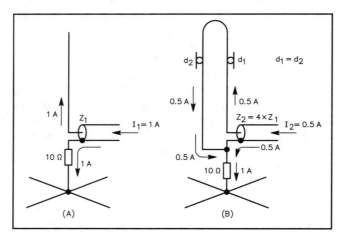

Fig 9-46—The same net current flows in the ground system, whether an open or a folded element is used. This is clearly illustrated for both cases. See text for details.

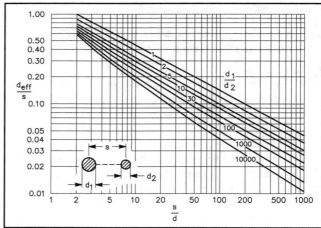

Fig 9-47—Normalized effective antenna diameter of a folded dipole using two conductors of unequal diameter, as a function of the individual conductor diameters d1 and d2, as well as the spacing between the two conductors (S). (After Gerd Janzen, *Kurze Antennen*.)

Z_{surge} is the characteristic impedance of the antenna seen as a short single-wire transmission line. The surge impedance is given by:

$$Z_{surge} = 60 \left[\ln \left(\frac{4h}{d} \right) - 1 \right] \qquad \text{(Eq 9-12)}$$

where

h = antenna height (length of transmission line)
d = antenna diameter (transmission-line diameter)
and where values for h and d are in the same units

The 3-dB bandwidth is given by:

$$BW_{3\,dB} = \frac{f}{Q} \qquad \text{(Eq 9-13)}$$

where f = the operating frequency.

Example:

Assume a top-loaded vertical measuring 30 m, with an effective diameter of 25 cm and a capacitance hat that resonates the vertical at 1.835 MHz.

Using Eq 9-12: $Z_{surge} = 310\ \Omega$

The electrical length of the vertical is

$$\frac{1.835}{300 \times 0.96} \times 30 \text{ m} \times 360° = 68.8°$$

Using Eq 9-7: $R_{rad} = 31.8\ \Omega$

Assume: $R_{ground} = 10\ \Omega$ (average ground system)

Using Eq 9-11: $Q = \frac{310}{31.8 + 10} = 7.42$

Using Eq 9-13: $BW_{3\,dB} = \frac{1.835}{7.42} = 0.247$ MHz

3.5.2. The 2:1 SWR bandwidth

A more practical way of knowing the SWR bandwidth performance is to model the antenna at different frequencies, using eg, *MININEC*. The Q of the vertical is a clear indicator of bandwidth. Antenna Q and SWR bandwidth are discussed in Chapter 5, Section 3.10.1.

Table 9-4 shows the results obtained by modeling full-size quarter-wave verticals of various conductor diameters. Both the perfect as well as the real-ground case are calculated. The vertical with a folded element clearly exhibits a larger SWR bandwidth than the single-wire vertical. Note that with a tower-size vertical (25-cm diameter), both the CW as well as the phone DX portions of the 80-meter band are well covered. If a wire vertical is planned (eg, suspended from trees), the folded version is to be preferred. Matching can easily be done with an L network.

It is evident that loaded verticals exhibit a much narrower bandwidth than their full-size $\frac{1}{4}$-λ counterparts. With the shorter verticals, the quality of the ground system (the equivalent loss resistance) plays a very important role in the bandwidth of the antenna. **Table 9-5** shows the calculated

Table 9-4
Quarter-Wave Verticals on 80 Meters

Z_t, SWR_t and Q_t indicate the theoretical figures assuming zero ground loss. Z_g, SWR_g and Q_g values include an equivalent ground resistance of 10 ohms.

diameter vertical		2 mm (0.08")	40 mm (1.6")	250 mm (10")
3.5 MHz	Z_t=	31.6 – j35.9	31.4 – j23.5	31.1 – j16.7
	Z_g=	41.6 – j35.9	41.4 – j23.5	41.1 – j16.7
	SWR_t=	2.8:1	2.0:1	1.7:1
	SWR_g=	2.2:1	1.7:1	1.5:1
3.65 MHz	Z_t=	35.9	35.9	35.9
	Z_g=	45.9	45.9	45.9
	SWR_t=	1:1	1:1	1:1
	SWR_g=	1:1	1:1	1:1
3.8 MHz	Z_t=	40.0 + j35.5	40.9 + j24.5	41.1 + j16.6
	Z_g=	50.0 + j35.5	50.9 + j24.5	51.1 + j16.6
	SWR_t=	2.5:1	1.9:1	1.6:1
	SWR_g=	2.1:1	1.7:1	1.4:1
	Q_t=	12.1	8.1	5.6
	Q_g=	9.5	6.4	4.4

Table 9-5
Verticals with 40-mm OD for 80 Meters

Z_t, SWR_t and Q_t are the values for a zero ohm ground resistance. Z_g, SWR_g and Q_g relate to an equivalent ground resistance of 10 ohms.

Frequency		$\frac{1}{8}$ wave long (9.9 m) (28.4 ft)	$\frac{3}{16}$ wave long (12.6 m) (41.3 ft)
3.5 MHz	Z_t =	5.37 – j340	9.3 – j237
	Z_g =	15.37 – j340	19.3 – j237
	SWR_t =	15.7:1	6.0:1
	SWR_g =	3.6:1	2.7:1
3.65 MHz	Z_t =	5.9 – j319	10.3 – j217
	Z_g=	10.5 – j319	20.3 – j217
	SWR_t =	1:1	1:1
	SWR_g =	1:1	1:1
3.8 MHz	Z_t =	6.47 – j299	11.4 – j198
	Z_g =	16.47 – j299	21.4 – j198
	SWR_t =	12.3:1	4.9:1
	SWR_g =	3.3:1	2.4:1
	Q_t =	42	23
	Q_g =	15	12

impedances and SWR values for short top-loaded verticals. The same equivalent ground resistance of 10 Ω (used in Table 9-4) has a very drastic influence on the bandwidth of the very short vertical. Note the drastic drop in Q and the increase in bandwidth with the 10-Ω ground resistance.

It is clear that two factors will definitely influence the

SWR bandwidth of a vertical of a given length: the conductor diameter, and the total loss resistance. It should also be clear that we only want to use the first parameter to increase the bandwidth. If we want to use the second parameter (loss resistance), we can use a dummy load for an antenna; that's the antenna with the largest SWR bandwidth (and the worst radiating efficiency).

If you use a coil for loading a vertical (center or top loading), you will understand that for a given antenna diameter, the bandwidth will decrease up to a point as the antenna is shortened and the missing part is partly or totally replaced by a loading coil. Then with more shortening, the bandwidth will begin to increase again as the influence of the equivalent resistive loss in the large coil begins to affect the bandwidth of the antenna.

If you measure an unusually broad bandwidth for a given vertical design, you should suspect a poor-quality loading coil or some other lossy element in the system (did you forget a ground system, or maybe did you forget to connect it?).

3.6. Designing Short Loaded Verticals

Let us review some practical designs of short loaded verticals (Ref 794).

3.6.1. Base coil loading

Assume a 24-m vertical having an effective diameter of 25 cm, which we use as a $^3/_8$-λ vertical on 80 meters. We want to resonate it on 160 meters using a base-mounted loading coil (**Fig 9-48**). The length of the vertical on 160 meters is 53.5°. Let us now calculate the surge impedance of the short vertical using Eq 9-12:

$$Z_{surge} = 60 \left[\ln\left(\frac{4 \times 2400}{25}\right) - 1 \right] = 297 \ \Omega$$

Calculation of the loading coil:
The capacitive reactance of the short vertical is:

$$X_C = \frac{Z_{surge}}{\tan t} \qquad \text{(Eq 9-14)}$$

where t = the electrical length of the vertical in degrees (24 m is 53.5°).

In this example, $X_C = \dfrac{297 \ \Omega}{\tan 53.5°} = 220 \ \Omega$

Since X_L must equal X_C,

$$L = \frac{X_L}{2\pi \times f} = \frac{220}{2\pi \times 1.83} = 19.1 \ \mu H$$

Let us assume a Q factor of 300, which is easily achievable:

$$R_{loss} = \frac{X_L}{Q} = \frac{220 \ \Omega}{300} = 0.73 \ \Omega$$

This value of loss resistance is reasonably low, espe-

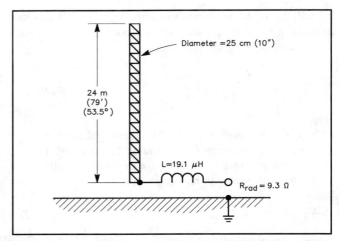

Fig 9-48—Base-loaded tower for 160 meters. See text for details on how to calculate the radiation resistance as well as the value of the loading coil. The loss resistance is effectively in series with the radiation resistance. With 60 $^1/_8$-λ radials over good ground, the feed-point impedance will be approximately 20 Ω and the radiation efficiency about 50%.

cially when you compare it with the value of R_{rad} (calculated using Eq 9-5):

$$R_{rad} = 36.6 \times \frac{(1 - \cos 53.5°)^2}{\sin^2 53.5°}$$

ELNEC also calculates R_{rad} as 9.3 Ω.

The radiation resistance is effectively in series with the ground-loss resistance. Assuming 60 $^1/_8$-λ radials over good ground, which yields an estimated equivalent loss resistance of about 10 Ω, the feed-point impedance will be approximately 20 Ω. The efficiency will be 50%.

3.6.2. Base linear loading

Section 3.1.5 explains why a linear-loading stub is better than a regular base-loading coil. The physical length of the linear-loading device required to bring the shortened vertical to resonance is typically 10 to 20% longer than the missing electrical length. A practical example is elaborated in Fig 9-45.

For 5 years I used such a vertical, where the linear-loading system was made of two 6-meter long aluminum tubes, 2 cm in diameter, spaced approx. 25 cm. This linear-loading device was spaced about 20 cm from the vertical. The easiest way to tune the system is to provide a movable shorting bar near the top of the device. Moving the bar changes the resonant frequency of the loaded vertical.

I. Payne, VE3DO, has reported great success with the linear loading on his 28.5-m vertical. **Fig 9-49** shows details of the professional work he has done on his linear-loaded vertical.

3.6.3. Capacitance-hat loading

Consider the design of a 30-m vertical that we want to load with a capacitance hat to resonate on 1.83 MHz. The electrical length of the 30-m vertical is 67°. We will replace the missing 23° with a capacitance hat (**Fig 9-50**).

First we calculate the surge impedance of the short vertical using Eq 9-12. Take a vertical diameter of 25 cm. The surge impedance is:

$$Z_{surge} = 60\left[\ln\left(\frac{4 \times 3000}{25}\right) - 1\right] = 310\ \Omega$$

Notice that the conductor diameter has a great influence on the surge impedance. The same vertical made of 5-cm tubing has a surge impedance of 407 Ω.

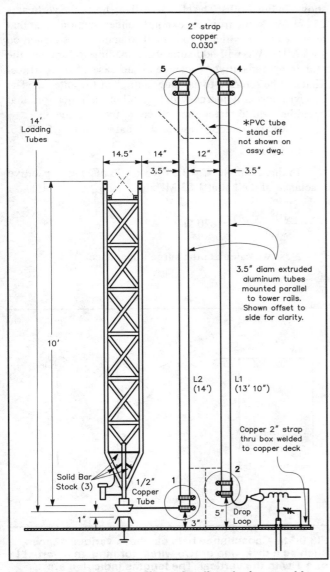

Fig 9-49—Sketch of the linear-loading device used by VE3DO on his 28.5 m vertical.

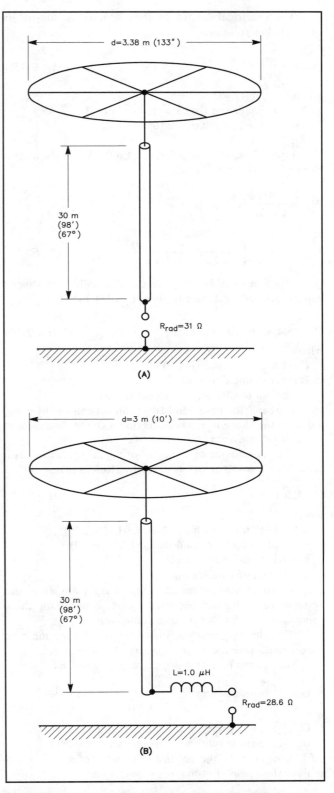

Fig 9-50—Examples of 160-meter verticals using capacitance hats. At A, the hat is dimensioned to tune the vertical to resonance of 1830 kHz. The antenna at B uses a capacitance hat of a given dimension, and resonance is achieved by using a small amount of base loading.

The electrical length of the capacitance top-hat is calculated as follows:

$$X_C = \frac{Z_{surge}}{\tan t} \qquad \text{(Eq 9-15)}$$

where
  X_C = reactance of the capacitance hat (Ω)
  t = electrical length of the top hat = 23°
  Z_{surge} = 310 Ω
  Eq 9-15 has the same form as Eq 9-14, but the definitions of terms are different.

$$X_C = \frac{310 \ \Omega}{\tan 23°} = 730 \ \Omega$$

$$C_{pF} = \frac{10^6}{2 \times \pi \times f \times X_C} = \frac{10^6}{2 \times \pi \times 1.82 \times 730} = 119 \ pF$$

The approximate capacitance of a solid-disk-shaped capacitive loading device is given by (Ref 7818):

$$C = 35.4 \times D \quad \text{(if } D > 2 \times h) \qquad \text{(Eq 9-16)}$$
where
  C = hat capacitance (in pF)
  D = hat diameter (in m)
  h = height of disk above ground (in m)
  The capacitance of a solid disk can be achieved by using a disk in the shape of a wheel, having 8 (large diameter) to 12 (small diameter) radial wires (Ref 7818).
  The capacitance of a single horizontal wire, used as a capacitive loading device is given by (Ref 7818):

$$C = k \times L \qquad \text{(Eq 9-17)}$$

where
  k = 10 pF/m for thick conductors (L/d < 200)
  k = 6 pF/m for thin conductors (L/d > 3000)
  C = hat capacitance (in pF)
  L = length of wire (in m)
  If two loading wires are used at right angles to the vertical, the k-factors become approx. 8 pF/m for thick conductors and 5 pF/m for thin conductors.
  If the loading wires are not horizontal, they must be longer to achieve the same capacitive loading effect.
  The capacitance of a sloping wire is given by:

$$C_{slope} = C_{horizontal} \times \cos \alpha \qquad \text{(Eq 9-18)}$$

where
  $C_{horizontal}$ = capacity of the horizontal wire
  α = slope angle (with a horizontal wire α = 0°)
  Using a disk, the required diameter of the disk (to achieve the 119 pF top-loading capacity) is:

$$D = \frac{119}{35.4} = 3.4 \ m$$

Using a wire, the total required length of the (thin) wire is:

$$L = \frac{119}{6} = 19.8 \ m$$

This wire can be in the shape of a single horizontal or gently sloping wire, can be the total length of the 2 legs of the T-shaped loading wire (horizontal or slightly sloping) or can be the total length of 4 wires as shown in **Fig 9-51**.

The disk of a capacitance hat has a large screening effect to what's above the disk. If there is a "whip" above a large disk, the lengthening effect of the whip may be largely undone. The same effect exists with towers loaded with Yagis. It is mainly the largest Yagi that will determine the capacity to ground. The capacitance hat makes one plate of a capacitor with air dielectric, the ground is the other one.

3.6.4. Capacitance hat with base loading

Consider the design of the same 30-m vertical with a 3-m diameter solid-disk capacitance hat for 1.83 MHz as shown in Fig 9-50B. The effective diameter of the vertical is 25 cm. We know from the example under Section 3.6.3 that this hat will be slightly too small to achieve resonance on 1.83 MHz. We will add some base loading to tune out the remaining capacitive reactance at the base of the vertical. That can be referred to as "fine tuning" of the antenna. The coil will normally merge with the coil of the L network that could be used to match the vertical to the feed line.

The capacitance of a solid-disk hat is given by Eq 9-16:

$$C = 35.4 \times D$$

In this example, C = 3.54 × 3 = 106 pF. The capacitive reactance of the hat at 1.83 MHz is:

$$\frac{10^6}{2 \times \pi \times 1.83 \times 85} = 820 \ \Omega$$

Next we calculate the surge impedance:

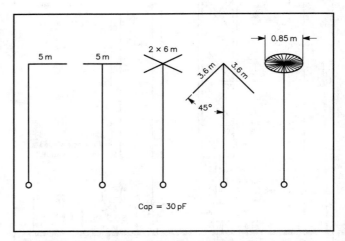

Fig 9-51—Capacitance hats can have various shapes, such as a disk, one or two wires, forming an inverted L or a T with the vertical. The lengths indicated are approximate values for a capacitance of 30 pF.

$$Z_{surge} = 60 \left[\ln \left(\frac{4 \times 3000}{25} \right) - 1 \right] = 310 \ \Omega$$

The electrical length of the capacitance top-hat is calculated using Eq 9-15, rewritten as:

$$\tan t = \frac{Z_{surge}}{X_C} \text{ or } t = \arctan \left(\frac{Z_{surge}}{X_C} \right)$$

$$t = \arctan \left(\frac{310}{820} \right) = 20.7°$$

For a thinner radiator, the electrical length of the hat would be higher as Z_{surge} would be greater.

The electrical length of our example vertical radiator is 67°, and the top-hat capacitance is 20.7°. Since the sum of the two is 87.7°, another 2.3° of loading is required to make a full 90°. Let us calculate the required loading coil for mounting at the base of the short vertical.

To do this calculation, we must first calculate the surge impedance of the vertical (with its capacitance top hat). The surge impedance was calculated above as 310 Ω. The capacitive reactance is calculated using Eq 9-14:

$$X_C = \frac{Z_{surge}}{\tan t} = \frac{310 \ \Omega}{\tan 87.7°} = 12.4 \ \Omega$$

Since X_L must equal X_C:

$$L = \frac{X_L}{2 \times \pi \times F} = \frac{12.4}{2 \times \pi \times 1.83} = 1.1 \ \mu H$$

The coil can be calculated using the program module available on the NEW LOW BAND SOFTWARE.

Let's see what the equivalent series loss resistance of the coil will be, in order to assess how much the base-loading coil influences the radiation efficiency of the system. We will assume a coil Q of 300. Using Eq 9-10 we calculate:

$$R_{loss} = \frac{X_L}{Q} = \frac{12 \ \Omega}{300} = 0.04 \ \Omega$$

This negligible loss resistance is effectively in series with the ground-loss resistance.

Calculation of radiation resistance using Eq 9-9:

$$R_{rad} = 36.6 \frac{(\sin t1 - \sin t2)^2}{\cos^2 t2}$$

$$= 36.6 \frac{(\sin 67° - \sin 2.3°)^2}{\cos^2 2.3°}$$

$$= 28.4 \ \Omega$$

With an equivalent ground resistance of 10 Ω, the efficiency of this system (Eq 9-3) is:

$$Eff = \frac{R_{rad}}{R_{rad} + R_{loss}} = \frac{28.4}{28.4 + 10 + 0.04} = 74\%$$

3.6.5. T-wire loading

If the vertical is attached at the center of the top-loading wire, the horizontal (high-angle) radiation from this top wire will be effectively canceled.

The capacitance of a top-loading wire of small diameter is approx. 6 pF/m for horizontal wires (see Chapter 8, Section 3.3.5). A fair approximation is to say that the total T-wire length is roughly twice the length of the missing portion of the vertical (to make it a $^1/_4$-λ antenna).

Fig 9-52 shows a typical configuration of a T antenna. Two existing supports, such as trees, are used to hold the flattop wire. Try to keep the vertical wire as far as possible away from the supports, as power will inevitably be lost in the supports if close coupling exists.

Fig 9-53 shows a design table that was derived using the *ELNEC* modeling program. The dimensions can easily be extrapolated to other design frequencies.

In practice the T-shaped loading wires will often be downward-sloping loading wires. In this case the radiation resistance will be slightly lower due to the vertical component from the downward-sloping current being in opposition with the current in the short vertical. Sloping loading wires will also be longer than horizontal ones, to achieve the same capacity (see Section 3.6.3, Eq 9-18).

3.6.6. Coil plus capacitance hat

Let's work out an example of a 1.8 MHz antenna using a 12-m mast, 5-cm OD, with a 1.2-m diameter capacitance hat above the loading coil (Fig 9-54). The length of the mast is 26.3°, and the capacitance of the top hat, by rearranging Eq 9-16, is:

$$C = \frac{D}{3.54} = \frac{1.2}{3.54} = 42.5 \ pF$$

$$X_C = \frac{10^6}{2 \times \pi \times 1.8 \times 42.5} = 2080 \ \Omega$$

The surge impedance of the vertical mast is calculated using Eq 9-12:

$$Z_{surge} = 60 \left[\ln \left(\frac{4 \times 1200}{5} \right) - 1 \right] = 352 \ \Omega$$

Let us look at the vertical as a short-circuited line having a characteristic impedance of 352 Ω. The input impedance of the short-circuited transmission line is given by:

$$Z = X_L = + j \ Z_0 \tan t \qquad \text{(Eq 9-19)}$$

where

    Z = input impedance of short-circuited line
    Z_0 = characteristic impedance of the line (352 Ω)
    t = line length in degrees

Thus,

$$Z = +j \ 352 \times \tan (26.3°) = +j \ 174 \ \Omega$$

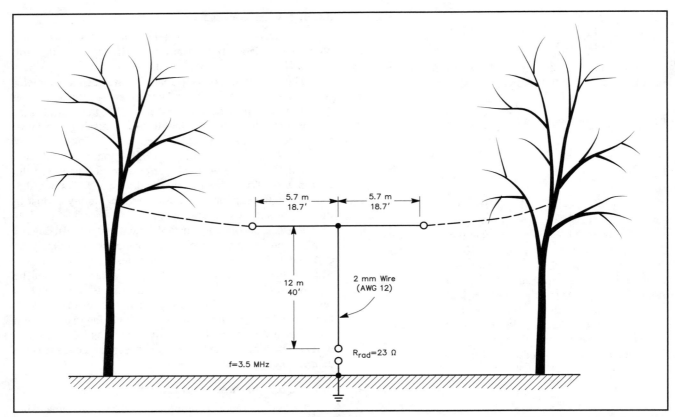

Fig 9-52—Typical setup of a current-fed T antenna for the low bands. Good-quality insulators should be used at both ends of the horizontal wire, as high voltages are present.

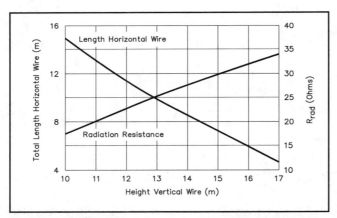

Fig 9-53—Design chart for a wire-type ¼-λ current-fed T antenna made of 2-mm OD wire (AWG 12) for a design frequency of 3.5 MHz. For 160 meters the dimension should be multiplied by a factor of 1.9.

This means that the mast, as seen from above, has an inductive reactance of 174 Ω at the top. The capacitive reactance from the top hat is 2080 Ω. This means that the loading coil, installed at the top of the mast, will need to have an inductive reactance of 2080 Ω – 174 Ω = 1906 Ω.

$$L = \frac{1906}{2 \times \pi \times 1.8} = 169 \ \mu H$$

Assuming you build a loading coil of such a high value with a Q of 200, the equivalent series loss resistance is:

$$R_{loss} = \frac{1906 \ \Omega}{200} = 9.5 \ \Omega$$

Using Eq 9-7, we calculate the radiation resistance of the 12-m long top-loaded vertical:

$$R_{rad} = 36.6 \times \sin^2 26.3° = 7.2 \ \Omega$$

Notice that if we want to use the loss resistance of the (top) loading coil for determining the efficiency (or the feed-point impedance) of the vertical, we must transpose the loss resistance to the base of the vertical. This can be done by multiplying the loss resistance of the coil times the square of the cosine of the height of the coil. In our example the loss resistance transposed to the base is:

$$Loss_{base} = Loss_{coil} \times \cos^2 h = 9.5 \times \cos^2 (26.3°) = 7.6 \ \Omega$$
(Eq 9-20)

Assuming a ground loss of 10 Ω, the efficiency of the antenna is:

$$Eff = \frac{7.6}{7.6 + 10 + 7.4} = 29\%$$

If there were no coil loss, the efficiency would be 42%. This brings us to the point of power-handling capability of the loading coil.

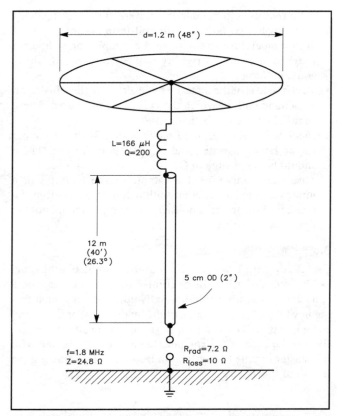

Fig 9-54—Top-loaded vertical for 160 meters, using a combination of a capacitance hat and a loading coil. See text for details.

3.6.6.1. Power dissipation of the loading coil

Let us calculate how much power is dissipated in the loading coil as calculated in Section 3.6.6 for an input power to the antenna of 1500 W. The base feed impedance is the sum of R_{rad}, R_{ground} and R_{coil}. This sum is = 7.2 + 10 + 7.6 = 24.8 Ω. The base current is:

$$I_{base} = \sqrt{\frac{1500}{24.8}} = 7.8 \text{ A}$$

The resistance loss of the loading coil is 7.6 Ω. The current at the position of the coil (26.3° above the feed point) is:

$$I_{coil} = 7.8 \times \cos 26.3° = 7 \text{ A}$$

The power dissipated in the coil is: $I_{coil}^2 \times R_{coil} = 7.0^2 \times 9.5 = 465 \text{ W}$

This is an extremely high figure, and it is unlikely that we can construct a coil that will be able to dissipate this amount of power without failing (melting!). In practice this means that we will have to do one of the following things if we want the loading coil to survive:

- Run lower power. For 100 W of RF, the power dissipated in the coil is 31 W, for 200 W it is 62 W, for 400 W it is 124 W. Let us assume that 150 W is the amount of power

that can safely be dissipated in a well-made very large size coil. In that case a maximum input power of 482 W can be allowed to the vertical (assumed coil Q = 200).

- Use a coil of lower inductance and use more capacitive loading (with a larger hat or longer T wires). To allow a power input of 1500 W, and assuming a ground loss of 10 Ω and a coil Q of 200, the maximum value of the loading coil for 150-W dissipation is 42.1 μH. This value is verified as follows (the intermediate results printed here are rounded):

The reactance of the coil is $X_L = 2 \times \pi \times 1.8 \times 42.1 = 476 \ \Omega$.

The R_{loss} of the coil is $\frac{476 \ \Omega}{200} = 2.4 \ \Omega$

Transposed to the base, $R_{loss} = 2.4 \times \cos^2 (26.3°) = 1.9 \ \Omega$

$$I_{base} = \sqrt{\frac{1500}{7.2 + 1 + 1.9}} = 12.2 \text{ A}$$

This current, transposed to the coil position, is $8.9 \times \cos 26.3° = 7.9 \text{ A}$.

$P_{coil} = 7.9^2 \times 2.4 = 150 \text{ W}$.

This is only about 20% of the value of the original 168-μH inductance needed to resonate the antenna at 1.8 MHz. This smaller coil will require a substantially larger capacitance hat to resonate the antenna on 160 meters. T wires would also be an appropriate way to tune the antenna to resonance.

- Make a coil with the largest possible Q. If we change the coil with a Q of 200 in the above example to 300 and run 1500 W, then the maximum coil inductance is 63.1 μH. The calculation procedure is identical to the above example:

The reactance of the coil is $X_L = 2 \times \pi \times 1.8 \times 63.1 = 714 \ \Omega$.

The R_{loss} of the coil is 714/300 = 2.4 Ω.

Transposed to the base, $R_{loss} = 2.4 \times \cos^2 (26.3°) = 1.9 \ \Omega$

$$I_{base} = \sqrt{\frac{1500}{7.2 + 10 + 1.9}} = 8.9 \text{ A}$$

This current, transposed to the coil position, is $8.9 \times \cos(26.3°) = 7.9 \text{ A}$.

$P_{coil} = 7.9^2 \times 2.4 = 150 \text{ W}$

This means that an increase of Q from 200 to 300 allows us to use a loading coil of 63.1 μH instead of 42.1 μH, resulting in the same power being dissipated in the coil. As you can see, the inductance is inversely proportional to the Q.

Notice that the ground loss resistance also has a great influence on the power dissipated in the loading coil. Staying with the same example as above (Q = 300, L = 63.1 μH), the power loss in the coil for a ground-loss resistance of 1.0 Ω (excellent ground system) is:

$$I_{base} = \sqrt{\frac{1500}{7.2 + 1 + 1.9}} = 12.2 \text{ A}$$

$$P_{coil} = (12.2 \times \cos 26.3°)^2 \times 2.4 = 284 \text{ W}$$

Conclusion: The better the ground system, the more power will be dissipated in the loading coil.

C. J. Michaels, W7XC, investigated the construction and the behavior of loading coils for 160 meters (Ref 797). In the above examples we have assumed Q factors of 200 and 300. The question is, how can we build loading coils having the highest possible Q? Michaels comes to the following conclusions:

• For coils with air dielectric, the L/D (length/diameter) ratio should not exceed 2:1.
• For coils wound on a coil form, this L/D ratio should be 1:1.
• Long, small-diameter coils are no good.
• The highest Q that can be achieved for a 150-µH loading coil for 160 meters is approximately 800. This can be done with a square coil (15-cm long by 15-cm diameter), using approx. 35 turns of AWG #7 (3.7 mm) wire (air wound), or a coil of 30 cm length by 15 cm diameter, wound with 55 turns of AWG #4 (5.1-mm diameter) wire.
• Coil diameters of 10 cm wound with AWG wire # 10 to 14 can yield Q factors of 600, while coil diameters of 5 cm wound with BSWG #20 to 22 will not yield Q factors higher than approximately 250. These smaller wire gauges should not be used for high-power applications.

You can use some common sense and simple test methods for selecting an acceptable plastic coil-form material:
• High-temperature strength: Boil a sample for ½ hour in water, and check its rigidity immediately after boiling, while still hot.
• Check the loss of the material by inserting a piece inside an air-wound coil, of which the Q is being measured. There should be little or no change in Q.
• Check water absorption of the material: Soak the sample for 24 hours in water, and repeat the above test. There should be no change in Q.
• Dissipation factor: Put a sample of the material in a microwave oven, together with a cup of water. Run the oven until the water is boiling. The sample should not get appreciably warm.

3.6.7. Coil with T wire

A coil with T-wire configuration is essentially the same as the one described in Section 3.6.6. In the case of a capacitance hat we would normally adjust the resonant frequency by pruning the value of the loading coil or by adding some reactance (inductor for positive or capacitor for negative) at the base of the antenna. In case of a T-loading wire it is easier to tune the vertical to resonance by adjusting the length of the T wire.

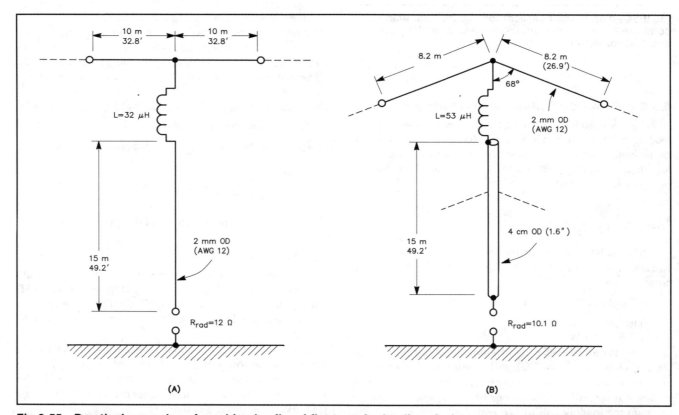

Fig 9-55—Practical examples of combined coil and flat-top wire loading. A shows a wire antenna with a loading coil at the top of the vertical section (no space for longer top-load wires). B shows a loaded vertical mast (4-cm OD) where two of the top guy wires, together with a loading coil, resonate the antenna at 1.835 MHz. The remaining guy wires are made of insulating material (eg, Kevlar, Phyllistran, etc.).

Fine tuning can also be done by changing the "slope" angle of the T wires. If the T wires are sloped downward the resonant frequency goes up, but also the radiation resistance will drop somewhat. **Fig 9-55** shows two examples of practical designs. In the example of the guyed vertical shown in Fig 9-55B, changing the slope angle by dropping the wires from 68° (end of T wires at 12-m height) to 43° (end at 9-m height) raises the resonant frequency of the antenna from 1.835 kHz to 1.860 kHz. Note, though, that with this change the radiation resistance drops from 10.1 Ω to 8.3 Ω!

The larger the value of the coil, the lower the efficiency will be. The equivalent loss resistance of the coil and the transposed loss resistance required to calculate the efficiency and the feed impedance of the vertical can be calculated as shown in Section 3.6.6. As explained in Section 3.6.6, we should avoid having a coil of more than approximately 75-μH inductance.

3.6.8. Coil with whip

Now we consider a vertical antenna loaded with a whip and a loading coil, as in **Fig 9-56**. Let's work out an example:

Mast length below the coil = 18.16 m = 40°
Mast length above the coil (whip) = 4.54 m = 10°
f_{design} = 1.835 MHz
Mast diameter = 5 cm
Whip diameter = 2 cm

Calculate the surge impedance of the bottom mast section using Eq 9-12:

$$Z_{surge} = 60 \left[\ln \left(\frac{4 \times 1816}{5} \right) - 1 \right] = 377 \ \Omega$$

Looking at the base section as a short-circuited line with an impedance of 377 Ω, we can calculate the reactance at the top of the base section using Eq 9-17:

$$Z = X_L = +j377 \times \tan 40° = +j316 \ \Omega$$

Calculate the surge impedance of the whip section, again using Eq 9-12:

$$Z_{surge} = 60 \left[\ln \left(\frac{4 \times 454}{2} \right) - 1 \right] = 349 \ \Omega$$

Let us look at the whip as an open-circuited line having a characteristic impedance of 349 Ω. The input impedance of the open-circuited transmission line is given by:

$$Z = X_C = - j \frac{Z_0}{\tan t} \qquad \text{(Eq 9-21)}$$

where

Z_0 = characteristic impedance (here = 349 Ω)
t = electrical length of whip (here = 10°)

The reactance of the whip is:

$$Z = X_C = - j \frac{349 \ \Omega}{\tan (10°)} = - j \ 1979 \ \Omega$$

Sum the reactances:

$$X_{tot} = +j316 \ \Omega - j1979 \ \Omega = -j1663 \ \Omega$$

This reactance is tuned out with a coil having a reactance of + j1663 Ω:

$$L = \frac{X_L}{2 \times \pi \times f} = \frac{1663}{2 \times \pi \times 1.835} = 144 \ \mu H$$

Assuming you build the loading coil with a Q of 300, the equivalent series loss resistance is

$$R_{loss} = \frac{X_L}{Q} = \frac{1663}{2 \times \pi \times 1.835} = 144 \ \mu H$$

The coil is placed at a height of 40°. Transpose this 5.5 Ω loss to the base using Eq 9-20:

$$R_{loss@base} = 5.5 \ \Omega \times \cos^2 (40°) = 3.2 \ \Omega$$

Calculate the radiation resistance using Eq 9-8:

$$R_{rad} = 36.6 \times (1 - \sin 80° + \sin 40°)^2 = 16 \ \Omega$$

Assuming a ground resistance of 10 Ω, the efficiency of this antenna is:

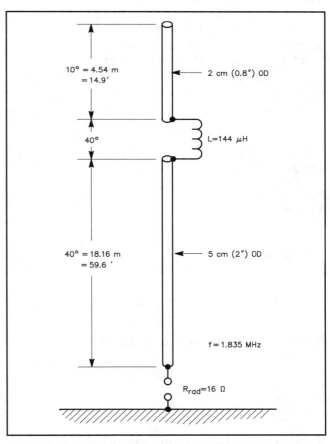

Fig 9-56—Practical example of a vertical loaded with a coil and whip. The length and diameter of the whip are kept within reasonable dimensions that can be realized on top of a loading coil without guying.

$$Eff = \frac{16}{16 + 10} = 55\%$$

I modeled the same configuration using *ELNEC* and found the following results:
Required coil = 1650 Ω reactance = 143 µH
R_{rad} = 20 Ω

The R_{rad} is 25% higher than what we found using Eq 9-8. This formula uses a few assumptions such as equal diameters for the mast section above and below the coil, which is not the case in our design. This is probably the reason for the difference in R_{rad}.

3.6.9. Comparing the different loading methods

In order to get a fair idea of how the different loading methods work, let's compare verticals of identical physical lengths over a relatively poor ground. Where one cannot erect a full-size vertical, it will often be the case that an elaborate radial system won't be possible either. That's why we'll use a rather high ground resistance in our comparative study.

The study is based on the following assumptions:
• Physical antenna length = 45° (1/8 λ).
• L = 20.5 m.
• Design frequency = 1.83 MHz.
• Antenna diameter = 0.1° on 160 meters = 4.55 cm
• Ground-system loss resistance = 15 Ω.

Quarter-wave full size:
R_{rad} = 36 Ω
R_{ground} = 15 Ω
$R_{ant\ loss}$ = 0 Ω
Z_{feed} = 51 Ω
Eff = 71%
Loss = 1.5 dB

Base loading:
R_{rad} = 6.2 Ω
R_{ground} = 15 Ω
Coil Q = 300
L_{coil} = 34 µH
$R_{coil\ loss}$ = 1.3 Ω
Z_{feed} = 22.5 Ω
Eff = 28%
Loss = 5.6 dB

Top-loaded vertical (capacitance hat or T wire):
R_{rad} = 18 Ω
R_{ground} = 15 Ω
Z_{feed} = 33 Ω
Eff = 55%
Loss = 2.6 dB

Top-loaded vertical (coil with capacitance hat at top):
R_{rad} = 18 Ω
R_{ground} = 15 Ω
Diameter of capacitance hat = 3 m
L_{coil} = 37 µH

Coil Q = 200
$R_{coil\ loss}$ = 2.1 Ω
$R_{coil\ loss}$ transposed to base = 1 Ω
Z_{feed} = 34 Ω
Eff = 53%
Loss = 2.8 dB

Top-loaded vertical (coil with whip):
R_{rad} = 12.7 Ω
R_{ground} = 15 Ω
Length of whip = 10° (4.55 m on 1.83 MHz)
L_{coil} = 150 µH
Coil Q = 200
$R_{coil\ loss}$ = 8.6 Ω
$R_{coil\ loss}$ transposed to base = 5.8 Ω
Z_{feed} = 33.5 Ω
Eff = 38%
Loss = 4.2 dB

Conclusions

With an average to poor ground system (15 Ω), a 1/8-λ vertical with capacitance top loading is only 1.1 dB down from a full-size 1/4-λ vertical. Over a better ground the difference is even less. If possible, stay away from loading schemes that use a coil.

■ 4. TALL VERTICALS

In this section verticals that are substantially longer than 1/4 λ are analyzed, especially their behavior over different types of ground. Is the very low wave angle, computed over ideal ground, ever realized in practice?

First of all there is the question if we really need the very low wave angles on the low bands. A very low incident angle grazes the ionosphere for a long distance increasing loss. More hops with less loss from a sharper angle can actually decrease propagation loss. We have learned in Chapter 1 that relatively high wave angles are actually a prerequisite to allow a "duct" to work on 160.

I will tackle the myth of voltage-fed antennas not requiring an elaborate ground system. Long verticals require an even better radial system and an even better ground quality in the Fresnel zone to achieve their low angle and gain potential.

In earlier sections of this chapter, short verticals are dealt with in detail, mostly for 160 meters. On higher frequencies, taller verticals are quite feasible. A full-size 1/4-λ radiator on 80 meters is only approximately 19.5 m in height. Long verticals are considered to be the 1/2-λ or 5/8-λ variety. Verticals that are slightly longer than a quarter-wave (up to 0.35 λ) do not fall in the *long vertical* category.

4.1. Vertical Radiation Angle

Fig 9-57 shows the vertical radiation patterns of two long verticals of different lengths. These are analyzed over an identical ground system consisting of average earth with 60 1/4-λ radials. A 1/4-λ vertical is included for comparison.

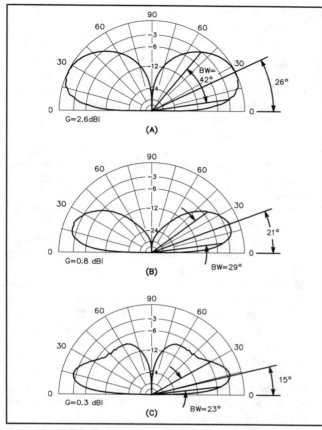

Fig 9-57—Vertical radiation patterns of diferent-length verticals over average ground, using 60 ¹/₄-λ radials. The 0-dB reference for all patterns is 2.6 dBi. A—¹/₄-λ vertical. B—¹/₂-λ. C—⁵/₈-λ.

Note that going from a ¹/₄-λ vertical to a ¹/₂-λ vertical drops the radiation angle from 26° to 21°. More important, however, is that the 3-dB vertical beamwidth drops from 42° to 29°. Going to a ⁵/₈-λ vertical drops the radiation angle to 15° with a 3-dB beamwidth of only 23°. But notice the high-angle lobe showing up with the ⁵/₈-λ vertical. If we make the vertical still longer, the low-angle lobe will disappear and be replaced by a high-angle lobe. A ³/₄-λ vertical has a radiation angle of 45°.

Whatever the quality of the ground is, the ⁵/₈-λ vertical will always produce a lower angle of radiation and also a more narrow vertical beamwidth. The story gets more complicated, though, when one compares the efficiency of the antennas.

4.2. Gain

I have modeled both a ¹/₄-λ as well as a ⁵/₈-λ vertical over different types of ground, in each case using a realistic number of 60 ¹/₄-λ radials. Fig 9-5 shows the patterns and the gains in dBi for the quarter-wave vertical, and **Fig 9-58** shows the results for the ⁵/₈-λ antenna.

Over perfect (ideal) ground, the ⁵/₈-λ vertical has 3.0 dB more gain than the ¹/₄-λ vertical at a 0° wave angle. Note the very narrow lobe width and the minor high-angle lobe (broken-line patterns in Fig 9-58).

Over sea-water the ⁵/₈ λ has lost 0.8 dB of its gain already, the ¹/₄ λ only 0.4 dB. The ⁵/₈-λ vertical has an extremely low wave angle of 5° and a vertical beamwidth of only 17°. The ¹/₄ λ has an 8° take off angle, but a 40° vertical beamwidth.

Over very good ground, the ⁵/₈-λ vertical has now lost 5.0 dB, the ¹/₄ λ only 1.9 dB. The actual gain of the ¹/₄ λ equals

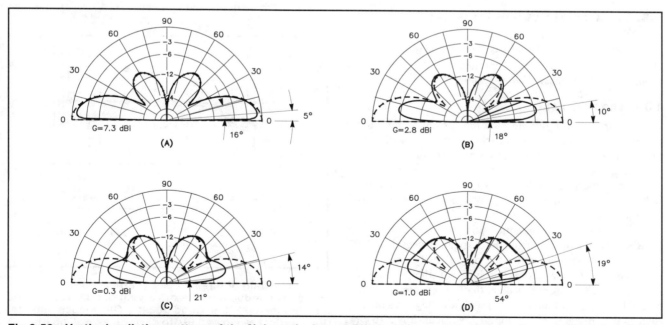

Fig 9-58—Vertical radiation pattern of the ⁵/₈-λ vertical over different types of ground. In all cases, 60 ¹/₄-λ radials were used. The theoretical perfect ground pattern is shown in each case as a reference (broken line, with a gain of 8.1 dBi). Compare with the patterns and gains of the ¹/₄-λ vertical, modeled under identical circumstances (Fig 9-5). A—Over sea. B—Over very good ground. C—Over average ground. D—Over very poor ground.

the gain of the $^5/_8$ λ! Note also that the high-angle lobe of the $^5/_8$ λ becomes more predominant as the quality of the ground decreases.

Over average ground the situation becomes really poor for the $^5/_8$-λ vertical. The gain has dropped 7.3 dB, and the secondary high-angle lobe is only 4 dB down from the low-angle lobe. The $^1/_4$-λ vertical has lost 2.6 dB versus ideal ground, and now shows 2.0 dB *more* gain than the $^5/_8$-λ vertical!

Over very poor ground the $^5/_8$-λ vertical has lost 6.6 dB from the perfect-reflector situation, and the $^1/_4$-λ vertical only 3.0 dB. Note that the $^5/_8$-λ vertical seems to pick up gain as compared to the situation over average ground. From Fig 9-58 it is clear that this is because the radiation at lower angles is now attenuated to such a degree that the radiation from the high-angle lobe (60°) becomes very dominant. Note also that the intensity of the high-angle lobe hardly changes from the perfect ground situation to the situation over very poor ground. This is because the reflection for this very high angle takes place right under the antenna, where the ground quality has been improved by the 60 $^1/_4$-λ radials.

I am sure this must come as a surprise to most. How can we explain this? An antenna that intrinsically produces a very low angle (as seen from the perfect ground model), relies on reflection at great distances from the antenna to produce the low angle. At these distances, radials of limited length do not play any role in improving the ground. With poor ground, a great deal of the power that is sent at a very low angle to the ground-reflection point is being absorbed in the ground rather than reflected (see also Section 1.1.2) . This means that from a Fresnel zone reflection point of view, the long vertical requires a better ground than the $^1/_4$-λ vertical in order to realize its full potential as a low angle radiator.

4.3. The Radial System for a Half-Wave Vertical

Here comes another surprise. A terrible misconception about voltage-fed verticals is that they do not require a good ground nor an extensive radial system.

4.3.1. The near field

It is true that if you measure the current going into the ground at the base of a $^1/_2$-λ vertical, this current will be very low (theoretically zero). With $^1/_4$-λ and shorter verticals, the current in the radials increases in value as you get closer to the base of the vertical. That's why, for a given amount of radial wire, it is better to use many short radials than just a few long ones.

With voltage-fed antennas, the earth current will increase as you move away from the vertical. Brown (Ref 7997) has calculated that the highest current density exists at approximately 0.35 λ from the base of the voltage-fed $^1/_2$-λ vertical. Therefore it is clear that it is even more important to have a good radial system with a voltage-fed antenna such as the voltage-fed T or the $^1/_2$-λ vertical. It is also clear that these verticals require longer radials to do their job efficiently than is the case with current-fed verticals.

4.3.2. The far field

In the far field, the requirement for a good ground with a long vertical is much more important than for the $^1/_4$-λ vertical. I have modeled the influence of the ground quality on the gain of a vertical by the following experiment.

- I compared three antennas: a $^1/_4$-λ vertical, a voltage-fed $^1/_4$-λ T (also called an inverted ground plane), and a $^1/_2$-λ vertical.
- I modeled all three antennas over average ground.
- I put them in the center of a disk of perfectly conducting material and changed the diameter of the disk to find out at what distance the Fresnel zone is situated for the three antennas.

The results of the experiment are shown in **Fig 9-59**. Let us analyze those results.

- With a disk $^1/_4$ λ in radius (equal to a large number of $^1/_4$-λ radials) the $^1/_4$-λ current-fed vertical is almost 2 dB better than the voltage-fed $^1/_4$-λ and the $^1/_2$-λ vertical.
- The $^1/_4$-λ vertical remains better than the other antennas up to a disk size of 1.5-λ diameter. This means that over good ground you must be able to put out radials at least 2-λ long with a $^1/_2$-λ vertical before the $^1/_2$-λ vertical will show any gain over the $^1/_4$-λ current-fed vertical.
- The voltage-fed $^1/_4$-λ vertical (voltage-fed T) equals the current-fed $^1/_4$-λ for a disk size of at least 2 λ in diameter. This is because the current maximum is at the top of the antenna, which means that for a given radiation angle (the wave angle), the Fresnel zone (the place where the main wave hits the ground to be reflected) is much farther away from the base of the vertical than is the case with a $^1/_4$-λ current-fed vertical. In other words, there is no advantage in using such a voltage-fed $^1/_4$-λ antenna.
- For both the voltage and the current-fed $^1/_4$-λ vertical, the Fresnel zone is situated up to approx. 4 λ away from the

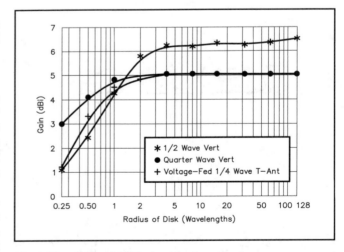

Fig 9-59—Gain of three types of verticals over a perfectly conducting disk of varying radius. The ground beyond the disk is of good quality. This means that the $^1/_2$-λ vertical requires 0.6-λ to perform as well as the $^1/_4$-λ vertical with $^1/_4$-λ radials. Be aware that the radiation angle of the $^1/_2$-λ vertical will be much lower, however.

vertical. For the $1/2$-λ vertical, the Fresnel zone stretches out to some 100 λ!

4.4. In Practice

On 40 meters, a height more than $1/4$ λ (10 m) should be easy to install in most places. In many cases it will be the same vertical that is used as a $1/4$ λ on 80 meters.

I have been using a $5/8$-λ vertical for 40 meters (equals $3/8$-λ wave on 80) for more than 20 years with good success. It does not compete with a Yagi at 100 feet or better, but with separate receiving antennas (Beverages), it has always been a relatively good performer. Now that I have been using a 3-element Yagi at 30 meters for a few years, I know, however, that the solution was far from ideal.

Earl Cunningham, K6SE's, experience confirms this: "I used a grounded $1/2$-λ vertical in Houston/Gulf Coast area where the soil conductivity is abnormally high. It was a super performer. The same vertical here in the desert (Palmdale, CA) was a ho-hum performer, even with a much more extensive ground radial system."

Another similar testimony comes from Tom Rauch, W8JI, who wrote "... I had the same results using BC arrays on 160 meters. The 250-ft to 300-ft verticals stunk, my $1/4$-λ vertical would beat them. I find the same effect on 80 meters."

A very low angle of radiation is certainly an advantage on the higher bands (10 through 20 meters) and is certainly a positive asset on 40, but I had the saddening experience that a $5/8$-λ vertical is too long a vertical radiator for the average DX paths on 80 meters.

The base resistance, $R_{rad(B)}$, and feed-point reactance for monopoles is given in Figs 9-8 and 9-11 as a function of the conductor diameter in degrees, and in Figs 9-9 and 9-12 as a function of the antenna length-to-diameter ratio. The graphs are accurate only for structures with rather large diameters (not for single-wire structures) and of *uniform* diameter. A conductor diameter of 1° equals 833/f (MHz) in mm.

■ 5. MODELING VERTICAL ANTENNAS

ELNEC as well as other versions of *MININEC* are well suited to do your own vertical antenna modeling. Be aware, however, that all *MININEC*-based antenna modeling programs assume a perfect ground for computing the impedance of the antenna. You cannot use these programs to assess the efficiency of the vertical, where we have defined efficiency as:

$$\text{Eff} = \frac{R_{rad}}{R_{rad} + R_{loss}}$$

MININEC will show the influence of the reflecting ground (in the far field) to make up the (low-angle) radiation pattern of the vertical antenna.

If you want to include the losses of the ground, you can insert a resistance at the feed point, having a value equivalent to the assumed loss resistance of the ground (see Table 9-1).

5.1. Wires and Segments

A wire is a straight conductor, and is part of the antenna

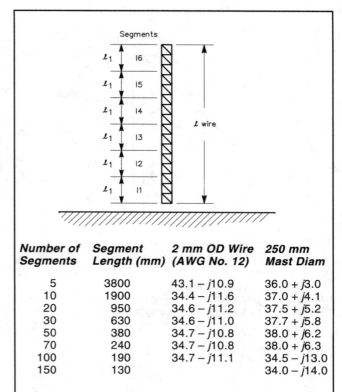

Number of Segments	Segment Length (mm)	2 mm OD Wire (AWG No. 12)	250 mm Mast Diam
5	3800	43.1 – j10.9	36.0 + j3.0
10	1900	34.4 – j11.6	37.0 + j4.1
20	950	34.6 – j11.2	37.5 + j5.2
30	630	34.6 – j11.0	37.7 + j5.8
50	380	34.7 – j10.8	38.0 + j6.2
70	240	34.7 – j10.8	38.0 + j6.3
100	190	34.7 – j11.1	34.5 – j13.0
150	130		34.0 – j14.0

Fig 9-60—*MININEC* analysis of a straight 19-m vertical antenna as shown in the drawing. The analysis frequency is 3.8 MHz. *MININEC* impedance results are shown as a function of the number of segments in the table. Note that for reliability with a "thick" (200-mm) vertical, the maximum number of segments (in this case segments=pulses) is 70. The *MININEC* documentation states that the segment length should be greater than 2.5 times the wire diameter (2.5 x 200 mm= 500 mm). In this particular case errors occur when the segment length is smaller than the wire diameter.

(or the total antenna). A segment is a part of the wire. Each wire can be broken up into several segments, all having the same length. Each segment has a different current. The more segments a wire has, the closer the current (pulse) distribution will come to the actual current distribution. There are limits, however.

• Many segments take a lot of computing time.
• Each segment should be at least 2.5 times the wire diameter (according to *MININEC* documentation).

There is no general rule as to the minimum number of segments that should be used on a wire. The only rule is the cut-and-try rule, whereby you gradually increase the number and look for the point where no further significant changes in the results are observed.

Fig 9-60 shows the example of a straight vertical for 80 meters (19-m long). This antenna consists of a single wire. In order to evaluate the effect of the wire length, I broke it up into several segments, going from 5 segments to 150 segments. Gain and pattern are very close to modeling with only

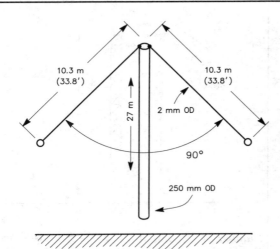

Fig 9-61—Impedances calculated by *MININEC* for a top-loaded 1.8-MHz vertical, using a 250-mm OD mast and two 2-mm OD slant loading wires. The segment lengths are indicated in mm. A large number of segments on all wires always gives more reliable results, provided the segment length is not very different. Judicious choice of segment length on the different wires can also yield very accurate results with a small number of total segments. In order to obtain accurate impedance results using *MININEC*, the wire sections near the acute-angle wire junctions must be short.

Vertical Mast			Slant Wires				
No. of Segments	Segment Length (min.)	Segment Length (max.)	No. of Segments	Segment Length (min.)	Segment Length (max.)	Total Pulses	Impedance
3	9000	9000	3	3765	3765	9	$9.0 - j184.0$
5	5400	5400	5	2260	2260	1	$14.2 - j107.0$
10	2700	2700	10	1130	1130	3	$16.2 - j83.5$
20	1350	1350	20	565	565	6	$16.6 - j87.1$
30	900	900	30	437	437	9	$16.7 - j77.0$
40	625	625	40	282	282	120	$16.6 - j76.5$
50	540	540	50	226	226	150	$16.8 - j76.3$
10	2700	2700	5	2260	2260	15	$16.8 - j78.6$
6	4500	4500	2	5650	5650	10	$16.8 - j80.4$
5	5400	5400	2	5650	5650	9	$16.7 - j80.7$
35	770	770	2	5650	5650	39	$14.9 - j103.0$

f = 1.8 MHz
length of vert mast = 27 m
slant wire = 10.3 m

5 segments. For impedance calculations at least 20 sections are required in order to obtain a reasonably accurate result. The table also shows an example of too many segments (for the vertical measuring 250 mm in diameter). As the segment length becomes very short in comparison to the wire diameter, the results become totally erroneous.

5.2. Modeling Antennas with Wire Connections

When the antenna consists of several straight conductors, things become more complicated. **Fig 9-61** shows the example of a 27-m vertical tower (250-mm OD), loaded by two sloping top-hat wires, measuring 2-mm OD (AWG #12).

The standard approach is to have three wires, one for each of the three antenna parts, and divide the three conductors into a number of segments (which are always of equal length inside each wire).

To obtain reliable results, one must make sure that the lengths of the segments near the junctions are similar. The table in Fig 9-61 shows the impedance obtained for the top-loaded vertical with different numbers of segments. A large number of segments on the vertical mast (eg, 35 segments, which results in a segment length of 770 mm), together with a small number of segments on the sloping wires, give an unreliable result, while a good result is obtained with a total of just 9 segments if the lengths are carefully matched. The segment *tapering* technique, as described in Chapter 14 on Yagis and quads, can also be used to minimize the number of segments and improve the accuracy of the results.

5.3. Modeling Verticals Including Radial Systems

MININEC does not analyze antenna systems with horizontal wires close to the ground. Therefore, modeling ground systems as part of the antenna requires the "professional" *NEC* software. *NEC-2*, or a software such as *EZNEC*, which uses the *NEC-2* engine, will model radials over ground. There seem, however, documented cases (models verified against real-world measurements) of *NEC-2* giving very optimistic results (sometimes up to nearly 6 dB too high gain). *NEC-3* and *NEC-4* will model buried radials (see Chapter 4, Section 2.2).

■ 6. PRACTICAL VERTICAL ANTENNAS

A number of practical designs of verticals for 40, 80 and 160 meters are covered in this section, as well as dual and triband systems. A number of practical matching cases are solved, and the component ratings for the elements are discussed. All the L networks have been calculated using the L-NETWORK DESIGN module from the NEW LOW BAND SOFTWARE.

6.1. Single-Band Quarter-Wave Vertical for 40, 80 or 160

Large diameter conductors are often used for various reasons such as increasing the bandwidth (by increasing the D/L ratio) or simply for mechanical reasons. The effective diameter of wire cages and flat multi-wire configurations is covered in Chapter 8 (Section 2.4.1).

Often, triangular tower sections are used to make vertical antennas. The effective equivalent diameter of a tower section is shown in **Fig 9-62**. A tower section measuring 25-cm wide, with vertical tubes measuring 2.5-cm diameter, has an equivalent diameter of $0.7 \times 25 = 17.5$ cm.

The length of a resonant full-size quarter-wave vertical depends on its physical diameter. **Fig 9-63** shows the physical shortening factor of a $1/4$-λ resonant antenna as a function of the ratio of antenna length to antenna diameter. The required physical length is given by:

$$L = \frac{74.95 \times p}{f_{MHz}}$$

where

L = length (height) of the vertical in m.
p = correction factor (from Fig 9-63).
f_{MHz} = design frequency in MHz.

Quarter-wave verticals are easy to match to 50-Ω coaxial feed lines. The radiation resistance plus the usual earth losses will produce a feed-point resistance close to 50 Ω.

If you don't mind using a matching network at the antenna base, and if you can manage a few more meters of antenna height, extra height will give you increased radiation resistance and higher efficiency. The feed-point impedance can be found in the charts of Figs 9-8, 9-9, 9-11 and 9-12.

Consider the following examples (see **Fig 9-64**):

Example 1:
Tower height = 27 m
Tower diameter = 25 cm
Design frequency = 3.8 MHz
From the appropriate charts or through modeling we find:
R = 185 Ω
X = + $j215$ Ω

Let's assume we have a pretty good ground radial system with an equivalent ground resistance of 5 Ω. We calculate the matching L network with the following values:
$Z_{in} = 190 + j215$ Ω
$Z_{out} = 50$ Ω

The values of the matching network were calculated for 3.8 MHz. The two matching-network alternates (low and high-pass) are shown in Fig 9-64A. The low-pass filter network gives a little additional harmonic suppression, while the high-pass assures a direct dc ground for the antenna, and some rejection of medium-wave broadcast signals.

Example 2:
This time we are setting out to build a vertical that is a little longer than a $1/4$ λ, so the resistive part of the feed-point impedance (at the design frequency) will be exactly 50 Ω. In this case the matching network will consist of a simple series capacitor to tune out the inductive reactance of the feed-point impedance. We use *ELNEC* to design two models:

• Vertical mast diameter = 5 cm, length = 20.5 m, Z = 50 + $j39$ Ω (see Fig 9-64B). The matching network consists of a series capacitor with a reactance of 39 Ω at 3.8 MHz. The value of the capacitor is:

$$C = \frac{10^6}{2 \times \pi \times 3.8 \times 39} = 1073 \text{ pF}$$

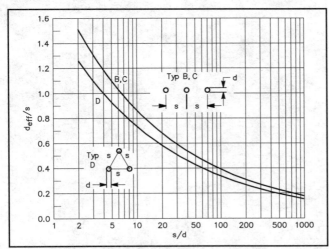

Fig 9-62—Normalized (round solid conductor) effective diameter of a triangular tower section as a function of the vertical tube diameter (d) and the tower width (s). The graph for three parallel conductors is also given (curve BC). (After Gerd Janzen, *Kurze Antennen*.)

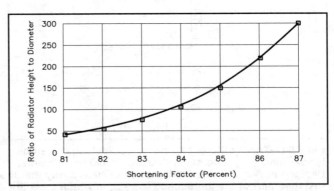

Fig 9-63—Graph showing the amount a $1/4$-λ vertical must be shortened for resonance as a function of the length-to-diameter ratio.

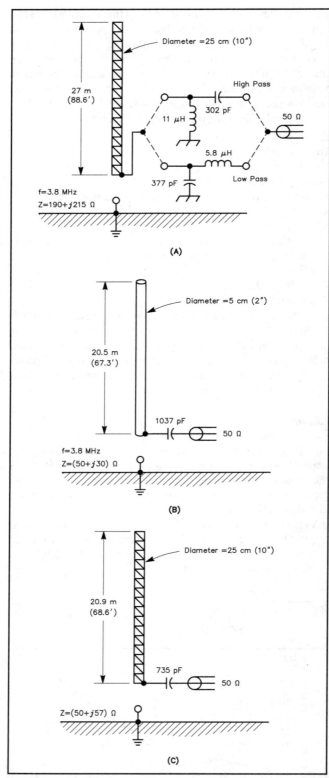

Fig 9-64—Three different 80-meter verticals that are no longer than ¼ λ, together with their matching networks. Designs at B and C are dimensioned in such a way that the resistive part of the impedance is 50 Ω, in which case the matching network consists of only a series capacitor. The difference between B and C is the diameter of the vertical. In all cases a perfect ground (zero loss) is assumed.

• Vertical mast diameter = 25 cm, length = 20.9 m, Z = 50 + j57 Ω (see Fig 9-64C). In this case the series-matching capacitor has a value of 735 pF.

Note that for the above examples we assumed a zero ground loss. The values of the series-matching capacitor can also be calculated using the SERIES IMPEDANCE module of the NEW LOW BAND SOFTWARE package.

6.1.1. Mechanical design

I don't want to give many detailed mechanical designs, listing materials, tubing diameter, etc as their availability is different in every country. Guy Hamblen, AA7ZQ/2 described an attractive 80/75-meter design that uses 12-ft long aluminum tubing sections ranging from 1.5-inch OD to 0.875-inch OD. He also describes the installation details (Ref 7819).

If you consider making a vertical with a rather long non-guyed top section, you can use the ELEMENT STRENGTH MODULE of the ON4UN YAGI DESIGN SOFTWARE. Using the software you can design a Yagi element with a length equal to twice the length you need for the non-guyed top section of the vertical. Because this top section, unlike the Yagi half-element, will not be loaded by its own weight (causing the sag in a Yagi element), the vertical section will have an added safety factor.

Fig 9-65 shows the design for an 80-meter vertical using 4 and 3-inch aluminum irrigation tubing, as designed by

Ninety ft irrigation tube vertical at W7LR in Montana, which is used for both 80 and 160 meters.

Steve Kelly, K7EM. The verticals are mounted on 6″ × 6″ pressure treated lumber. The total length of each post is 12 feet of which 4 feet is in the ground. The arms that hold the verticals in place are made from 2″ × 6″ lumber. Steve used ³/₈″ threaded rod to bind the arms to the posts and a ¹/₂″ threaded rod goes through the base of each vertical (see **Fig 9-66**). The ¹/₂″ rod acts as a hinge for raising and lowering and is insulated from the vertical with PVC tubing. Steve recently replaced the 2″ × 6″ lumber with ¹/₂″ thick Plexiglas sheet. The bottom ends of the 4″ tubing are insulated by 4″ (inside diameter) PVC pipe. Steve mentions splitting this pipe lengthwise, heating it with a special PVC bending blanket and then sliding it over the 4″ irrigation tubing.

To construct his 160 plus 80-meter vertical, Bob Leo, W7LR, uses 4-inch OD irrigation pipe, which comes in 40-foot lengths (see photo). At the joints, he uses a round wooden fence post inside to stiffen the joint. He then adds a piece of the same kind of pipe that has been cut lengthwise around the joint. He then bolts that and adds hose clamps around it. Bob uses nylon guy ropes at each 15-foot point, out in four directions. The rope is fastened to oval shaped hardware, with those ovals fastened with two hose clamps at each point. At the top, Bob uses four 12.8-m long top loading wires (#12 copper wire), sloping at a 45° angle. These loading wires are attached to the top with eye bolts that go into a metal top flange. Bob put a plastic cap over the top end of that flange so birds don't fall down inside the tubing.

6.2. Top-Loaded Vertical

The design of loaded verticals has been covered in great detail in Section 3.6. Capacitive top loading using wires (horizontal or sloping) are quite easily realized from a mechanical point of view. It is more difficult to insert a husky loading coil in a vertical antenna. In addition, because of their intrinsic losses, loading coils are always a second choice when it comes to loading a vertical.

A wire-loaded vertical for 160 meters is described in Section 6.5 as part of an 80/160-meter duoband system.

Inverted-L antennas, which are a specific form of top-loaded verticals, are the subject of Section 6.

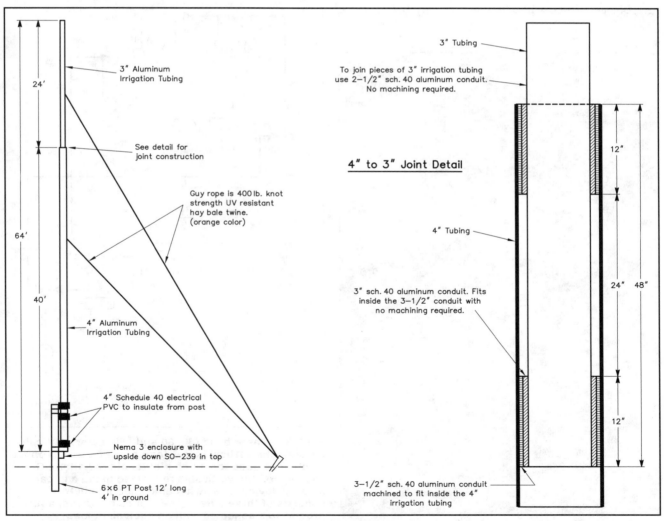

Fig 9-65—Construction details for 80-meter ¹/₄-λ elements made of 4-inch and 3-inch irrigation tubing, as designed by K7EM.

6.3. Three-Band Vertical for 40, 80 and 160 Meters

A $^5/_8$-λ vertical for 40 meters produces a very low angle of radiation, which can be used advantageously when working long-haul DX on that band. The condition for producing that very low angle, and for being an efficient radiator, is to erect the antenna over excellent ground.

The length of a $^5/_8$-λ vertical made of a 25-cm (effective) diameter mast on 7.05 MHz is 24 meters. On 3.65 MHz, this antenna is approximately 0.3-λ long (3.65 MHz was chosen because this antenna will easily cover 3.5-3.8 MHz with less than 2:1 SWR). On 160 meters the antenna is only about 0.15-λ long.

The feed-point impedances can again be found in the charts of Figs 9-8, 9-9, 9-11 and 9-12. I also modeled the antenna with *MININEC*. The impedance of the 24-m vertical (25-cm tower diameter) on the different bands is:

40 meters: $(99 - j\,255)$ Ω
80 meters: $(77 + j\,98)$ Ω
160 meters: $(9.3 - j\,255)$ Ω

Linear loading will add approximately 0.7 Ω to the

Fig 9-66—Base of one of the irrigation-pipe 80-meter verticals as constructed by K7EM.

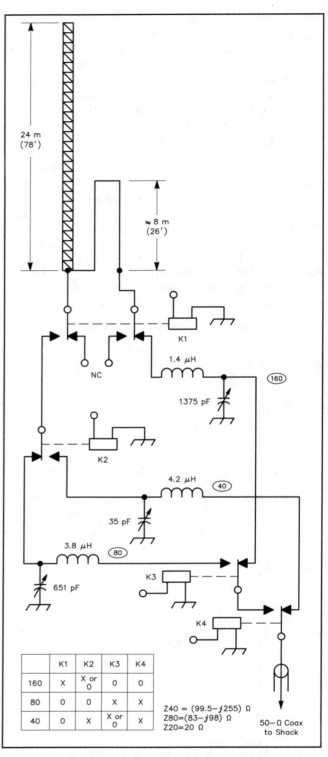

	K1	K2	K3	K4
160	X	X or 0	0	0
80	0	0	X	X
40	0	X	X or 0	X

Z40 = $(99.5 - j\,255)$ Ω
Z80 = $(83 - j\,98)$ Ω
Z20 = 20 Ω

50-Ω Coax to Shack

Fig 9-67—Three-band (40, 80 and 160-meter) vertical system. The vertical is $^5/_8$-λ long on 40 meters. (Longer would introduce too much high-angle radiation.) On 160 meters, the vertical is resonated using a linear-loading device. The switchable matching network consists of three L networks. For calculations, a real ground was assumed with an equivalent loss resistance of 10, 6 and 0.5 Ω respectively on 160, 80 and 40 meters.

radiation resistance on 160 meters. The feed-point impedance becomes: $10 + j\,0\ \Omega$.

For calculating the three-band switchable matching networks, we assume using radials of 20 m length. The equivalent ground-loss resistance for this radial system is estimated at 10 Ω for 160 meters, 6 Ω for 80 meters and 0.5 Ω on 40 meters (see Table 9-1). The antenna with its three-band switchable matching system, comprising three L networks, is shown in **Fig 9-67**.

6.4. Linear-Loaded Duo-bander for 80/160 Meters

Full-size, $\frac{1}{4}$-λ verticals (40-m tall on 160 meters) are out of reach for most amateurs. Often an 80/160-meter duo-band vertical will be limited to a height of around 30 m. This represents an electrical length of 140° at 3.65 MHz and 70° on 160 meters. We can determine R and X from Figs 9-9 and 9-11 or through modeling:

80 meters: $Z = (280 + j\,278)\ \Omega$
160 meters: $Z = (17 - j\,102)\ \Omega$

Fig 9-68 shows the antenna configuration together with the switchable matching system. For calculating the L network for 160 meters, we assume that the linear-loading device will add 1 Ω to the radiation resistance. The assumed ground-loss resistance is 10 Ω on 160 and 5 Ω on 80 meters.

Linear loading provides a higher radiation resistance and

ensures low coil loss, as no large-value inductors are required in the matching network. The exact length of the linear-loading device must be found experimentally, but a good starting point is to make the loading device as long as the antenna would have to be extended to make it a full $\frac{1}{4}$-λ long. An Antennascope or one of the now popular antenna analyzers are good instruments to prune the length of a linear-loading device. An Antennascope will only produce a full null when the measured impedance is purely resistive. Adjust the length of the loading device until a complete null is obtained on the Antennascope or minimum SWR on the antenna analyzer (or 0-Ω reactive impedance component if you are using one of the newer analyzers which allows you to directly measure complex antenna impedances. At the same time you will be able to read the radiation resistance of the antenna on 160 meters from the scale of the instrument. Deduct the theoretical value of the vertical (17 + 1 Ω) and you will have the value of the equivalent ground-loss resistance.

Fig 9-68 shows the two-band L-network matching device for the 160/80 meter antenna. If the vertical is made from lattice-type tower sections, the linear-loading device can be made of two 2.5 cm aluminum tubes spaced about 30 cm apart and about the same distance from the tower. Fig 9-49 shows the linear-loading device on VE3DO's 27-m vertical. A similar 160/80-meter duo-band vertical, using linear loading plus a capacitive top-hat with its switchable tuning network was described by Gale Stewart, K3ND, (Ref 7829).

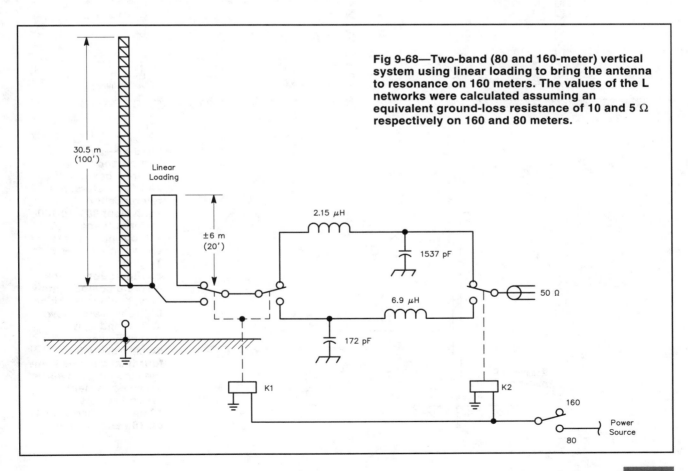

Fig 9-68—Two-band (80 and 160-meter) vertical system using linear loading to bring the antenna to resonance on 160 meters. The values of the L networks were calculated assuming an equivalent ground-loss resistance of 10 and 5 Ω respectively on 160 and 80 meters.

30.5 m (100')

Linear Loading

±6 m (20')

2.15 μH

1537 pF

6.9 μH

50 Ω

172 pF

K1

K2

160

80

Power Source

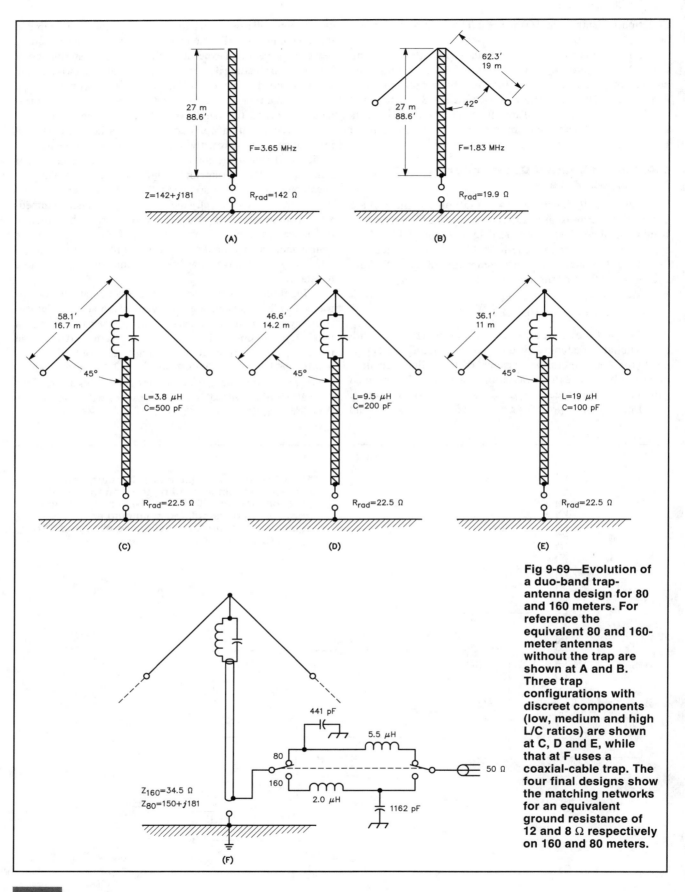

Fig 9-69—Evolution of a duo-band trap-antenna design for 80 and 160 meters. For reference the equivalent 80 and 160-meter antennas without the trap are shown at A and B. Three trap configurations with discreet components (low, medium and high L/C ratios) are shown at C, D and E, while that at F uses a coaxial-cable trap. The four final designs show the matching networks for an equivalent ground resistance of 12 and 8 Ω respectively on 160 and 80 meters.

6.5. 80/160 Top-Loaded Vertical with Trap

Traps are frequency-selective insulating devices, and are incorporated in radiating elements to adapt the electrical length of the element depending on the frequency for which the element is being used.

Commercial multiband antennas make frequent use of traps. Home-made antennas use the technique much more infrequently. There are two types of commonly used traps:
- Isolating traps
- Shortening/lengthening traps

6.5.1. Isolating traps

An isolating trap is a parallel-tuned circuit that represents a high impedance at the design frequency, effectively decoupling (by its high impedance) the "outer" section of the radiator from the "inner" section.

A good isolating trap therefore will meet the following specifications:
- It represents a high impedance on the design frequency.
- It represents as low a Q as possible, together with the high impedance.
- It represents as low a series inductance as possible on the frequencies where the trap is not resonant (minimize inductive loading).

An effective way of making one's own traps is to build a trap of coaxial cable. Several articles describe in detail how to make these traps (Ref 684, 662 and 689). The Battle Creek Special (Section 8.7) uses the coax-cable trap very successfully to make an outstanding three-band (40, 80, 160-meter) compact inverted L.

The example in **Fig 9-69** shows a 27-m vertical mast, measuring 25 cm in effective diameter. We want to use this mast on 80 meters and load it to resonance on 160 meters using two flat-top wires. The trap at the top of the vertical will isolate the loading wires from the mast when operating on 80 meters; it will have to be resonant on 80 meters. Let us design a system that covers 3.5 to 3.8 MHz and see what the performance will be when compared to two monoband systems using basically the same configuration.

Resonance of the trap at 3.65 MHz can be obtained by an unlimited number of L/C combinations:

$$F_{res} = \frac{10^3}{2 \pi \sqrt{L \times C}}$$

where f is in MHz, L in µH and C in pF.

Let us evaluate three different L/C ratios that resonate at 3.65 MHz:

L = 3.8 µH, C = 500 pF (X = 87 Ω)
L = 9.5 µH, C = 200 pF (X = 218 Ω)
L = 19 µH, C =100 pF (X = 436 Ω)

As standards of comparison we'll use the stand-alone 27 m tower (no trap, no flat-top wires) on 80 meters, and the same 27-m tower with two sloping flat-top wires on 160 meters. The dimensions of the 5 different configurations are shown in Fig 9-69.

The impedance at 1.835 MHz (resonance with the loading wires, Fig 9-69B) is 22.5 Ω. At 3.65 MHz (Fig 9-69A) the R_{rad} is 142 Ω (Z = 142 + j 181 Ω). Here we see the advantage of a vertical that is slightly longer than $^1/_4$ λ: improved efficiency. The estimated ground loss for 60 $^1/_4$-λ radials (8 Ω on 80 meters) hardly influences the efficiency:

Fig 9-70 shows the influence of the slope angle of the top wires on the resonant frequency. The values are for a 27-m vertical with two sloping loading wires, 19-m long.

On 160 meters the ground loss of 60 $^1/_8$-λ radials (estimated at 12 Ω) will have a substantial influence in the efficiency:

$$Eff = \frac{22.5}{22.5 + 12} = 65\%$$

The three different trap solutions have a significant influence on the SWR-bandwidth behavior of the antenna. The influence of the L/C ratio is the opposite on 80 meters from what it is on 160. The low-L, high-C solution (3.8 µH and 500 pF) yields the highest bandwidth on 160, and the lowest bandwidth on 80 meters. The opposite is also true. With L = 19 µH and C = 100 pF, the SWR curve on 80 is almost as flat as for the reference antenna (just the 27 m vertical with no trap nor loading wire).

The bandwidth results for the different designs are shown in **Fig 9-71** for both 80 and 160 meters. I have calculated the theoretical bandwidth, excluding the ground losses, as well as the practical bandwidth, including ground losses.

Solution two (9.5 µH and 200 pF, Fig 9-69D) is certainly an excellent compromise if both bands are to be treated with equal attention. The reactance of the circuit for these LC values is X = 218 Ω. Assuming an equivalent series loss resistance of 1 Ω, we end up with a Q factor of 218/1 = 218.

For the parallel capacitor we must use a high-voltage transmitting type (eg, doorknob-type) ceramic capacitor. The winding data for the coil can be calculated using the COIL module of the NEW LOW BAND SOFTWARE.

Example: The 9.5-µH coil can be wound on a 5-cm coil form, with close-wound turns of 3-mm OD double-enameled wire, giving a coil length of 5 cm. The square form will yield a good Q.

In practice you will need slightly less inductance to establish resonance of the trap at 3.65 MHz because of the self-capacitance of the coil. After winding the coil, connect the 200 pF doorknob in parallel (it can be inserted inside the coil form), and grid dip the parallel circuit. Use a receiver to monitor the grid-dip frequency accurately. Remove turns until you obtain resonance at the desired frequency. When all is done, coat the assembly generously with Q-dope.

In recent years coaxial cable traps have become popular. These traps rely upon the inherent capacitance in coaxial cable, along with the inductance that results when the proper number of cable turns are wound on a coil form. Such traps were described by C. Sommer, N4UU (Ref 662). In his design we require a reactance of 173 Ω using RG-58 cable to make the coil. Larry East, W1HUE, published a basic computer program for designing coaxial traps (Ref 692). VE3ERP made a computer program (*HAMCALC 9.4* available for

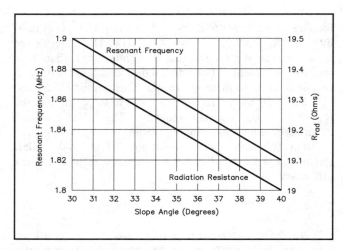

Fig 9-70—This chart shows the variation in resonant frequency and radiation resistance for a 27-m vertical with 19-m long sloping top-load wires, 30 to 40° shifts the resonant frequency by 80 kHz.

$5 from George Murphy, VE3ERP, 77 McKenzie St, Orillia, ON L3V 6A6, Canada) for designing coaxial traps. The detailed design is beyond the scope of this book.

For this example, the inductance of the trap is:

$$L = \frac{173}{2 \times \pi \times 3.65} = 7.5 \ \mu H$$

The equivalent capacitance of the coax trap will be 252 pF. In practice, 14 close-wound turns of RG-58 cable on a 5-cm-diameter coil form, connected as shown in **Fig 9-72**, will yield a trap with resonance near 3.65 MHz.

RG-58 cable has a polyethylene inner-conductor insulation. Polyethylene is a fairly low-temperature material, and the heating of the cable due to $I^2 R$ losses may cause the inner conductor to move through the heated PE and short with the outer shield. Therefore traps using RG-58 or RG-59 cable should only be used with medium power (typically 600 W).

One can build equally compact traps using 50-Ω coaxial cable of the same dimension of RG-58, but using Teflon

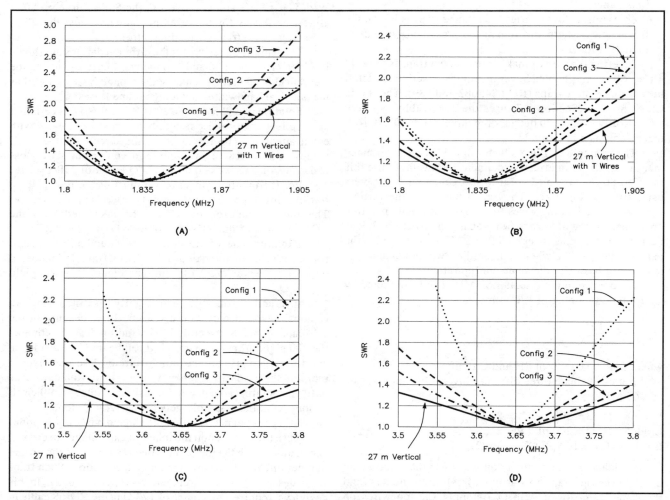

Fig 9-71—SWR curves for 80 and 160 meters for the different antenna configurations of Fig 9-69. Config 1 refers to Fig 9-69C, Config 2 ro Fig 9-69D, and Config 3 to Fig 9-69E. The curves are plotted for perfect ground (at A and C) as well as for real ground (at B and D). Note the increased bandwidth on 160 meters caused by the ground losses and the low R_{rad}. On 80 meters the ground losses have almost no influence due to the high R_{rad}.

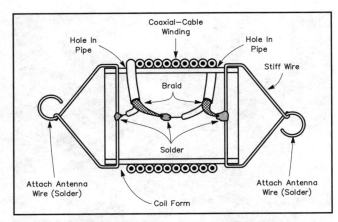

Fig 9-72—Cutaway view of the coaxial-cable trap wound on plastic tubing. At the position of the last turn you can drill four holes around the circumference of the coil form. That will help you in pruning the trap to the desired resonant frequency.

insulation (eg, RG-141 or RG-142). The velocity factor of these cables is slightly higher (0.70) than for RG-58 (0.66), which means that the turns data may be slightly different from the RG-58 version. A trap made with this Teflon (PTFE) insulated coax will easily stand a couple of kW. Make sure you cover the assembly with a generous coat of black vinyl tape or with black colored heat shrink tube (eg, Raychem) to protect the PTFE cable from UV light.

A detailed description on how to make your own high-power coaxial traps is given in Section 5.7.1.

Matching networks

The impedances to be matched are:
80 meters (3.65 MHz): $(150 + j\ 181)\ \Omega$
160 meters (1.835 MHz): $(34.5 + j\ 0)\ \Omega$

The values and the wiring of the switchable network are shown in Fig 9-69F.

6.5.2. Shortening/lengthening traps

If the isolating trap principle were to be used on a triband antenna, it would require two isolating traps. Three-band trap Yagis of the early sixties indeed used two traps on each element half, the inner one being resonant on the highest band, the outer one on the middle band. Modern trap-design Yagis only use a single trap in each element half to achieve the same purpose. Y. Beers, WØJF, wrote an excellent article covering the design of these traps (Ref 680). In this design, the trap is not resonant on the high-band frequency, but somewhere in between the low and the high band. In the balanced design described by Y. Beers, the frequency at which the trap is resonant is the geometrical mean of the two operating frequencies (equal to the square root of the product of the two operating frequencies). For an 80/160-meter vertical, the trap would be resonant at:

$$f = \sqrt{1.83 \times 3.85} = 2.65\ \text{MHz}$$

The principle is that, on a frequency below the trap resonant frequency, the trap will show a positive reactance

(acts as an inductor), while above the resonant frequency the trap acts as a capacitor. A single parallel-tuned circuit can be designed, which inserts the necessary positive reactance (at the lowest frequency) and negative reactance (at the highest frequency). In the balanced design the absolute value of the reactances is identical for the two bands; only the sign is different. There are five variables involved in the design of such a trap system: the two operating frequencies, the trap resonant frequency, the total length, and the L/C ratio used in the trap parallel circuit. The design procedure and the mathematics are covered in detail in the above-mentioned article.

6.6. The Self-Supporting Full-Size 160-M Vertical at ON4UN

A full-size $1/4$-λ vertical antenna for 160 meters is just about the best transmitting antenna you can have on that band, with the exception of an array made of full-size or top-loaded verticals. I use a 32-m triangular self-supporting tower, measuring 1.8 m across at the base, and tapering to 20 cm at the top. I knew that the taper would make the tower electrically shorter than if it had a constant diameter, so that had to be accounted for. On top of the tower I mounted a 7-m-long mast. It is steel at the bottom and aluminum at the top, tapering from 50-mm OD to 12 mm at the top.

In order to make up for the shortening due to the tower taper, I knew I had to install a capacitance hat somewhere near the top of the tower. The highest point I could do this was at 32.5 m. I decided to try a disk with a diameter of 6 m, because I had 6-m-long aluminum tubing available. Two aluminum tubes were mounted at right angles, the ends being connected by copper wire to make a square. **Fig 9-73** shows the vertical.

I hoped I would come close to an electrical $1/4$-λ on 160 meters, and fortunately the antenna resonated on exactly 1830 kHz. In the beginning I had the tower insulated at the base, and was able to measure its impedance, approximately 20 Ω, where 36 Ω would be expected with a 0-Ω earth-system resistance. Such a low radiation resistance has been reported in the literature, and must be due to the large tower cross-section. Originally I suspected mutual coupling with one of two other towers (or both), but decoupling or detuning those towers did not change anything.

Fig 9-74 shows the radiation resistance of a $1/4$-λ vertical over a radial system consisting of 60 $1/4$-λ radials, measured as a function of the diameter of the vertical. From the graph, we see that for a height/diameter ratio of 44 (eg, a self-supporting tower with a diameter of 1 m operating at 1.83 MHz) it shows a radiation resistance of approximately 20 Ω. The classic 36 Ω applies for a very thin conductor!

After a series of unsuccessful attempts to use the vertical on 80 meters, I grounded the tower and shunt fed it using a gamma match. A tap at 8-m height, and a 500-pF series capacitor provided a 1:1 SWR on top band, and a 2:1 SWR bandwidth of 175 kHz. The gamma wire is approximately 1.5 m from he tower. This vertical really plays extremely well. I use quite an extensive radial system, consisting of approximately 250 radials ranging from 18 m to 75 m in length. The tower now also supports a 4-square sloping $1/4$-λ vertical array as described in the chapter on vertical arrays.

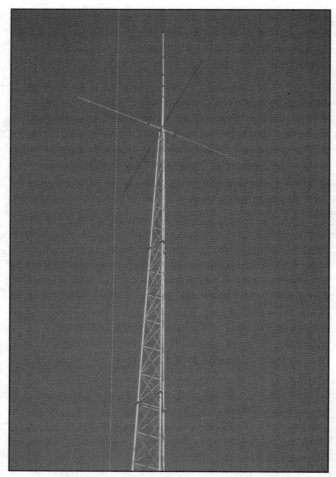

Fig 9-73—Self-supporting 39.5-m ¹/₄-λ vertical for 160 meters at ON4UN. The base is 1.8-m wide and the tower tapers to just a few inches at the top. The tower is shunt fed with a gamma match and also serves as a support for an 80-meter 4-element square array made of ¹/₄-λ verticals, supported from sloping catenary lines running from the 160-meter tower.

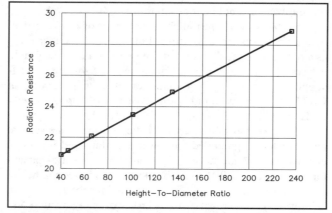

Fig 9-74—The feed-point resistance of a resonant ¹/₄-λ vertical over 60 ¹/₄-λ radials, as a function of the conductor diameter. Verticals made of a large-diameter conductor, such as a tower, exhibit much lower feed-point resistances than encountered with wire verticals.

Fig 9-75—Giving out new countries and chasing new countries on 160 meters are not the only hobbies for Rudi, DK7PE (left) and ON4UN, who are ready to go on a bike trip. In the background is the base of the ON4UN 160-meter vertical showing the cabinet that houses the matching circuitry for the 160-meter vertical and the 4-square 80-meter system.

Fig 9-76—R. Vermet, ON6WU, with his professional antenna measuring setup, tuning the new vertical at ON4UN. An HP Network Analyzer is used, which directly produces a Smith Chart. Although such sophisticated test equipment is not necessary to tune a vertical, it is very instructive to know the impedance (admittance) curve.

Fig 9-76 shows the base of the vertical and the cabinet housing the series capacitor for the 160-m gamma match as well as the hybrid coupler for the 80-meter 4-square array. Information about the ON4UN vertical was obtained with the assistance of ON6WU and his professional-grade test equipment (HP Network Analyzer).

6.7. The Battle Creek Special Antenna

Everyone familiar with DX operating on 160 meters has heard about the Battle Creek Special and its predecessor, the Minooka Special. These antennas are transportable verticals for operating on the low bands. The Minooka Special (Ref 761) was designed by B. Boothe, W9UCW, for B. Walsh, WA8MOA, to take on his trips to Mellish Reef and Heard Island many years ago.

Basically the antennas were designed to complement a triband Yagi on DXpeditions to provide excellent six-band coverage for the serious DXpeditioner. The original Minooka was a 40 through 160-meter antenna, using an L network for matching and an impressively long 160-meter loading coil near the top. WØCD built a very rugged and easily transportable version of the Minooka Special, but soon found out that the slender loading coil simply melted when the antenna was taking high power for longer than a few seconds. No wonder! It was more than 100-cm long with a diameter of only 27 mm. Michaels, W7XR, later calculated the Q factor of the coil to be around 20! That's an equivalent loss resistance of 100 Ω!

WØCD improved the antenna both mechanically and electrically. Instead of developing a better loading coil, he simply did away with the delicate part, and replaced the loading coil with a loading wire. His design uses two sloping wires, one for 80 meters and one for 160, which now makes it really an inverted L, but nothing would prevent you from using a T-shaped loading wire as described in Section 3.6.5.

The new design, named the Battle Creek Special, takes 1.5 kW of RF on SSB or CW without any problem for several minutes. For continuous-duty digital modes the RF output should not exceed 600 W. An 80-meter trap isolates the loading wires for 80 and 160. The Battle Creek Special has been reported by the 3Y5X DXpedition operators to outperform commercially available verticals such as the Butternut with the 160-meter option by a solid two S units!

The section below the 40-meter trap is 9.75 m long, which makes it a full-size quarter-wave on that band. The SWR bandwidth is less than 2:1 from 7 to 7.3 MHz.

On 80 meters the 15 m of tubing below the 80-meter trap, together with the loading wire, make it an inverted L. The antenna will cover 3.5 to 3.6 MHz with an SWR of less than 2:1. On 3.8 MHz the antenna is "too long," but a simple series capacitor of 200 to 250 pF will reduce the SWR to a very acceptable level (typically 1.3:1).

On 160 meters the entire vertical antenna plus the top-loading wire make it a $^1/_4$-λ L antenna. The SWR is typically 2:1 over 20 kHz, indicating a feed-point impedance of approximately 25 Ω (depending to a large extent on the quality of the radial system).

There are several ways to obtain a better match to the feed line. WØCD uses an unun with a 2:1 impedance ratio (see also Chapter 6 on Feed Lines and Matching), which is switched in the circuit on 160 meters, and out of the circuit on 80 and 40 meters. The unun is an unbalanced-to-unbalanced wideband toroidal transformer (Ref 1521 and 1522). WØCD actually built a 9:4 (2.25:1) balun, and removed the top turn to get an exact 2:1 ratio.

Another alternative is to use an L network. A simple tunable L network that has been especially designed for matching "short" 160-meter loaded verticals is shown in **Figs 9-77** and **9-78**. The L network was made by ON7TK and has been traveling around the world on several DXpeditions (A61, 9K2, FOØC, etc).

The Battle Creek Special uses high strength aluminum tubing, 6061-T6 alloy, in sizes ranging from 2 inches to 1 inch (5 to 2.5 cm). The guy lines are 2.4 mm Dacron double-braided rope with a rating of 118 kg breaking strength. Wind survival rating is 160 k/hr assuming proper guy-rope anchors. It is guyed four ways at three levels so the side guy ropes act as a hinge allowing it to be "walked up" by one person.

Original traps were coaxial traps (see Section 6.5.1) but using RG-58 they ran too hot with power levels over 800 W. Instead of going for the Teflon coax the designers decided to switch to regular L/C traps with the "L" being #10 wire and the "C" made from some lengths of RG-213 approx. 100 pF/m. The coaxial capacitors fit inside the aluminum mast sections. A single open-ended coax stub of approx. 90 cm length (90 pF) is used for the 40-meter trap and two parallel-connected pieces of coax of approx. 120-cm length

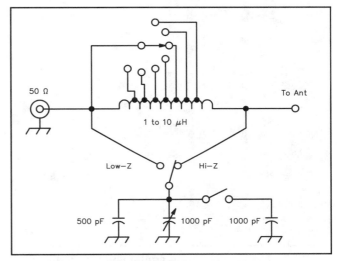

Fig 9-77—L network to be used with Inverted-L antennas and other loaded 160-meter verticals. With the component values shown, impedances in the range 20 – *j*100 to 100 + *j*100 can easily be matched on 160 meters.

Fig 9-78—The L network of Fig 9-77 is contained in a small plastic housing. This particular unit was built by ON7TK and used on several DXpeditions (A61, 9K2, FOØC).

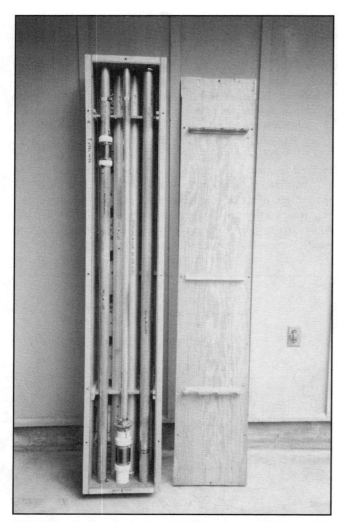

Fig 9-79—The wooden crate containing the Battle Creek Special antenna, a three-band (40, 80 and 160-meter) vertical. The wooden crate is especially designed to ensure safe transportation of the antenna to the most remote parts of the world. It contains all the antenna parts and accessories, such as guy ropes, anchors, hinged base plate, radial wires, etc.

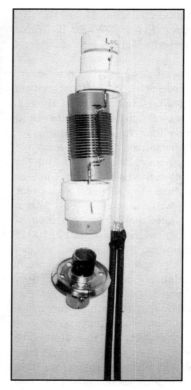

Fig 9-80—The new 80-meter trap for the Battle Creek Special. The original coaxial-cable traps have been replaced with regular LC traps, the capacitor being made by short pieces of coaxial cable. See text for details.

Fig 9-79 shows the wooden crate containing the Battle Creek Special. **Fig 9-80** shows the 80-meter coax-cable trap. The 40-meter trap is 13 turns of 10 gauge copper wire, 4.8-cm ID by 6.35-cm long. The 40-meter trap is tuned to 7050 kHz by using a grid dip meter, starting with a longer than theoretical piece of RG-213 connected across the coil, and trimming the shield to get the capacity to resonate the coil properly. The 80-meter trap uses 17 turns, 10 gauge wire, 6-cm ID by 7.6-cm long. The 80-meter trap is tuned/dipped at 3.525 kHz by pruning the shield on the coaxial cable capacitor(s) the same way.

6.7.1. The BC Trapper

Jeff Briggs, K1ZM, eminent 160-meter DXer and author of *DXing on the Edge* (Ref 511) developed a very husky version of the coaxial-cable traps, using RG-213 coax, and provided a detailed description on how to make your own traps, and how to set up and tune your wire-type Battle Creek Special, which he calls the BC Trapper.

Because the Battle Creek Special is *not* a commercially available product, because of its repeated proof of quality, and because of its ever growing popularity, I want to include a detailed description in this book of how to make the traps and how to assemble and tune the BC Trapper (thanks Jeff, K1ZM). I have made one change, though; that is to use two 160-meter top-loading wires, which makes this a T-loading configuration, which is slightly superior to the inverted-L-loading configuration. The wire version is intended to be used in places where an adequate support is available (eg, a tree).

The BC Trapper is slightly different from the Battle Creek Special as the vertical part is slightly longer (approx.

each (240 pF total) are used for the 80-meter trap.

WØCD recommends using at least 30 radials, each of 20-m length. I would consider this a bare minimum. The Battle Creek Special is not for sale, but is available for loan to DXpeditions to rare countries. Interested and qualified DXpeditioners should contact W8UVZ for further details. The antenna was used at Bouvet on 80 and 160 meters in 1989/90, and during the DXpeditions to ZSØZ, 7P8EN, 7P8BH, G4FAM/3DA, 3Y5X, 5X4F, ZS8IR, XRØY, VP8SGP, YKØA, 8Q7AJ, VKØIR, ZS9Z, V51Z, P40GG, CY9AA, ZS6EX, ZS6NW, AHØ/AC8W, AL7EL/KH9, XF4DX, AH1A, 3YØPI, 9MØC, J37XT, K5VT/JT, VK9LX, ZK1XXP, 3B7RF and many other locations with great success.

The entire antenna, with its base, guy-wires and radials is packed in a strong wooden case for safe transport to the remotest DXpedition spot. The package weighs 30 kg (66 lb).

19 m) which makes it unnecessary to add a (short) loading wire below the 80-meter trap. This extra length (it is approx. 4 m longer than the Battle Creek Special) makes this an even better performer on 160 meters.

Even if you do not have a 19-m tall support you can install the BC Trapper in a slightly sloping fashion, or you may want to add a short loading wire just below the 80-meter trap as shown in **Fig 9-81**. A minimum support height of 15 m is recommended.

Parts list for the 80-meter trap

- 18-ft RG-213 (PE dielectric) 50-Ω coax
- About 2 feet (60 cm) of schedule 40, 4" (10-cm) diameter PVC septic line (schedule 40 is relatively thin wall and lighter in weight than schedule 80 pipe)
- Four end-caps for 4" diameter PVC
- Four ft (120 cm) of #12 THHN stranded copper wire
- Four rolls of PVC electrical tape
- One tube of GE silicone sealant—the "Home Line"
- Four $^1/_4$" eye hooks, eight $^1/_4$" flat washers, four $^1/_4$" nuts, four $^1/_4$" lock washers (6 mm hardware can be used as well)
- A small can of PVC pipe cement

Trap assembly

- Prepare two lengths of RG-213 as follows: Cut two lengths, one to 11 feet (335 cm) and the other to 6′ 3.5″ (192 cm). Separate the braid from the PE center dielectric in order to create "pigtails," 2.75″ (7 cm) long at each of the four ends of coax. The 11 ft (335 cm) length will be used for the 80-meter trap and the 6′ 3.5″ (192 cm) length will be used for the 40-meter trap.
- Place a PVC end cap temporarily on one end of the 4″ PVC tube.
- Size the actual length of PVC pipe required by temporarily winding the 11 ft (335 cm) length of RG-213 around the PVC tube. In order to wind up with the smallest required amount of PVC, start the coil about $^1/_2$ inch (13 mm) below the point on the PVC where the end cap ends.
- The RG-213 should be tightly wound with each turn directly adjacent to the preceding one. It may be necessary to tape the first couple of turns temporarily to the PVC in order to effect this.
- When all turns have been wound, estimate the additional length of PVC required to accommodate the lower PVC end cap. Now remove the RG-213 and cut the PVC pipe to its final length.
- With one end cap still in place, drill a hole in the PVC at the point at which you started winding the coil. This hole should be just large enough to offer a snug fit for the RG-213 that will be threaded through it later.
- When the hole size is right, thread one end of the coax pigtail into the PVC (from the outside of the coil) such that all of the 2.75″ (7 cm) of the coax pigtail plus about $^3/_4$″ (2 cm) of black coax outer jacket fit inside the PVC tube.
- Now wind the coil as tightly as you can, and use electrical tape to hold the turns in place if required. Some dexterity is required.

- When you are nearing the bottom of the coil, you will need to estimate where to drill the lower hole for the coax to enter the PVC tube.
- Be sure to allow sufficient length of the lower coil coax end to enter the PVC tube so that the pigtails can reach each other inside the coil! This is necessary in order to make the internal connections inside the PVC pipe. If the lower pigtail needs to be increased in length to achieve this, separate a wee bit more braid from the PE insulation.
- When you are sure you have estimated it correctly, drill your lower hole, insert the lower end of the coax into the PVC and test the pigtail lengths to ensure they will actually reach each other comfortably inside the PVC. Now tape your lower turns snugly to the PVC in order to hold them in place while you make your connections.

Making the connections

- To form the coaxial capacitor inside the coil, connect the braid from one pigtail to the center conductor of the other pigtail. Do not connect braid to braid or center conductor to center conductor!
- Initially, this connection should not be soldered to allow for grid dip meter testing of the trap's resonant frequency. I have found that these dimensions typically resonate anywhere between 3.490 and 3.525 MHz.
- Should they resonate significantly *lower*, then the coax can be trimmed and a new hole drilled for the lower coil pigtail entry. If they resonate much higher, you are in bad luck.
- When you are happy with the trap's resonant frequency, twist these wires firmly together and solder them well! Do your very best to provide as much tension as possible between the lower and upper pigtails as you do this. In other words, do not leave any slack inside the PVC tube. If you get it just right, there will actually be some tension inside the PVC holding the outer coil windings tight.
- This leaves you with two leftover ends of the pigtails—one braid and one center conductor. One should be angled toward one end of the PVC and the other toward the opposite end of the PVC.
- Now solder a 12″ (30-cm) piece of #12 (2-mm) THHN stranded wire to each of these conductors (the braid and the center conductor). Twist these wires firmly together and solder them well as they will never be accessible once you close up the PVC tube—and they will be handling 1.5 kW!
- Liberally seal the holes where the coax enters the PVC from the inside, using GE silicone glue/sealant. It's also a good idea to cover the soldered connections inside the PVC with the sealant as well to ensure they don't float around inside somehow—although proper tension on these leads should not allow this to occur.

Preparing the end caps

- Drill a hole in the center of each end cap just large enough for the $^1/_4$" (6-mm) eye hooks to fit through. Mount the eye hooks to the end caps using one flat washer on the inside and one flat washer on the outside of the end cap.

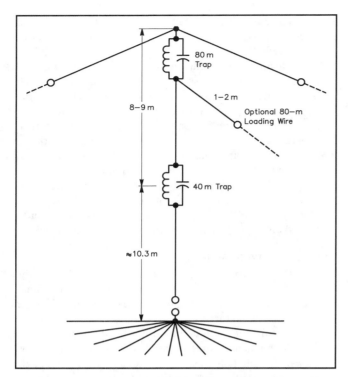

Fig 9-81—This version of the BC Trapper is about 18 to 19-m long. If resonance on 3.5 MHz cannot be established (support not high enough) add an 80-meter loading wire as shown. It would be typically 1 to 2-m long. Slope this wire at a right angle to the 160-meter loading wires. The slope angle of the 160-meter T loading wires greatly influences the resonance of the antenna. You can actually use this to fine-tune the antenna on Top Band.

Tighten the eye hook in place using the ¹/₄" (6-mm) nut inside the coil. Get this tight inside and use a lock washer if in doubt! These eye hooks are going to hold the weight of the traps and the weight of the wire as well, so they need to be tightly affixed inside the PVC tube. They too will be inaccessible later!

• Now drill a hole somewhere in the PVC end cap just large enough for the THHN #12 (2-mm) wire to snugly pass through.

• When both PVC end caps have been so prepared, it's time to glue them onto the PVC. This requires some dexterity, as you must first pass the open end of the #12 wire through the hole in the end cap and spread PVC cement on the inside of the end cap and the top of the PVC as well—all at the same time.

• Since PVC cement dries in seconds, this needs to be done quickly. Push the PVC end cap onto the PVC as far as it will go and pull the #12 THHN lead snug as you do this. You should end up with about 8" (10 cm) of THHN wire outside the coil when completed.

Finishing off the trap

• You should end up with a tightly wound coaxial coil around the PVC and the GE sealant should keep water out of the

coil. However, to be doubly sure, I now tape the entire outer surface of the coil with PVC electrical tape, wrapping it tightly. Usually I expend 1-2 rolls of tape per trap to keep the coil turns snug and to keep water from ever seeping into the assembly.

• I also seal the end caps liberally around the eye hook flat washer and around the eye hook where it exits the end cap. Also liberally seal the point where the #12 (2-mm) THHN wire exits the PVC tube through the end cap. This I do several times—at about 6 hour intervals to allow the GE silicone sealer to set properly between each coating.

The 40-meter trap

The 40-meter trap should be constructed exactly the same way, but using the 6' 3.5" (192 cm) piece of RG-213. This length coax should yield a 40-meter trap resonant around 7.000 MHz, plus or minus. Again, trim to final length with a grid dip meter, if necessary.

Assembling the BC Trapper

• It is essential to use rather heavy gauge wire for the "ground to 40-meter trap" section, as this is where the highest current flows. For the remainder of the antenna, smaller gauges, such as #14 or #16 can be used (the issue here is mechanical strength rather than current carrying capability). The dimensions of the three required THHN wires are as follows:
 1. Ground level to 40-meter trap: 10.3 m. Cut to 12 m initially!
 2. 40-meter trap to 80-meter trap: 8 to 9 m. Cut to 9.5 m initially!
 3. 80-meter trap to end insulator: two pieces of approx. 11 m. Cut to 13 m initially!

• The various sections of wire should be tied and knotted around the ¹/₄" (6-mm) eye hooks on the ends of the traps. Allow enough wire on the dead ends to connect to the pigtail lead of #12 (2-mm) THHN exiting the PVC end caps for making your connections. Initially do not solder these connections, just twist them well!

• Mount the BC Trapper (eg, in a tree), and make sure to pull the tails of the T wires (160-meter loading wires) out into their approximate final position when doing all tuning, testing and trimming. If you let them dangle, you will be sorry later as it makes a real difference in the final tuning of the 160-meter section of the T wires!

• Start with 40-meters and trim the 10.3-m section to final length on 7.0 MHz, as indicated by minimum SWR. One of the popular antenna analyzers is a handy tool for this, but you can also do it with your transmitter and an SWR bridge.

• Next comes the 80-meter section. If your support is high enough, try to resonate the antenna without the use of a loading wire. Trim the 9.5-m long wire for resonance on 3.500 MHz. Trimming this wire length should not impact the resonant frequency on 40 meters if the 40-meter trap is working properly. If your support is not high enough, and you cannot obtain resonance on 3.5 MHz, you will have to add a short wire (1 to 2-m long) at the top of this wire, just below the 80-meter trap. Slope this wire to the ground at an angle of approx. 45° in the direction opposite to the 160-meter loading wire.

- Now move to 160 meters and trim the T wires (as symmetrically as possible) to achieve resonance on 1.830 MHz (or your preferred operating frequency). Trimming these wires should not impact the resonant frequency on 40 or 80 meters if both traps are working properly.
- Now go back and recheck 40, 80 and again 160. When you do this, be sure to hang your T-loading wires in their final elevated position.
- When you are happy with the results, solder all connections and tape these points liberally with PVC tape.

Radial system

A minimum of thirty 20-m long (or longer) radials on, or in, the ground are recommended (the more, the better, though). In fact the N7CL rule (see Section 2.1.4) applies. This means that with 20-m long radials, and considering both 80 and 160, you will be almost 0.5 dB better with 52 radials than with half of them! Thirty is indeed a strict minimum.

You can also use an elevated radial system, and as few as *one* radial per band will do fine in collecting the antenna return currents. Do *not*, I repeat *not* use just one or two radials over "any" ground, you will be trading in about 5 or 6 dB of signal strength. A few elevated radials are perfect *over salt water*, or over a large ground screen only (see Section 2.2). If you use such a ground screen (in the shape of radial wires, chicken wire, etc) do not connect these radials or nets anywhere to the antenna or antenna feed line (see Section 2.2.12 and 2.2.13).

Using one or two elevated radials (per band) also makes changing from the CW to the SSB portion of the band very easy.

This system, however. makes the tuning procedure of the vertical element somewhat more complicated, as the resonant frequency on the different bands is now not only determined by the length of the vertical antenna but also by the length of the radials (although it is *not* important that the vertical radiator be exactly ¼ λ, it is the radiator and the radial system together that have to be resonant for ease of matching). Therefore, even when an elevated radial system is contemplated, first tune the antenna against ground (which is nonresonant!), using a minimum of a few radials or a chicken wire net.

SWR values

It is likely that on both 40 and 80 meters the impedance of the antenna will be fairly close to 50 Ω, depending on the equivalent loss resistance of the ground system. On 160 meters the minimum SWR will likely be between 1.5:1 and 2:1 again depending on the ground system losses. In this case an appropriate unun transformer should be used to obtain a lower SWR (Ref 1522).

The BC Trapper antenna, as described, is resonant in the CW part of 40 and 80 meters. The easiest way to tune the antenna for the phone end of these bands is to connect a series

capacitor between the base of the antenna and the feed line on SSB. As the antenna, cut for the low end of the band, will be too long, which means that it will show an inductive-reactance component in its impedance, this inductive reactance can be tuned out by connecting a capacitor in series, having the same reactance (but opposite sign).

Table 9-6 shows the estimated values for switching from CW to SSB. These capacitors should be transmitter-type capacitors (high current, high voltage ceramic). With the antenna tuned to 7.0 MHz it is impossible to obtain a perfect match on 7.2 MHz with just a series capacitor. But a SWR of better than 1.4:1 should be possible all the way up to 7.1 MHz by inserting a series capacitor with a value of approx. 250-300 pF.

On 80 meters a min. SWR of approx. 1.6:1 should be obtainable on 3.8 MHz using a series capacitor of approx. 100 pF. Watch out—this should be a fairly high voltage capacitor! To cover both bands, and all frequencies, one could use a 0-500 pF or 0-1000 pF vacuum variable (5 kV min.), which could be adjusted for min. SWR anywhere in the band (both on 40 and 80/75 meters). On the CW end this capacitor should be short circuited.

Note that for the Battle Creek Special, using tubing for the vertical conductor (lower Q), the capacitor value needed to tune out the inductive reactance on 3.8 MHz will be larger (typically 200 pF).

One could also simply use an antenna tuner, but in that case a high SWR will exist on the feed line. While RG-58 can be used on 160 meters with 1500 W (intermittent duty), this holds only true for a line with low SWR. This consideration can be important for those who want to use such antennas on a DXpedition (use of RG-58 dictated by weight limitation).

If you use two elevated radials (per band), the change from CW to SSB becomes very simple: just shorten the radial wires an equal amount on both sides. This will perfectly shift the resonant frequency of the antenna. In a temporary set-up (eg, a DXpedition), this can be done by connecting, or disconnecting, the outer ends of the radials using alligator clips. As these points are high voltage points, current capability concerns using alligator clips are out of order.

Table 9-6

SWR and Capacitor Values for Resonating the BC Trapper for Phone Operation

Frequency	SWR without Series C	Capacitor Value	SWR with Series C
3.5 MHz	1.2:1	none (short)	1.2:1
3.6 MHz	11:1	≈300 pF	1.0:1
3.7 MHz	32:1	≈140 pF	1.3:1
3.8 MHz	61:1	≈100 pF	1.6:1
7.0 MHz	1.1:1	none (short)	1.1:1
7.080 MHz	2.2:1	≈300 pF	1.2:1

Fig 9-82—BC Trapper antenna, made by K1ZM, ready to go up. The 40 and 80-meter traps use RG-213 coaxial cable to make the traps that will withstand a couple kW with no problem.

Using smaller Teflon coax

The same construction method as outlined above, could of course be used for a design with RG-141 or RG-142 Teflon insulated coax (diameter approx. 5 mm). The capacitance per unit of length of this cable is about 5% less than for RG-213, which means that for the same resonant frequency slightly more inductance (coax length) will be required. The power rating of such traps will be the same as for traps made with RG-213, but the weight will be significantly less. To be on the safe side and to prevent possible cold flow of the central conductor, it is best to make the diameter of the trap not less than 7.5 cm. The higher quality RG-400 (M17/128-RG-400) uses 19 strands of silver covered copper as a central conductor, as opposed to the RG-141 and RG-142 cables which use silver covered copper-clad steel. This makes the RG-400 less springy and suitable for a trap with a much smaller diameter (eg, 5 cm).

A word of caution

Wire antennas such as the BC Trapper will often be erected using trees as supports. Be aware that high voltages are present on the outer ends of any antenna. Bare antenna conductors, when touching dry branches or leaves can set them on fire. It is not a bad idea to use Teflon coated wires when the antenna is likely to touch branches or leaves. This should certainly give an additional level of protection.

6.8. Using the Beam Tower as a Low-Band Vertical

The tower supporting the HF antennas can make a very good loaded vertical for 160 meters. A 24-m tower with a triband or monoband Yagi, or a stack of Yagis, will exhibit an electrical length between 90° and 150° on 160 meters. These are lengths that are very attractive for low-angle work on 160.

6.8.1. The electrical length of a loaded tower

Fig 9-83 can be used to assess the electrical length of a tower loaded with a Yagi antenna. The chart shows the situation for a tower, with an effective diameter of 30 cm, loaded with five different Yagis, ranging from a 3-element, 20-meter Yagi to a 3-element 40-meter full-size Yagi. These figures are for Yagis that have there elements electrically connected to the boom! A KT-34 will show little capacity loading, because all elements are insulated from the boom.

A 24-m tower, loaded with a 5-element, 20-meter Yagi, will have an electrical length of 103° on 1.825 MHz. The effect of capacitive top loading depends to a great extent on the diameter of the tower under the capacitance hat. The capacitance hat (the Yagis) will have a greater influence with "slim" towers than with towers having a large diameter. If we increase the tower diameter to 60 cm, this will shorten the electrical length between 4° and 7° (4° for the tower loaded with the 3-element, 20-meter Yagi, and 7° for the tower loaded with the 40-meter, 3-element full-size Yagi).

W. J. Schultz, K3OQF, published the mathematical derivation of the shunt fed top-loaded vertical (Ref 7995).

There are no data nor formulas available for calculating the electrical length of a tower loaded with multiple Yagis. The best way to find out is to attach a drop wire to the very top of the tower (turn it into a folded element) and grid dip the entire structure as shown in **Fig 9-84**.

If your tower, top loaded with a Yagi, is still a little short, you may want to add some extra wires from the top of the tower sloping down to increase the top loading (maybe use part of the top set of guy wires).

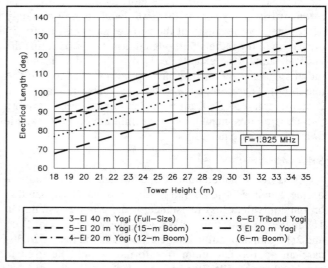

Fig 9-83—Electrical length of a tower loaded with a Yagi antenna. The chart is valid for 160 meters (1.825 MHz) and a tower diameter of 30 cm. For a larger tower diameter, the electrical length will be shorter (4 to 7° for a tower measuring 60 cm in diameter).

6.8.2. Measuring the electrical length

A second and very practical method of determining the resonant length of a tower system was given by DeMaw, W1FB (Ref 774). A shunt-feed wire is dropped from the top of the tower to ground level. What you do is turn a grounded single-conductor vertical into a folded-element vertical, where you now can easily do measurements in the drop wire. Attach a small 2-turn loop between the end of the wire and ground and couple this loop to the grid dip meter (Fig 9-84).

The lowest dip found then is the resonant frequency of the tower/beam. The electrical length at the design frequency is given by:

$$L \text{ (in degrees)} = 90° \frac{f_{design}}{f_{resonant}}$$

Therefore, if $f_{resonant} = 1.6$ MHz and $f_{design} = 1.8$ MHz, then L = 101°.

6.8.3. Gamma and omega matching

There are many approaches to matching a loaded, grounded tower. Three popular methods are
- Slant-wire shunt feeding (Section 6.8.4)
- Folded monopole feeding (Section 6.8.5)
- Gamma or omega-match shunt feeding.

Gamma and omega-matching techniques are most widely used on loaded towers. The design of gamma matches has frequently been described in the literature (Ref 1401, 1414, 1421, 1426 and 1441).

Fig 9-85 shows the height of the gamma-match tap as well as the value of the gamma capacitor for a range of antenna lengths varying from 60° to 180°. The chart was developed using a gamma wire of 10-mm diameter. There are three sets of graphs, for three different wire spacings.

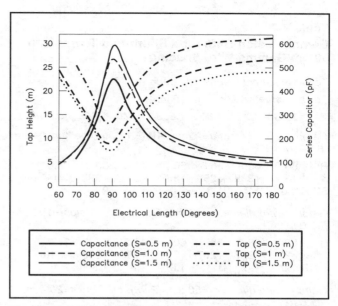

Fig 9-85—Tap height and values of the gamma series capacitor for a shunt-fed tower at 1.835 MHz. The tower diameter is 250 mm, and the gamma wire has a diameter of 10 mm. Three sets of curves are shown, for three spacings (S). The spacing is the distance from the wire to the tower center.

6.8.3.1. Close spacing versus wide spacing

The wider the spacing, the shorter the gamma wire needs to be. Shorter gamma wires will logically show less inductive reactance, which means that the series capacitor will be larger in value.

Electrically very long verticals will require a tap which is 20 to 30 m up on the tower. The required series capacitor will be small in value (typically 100 to 150 pF). There will be a very high voltage across capacitors of such small value.

In case the required gamma-wire length is longer than the physical length of the tower, an omega match will be required (see Section 6.8.3.5).

6.8.3.2. Influence of gamma-wire diameter

The gamma-wire diameter has little influence on the length of the gamma wire (position of the tap on the tower). A larger diameter wire will require a somewhat shorter gamma wire. The wire diameter has a pronounced influence on the required gamma capacitor. It also has some influence on the SWR bandwidth of the antenna system, but less than most believe.

6.8.3.3. SWR bandwidth

Tables 9-7 and **9-8** show the feed-point impedance and the SWR versus frequency for a vertical of 100° electrical length, fed with a gamma match. A spacing of 50 cm is used in Table 9-7, and 150-cm spacing in Table 9-8. Wire diameters of 2 mm (AWG 12), 10 mm, 50 mm and 250 mm are included. The 2-mm (AWG 12) wire is certainly not responsible for a narrow bandwidth. It does not seem worth using

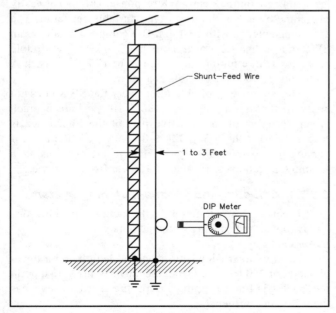

Fig 9-84—A method of "dipping" a tower with a shunt-feed wire connected to the top.

Table 9-7
Gamma-Match Data for a Shunt-Fed Tower with 50-cm Gamma-Wire Spacing

Tower electrical height=100 degrees

Tower diameter=250 mm (10 in.)

	1.730	1.765	1.800	1.835	1.870	1.905	1.940 MHz

Gamma-wire diameter=2 mm (AWG 12); tap height=19.5 m (64.0 ft)

	1.730	1.765	1.800	1.835	1.870	1.905	1.940
R	80.6	66.8	56.5	50.0	43.3	39.2	36.0
X	+330	+338	+350	+363	+377	+392	+407
SWR	2.0	1.7	1.3	1.0	1.4	2.0	2.8

Gamma-wire diameter=10 mm (0.4 in.); tap height=19.8 m (65.0 ft)

	1.730	1.765	1.800	1.835	1.870	1.905	1.940
R	82.9	68.6	58	50	44.8	40.6	37.4
X	+250	+257	+267	+278	+291	+303	+316
SWR	1.9	1.6	1.3	1.0	1.3	1.8	2.4

Gamma-wire diameter=50 mm (2 in.); tap height=20.0 m (65.6 ft)

	1.730	1.765	1.800	1.835	1.870	1.905	1.940
R	80.8	66.9	56.9	50.0	44.3	40.3	37.3
X	+164	+171	+179	+188	+198	+208	+218
SWR	1.8	1.5	1.2	1.0	1.3	1.6	2.1

Gamma-wire diameter=250 mm (10 in.); tap height=20.2 m (66.3 ft)

	1.730	1.765	1.800	1.835	1.870	1.905	1.940
R	78.8	65.5	56	50	44	41	38.3
X	+75	+82	+90	+98	+105	+113	+121
SWR	1.8	1.5	1.2	1.0	1.2	1.5	1.8

Table 9-8
Gamma-Match Data for a Shunt-Fed Tower with 150-cm Gamma-Wire Spacing

Tower electrical height=100 degrees

Tower diameter=250 mm (10 in.)

	1.730	1.765	1.800	1.835	1.870	1.905	1.940 MHz

Gamma-wire diameter=2 mm (AWG 12); tap height=11.9 m (39.0 ft)

	1.730	1.765	1.800	1.835	1.870	1.905	1.940
R	86.8	71.0	59.0	50.0	43.7	38.8	35.0
X	+226	+229	+36	+244	+53	+262	+272
SWR	1.8	1.5	132	1.0	1.2	1.6	2.1

Gamma-wire diameter=10 mm (0.4 in.); tap height=12.0 m (39.4 ft)

	1.730	1.765	1.800	1.835	1.870	1.905	1.940
R	87.8	71.7	59.8	50.0	44.4	39.5	35.7
X	+179	+181	+187	+195	+203	+212	+220
SWR	1.8	1.5	1.3	1.0	1.2	1.6	2.0

Gamma-wire diameter=50 mm (2 in.); tap height=12.0 m (39.4 ft)

	1.730	1.765	1.800	1.835	1.870	1.905	1.940
R	87.0	71.0	59.0	50.0	44.3	39.5	35.8
X	+1230	+132	+137	+44	+152	+159	+166
SWR	1.8	1.5	1.2	1.0	1.2	1.5	1.8

Gamma-wire diameter=250 mm (10 in.); tap height=11.9 m (39.0 ft)

	1.730	1.765	1.800	1.835	1.870	1.905	1.940
R	85.2	69.6	58.3	50	44	39.4	36
X	+79	+82	+87	+93	+100	+106	+112
SWR	1.8	1.5	1.2	1.0	1.2	1.5	1.7

Table 9-9
Gamma-Match Data for a Shunt-Fed Tower with 150-cm Gamma-Wire Spacing

Tower electrical height=100 degrees

Tower diameter=250 mm (10 in.)

	1.730	1.765	1.800	1.835	1.870	1.905	1.940 MHz

Gamma-wire diameter=10 mm (0.4 in.); tap height=25.9 m (65.0 ft)

	1.730	1.765	1.800	1.835	1.870	1.905	1.940
R	43.1	45.2	47.7	50.0	54	58	62.7
X	+567	+597	+628	+661	+697	+736	+778
SWR	6.0	3.5	1.9	1.0	2.0	3.7	6.3

Gamma-wire diameter=250 mm (10 in.); tap height=24.8 m (81.4 ft)

	1.730	1.765	1.800	1.835	1.870	1.905	1.940
R	41.5	43.8	46.6	50	54	58.54	64.0
X	+286	+303	+320	+340	+362	+384	+409
SWR	3.2	2.2	1.5	1.0	1.5	2.2	3.2

a "wire cage" gamma-wire to improve the bandwidth.

For loaded towers that are much longer, the bandwidth behavior is quite different. The longer the electrical length of the vertical, the narrower the SWR bandwidth. **Table 9-9** shows the feed-point impedance and the SWR for a vertical of 150° electrical length, fed with a gamma-match and a gamma-wire of both 10 mm and 250 mm OD. In contrast with the effect on the shorter vertical (100°), the wire diameter now has a pronounced influence on the bandwidth. The 10-mm wire yields a 70-kHz bandwidth, the 250-mm wire cage almost 130 kHz. As can be seen from the impedance values listed in Table 9-9, it is the large variation in reactance that is responsible for the steep SWR response. This can be easily overcome by using a motor-driven variable capacitor. The 150° long antenna with a 10-mm-OD gamma wire shows an SWR of less than 1.3:1 over more than 200 kHz, if a variable capacitor with a tuning range of 100 to 175 pF is used. A high-voltage vacuum variable is a must.

This simple way of obtaining a very flat SWR does not apply to the shorter verticals (90°-110°), where a much larger variation of the resistive part of the impedance is responsible for the SWR. **Fig 9-86** shows SWR plots for gamma-fed towers of varying electrical length, using a 10-mm OD gamma wire, spaced 150 cm from the tower.

6.8.3.4. Adjusting the gamma-matching system

The easiest way to fine tune the gamma-matching system is to vary the spacing of the gamma wire.
Example:

For a vertical with 100° electrical length and a tower diameter of 250 mm, we install the tap at 14 m. At that point the spacing is 1 m. Changing the spacing at ground level has the following influence:

Spacing = 0.5 m: Z = (38 + j 206) Ω
Spacing = 0.75 m: Z = (44.8 + j 298) Ω

Spacing = 1.0 m: Z = (49.3 + j 211) Ω
Spacing = 1.25 m: Z = (56.4 + j 213) Ω
Spacing = 1.5 m: Z = (61.5 + j 214) Ω

This demonstrates how fine tuning can easily be done on the gamma-matching system.

6.8.3.5. Using the omega matching system

If you can use a gamma, I would not advise an omega-system. The omega match requires one more component, which means additional losses and additional chances for a component breakdown. It is possible, however, to use a gamma-rod (wire) length that is up to 50% shorter than the length shown in Fig 9-85 when an omega match is employed. In this case a parallel capacitor will be required between the bottom end of the gamma wire and ground.

The 100° long vertical requires a 14-m long gamma wire, with 100-cm gamma-wire (OD 10 mm) spacing. If we shorten the gamma wire to 8 m, the impedance becomes (14.1 + j 127) Ω. This can be matched to 50 Ω using an L network. One of the solutions of this L network consists of two capacitors: the well-known parallel and series capacitor of the omega-matching system.

To calculate the omega-system, the following procedure should be used:

- Model the vertical with the "short" gamma rod. Make sure you use enough segments (pulses). For 160 meters, segment lengths of 100 cm give good results. Note the input impedance, which will be lower than 50 Ω, and inductive).
- Use the L NETWORK module of the NEW LOW BAND SOFTWARE to calculate the capacitance of the parallel and the series capacitor.

In our example above, the 8-m-long gamma wire requires a parallel capacitor of 369 pF and a series capacitor of 323 pF.

If you have a physically short tower with a lot of loading, it may be that the required tap height is greater than your tower height. In this case an omega match is the only solution (if you have already tried a larger spacing).
Example: See **Fig 9-87**. The tower was "dipped" and the electrical length turned out to be 140°. The physical height is 24 m. Fig 9-85 shows a required gamma-wire length of 30 m for a 2-mm OD gamma wire and a 50-cm spacing. In this case we will connect the gamma wire at the top of the tower (h = 24 m). Using *MININEC*, we calculate the feed-point impedance as Z = 17.2 + j 579 Ω. From the L NET-WORK software module, the capacitor values are calculated as C_{par} = 62 pF, C_{series} = 88 pF. Note that these very low-value capacitors will carry very high voltages across their terminals with high power.

This L network is a very high-Q network. **Table 9-10** lists the impedances at the end of the gamma wire before and after transformation by the capacitors of the omega-match system (an L network using two capacitors). Note the very narrow bandwidth of this extremely high-Q matching system. If we adjust the omega-capacitors for a 1:1 SWR on 1835 kHz, the 2:1 bandwidth will be typically 20 kHz! If we make the series capacitor adjustable (60 to 120 pF), we can tune the antenna to an SWR of less than 1.5:1 over more than 200 kHz.

6.8.3.6. Conclusion

If you have an electrically long vertical, it pays to use a large-diameter cage-type gamma wire and a large wire-to-tower spacing. Making the series capacitor remotely tunable will certainly make the antenna much more broadbanded. Do not shorten the gamma wire unless required because of the

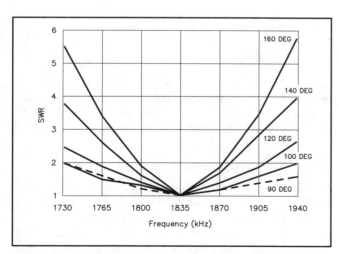

Fig 9-86—SWR curves for gamma-fed towers using a 10-mm OD gamma wire and a spacing of 50 cm, for electrical tower lengths varying from 90° to 160°. The SWR bandwidth of the longer vertical can be "tuned" to a very low SWR over a wide bandwidth by using a motor-driven variable series capacitor.

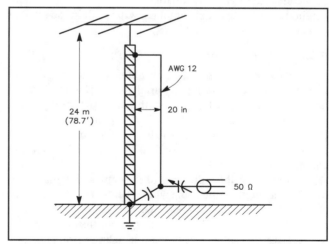

Fig 9-87—A shunt-fed tower using an omega-matching system. The tower is electrically 140° long. An omega match is required, as the tower is physically too short to accommodate a gamma match with a 2-mm gamma wire. Table 9-10 lists the impedances at the end of the gamma wire before and after transformation by the capacitors of the omega-match system.

Table 9-10
Omega-Match Data for a Shunt-Fed Tower with 50-cm Gamma-Wire Spacing

Tower electrical height=140 degrees

Omega-wire diameter=250 mm (10 in.)

	1.730	1.765	1.800	1.835	1.870	1.905	1.940 MHz

Gamma-wire diameter=2 mm (AWG 12); tap height=24.0 m (78.7 ft)

R	16.0	16.3	16.7	17.2	17.9	18.6	19.3
X	+514	+535	+552	+579	+603	+629	+650

With parallel capacitor of 62 pF added

R	37.4	40.7	44.8	50.0	56.8	65.3	74.7
X	+785	+845	+910	+986	+1073	+1178	+1287

With fixed series capacitor of 88 pF added

R	37.4	40.7	44.8	50.0	56.8	65.3	74.7
X	−261	−180	−95	0	+106	+228	+348
SWR	38.0	17.9	5.9	1.0	5.8	17.9	35

With variable series capacitor, 50 to 125 pF (adjusted to cancel inductive reactance)

R	37.4	40.7	44.8	50.0	56.8	65.3	74.7
X	0	0	0	0	0	0	0
SWR	1.3	1.2	1.1	1.0	1.1	1.3	1.5

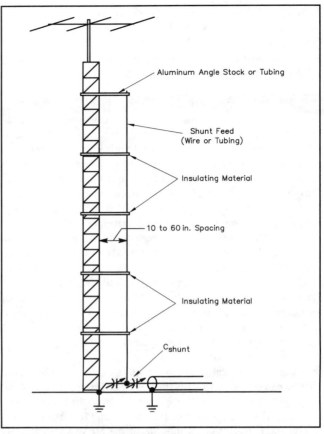

Fig 9-88—The omega-matching system (a gamma match with an additional shunt capacitor) adds a great deal of flexibility to the shunt-fed-tower arrangement. In order to maintain maximum bandwidth, make the gamma wire as long as possible. If the antenna is electrically longer than 120°, a variable series capacitor will make it possible to obtain a very low SWR over a very wide bandwidth.

physical length of the tower.

Fig 9-88 shows the correct wiring of both the gamma and omega-matching networks on a loaded tower. Notice the correct connection of the shunt capacitor in the case of the omega match.

The same principles can, of course, be applied to 80 meters, although it is probable that a tower of reasonable height, loaded with a Yagi antenna, will result in too long an antenna for operation on 80 meters.

6.8.3.7. Practical hints

All cables leading to the tower and up to the rotator and antennas should be firmly secured to a tower leg, on the inside of the tower. All leads from the shack to the tower base should be buried underground in order to provide sufficient RF decoupling. A stack of ferrite beads (similar to those used on current baluns) can be used to decouple RF from any conductors.

If there is still RF on some of the cables, you may wish to coil up a length of the cable where it enters the shack. Care should be taken to ensure good electrical continuity between the tower sections, and between the rotator, the mast, and the tower. Large braid (such as the flattened braid from old coax or a piece of car-battery cable) can provide the necessary electrical contact and physical flexibility. The gamma rod can be supported with sections of plastic pipe, attached to the tower with U bolts or stainless-steel radiator hose clamps. If the tower is a crank-up type, heavy, insulated copper wire can be used for the gamma element. As with any vertical, this system requires the best possible ground system for optimum low-angle radiation and efficiency. The longer the electrical length of the loaded tower, the better the quality of the ground will need to be to achieve the potential low-angle radiation.

6.8.4. The slant-wire feed system

The slant-wire feed system is very similar to the gamma feed system. The feed wire is attached at a certain height on the tower and slopes at an angle to the ground, where a series capacitor tunes out the reactance. The advantage of this system is that a match can be obtained with a lower tap point, which makes it possible to avoid using an omega match on physically short towers. The disadvantage is that the slant-wire feed also radiates a horizontally polarized component.

The slant-wire feed system can easily be modeled using *MININEC*, just as the gamma and omega-matching systems.

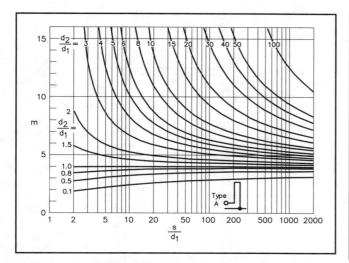

Fig 9-89—Transformation ratio (m) of a two-wire folded monopole, as function of the wire spacing (S/d1) and the ratio of the conductor diameter (d2= diameter of the grounded conductor, d1=diameter of the fed conductor). (After Gerd Janzen, *Kurze Antennen*.)

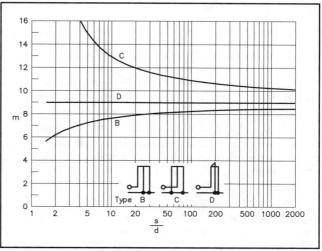

Fig 9-90—Transformation ratio (m) of a three-wire folded monopole, as function of the spacing between the wire (s/d) and the configuration B, C or D. In this case three conductors of equal diameter are assumed. (After Gerd Janzen, *Kurze Antennen*.)

6.8.5. Folded monopoles

Folded antennas have the following advantages:
• Higher bandwidth due to a larger effective antenna diameter.
• Higher feed-point impedance.
• Grounded element (advantage for lightning protection).

Fig 9-89 shows how we can manipulate the wire diameter (ratios) and spacing to obtain up-transformation ratios ranging from 2 to well over 10. (source: *Kurze Antennen*, by Gerd Janzen, ISBN 3-440-05469-1). The configuration of the two-wire folded monopole is shown in the same figure. One leg is grounded, while the antenna is fed between the bottom of the other leg and ground.

The effective diameter of multi-wire elements can be calculated from the chart shown in Fig 9-47.

Three-wire folded-element configurations allow even higher transformation ratios as can be seen in **Fig 9-90**. The effective antenna diameter (which determines the bandwidth of the antenna) is given in Fig 9-62 for the various configurations. The configurations are also shown in the same figure.

6.8.6 Modeling shunt-fed towers

MININEC can be used for modeling the gamma, omega and slant-wire matching systems on shunt-fed grounded towers. Satisfactory results are obtained using the following guidelines:
• The horizontal wire connecting the gamma wire to the tower has one segment (the length of the segment is the spacing from the gamma wire to the tower).
• Use approximately the same segment lengths on all wires of the antenna.
• Do not try to model the capacitance top load. It is much

easier to first "dip" the tower (see Fig 9-84), calculate the electrical length of the loaded tower, and then use an equivalent straight tower to do the gamma-match modeling.

Example:

A tower dips at 1.42 MHz. The required operating frequency is 1.835 MHz. The electrical length is:

$$L \text{ (in degrees)} = 90° \frac{1.835}{1.42} = 116°$$

The physical length of a $^{1}/_{4}$-λ tower (equivalent diameter = 250 mm) is 39 m. The equivalent tower length for 116° is:

$$39 \text{ m} \times \frac{116°}{90°} = 50.3 \text{ m}$$

Now model a vertical with a diameter of 250 mm and of 50.3 m length. According to Fig 9-85, the tap will be at a height of between 17 and 25 m, depending on the wire spacing.

6.8.7. Decoupling antennas at high impedance points

When shunt feeding a tower that supports various antennas, including wire antennas, one may wish to decouple the wire antennas from the vertical radiator. Otherwise, the wire antennas will act as top loading to the vertical. For relatively short towers, the extra loading may be welcome, but in other cases the loaded vertical may become too long with the additional loading of large wire antennas, such as eg, a 160-meter or 80-meter inverted V.

These wire antennas are usually installed at the top of the tower, at a high impedance point. This makes decoupling of the wire antennas more difficult. Conventional current baluns are not suitable as they do not introduce enough impedance to effectively decouple the antenna. Such baluns, installed at the high-impedance points will heat and may even cause the coaxial cable to melt. Problems with such baluns will also lead to unusual and unexpected changes in the feed-point impedance of the loaded vertical, while transmitting. Upon first transmitting the SWR will be normal, but soon the ferrite material used in the current balun will heat to the point where the Curie temperature is reached, resulting in a sudden drop in magnetic susceptibility of the ferrite material. The balun will no longer represent enough impedance, causing the dipole to load the tower, with a change in SWR as a result.

For effective decoupling in this application, a high-impedance balun is required. This balun is rather like a trap, a parallel-tuned resonant circuit, tuned to the frequency of the loaded vertical. The RF currents that flow from the vertical to the dipole (which we want to decouple) are common-mode currents, which means they flow only on the outside of the coaxial feed line of the inverted V dipole.

The trap is made by winding a single-layer coil of coax onto a suitable form, and resonating the coil with a suitable capacitor. Jim Jorgenson, K9RJ, made such a trap for 160 meters consisting of 21 turns of RG-213 wound close spaced on a PVC pipe of approx. 10-cm diameter. The coil is approx. 33-cm long, and the measured inductance of this coil is 33 μH. The coil is held in place on the form by drilling close-fitting holes on an angle through the PVC pipe and passing the coax through these holes into the interior of the pipe at both ends. At one end the shield and the inner connector are separated and connected to stainless steel eye bolts that are used to connect the two legs of the inverted V antenna. At the other (bottom) end the coax passes out through a standard PVC end cap and a PL-259 connector is attached at that end.

The capacitance to resonate this coil on 1830 kHz is about 200 pF. One could use a quality transmitting-type ceramic capacitor, but a suitable capacitor can be made from a short piece of coax. RG-213 coax has a capacitance of 100 pF/m, which means that an open-ended piece of RG-213, measuring 2-m long will resonate the coil on 160 meters. The resonant frequency of the trap can easily be measured using a grid-dip meter. **Fig 9-91** shows the layout of the trap. To tune the trap, one can deliberately make the coaxial-line capacitor too long, and then cut small pieces at a time until resonance is obtained at the desired frequency. The stub capacitor can then be folded inside the PVC tube before putting in the bottom end cap. Jim reports that since he has been using this balun there has been no change of the shunt-fed tower impedance, with the 160-meter inverted V attached. This proves that the trap is now fully decoupling the inverted V from the vertical.

■ 7. INVERTED-L ANTENNA

The ever-so popular inverted L is analyzed in this section and a few practical designs, such as the well-known AKI Special, are given particular attention.

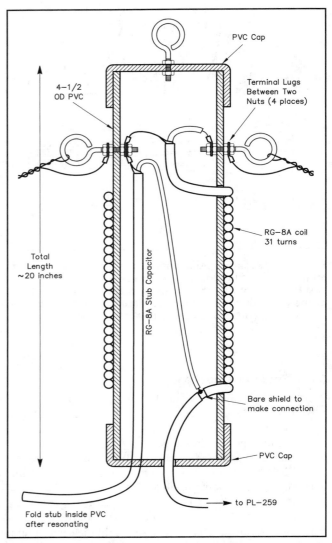

Fig 9-91—Construction details of the K9RJ decoupling trap.

The inverted L is a popular antenna, especially on 160 meters. These antennas are not truly verticals, as part of the antenna is horizontal and radiates a horizontally polarized wave. Most inverted Ls are of the $1/4$-λ variety, although this does not necessarily need to be the case. The vertical portion of an inverted L can be put up alongside a tower supporting HF antennas. In such a setup one must take care that the tower (plus antenna) does not resonate near the design frequency of the inverted L. Grid dip your supporting tower using the method shown in Fig 9-84. If it dips anywhere near the operating frequency, maybe you should use the tower (shunt fed) instead of using it as a support for the inverted L.

If you choose not to do so, you can detune the tower to make sure the smallest possible current flows in the structure. The easiest way I found to do this is to use the drop wire, which you used to dip the tower, and terminate this wire via a 0-2000-pF variable capacitor to ground. Next use a current

Fig 9-92—The K9RJ coaxial trap feeds a 160-meter inverted V hanging from the top of a tower, which is also used as a shunt-fed vertical for 160 meters. The entire construction looks somewhat like an oversized center insulator or balun.

probe (eg, as shown in Fig 11-17 in the chapter on vertical arrays) and adjust the variable capacitor for minimum current. The capacitor can be replaced with a fixed one (parallel combination of several values, if necessary) having the same value. This procedure will guarantee minimum mutual coupling between the inverted L and the supporting tower.

The longer the vertical part of the antenna, the better the low-angle radiation characteristics of the antenna and the higher the radiation resistance (see **Fig 9-93**). The horizontal part of the antenna accounts for the high-angle radiation that the antenna produces. If you are looking for an antenna that radiates reasonably well at both low and high angles, an inverted L may be an excellent choice for you. Since it is a loaded monopole, an inverted L requires a good ground system for optimum low-angle radiation.

Fig 9-94 shows the vertical and horizontal radiation patterns for a practical design of an inverted-L antenna for 3.5 MHz, one having a 12-m vertical mast. Notice how the vertical part of the antenna takes care of the low-angle radiation, while the horizontal part assures high-angle output. The radiation pattern shown is for the direction perpendicular to the plane of the inverted L.

An inverted L is also an attractive solution for the operator

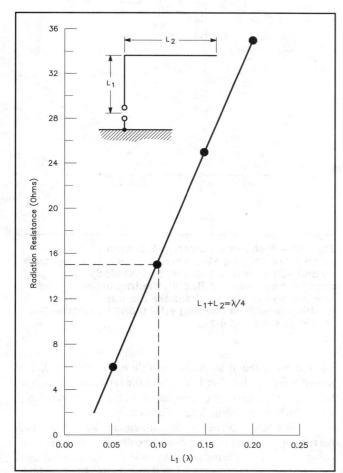

Fig 9-93—Radiation resistance of an inverted-L antenna as a function of the lengths of the horizontal wire versus the vertical conductor.

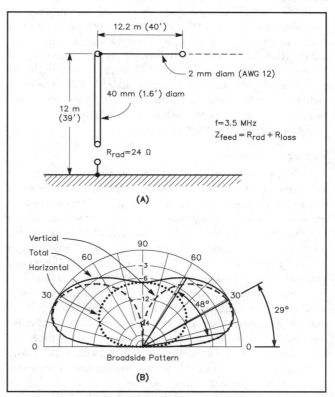

Fig 9-94—At A is a 3.5-MHz inverted L with a 12-m vertical mast. The vertical radiation pattern is shown at B. The pattern has both vertically and horizontally polarized components, and these components are also plotted at B. The pattern is generated over average ground, using 60 $1/4$-λ radials. Note that the angle of maximum radiation is 29°, not bad for a DX antenna.

who needs to use an 80-meter vertical antenna as a support for a 160-meter antenna (**Fig 9-95A**). The easiest solution is to insert a trap at the top of the 80-meter vertical. The exact L/C ratio is not important, but influences the length of the loading wire and the SWR behavior of the antenna on both 80 and 160 meters. See also Figs 9-69 and 9-71.

A second alternative, shown in Fig 9-95B, uses an 80-meter trap to isolate the horizontal part of the 160-meter inverted-L antenna when operating on 80 meters (Ref 659). The trap can be a coaxial-cable trap as explained in Section 6.5.

The inverted L has been extensively described in amateur literature as the better antenna for producing a low-angle signal on top-band (Ref 798 and 7994). The Battle Creek Special and the BC Trapper, described in Sections 6.7 and 6.7.1 are excellent examples of inverted Ls (on 80 and 160 meters).

The AKI Special is another DXpedition-style inverted L, as used by Aki Nago, JA5DQH, during his operation on 160 meters from several rare DX spots. From Kingman Reef (May 1988), Nago used the inverted L as shown in **Fig 9-96**. The vertical part is made of a 12-m aluminum mast, which is extended by an 8-m-long fiberglass fishing rod, to which a copper wire has been attached. From the tip of the (bent) fishing rod, the sloping wire extends another 23.5 m, to be terminated with a fishing line supported by a 3-m pole at some distance. Aki used about 800 m of radials running into the Pacific Ocean. A very similar 160-meter antenna was successfully used from Palmyra during the same DXpedition trip in 1988, and during a more recent DXpedition to Ogasawara by JA5AUC. The calculated radiation resistance of this antenna is approximately 14 Ω. The main radiation angle (over sea water) is 10°, but due to the relatively long horizontal (sloping) wire, the radiation at higher angles is only slightly suppressed.

Tuning procedure

When cutting the length of the sloping wire, cut it at first a little long (2 m too long). Put up the antenna, and connect one of the popular antenna analyzers (MFJ, AEA or Autek) between the bottom of the antenna and the ground system. Adjust the length of the sloping top wire for minimum SWR. Now read the resistance value off the scale of your analyzer. If it is between 35 Ω and 70 Ω, the SWR will be pretty acceptable (1.5:1) and you may want to feed the antenna directly with 50-Ω feed line. From the difference between the R value and the calculated 14-Ω radiation resistance, you can calculate the effective ground-loss resistance of the ground (radial) system. If the feed-point impedance is above 50 Ω, you really need to improve the radial system. For 50 Ω the efficiency would be 14/50 = 28%. Any value higher than 50 Ω would indicate an even lower efficiency. If you want a perfect match you can use an L network, as described in Section 5.7 or an unun (Ref 1522).

Inverted Ls longer than ¹/₄ λ

Some have advocated the use of a total length that is greater than ¹/₄ λ electrically.

If the vertical section is at least ¹/₈-λ long you may

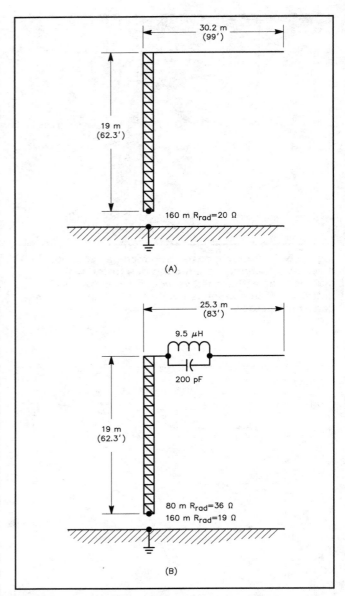

Fig 9-95—A shows an inverted-L antenna for 160 meters, using a 19-m vertical tower. To cover both 80 and 160 meters, a trap can be installed at the top of the tower as shown at B. With the trap installed, the loading wire is shorter, because the trap shows a positive reactance (loading effect) on 160 meters. See also Figs 9-69 and 9-71.

want to make the total antenna a little longer than ¹/₄ λ, and install a series capacitor to tune out the inductive reactance. this may also give you a better match to a 50 Ω cable, depending on the earth-loss resistance.

Don't forget however, that maximum FS occurs with maximum current integrated over the entire length of the *vertical* structure. Don't bring a current node in the horizontal wire, unless of course you want a lot of horizontally polarized high-angle radiation! This means that a ³/₄-λ long inverted L is not what you want. Current moved into the flat top represents higher

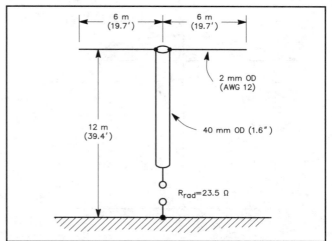

Fig 9-97—T-wire loaded current-fed ¼-λ antenna for 3.5 MHz.

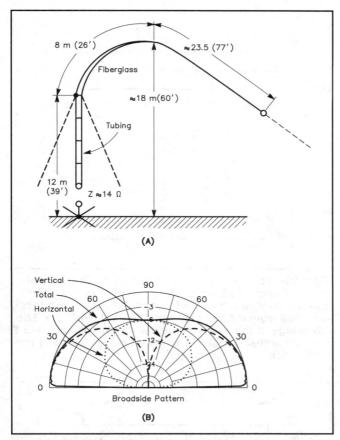

Fig 9-96—The AKI Special, a typical DXpedition type 160-meter inverted L. A collapsible fiberglass fishing rod (available in Europe in lengths of up to 12 m) is used on top of a 12-m aluminum mast. A #12 wire is attached to the rod, and slopes to a distant point to make the sloping (horizontal) part of the antenna. The radiation pattern is over salt water. (That's where the island DXpeditioners put these antennas.)

radiation resistance, but unfortunately the radiation is straight up. The effect is one of robbing current from the vertical section of the antenna, and it (unless the ground system is *very* poor) will decrease vertical radiation.

■ 8. THE T ANTENNA

The current-fed T antenna is a top-loaded short vertical, as covered earlier in this chapter. The voltage-fed T antenna is given special attention here, as well as the different top-loading structures.

8.1. Current-Fed T Antennas

T-wire loading (flat-top wire) is covered in detail in Sections 3.1.2 and 3.6.5 when dealing with top loading of short verticals. The advantage of the horizontal T-wire loading system over the inverted-L system is that the top-wire does not contribute to the total radiation pattern. **Fig 9-97** shows a practical design where a 12-m long vertical is loaded with a horizontal top-load wire to achieve resonance at 3.5 MHz. The

The photo shows an 80-meter T-loaded quarter-wave vertical that I developed for use on Heard Island (VKØIR). The bottom part consists of several sections of aluminum tubing, measuring approx. 9 m in total length. The top part is a 7 m long fiberglass fishing rod, with a wire running inside. From the top of the wire a sloping horizontal section works as a capacitance hat and resonates the antenna on 80 meters. This antenna was primarily used on 80 meter SSB. It used two elevated quarter-wave radials, plus a good ground screen, made of thousands of meters of copper wire, lying on the ground but not connected to the elevated radial system. This antenna was on a little platform, so there was no salt water under it, but it has a clear view toward the ocean for Fresnel reflection. This antenna appeared to have excellent efficiency, as the signals on 80 meter SSB were consistently 3 to 4 S-units louder than the CW signals that used a 4-square with a commercial short, center-loaded vertical, standing in salt water right near the beach.

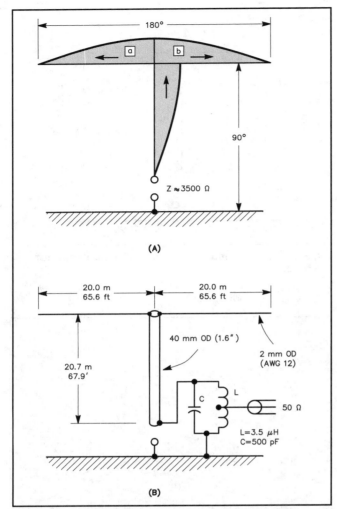

(A)

(B)

Fig 9-98—Voltage-fed 80-meter ¹/₄-λ vertical (also called inverted vertical), using a ¹/₂-λ long top-loading wire. The T wire has a twofold function—providing a low impedance at the top of the vertical, and having a configuration whereby horizontally polarized radiation is essentially canceled (area A = area B, hence no radiation). C—500 pF. L—3.5 μH.

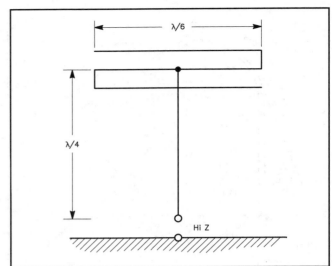

Fig 9-99—Voltage-fed ¹/₄-λ T antenna with the ¹/₂-λ flat-top wire folded to have a total span of only ¹/₆ λ. The current distribution in the folded top-load is such that radiation from the top-load is effectively canceled. The advantage of this design over the original voltage-fed T antenna is that it requires a much smaller (shorter) top-load space.

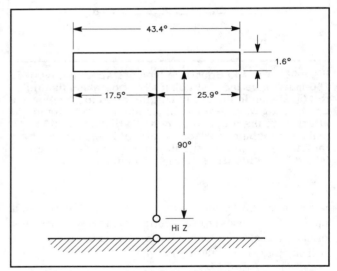

Fig 9-100—Voltage-fed T antenna with a ¹/₄-λ long top load, arranged in such a way that there is no radiation from the flat-top section.

R_{rad} of this design is approximately 23.5 Ω.

Fig 9-53 gives a design chart for ¹/₄-λ T antennas. If there is not enough room for a single flat-top wire, two wires at right angles can be used. In this case the length of the wire will be approx. 60% of the length of a single wire.

8.2. Voltage-Fed T Antennas

Voltage-fed T antennas are loaded vertical antennas with a current minimum at ground level. A specific case consists of a quarter-wave vertical, loaded with a half-wave top wire.

Fig 9-98 shows the configuration of this antenna and the current distribution. In this case, the impedance at the base of the antenna is high and purely resistive. The current maximum is at the antenna top. The antenna is sometimes called an inverted

vertical, as it has its current maximum at the top. In theory, the current in both halves of the flat-top wire is such that radiation from that wire is zero. (In practice there is a very small amount of horizontal radiation.) The disadvantage of this construction is that the antenna requires a very long flat-top wire. Fig 9-98 also shows the dimensions for such a vertical for a practical design on 3.5 MHz.

Hille, DL1VU, dramatically improved the T antenna by folding the ¹/₂-λ flat-top section in such a way that the radiation

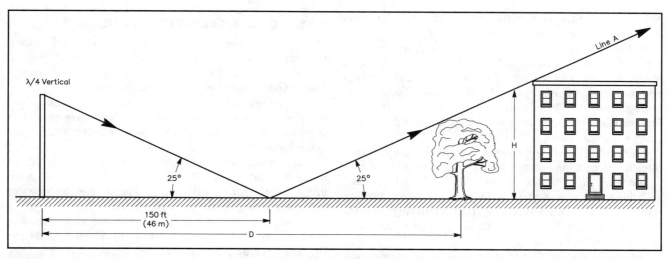

Fig 9-101—Clearance required for a good layout of a vertical antenna. The dimensions are given for 3.5 MHz. For 160 meters, all dimensions should be multiplied by 1.9. All neighboring structures should fall below line A. See text for discussion.

from the flat-top section is effectively suppressed. **Fig 9-99** shows the configuration of this antenna. It can easily be proved that the area under the current distribution line for the central part (which is $^1/_{12}$-λ long) is the same as the area for the remaining part of the loading device (which is $^1/_6$-λ long). Because of the way the wires are folded, the radiation from the horizontal loading device is effectively canceled.

The latest design of a T-type top-load by Hille requires only a single $^1/_4$-λ flat top. In order to cancel all possible horizontal radiation from this flat-top wire, the $^1/_4$ λ is folded back as shown in **Fig 9-100**. Notice that the top load is asymmetrical.

A single quarter-wave flat top acts as a short circuit at the top of the vertical, the same way as radials provide a low-impedance attachment point for the outer conductor of the coax feed line in the case of a ground plane (vertical) antenna.

Hille also described a vertical with a physical length of only 0.39 λ, using the $^1/_4$-λ-long top-load wire configuration as described above (Ref 7991). This antenna produces the same field strength as a $^5/_8$-λ (0.64-λ) vertical antenna.

The T antenna can also be seen as a bobtail curtain antenna with the two vertical end sections missing (see Fig 12-18). As such, this antenna is a poor performer with respect to the bobtail antenna, where the directivity and gain is obtained through the use of three vertical elements.

8.2.1. Feeding the antenna

The voltage-fed T antenna can best be fed by means of a parallel tuned circuit (see Fig 9-98). You can either tap the coax on the coil for the lowest SWR point or tap the antenna near the top of the coil. Either method is valid.

8.2.2. The required ground and radial system

The ground and radial requirements are identical to those required for a $^1/_2$-λ vertical (see Section 4.3).

■ 9. LOCATION OF THE VERTICAL ANTENNA

Let's tackle the so-often-asked question, "Will a vertical work in my particular location?" Verticals for working DX on the low bands are certainly not space-saving antennas but to the contrary, require a lot of space and a good ground. Many low-band DXers have wondered why some verticals don't work well at all, while others work "like gangbusters." The poor performers generally have the poor locations. To repeat, a vertical is not a space-saving antenna! A good vertical takes a lot of real estate. In addition, it must be real estate with a good RF ground!

The standard for buried radials is that for best radiation efficiency you need 120 $^1/_2$-λ radials. This means that for 80 meters, you need about an 80 × 80-meter lot in which to place all the radials. The radials are there to provide a low-resistance return path for the antenna current in order to achieve a good efficiency. You can do as well with just a few elevated $^1/_4$-λ radials, as far as the radiation efficiency is concerned.

The area beyond the ends of the radials is at least as important, because that's where the low-angle reflection at ground level takes place (the Fresnel zone). This is where the reflection efficiency is determined.

Fig 9-101 shows how much clearance a current-fed $^1/_4$-λ vertical should have for adequate performance. Assuming a $^1/_4$-λ vertical with an excellent ground system, the wave angle can be as low as 20° to 25°. RF radiated from the top of the $^1/_4$-λ vertical at an angle of 25° will hit the ground about $^1/_2$ λ away from the base, reflecting at an angle of 25° (the main wave angle). Here you can see that the 1-λ long radials start having an influence on the very low-angle radiation behavior of verticals, but for the really low angles, it all happens much farther away!

Up to $^1/_2$ λ away from the vertical, most of the reflection will take place that is responsible for the 25° radiation (main angle) of a $^1/_4$-λ vertical. Therefore, beyond this point, a clear path should be available for these low-angle rays in order to obtain maximum low-angle radiation. It is clear that for even lower angles of radiation, the ground at even greater distances becomes important. As explained earlier, this is of course very much more so with "long" verticals (eg, $^1/_2$-λ vertical) where the Fresnel reflection takes place up to 100 λ away from the antenna (for wave angles down to 0.25°).

From Fig 9-101 you can see that this means no structures taller than the antenna should be closer than 1 wavelength away from a $^1/_4$-λ vertical. Smaller interfering and absorbing obstacles can be a little closer, as long as the size remains small enough to refrain from interfering with the low-angle energy reflected from the ground within $^1/_2$ λ from the base of the vertical.

For a $^1/_4$-λ vertical, the maximum height of a neighboring obstacle can be calculated with the following formula:

$$h_{max} \text{ (in meters)} = \left(D - \frac{468}{f}\right) \tan \alpha$$

where

    D = distance of the obstacle from the antenna base, same units as h

    f = frequency, MHz

    α = wave angle in degrees (25° is a good rule of thumb)

This means, for instance, that at a point 60 m from a 3.5-MHz antenna, the maximum height of a structure should be limited to 9.1 m. What about trees closer in? Trees can be reasonably good conductors and can be very lossy elements in the near field of a radiator. A case has been reported in the literature where a $^1/_4$-λ vertical with an excellent ground system showed a much lower radiation resistance than expected. It was found that trees in the immediate area were coupling heavily with the vertical and were causing the radiation resistance of the vertical to be very low. Under such circumstances of uncontrolled coupling into very lossy elements, far from optimum performance can be expected. Of course if the trees are short in relation to the (quarter) wavelength, it is reasonable to assume that the result of such coupling will be minimal.

Even though neighboring (lossy) structures such as trees may not be resonant, they will always absorb some RF to an unknown degree. Other objects that are very likely to affect the performance of a vertical are nearby antennas and towers. Mutual coupling can be considered the culprit if the radiation resistance of the vertical is lower than expected. Another way of checking for coupling with other antennas is to alternately open and short-circuit the suspected antenna feed lines while watching the SWR or the radiation resistance of the vertical antenna. If there is any change, you are in trouble. Checking for resonance of towers has been described in Section 7.

It may come as a surprise that a vertical is so demanding of space. Most amateur verticals are not anywhere near ideal, and good performance can still be obtained from practical setups, but the builder of a vertical should understand which factors are important for optimum performance, and why.

■ 10. 160-M DXPEDITION ANTENNAS

I have talked at great length with well-known DXpeditioners who have been especially successful on the low bands. I'd like to share the following rules with candidate DXpeditioners with respect to the low bands.

If you're on an island, erect the station on that side of the island where you will have the most difficult path or where you are facing the most stations (eg, if you are on an island in the South Indian Ocean, try a shore on the northwest side of the island, looking into both Europe and North America). By all means erect the antenna very close to the salt water, or over (or in) the salt water. This will help you lower the pseudo-Brewster angle, and assure a good low-angle take off.

Unless you have a very tall support of at least 30 meters, use a vertical. Good choices are the Battle Creek Special, the BC Trapper, the AKI Special, the Titanex V160E or any inverted L, for which you should try to make the vertical part as long as possible. The vertical section should be at least 15 m tall, 12 m being an absolute minimum for 160 meters. If there are some trees, you may try to climb a tall tree, and use a collapsible fiberglass fishing rod (they exist in 12 m lengths) to extend the effective support height. Use as many radials as you can, and let them run into the salt water. Very thin wire is just fine if you use many (current is shared by the many wires). A small spool of #28 enameled copper wire (magnet wire) can hold a lot of wire and takes little space. Equally as good is to use two in-line elevated radials and it makes switching from the CW to the phone band very simple by merely adjusting the length of the two radials (see also Section 6.7.1).

The Titanex verticals are very special in that they are made of an aluminum-titanium alloy that is very strong and extremely lightweight. The model V160E vertical is a 26.7-m long vertical, that weighs only 7.5 kg. The maximum section length is only 2.1 m, and the total antenna can easily be erected by two to three persons. This, as well as the low weight, makes this a very attractive antenna for DXpeditions. The guy wires are 2-mm Kevlar, and guys are placed at 6 m, 9 m, 12 m, 15 m and 18 m. The upper 8 m of the vertical swings freely in the wind. With a total length of 26 m, this antenna has a very respectable radiation resistance of approx. 12 Ω, which is 50% higher than that of the Battle Creek Special (which is of course 10 m shorter!). The antenna is $^3/_8$ λ on 80, and $^5/_8$ λ on 40. Also on 40 this should make it a killer antenna if erected over salt water. The V80E vertical measures "only" 20-m tall, which is good for a R_{rad} of approx. 8.5 Ω, which is similar to the R_{rad} of the Battle Creek Special. The V80E (See **Fig 9-102.**) weighs only 6.5 kg! Titanex also provides a three-band relay-switched matchbox providing a 1:1 SWR on the three low bands. More info at **http://www.qth.com/titanex**. The Titanex antennas are expensive mechanical marvels and have been used extremely successfully during a number of expeditions eg, VK9CR, VK9XY, C56CW, FW2OI, S21XX, P29VXX, DL7FD/HR3, K7K, K4M, T31BB, 9MØC, TJ1GB, ZL7DK, YJØADJ, FOØFI, FOØFR, and 3B7RF.

Don't bother putting up a Beverage near the sea; it

Fig 9-102—The Titanex V160E antenna on the "beach" on VK9CR, surrounded by beautiful coconut trees. This 26.8-m long special DXpedition vertical weighs only 7.5 kg and disassembles into 2.1-m long sections, ideal for traveling!

won't work well. The VKØIR guys did not believe me. They put one up; it never worked. Anyhow, it's unlikely you will have to deal with a lot of local QRM or man-made noise, which makes the use of directive antennas pretty senseless. If you do need directivity, try an EWE or a K9AY loop, which are receiving antennas that actually work better over good ground.

If there is a tall support, you may want to use a sloping half-wave vertical, especially *if* you are near the sea (see Chapter 8 on dipole antennas). The sloping vertical builds up its image as far as 100 wavelengths away from the antenna. If there is no salt water nearby and ground conductivity is poor, use a high support for an inverted-V dipole. Don't try an inverted V or any other horizontally polarized antenna at a height of 15 meters or less. All you will get is very high-angle radiation.

Here is a hint I got from Rudi, DK7PE: If you are on a DXpedition in a country with a substantial tourist business, choose the tallest hotel (Hilton, Sheraton or Intercontinental hotels are usually doing well in this respect). Slope a dipole from the top of the building to some distant point and let the feed line come to your room, which can be a few stories below the roof. Make the dipole as vertical as possible. This

is by far the best antenna if you are in such a situation.

DK7PE proved it during his operation from D2CW (August '92) where he had his sloping dipole attached some 60 meters above street level, facing north, and within 1 λ of the South Atlantic Ocean. Rudi's signals were always S9 in Europe on 160 meters. During his more recent operation from Ethiopia and Eritrea (9F2CW), he proved it again. Rudi's total antenna system for his DXpeditions (covering 160 through 10 meters) can be packed in a very small handbag. The RG-58 cable takes up 80% of the volume. The antenna consists of precut lengths of flexible insulated wire, with small insulators and a variety of alligator clips that let him change bands. On the higher bands he can configure the wire into a 2-element Yagi.

R. E. Tanaka, 9M2AX, well-known 160-meter operator from the Far East, sent me the sketches of the antennas he is using in 9M2 as well as when he operated from 9M8AX. The antennas Ross was using can be put up at any tall hotel, and should be excellent suggestions for 160-meter DXpeditioners. **Figs 9-103** and **9-104** show the layouts of the two antenna setups and their radiation patterns. The radial system covering only one quadrant (90°) results in a significant high-angle radiation component with the 9M2AX version. The low-angle radiation is very pronounced as well. From modeling, the "inverted sloping wire vertical" from the 9M2 QTH has a feed-point impedance of approx. 75 Ω. The 9M8AX configuration is an inverted L with a sloping flat-top. The calculated impedance from modeling is nearly 60 Ω. This antenna has better low-angle radiation than the 9M2AX version, which is normal. In order to eliminate the high-angle radiation for the 9M2AX version, it would be necessary to install just two radials (in line with one another), so that the radiation from these wires would be canceled. The radials are *not* there to provide a ground plane, but are merely serving to provide a low-impedance point to which to connect the outer shield of the feed line. One ¼-λ radial would serve that purpose, but would radiate a lot of horizontal component. Two radials in line would provide a low impedance point just as well, but not radiate any high-angle horizontal component.

■ 11. BUYING A COMMERCIAL VERTICAL

I sincerely hope that this chapter on verticals has incited you to build your own antenna. You cannot believe how much more satisfaction you get out of using something you made or designed yourself, rather than going to the store, opening your wallet and then play the appliance-type ham.

Anyhow, if you choose not to make your own (you don't know what you miss!), here are a few rules to help you select your new low-band commercial vertical:

1. Most, if not all companies advertising their products, largely exaggerate the performance.
2. A short vertical with a large bandwidth means there are a lot of losses. With short antennas a large bandwidth is a direct measure of its poor efficiency (lots of losses).
3. The efficiency of a vertical is in the first place determined by the physical length of the vertical.
4. Only top loading is efficient.
5. Verticals with coil loading are bound to be inefficient. An

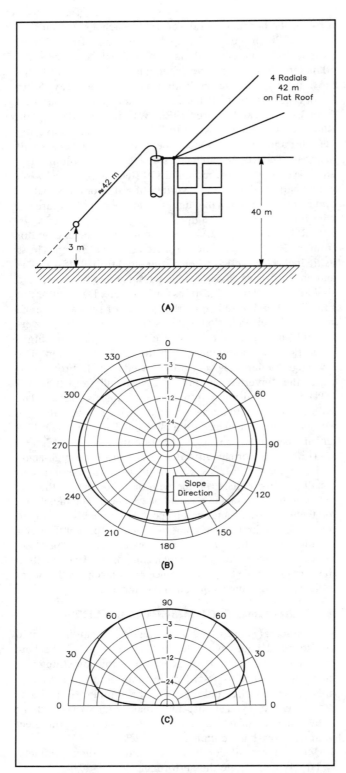

Fig 9-103—Configuration and radiation patterns of the inverted $^1/_4$-λ sloper antenna used by 9M2AX. The azimuth pattern is shown at B for an elevation angle of 30°, and at C is the elevation pattern. (The elevation pattern is taken in the 90-270° direction as displayed in the azimuth pattern.) Note the relative high amount of high-angle radiation. Using just two radials in line would improve this situation considerably.

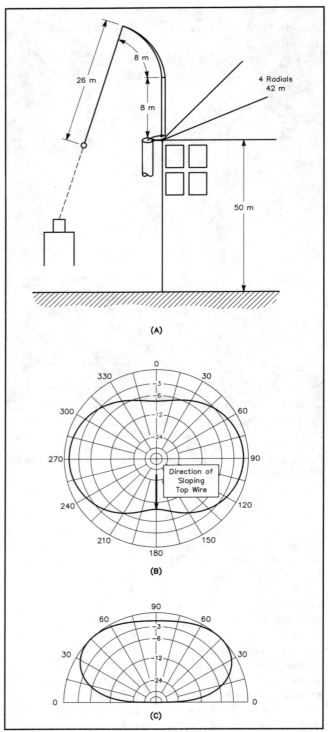

Fig 9-104—Configuration and radiation patterns of the inverted $^1/_4$-λ sloper antenna used by 9M2AX during his expedition from East Malaysia (Sarawak) as 9M8AX. The azimuth pattern is shown at B, and the elevation pattern at C. (The elevation pattern is taken in the 90-270° direction as displayed in the azimuth pattern.) The antenna was installed on the edge of a 50-m high flat roof. Four $^1/_4$-λ long radials were laid on the roof. The metal mast plus the fiberglass rod are 16-m long. The sloping wire was adjusted for minimum SWR at resonance.

8-m long vertical with center (coil) loading is bound to be a very poor performer on 160 meters.

6. To be a reasonable performer a minimum physical length of approx. 14 m is needed on 160 meters.

7. Good hardware (stainless steel, good finishing, etc) are no guarantee for a good antenna.

8. A fancy feed system, or folded elements that claims to reduce losses and increase efficiency are a total fallacy.

9. A producer of a 160-meter vertical who prescribes using a few 10-m-long radials does not know what he is talking about.

10. Advertisers bragging that their product is bought by government agencies are not proving anything. Remember the Maxcom "dummy load" antenna-matching network used extensively by the armed forces?

11. An advertiser specifying his 8-m long 160-meter vertical has 75% efficiency, without specifying the ground (radial) system is telling you stories.

12. Advertisers selling their product by telling how many new countries one of their customers has worked with it, are. . . Well, you know. Maybe, with a good home-made vertical (an inverted L, a BC Trapper or a shunt-fed 60-ft tower with a Yagi on it), he would have worked double the number of new countries. Not very scientific advertising, anyhow.

It is a shame that some Amateur Radio publications continue to make product reviews that are not at all technical, but merely a sales promotion leaflet.

Spending nearly $500 for a 9-m long radiator is a heck of a lot of money. You could buy some simple aluminum tubing (TV-type push-up mast, approx. $70) some copper wire to make a number of top-loading wires (add another $10), some nylon guy rope (another $10), maybe an (empty) Coke bottle for an insulator (free), and you have exactly the same for maybe 20% of the price of the commercial thing. It won't work any better, but at least you won't feel like you've been robbed. And spending nearly $400 for an 8-m long 160-meter vertical, with a slim (and thus very lossy) loading coil, is even worse, of course.

Amateur Radio is a technical hobby. It is true that the progress of micro-electronics has made it very difficult for the average ham to do much home designing and home building in the field of receivers and transmitters. Antennas is one of the few fields where we can, ourselves, through our own knowledge, understanding and expertise, do as well and usually much better than the commercial companies. Let's grab this opportunity with both hands, and build our own vertical for the low bands. This will give you the ultimate kick, I promise you!

LARGE LOOP ANTENNAS

10 LARGE LOOP ANTENNAS

The delta loop antenna is a superb example of a high-performance compromise antenna. The single-element loop antenna is almost exclusively used on the low bands, where it can produce low-angle radiation, requiring only a single quarter-wave high support. We will see that a vertically polarized loop is really an array of two phased verticals, and that the ground requirements are the same as for any other vertically polarized antenna.

This means that with low delta loops, the horizontal wire will couple heavily to the lossy ground and induce significant losses, unless we have improved the ground by putting a ground screen under the antenna. (See Chapter 9, Section 1.3.3 and Section 2.) I have seen it stated in various places that delta loops don't require a good ground system. This is as true as saying that verticals with a single elevated radial don't require a good ground system.

Loop antennas have been popular with 80-meter DXers for the last 25 years or so. Resonant loop antennas have a circumference of 1 λ. The exact shape of the loop is not particularly important. In free space, the loop with the highest gain, however, is the loop with the shape that encloses the largest area for a given circumference. This is a circular loop, which is difficult to construct. Second best is the square loop (quad), and in third place comes the equilateral triangle (delta) loop (Ref 677).

The maximum gain of a 1-λ loop over a ¹/₂-λ dipole in free space is approximately 1.35 dB. Delta loops are used extensively on the low bands at apex heights of ¹/₄ to ³/₈ λ above ground. At such heights the vertically polarized loops far outperform dipoles or inverted-V dipoles for low-angle DXing, assuming good ground conductivity.

Loops are generally erected with the plane of the loop perpendicular to the ground. Whether or not the loop produces a vertically or a horizontally polarized signal (or a combination of both) depends only on how (or on which side) the loop is being fed.

Sometimes we hear about horizontal loops. These are antennas with the plane of the loop parallel to the ground. Such horizontal loops have the reputation of being excellent low-noise receiving antennas. On transmit they produce exclusively high-angle radiation.

■ 1. QUAD LOOPS

Belcher, WA4JVE, Casper, K4HKX (Ref 1128), and Dietrich, WAØRDX (Ref 677), have published studies com-paring the horizontally polarized quad loop with a dipole. A horizontally polarized quad loop antenna (**Fig 10-1A**) can be seen as two short, end-loaded dipoles, stacked ¹/₄ λ apart, with the top antenna at ¹/₄ λ and the bottom one just above ground level. There is no broadside radiation from the vertical wires of the quad because of the current opposition in the vertical members. In a similar manner, the vertically polarized quad loop (Fig 10-1B) consists of two top-loaded, ¹/₄-λ vertical dipoles, spaced ¹/₄ λ apart. Fig 10-1 shows how the current distribution along the elements produces cancel-

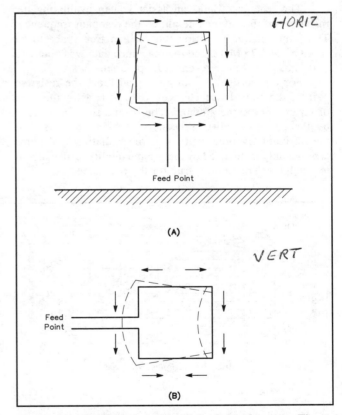

Fig 10-1 — Quad loops with a 1-λ circumference. The current distribution is shown for (A) horizontal and (B) vertical polarization. Note how the opposing currents in the two legs results in cancellation of the radiation in the plane of those legs, while the currents in the other legs are in phase and reinforce one another in the broadside direction (perpendicular to the plane of the antenna).

lation of radiation from certain parts of the antenna, while radiation from other parts (the horizontally or vertically stacked short dipoles) is reinforced.

The square quad can be fed for either horizontal or vertical polarization merely by placing the feed point at the center of a horizontal arm or at the center of a vertical arm. At the higher frequencies in the HF range, where the quads are typically half to several wavelengths high, quad loops are usually fed to produce horizontal polarization, although there is no specific reason for that except maybe from a mechanical standpoint. Polarization by itself is of little importance at HF, because it becomes random after ionospheric reflection.

1.1. Impedance

The radiation resistance of an equilateral quad loop in free space is approximately 120 Ω. The radiation resistance for quad loops as a function of their height above ground is given in **Fig 10-2**. The impedance data were obtained by modeling an equilateral quad loop over three types of ground (very good, average, and very poor ground) using *NEC. MININEC* cannot be used for calculating loop impedances at low heights (see Section 2.7).

The reactance data can assist you in evaluating the influence of the antenna height on the resonant frequency. The loop antenna was first modeled in free space to be resonant at 3.75 MHz, and the reactance data was obtained with those free-space resonant loop dimensions.

For the vertically polarized quad loop, the resistive part of the impedance changes very little with the type of ground under the antenna. The reactance is influenced by the ground quality, especially at lower heights. For the horizontally polarized loop, the radiation resistance is noticeably influenced by the ground quality, especially at low heights. The same is true for the reactance.

1.2. Square Loop Patterns

1.2.1. Vertical polarization

The vertically polarized quad loop, Fig 10-1B, can be considered as two shortened top-loaded vertical dipoles, spaced $1/4$ λ apart. Broadside radiation from the horizontal elements of the quad is canceled, because of the opposition of currents in the vertical legs. The wave angle in the broadside direction will be essentially the same as for either of the vertical members. The resulting radiation angle will depend on the quality of the ground up to several wavelengths away from the antenna, as is the case with all vertically polarized antennas.

The quality of the reflecting ground will also influence the gain of the vertically polarized loop to a great extent. The quality of the ground is as important as it is for any other vertical antenna. This means that vertically polarized loops close to the ground will not work well over poor soil.

Fig 10-3 shows both the azimuth and elevation radiation patterns of a vertically polarized quad loop with a top height of 0.3 λ (bottom wire at approximately 0.04 λ). This is a very realistic situation, especially on 80 meters. The loop radiates an excellent low-angle wave (lobe peak at approximately 21°) when operated over average ground. Over poorer ground, the wave angle would be closer to 30°. The horizontal directivity, Fig 10-3C, is rather poor, and amounts to approximately 3.3 dB of side rejection at any wave angle.

1.2.2. Horizontal polarization

A horizontally polarized quad loop antenna (two stacked short dipoles) produces a wave angle that is dependent on the height of the loop. The low horizontally polarized quad (top at 0.3 λ) radiates most of its energy right at or near zenith.

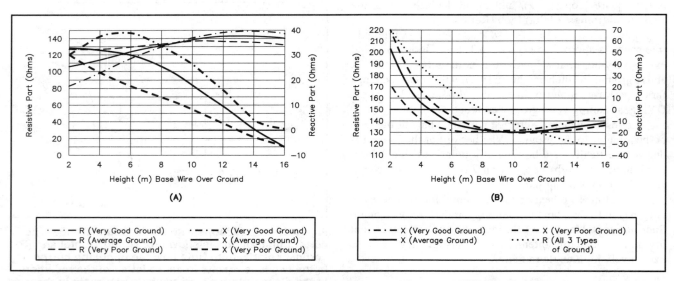

Fig 10-2 — Radiation resistance and feed-point resistance for square loops at different heights above real ground. The loop was first dimensioned to be resonant in free space (reactance equal to zero), and those dimensions were used for calculating the impedance over ground. At A, for horizontal polarization, and at B, for vertical polarization. Analysis was with *NEC* at 3.75 MHz.

Fig 10-4 shows directivity patterns for a horizontally polarized loop. The horizontal pattern, Fig 10-4C, is plotted for a wave angle of 30°. At low wave angles (20° to 45°), the horizontally polarized loop shows more front-to-side ratio (5 to 10 dB) than the vertically polarized rectangular loop.

1.2.3. Vertical versus horizontal polarization

Vertically polarized loops should be used only where very good ground conductivity is available. Installing radials under the loop does not pay off unless they are many wavelengths long. From **Fig 10-5A** we learn that the gain of

the vertically polarized quad loop, as well as the wave angle, does not change very much as a function of the antenna height. This makes sense, as the vertically polarized loop is in the first place two phased verticals, each with its own radial. However, the gain is drastically influenced by the quality of the ground. At low heights, the gain difference between very poor ground and very good ground is a solid 5 dB! The wave angle for the vertically polarized quad loop at a low height (bottom wire at 0.03 λ) varies from 25° over very poor ground to 17° over very good ground.

I have frequently read in Internet messages that a delta

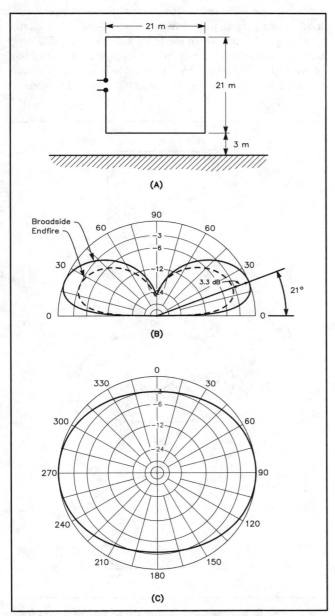

Fig 10-3 — Shown at A is a square loop, with its elevation-plane pattern at B and azimuth pattern at C. The patterns are generated for good ground. The bottom wire is 0.0375 λ above ground (3 m or 10 ft on 80 meters). At C, the pattern is for a wave angle of 21°.

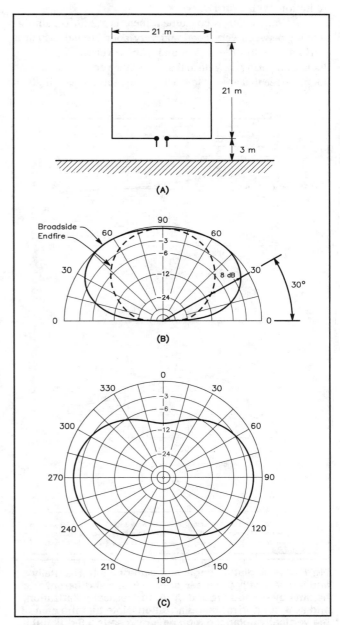

Fig 10-4 — Azimuth and elevation patterns of the horizontally polarized quad loop at low height (bottom wire is 0.0375 λ above ground). At a wave angle of 30°, the loop has a front-to-side ratio of approximately 8 dB.

loop has certain advantages over vertical antennas (or arrays of vertical antennas) as the loop antenna does not require any radials. This statement is really quite misleading, and is like saying that a vertical *with* elevated radials does not require any radials. Indeed, in a delta loop (and a quad loop), the "element" that takes care of the return current is part of the antenna itself, just like with a dipole!

With a horizontally polarized quad loop the wave angle is very dependent on the antenna height, but not so much by the quality of the ground. At very low heights, the main wave angle varies between 50° and 60° (but is rather constant all the way up to 90°), but these are rather useless radiation angles for DX work.

As far as gain is concerned, there is a 2.5-dB gain difference between very good and very poor ground, which is only half the difference we found with the vertically polarized loop. Comparing the gain to the gain of the vertically polarized loop, we see that at very low antenna heights the gain is about

3-dB better than for the vertically polarized loop. But this gain exists at a high wave angle (50° to 90°), while the vertically polarized loop at very low heights radiates at 17° to 25°.

Fig 10-6 shows the vertical-plane radiation patterns for both types of quad loops over very poor ground and over very good ground on the same dB scale. For more details see Section 2.3.

1.3. A Rectangular Quad Loop

A rectangular quad loop, with unequal side dimensions, can be used with very good results on the low bands. An impressive signal used to be generated by 5NØMVE from Nigeria with such a loop antenna. The single quad-loop element is strung between two 30-m high coconut trees, some 57 m apart in the bushes of Nigeria. 5NØMVE used this configuration and fed the loop in the center of one of the vertical members. He first tried to feed it for horizontal polarization but he says it did not work well. The vertical

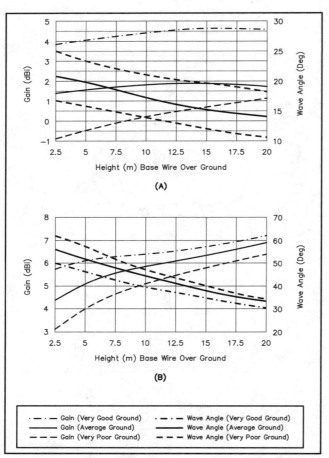

Fig 10-5 — Radiation angle and gain of the horizontally and the vertically polarized square loops at different heights over good ground. At A, for vertical polarization, and at B, for horizontal polarization. Note that the gain of the vertically polarized loop never exceeds 4.6 dBi, but its wave angle is low for any height (14 to 20°). The horizontally polarized loop can exhibit a much higher gain provided the loop is very high. Modeling was done over average ground for a frequency of 3.75 MHz, using *NEC*.

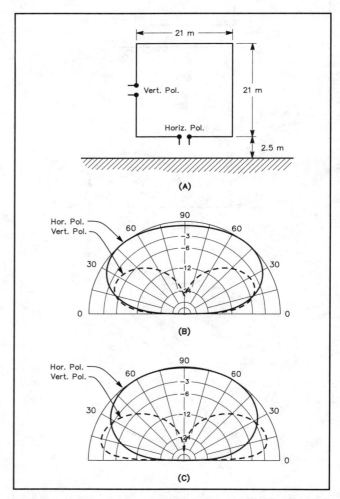

Fig 10-6 — Superimposed (same dB scale) patterns for horizontally and vertically polarized square quad loops (shown at A) over very poor ground (B) and very good ground (C). In the vertical polarization mode the ground quality is of utmost importance, as it is with all verticals. See also Fig 10-14.

and the horizontal radiation patterns for this quad loop over good ground are shown in **Fig 10-7**. The horizontal directivity is approximately 6 dB (front-to-side ratio).

I have analyzed the antenna. It is significant that even in free space, the impedance of the two varieties of this rectangular loop are not the same. When fed in the center of a short (27-m) side, the radiation resistance at resonance is 44 Ω. When fed in the center of one of the long (57-m) sides, the resistance is 215 Ω. Over real ground the feed-point impedance is different in both configurations as well; depending on the quality of the ground, the impedance varies between 40 and 90 Ω.

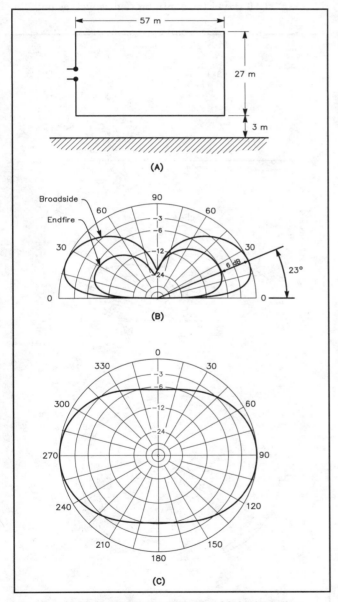

Fig 10-7 — At A, a rectangular loop with its baseline approximately twice as long as the vertical height. At B and C, the vertical and horizontal radiation patterns, generated over good ground. The loop was dimensioned to be resonant at 1.83 MHz. The azimuth pattern at C is taken at a 23° wave angle.

1.4. Loop Dimensions

The total length for a resonant loop is approximately 5 to 6% longer than the free-space wavelength.

1.5. Feeding the Quad Loop

The quad loop feed point is symmetrical, whether you feed the quad in the middle of the vertical or the horizontal wire. A balun must be used. Baluns are described in Chapter 6 on matching and feed lines.

Alternatively one could use open-wire feeders (450-Ω line). The open-wire-feeder alternative has the advantage of being a lightweight solution. With a tuner you will be able to cover a wide frequency spectrum with no compromises.

■ 2. DELTA LOOPS

Just as the inverted-V dipole has been described as the poor man's dipole, the delta loop can be called the poor man's quad loop. Because of its shape, the delta loop with the apex on top is a very popular antenna for the low bands; it needs only one support.

In free space the equilateral triangle produces the highest gain and the highest radiation resistance for a three-sided loop configuration. As we deviate from an equilateral triangle toward a triangle with a long baseline, the effective gain and the radiation resistance of the loop will decrease for a bottom-corner-fed delta loop. In the extreme case (where the height of the triangle is reduced to zero), the loop has become a half-wavelength-long transmission line that is shorted at the end, which shows a zero-ohm input impedance (radiation resistance), and thus zero radiation (well-balanced open-wire line does not radiate).

Just as with the quad loop, we can switch from horizontal to vertical polarization by changing the position of the feed point on the loop. For horizontal polarization the loop is fed either at the center of the baseline or at the top of the loop. For vertical polarization the loop should be fed on one of the sloping sides, at ¹/₄ λ from the apex of the delta. **Fig 10-8** shows the current distribution in both cases.

2.1. Vertical Polarization

2.1.1. How it works

Refer to **Fig 10-9**. In the vertical-polarization mode the delta loop can be seen as two sloping quarter-wave verticals

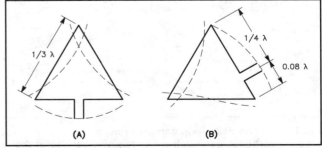

Fig 10-8 — Current distribution for equilateral delta loops fed for (A) horizontal and (B) vertical polarization.

(their apex touching at the top of the support), while the baseline (and the part of the sloper under the feed point) takes care of feeding the "other" sloper with the correct phase. The second sloping vertical is also fed via the top (both sloping verticals are connected at the top). This top connection can be left open without changing anything about the operation of the delta loop. The same is true for the baseline, where the middle of the baseline can be opened without changing anything. These two points are the high-impedance points of the antenna. Either the apex or the center of the baseline must be "shorted," however, in order to provide feed voltage to the "other half" of the antenna. Normally we use a fully closed loop in the standard delta loop, although for single-band operation this is not strictly necessary.

Assume we construct the antenna with the center of the horizontal bottom wire open. Now we can see the two half baselines as two $1/4$-λ radials, one of which provides the necessary low-impedance point for connecting the shield of the coax. The other "radial" is connected to the bottom of the second sloping vertical, which is the other sloping wire of the delta loop.

This is similar to the situation encountered with a $1/4$-λ vertical using a single elevated radial (see Chapter 9 on vertical antennas). The current distribution in the two quarter-wave radials is such that all radiation from these radials is effectively canceled. The same situation exists with the voltage-fed T antenna (see Chapter 9), where we use a half-wave flat-top (equals two $1/4$-λ radials) to provide the necessary low-impedance point to raise the current maximum to the top of the T antenna.

The vertically polarized delta loop is really an array of two $1/4$-λ verticals, with the high-current points spaced 0.25 λ to 0.3 λ, and operating in phase. The fact that the tops of the verticals are close together does not influence the performance to a large degree. The reason is that the current near the apex of the delta is at a minimum (it is *current* that takes care of radiation!).

Considering a pair of phased verticals, we know from the study on verticals that the quality of the ground will be very important as to the efficient operation of the antenna:

- This does not mean that the delta loop requires radials. It has two elevated radials that are an integral part of the loop and take care of the return currents (radiation efficiency).
- As with all vertically polarized antennas, however, the quality of the ground within a radius of several wavelengths will determine the low-angle radiation of the loop antenna (reflection efficiency).

Refer to Chapter 9 on verticals for more details on this topic (see pseudo-Brewster angle, Section 1.1.2.2).

2.1.2. Radiation patterns

2.1.2.1. The equilateral triangle

Fig 10-10 shows the configuration as well as both the

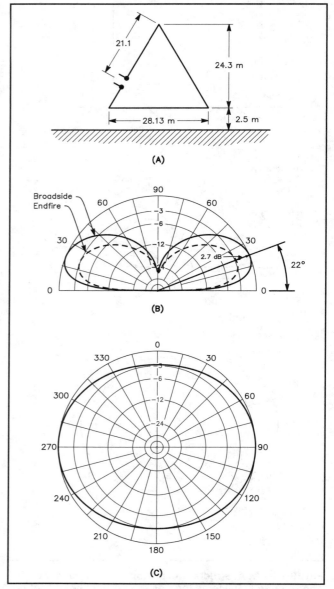

Fig 10-10 — Configuration and radiation patterns for a vertically polarized equilateral delta loop antenna. The model was calculated over good ground, for a frequency of 3.8 MHz. The wave angle for the azimuth pattern at C is 22°.

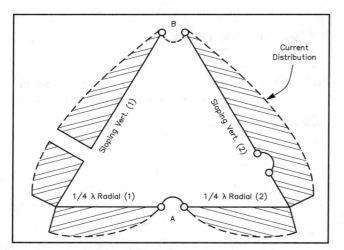

Fig 10-9 — The delta loop can be seen as two $1/4$-λ sloping verticals, each using one radial. Because of the current distribution in the radials, the radiation from the radials is effectively canceled.

broadside and the end-fire vertical radiation patterns of the vertically polarized equilateral-triangle delta loop antenna. The model was constructed for a frequency of 3.75 MHz. The baseline is 2.5 m above ground, which puts the apex at 26.83 m. The model was made over good ground. The delta loop shows nearly 3 dB front-to-side ratio at the main wave angle of 22°. With average ground the gain is 1.3 dBi.

2.1.2.2. The compressed delta loop

Fig 10-11 shows an 80-meter delta loop with the apex at 24 m and the baseline at 3 m. This delta loop has a long baseline

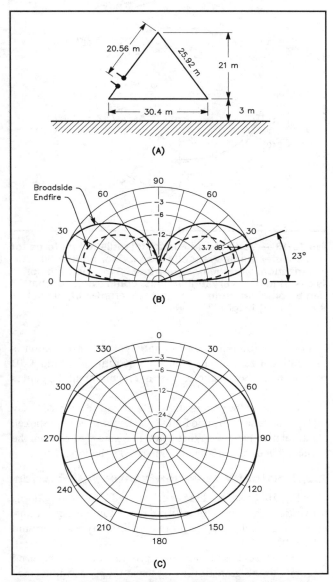

Fig 10-11 — Configuration and radiation patterns for the "compressed" delta loop, which has a baseline slightly longer than the sloping wires. The model was dimensioned for 3.8 MHz to have an apex height of 24 m and a bottom wire height of 3 m. Calculations are done over good ground at a frequency of 3.8 MHz. The azimuth pattern at C is for a wave angle of 23°. Note that the correct feed point remains at ¹/₄ λ from the apex of the loop.

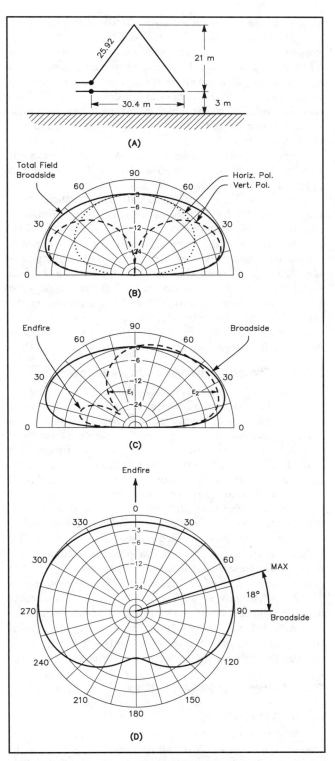

Fig 10-12 — Configuration and radiation patterns for the compressed delta loop of Fig 10-11 when fed in one of the bottom corners at a frequency of 3.75 MHz. Improper cancellation of radiation from the horizontal wire produces a very strong high-angle horizontally polarized component. The delta loop now also shows a strange horizontal directivity pattern (at D), the shape of which is very sensitive to slight frequency deviations. This pattern is for a wave angle of 29°.

of 30.4 m. The feed point is again located $\frac{1}{4}\lambda$ from the apex.

The front-to-side ratio is 3.8 dB. The gain with average ground is 1.6 dBi. In free space the equilateral triangle gives a higher gain than the "flat" delta. Over real ground and in the vertically polarized mode, the gain of the flat delta loop is 0.3 dB better than the equilateral delta, however. This must be explained by the fact that the longer baseline yields a wider separation of the two "sloping" verticals, yielding a slightly higher gain.

For a 100-kHz bandwidth (on 80 meters) the SWR rises to 1.4:1 at the edges. The 2:1-SWR bandwidth is approximately 175 kHz.

Bill, W4ZV, used what he calls a "squashed" delta loop very successfully on 160 meters. The apex is 36-m high and Bill claims that this configuration actually has improved gain over the equilateral delta loop, which can indeed be verified by accurate modeling. The antenna is also fed a $\frac{1}{4}\lambda$ from the apex, using a $\frac{1}{4}$-wave 75-Ω matching stub. Bill says that this loop can actually be installed on a 27-m tower by pulling the base away from the tower. By pulling the base away about 8 m from a tower, one can actually use a full-wave delta on a 24-m high tower, with very little trade-off.

2.1.2.3. The bottom-corner-fed delta loop

Fig 10-12 shows the layout of the delta loop being fed at one of the two bottom corners. The antenna has the same apex and baseline height as the loop described in Section 2.1.2.2. Because of the "incorrect" location of the feed point, cancellation of radiation from the base wire (the two "radials") is not 100% effective, resulting in a significant horizontally polarized radiation component. The total field has a very uniform gain coverage (within 1 dB) from 25° to 90°. This may be a disadvantage for the rejection of high-angle signals when operating DX at low wave angles.

Due to the incorrect feed-point location, the end-fire radiation (radiation in line with the loop) has become asymmetrical. The horizontal radiation pattern shown in Fig 10-12D is for a wave angle of 29°. Note the deep side null (nearly 12 dB) at that wave angle. The loop actually radiates maximum signal approximately 18° off the broadside direction.

All this is to explain that this feed-point configuration (in the corner of the compressed loop) is to be avoided, as it really deteriorates the performance of the antenna.

2.2. Horizontal Polarization

2.2.1. How it works

In the horizontal polarization mode, the delta loop can be seen as an inverted-V dipole on top of a very low dipole with its ends bent upward to connect to the tips of the inverted V. The loop will act as any horizontally polarized antenna over real ground; its wave angle will depend on the height of the antenna over the ground.

2.2.2. Radiation patterns

Fig 10-13 shows the vertical and the horizontal radiation patterns for an equilateral-triangle delta loop, fed at the

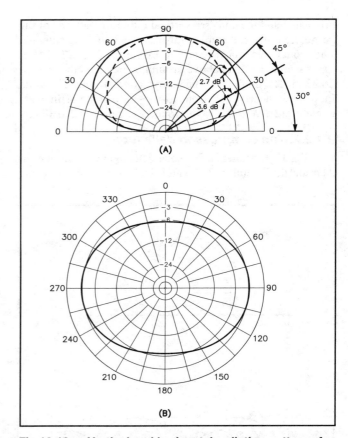

Fig 10-13 — Vertical and horizontal radiation patterns for an 80-meter equilateral delta loop fed for horizontal polarization, with the bottom wire at 3 m. The radiation is essentially at very high angles, comparable to what can be obtained from a dipole or inverted-V dipole at the same (apex) height.

center of the bottom wire. As anticipated, the radiation is maximum at zenith. The front-to-side ratio is around 3 dB for a 15 to 45° wave angle. Over average ground the gain is 2.5 dBi.

Looking at the pattern shape, one would be tempted to say that this antenna is no good for DX. So far we have only spoken about relative patterns. What about real gain figures from the vertically and the horizontally polarized delta loops?

2.3. Vertical Versus Horizontal Polarization

Fig 10-14 shows the superimposed elevation patterns for vertically and horizontally polarized low-height equilateral-triangle delta loops over two different types of ground (same dB scale).

MININEC-based modeling programs cannot be used to compute the gain figures of these loops, as impedance and gain figures are incorrect for very low antenna heights.

Over very poor ground

The horizontally polarized delta loop is better than the vertically polarized loop for all wave angles above 35°. Below 35° the vertically polarized loop takes over, but quite marginally. The maximum gain of the vertically and the horizontally polarized loops differs by only 2 dB, but the big difference is that for the horizontally polarized loop, the gain occurs at almost 90°, while for the vertically polarized loop it occurs at 25°.

One might argue that for a 30° wave angle, the horizontally polarized loop is as good as the vertically polarized loop. It is clear, however, that the vertically polarized antenna gives good high-angle rejection (rejection against local signals), while the horizontally polarized loop will not.

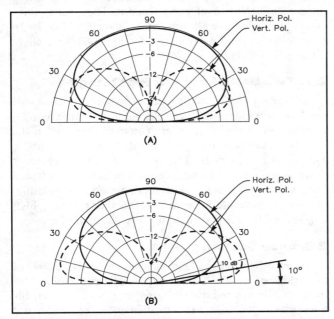

(A)

(B)

Fig 10-14 — Radiation patterns of vertically and horizontally polarized delta loops on the same dB scale. At A, over very poor ground, and at B, over very good ground. These patterns illustrate the tremendous importance of ground conductivity with vertically polarized antennas. Over better ground, the vertically polarized loop performs much better at low radiation angles, while over both good and poor ground the vertically polarized loop gives good discrimination against high-angle radiation. This is not the case for the horizontally polarized loop.

Over very good ground

The same thing that happens with any vertical happens with our vertically polarized delta: The performance at low angles is greatly improved with good ground. The vertically polarized loop is still better at any wave angle under 30° than when horizontally polarized. At a 10° radiation angle, the difference is as high as 10 dB.

Conclusion

Over very poor ground, the vertically polarized loops do not provide much better low-angle radiation when compared to the horizontally polarized loops. They have the advantage of giving substantial rejection at high angles, however.

Over good ground, the vertically polarized loop will give up to 10-dB gain at low radiation angles as compared to the horizontally polarized loop, in addition to its high-angle rejection. See Fig 10-14B.

2.4. Dimensions

The length of the resonant delta loop is approximately 1.05 to 1.06 λ. When putting up a loop, cut the wire at 1.06 λ, check the frequency of minimum SWR (it is always the resonant frequency), and trim the length.

The wavelength is given by

$$\lambda\,(m) = \frac{299.8}{f\,(MHz)}$$

2.5. Feeding The Delta Loop

The feed point of the delta loop in free space is symmetrical. At high heights above ground the loop feed point is to be considered as symmetrical, especially when we feed the loop in the center of the bottom line (or at the apex), because of its full symmetry with respect to the ground.

Fig 10-15 shows the radiation resistance and reactance for both the horizontally and the vertically polarized equilateral delta loops as a function of height above ground. At low heights, when fed for vertical polarization, the feed point is to

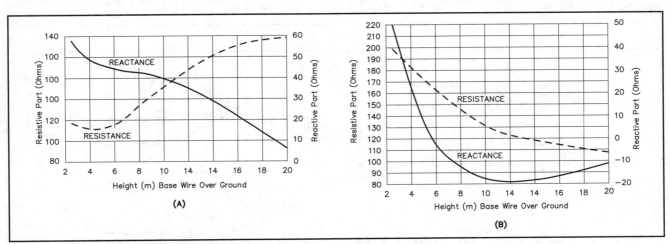

(A)

(B)

Fig 10-15 — Radiation resistance of (A) horizontally and (B) vertically polarized equilateral delta loops as a function of height above average ground. The delta loop was first dimensioned to be resonant in free space (reactance equals zero). Those dimensions were then used for calculating the impedance over real ground. Modeling was done at 3.75 MHz over good ground, using *NEC*.

be considered as asymmetric, whereby the "cold" point is the point to which the "radials" are connected. The center conductor of a coax feed line goes to the sloping vertical section. Many users have, however, used (symmetric) open-wire line to feed the vertically polarized loop (eg, 450-Ω line).

Most practical delta loops show a feed-point impedance between 50 and 100 Ω, depending on the exact geometry and coupling to other antennas. In most cases the feed point can be reached, so it is quite easy to measure the feed-point impedance using, for example, a good-quality noise bridge connected directly to the antenna terminals. If the impedance is much higher than 100 Ω (equilateral triangle), feeding via a 450-Ω open-wire feeder may be warranted. Alternatively, one could use an *unun* (unbalanced-to-unbalanced) transformer, which can be made to cover a very wide range of impedance ratios (see Chapter 6 on feed lines and matching). With somewhat compressed delta loops, the feed-point impedance is usually between 50 and 100 Ω. Feeding can be done directly with a 50 or 70-Ω coaxial cable, or with a 50-Ω cable via a 70-Ω quarter-wave transformer (Z_{ant} = 100 Ω).

To keep RF off the feed line it is best to use a balun, although the feed point of the vertically polarized delta loop is not strictly symmetrical. In this case, however, we want to keep any RF current from flowing on the outside of the coaxial feed line, as these parasitic currents could upset the radiation pattern of the delta loop. A stack of toroidal cores on the feed line near the feed point, or a coiled-up length of transmission line (making an RF choke) will also be useful. For more details refer to Section 7 of Chapter 6 on feed lines and matching.

2.6. Gain and Radiation Angle

Fig 10-16 shows the gain and the main-lobe radiation angle for the equilateral delta loop at different heights. The values were obtained by modeling a 3.8-MHz loop over average ground using *NEC*.

Earl Cunningham, K6SE, investigated different configurations of single element loops for 160 meters, and came up with the results listed in **Table 10-1** (modeling done with *EZNEC* over good ground). These data correspond surprisingly well with those shown in Fig 10-16 (where the ground was average), which explains the slight difference in gain.

2.7. Two Delta Loops at Right Angles

If the 4 to 5 dB front-to-side ratio bothers you, and if you have sufficient space, you can put up two delta loops at right angles on the same tower. You must, however, make provisions to open up the feed point of the antenna not in use, as well as its apex. This results in two non-resonant wires that do not influence the loop in use. If you would leave the unused delta loop in its connected configuration, the two loops would influence one another to a very high degree, and the results would be very disappointing.

2.8. Loop Supports

Vertically polarized loop antennas are really an array of two (sloping) verticals, each with an elevated radial. This means that if you support the delta loop from a metal tower, this tower may well influence the radiation pattern of the loop if it

Table 10-1
Loop Antennas for 160 Meters

Description	Feeding Method	Gain dBi	Radiation Angle
Diamond loop, bottom 2.5 m high	In side corner	2.15 dBi	18.0°
Square loop, bottom 2.5 m high	In center of one vertical wire	2.06 dBi	20.5°
Inverted-equilateral delta loop (flat wire on top)	Fed ¼ λ from bottom	1.91 dBi	20.9°
Regular equilateral delta loop	Fed ¼ λ from top	1.90 dBi	18.1°

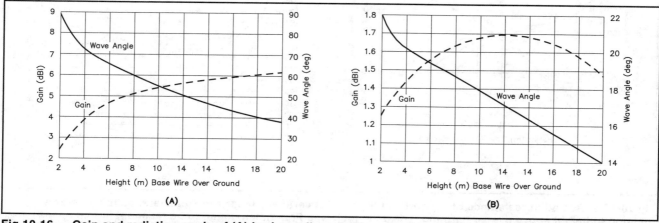

Fig 10-16 — Gain and radiation angle of (A) horizontally and (B) vertically polarized equilateral delta loops as a function of the height above ground. Modeling was done at 3.75 MHz over average ground, using *NEC*.

resonated anywhere in the vicinity of the loop. You can investigate this by modeling, but when the tower is loaded with Yagis, it is often difficult to exactly model the Yagis and their influence on the electrical length of the tower.

The safest thing you can do is to detune the tower to make sure the smallest possible current flows in the structure. The easiest way I found to do this is to drop a wire from the top of the tower, parallel with the tower (at 0.5 to 1.0 m distance) and terminate this wire via a 0 to 2000 pF variable capacitor to ground. Next use a current probe (such as is shown in Fig 11-17 in the chapter on Vertical Arrays) and adjust the variable capacitor for minimum current. The capacitor can be replaced with a fixed one (using a parallel combination of several values, if necessary) having the same value. This procedure will guarantee minimum mutual coupling between the loop and the supporting tower.

2.9. Modeling Loops

Loops can be modeled with *MININEC* when it comes to radiation pattern generation. Because of the acute angles at the corners of the delta loop, special attention must be paid to the length of the wire segments near the corners. Wire segments that are too long near wire junctions with acute angles will cause pulse overlap (the total conductor will look shorter than it actually is). The wire segments need to be short enough in order to obtain reliable impedance results. Wire segments of 20 cm length are in order for an 80-meter delta loop if accurate results are required. To limit the total number of pulses, the segment-length tapering technique can be used: The segments are shortest near the wire junction, and get gradually longer away from the junction. *ELNEC* as well as *EZNEC* have a special provision that automatically generates tapered wire segments (Ref 678).

At low heights (bottom of the antenna below approx. 0.2 λ), the gain and impedance figures obtained with a *MININEC*-based program are incorrect. The gain is too high, and the impedance too low. This is because *MININEC* calculates using a perfect ground under the antenna. Correct gain and impedance calculations at such low heights require modeling with a *NEC*-based program, such as *EZNEC*. All gain and impedance data listed in this chapter were obtained by using such a *NEC*-based modeling program.

■ 3. LOADED LOOPS

3.1. CW and SSB 80-Meter Coverage

An 80-meter delta loop or quad loop will not cover 3.5 through 3.8 MHz with an SWR below 2:1. There are two ways to achieve a wide-band coverage:
1) Feed the loop with an open-wire line (450 Ω to a matching network).
2) Use inductive or capacitive loading on the loop to lower its resonant frequency.

3.1.1. Inductive loading

For more information about inductive loading, you can refer to the detailed treatment of short verticals in Chapter 9 on vertical antennas.

There are three principles:
- The required inductance of the loading devices (coils or stubs) to achieve a given downward shift of the resonant frequency will be minimum if the devices are inserted at the maximum current point (similar to base loading with a vertical). At the minimum current point the inductive loading devices will not have any influence. This means that for a vertically polarized delta loop, the loading coil (or stub) cannot be inserted at the apex of the loop, nor in the middle of the bottom wire.
- Do not insert the loading devices in the radiating parts of the loop. Insert them in the part where the radiation is canceled. For example, in a vertically polarized delta loop, the loading devices should be inserted in the bottom (horizontal) wire near the corners.
- Always keep the symmetry of the loop intact, including after having added a loading device.

From a practical (mechanical) point of view it is convenient to insert the loading coil (stub) in one of the bottom corners. **Fig 10-17A** shows the loaded, compressed delta loop (with the same physical dimensions as the loop shown in Fig 10-11), where we have inserted a loading inductance in the bottom corner near the feed point.

A coil with a reactance of 240 Ω (on 3.5 MHz) or an inductance of 10.9 μH will resonate the delta on 3.5 MHz. The 100-kHz SWR bandwidth is 1.5:1. Note again the high-angle fill in the broadside pattern (no longer symmetrical baseline configuration), as well as the asymmetrical front-to-side ratio of the loop. The 10.9-μH coil can be replaced with a shorted stub. The inductive reactance of the closed stub is given by:

$$X_L = Z \tan \ell$$

where
- Z = characteristic impedance of stub (transmission line)
- ℓ = length of line, degrees
- X_L = inductive reactance

From this,

$$\ell = \arctan\left(\frac{X_L}{Z}\right)$$

In our example:
- $X_L = 240$ Ω
- $Z = 450$ Ω

$$\ell = \arctan\left(\frac{240}{450}\right) = 28°$$

Assuming a 95% velocity factor for the transmission line, we can calculate the physical length of the stub as follows:

$$\text{Wavelength} = \frac{299.8}{3.5} = 85.66 \text{ m} \left(\text{for } 360°\right)$$

$$\text{Physical Length} = 85.66 \text{ m} \times 0.95 \times \frac{28°}{360°} = 6.33 \text{ m}$$

Parts B and C of Fig 10-17 show the radiation patterns resulting from the insertion of a single stub (or coil) in one of the bottom corners of the delta loop. The insertion of the single loading device has broken the symmetry in the loop, and the bottom wire now radiates as well, upsetting the pattern of the loop.

This can be avoided by using two loading coils or stubs, located symmetrically about the center of the baseline. The example in **Fig 10-18A** shows two stubs, one located in each bottom corner of the loop. Each loading device has an inductive reactance of 142 Ω. For 3.5 MHz this is

$$\frac{142}{2\pi \times 3.5} = 6.46 \ \mu H$$

A 450-Ω short circuited line is 3.96 m long (see calculation method above). The corresponding radiation patterns in Fig 10-18 are now fully symmetrical, and the annoying high-angle radiation is totally gone.

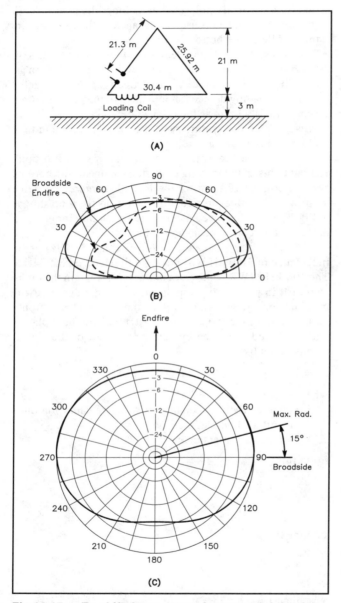

Fig 10-17 — To shift the resonant frequency of the delta loop from 3.75 MHz to 3.55 MHz, a loading coil (or stub) is inserted in one bottom corner of the loop, near the feed point (A). This has eliminated the reactive component, but has also upset the symmetrical current distribution in the bottom wire. Vertical patterns are shown at B, and the horizontal pattern is shown at C for a 27° wave angle. As with the loop shown in Fig 10-12, high-angle radiation (horizontal component) has appeared, and the horizontal pattern exhibits a notch in the endfire direction. Maximum radiation is again slightly off from the broadside direction.

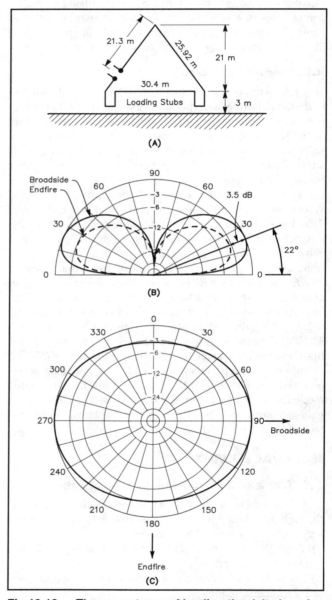

Fig 10-18 — The correct way of loading the delta loop is to insert two loading coils (or stubs), one in each bottom corner. This keeps the current distribution in the baseline symmetrical, and preserves a "clean" radiation pattern in the horizontal as well as the vertical plane. The horizontal pattern at C is for a wave angle of 22°.

The 100-kHz SWR bandwidth is 1.45:1. The 2:1 SWR bandwidth is 170 kHz.

Fig 10-19 shows the practical arrangement that can be used for installing the switchable stubs at the two delta-loop bottom corners. A small plastic box is mounted on a piece of epoxy printed-circuit-board material that is also part of the guying system. In the high-frequency position the stub should be completely isolated from the loop. Use a good-quality open-wire line and DPDT relay with ceramic insulation. The stub can be attached to the guy lines, which must be made of insulating material.

3.1.2. Capacitive loading

We can also use capacitive loading in the same way that we employ capacitive loading on a vertical. Capacitive loading has the most effect when applied at a voltage antinode (also called a voltage point).

This capacitive loading is much easier to install than the inductive loading, and requires only a single-pole (high-voltage!) relay to switch the capacitance wires in or out of the circuit. **Keep the ends of the wires out of reach of people and animals, as extremely high voltage is present.**

Fig 10-20 shows different possibilities for capacitive loading on both horizontally and vertically polarized loops. If installed at the top of the delta loop as in Fig 10-20C, a 9-m long wire inside the loop will shift the 3.8-MHz loop (from Fig 10-11) to resonance at 3.5 MHz. For installation at the center of the baseline, you can use a single wire (Fig 10-20D), or two wires in the configuration of an inverted V (Fig 10-20E). Several wires can be connected in parallel to increase the capacitance. **(Watch out, as there is very high voltage on those wires while transmitting!)**

The same symmetry guidelines should be applied as explained in Section 3.1 in order to preserve symmetrical current distribution.

Adjustment

Once the loop has been trimmed for resonance at the high-frequency end of the band, just attach a length of wire with a clip at the voltage point and check the SWR to see

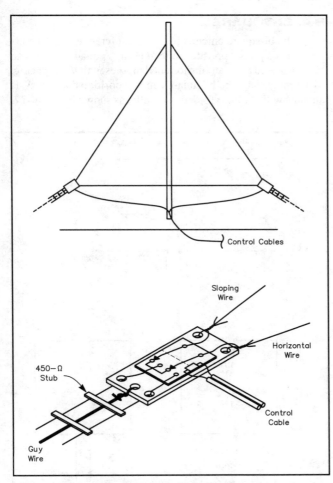

Fig 10-19 — Small plastic boxes, mounted on a piece of glass-epoxy board, are mounted at both bottom corners of the loop, and house DPDT relays for switching the stubs in and out of the circuit. The stubs can be routed along the guy lines (guy lines must be made of insulating material). The control-voltage lines for the relays can be run to a post at the center of the baseline and from there to the shack. Do not install the control lines parallel to the stubs.

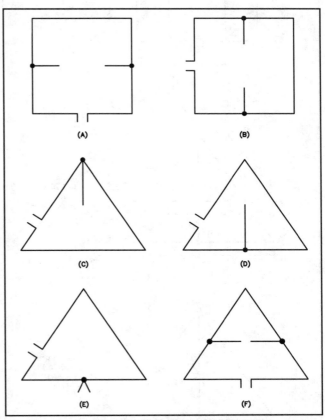

Fig 10-20 — Various loop configurations and possible capacitive loading alternatives. Capacitive loading must be applied at the voltage maximum points of the loops to have maximum effectiveness. The loading wires carry very high voltages, and good insulators should be used in their installation.

how much the resonant frequency has been lowered. It should not take you more than a few iterations to determine the correct wire length. If a single wire turns out to require too much length, connect two or more wires in parallel, and fan out the wire ends to create a higher capacitance.

3.1.3. Bandwidth

By using one of the above-mentioned loading methods and a switching arrangement, a loop can be made that covers the entire 80-meter band with an SWR below 2:1.

3.2. Reduced-Size Loops

Reduced-size loops have been described in amateur literature (Refs 1115, 1116, 1121, 1129). **Fig 10-21** shows some of the possibilities of applying capacitive loading to loops, whereby a substantial shift in frequency can be obtained. G3FPQ uses a reduced-size 2-element 80-meter quad that makes use of capacitive-loaded square elements as shown in Fig 10-21A. The loading wires are supported by the fiberglass spreaders of the quad.

It is possible to lower the frequency by a factor of 1.5 with this method, without lowering the radiation resistance to an unacceptable value (a loop dimensioned for 5.7 MHz can be loaded down to 3.8 MHz).

The triangular loop can also be loaded in the same way, although the mechanical construction may be more complicated than with the square loop. See Fig 10-21B.

In principle, we can replace the parallel wires with a (variable) capacitor. This would allow us to tune the loop. The example in Fig 10-21C requires approximately 30 pF to shift the antenna from 5.7 to 3.8 MHz. Beware, however, that extremely high voltages exist across the capacitor. It would certainly not be over-engineering to use a 50 kV or higher capacitor for the application.

■ 4. BI-SQUARE

The bi-square antenna has a circumference of 2 λ and is opened at a point opposite the feed point. A quad antenna can be considered as a pair of shortened dipoles with ¹/₄-λ spacing. In a similar way, the bi-square can be considered as a lazy-H antenna with the ends folded vertically, as shown in **Fig 10-22**.

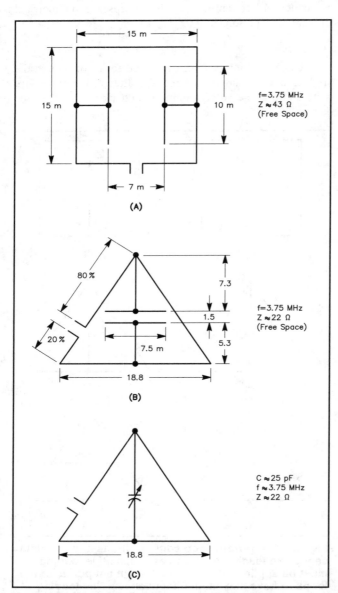

Fig 10-21 — Capacitive loading can be used on loops of approximately ²/₃ full size. See text for details.

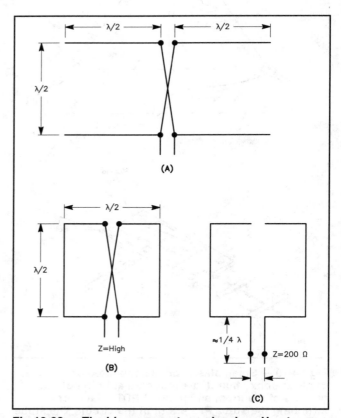

Fig 10-22 — The bi-square antenna is a lazy-H antenna (two ¹/₂-λ collinear dipoles, stacked ¹/₂ λ apart and fed in phase), with the ends of the dipoles bent down (or up) and connected. The feed-point impedance is high and the array can best be fed via a ¹/₄-λ stub arrangement.

Not many people are able to erect a bi-square antenna, as the dimensions involved on the low bands are quite large.

In free space the bi-square has 3-dB gain over two $\frac{1}{2}$-λ dipoles in phase (collinear), and almost 5 dB over a single $\frac{1}{2}$-λ dipole. Over real ground, with the bottom wire $\frac{1}{8}$ λ above ground (10 m for an 80-meter bi-square), the gain of the bi-square is the same as for the two $\frac{1}{2}$-λ dipoles in phase. The bottom two $\frac{1}{2}$-λ sections do not contribute to low-angle radiation of the antenna.

The bi-square has the advantage over two half-waves in phase that the antenna does not exhibit the major high-angle sidelobe that is present with the collinear antenna when the height is over $\frac{1}{2}$ λ. **Fig 10-23** shows the radiation patterns of the bi-square and the collinear with the top of the antenna $\frac{5}{8}$-λ high. Notice the cleaner low-angle pattern of the bi-square. Of course one could obtain almost the same result by lowering the collinear from $\frac{5}{8}$ to $\frac{1}{2}$-λ high!

The bi-square can be raised even higher in order to further reduce the wave angle without introducing high-angle lobes, up to a top height of 2 λ. At that height the wave angle is 14°, without any secondary high-angle lobe. With the top at $\frac{5}{8}$ λ, the wave angle is 26°.

In order to exploit the advantages of the bi-square antenna, you need quite impressive heights on the low bands. N7UA is one of the few stations using such an antenna, and he is producing a most impressive signal on the long path into Europe on 80 meters. With a proper switching arrangement, the antenna can be made to operate as a full-wave loop on half the frequency (eg, 160 m for an 80-meter bi-square).

The feed-point impedance is high (a few thousand ohms), and the recommended feed system consists of 600-Ω line with a stub to obtain a 200-Ω feed point. By using a 4:1 balun, a coaxial cable can be run from that point to the shack. Another alternative is to run the 600-Ω line all the way to the shack into an open-wire antenna tuner.

■ 5. THE HALF LOOP

The half loop was first described by Belrose, VE2CV (Ref 1120 and 1130). This antenna, unlike the half sloper, cannot be mounted on a tall tower supporting a quad or Yagi. If this was done, the half loop would shunt-feed RF to the tower and the radiation pattern would be upset. This can be avoided by decoupling the tower using a $\frac{1}{4}$-λ stub (Ref 1130). The half loop as shown in **Fig 10-24** can be fed in different ways.

5.1. The Low-Angle Half Loop

For low-angle radiation, the feed point can be at the end of the sloping wire (with the tower grounded), or else at the

Fig 10-23 — The bi-square antenna (A) and its radiation patterns (B and C). The azimuth pattern at B is for a wave angle of 25°. At D, two half waves in phase and at E, its radiation pattern. Note that for a top-wire height of $\frac{5}{8}$ λ, the bi-square does not exhibit the annoying high-angle lobe of the collinear antenna.

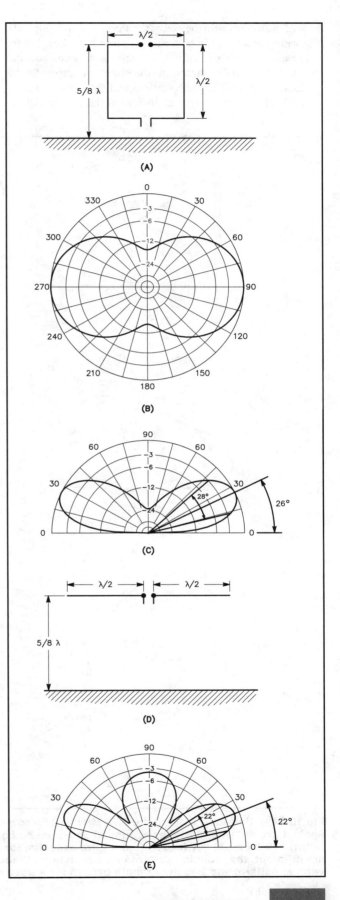

base of the tower (with the end of the sloping wire grounded). The radiation pattern in both cases is identical. The front-to-side ratio is approximately 3 dB, and the antenna radiates best in the broadside direction (the direction perpendicular to the plane containing the vertical and the sloping wire).

There is some pattern distortion in the end-fire direc-

tion, but the horizontal radiation pattern is fairly omnidirectional. Most of the radiation is vertically polarized, so the antenna requires a good ground and radial system, as for any vertical antenna. As such, the half loop does not really belong to the family of large loop antennas, but as it is derived from the full-size loop, it is treated in this chapter

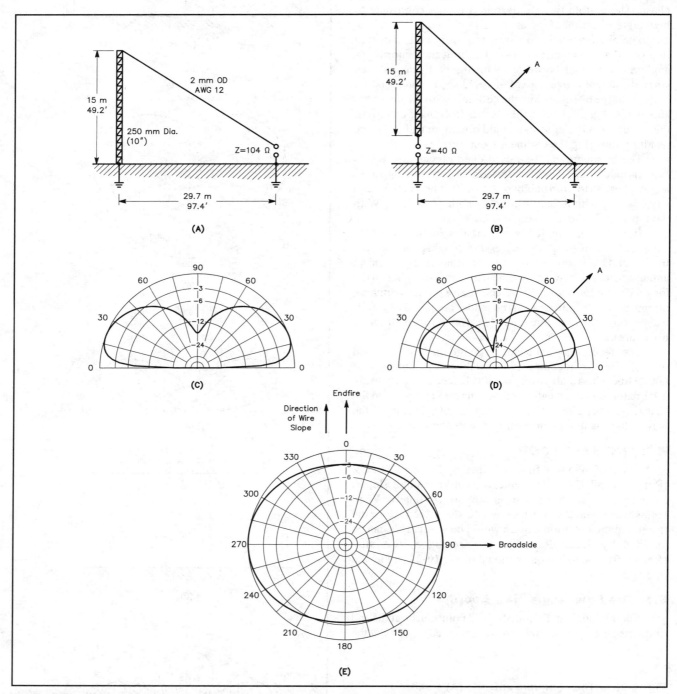

Fig 10-24 — Half-loop antenna for 3.75 MHz, fed for low-angle radiation. The antenna can be fed at either end against ground (A and B). The grounded end must be connected to a good ground system, as must the ground-return conductor of the feeder. Radials are essential for proper operation. Note that while the feed-point locations are different, the radiation patterns do not change. C shows the broadside vertical pattern, D is the endfire vertical pattern, and E is the azimuth pattern for a wave angle of 20°.

rather than as a top-loaded short vertical.

The exact resonant frequency depends to a great extent on the ratio of the diameter of the vertical mast to the slant wire. The dimensions shown in Fig 10-24 are only indicative. Fine-tuning the dimensions will have to be done in the field.

5.2. The High-Angle Half Loop

The half delta loop antenna can also be used as a high-angle antenna. In that case you must isolate the tower section from the ground (use a good insulator because it now will be at a high-impedance point) and feed the end of the sloping wire. Alternatively, you can feed the antenna between the end of the sloping wire and ground, while insulating the bottom of the tower from ground. Using the same dimensions that made the low-angle version resonant no longer produces resonance in these configurations.

Fig 10-25 shows the low-angle configurations with the radiation patterns. Note that the alternative where the end of the slant wire is fed against ground produces much more high-angle radiation than the alternative where the bottom end of the tower is fed. In both cases, the other end of the

aerial is left floating (not connected to ground).

Dimensional configurations other than those shown in the relevant figures can be used as well, such as with a higher tower section and a shorter slant wire. If you move the end of the sloping wire farther away from the tower, you will need to decrease the height of the tower to keep resonance, and the radiation resistance will decrease. This will, of course, adversely influence the efficiency of the antenna. If the bottom of the sloping wire is moved toward the tower, the length of the vertical will have to be increased to preserve resonance. When the end of the sloping wire has been moved all the way to the base of the tower we have a $^1/_4$-λ vertical with a folded feed system. The feed-point impedance will depend on the spacing and the ratio of the tower diameter to the feed-wire diameter.

Compared to a loaded vertical, this antenna has the advantage of giving the added possibility for switching to a high-angle configuration. For a given height, the radiation resistance is slightly higher than for the top-loaded vertical, whereby there is no radiation from the top load. The sloping wire in this half-loop configuration adds somewhat to the

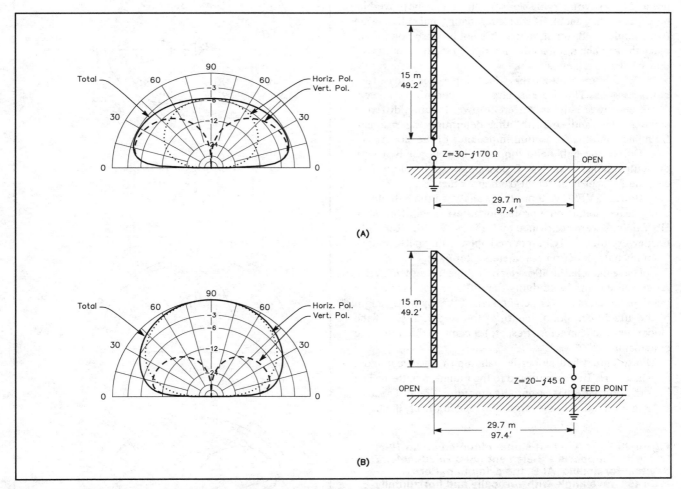

Fig 10-25 — High-angle versions of the half delta loop antenna for 3.75 MHz. As with the low-angle version, the antenna can be fed at either end (against ground). The other end, however, must be left floating. The two different feed points produce different high-angle patterns as well as different feed-point impedances.

vertical radiation, hence the increase (10% to 15%) in R_{rad}.

Being able to feed the antenna at the end of the sloping wire may also be an advantage: This point may be located at the transmitter location, so the sloping wire can be directly connected to an antenna tuner. This would enable wide-band coverage by simply retuning the antenna tuner. Switching from a high to a low-angle antenna in that case consists of shorting the base of the tower to ground (for low-angle radiation).

■ 6. THE HALF SLOPER

Although the so-called half sloper of **Fig 10-26A** may look like a half delta, it really does not belong with the loop antennas. As we will see, it is rather a loaded vertical with a specific matching system and current distribution.

Quarter-wave slopers are the typical result of ham ingenuity and inventiveness. Many DXers, short of space for putting up large, proven low-angle radiators, have found their half slopers to be good performers. Of course they don't know how much better other antennas might be, as they have no room to try them. Others have reported that they could not get their half sloper to resonate on the desired frequency (that's because they gave up trying before having found the proverbial needle in the haystack). Of course resonating and radiating are two completely different things. It's not because you cannot make the antenna resonant that it will not radiate well. Maybe they need a matching network?

To make a long story short, half slopers seem to be very unpredictable. There are a large number of parameters (different tower heights, different tower loading, different slope angles, and so forth) that determine the resonant frequency and the feed-point impedance of the sloper.

Unlike the half delta loop, the half sloper is a very difficult antenna to analyze from a generic point of view, as each half sloper is different from any other.

Belrose, VE2CV, thoroughly analyzed the half sloper using scale models on a professional test range (Ref 647). His findings were confirmed by DeMaw, W1FB (Ref 650). Earlier, Atchley, W1CF, reported outstanding performance from his half sloper on 160 meters (Ref 645).

I have modeled an 80-meter half sloper using *MININEC*. After many hours of studying the influence of varying the many parameters (tower height, size of the top load, height of the attachment point, length of the sloper, angle of the sloper, ground characteristics, etc), I came to the following conclusion:

- The so-called half sloper is made up of a vertical and a slant wire. Both contribute to the radiation pattern. The radiation pattern is essentially omnidirectional. The low-angle radiation comes from the loaded tower, the high-

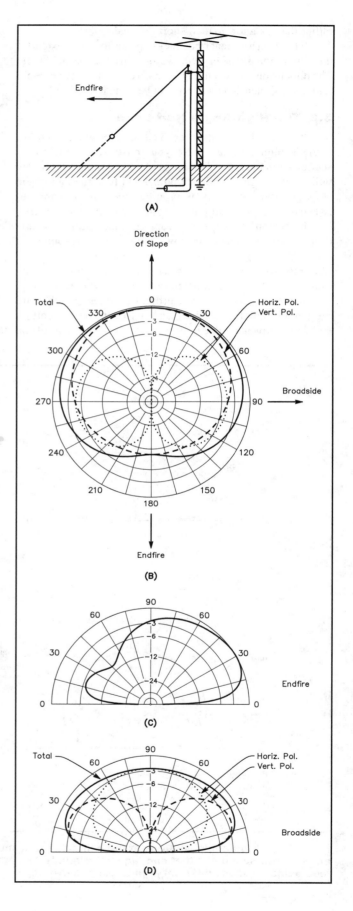

Fig 10-26 — At A, a half-sloper mounted on an 18-m tower that supports a 3-element full-size 20-meter Yagi. See text for details. At B, the azimuth pattern for a 45° wave angle with vertically and horizontally polarized components, and at C and D, elevation patterns. The antenna shows a modest F/B ratio in the end-fire direction at a 45° wave angle.

angle radiation from the horizontal component of the slant wire. The antenna radiates a lot of high-angle signal (coming from the slant wire).

- Over poor ground the antenna has some front-to-back advantage in the direction of the slope, ranging from 10 to 15 dB at certain wave angles.
- Over good and excellent ground the F/B ratio is not more than a few dB.

An interesting testimony was sent on Internet by Rys, SP5EWY, who wrote "Well, I had previously used my tower without radials and the half sloper favored the South, with the wire sloping in that direction, by at least 1 S-unit. Later I added 20 radials and since then it seems to radiate equally well in all directions."

In essence the half sloper is a top-loaded vertical, which is fed at a point along the tower where the combination of the tower impedance and the impedance presented by the sloping wire combine to a 50-Ω impedance (at least that's what we want). The sloping wire also acts as a sort of "radial" to which the "other" conductor of the feed line is connected (like radials on a vertical). In other words, the sloping wire is only a minor part of the antenna, a part that helps to create resonance as well as to match the feed line. Belrose (Ref 647) also recognized that the half sloper is effectively a top-loaded vertical. Fig 10-26 (B through D) shows the typical radiation patterns obtained with a half-sloper antenna.

While modeling the antenna, it was very critical to find a point on the tower and a sloper length and angle that give a good match to a 50-Ω line. The attachment point on the tower need not be at the top. It is not important how high it is, as you are not really interested in the radiation from the slant wire.

Changing the attachment point and the sloper length does not appreciably change the radiation pattern. This indicates that it is the tower (capacitively loaded with the Yagi) that does the bulk of the radiating. As the antenna mainly produces a vertically polarized wave, it requires a good ground system, at least as far as its performance as a low-angle radiator is concerned.

From my experience in spending a few nights modeling half slopers, I would highly recommend any prospective user to first model the antenna on MININEC. (ELNEC is just great for such a purpose, as it has the most user-friendly interface for multiple iterations.)

There is an interesting analysis by D. DeMaw (Ref 650). DeMaw correctly points out that the antenna requires a metal support, and that a tree or a wooden mast will not do. But he does not emphasize anywhere in his study that it is the metal support that is responsible for most of the desirable low-angle radiation. DeMaw, however, recognizes the necessity of a good ground system on the tower, which implicitly admits that the tower does the radiating. DeMaw also says, "The antenna is not resonant at the operating frequency," by which he means that the slant wire is not a quarter-wave long. This is again very confusing, as it seems to indicate that the slant wire is the antenna, which it is not. Describing the on-the-air results, DeMaw confirms what we have modeled: Due to the presence of high-angle radiation, it outperforms the vertical for short and medium-range contacts, while the vertical takes over at low angles for real DX contacts.

To summarize the performance of half slopers, it is worthwhile to note Belrose's comment, "If I had a single quarter-wave tower, I'd employ a full-wave delta loop, apex up, lower-corner fed, the best DX-type antenna I have modeled."

Of course, a delta loop still has a baseline of approximately 100 feet (on the 80-meter band), which is not the case with the half sloper. But the half sloper, like any vertical, requires radials in order to work well. It may look like the half sloper has a space advantage over many other low-band antennas, but this is only as true as for any vertical.

PHASED ARRAYS

11 PHASED ARRAYS

THANKS WØUN

Hardly any active ham needs an introduction to John Brosnahan, WØUN. Another American antenna guru, John lives on the high plains of Colorado. His 160 acre (64 hectare) playground is all one can dream of to realize the most exotic antenna dreams. The main thing John lacks is time to realize these dreams. John is the president and technical director of Alpha/Power, Inc., the Colorado-based manufacturer of top-notch power amplifiers. Although I had met John eye to eye on a number of occasions at the Dayton Hamvention over the years it was during WRTC 1996 in San Francisco that we got to know each other better, and I have followed his moves in Amateur Radio ever since.

John is a research physicist by education who has spent his career on the electrical engineering side of remote-sensing instrumentation. From 1973 to 1978 he was the engineer for the University of Colorado's radio astronomy observatory, designing receivers and antenna arrays for HF and VHF radio astronomy. Since then he has been founder and president of two companies that design and build HF and VHF radar systems for remote sensing of the atmosphere and ionosphere. He has designed and built arrays all the way from 80 dipoles at 2.66 MHz, which covered 40 acres (16 hectares), to a 12,288-dipole array at 49 MHz. He has also built numerous Yagi arrays, including a 768-element array at 52 MHz to a 500-element array at 404 MHz.

When asked, John immediately volunteered to review the two chapters on arrays for the low bands. Thank you, John, for your help, your input, and also your friendship!

If you want gain and directivity on one of the low bands and if you live in an area with good or excellent ground, an array made of vertical elements may be the answer, provided you have room for it. If you are in the desert or live in a rocky area with poor electrical ground properties, you probably would do better looking at an array made of horizontally polarized elements or maybe a (shortened) Yagi or quad antenna if you want gain and efficiency. The reason is that reflection of radio waves on imperfect ground behaves very different with horizontally polarized signals than with vertically polarized signals. (See Chapter 8, Sections 1.1.2 and 1.2.1.4.)

Arrays made with vertical elements have the same requirements as single vertical antennas as far as ground quality is concerned. If you can take care of a good ground system for collecting element return currents (that is usually not a problem) and have enough ground screen to hide the lossy ground from the vertical antennas (or maybe have an elevated radial system that's at least $1/4$-λ long) and if you are fortunate to have very good electrical ground in your area (that's harder if not impossible to control), then you are in the game for a well-performing vertical array. Before you decide to put one up, take the time to understand the mechanism of an array with all-fed elements.

Dipoles, monopoles and other antennas can be combined to form an array in order to obtain increased directivity (and gain) in either the horizontal or the vertical plane (or both). Horizontal directivity is commonly sought on the HF bands, while at VHF, vertical directivity is often needed to concentrate RF energy close to the horizon (as with repeater antennas).

In this chapter I cover the subject of arrays made of elements that, by themselves, have an omnidirectional horizontal radiation pattern: vertical antennas.

■ 1. RADIATION PATTERNS

How the pattern is formed

Let us consider two verticals with spacing D and fed with a phase difference of a'. Fig **11-1** shows two vertical antennas (A and B). The paper represents the ground and both radiators are omnidirectional by definition. Rays a and a' from antenna A and B have a phase difference that depends on four factors:

1) Spacing.
2) The phase difference with which RF current is applied at the feed point of the antenna.
3) Angle of ray a and a' with respect to line AB.
4) The current magnitude in each element.

Consider the specific case where the spacing is $1/4$ λ and the phase difference is 90°, as shown in Fig 11-1. Rays b and b' are clearly in phase ($1/4$ λ due to spacing minus $1/4$ λ due to phase difference of 90°). Similarly, d and d' are 180° out of phase. Rays a and a' will reinforce one another, but c and c' will complement each other to a much lesser degree. The resulting radiation pattern is called a cardioid.

Directivity over perfect ground

Fig **11-2** shows a range of radiation patterns obtained by different combinations of two monopoles over perfect ground and at zero wave angle. These directivity patterns are classics

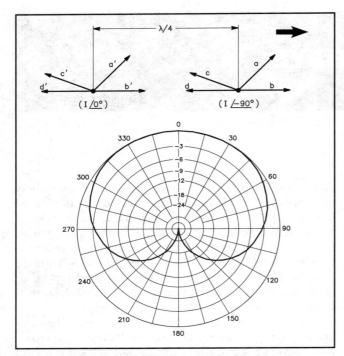

Fig 11-1—Antennas A and B are fed with the same current, but antenna A leads in phase by 90°. Graphic analysis of a few rays shows that the array will radiate most power in the direction of rays b and b′ where these rays, because of the physical separation and phase relationship between the elements, will show maximum reinforcement. The resultant radiation pattern is calculated at a 0° wave angle over ideal ground.

in every good antenna handbook. These are patterns for a theoretical zero wave angle.

Directivity over real ground

Over real ground there is no radiation at zero wave angle. All the effects of real ground, which were described in detail in the chapter on verticals, apply to arrays of verticals.

Direction of firing

The rule is simple: An array always fires in the direction of the element with the lagging feed current.

■ 2. ARRAY ELEMENTS

In principle, the whole range of verticals described in the chapter on vertical antennas can be used as elements for a vertical array. Quarter-wave elements have gained a reputation of giving a reasonable match to a 50-Ω line, which is certainly true for single vertical antennas. In this chapter we will learn the reason why quarter-wave resonant verticals do not have a resistive 36-Ω feed-point impedance when operated in arrays (even assuming a perfect ground). Quarter-wave elements still remain a good choice as they have a reasonably high radiation resistance, which ensures good overall efficiency. On 160 meters, the elements could be top-loaded verticals as described in the chapter on verticals.

The design methodology for arrays given in Section 3, as well as all the designs described in Section 4, assume that all the array elements are physically identical elements, with a current distribution that is the same on each element. In practice this means that only elements with a length of up to $1/4 \lambda$ should be used. The patterns given in Section 4, do not apply if you use elements much longer than $1/4 \lambda$. They certainly do *not* apply for elements that are $1/2 \lambda$ or $5/8-\lambda$ long. If you want to use "long" elements, you will have to model the design using the particular element lengths (Ref 959). This may be a problem if you want to use shunt-fed towers, carrying HF beams as elements for an array. With their top load, these towers are often much longer than $1/4 \lambda$.

■ 3. DESIGNING AN ARRAY

The radiation patterns as shown in Fig 11-2 give a good idea what can be obtained with different spacings and different current phase delays for a 2-element array. For arrays with more elements there are a number of popular classic designs. Many of those are covered in detail in this chapter.

A good array should meet the following specifications:
• High gain.
• Good front-to-back ratio, also at high elevation angles.
• Ease of feeding.
• Ease of direction switching.

3.1. Modeling Arrays

The original *MININEC* was not directly suited for modeling arrays. Excitation of elements in *MININEC* was by voltage. In arrays we normally define currents for the RF sources. It is the current (magnitude and phase) in each element that determines the radiation pattern of the array. Therefore currents, rather than voltages, must be specified.

ELNEC, which is based on *MININEC*, has overcome this problem, and the user has the choice of defining the sources as voltage or current sources.

It is only if you want to do some modeling that includes the influence of a radial system plus the influence of a "poor" ground that the full-blown *NEC* program is required. Studies on elevated radial systems and on buried radials require *NEC* (*NEC-3* if buried radials are involved).

In this chapter we will compare arrays against a single vertical, over an identical ground. The influence of the radial system has been included in the form of an equivalent loss resistance in series with each element feed point.

3.2. Getting the Right Current Magnitude and Phase

There is a world of difference between designing an array on paper or with a computer modeling program and realizing it in real life. With single-element antennas (a single vertical, a dipole, etc) we do not have to bother about the feed current (magnitude and phase), as there is only one feed point anyway. With phased arrays things are vastly different.

First we will need to make up our mind which array to build. Once this is done, the problem will be how to achieve the right feed currents in all the elements (magnitude and phase angle).

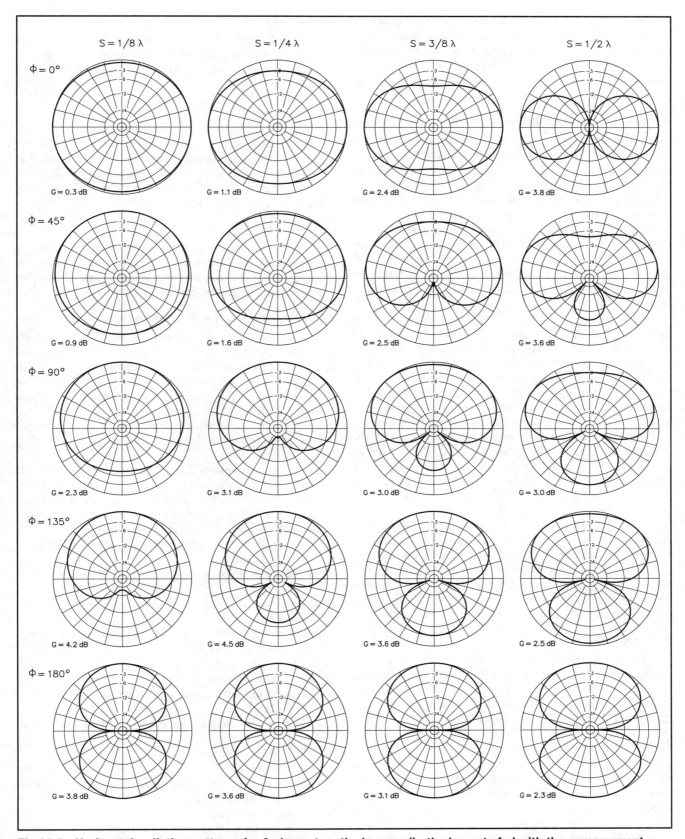

Fig 11-2—Horizontal radiation patterns for 2-element vertical arrays (both elements fed with the same current magnitude). The elements are in the vertical axis, and the top element is the one with the lagging phase angle. Patterns are for 0° wave angle over ideal ground. *(After The ARRL Antenna Book)*

When we analyze an array with a modeling program, we notice that the feed-point impedances of the elements change from the value for a single element. If the feed-point impedance of a single quarter-wave vertical is 36 Ω over perfect ground, it is always different from that value in an array because of mutual coupling.

3.2.1. The effects of mutual coupling

Few articles in Amateur Radio publications have addressed the problems associated with mutual coupling in designing a phased array and in making it work as it should. Gehrke, K2BT, wrote an outstanding series of articles on the design of phased arrays (Ref 921-925, 927). These are highly recommended for anyone who is considering putting up phased arrays of verticals. Another excellent article by Christman, K3LC (ex KB8I) (Ref 929), covers the same subject. The subject has been very well covered in the 15th and later editions of *The ARRL Antenna Book*, where R. Lewallen, W7EL, wrote a comprehensive contribution on arrays.

If we bring two (nearly) resonant circuits into the vicinity of each other, mutual coupling will occur. This is the reason that antennas with parasitic elements work as they do. Horizontally polarized antennas with parasitically excited elements are widely used on the higher bands. On the low bands the proximity of the ground limits the amount of control the designer has on the current in each of the elements. Arrays of (vertical) antennas, where each element is fed, overcome this limitation, and in principle the designer has an unlimited control over all the design parameters. With so-called phased arrays, all elements are individually and physically excited by applying power to the elements via individual feed lines. Each feed line supplies current of the correct magnitude and phase.

There is one frequently overlooked major problem with arrays. As we have made up our minds to feed all elements, we too often assume (incorrectly) there is no mutual coupling or that it is so small that we can ignore it. Taking mutual coupling into account complicates life, as we now have two sources of applied power to the elements of the array: parasitic coupling plus direct feeding.

Self-impedance

If a single quarter-wave vertical is erected, we know that the feed-point impedance will be 36 + *j*0 Ω, assuming resonance, a perfect ground system, and a reasonably thin conductor diameter. In the context of our array we will call this the self-impedance of the element.

Coupled impedance

If other elements are closely coupled to the original element, the impedance of the original element will change. Each of the other elements will couple energy into the original element and vice versa. The coupled impedance is the impedance of an element being influenced by one other element. Coupling from the coupled element results in a coupled impedance being totally different from the self-impedance in most cases.

Mutual coupling

Mutual coupling is the phenomenon that relates to the interaction of closely spaced elements in an array, causing the coupled impedances of the elements to be different from the self-impedance.

Mutual impedance

The mutual impedance is a term that defines unambiguously the effect of mutual coupling between a set of two antennas (elements). Mutual impedance is an impedance that cannot be measured. It can only be calculated. The calculated mutual impedances and driving impedances have been extensively covered by Gehrke, K2BT (Ref 923).

Drive impedance

In order to design a correct feed-system for an array, we must know the drive impedances of each of the elements, the elements being fed as required to produce the wanted radiation pattern (correct current magnitude and current angle).

3.2.2. Calculating the drive impedances

You cannot measure mutual impedance. It must be calculated. Mutual impedances are calculated from measured self-impedances and drive impedances. Here is an example: We are constructing an array with three ¼-λ elements in a triangle, spaced ¼-λ apart. We erect the three elements and install the final ground system. Make the ground system as symmetrical as possible. Where the buried radials cross, terminate them in a bus. Then the following steps are carried out:

1) Open-circuit elements 2 and 3 (opening an element will effectively isolate it from the other elements in the case of quarter-wave elements; when using half-wave elements the elements must be grounded for maximum isolation, and open-circuited for maximum coupling).
2) Measure the self-impedance of element 1 (= Z11).
3) Ground element 2.
4) Measure the coupled impedance of element 1 with element 2 coupled (= Z1,2).
5) Open-circuit element 2.
6) Ground element 3.
7) Measure the coupled impedance of element 1 with element 3 coupled (= Z1,3).
8) Open-circuit element 3.
9) Open-circuit element 1.
10) Measure the self-impedance of element 2 (= Z22).
11) Ground element 3.
12) Measure the coupled impedance of element 2 with element 3 coupled (= Z2,3).
13) Open-circuit element 3.
14) Ground element 1.
15) Measure the coupled impedance of element 2 with element 1 coupled (= Z2,1).
16) Open-circuit element 1.
17) Open-circuit element 2.
18) Measure the self-impedance of element 3 (= Z33).
19) Ground element 2.

20) Measure the coupled impedance of element 3 with element 2 coupled (= Z3,2).
21) Open-circuit element 2.
22) Ground element 1.
23) Measure the coupled impedance of element 3 with element 1 coupled (= Z3,1).

This is the procedure for an array with three elements. The procedures for 2 and 4-element arrays can be derived from the above.

As you can see, measurement of coupling is done by pairs of elements. At step 15, we are measuring the effect of mutual coupling between elements 2 and 1, and it may be argued that this has already been done in step 4. It is useful, however, to make these measurements again in order to recheck the previous measurements and calculations. Calculated mutual couplings Z12 and Z21 (see below) using the Z1,2 and Z2,1 inputs should in theory be identical, and in practice should be within an ohm or so.

The self-impedances and the driving impedances of the different elements should match closely if the array is to be made switchable.

The mutual impedances can be calculated as follows:

$$Z12 = \pm\sqrt{Z22 \times (Z11 - Z1,2)}$$

$$Z21 = \pm\sqrt{Z11 \times (Z22 - Z2,1)}$$

$$Z13 = \pm\sqrt{Z33 \times (Z11 - Z1,3)}$$

$$Z31 = \pm\sqrt{Z11 \times (Z33 - Z3,1)}$$

$$Z23 = \pm\sqrt{Z33 \times (Z22 - Z2,3)}$$

$$Z32 = \pm\sqrt{Z22 \times (Z33 - Z3,2)}$$

It is obvious that if Z11 = Z22 and Z1,2 = Z2,1, then Z12 = Z21.

If the array is perfectly symmetrical (such as in a 2-element array or in a 3-element array with the elements in an equilateral triangle), all self-impedances will be identical (Z11 = Z22 = Z33), and all driving impedances as well (Z2,1 = Z1,2 = Z3,1 = Z1,3 = Z2,3 = Z3,2). Consequently, all mutual impedances will be identical as well (Z12 = Z21 = Z31 = Z13 = Z23 = Z32). In practice, the values of the mutual impedances will vary slightly, even when good care is taken to obtain maximum symmetry.

Because all impedances are complex values (having real and imaginary components), the mathematics involved are difficult.

The MUTUAL IMPEDANCE AND DRIVING IMPEDANCE software module of the NEW LOW BAND SOFTWARE will do all the calculations in seconds. No need to bother with complex algebra. Just answer the questions on the screen.

Fig 11-3 shows the mutual impedance to be expected for quarter-wave elements at spacings from 0 to 1.0 λ. The resistance and reactance values vary with element separation as a damped sine wave, starting at zero separation with both signs positive. At about 0.10 to 0.15 λ spacing, the reactance sign changes from + to −. This is important to know in order

to assign the correct sign to the reactive value (obtained via a square root).

Gehrke, K2BT, emphasizes that the designer should actually measure the impedances and not take them from tables. Some methods of doing this are described in Ref 923. The published tables show ballpark figures, enabling you to verify the square-root sign of your calculated results.

After calculating the mutual impedances, the drive impedances can be calculated, taking into account the drive current (amplitude and phase). The driving-point impedances are given by

$$Zn = \frac{I1}{In} \times Zn1 + \frac{I2}{In} \times Zn2 + \frac{I3}{In} \times Zn3... + \frac{In}{In} \times Znn$$

where n is the total number of elements. The number of equations is n. The above formula is for the nth element. Note also that Z12 = Z21 and Z13 = Z31, etc.

The above-mentioned program module performs the rather complex driving-point impedance calculations for arrays with up to 4 elements. The required inputs are:
1) The number of elements.
2) The driving current and phase for each element.
3) The mutual impedances for all element pairs.

The outputs are the driving-point impedances Z1 through Zn.

Design example

Let us take the example of an array consisting of two $\frac{1}{4}$-λ long verticals, spaced $\frac{1}{4}$ λ apart and fed with equal magnitude currents, with the current in element 2 lagging the current in element 1 by 90°. This is the most common end-fire configuration with a cardioid pattern.

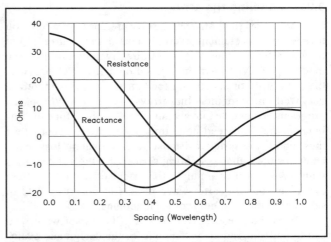

Fig 11-3—Mutual impedance for two $\frac{1}{4}$-λ elements. For shorter vertical elements (length between 0.1 and 0.25 λ), one can calculate the mutual impedance by multiplying the figures from the graph by the ratio R_{rad} / 36.6 where R_{rad} = the radiation resistance of the short vertical.

Self impedance

The quarter-wave long elements of such an array are assumed to have a self-impedance of 36.4 Ω over a perfect ground. A nearly perfect ground system consists of at least 120 half-wave radials (see Chapter 9, Vertical Antennas). For example, a system with "only" 60 radials may (depending on the ground quality) show a self-impedance on the order of 40 Ω.

Coupled impedance

We measured $37.5 + j15.2$ Ω.

Mutual impedance

The mutual impedances were calculated with the above-mentioned computer program: $Z12 = Z21 = 19.76 - j15.18$ Ω. From the mutual impedance curves in Fig 11-3 it is clear that the minus sign is the correct sign for the reactive part of the impedance.

Drive impedance

The same software module calculates the drive impedances (also called feed-point impedances) of the two elements.

$Z1 = 55.8 + j19.8$ Ω for the $-90°$ element.

$Z2 = 24.8 - j19.8$ Ω for the $0°$ element

We have now calculated the impedance of each element of the array, the array being fed with the current (magnitude and phase) as set out. We have used impedances that we have measured; we are not working with theoretical impedances.

The 2 EL AND 4 EL VERTICAL ARRAYS module of the NEW LOW BAND SOFTWARE is a perfect tool to guide you along the design of an array. You can enter your own values or just work your way through using a standard set of values.

3.2.3. Modeling the array

As the original *MININEC* software does not allow you to specify the element excitation as a feed current (but only as a feed voltage), it is not directly suited to do array modeling. *ELNEC*, which is based on *MININEC*, provides the possibility of specifying feed *currents* and is therefore very well suited for modeling arrays. As the *MININEC* based programs do not take into account "real ground" for calculating near-field and feed-point impedances, to be accurate we must add some series resistance (the equivalent loss resistance for the radial system) in series with the antenna feed point. We can, eg, simulate the effect of a radial system consisting of 60 quarter-wave radials by inserting 4 Ω in series with the feed point of each antenna.

With the latest *NEC-3* or *NEC-4* based software, you can include the radial system, but for the design and evaluation of arrays, a *MININEC*-based modeling program will do, as long as we realize that we must add some equivalent series resistance to account for the ground losses of the radial system.

Modeling the cardioid antenna over a perfect ground

system, *ELNEC* comes up with the following impedances:

$Z1 = 50.9 + j18.4$ Ω

$Z2 = 19.2 - j18.4$ Ω

These are close to the values worked out with the NEW LOW BAND SOFTWARE, which were based on measured values of coupled and self impedances. The difference seems to indicate the actual ground losses are larger than those assumed in the modeled example (R-part being larger). The vertical and the horizontal radiation patterns for the 2-element cardioid array are shown in **Fig 11-4**.

3.3. Designing a Feed System

The challenge now is to design a feed system that will supply the right current to each of the array elements. As we now know the current requirements as well as the drive-

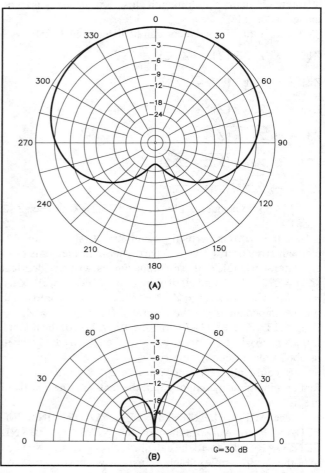

Fig 11-4—Vertical and horizontal radiation patterns for the 2-element cardioid array, spaced 90° and fed with 90° phase difference. The pattern was calculated for very good ground with a radial system consisting of 120 radials, each 0.4 λ long (the equivalent ground resistance is 2 Ω). The gain is 3.0 dB as compared to a single vertical over the same ground and radial system. The horizontal pattern at A is for an elevation angle of 19°.

impedance data for each element of the array, we have all the required inputs to design a correct feed system.

Each element will need to be supplied power with its own feed line. In a driven array each element gets power or possibly delivers power. During calculations we will sometimes encounter a negative feed-point impedance, which means the element is actually delivering power into the feed network. If the element impedance is zero, this means that the element can be shorted to ground. It then acts as a parasitic element.

Eventually all the feed lines will be connected to a common point, which will be the common feed point for the entire array. You can only connect feed lines in parallel if the voltages on the feed lines (at that point) are identical (in magnitude and phase)—the same as with ac power!

Designing a feed system consists of calculating the feed lines (impedance, length) as well as the component values of networks used in the feed system, so that the voltages at the input ends of the lines are identical. It is as simple as that.

The ARRL has published the original (1982) work by Lewallen, W7EL, in the latest editions of *The ARRL Antenna Book*. This material is a must for every potential array builder. However, there are other feed methods than the Lewallen method.

3.3.1. The wrong way

In just about all cases, the drive impedance of each element will be different from the characteristic impedance of the feed line. This means that there will be standing waves on the line. This has the following consequences:

• The impedance, voltage and current will be different in each point of the feed line.
• The current and voltage phase shift is not proportional to the feed line length, except for a few special cases (eg, a half-wave-long feed line).

This means that if we feed these elements with 50-Ω coaxial cable, we cannot simply use lengths of feed line as delay lines by making the line length in degrees equal to the desired delay in degrees. In the past we have seen arrays where a 90° long coax line was inserted in one of the feed lines to an element to create a 90° antenna current phase shift. Let us take the example of the 2-element cardioid array (as described above) and see what happens (see **Fig 11-5**).

We run two 90° long coax cables to a common point. Using the COAX TRANSFORMER/SMITH CHART software module of the NEW LOW BAND SOFTWARE, we calculate the impedances at the end of those lines (I took RG-213 with 0.35 dB/100 ft. attenuation at 3.5 MHz). The array element feed impedances (including 2 Ω of equivalent ground loss resistance) are (let's use round figures):

$$Z1 = 51 + j20 \ \Omega$$

$$I1 = 1 \ A, \underline{/-90°}$$

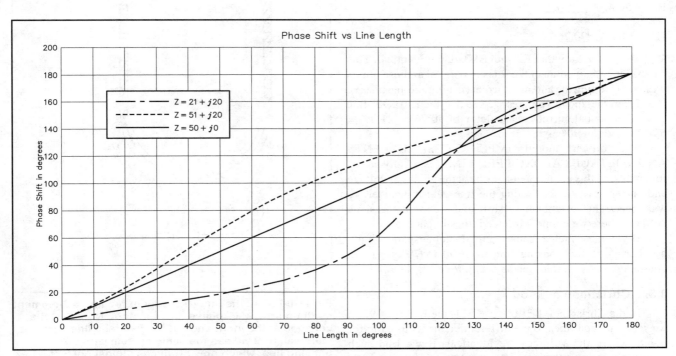

Fig 11-5—Graph showing the current phase shift in a 50-Ω line (RG-213, on 80 meters) as a function of the load impedance. The loads shown are those for a 2-element cardioid array as used in the text. Note that the phase shift does not equal line length, except when the line is terminated in its own characteristic impedance!

From E = Z / I we can calculate (don't worry that software does it for you):

E1 = 54.8 V, $\underline{/-68.6°}$

and

Z2 = 21 – j20 Ω

I2 = 1 A, $\underline{/0°}$

E2 = 29 V, $\underline{/-43.6°}$

At the end of the 90° long RG-213 feed lines the impedances (and voltages) become:

Z1′ = 42.81 – j16.18 Ω

I1′ = 50.89 A, $\underline{/0.39°}$

E1′ = 1.11 V, $\underline{/21.09°}$

and

Z2′ = 63.1 + j56.94 Ω

I2′ = 50.37 A, $\underline{/89.61°}$

E2′ = 0.59 V, $\underline{/47.54°}$

If we make the line to the lagging element 180° long (plus the extra 90° for obtaining an extra 90° phase shift), we end up with:

Z1″ = 51.18 + j18.64 Ω

E1″ = 56.42 V, $\underline{/110.77°}$

I1″ = 1.04 A, $\underline{/90.76°}$

We see already that E2′ and E1″ are not identical. This means we cannot connect the lines in parallel at those points without upsetting the antenna current (magnitude and phase).

From the above voltages we see that the extra 90° line created an actual current phase delay of 90.76° – 21.09° = 68.67°, and *not* 90° as required.

The software module IMPEDANCES, CURRENTS AND VOLTAGES ALONG FEED LINES is ideally suited for analyzing this phenomenon. Look at the values of voltage and current as you scan along the line, and remember we want the right current phase shift and we want the same voltage where we connect the feed lines in parallel.

If you have such a feed system, do not despair. Simply by shortening the phasing line from 90 to 70°, you can obtain an almost perfect feed system. (See **Fig 11-6**.)

3.3.2. Christman method

In the Christman, K3LC (ex KB8I) method (Ref 929), we scan the feed lines to the different elements looking for points where the voltages are identical. If we find such points, we connect them together, and we are all done! It's really as simple as that. Whatever the length of the lines are, provided you have the right current magnitude and phase at

the input ends of the lines, you can always connect two points with identical voltages in parallel. That's also where you feed the entire array.

Christman makes very clever use of the transformation characteristics of the feed lines. We know that on a feed line

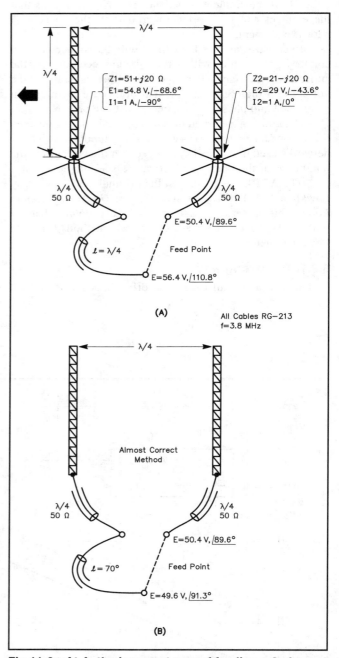

(A)

All Cables RG-213
f=3.8 MHz

(B)

Fig 11-6—At A, the incorrect way of feeding a 2-element cardioid array (90° phase, 90° spacing). Note that the voltages at the input ends of the two feed lines are not identical. In B we see the same system with a 70° phasing line, which now produces almost correct voltages. The F/B ratio of existing installations will jump by 10 or 15 dB, just by changing the line length from 90 to 73°.

with SWR, voltage, current and impedance are different in every point of the line.

The question is now, "Are there points with identical voltage to be found on all of the feed lines?" and "Are the points located conveniently; in other words are the feed lines long enough to be joined?" This has to be examined case by case.

It must be said that we cannot apply the Christman method in all cases. I have encountered situations where identical voltage points along the feed lines could not be found.

The software module, IMPEDANCE, CURRENT AND VOLTAGE ALONG FEED LINES, which is part of the NEW LOW BAND SOFTWARE, can provide a printout of the voltages along the feed lines. The required inputs are:

- Feed-line impedance.
- Driving-point impedances (R and X).
- Current magnitude and phase.

Continuing with the above example of a 2-element configuration (90° spacing, 90° phase difference, equal currents, cardioid pattern), we find:

E1 = (155° from the antenna element) = 47.28 V, $\underline{/86.1°}$

E2 = (84° from the antenna element) = 47.27 V, $\underline{/85.9°}$

Notice on the printout that the voltages at the 180° point on line 1 and at the 90° point on line 2 are not identical (see Section 3.3.1), which means that if you connect the lines in parallel in those points, you will not have the proper current in the antennas.

All we need to do now is to connect the two feed lines together where the voltages are identical. If you want to make the array switchable, run two 84° long feed lines to a switch box, and insert a 155° – 84° = 71° long phasing line, which will give you the required 90° antenna-current phase shift. **Fig 11-7** shows the Christman feed method.

Of course the impedance at the junction of the two feed lines is not 50 Ω. Using the COAX TRANSFORMER/SMITH CHART software module, we calculate the impedances at the input ends of the two lines we are connecting in parallel.

$Z1_{end} = 39 + j12$ Ω

$Z2_{end} = 50 + j52$ Ω

The software module PARALLEL IMPEDANCES calculates the parallel impedance as 23.8 + j12.4 Ω. This is the feed-point impedance of the array. You can use an L network, or any other appropriate matching system to obtain a more convenient SWR on the 50-Ω feed line.

3.3.3. Lewallen method

Roy Lewallen, W7EL, uses a method that takes advantage of the specific properties of quarter-wave feed lines (Lewallen calls it "current-forcing"). This method is covered in great detail by W7EL in recent editions of *The ARRL Antenna Book*.

A quarter-wave feed line has the following wonderful property, which is put at work with this particular feed method:

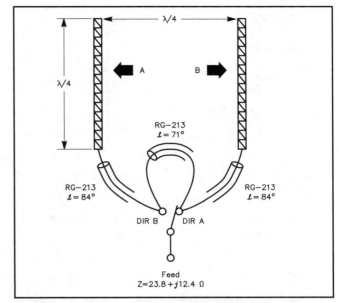

Fig 11-7—Feed system for the 2-element ¼-λ spaced cardioid array fed 90° out of phase. Note that the two feed lines are 84° long (not 90°), and that the "90° phasing line" is actually 71 electrical degrees in length. The impedance at the connection point of the two lines is 23.8 + j12.4 Ω (representing an SWR of 2.3:1 in 50-Ω line), so some form of matching network is desirable.

The magnitude of the input current of a ¼-λ transmission line is equal to the output voltage divided by the characteristic impedance of the line, and it is independent of the load impedance. In addition, the input current lags the output voltage by 90° and is also independent of the load impedance.

There are very slight deviations from this statement with real cables, as the above statement is 100% correct only if there are no losses in the cables. The Lewallen method is a feed method that can only be applied to antennas fed in quadrature, which means antennas where the elements are fed with phase differences that are a multiple of 90°.

Fig 11-8 shows the application of the Lewallen method for our 2-element cardioid array. From Section 3.3.1 we know the voltages at the end of the 90° long "real" feed lines (quarter-wave RG-213 with 0.35 dB loss per 100 ft on 80 meters). Notice that even with the losses, the real values are very close to the theoretical values (50 V, 90° and 0°):

E1′ = 50.9 V, $\underline{/0.39°}$

Z1′ = 42.81 – j16.18 Ω

E2′ = 50.4 V, $\underline{/89.6°}$

Z2′ = 63.1 + j56.94 Ω

If the feed line had no losses, E1′ would be 50 V, $\underline{/0°}$, and E2′ would be 50 V, $\underline{/+ 90°}$.

Remember, we said that in order to be able to connect the feed lines in parallel, the voltages need to be identical.

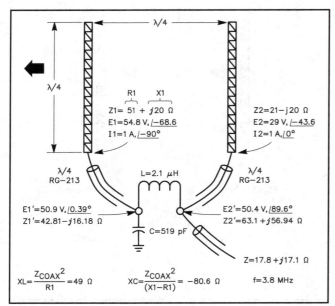

Fig 11-8—The Lewallen, W7EL, feed method for quadrature-fed arrays as applied to the 2-element cardioid (90° spacing, 90° phasing). Note the voltage magnitudes and phase angles at the input ends of the 90° long feed lines. The 50 V comes from 50 Ω and 1 A antenna current. The L network provides a 90° phase shift without changing the voltage magnitude. See text for details.

The magnitudes are almost the same but there still is a 90° phase difference.

In a Lewallen feed system, the array feed point is always at the end of the quarter-wave feed line that goes to the element(s) with the leading current angle. A network must be installed between the end of the feed lines going to the element(s) with the lagging feed current. The L network of Fig 11-8 provides the required 90° voltage-angle transformation, without changing the voltage magnitude.

Calculation for 2-element arrays

The elements of the L network can easily be calculated (this formula is valid only for 2-element arrays):

$$X_S = \frac{Z_{coax}^2}{R1}$$

$$X_P = \frac{Z_{coax}^2}{X1 - R1}$$

where:

X_S = reactance of the series element of the L network
X_P = reactance of the parallel element of the L network
Z_{coax} = characteristic impedance of the quarter-wave feed lines
R1 = feed-point resistance of the antenna element with the lagging feed current

X1 = feed-point reactance of the antenna element with the lagging feed current

These formulas are somewhat approximate as they assume lossless cables.

The impedance of the element with the lagging current is:

$$Z1 = 51 + j\,20\ \Omega\ \text{(with I1 = 1 A, } \angle{-90°}\ )$$

For our 2-element cardioid array we calculate the component reactances as follows:

$$X_S = \frac{50^2}{51} = j49\ \Omega$$

$$X_P = \frac{50^2}{(20 - 51)} = -\,j80.6\ \Omega$$

The values of the components can be calculated as follows:

If the reactance is positive, $L = \dfrac{X}{2 \times \pi \times f}$

If the reactance is negative, $C = \dfrac{10^6}{2 \times \pi \times f}$

where:

L = inductance, μH
C = capacitance, pF
f = frequency, MHz

In our case and for a design frequency of 3.8 MHz, the series element is a coil with an inductance of 2.1 μH; the shunt element is a capacitor with a capacitance of 519 pF.

Calculation of array feed impedance

Let's calculate the resulting feed impedance of the array. The impedance at the input end of the line running to the element with the lagging current is transformed by the L network. We can use the SHUNT/SERIES IMPEDANCE NETWORK to calculate the transformed impedance:

$-j80.6\ \Omega$ in parallel with $42.81 - j16.18\ \Omega$
$= 24.83 - j24.46\ \Omega$.

This, in series with $+j49\ \Omega = 24.83 + j24.54\ \Omega$.

Now we connect this impedance in parallel with $63.1 + j\,56.94\ \Omega$. The result is $17.8 + j17.1\ \Omega$.

The 2 EL AND 4 EL VERTICAL ARRAYS module of the NEW LOW BAND SOFTWARE is a tutorial and engineering program that takes you step by step through the design of a 2-element cardioid type phased array (and also the famous 4-element square array, which is described later). The results as displayed in that program will be slightly different from the results shown here, as the software uses ideal (lossless) feed lines. It is interesting, however, to compare the figures from this paragraph with the figures from the software program to assess the error caused by using lossless cables. **Figs 11-9** and **11-10** show the screen print of two pages from the 2-element design case as covered by the tutorial program.

```
  ed920502 ════════ THE 2 ELEMENT CARDIOID ARRAY ═════════ by on4un
```

	EL # 1		EL # 2	
	R	X	R	X
SELF IMPEDANCE	36.0	0.0	36.0	0.0
MUTUAL Z TO EL # 1			20.0	-15.0
MUTUAL Z TO EL # 2	20.0	-15.0		
ANTENNA CURRENT	1		1	
PHASE (DEG)	-90		0	

```
                                           R         X
                                         -----     -----
                   FEED IMPEDANCE EL #  1  ->  51.0      20.0
                   FEED IMPEDANCE EL #  2  ->  21.0     -20.0


   Here we have the feedpoint impedances of the 2 elements of our 2-element
   cardioid array. We will need those for futher calculations.

   ═ >RETURN< = CONTINUE ═════ >B< BACK 1 PAGE ═════ >G< GRAPH ═══ >X< EXIT ═
```

Fig 11-9—Screen-print of a worksheet from the VERTICAL ARRAY TUTORIAL software program showing how the element drive impedances are calculated from the self-impedances, the mutual impedances and the antenna current. The program can be used in a tutorial mode (showing a classic design case) or as a calculator to calculate a specific case with data entered from the keyboard.

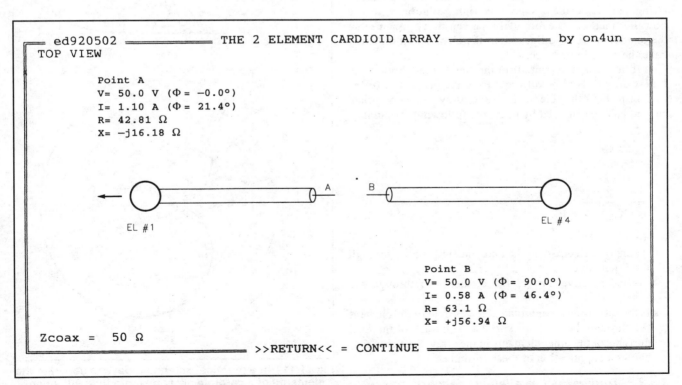

Fig 11-10—The CALCULATION/TUTORIAL program makes use of many graphic representations to clearly indicate where in the array we have certain impedances, voltages, currents, etc. Every design step is covered by a graphic representation.

3.3.3.1. Calculation for arrays with more elements

If more than one element is fed via the L network, and all these elements are fed via feed lines of the same impedance (Z_{coax}), the formulas for X_S and X_P are:

$$X_S = \frac{Z_{coax}^2}{\Sigma\,(R)}$$

$$X_P = \frac{Z_{coax}^2}{\left[\Sigma\,(X) - \Sigma\,(R)\right]}$$

where:

$\Sigma(R)$ = The sum of the feed-point resistances of all elements connected to the output side of the L network (the elements with the lagging current). In our case, $\Sigma(R) = 2 \times R2$

$\Sigma(X)$ = The sum of the feed-point reactances of all elements connected to the output side of the L network (the elements with the lagging current). Here, $\Sigma(X) = 2 \times X2$

This formula is applicable only if all the elements are fed with $1/4$-λ feed lines of the same impedance. This is the formula to be used for designing the feed system in several quadrature-fed arrays with more than 2 elements (see Section 4.9.2.1, Section 4.10.1, Section 4.10.3, 4.11.1.1 and 4.11.1.2).

3.3.3.2. Using the Lewallen method for arrays with different feed-current magnitudes

One can also use the Lewallen feed system in arrays using different current magnitudes on each of the elements. Using parallel cables is not the right solution with the Lewallen method. The formulas, as given above, assume that the quarter-wave feed lines to all the elements in the array have the same impedance.

If the current magnitude of the element fed through the L network needs to be different from the magnitude of the current to the other elements in the array, the appropriate current can be achieved by using the following formula:

$$X_S = \frac{Z_{coax}^2}{\left(k \times R\right)}$$

$$X_P = \frac{Z_{coax}^2}{\left[X - \left(k \times R\right)\right]}$$

where:

R is the resistive part of the antenna element fed through the L network

X is the reactive part of the element fed through the L network

k is the ratio of the magnitude of the current of the element fed through the L network to the magnitude of the feed currents of the other elements in the array. Section 4.6.1. shows an application of these formulas.

3.3.3.3. Tuning the Lewallen LC network

In Section 3.3.8.7, I will cover some measurement methods for adjusting quadrature-fed arrays. Using one of the phase meters described one can tune the Lewallen LC network for a perfect 90° voltage phase shift at the output of the networks (which, using $1/4$-λ transmission lines, corresponds to 90° current phase shift at the antenna feed points). Once the network is adjusted for 90°, one can fine tune the network for optimum F/B using a field strength meter. One should be aware however, that the array is not necessarily designed for a maximum F/B (at 90° phase shift) for a direction 180° off the back of the array.

Take the example of a 2-element end-fire cardioid array, where we feed the 2 elements with 90° phase difference. If the 2 elements are spaced $1/4$ λ, then infinite rejection will occur at a 0° wave angle. This is not necessarily what we want. A perfectly valid strategy can be to design the array for a maximum rejection at a wave angle of eg, 30°. In this case we will have a better rejection at higher wave angles, and somewhat less at zero-wave angle, which is of no consequence, as we have no zero-wave angle signals anyhow! In that case we will have to separate the two verticals by (90°) × cos α, where α = wave angle. In our example case the

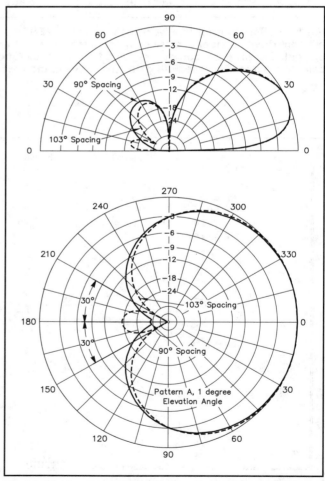

Fig 11-11—In a practical antenna it pays to separate the 2 elements of a cardioid array by more than 90° when the elements are fed 90° out of phase. This achieves maximum rejection at a practical wave angle such as 30° rather than a wave angle of 0°.

separation will be 90° / cos(30°) = 103°. In that case we will have to put our FS meter at an angle of 30° vs the geometric back of the antenna to do the final F/B adjustment (ie at either 60 or 120°), when doing the measurement at near ground level (eg, 1° wave angle), as shown in **Fig 11-11**.

3.3.4. The modified Lewallen method

The L network described by Lewallen achieves a 90° voltage phase and is applicable only to quadrature-fed arrays. Let's take the case where we change our 2-element cardioid slightly from a 90° phase shift to a 110° phase shift. This increases the gain of the array by 0.6 dB!

Refer to **Fig 11-12**. The data for the two elements are now:

Z1 = 43.3 + j23.9 Ω

I1 = 1 A, $\underline{/- 110°}$

E1 = 49.46 V, $\underline{/- 81.1°}$

Z2 = 15.1 – j13.7 Ω

I2 = 1 A, $\underline{/0°}$

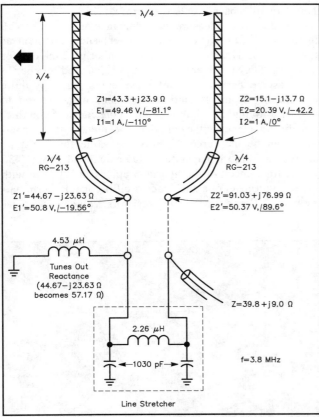

Fig 11-12—The modified Lewallen method can be used for arrays that are not quadrature fed. The combination of a shunt element and a pi or T network makes it possible to insert a voltage phase shift for any angle. See text for details.

E2 = 20.39 V, $\underline{/- 42.2°}$

At the end of the 90° long RG-213 feed lines the impedances (and voltage) become:

Z1' = 44.67 – j23.63 Ω

E1' = 50.8 V, $\underline{/- 19.56}$

Z2' = 91.03 + j76.99 Ω

E2' = 50.37 V, $\underline{/89.6°}$

This time we need a network between the ends of the two cables that takes care of a phase shift of 89.6° – (–19.56°) = 109.16°.

To do this we can use a constant-impedance pi or T phase-shift network. The pi and T networks are designed to provide the correct phase shift in a system impedance that has no reactive part. This means we will first have to add a shunt element to the end of the feed line going to the element with the lagging feed current, in order to make the impedance real. We cannot use a series element as that would change the voltage magnitude as well as the angle. We can use the SHUNT/ SERIES IMPEDANCE NETWORK module to do that.

Connecting an inductor with a reactance of 108.1 Ω in parallel with the feed line to Z1 brings the impedance to 57.17 Ω. At 3.8 MHz the required inductance is 4.53 µH.

Next we use the LINE STRETCHER (pi or T) module to calculate the values of the phase-shift network. The output phase angle must be specified as –19.56°, the input phase angle as 89.6°, and the system impedance as 57.17 Ω. The resulting pi-network values are:

X_S = + j54 Ω

X_P = – j40.66 Ω

For f = 3.8 MHz the values are:
  Series coil: 2.26 µH
  Parallel capacitors: 1030 pF

The final feed system is shown in Fig 11-12. The impedance in the junction is the parallel equivalent for:

Z1″ = 57.17 Ω

Z2′ = 91.03 + j76.99 Ω

The software module PARALLEL IMPEDANCES calculates the parallel impedance as 39.8 + j9.0 Ω. This is the feed-point impedance of the array. You can use an L network, or any other appropriate matching system to obtain a more convenient SWR on the 50-Ω feed line.

Conclusion: The modified Lewallen method allows you to use the "current-forcing" method with arrays that are not quadrature-fed. The network that must provide the correct voltage phase shift is no longer a simple L network, however.

Remark: In the above example we have inserted the phase-adjusting network in the leg going to the element with the lagging current angle, while the array feed line is connected directly to the quarter-wave line going to the element with the leading current. The other alternative is equally valid, connecting the feed line directly to the quarter-wave line going to the element with the current lagging. Make sure you specify input and output phase angles correctly when using the PI or T LINE-STRETCHER software module. The choice, as well as the choice between pi and T-network stretchers, should in practice be determined by practical aspects such as impedances and component values (too low impedance values result in too high currents; too high impedance values result in too high voltages).

3.3.5. Collins method

The Collins, W1FC, feed system is similar to the Lewallen system in that it uses "current-forcing" $1/4$-λ feed lines to the individual elements. There is one difference, however. Instead of using an L network, Collins uses a quadrature hybrid coupler, shown in **Fig 11-13**.

The hybrid coupler divides the input power (at port no. 1) equally between port nos. 2 and 4, with theoretically no power output at port 3 if all four port impedances are the same. When the output impedances are not the same, power will be dissipated in the load resistor connected to port no. 4. In addition, the phase difference between the signal at port nos. 2 and 4 will be different by 90° if the load impedance of these ports is not real or, if complex, they do not have an identical reactive part. We will examine whether or not this characteristic of the hybrid coupler is important as to its application as a feed system for a quadrature-fed array.

Hybrid coupler construction

The values of the hybrid coupler components are:

$X_{L1} = X_{L2} = 50\ \Omega$ (system impedance)

$X_{C1} = X_{C2} = 2 \times 50 = 100\ \Omega$

For 3.65 MHz the component values are:

$$L1 = L2 = \frac{X_L}{2\pi f} = \frac{50}{2\pi \times 3.65} = 2.18\ \mu H$$

$$C1 = C2 = \frac{10^6}{2\pi f X_C} = \frac{10^6}{2\pi \times 3.65 \times 100} = 436\ pF$$

When constructing the coupler, one should take into account the capacitance between the wires of the inductors, L1 and L2, which can be as high as 10% of the required total value for C1 and C2. The correct procedure is to first wind the tightly coupled coils L1 and L2, then measure the interwinding capacitance and deduct that value from the theoretical value of C1 and C2 to determine the required capacitor value. For best coupling, the coils should be wound on powdered-iron toroidal cores. The T225-2 (μ = 10) cores from Amidon are a good choice for power levels well in excess of 2 kW. The larger the core, the higher the power-handling capability. Consult Table 6-1 in Chapter 6 for core data. The T225-2 core has an A_L factor of 120. The required number of turns is calculated as

$$N = 100\sqrt{\frac{2.18}{120}} = 13.4\ turns$$

The coils can be wound with AWG 14 or AWG 16 multistrand Teflon-covered wire. The two coils can be wound with the turns of both coils wound adjacent to one another, or the two wires of the two coils can be twisted together at a rate of 5 to 7 turns per inch before winding them (equally spaced) onto the core.

At this point, measure the inductance of the coils (with an impedance bridge or an LC meter) and trim them as closely as possible to the required value of 2.09 μH for each coil. Do NOT merely go by the calculated number of turns, as the permeability of these cores can vary quite significantly from production lot or one manufacturer to another. Moving the windings on the core can help you fine-tune the inductance of the coil. Now the interwinding capacitance can be measured. This is the value that must be subtracted from the capacitor value calculated above (436 pF). A final check of the hybrid coupler can be made with a vector voltmeter or a dual-trace oscilloscope. By terminating ports 2, 3 and 4 with 50-Ω resistors, you can now fine-

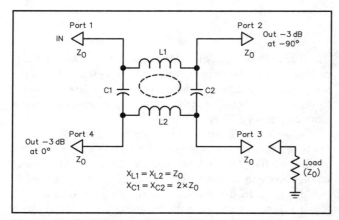

Fig 11-13—Hybrid coupler providing two —3 dB outputs with a phase difference of 90°. L1 and L2 are closely coupled. See text for construction details.

Eighty-meter home-made hybrid coupler with vacuum direction switching relays, made by JA6BJT.

tune the hybrid for an exact 90° phase shift between ports 2 and 4. The output voltage amplitudes should be equal.

Performance of the hybrid coupler

I have tested the performance of a commercially made hybrid coupler (ComTek, Ref Section 3.3.7). First the coupler was tested with the two load ports (port nos. 2 and 4) terminated in a 50-Ω load resistor. Under those conditions the power dissipated in the 50-Ω dummy resistor (port no. 3) was 21 dB down from the input power level. This means that the coupler has a directivity of 21 dB under ideal loading conditions (equivalent to 12 W dissipated in the dummy load for a 1500-W input). The results were identical for both 3.5 and 3.8 MHz. The input SWR under the same test conditions was approximately 1.1:1 (25-dB return loss).

I also checked the hybrid coupler for its ability to provide a 3-dB signal split with a 90° phase-angle difference. When the two hybrid ports were terminated in a 50-Ω load I measured a difference in voltage magnitude between the two output ports of 1.7 dB, with a phase-angle difference of 88° at 3.8 MHz. At 3.5 MHz the phase-angle difference remained 88°, but the difference in magnitude was down to 1.2 dB (theoretically the difference should be 0 dB and 90°).

The commercially available hybrid coupler system from ComTek Systems (**comtek4@juno.com**) uses a toroidal-wound transmission line to achieve a 180° phase shift over a wide band-width. The phase transformer consists of a bifilar-wound conductor pair, where wire A is grounded on one end and wire B on the other end of the coil. The other two ends are the input and output connections, whereby the voltages are shifted 180° in phase. This approach eliminates the long ($1/2$ λ) coax that is otherwise required for achieving the 180° phase shift, and it is broadbanded as well.

I have measured the performance of this "compressed" delay line. Using a 50-Ω load, the delay was 168°, with an insertion loss of 0.8 dB. With a complex-impedance load the phase shift varied between 160 and 178°. Measurements were done with a Hewlett-Packard vector voltmeter. The hybrid coupler was also evaluated using "real" loads in a four-square array configuration. See Section 5.3 for details.

After investigating the components of the ComTek hybrid-coupler system, I evaluated the performance of the coupler (without the delay line), using impedances as found at the input ends of the $1/4$-λ feed lines in real arrays as load impedances for ports 2 and 4 of the coupler. Let us examine the facts and figures for our 2-element end-fire cardioid array.

The SWR on the quarter-wave feed lines to the two elements (in the cardioid-pattern configuration) is not 1:1. Therefore, the impedance at the ends of the quarter-wave feed lines will depend on the element impedances and the characteristic impedances of the feed lines. We want to choose the feed-line impedances such that a minimum amount of power is dissipated in the port 3 terminating resistor.

The impedances at the end of the 90° long "real" feed lines ($1/4$-λ RG-213 with 0.35 dB loss per 100 ft on 80 meters) are:

$$Z1' = 42.81 - j16.18 \ \Omega$$

$$Z2' = 63.1 - j56.94 \ \Omega$$

These values are reasonably close to the 50-Ω design impedance of our commercial hybrid coupler. With 75-Ω feed lines the impedance would be

$$Z1' = 95.11 - j35.88 \ \Omega$$

$$Z2' = 141.05 - j125.4 \ \Omega$$

It is obvious that for our 2-element cardioid array, 50 Ω is the logical choice for the feed-line impedance. This can be different for other types of arrays. The basic 4-element four-square array, with $1/4$-λ spacing and quadrature-fed, is covered in detail in Section 4.9.2.2. A special version of the four-square array is analyzed in detail in Section 5.3.

Array performance

Although the voltage magnitudes and phase at the ends of the two quarter-wave feed lines are not exactly what is needed for a perfect quadrature feed, it turns out that the array only suffers slightly from the minor difference. The incorrect phase angle will likely deteriorate the F/B, but the gain will remain almost the same as with the ideal driving conditions (see also Fig 11-16).

Different design impedance

We can also design the hybrid coupler with an impedance that is different from the 50-Ω quarter-wave feed-line impedance in order to realize a lower SWR at ports 2 and 4 of the coupler. The load resistor at port 3 must of course have the same ohmic value as the hybrid design impedance. Alternatively we can use a standard 50-Ω dummy load with a small L network connected between the load and the output of the hybrid coupler.

With the aid of the software module SWR ITERATION, one can scan the SWR values at ports 2 and 4 for a range of design impedances. The results can be cross-checked by measuring the power in the terminating resistor and alternately connecting 50 Ω and 75-Ω quarter-wave feed lines to the elements. A practical design case is illustrated in Section 4.9.2.2.

By choosing the most appropriate feed-line impedance as well as the optimum hybrid-coupler design impedance, it is possible to reduce the power dissipated in the load resistor to 2% to 5% of the input power. Whether or not reducing the lost power to such a "low" degree is worth all the effort may be questionable, but covering the issue in detail will certainly help in better understanding the hybrid coupler and its operation as a feed system for a phased array with elements fed in quadrature.

3.3.6. Gehrke method

Gehrke, K2BT, has developed a technique that is fairly standard in the broadcast world. The elements of the array are fed with randomly selected lengths of feed line, and the required feed currents at each element are obtained by the insertion of discrete component (lumped-constant) networks

The ComTek hybrid coupler unit as installed at K4ZW. The double-sided PC board is mounted on anodized metal frame using a Plexiglas cover, which makes it possible to inspect the unit without having to open the box.

in the feed system. He makes use of L networks and constant-impedance T or pi delay networks. The detailed description of this procedure is given in Ref 924.

The Gehrke method consists of selecting equal lengths (not necessarily 90° lengths) for the feed lines running from the elements to a common point where the array switching and matching are done. With this method, the length of the feed linescan be chosen by the designer to suit any physical requirements of the particular installation. The cables should be long enough to reach a common point, such as the middle of the triangle in the case of a traingle-shaped array.

As an example, we will work out a 2-element end-fire array with a cardioid pattern (fed with equal current magnitudes, but 90° out of phase). All the calculations are done assuming we use RG-213 coax (0.35 dB loss / 100 ft at 3.8 MHz). We'll use the software for "real" feed lines, which takes into account the effect of cable loss.

Next, the required impedances at the T junction must be determined. Continuing with the above example, the element feed-point impedances are:

$Z1 = 51 + j20 \, \Omega$ for the $-90°$ element

$Z2 = 21 - j20 \, \Omega$ for the 0° element

As the current in both antennas is 1 A (arbitrarily chosen), the power levels are:

$P1 = I^2 \times 51 = 51 \, W$

$P2 = I^2 \times 21 = 21 \, W$

$P_{tot} = 21 + 51 = 72 \, W$

If you want to work with real powers, voltages, and currents, divide the transmitter output power by P_{tot}, and replace the 1 A of current with the square root of the output power divided by the total power. Example: If $P_{out} = 1500 \, W$, then

$$I = \sqrt{\frac{1500}{72}} = 4.56 \, A$$

Now you can read real voltages and real currents, which can be an advantage when determining components for the circuit.

At the T junction (assuming we want an SWR of 1:1 on the 50-Ω line to the shack), the voltage will be:

$$E = \sqrt{R \times P} = \sqrt{50 \times 72} = 60 \, V$$

Looking toward element 1, the impedance at the T junction must be:

$$Z = \frac{E^2}{P} = \frac{60^2}{51} = 70.6 \, \Omega$$

For element 2, the impedance at the T point must be 171 Ω. A simple check can be made to verify that the resulting impedance will be 50 Ω:

$$\frac{1}{70.6} + \frac{1}{171} = \frac{1}{50}$$

The impedance, current and voltage at the element feed points are (see Section 3.3.1):

$Z1 = 1 + j20 \, \Omega$

$I1 = 1 \, A, \; \underline{/-90°}$

$E1 = 54.8 \, V, \; \underline{/-68.6°}$

$Z2 = 1 - j20 \, \Omega$

$I2 = 1 \, A, \; \underline{/0°}$

$E2 = 29 \, V, \; \underline{/-43.6°}$

E1 and E2 are the voltages (magnitude and phase) that must be present at the feed points of the elements in order to produce the pattern resulting from quarter-wave spacing, equal current magnitude, and a 90° phase delay. **Fig 11-14** shows the impedance, current and voltage values, and where they are present in the array.

The values of I, E, and Z at the element feed points must now be transformed to the input ends of the coaxial feed lines. If we decide to make the direction of the array switchable, it is recommended that all feed lines be cut to the same electrical length, unless one element remains in the same position (such as the center element in a 3-element in-line array), because the direction-switching harness will be easier to construct. Otherwise, there is no reason

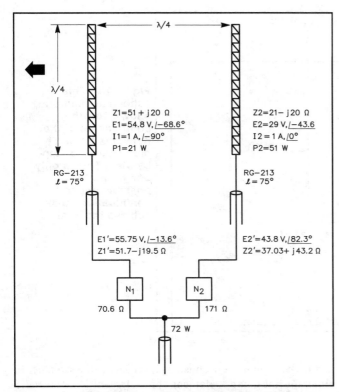

Fig 11-14—Schematic representation of the Gehrke approach to feeding the elements of an array. The networks N1 and N2 transform the complex impedances at the end of the feed lines to resistive values at the T junction and provide voltages at the junction that are identical in magnitude and phase. See text for details.

why the feed lines must be of equal length. Let's take two cables each 75° long.

We calculate the impedance at the input end of the two coaxial cables using the software module COAX TRANS-FORMER/SMITH CHART. The result of the transformation via the 75° long 50-Ω lines is:

$Z1' = 51.7 - j19.5\ \Omega$

$E1' = 55.75$ A, $\underline{/-13.6°}$

$I1' = 1.01$ V, $\underline{/7.05°}$

$Z2' = 37.03 + j43.2\ \Omega$

$E2' = 43.8$ V, $\underline{/82.3°}$

$I2' = 0.77$ A, $\underline{/32.88°}$

These are the values of Z, E and I that must be present at the ends of the coaxial cables in order for the array to perform as it was designed to.

Next we have to calculate the lumped-constant networks. The networks must be designed to transform the impedances at the end of the cables to the resistive values

(determined above as 70.6 Ω and 171 Ω). This transformation can be done with a shunt-input L network, as the output impedance is lower than the input impedance. The only requirement for the network is the impedance transformation. L networks have a phase delay that is inherently coupled to the transformation ratio, which means that we have no separate control over the phase delay. The phase delays in current and voltage can be calculated using matrix algebra (Ref 925). The software module L-NETWORK does the job in seconds. Each L-network design yields at least two solutions, where the input voltage magnitudes are the same, but the phase angles are different.

L Network for Z1 (Element with the Lagging Current)
Input data:

$Z_{out} = 70.6\ \Omega$

$Z1' = 51.7 - j19.5\ \Omega$

$E1' = 55.85$ V, $\underline{/-13.6°}$

$I1' = 1.01$ A, $\underline{/7.05°}$

Solution 1

$X_S = -j11.76\ \Omega$ (the minus sign indicates capacitive reactance)

$X_P = j116.77\ \Omega$

Solution 2

$X_S = j50.76\ \Omega$

$X_P = -j116.77\ \Omega$ (capacitive reactance)

L Network for Z2 (Element with the Leading Current)
Input data:

$Z_{out} = 171\ \Omega$

$Z2' = 37.03 + j43.2\ \Omega$

Solution 3

$X_S = -j113.63\ \Omega$

$X_P = j89.9\ \Omega$

Solution 4

$X_S = j27.23\ \Omega$

$X_P = -j89.9\ \Omega$

For each solution the voltage and current are in phase, as the impedance is a pure resistance. The networks can now be inserted in the feed lines to the elements. See **Fig 11-15**. First we must decide which L networks to use. Our choice will depend on the values of the components involved, as well as on the phase difference that will have to be compensated for with the line-stretcher network. Let's assume we have chosen solutions 2 and 3.

Looking at the voltage inputs to the above two net-

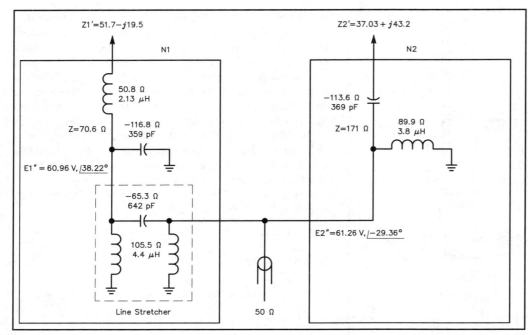

Fig 11-15—This schematic diagram shows how the components of the networks N1 and N2 are calculated step-by-step. This is only one of the 16 possible network combinations that achieve the goal.

works, E1″ and E2″, we can see that the magnitudes are practically the same but the phase relationships are different: + 38.22° for element 1 and –29.36° for element 2. In this case, an additional phase delay will have to be accomplished in a line stretcher in order to make up for the phase difference. This is done with a symmetrical pi or T network designed around an input (and output) impedance of 70.6 Ω or 171 Ω and delivering the required phase shift.

This network can be designed using the software module LINE STRETCHER (pi or T). If you use the software make sure you specify input and output angles correctly. One possible solution is a pi network with the branch data that follows.

Let's put the stretcher in the 70.6-Ω branch: The output phase angle is 38.22°, and the input phase angle –29.36°. Make sure you specify input phase and output phase correctly: Looking from the input of the network toward the antenna loads you first have the input phase; the output phase is the phase toward the load.

The values of the pi network are:

$$X_S = -j65.26 \ \Omega \ (642 \ pF \ at \ 3.8 \ MHz)$$

$$X_P = j105.5 \ \Omega \ (4.4 \ \mu H \ at \ 3.8 \ MHz)$$

Ahead of the line stretcher, the voltages are identical in both amplitude and phase, and the input terminals of networks N1 and N2 can be connected in parallel to obtain the required feed-system impedance.

The input impedance of the array is 1 / (1 / 171 + 1 / 70.6) = 50 Ω. Remember, that's how we started the design procedure. Fig 11-14 shows the resulting feed configuration as calculated above.

There are a large number of solutions that can be designed. For each of the L networks we have two solutions,

and we can connect the constant-impedance line stretcher in either the line going to element 1 or to element 2. All are valid solutions. For this 2-element cardioid array there are a total of 16 possible L-network line-stretcher combinations that will meet our design criteria.

3.3.7. Choosing the best feed system

Until Gehrke published his excellent series on vertical arrays, it was general practice to simply use feed lines as delay lines, and to equate electrical line length to phase delay under all circumstances. We now know that there are better ways of accomplishing the same goal (Ref Section 3.3.1).

Fortunately, as Gehrke states, these vertical arrays are relatively easy to get working. **Fig 11-16** shows the results of an analysis of the 2-element cardioid array with deviating feed currents. The feed-current magnitude ratio as well as the phase angle are quite forgiving as far as gain is concerned. As a matter of fact, a greater phase delay (eg, 100 versus 90°) will increase the gain by about 0.3 dB. The picture is totally different as far as F/B ratio is concerned. To achieve an F/B of better than 20 dB, the current magnitude as well as the phase angle need to be tightly controlled. But even with a "way off" feed system it looks like you always get between 8 and 12 dB of F/B ratio, which is indeed what we used to see from arrays that were incorrectly fed with coaxial phasing lines having the electrical length of the required phase shift.

Current-forcing methods (Lewallen, modified Lewallen and Collins) can be used if the current magnitudes are all the same or if the correct current magnitude can be provided by the correct choice of coaxial feed-line impedance.

Let's take an example. A 4-element array, where relative current magnitudes of 1, 1.5, 2 and 3 are required on the four elements, can be fed using coaxial feed lines of 75 Ω, 50 Ω, 37.5 Ω (two 75-Ω lines in parallel) and 25 Ω (two

50-Ω lines in parallel). At the end of each of the quarter-wave feed lines the voltage magnitude will be 75 V (1 × 75 = 75, 2 × 37.5 = 75, 1.5 × 50 = 75 and 3 × 25 = 75).

If the current angles are in a quadrature relationship (multiples of 90°) either the Lewallen or the Collins feed method can be applied. The big advantage of the Collins method is that you don't need to measure anything. The Collins system requires very few discrete components.

With the Collins system, essentially the same front-to-back ratio can be achieved in the range from 3.5 to 4.0 MHz, which makes this approach very attractive for those who want broadband performance. It does not make sense, however, to judge the operational bandwidth of the hybrid-coupler system by measuring the SWR curve at the input of the coupler. The coupler will show a very flat SWR curve (typically less than 1.3:1) under ALL circumstances, even from 3.5 to 4 MHz or from 1.8 MHz to 1.9 MHz on 160 meters. The reason is that, away from its design frequency, the impedances on the hybrid ports will be extremely reactive, resulting in the fact that nearly all power fed into the system will be dissipated in the dummy resistor. It is typical that an array tuned for element resonance at 3.8 MHz will dissipate 50 to 80% of its input power in the dummy load when operating at 3.5 MHz (the exact amount will depend on the Q factor of the elements). On receive, the same array will still exhibit excellent directivity on 3.5 MHz, but its gain will be down by 3 to 7 dB from the gain at 3.8 MHz (as we are wasting 50 to 80% of the received signal as well into the dummy resistor). It is clear that the only bandwidth-determining parameter is the power wasted in the load resistor. So stop bragging about your SWR curves, but let's see your dummy-load power instead!

Compared to the other current-forcing methods (Lewallen and modified Lewallen methods), the hybrid coupler has the disadvantage of wasting some of the transmitter power (and receive power as well, but that's probably much less relevant) in the dummy-load resistor. As long as the power is not more than 10%, the loss is really negligible from a practical point of view. The amount of wasted power can be reduced by the correct choice of the impedance of the quarter-wave feed lines, as well as of the hybrid-coupler design impedance (see Section 3.3.5 and 4.9.2.2).

Another relative disadvantage of the hybrid-coupler system is that the coupler does not produce the exact phase-quadrature phase shift unless some very specific load conditions exist (resistive loading or loading with identical reactive components on both ports). Most of the quadrature-fed arrays, however, are quite lenient, tolerating a certain degree of deviation from the perfect quadrature condition. The Collins-type coupler can only be used with an array that has quadrature-fed elements (current phase angles in multiples of 90°).

It is clear that over the years the Collins method has become the most popular feed method, clearly because it is an appliance-operator type solution (plug it in, no questions asked, and it works —most of the time)!

The Christman method makes maximum use of the transformation characteristics of coaxial feed lines, thus minimizing the number of discrete components required in the feed network. This is an attractive solution, and should not scare off potential array builders. For a 2-element cardioid array this is certainly the way to go. Of course you need to go through the trouble of measuring the impedances.

The Christman method is very useful for 2-element arrays. With arrays of more elements, it is likely that identical voltages will only be found on two lines. For the third line, lumped-constant networks will have to be added. In such case the Lewallen or modified Lewallen method is preferred.

The Collins and Lewallen system of "current-forcing" have the advantage that they are not as sensitive to less-than-perfect symmetry in the array in the case of direction switch-

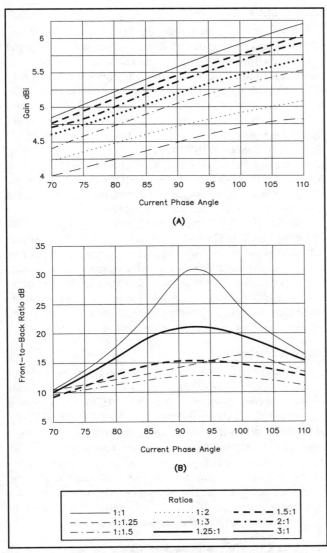

(A)

(B)

Ratios		
——— 1:1	·········· 1:2	– – – 1.5:1
– – – 1:1.25	– · — 1:3	– · – · – 2:1
– · — 1:1.5	——— 1.25:1	——— 3:1

Fig 11-16—Calculated gain and the front-to-back ratio of a 2-element cardioid array versus current magnitudes and phase shifts. Calculations are for very good ground at the main wave angle. The array tolerates large variations as far as gain is concerned, but is very sensitive as far as front-to-back ratio is concerned.

ing, as the other systems. This can be seen in the more constant front-to-back ratios in different directions with these systems.

For the Lewallen method, one only needs to know the impedance of the element with the lagging current. I don't know if this is an advantage. If you measure one element, you can just as well measure all, and if you want to switch directions, you have to, anyhow. With the modified Lewallen method, non-quadrature-fed arrays can also be handled. Lewallen has published a number of L-network values for the 2-element cardioid and the 4-element square arrays, which a builder can use for building the L network without doing any measuring. In that case (in any case) some performance measuring will have to be done to come to the right component values (see Section 3.3.7). The Lewallen method will not yield a good F/B over the entire 80-meter band, but it is entirely feasible to make two switchable L networks, one for 3.8 and one for 3.5 MHz.

The Gehrke design is rather impractical for amateur applications as it uses a large number of networks. The Gehrke approach is nevertheless given, as it helps in understanding how feed systems for arrays can be designed. The adjustment is very complex. Also, the system is narrow banded. Typically, a Gehrke system designed for 3.8 MHz would show very little F/B ratio at 3.5 MHz, although the gain would remain essentially unchanged.

3.3.8. Tuning and measuring vertical arrays

None of the arrays described in this chapter can be built or set-up without any measuring. The most simple array uses a quadrature configuration, which makes it possible to use a hybrid coupler for obtaining the required phase shift. Even in that case, the elements of the array will have to be tuned to proper resonance.

3.3.8.1. Measuring antenna resonance

In order to obtain maximum directivity from an array, it is essential that the self-impedances of the elements be identical. Measurement of these impedances also requires a good impedance or noise bridge. Equalizing the resonant frequency can be done by changing the radiator lengths, while equalizing the self-impedance can be done by changing the number of radials used. If you start putting down perfectly identical and symmetrical radial systems, you will likely get very similar values for the resistive part of the various elements. If you cannot easily get equal impedances, you will have to suspect that one or more of the array elements are coupling into another antenna or conducting structure. Take down all other antennas that are within $1/2$ λ from the array to be erected. Do not change the length of one of the radiators to get the equal values for the resistive parts of the elements. The elements should all have the same physical height (within a few percent).

You can measure the resonant frequency by using the simple built-in SWR meter of a modern transceiver. One by one, while the other elements are fully decoupled (left floating in case of $1/4$-λ elements, the elements will be trimmed for minimum SWR on the design frequency. It is not strictly so that the value of minimum SWR is the true resonant frequency, but for this purpose, the approximation is sufficient (see also Chapter 7, Section 1.4.4).

3.3.8.2. Measuring antenna impedance

Most of the arrays described in this book, with the exception of those using the Collins, W1FC, Hybrid Coupler, will require some impedance measuring in order to be able to calculate the values of the components required to build the necessary phasing network(s).

Access to accurate impedance measuring equipment is essential. This equipment will be needed for measurement of element impedances and, as well as for measuring the reactances of the components that make up the phasing networks.

Noise bridges

Commercially available noise bridges will almost certainly not give the required degree of accuracy, as rather small deviations in resistance and reactance must be accurately recorded. A genuine impedance bridge is more suitable, but with care, a well-constructed and carefully calibrated noise bridge may be used. Several excellent articles covering noise bridge design and construction were published: one written by Hubbs, W6BXI; Doting, W6NKU (Ref 1607); Gehrke, K2BT (Ref 1610); and J. Grebenkemper, KI6WX (Ref 1623); D. DeMaw, W1FB (Ref 1620); and J. Belrose, VE2CV (Ref 1621). These articles are recommended reading material for anyone considering using a noise bridge in array design and measurement work. The software module RC/RL TRANSFORMATION part of the NEW LOW BAND SOFTWARE is very handy for transforming the value of the noise-bridge capacitor, connected in parallel with either the variable resistor or the unknown impedance, first to a parallel reactance value and then to an equivalent reactance value for a series LC circuit. This enables the immediate computation of the real and imaginary parts of the series impedance equivalent, expressed in "A + jB" form.

Network analyzers

Network analyzers are, in principle, the ideal tools for measuring impedances. There are various types on the market, and second hand you may be able to get a system comprising the analyzer and the generator for between $1000 and $2000. When measuring antennas on 80 and 160, it is important that you do these measurements during day time, as during the night the average signal power on the band is so great that this background noise will cause erroneous readings on the equipment. With good quality equipment, one can adjust the generator power level to overcome this problem to a certain degree.

The AEA CIA-HF Antenna Analyzer

The AEA-CIA-HF Analyzer is a one-port network analyzer (with limited capabilities), which measures impedances (and of course SWR, etc) by a swept frequency method, that you can set between 0.4 and 54 MHz. The nice thing is that it is portable, and can operate from built-in batteries. This equipment is only a single-port analyzer, it cannot measure gain/phase through a component, such as filters. When using it to measure antenna impedances, I have found it quite useful on all bands, down to 40 meters, and sometimes 80 meters. On 160, signals picked up from the broadcast band are too strong and mess up the readings, even during day-time. the challenge will be for someone to come up with filters that will eliminate the BC interference, without causing any impedance transformation in the measuring range. Quite a challenge. Another solution would be to have a higher output, but that may conflict with the FCC regulations on this subject. **Fig 11-17** shows the AEA CIA-HF showing the SWR curve of my 160-meter vertical, which is really 1:1 on 1830, but shows a much higher value due to spurious signal pick up. Note that this reading was taken in the middle of the day, to minimize strong signals via sky-wave. The closest medium wave BC transmitter is about 35 km from my QTH.

The MFJ-259 Antenna Analyzer

The MFJ-259B Antenna Analyzer is different from the older MFJ-259. It uses a microprocessor and four voltage detectors in a bridge to directly measure reactance, resistance, and SWR. With so much information available, uses are limited mostly by your imagination and technical knowledge.

One main application for antenna builders is its capability of measuring SWR (also in terms of reflection loss). It will also measure the resistive part and the absolute value of the reactive part of a complex impedance. The MFJ-259B isn't smart enough to give the sign of the reactive part, without some minor help. You must vary the frequency slightly and watch the reactance change to determine the sign of the reactance (and the type of component required to resonate the system). If adjusting the frequency slightly higher increases reactance (X), the load is inductive and requires a series

capacitance for resonance. If increasing frequency slightly reduces reactance, the load is capacitive and requires a series inductance for resonance. This general rule works with most antennas, but not necessarily all of them.

The designers of the unit have added a "transparent filter" to cope with the problems of strong signals messing up low-level reflected-power readings in the vicinity of broadcast transmitters or during night time measurements on the low bands. This accessory includes an adjustable notch filter and selective band-pass filter. This handy accessory allows the MFJ-259B to be used on large low band antennas, even if the antenna is located in the area of a broadcast transmitter.

At first sight the major difference between the MFJ unit and both the AEA and the Alpha Power units is the fact that this unit does not generate a spectral display of the units measured. It is basically a point to point measurement system.

The older type of antenna analyzer such as the MFJ 259 can also easily be adapted for doing a full range of RF measurements, by using some extra outboard components. Gerd Janzen, DF6SJ/VK2BJZ wrote an extremely interesting booklet (in German language) on this subject (Ref 1626). Hopefully this will be translated into English so that more amateurs can benefit from it.

The Alpha/Power Analyzer

The Alpha/Power 2-port analyzer is a different breed really. It is a genuine, full-fledged network analyzer that allows you to measure complex impedance in single point or swept frequency modes. It also allows you to do two-port gain and phase measurements. The unit (**Fig 11-18**), operates from 1 to 30 MHz, and works with any computer running under Windows (3.x, 95 or 98).

This equipment is extremely accurate, and will, even during the night, measure accurate impedances on a big 160-meter antenna. This is due to the use of very efficient

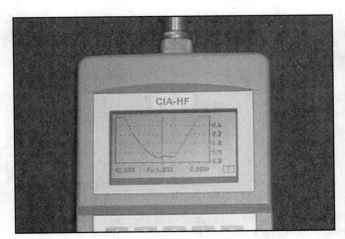

Fig 11-17—Picture of the AEA CIA-HF Antenna Analyzer, showing the SWR curve for the 160-meter vertical at ON4UN. Note that the SWR curve does not drop completely to the 1:1 line. This is due to signals and noise on the band that are stronger than the level of the reflected test signal.

Fig 11-18—Together with a notebook PC, the battery-powered Alpha/Power Analyzer makes a portable, high-performance test instrument for antenna and network measurements.

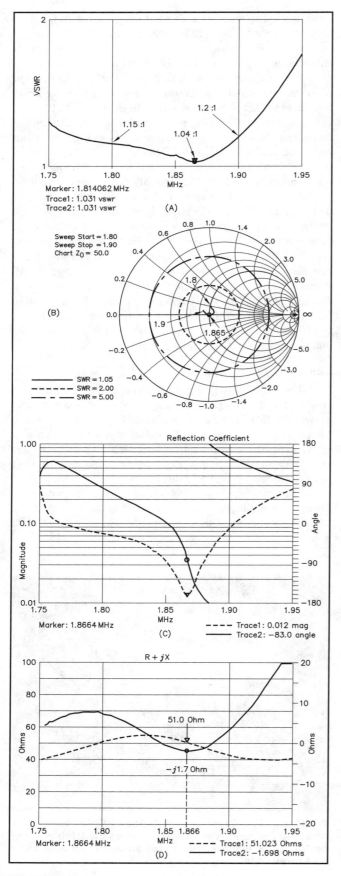

Marker: 1.814062 MHz
Trace1: 1.031 vswr
Trace2: 1.031 vswr

(A)

Sweep Start = 1.80
Sweep Stop = 1.90
Chart Z₀ = 50.0

(B)

——— SWR = 1.05
— — — SWR = 2.00
— ·· — SWR = 5.00

Reflection Coefficient

Marker: 1.8664 MHz

(C)

— — — Trace1: 0.012 mag
——— Trace2: −83.0 angle

R + jX

Marker: 1.8664 MHz

(D)

— — — Trace1: 51.023 Ohms
——— Trace2: −1.698 Ohms

software averaging and smoothing algorithms, which compensate for the effects of received power not generated by the forward power of the measuring device. **Fig 11-19** shows several screen prints of measurements done on the author's 160-meter $1/4$-λ vertical. Note that these measurements were done at night, when the signals on top band are strong. Using frequency markers, which can be moved with the help of the PC mouse, exact frequencies can be very accurately determined.

Impedance bridge

All the above mentioned equipment except the Alpha/Power Analyzer has the same intrinsic problem of suffering from alien signal overload when measuring large antennas on the low bands, especially 160, where BC signals are likely to cause false readings unless clever computer algorithms are used to compensate for this phenomenon. The only way to overcome this problem is to measure with more power, which, to a degree, is possible with professional-grade network analyzers, or in the worst of cases one will have to resort to a "good old" General Radio (or similar) bridge, driven by a signal source of sufficient level. This method, of course, lacks the flexibility of a real frequency-sweeping network analyzer.

3.3.8.3. Cutting feed-line cable lengths

Never go by the published velocity-factor figures, certainly not when you are dealing with foam coax. There are several valid methods for cutting $1/4$-λ or $1/2$-λ cable lengths. *The ARRL Antenna Book* describes the professional way of doing it in its chapter on driven arrays (see **Fig 11-20**).

With the method described, when adjusting $1/4$-λ stubs the far end should be open. With $1/2$-λ stubs the far end of the cable should be short-circuited. Odd lengths can also be adjusted this way after a little calculating. Assume you need a 72° long line at 3.65 MHz. This line will be 90° long at 3.65 × 90 / 72 = 4.56 MHz. The cable can now be cut to $1/4$ λ on 4.56 MHz using the method described above.

Using your transmitter, a dummy load and a good SWR bridge

If you do not have access to the test equipment described above, you can do quite well using your transceiver, a good SWR bridge and a good dummy load. Connect your transmitter through a good SWR meter (a Bird 43 is a good choice) to a 50-Ω dummy load. Insert a coaxial T connector

Fig 11-19—Screen dumps of the measurements taken of the 160-meter vertical using the Alpha Analyzer. At A the classical SWR curve which goes down to 1:1, although this measurement was done at night, indicating an excellent rejection of signals received off the air. At B the Smith Chart diagram, at C the reflection coefficient (magnitude and phase) and at D the value of R and X of the impedance. Values shown are for a frequency span going from 1.75 MHz to 1.95 MHz. Plots generated by the Alpha/Power Analyzer.

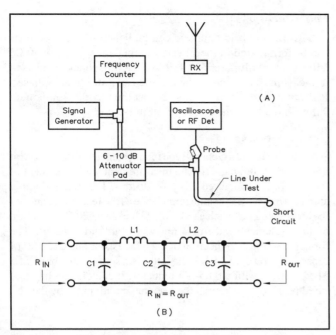

Fig 11-20—At A, the setup for measurement of the electrical length of a transmission line. The receiver may be used in place of the frequency counter to determine the frequency of the signal generator. The signal generator output must be free of harmonics; the half-wave harmonic filter at B may be used outboard if there is any doubt. It must be constructed for the frequency band of operation. Connect the filter between the signal generator and the attenuator pad.

C1,C3—Value to have a capacitive reactance = R_{IN}
C2—Value to have a capacitive reactance = $\frac{1}{2} R_{IN}$
L1,L2—Value to have an inductive reactance=R_{IN}

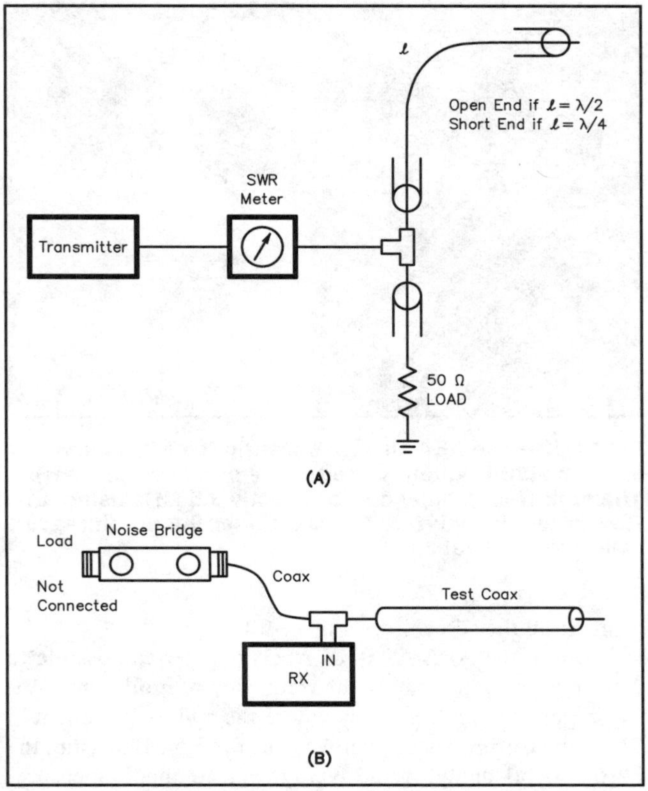

Fig 11-21—Very precise trimming of $\frac{1}{4}$-λ and $\frac{1}{2}$-λ lines can be done by connecting the line under test in parallel with a 50-W dummy load, shown at A. Watch the SWR meter while the line length or the transmitter frequency is being changed. The alternative method at B uses a noise bridge and a receiver. See text for details.

at the output of the SWR bridge. See **Fig 11-21**.

If you need to cut a quarter-wave line (or odd multiple of quarter waves), short the end of the coax. Make sure it is a good short, not a short with a lot of inductance. Insert the cable in the T connector. If the cable is a quarter-wave long, the cable end at the T connector will show as an infinite impedance and there will be no change at all in SWR (will remain 1:1). If you change the frequency of the transmitter you will see that on both sides of the resonant frequency of the line, the SWR will rise rather sharply. For fine tuning you can use high power (eg, 1 kW) and use a sensitive meter position for measuring the reflected power. I have found this method very accurate, and the cable lengths can be trimmed very precisely.

Using a noise bridge

If you have a noise bridge there is another simple method, whereby you use the noise bridge only as a wideband noise source, without using the internal bridge. Instead you will connect the line to be trimmed across the output of the noise bridge, and trim the length until the noise level on the receiver will be reduced to zero. Switch off the receiver AGC to make the final adjustments (see Fig 11-16B). Tuning

the receiver back and forth across the frequency makes it possible to determine the frequency of maximum rejection quite accurately. in this method the $\frac{1}{4}$-λ lines should be open circuited at the end, and $\frac{1}{2}$-λ lines should be short circuited.

Using a grid-dip oscillator

Using a grid-dip oscillator is not a very accurate way of trimming precise cable lengths. The pick-up loop will typically account for a 2° to 3° error when dipping a quarter-wave cable length. This method is to be avoided by all means.

Using one of the popular antenna analyzers

All of the popular antenna analyzers can be used as well. The method consists of connecting a 50-Ω dummy load to the analyzer via a T connector. The transmission line or stub is connected in parallel with a precision 50-Ω dummy load. The antenna analyzer is then adjusted for the frequency with the lowest SWR ratio. For an open ended cable this is at the frequencies where the cable is $\frac{1}{2}$ λ or multiple thereof. For a shorted stub this is for a length of $\frac{1}{4}$ λ or any odd multiple thereof. The AEA CIA-HF Analyzer has a nice feature whereby it calculates the frequency of lowest SWR and

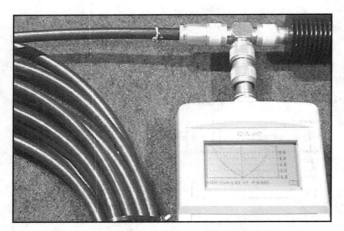

Fig 11-22—The AEA CIA-HF showing the SWR curve and the stub frequency (frequency of minimum SWR). The stub was initially cut for exactly 3.5 MHz using an R&S network analyzer. The text under the graph reads "MIN SWR 1.01 at 3.500."

prints it on the screen (see **Fig 11-22**).

Using the AEA CIA-HF Analyzer, it was possible to determine very accurately the frequency of minimum SWR (which equals the frequency where the stub is "resonant"). When measuring a stub that was cut for 3.5 MHz using the R&S network analyzer, the average of a number of measurements gave 3.492 MHz for stub resonant frequency, which is within 0.2%, which is excellent.

The new MFJ-259B Antenna Analyzer can also be used to do the job, although I have personally not been able to test this unit's accuracy. It will directly display electrical length of a transmission line in feet but not in degrees, which is a pity. Of particular importance to low band DX'ers are functions describing R and X and electrical lengths. I hope they will be added with later software versions.

Using the Alpha/Power Analyzer

This equipment (see also Section 3.3.8.2), which is really a network analyzer, is ideally suited for cutting stubs. When using the same method as described above (stub connected across a 50-Ω dummy load), identical results can be obtained as compared to the AEA CIA-HF Analyzer (see **Fig 11-23**). There are other methods that can be used which are more accurate and allow the stubs to be cut with an accuracy of typically 1 kHz! When using the rectangular display (R + jX), the reactance goes through zero for 3.495 kHz. Using the Z, $\underline{/\theta}$ display (magnitude and angle of the impedance) the stub frequency appears to be 3.497 kHz. This result is within 1% of the value measured on an R&S network analyzer, which is truly outstanding!

3.3.8.4. Network measurements

Lewallen networks

Lewallen LC networks require little testing. There is no on-the-bench testing possible, because the networks have to be terminated into the antennas to do their proper phasing job.

Collins networks

Collins hybrid couplers can be bench tested. All ports can be terminated in their system impedance (usually 50 Ω) and a vector voltmeter or a suitable oscilloscope can be used to measure the output at the 2 ports (magnitude and phase). It is not necessary to do this test, as long as the component values have been measured accurately.

Gehrke networks

Constructing a Gehrke-type feed system requires quite a bit of network designing. Off-site measurement and adjustment of the networks is required also. To do this, first make dummy antennas that have the same impedance as the feed impedance of the array elements (such as 21 − j20 Ω and 51 + j20 Ω in the case of the 2-element end-fire array with cardioid pattern). The former is a series connection of a 21-Ω resistor and 2122 pF of capacitance (at 3.75 MHz); the latter is a series connection of a 51-Ω resistor with a 0.85-μH inductor. Inductors can be measured precisely by dipping the resonant circuit made by the coil

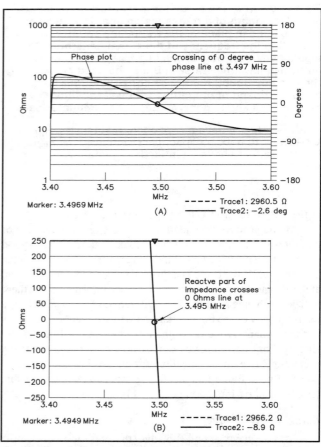

Fig 11-23—At A, the Z, $\underline{/\theta}$ display showing Z as >1000 Ω, and the phase angle (θ) going through zero at exactly 3497 kHz. At B the R and J X display, with R >250 Ω and X going through zero at 3495 kHz. The curves are extremely sharp and the frequency can be read easily within 1 kHz. These displays are generated with the Alpha/Power Analyzer.

and a standard-value capacitor. Alternatively, one may measure air-wound or toroidal coils using an impedance bridge. Do *not* measure toroidal cores using a Digital L-C meter. These meters work on a fixed frequency (often 1 MHz) and the exact inductance of a toroidal core is often frequency dependent (always with ferrite materials). The various sub-units of the feed system can now be tested in the lab. Voltages can be measured using a high-quality oscilloscope. If a dual-trace scope is used, the phase difference between two points of interest can also be measured.

Suitable current amplitude and phase probes have been described by Gehrke (Ref 927 and *The ARRL Antenna Book*). When measuring the impedances of the building blocks of a total feed system, make sure you know exactly what you are measuring. All impedances mentioned during the design were impedances looking from the generator toward the load. If a network is measured in the reverse direction (terminating the input of the network with a resistive impedance to measure the output impedance), you will measure the conjugate value. (A conjugate value has the opposite reactance sign; the conjugate of $A + jB$ is $A - jB$). All tuning and measurement procedures are adequately covered by Gehrke in Ref 927. The correct coaxial cable lengths can be cut using the methods described in Chapter 6 (The Feed Line and the Antenna), Section 4.1.

3.3.8.5. Antenna feed-current magnitude and phase measurement

It is essential to be able to measure the feed current in order to assess the correct operation of the array. A good-quality RF ammeter is used for element-current magnitude measurements and a good dual-trace oscilloscope to measure the phase difference. The two inputs to the oscilloscope will have to be fed via identical lengths of coaxial cable. **Fig 11-24** shows the schematic diagram of the RF current probes for current amplitude and phase-angle measurement. Details of the devices can be found in Ref 927. D. M. Malozzi, N1DM, pointed out that it is important that the secondary of the toroidal transformer always sees its load resistor, as otherwise the voltage on the secondary can rise to extremely high values and destroy components and also the input of an oscilloscope if the probe is to be used with a scope. He also pointed out that it is best to connect two identical resistors at each end of the coax connecting the probe to the oscilloscope. Both resistors should have the impedance of the coax. Make sure the resistors are non-inductive, and of adequate power rating. It is not necessary to do your measurement with high power (nor advisable from a safety point of view).

3.3.8.6. Quadrature-fed arrays

If you are designing an array with quadrature-fed elements, it is possible to check the element current magnitudes and phase relationships by using the simple piece of test equipment described below. Remember that the antenna feed current (magnitude and phase) is perfectly reflected in the voltage at the end of the $1/4$-λ feed lines. The magnitude of the input current of a $1/4$-λ transmission line is equal to the output

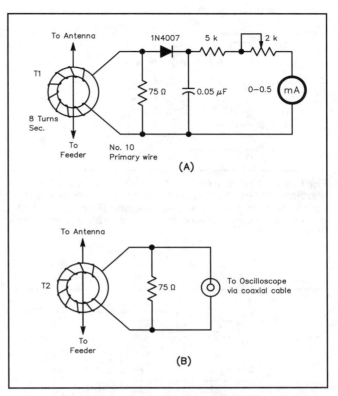

Fig 11-24—Current amplitude probe (at A) and phase probe (at B) for measuring the exact current at the feed point of each array element. See text for details.

T1,T2—Primary, single wire passing through center of core; secondary, 8 turns evenly spaced. Core is $1/2$-in. diameter ferrite, $\mu_i = 125$ (Amidon FT-5061 or equiv.)

voltage divided by the characteristic impedance of the line, and it is independent of the load impedance. In addition, the input current lags the output voltage by 90° and is also independent of the load impedance. This means we can measure the voltage at the end of the feed lines to know all about the current at the element feed points. We can of course use a vector voltmeter, or suitable oscilloscope, but we can also build a simple piece of test equipment to do the same.

An analog phase meter

The heart of this test equipment's circuit is a wideband, single-ended, push-pull transformer. When fed as shown in **Fig 11-25**, the secondary (C-D) should be completely out of phase. To test this, you can rectify the RF and measure the sum of the two voltages. If you connect points A and B to the end of the quarter-wave feed lines (feeding the elements with a current phase difference of 90°), the voltages at C and D will be 180° out of phase if the voltages at A and B are 90° out of phase. The transformer can be wound on a ferrite core, such as an Indiana General BBR-7731 (also used for feeding Beverage antennas; see Chapter 7, Special Receiving Antennas). The winding consists of four turns of AWG 20 to 26

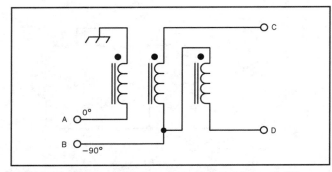

Fig 11-25—This quadrature transformer has a trifilar winding. Notice the phasing dots; the windings must be correctly phased. If voltages A and B are 90° out of phase, the voltage from B to C will be 180° out of phase.

enameled copper wire, trifilar wound. Make sure the phasing is as shown in Fig 11-25.

Fig 11-26 shows the schematic diagram of this piece of equipment. It can be built in a small, well-shielded box. The circuit contains four voltage-doubling rectifier arrangements with separate sensitivity adjustment potentiometers. All modules should be adjusted for the same sensitivity with the output potentiometers.

The measuring circuit is a straightforward differential FET voltmeter that is powered with a small 9-V battery. The ganged potentiometers in the source circuit of the FETs can be used to adjust the overall circuit sensitivity. The three-position input switch has a balancing position (position 1) for adjustment of the 50-kΩ balance potentiometer. With the switch in position 2, the circuit measures the difference between the magnitudes of the voltages at the two inputs. Identical voltage inputs are indicated by a zero meter reading.

With the switch in position 3, the circuit checks for a 90° phase difference between the two inputs. In order to make a nonzero phase-shift reading more meaningful, the meter can be calibrated with the setup shown in **Fig 11-27**. The phase calibrator consists of a ¹/₄-λ, 50-Ω delay line. At the frequency where the delay line is exactly ¹/₄-λ long, the voltages at A and B will be equal in magnitude but 90° out of phase. The meter can be calibrated by varying the transmit frequency by 5% down and noting the meter reading, then moving the transmit frequency 5% up and making sure that the meter reading is the same. The test circuit can be checked for symmetry by inverting the inputs. The meter should read the same value for the normal or inverted mode. Note that for actual testing in an array, the device must be installed at the ends of the quarter-wave feed lines to the elements (where the direction switching is done).

This device is a very handy tool for making fine adjustments in a Lewallen type of feed system. The meter could be mounted in the shack for convenience. If you are building a Collins-type feed system, you can use this device to check

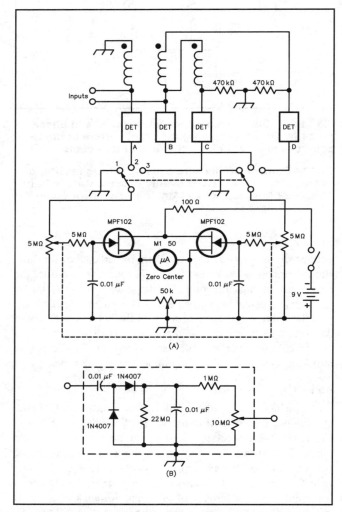

Fig 11-26—The simple quadrature tester consists of a transformer, four detector modules, a mode switch and a differential voltmeter. The diagram for the detector modules is shown at B. Note that the two 5-MΩ potentiometers of the tester are ganged. This is the sensitivity control.

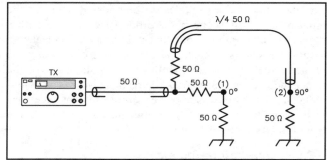

Fig 11-27—In this phase-calibration system for the quadrature tester, RF voltage from the transmitter is divided down with two 50-Ω series resistors (to ensure a 1:1 SWR), routed directly to a 50-Ω lead, and through a 90°-long 50-Ω line (RG-58) to the second 50-Ω load. For a frequency of 3.65 MHz, the cable has a nominal length of 44.49 feet (13.56 m). The cable length should be tuned using the method described in Chapter 6 on feed lines and matching.

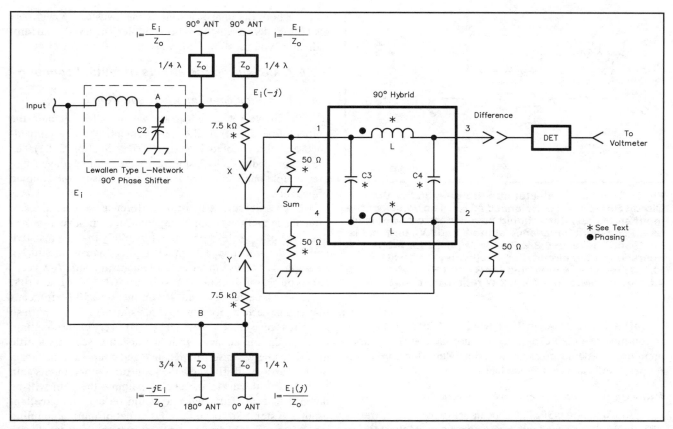

Fig 11-28—The W1MK phase measuring setup for quadrature-fed arrays. The unit employs a hybrid coupler as used in the Collins feed system for arrays. The unit can be left permanently in the circuit if the voltage dividing resistors are of adequate wattage. See text for details.

the performance of the hybrid coupler by terminating all three output ports of the coupler with resistors equal to the design impedance of the coupler and then connecting the quadrature checker between the ports that have a 90° phase shift between them. If the coupler is working properly, the meter should read zero.

A hybrid coupler/monitor circuit

R. Lahlum, W1MK (Ref 968), described a hybrid coupler (identical to the coupler used in the Collins feed system) for monitoring the exact 90° phase relationship in quadrature-fed arrays (see **Fig 11-28**). If high-wattage resistors are used (15-W non-inductive resistors are required for the 7.5-kΩ resistors!), the hybrid coupler can be left connected to the feed system at all times. The voltage from the HF detector at the output port of the coupler can be routed to the shack via a (long) shielded cable.

Both elements (C and L) of the L network for the Lewallen-type feed system can be made continuously variable by a little trick. If the networks require eg, a coil with a reactance of 50 Ω, make a coil with double the reactance (100 Ω or 4.2 μH at 3.8 MHz) and connect in series a variable capacitor with (at maximum capacitance) a reactance of − 50 Ω or less. If you use −25 Ω (1675 pF at 3.8 MHz), the series connection of the two elements will now yield a continuously variable reactance (at

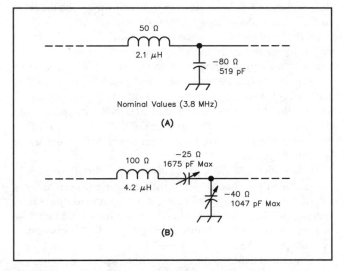

Fig 11-29—To make the Lewallen L network continuously adjustable, replace the coil with a coil of twice the required value and connect a capacitor in series. The net result will be a continuously variable reactance. With the values shown, the nominal +50-Ω reactance is adjustable from +75 to +25 Ω (and less). The two capacitors can be motor driven to make the phase-shift network remotely controllable.

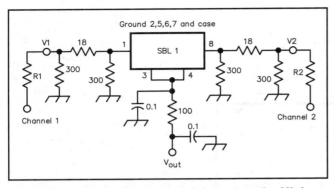

Fig 11-30—Phase detector circuit based on the Mini-Circuit SBL-1 Double Balanced Mixer. The values of the input series resistors should be adapted to power level used. The input impedance at the V1 and V2 terminal is 50 Ω, and a level of 0.3 V RMS is required for proper operation of the circuit. For a 100 W level a 10 kΩ (0.5 W) resistor is indicated for R1 and R2, for 1.5 kW, a resistor of 47 kΩ, 2 W rating, is required.

3.8 MHz) of 25 (or less) to 75 Ω. See **Fig 11-29** for a diagram. If you motorize both capacitors and use the above-described monitoring system, you can now remotely tune the L networks for peak performance over the band.

Using an integrated circuit product detector

Bob Whelan, G3PJT, has been using the Mini-Circuits SBL-1 double balanced mixer as a phase detector/meter. For the Lewallen and Collins "current forcing" methods, the voltages at the ends of the $1/4$-λ feed lines are in phase with the current into the elements. Therefore, all that is necessary is to measure the voltages at the switch box with a phase detector and an RF voltmeter. **Fig 11-30** shows the diagram of the SBL-1 phase meter. The output of the circuit will be zero if the phase difference is 90°. G3PJT uses the DBM phase meter to adjust the LC components of the Lewallen network for zero output. Then he uses an RF detector and a remote source off the back of the array (see Section 3.3.3.3) and carefully adjusts the network L and C values "slightly" around the 90° settings, to obtain best F/B. Bob points out that it is almost impossible to tune the L-C networks for best F/B without first having set the theoretical 90° values using the DBM circuit.

As with the W1MK phase monitoring circuit, one could leave the SBL-1 based phase monitor circuit permanently wired into the array switching box, provided the series resistor of 2 kΩ (Fig 11-30) is replaced with a 7.5-kΩ, 15-W non-inductive resistor if monitoring of a 1.5-kW transmitter is anticipated. The output of the SBL-1 circuit could be fed via a shielded cable to the shack, where a permanent indication would be available, using a zero-center 0.5-mA moving coil meter.

If a Lewallen L network is used, the SWR meter and of course the phase monitor will be excellent fault indicators.

If a Collins hybrid coupler is used, one could use a small relay and feed either the output of the SBL-1 circuit or the output of an RF-detector on the dummy load to the shack, to

monitor both the phase as well as the "wasted" power (see also Fig 11-89). With the hybrid coupler, the SWR is no fault indicator, you must monitor wasted power and/or phase.

3.3.8.7. Measuring the effects of mutual coupling

Too little mutual coupling where you want it

When we set up an array, we need to calculate the mutual impedance from the measurements of the self impedance and the coupled impedance (Ref Section 3.2.1). The normal procedure is to first measure the self impedance, and then couple one element at a time, and measure the coupled impedance.

If you measure little or no difference between the self impedance and the coupled impedance, then have a look at the value of the self impedance. It is likely that the resistive part of the impedance is much higher than it should be. Example: if you use inverted-L elements that are $1/8$-λ vertical, you should expect a self impedance of approx. 17 Ω over a perfect ground. If you measure 50 Ω, it means that you have an equivalent loss resistance of 33 Ω! With so much loss resistance you will, even with very close coupling, as in the case in an array with $1/8$ λ spacing, see only a little difference between self impedance and coupled impedance. Such an array will still show the same directivity, but its gain will be way down. In the above example the gain will be down 4 to 5 dB from what it would be over an excellent ground system. So, if you see no effect of mutual coupling where you should see it, suspect you have large losses involved somewhere.

Unwanted mutual coupling

There are cases where you don't want to see the effect of mutual coupling. But they are there, and you want to control them. If you happen to have towers (or other metal structures or antennas) within $1/4$ λ of one of the elements of the array, it is possible that you will induce a lot of current into that tower by mutual coupling. The tower will act as a parasitic element, which will upset the radiation pattern of the array and also change the feed impedances of the elements and the array.

To eliminate the unwanted effect from the parasitic coupling proceed as follows:

- Decouple all the elements of the array with the exception of the element closest to the suspect parasitic tower. For quarter-wave element decoupling, this means lifting the elements from ground (see Section 3.2.2).
- Measure the feed-point impedance of the vertical under investigation.
- If a suspect tower is heavily coupled to one of the elements of the array, a substantial current will flow in it. Probe the current by one of the methods described by D. DeMaw, W1FB (Ref *W1FB's Antenna Notebook*, ARRL publication, 1987, p 121) and shown in **Fig 11-31**. If there is an appreciable current, you will have to "detune" the tower.
- Attach a shunt-fed wire at the top of the suspect tower (see Fig 9-84) and connect the bottom end to ground through a 1000-pF variable capacitor (a broadcast variable will do).

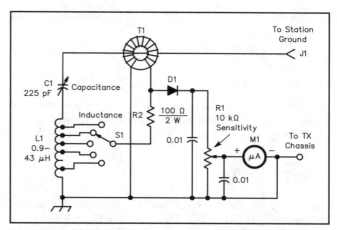

Fig 11-32—Schematic diagram of the MFJ-931 artificial RF ground. The unit includes a current probe (toroidal transformer), a detector, and a series tuning arrangement (L and C). When used to decouple a tower, L and C should be varied until *minimum* current is obtained. See text for details.

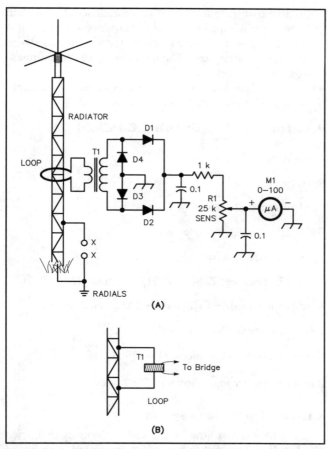

Fig 11-31—Current-sampling methods for use with vertical antennas, as described by DeMaw, W1FB. Method A requires a single-turn loop of insulated wire around the tower. The loop is connected to a broadband transformer, T1. A high-mu ferrite toroid, as used with Beverage receiving antennas (see Chapter 7 on special receiving antennas), can be used with a 2-turn primary and 2 to 10-turn secondary, depending on the power level used for testing.

Before closing the circuit, pass the vertical wire through the core of a current probe (see Fig 11-24).

• While applying power to the single vertical, tune the BC variable for minimum current in the "parasitic" tower. You may have to choose another attachment point on the tower to bring the current down to zero or a very small value. Note the value of the tuning capacitor.

• Measure the feed-point impedance of the vertical again. If you have properly detuned the parasitic tower, you will likely see a rise in impedance and a shift in resonant frequency.

• Reconnect the whole array and fire in the direction of the "parasitic" tower.

• Check the current in the parasitic tower, and if necessary make final adjustments to minimize the current in the tower. You can use high power now in order to be able to tune the tower very sharply. In general the tuning will be quite broad, however.

• Replace the BC variable with a fixed capacitor of the same value.

You now have made the offending tower invisible to your array.

The MFJ-931 (artificial RF ground) is a useful instrument for detuning a tower. **Fig 11-32** shows the schematic of the unit. The inductance/capacitance can be varied until minimum current flows in the circuit.

3.3.8.8. Performance measurement

Measuring gain is something that is out of reach for all but a few of those lucky enough to have access to an antenna test range. Front-to-back ratio, however, can be measured fairly easily with the use of a second antenna mounted several wavelengths away from the antenna under test, or by the help of a neighbor ham (he must be located quite accurately in the back of the array in order to optimize F/B).

The directivity of most of the arrays described in this chapter is very high at low angles. Some designs show major back lobes at high angles (60 or even 90°). Even the best array will show very little directivity at high angles, because it also radiates very little power at high angles in the forward direction. All this is to say that in practice you must evaluate the array with DX signals. Properly fed 4-square arrays show an operational F/B of 25 to 35 dB in practice. In many instances it is like going from an S9 signal to "inaudible." On signals within 1500 km (900 miles) the directivity of the array will be mediocre at best.

3.3.9. Network component dimensioning

When designing array feed networks using the computer modules from the NEW LOW BAND SOFTWARE, you can use absolute currents instead of relative currents.

The feed currents for the 2-element cardioid array (used so far as a design example) have so far been specified as

I1 = 1A, $\underline{/-90°}$ and I2 = 1A, $\underline{/0°}$. The feed-point impedances of the array are:

Z1 = 51 + j20 Ω

Z2 = 21 − j20 Ω

With 1 A antenna current in each element, the total power taken by the array is 51 + 21 = 72 W. If the power is 1500 W, the true current in each of the elements will be

$$I = \sqrt{\frac{1500}{72}} = 4.56 \text{ A}$$

Using this current magnitude in the relevant computer program module COAXIAL TRANSFORMER will now show the user the real current and voltage information all through the network design phase. The components can be chosen according to the current and voltage information shown.

If there is any question as to the voltage rating of any of the feed lines that are used as transformers in our designs (all have an SWR greater than 1), the program FEED LINE VOLTAGE with the real current as an input can be used to calculate what the highest voltage is at any point on the line. We find that for the 2-element array with a cardioid pattern (fed according to the Christman method), the highest voltage on a feed line of any length to element 1 (which has a 1.48:1 SWR) is only 397 V with 3 kW applied. For feed line 2, the maximum voltage is 352 V. For the 4-square array with $^1/_4$ λ spacing, the feed-line-voltage values are 234 V, 253 V, 253 V and 391 V. This should not represent any problem with good-quality RG-213 cable. In a similar fashion the voltages across capacitors or currents through capacitors in the lumped-constant networks can be determined. When evaluating coils, use the following guidelines: For up to 5 μH, it is advisable to use air-wound coils. The best Q factors are achieved with coils having a length-to-diameter ratio of 1. For higher values, use powdered-iron toroidal cores if necessary (never use ferrite material for these applications). Information on this subject as well as on the subject of dimensioning capacitors in a network is given in Chapter 6, on feed lines and matching. The computer module COIL CALCULATION of the NEW LOW BAND SOFTWARE may be helpful in designing the coils.

■ 4. POPULAR ARRAYS

This section briefly describes the most popular arrays and a few feed systems. I will not systematically cover all the possible feed systems.

The gains quoted in the following paragraphs are for arrays made of quarter-wave verticals and are quoted over a single $^1/_4$-λ vertical over the same ground. All arrays are modeled over very good ground, with an extensive radial system that accounts for an equivalent series loss resistance of 2 Ω (for each element). The element feed-point impedances shown include this 2 Ω of loss resistance. If you want to calculate your feed system for different equivalent ground loss resistances, apply the following procedure:
• Take the values from the array data (see below). The

resistive part includes 2 Ω of loss resistance. If you want the feed-point impedance with 10 Ω of loss resistance, just add 8 Ω to the resistive part of the feed-point impedance shown in the array data. The imaginary part of the impedance remains unchanged.
• Follow the feed-system design criteria as shown, but apply the new feed-point impedance values.

4.1. The 90°, $^1/_4$-λ Spacing Cardioid

Array data

Spacing: $^1/_4$ λ

Feed currents: I1=1 A, $\underline{/0°}$, I2=1 A, $\underline{/-90°}$

Gain: 3.1 dB over a single vertical

3 dB beamwidth: 176°

Mutual Impedance: Z(12) = Z(21) = approx. 15 + j20 Ω

Feed-point impedance (including 2 Ω ground loss):

Z (lagging phase element) = 51 + j20 Ω

Z (leading phase element) = 21 − j20 Ω

Radiation patterns are shown in Fig 11-4.

4.1.1. Christman feed system

Fig 11-7 shows how we can switch the array in the two end-fire directions. When both elements are fed in phase the array will have a bi-directional broadside pattern with a gain of 1 dB over a single vertical. The front-to-side ratio is only 3 dB. The feed impedance of two quarter-wave-spaced elements fed in phase is approximately 57 − j15 Ω, assuming an almost-perfect ground system with 2 Ω equivalent ground loss resistance. Notice that both elements have the same impedance, which is logical as they are fed in phase.

We can easily add the broadside direction (both elements fed in phase) by adding a switch (relay) that shorts the 71° long phasing line, as shown in **Fig 11-33**. L networks can be designed to match the array output impedance to the feed line. Don't forget that you need to measure impedances in order to calculate the line lengths that will give you the required phase shifts. Merely going by published figures will not get you optimum performance!

4.1.2. Lewallen feed system

When using the Lewallen system, one can also add the broadside bi-directional configuration. Fig 11-33 shows the wiring of the array with the Lewallen feed method, including the switching harness.

Assuming a feed impedance of 57 − j15 Ω and RG-213 coax with 0.35 dB loss per 100 ft at 3.8 MHz, the impedance at the end of the 90° feed lines (calculated with the COAXIAL TRANSFORMER software module) is Z = 41.3 + j0.5 Ω

Paralleling the two feed lines yields an impedance of

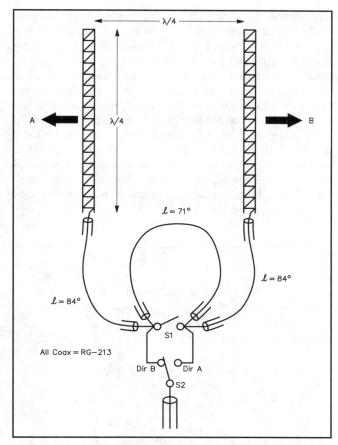

Fig 11-33—The 2-element vertical array (¹/₄-λ spacing) can be fed in phase to cover the broadside directions. I added switch S1 to the Christman feed system as described in Fig 11-7. When S1 is closed, both antennas are fed in phase, resulting in bi-directional broadside radiation.

$20.65 + j5.25 \ \Omega$. This impedance can be matched to the feed-line impedance with an L network, as shown in **Fig 11-34A**.

4.1.3. Using the Collins hybrid coupler

When you buy a commercial hybrid coupler, you don't really need to do any impedance measurements. All you will have to do is trim the elements to resonance (decoupled from one another!). Commercial hybrid couplers are made to accommodate 4-square arrays, and normally use 4 relays to do the direction switching. For a 2-element end-fire array, a much simpler switching system, using a single DPDT relay will do the job if only the two end-fire directions are required. In this case you can delete K1 and its associated wiring from the schematic shown in Fig 11-34. On the low-bands any 10 A relay will do. If you want the bi-directional broadside pattern as well, two relays and an L-C network are used as shown in Fig 11-34.

4.1.3. How to prevent hot switching

A potential problem with any antenna switching, is that one inadvertently switches while transmitting. This guaran-

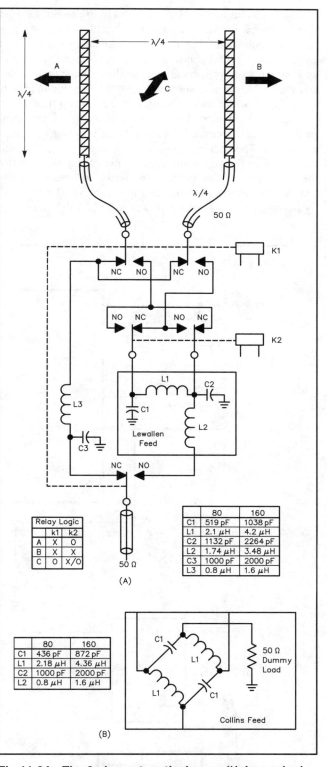

Fig 11-34—The 2-element vertical array (¹/₄-λ spacing) can be fed in phase to cover the broadside directions. Two additional feed methods are shown. At A the wiring for the Lewallen feed method, and at B the Collins hybrid-coupler method. In both these circuits relay K1 chooses between the end-fire and the broadside configuration, relay K2 switches directions in the end-fire position.

tees burnt contacts, unless you happen to use husky vacuum relays. This is not so much a problem with regular antenna band selection switches (relay or manual), as you don't switch those around continuously. It is not uncommon that you have one of your hands on the direction switch of the array while doing some listening. If however, you inadvertently switch while transmitting, you are in deep trouble.

Frank Veldeman, ON1ACV, a member of my local radio club, developed a little circuit that prevents this from happening. The principle is as follows. The line(s) that go to the relays go through a little electronic device that I call a memory system. The circuit also senses the status of the transmitter (whether it is on transmit or receive). This is done by passing the relay-out line (used to switch the amplifier)

through the little circuit. When that line is low (grounded), the electronic circuit will temporarily hold in memory any change you make in direction switching, until the transmitter status line goes high (that means the transceiver is on receive). Within a fraction of a second the unit will then switch to the direction you actually selected while transmitting (see **Fig 11-35**). This unit is *not* fast enough to be used with full break-in CW, but works fine on semi break-in CW and SSB/VOX control. It is utterly important that the transmitter-sense line is well connected, and is connected to the transceiver you are actually using on the antenna you are protecting with this circuit. (I have various rigs in the shack, and I once had it connected to the wrong radio, and I kept burning antenna relay contacts—no wonder!)

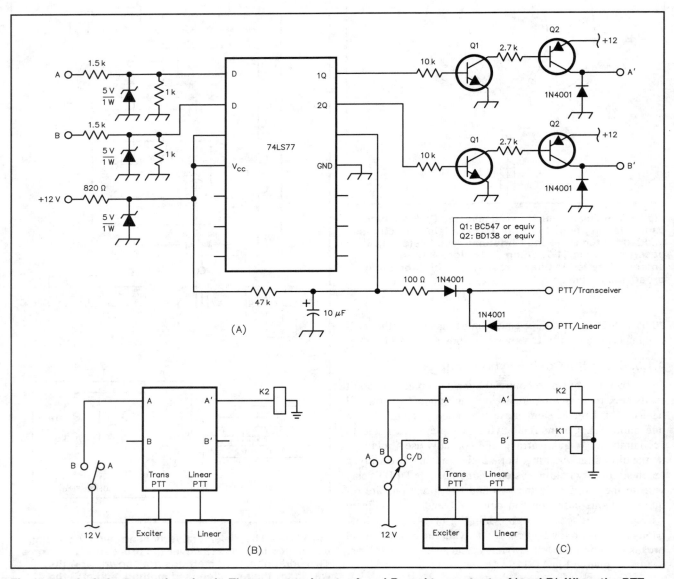

Fig 11-35—At A the protective circuit. There are two inputs, A and B, and two outputs, A' and B'. When the PTT (Transceiver) line is high (floating), A' and B' follow A and B without delay. When the PTT (Transceiver) line is low, then A' and B' will not follow A or B until the PTT will go high again. At B the wiring for switching the end-fire array from Fig 11-33 in two directions. At C the circuit for switching three directions, as shown in Fig 11-34.

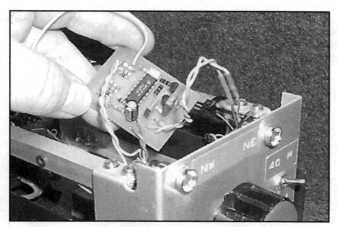

Fig 11-36—The protective circuit is built into the control box of the 40 as well as the 80-meter 4-square switching box at ON4UN. (The 40-m unit is shown.)

The circuit can handle two control lines, which makes it suitable for use with a control box for 4-square arrays as well (see Section 4.9.1, Fig 11-43). A fully tested circuit, on a 40×50 mm printed board, is available from the author ($50, including air-mail shipping world wide). See **Fig 11-36**.

4.2. The 135°, ¹/₈-λ Spacing Cardioid

This array is known as the ZL-special or the HB9CV array.

Array data

Spacing: ¹/₈ λ

$I1 = 1$ A, /0° ; $I2 = 1$ A, /–135°

Gain: 3.7 dB over a single vertical

3 dB beamwidth: 142°

The feed-point impedance (including 2 Ω ground loss):

Z (leading phase element) = $13 - j21$ Ω

Z (lagging phase element) = $18 + j23$ Ω

Radiation patterns are shown in **Fig 11-37**. The horizontal pattern also shows the pattern for the ¹/₄ λ spacing, 90° out of phase, for comparison.

Because the elements are not fed in quadrature, the Collins and the Lewallen feed methods cannot be used. The modified Lewallen, the Christman and the Gehrke methods can be used, however.

4.2.1. The Modified Lewallen feed system

Using the COAX TRANSFORMER program from the NEW LOW BAND SOFTWARE, we first calculate Z, I, and

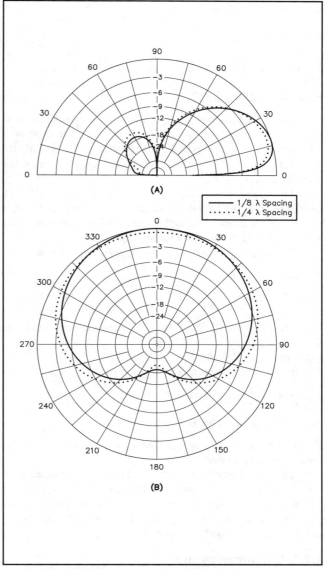

Fig 11-37—Radiation patterns for the 2-element cardioid array with ¹/₈-λ spacing when fed 135° out of phase. The patterns of the ¹/₄-λ spaced (90° out of phase) array are superimposed for comparison. The gain is calculated over very good ground, using a radial system with an equivalent loss resistance of 2 Ω.

E at the ends of the four quarter-wave feed lines.

Input data

$I1 = 1$ A, /0°

$Z1 = 13 - j21$ Ω

$I2 = 1$ A, /–135°

$IZ2 = 18 + j23$ Ω

At the end of the quarter-wave feed lines we find:

$Z1' = 55.6 + j82.9\ \Omega$

$E1 = 50.23\ V,\ \underline{/89.6°}$

$Z2' = 54.14 - j65\ \Omega$

$E2 = 50.23\ V,\ \underline{/-44.55°}$

Let us feed the array at the end of the line to the element with the leading phase (we can choose either). We must now add a shunt reactance to make the input impedance real (we will use a constant-impedance line stretcher to provide the additional phase shift).

The shunt impedance (calculated with the SHUNT IMPEDANCE NETWORK module) required to achieve a resistive impedance at the end of the $1/4$-λ feed line to element 1 is $-j120.2\ \Omega$ (this is 348 pF at 3.8 MHz).

With the capacitor in parallel the values have become:

$Z1'' = 179.2\ \Omega$

$E1'' = 50.23\ V,\ \underline{/89.59}$

Now we calculate the constant-impedance π-line stretcher:
Output phase angle: 89.59°
Input phase angle: −44.55°
The pi-filter components are

$X_S = -j128.6\ \Omega\ (= 326\ pF\ at\ 3.8\ MHz)$

$X_P = j75.8\ \Omega\ (= 3.18\ \mu H\ at\ 3.8\ MHz)$

Fig 11-38 shows the final layout of the array with all the feed-system components.

The impedance at the array feed point can be calculated by paralleling 179.2 Ω and $54.14 - j\ 64.97\ \Omega$. The result is $51.48 - j35.6\ \Omega$. This impedance can be matched almost perfectly to a 50-Ω line by merely connecting a 1.5-µH inductor in series with the feed line.

4.2.2. Christman method

Fig 11-39 shows the Christman feed method for this array.

Input data

$Z1 = 13 - j21\ \Omega$

$I1 = 1\ A,\ \underline{/0°}$

$Z2 = 18 + j23\ \Omega$

$I2 = 1\ A,\ \underline{/-135°}$

Using the program FEED LINE VOLTAGE, one can scan both 50-Ω feed lines for points where the voltages are the same on both lines. Two such sets of points are found in this case.

- One is 157° from element 1 (the element with the leading current) and 196° from element 2. Note that the 135° current phase shift is accomplished with a feed-line-length

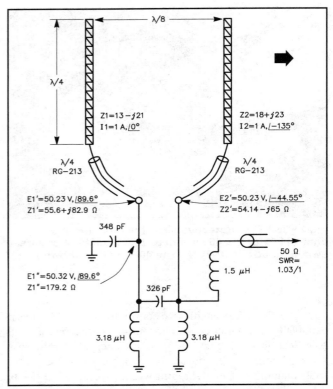

Fig 11-38—Modified Lewallen feed system for the 2-element cardioid array ($1/8$-λ spacing, 135° out of phase). The modified Lewallen network uses a pi (or T) filter plus a shunt reactance (coil or capacitor) instead of an L network, but can handle any phase shift, whereas the Lewallen network handles only 90° phase shifts.

difference of only 39°.
- The second one is 38° from element 1 and 158° from element 2.

There's a clear choice here because the second set of points occurs at areas of high variation of feed-line voltage for small variations in distance, making these points a poor choice. Using the second set of points would also decrease the bandwidth of the array. The location of one point at only 38° from the element would make it impossible to make the array switchable in direction. Neither of these problems are encountered with the first set of points. The first solution requires a fair amount of feed line, however.

The feed lines can now simply be parallel-connected at these points. The impedances can be calculated using the COAX TRANSFORMER software module:

$Z1' = 24.9 - j47\ \Omega$

$E1' = 41\ V,\ \underline{/108°}$

$Z2' = 28 + j38.4\ \Omega$

$E2' = 41\ V,\ \underline{/108°}$

The combined impedance can then be calculated with the T-JUNCTION module. The impedance for this array

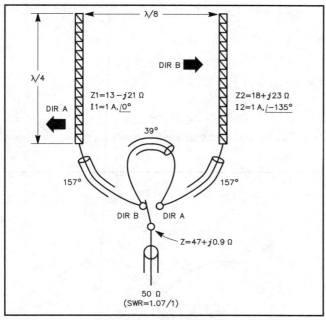

Fig 11-39—Christman feed method for the 1/8-λ-spaced cardioid pattern array. This system turns out to be very simple, and it yields an almost perfect 1:1 SWR on a 50-Ω feed line without the need for a matching network. The direction switching is done with an SPDT relay.

comes out to $47.2 + j0.9 \ \Omega$, which yields an SWR of 1.07:1 when connected to a 50-Ω cable, and will require no further matching.

4.3. Two Elements in Phase with $\frac{1}{2}$-λ Spacing

Array data

Spacing: $\frac{1}{2} \lambda$

Feed currents: I1 = 1 A, $\underline{/0°}$; I2 = 1 A, $\underline{/0°}$

Gain: 3.8 dB over a single vertical

3 dB beamwidth: 62°

Radiation: Bi-directional, broadside

Mutual impedance: Z(12) = Z(21) = approx. $-9 - j13 \ \Omega$

The feed-point impedance (including 2 Ω ground loss resistance) is Z1 = Z2 = 31 − j14 Ω

Fig 11-40 shows the radiation patterns over very good ground.

4.3.1. Feed system

In principle the array can be fed with two feed lines of equal lengths. Feeding via $\frac{1}{4}$-λ (or $\frac{3}{4}$-λ) feed lines, however, has the advantage of "forcing" equal currents in both elements, whatever the difference in element impedances might be. It is therefore advised to feed the array via two $\frac{3}{4}$-λ feed lines. Quarter-wave feed lines are too short (due to the velocity factor) to reach the center of the array.

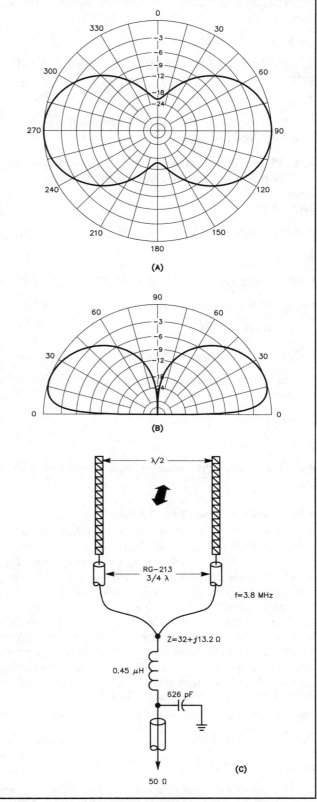

Fig 11-40—A, horizontal and B, vertical radiation patterns for an array made of two verticals spaced $\frac{1}{2}$-λ apart and fed in phase. It is best to feed the elements via two $\frac{3}{4}$-λ long feed lines, as shown at C.

The impedance at the end of the 270° long RG-213 feed line (attenuation = 0.35 dB/100 ft at 3.8 MHz) is:

$Z = 65.84 + j26.37\ \Omega$

The two feed lines in parallel result in an array impedance of:

$Z_{array} = 32.0 + j13.2\ \Omega$

The feed system and an L network providing a 1:1 SW to a 50-Ω feed line are shown in Fig 11-40.

4.4. Three Elements in Phase, ¹/₂-λ Spacing, Binomial Current Distribution

Array data

Number of elements: 3

Spacing: ¹/₂ λ

Feed currents: I1 = 1 A, /0°; I2 = 2 A, /0°; I3 = 1 A, /0°

Gain: 5.2 dB over a single vertical

3-dB beamwidth: 46°

Radiation: bi-directional, broadside

The feed-point impedances (including 2 Ω ground loss resistance) are:

$Z1 = Z3 = 25.0 - j19.4\ \Omega$

$Z2 = 30.7 - j13.7\ \Omega$

Fig 11-41 shows the radiation patterns over very good ground.

4.4.1. Current-forcing feed system

For reasons explained in Section 3.3.7, it is always best to try to feed the elements with feed lines that are an odd multiple of ¹/₄ λ long. Three-quarter-wavelength-long feed lines to elements 1 and 3 will just reach the center of the array if they are made of solid PE coaxial cables (0.75-λ long feed line × 0.66 velocity factor = 0.5 λ).

We will run ³/₄-λ-long feed lines to all elements. The feed line to the center element will require half the impedance of the cables to the outer elements to obtain the proper current magnitude. We use two 50-Ω feed lines in parallel (Z = 25 Ω). All calculations are done with RG-213 coax (0.35 dB/100 ft attenuation at 3.8 MHz).

The element impedances are:

$Z1 = Z3 = 25.0 - j19.4\ \Omega$

$Z2 = 30.7 - j13.7\ \Omega$

The impedances and voltages at the end of the feed lines are calculated using the software module COAXIAL CABLE TRANSFORMER/SMITH CHART. Elements 1 and 3 are fed via a 270°-long, 50-Ω line:

$Z1' = Z3' = 63.1 + j42.6\ \Omega$

E1′ = E3′ = 51.37 V, /– 91.1°

Element 2 is fed via a 270°-long, 25-Ω line:

$Z2' = 7.41 + j7.3\ \Omega$

E1′ = 51.6 V, /– 90.8°

The voltages are essentially identical, so the feed lines

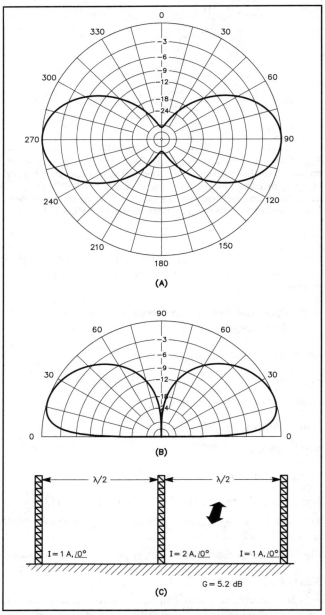

Fig 11-41—A, horizontal and B, vertical radiation pattern for the three-in-line array with ¹/₂-λ spacing. The element feed currents are in phase but the center element gets twice as much current as the outer ones (binomial current distribution). Notice the rather narrow lobe beamwidth in the azimuth pattern at A.

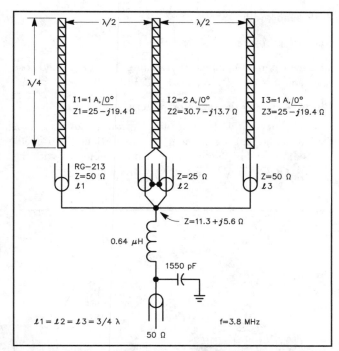

Fig 11-42—Feed system for the three verticals in line with binomial current distribution. The center element is fed via two parallel 50-Ω feed lines in order to obtain the double current. The current-forcing method ensures that variations in element self-impedances have a minimum impact on the performance of the array.

can simply be connected in parallel. The array impedance is made up by the parallel connection of the impedances at the end of the three feed lines:

$$Z_{array} = 11.3 + j5.6 \; \Omega$$

We can design an L network to match this impedance to the 50-Ω feed line. The layout with all the values is shown in **Fig 11-42**.

4.5. Two-Element Bidirectional End-Fire Array, ¹/₂-λ Spacing, 180° Out of Phase

Array data

Number of elements: 2

Spacing: ¹/₂ λ

Feed currents: I1 = 1 A, $\underline{/1°}$; I2 = 1 A, $\underline{/-180°}$

Gain: 2.4 dB over a single vertical

3-dB beamwidth: 116°

Radiation: bi-directional, end-fire

Mutual impedance: Z(12) = Z(21) = approx. − 9 − j13 Ω

The feed-point impedance (including 2 Ω ground loss resistance): Z1 = Z2 = 45.4 + j14 Ω.

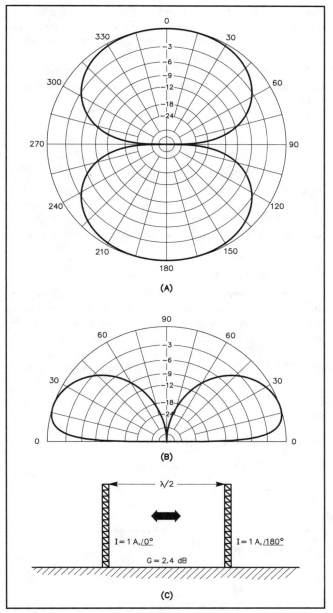

Fig 11-43—Radiation patterns for the ¹/₂-λ spaced bidirectional end-fire array. The two elements are fed in phase opposition (180° out of phase).

Fig 11-43 shows the radiation patterns over very good ground.

4.5.1. Current-forcing feed system.

We will run a 270°-long feed line to the element with the leading current, and a 450°-long feed line to the element with the lagging feed current. With the lines being odd multiples of ¹/₄-λ long, we enhance the "current-forcing" principle (currents will be equal in magnitude even though element impedances may be slightly different). A 90° and a 270°-long feed line are too short for the array, as the elements are spaced $^{1}/_{2}\lambda$. To preserve symmetry, the T junction, where the lines to the elements join, must be located at the center of the array. The element with the leading current is fed via a $^{3}/_{4}$-λ feed line, the lagging element via a $^{5}/_{4}$-λ line.

The impedances at the end of the feed lines can be calculated with the COAX TRANSFORMER software module.

$Z1' = 50.5 - j14\ \Omega$

$Z2' = 59.6 - j13\ \Omega$

The combined impedance can then be calculated with the T-JUNCTION module.

The impedance for this array comes out to $27.4 - j6.8\ \Omega$. An L network can be designed for matching this impedance to a 50-Ω line. The feed system is shown in **Fig 11-44**.

4.5.2. Closer spacing

An alternative where you can use a 90° and a 270° feed line is to use foam dielectric line (VF ≈ 80%) and move the two radiators closer together. As you reduce the spacing the gain increases, and reaches a maximum at 0.3 λ separation. At this spacing, feed lines with solid PE dielectric (eg, RG-213) will reach. We will analyze this configuration.

Array data

Spacing: 0.3 λ

Feed currents: I1 = 1 A, /0°; I2 = 1 A, /–180°

Gain: 3.1 dB over a single vertical

3-dB beamwidth: 98°

Radiation: bi-directional, end-fire

The feed-point impedance (including 2 Ω of ground loss resistance) is Z1 = Z2 = 24.5 + j18.3 Ω

Fig 11-45 shows the radiation pattern.

The element with the leading current is fed via a $^{1}/_{4}$-λ feed line, the lagging element via a $^{3}/_{2}$-λ line. The impedances at the end of the feed lines can be calculated with the COAX TRANSFORMER software module:

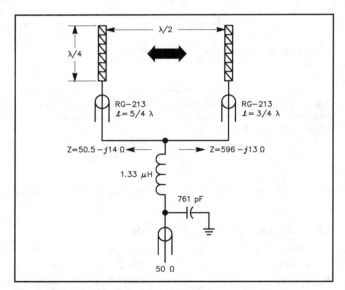

Fig 11-44—Current-forcing feed system for the 2-element bidirectional end-fire array. $^{3}/_{4}$-λ long feed lines are used because $^{1}/_{4}$-λ-long feed lines would not reach to the center of the array.

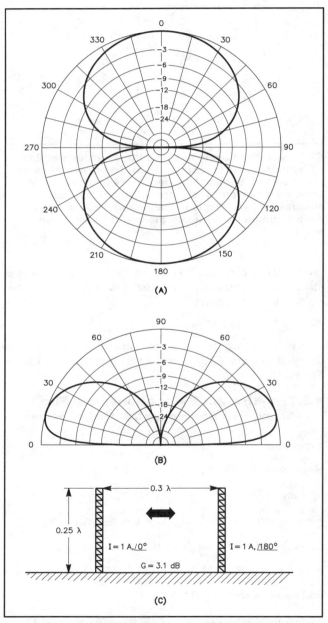

Fig 11-45—Radiation patterns for the 2-element out-of-phase end-fire array using reduced spacing (0.3 λ). The reduced spacing results in superior gain.

$Z1' = 65.7 - j46.8 \ \Omega$

$Z2' = 65.8 - j42.7 \ \Omega$

The two impedances would have been identical if the two feed lines were lossless.

The combined impedance can then be calculated with the T-JUNCTION module. The impedance for this array comes out to $32.9 - j22.4 \ \Omega$. An L network can be designed for matching this impedance to a 50-Ω line (**Fig 11-46**).

4.6. Three-Element $^{1}/_{4}$-λ-Spaced Quadrature-Fed End-Fire Array

A very effective 3-element end-fire array uses quarter-wave spacing and 90° of phase shift between adjacent elements. The center element is supplied with twice as much current as the outer ones.

Array data

Number of elements: 3

Spacing: 0.25 λ

Feed currents: I1= 1 A, $\underline{/-90°}$; I2 = 2 A, $\underline{/0°}$; I3 = 1 A, $\underline{/90°}$

Gain: 4.1 dB over a single vertical

3-dB beamwidth: 142°

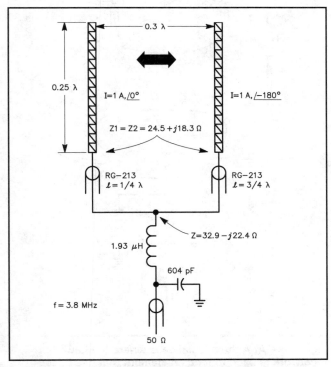

Fig 11-46—Current-forcing feed method for the 2-element out-of-phase end-fire array using reduced spacing (0.3 λ). The reduced spacing allows the use of $^{1}/_{4}$-λ long feed lines.

Radiation: Unidirectional, end-fire

Mutual impedances (approx. values):

$Z(12) = 15 - j15 \ \Omega$

$Z(13) = -9 - j13 \ \Omega$

$Z(23) = Z(12)$

The feed-point impedances (including 2 Ω ground loss resistance) are:

$Z1 = 76.1 + j51 \ \Omega \ (-90°)$

$Z2 = 26.3 - j0.4 \ \Omega$

$Z3 = 15 - j22.6 \ \Omega \ (+90°)$

Fig 11-47 shows the radiation patterns over very good ground.

As this antenna is quadrature-fed, it is obvious that the Lewallen or the Collins feed methods are the logical choices for the feed system. Quarter-wave feed lines can only be used to feed the elements if the array is not to be made switchable. If the array is to be made switchable, $^{3}/_{4}$-λ long feed lines will have to be used.

4.6.1. The Lewallen feed system

Fig 11-48 shows the schematic of the Lewallen feed system. A half-wave line goes from A to C, taking care of the 180° phase shift between these 2 points. Point B will require a 90° delay versus point A, so we will insert the L network between A and B. In this case however the element at the end of the $^{3}/_{4}$-λ feed line going to point B requires twice the current magnitude as compared to the elements A and C.

In this case the following formulas from Section 3.3.3.2 must be applied:

$$X_S = \frac{Z_{coax}^{\ 2}}{k \times R}$$

$$X_P = \frac{Z_{coax}^{\ 2}}{X - (k \times R)}$$

where:
R is the resistive part of the antenna element fed through the L network

X is the reactive part of the element fed through the L network

k is the ratio of the magnitude of the current of the element fed through the L network to the magnitude of the feed currents of the other elements in the array.

The impedance of the center element is:

$Z = 26.3 - j0.4 \ \Omega$

From this:

R = 26.3

$X = -j0.4$

k = 2 (magnitude current center element is twice the magnitude of the current in the outer elements)

$$X_S = \frac{2500}{2 \times 26.3} = j47.5 \; \Omega$$

$$X_P = \frac{2500}{\left[-0.4 - \left(2 \times 26.3\right)\right]} = -j47.2 \; \Omega$$

For a frequency of 3.8 MHz the components are:

$$\text{Parallel capacitor} = \frac{10^6}{2 \times \pi \times f \times 47.2} = 888 \text{ pF}$$

$$\text{Series coil: } \frac{47.5}{2 \times \pi \times f} = 1.99 \; \mu H$$

Calculation of the impedance of the array

The impedances at the end of the $^3/_4$-λ feed lines were calculated using the COAX TRANSFORMER software module (all elements are fed with RG-213 50-Ω cables with 0.35 dB/100 ft loss at 3.8 MHz):

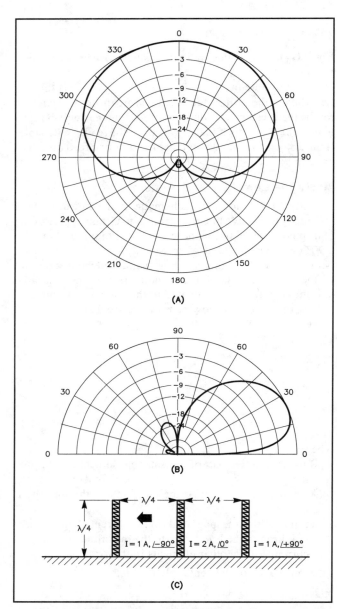

(A)

(B)

(C)

Fig 11-47—Radiation pattern of the 3-element in-line array, with elements spaced $^1/_4$ λ and with quadrature phasing conditions. The center element gets twice as much current as the outer elements.

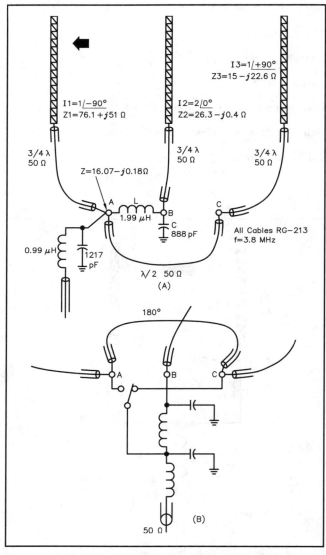

Fig 11-48—At A the special case where the Lewallen L-network is used to feed an element that has a feed-current magnitude that is different from the feed-current magnitude of the other elements in the array. See text for details. At B the direction switching applied to this feed system. One SPDT relay does the job.

Element 1 is fed via a 270° (³/₄-λ long) 50-Ω cable:

$Z1' = 24.9 - j14.46 \ \Omega$

$E1' = 54.06 \ V, \underline{/-177.21°}$

Element 2 is fed via a 270°-long 50-Ω cable:

$Z2' = 88.9 + j1.2 \ \Omega$

$E2' = 102.85 \ V, \underline{/-90.02°}$

The third element is fed via an extra 180° long line; the total feed line is 450° long:

$Z3' = 58.5 + j63.5 \ \Omega$

$E3' = 52.5 \ V, \underline{/177.8°}$

The end of the feed lines to elements 1 and 3 can be connected at point A. (Apart from some influence of cable losses, the voltages at the end of these cables are identical.) Their parallel impedance is:

$Z(1,3) = 25.01 - j5.9 \ \Omega$

The impedance at the end of the feed line to the center element (2) is $Z2' = 88.9 + j1.2 \ \Omega$. In point B this impedance is shunted with a reactance of $-j47.2 \ \Omega$. The combined impedance is $Z'(B) = 19.77 - j36.97 \ \Omega$. In this point we have a inductive reactance of $j47.5 \ \Omega$ in series, which turns the impedance at the other side of the inductor into $Z''(B) = 19.77 + j10.53 \ \Omega$. In this point this impedance is in parallel with $Z(1,3) = 25.01 - j5.9 \ \Omega$. The final impedance of the array becomes:

$Z_{array} = 2.63 + j1.97 \ \Omega$.

This impedance can be matched to a 50-Ω line by an appropriate L-network (see Fig 11-48).

4.6.2. The Collins feed system

In the case of the Collins hybrid feed system, if we need double the current in one of the elements of an array, all we need to do is to run a coaxial cable with half the impedance of the coax feeding the other elements. In other words, the feed line to the center element will consist of two parallel-connected feed lines, as shown in **Fig 11-49**.

We have calculated the impedances at the end of the feed lines to the outer elements (270° and 450° long) Z1' and Z3' in Section 4.6.1.

The transformed data for the center element (now being fed via a 270° long 25-Ω line) are:

$Z2' = 23.8 + j0.3 \ \Omega$

$E2' = 51.4 \ V, \underline{/-90°}$

The impedance at the parallel connection of the feed lines to elements 1 and 3 was calculated in Section 4.6.1. as:

$Z(1,3) = 25 - j5.9 \ \Omega$.

The impedance at the end of the two parallel 50-Ω feed lines to element 2 is:

$Z2' = 23.8 + j0.3 \ \Omega$.

Notice that both impedances result in a very low SWR in a 25-Ω system. The performance of the coupler will be very good if we design the hybrid coupler with a nominal impedance of 25 Ω.

The values of the coupler components are (see Section 3.3.5):

$X_{L1} = X_{L2} = 25 \ \Omega$ (system impedance)

$X_{C1} = X_{C2} = 2 \times 25 = 50 \ \Omega$

For 3.65 MHz the component values are

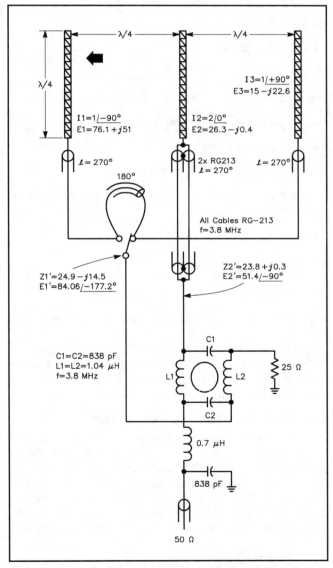

Fig 11-49—Collins feed method for the 3-element in-line end-fire array. Direction switching is very simple and requires only one SPDT relay.

$$L1 = L2 = \frac{X_L}{2 \times \pi \times f} = \frac{25}{2 \times \pi \times 3.65} = 1.09 \ \mu H$$

$$C1 = C2 = \frac{10^6}{2 \times \pi \times f \times X_C} = \frac{10^6}{2 \times \pi \times 3.65 \times 50} = 872 \ pF$$

For 160 meters these values must be doubled. Fig 11-49 shows the feed system for the array.

4.7. Three-Element ¹/₄-λ-Spaced End-Fire Array, Nonquadrature Fed

A. Christman, K3LC (ex KB8I), published a configuration of a 3-element in-line array (Ref 963), with a non-quadrature-fed current distribution, which results in superior gain and front-to-back ratio as compared to the quadrature-fed array described in Section 4.6.

Array data

Number of elements: 3

Spacing: 0.25 λ

Feed currents: I1 = 1 A, $\underline{/-116°}$; I2 = 2 A, $\underline{/0°}$; I3 = 1A, $\underline{/117°}$

Gain: 5.2 dB over a single vertical

3-dB beamwidth: 114°

Radiation: Unidirectional, end-fire

The feed-point impedances (including 2 Ω ground loss resistance) are:

Z1 = 42.3 + j61.5 Ω (− 116°)

Z2 = 28 + j6.3 Ω

Z3 = 9.6 − j17 Ω (+117°)

Fig 11-50 shows the radiation patterns over very good ground.

As this is not a quadrature-fed array, this array can obviously not be fed using the Collins or the Lewallen methods. We will calculate a modified Christman and a current-forcing system (modified Lewallen feed system).

4.7.1. Modified Christman feed system

First we'll design a feed system using the modified Christman method (Section 3.3.2). We use the software program VOLTAGE ALONG FEED LINES and see if we can find points on the three feed lines where the voltages are identical. Don't forget to specify cable losses per wavelength (not the usual dB per 100 ft!). Such points can be found on the lines to element 1 and element 3, as follows.

Line 1, at 304.5° from the load:

E1′ = 29.8 V, $\underline{/-132.4°}$

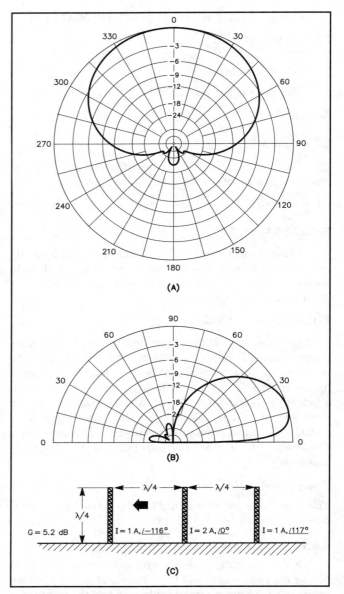

Fig 11-50—Radiation patterns for the 3-element in-line array using ¹/₄-λ spacing and optimized current angles (after Christman). Notice the superior high-angle F/B performance as compared to the quadrature-fed array from Fig 11-47.

Z1′ = 16.8 + j2.15 Ω

Line 3, at 167° from the load:

E2′ = 29.9 V, $\underline{/-131.8°}$

Z2′ = 13.95 − j29.9 Ω

Note from the listing that the length of line 1 is very critical! For all practical purposes we can consider these values as identical. This means we can connect the end of those two 50-Ω feed lines together. I tried several feed-line impedances (25 Ω, 50 Ω and 75 Ω) but could not find a point on a feed line running to the central element that had the same voltage.

The solution is to use an L network (Ref the Gehrke system, Section 3.3.5). With the software module SERIES/SHUNT INPUT L-NETWORK ITERATION we can design an L network that will transform the impedance at the base of the central element (Z2 = 28 + j6.3 Ω) so that the voltage magnitude at the input of the filter (the resistive impedance end) is 29.9 V. There are two solutions to the problem.

Solution 1

$X_S = j29.9$ Ω (1.25 μH at 3.8 MHz)

$X_P = j20.8$ Ω (0.87 μH at 3.8 MHz)

$Z_{in} = 8$ Ω

$E_{in} = 29.9$ V, $\underline{/-45.9°}$

Solution 2

$X_S = j13.1$ Ω (0.55 μH at 3.8 MHz)

$X_P = -j15.8$ Ω (2652 pF at 3.8 MHz)

$Z_{in} = 8$ Ω

$E_{in} = 29.9$ V, $\underline{/71.2°}$

Now that the voltage magnitude is correct we need to shift the phase to −132.5°. Let us take solution 1.

We use a constant-impedance T or pi network to do the phase shifting. The software module LINE STRETCHER (PI OR T) does the calculations in seconds. Let us take a pi network. The solution is

Z_{filter}: 8 Ω

Voltage input phase angle: −132.5°

Voltage output phase angle: −45.9°

$X_S = -j8$ Ω (5244 pF at 3.8 MHz)

$Z_P = j8.5$ Ω (0.36 μH at 3.8 MHz)

The final network is shown in **Fig 11-51**. The input impedance of the array is given by the parallel connection of three impedances:

Z1 = 16.8 + j2.15 Ω

Z2 = 8 Ω

Z3 = 13.95 − j29.9 Ω

The parallel combination of these three impedances is Z_{array} = 5.0 − j0.51 Ω. This is a very low value. Very high currents will be circulating in the components of the network, so care must be taken in the construction. An L network seems to be the logical choice for matching this impedance to the feed line.

I'd like to comment on this solution. In order to be able to draw double the current in the central element (as com-

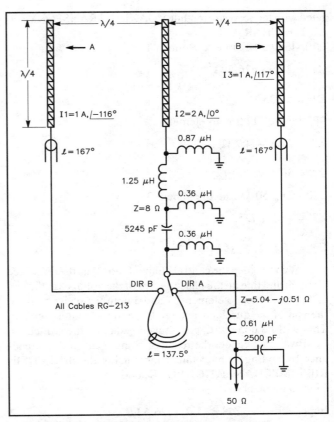

Fig 11-51—Modified Christman feed system for the three-in-line end-fire array ($^1/_4$-λ spacing) with optimized current phase angles. See text for details.

pared to the current in the outer elements), the impedances involved are very low. This is because the impedances at the end of the two feed lines are already low (16.8 + j21.5 Ω and 14 − j30 Ω) to start with. This results in components in the phasing system that have very low impedances. If you would like to construct this system, I would strongly advise you to take extreme care when making the phasing network, as extremely high currents will circulate in the components with high power.

4.7.2. Current-forcing system

We will run $^3/_4$-λ feed lines from the outer elements to our switch-box and a $^1/_4$-λ feed line of paralleled 50-Ω cables from the center element. This combination guarantees us "current-forcing." The magnitudes of the voltages at the end of the three lines will be identical (50 V). All we will then do is use constant-impedance line stretchers to obtain the proper phase angle.

The element feed-point impedances are:

Z1 = 42.3 + j61.5 Ω (−116°)

Z2 = 28 + j6.3 Ω

Z3 = 9.9 − j17 Ω (+ 117°)

The impedances at the end of the feed lines are calculated with the software module COAXIAL TRANSFORMER/SMITH CHART.

270°-long 50-Ω line to element 1:

$Z1' = 21.9 - j26.5\ \Omega$

$E1' = 52.35\ V,\ \underline{/157.5°}$

90°-long 25-Ω line to element 2:

$Z2' = 21.3 - j4.7\ \Omega$

$E2' = 50.5\ V,\ \underline{/90.12°}$

270°-long 50-Ω line to element 3:

$Z3' = 72.5 + j95.3\ \Omega$

$E3' = 50.6\ V,\ \underline{/- 153°}$

We will keep the central element (no. 2) as the reference element and line up the phase angles of the voltages at the end of the feed lines to elements 1 and 3 with the phase angle at the end of the quarter-wave feed line to the center element. This will be done with constant-impedance line stretchers. First we must cancel the reactive component in the impedance by adding a shunt impedance, using the SHUNT/SERIES IMPEDANCE NETWORK module.

Line to element 1:

Impedance was $Z1' = 21.9 - j26.5\ \Omega$

Adding a shunt reactance of $j44.6\ \Omega$ (1.9 μH coil at 3.8 MHz) tunes out the reactive part of the impedance. The impedance now is:

$Z1'' = 54\ \Omega$

Line to element 3:

Impedance was: $Z1' = 72.5 + j95.3\ \Omega$

Adding a shunt reactance of $-j150.5\ \Omega$ (278 pF at 3.8 MHz) tunes out the reactive part of the impedance. The impedance now is

$Z1'' = 198\ \Omega$

Now we will design the pi-network line stretchers (T would also be possible) using the LINE STRETCHER (PI or T) software module:

Line stretcher to element 1:

Impedance: 54 Ω

Voltage input phase angle: 90.12°

Voltage output phase angle: 157.5°

$X_S = - j49.8\ \Omega$ (840 pF at 3.8 MHz)

$X_P = j81\ \Omega$ (3.4 μH at 3.8 MHz)

Line stretcher to element 3:

Impedance: 198 Ω

Voltage input phase angle: 90.12°

Voltage output phase angle: $-153° = +207°$

$X_S = - j177\ \Omega$ (237 pF at 3.8 MHz)

$X_P = j121.6\ \Omega$ (5.1 μH at 3.8 MHz)

The output of the line stretchers and the line to element 2 can now be connected in parallel, as the voltages are identical. The impedance in that point is the parallel of:

$Z1'' = 54\ \Omega$

$Z2 = 21.3 - j4.7\ \Omega$

$Z3 = 198\ \Omega$

The resulting impedance is the array feed-impedance (calculated with the PARALLEL IMPEDANCES module):

$Z_{array} = 14.3 - j2.1\ \Omega$

An appropriate L network will ensure a 1:1 SWR into a 50-Ω feed line. This feed system is shown in **Fig 11-52**. Comparing this solution with the solution described in Section 4.7.1, it is obvious that this solution is far superior. Component values are more "normal" because "normal" impedances are involved. This feed system is obviously the better choice.

4.8. Three-Element ¹/₈-λ Spaced End-Fire Array, Non-quadrature Fed

Another design by A. Christman, K3LC (ex KB8I), uses three ¹/₈-λ spaced elements in line to obtain an excellent F/B behavior and a gain comparable to that of the ¹/₄-λ spaced array.

Array data

Number of elements: 3

Spacing: 0.125 λ

Feed Currents: I1 = 1 A, $\underline{/-149°}$; I2 = 2 A, $\underline{/0°}$; I3 = 1 A, $\underline{/146°}$

Gain: 4.1 dB over a single vertical

3-dB beamwidth: 94°

Radiation: Unidirectional, end-fire

The feed-point impedances (including 2 Ω ground loss resistance) are:

$Z1 = 4.6 + j19\ \Omega\ (-149°)$

$Z2 = 12 + j1.6\ \Omega$

$Z3 = - 19 + j13.4\ \Omega\ (+146°)$

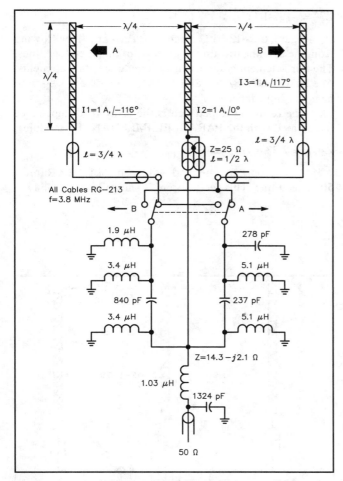

Fig 11-52—Current-forcing feed system for the three-in-line end-fire array (¹/₄-λ spacing) with optimized feed current angles. The system requires a DPDT relay for direction switching but is nevertheless the logical choice as compared to the feed system from Fig 11-39. See text for details.

Fig 11-53 shows the radiation patterns over very good ground.

In view of the very low impedances involved with ¹/₈-λ spacing, I highly recommend using a "current-forcing" feed system, as the system is more lenient with changes in impedance of the individual elements than other systems. Note the negative impedance of the third element, which means that this element is actually returning current to the feed system.

4.8.1. Current-forcing feed system

We will run ¹/₄-λ feed lines from the outer elements to our switch-box and a ¹/₄-λ feed line of parallel-connected 50-Ω cables from the center element. These combinations guarantee us "current-forcing." The magnitudes of the voltages at the end of the three lines will be identical (50 V). All we will then do, is use constant-impedance line stretchers to obtain the proper phase angle.

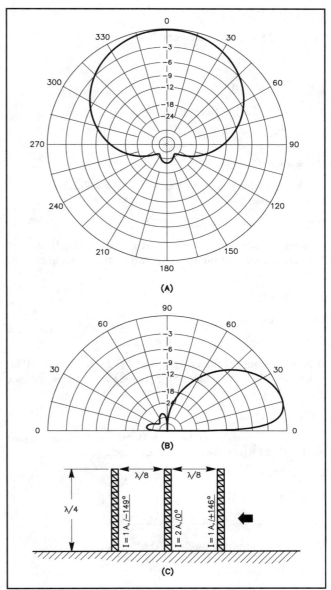

Fig 11-53—Radiation patterns for the three-in-line end-fire array with ¹/₈-λ element spacing and optimized feed current angle distribution (−149, 0 and +146°).

The element feed-point impedances are:

$Z1 = 4.6 + j19 \ \Omega \ (-149°)$

$Z2 = 12 + j1.6 \ \Omega$

$Z3 = -19 + j13.4 \ \Omega \ (+ 146°)$

The impedances at the end of the feed lines are calculated with the software module COAXIAL TRANSFORMER/SMITH CHART.
90°-long 50-Ω line to element 1:

$Z1' = 35.7 - j121.5 \ \Omega$

E1′ = 50.1 V, $\underline{/-50.6°}$

90°-long 25-Ω line to element 2:

Z2′ = 50.5 − j6.6 Ω

E2′ = 50.2 V, $\underline{/90.3°}$

90°-long 50-Ω line to element 3:

Z3′ = −88.3 − j65.8 Ω

E3′ = 49.7 V, $\underline{/-123.7°}$

The methodology is described in Section 4.7.2.
Line to element 1:

Impedance was Z1′ = 35.7 − j121.5 Ω

Adding a shunt reactance of j132 Ω (5.53-μH coil at 3.8 MHz) tunes out the reactive part of the impedance. The impedance now is:

Z1″ = 449 Ω

Line to element 3:

Impedance was Z3′ = −88.3 − j65.8 Ω

Adding a shunt reactance of + j184 Ω (7.7 μH at 3.8 MHz) tunes out the reactive part of the impedance. The impedance now is:

Z3″ = −137.1 Ω

Now we will design the pi-network line stretchers using the LINE STRETCHERS software module.

Line stretcher to element 1:

Impedance: 449 Ω

Voltage input phase angle: 90.03°

Voltage output phase angle: −50.6°

X_S = −j285 Ω (147 pF at 3.8 MHz)

X_P = j161 Ω (6.7 μH at 3.8 MHz)

Line stretcher to element 3:

Impedance: 198 Ω

Voltage input phase angle: 90.03°

Voltage output phase angle: −123.7° = +236.3°

X_S = j110 Ω (4.6 μH at 3.8 MHz)

X_P = − j60 Ω (698 pF at 3.8 MHz)

The output of the line stretchers and the line to element 2 can now be connected in parallel, as the voltages are identical. The impedance in that point is the parallel of

Z1″ = 449 Ω

Z2″ = 50.5 − j6.6 Ω

Z3″ = −137.1 Ω

Again, the −137.1 Ω means that current will be flowing from the antenna toward the junction of the three feed lines. The third element is supplying current to the other elements. The NEW LOW BAND SOFTWARE copes very well with negative impedances.

The resulting impedance is the array feed impedance (calculated with the PARALLEL IMPEDANCES module):

Z_{array} = 7.33 − j11.9 Ω

An appropriate L network will ensure a 1:1 SWR into a 50-Ω feed line. This feed system is shown in **Fig 11-54**.

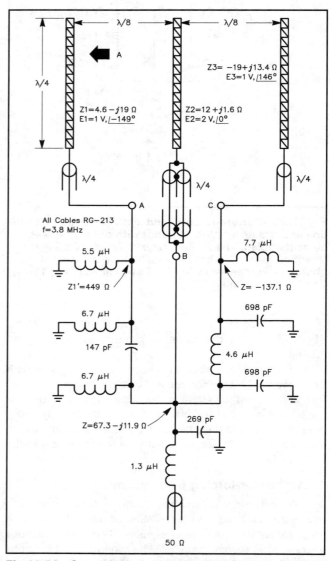

Fig 11-54—Current-forcing method for the $^1/_8$-λ spaced end-fire three-in-line array with optimized feed current angle distribution.

4.9. Four-Square Arrays

In 1965 D. Atchley, W1CF (then W1HKK), described two arrays that were computer modeled, and later built and tested with good success (Ref 930, 941). Although the theoretical benefits of the 4-square were well understood, it took a while before the correct feed methods were developed that could guarantee performance on a par with the paperwork.

The 4-square can be switched in 4 quadrants. Atchley also developed a switching arrangement that made it possible to switch the array directivity in increments of 45°. The second configuration consists of two side by side cardioid arrays. This antenna is discussed in detail in Section 4.10.3.

The practical advantage of the extra directivity steps, however, does not seem to be worth the effort required to design the much more complicated feeding and switching system, as the forward lobe is so broad that switching in 45° steps makes very little difference. It is also important to keep in mind that the more complicated a system is, the more failure-prone it is.

4.9.1. Pattern switching for the 4-square arrays

Fig 11-55 shows a direction-switching system that can be used with all the 4-square arrays, on condition that the two diagonal (central) elements are fed in phase. The "front" element (in the direction of firing) will of course be fed with the lagging feed angle, the back element with the leading feed angle.

While designing the feed systems of the individual arrays, I will simply show the "black box" containing the feed system. The box has four terminals:
1) 50-Ω input to transmitter.
2) Two outputs to the central elements (phase angle zero).
3) One output to the front element (lagging phase angle).
4) One output to the back element (leading phase angle).

The protective circuit described in Section 4.1.3 can of course be used with this 4-square switching system. There are two logic lines: one line controls K1 and K2, the second line controls K3.

4.9.2. Quarter-wave-spaced square, quadrature-fed

Array data

Number of elements: 4

Placement of elements: In a square, spaced 0.25 λ per side

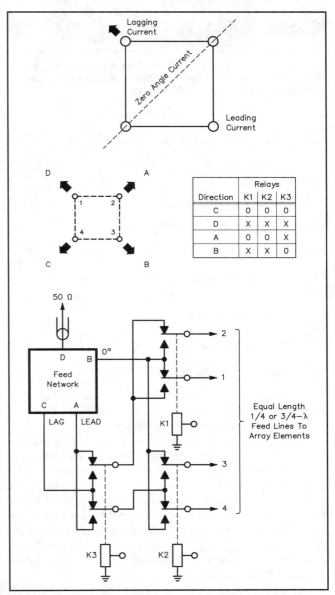

Direction	Relays		
	K1	K2	K3
C	0	0	0
D	X	X	X
A	0	0	X
B	X	X	0

Fig 11-55—Universal direction-switching system that can be used with any of the 4-square arrays. The condition is that the two "central" elements be fed with the reference (zero-angle) current, the front element with the lagging current, and the back element with the leading current. From the switching network, four lines of equal length (¼ or ¾-λ long) go to the individual elements. The lengths and specifications (impedance, type) are different for every type of square array.

Eighty-meter 4-square at K9JF/7. The verticals use an elevated radial system approx. 2 m above ground.

Eighty-meter 4-square at K7EM. The elements are made out of 4 and 3-inch irrigation tubing. For element construction, see Chapter 9, Section 6.1.1.

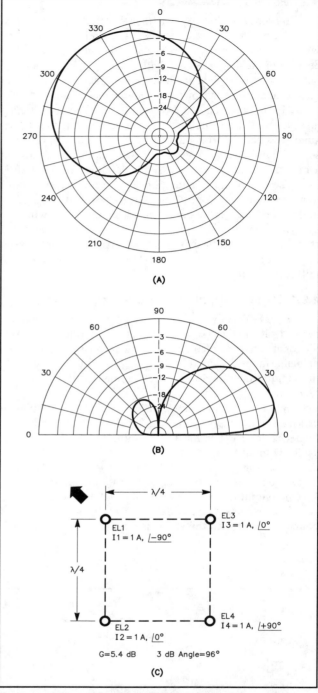

Fig 11-56—Radiation patterns for the 4-square array with $\frac{1}{4}$-λ element spacing.

Elements 2 and 3 = a diagonal of the square

Feed currents: I1 = 1 A, $\underline{/-90°}$; I2 = 1 A, $\underline{/0°}$; I3 = 1 A, $\underline{/0°}$, I4 = 1 A, $\underline{/+90°}$

Gain: 5.5 dB over a single vertical

3-dB beamwidth: 96°

Radiation: Unidirectional

The feed-point impedances (including 2 Ω ground loss resistance) are

Z1 = 61.7 + j59.4 Ω (−90°)

Z2 = Z3 = 41 − j19.3 Ω

Z4 = −0.4 − j15.4 Ω (+90°)

These impedance figures were obtained by modeling, and may differ slightly from those published elsewhere. **Fig 11-56** shows the radiation patterns over very good ground.

The four elements of the antenna are positioned in a square, with $\frac{1}{4}$-λ spacing between adjacent elements. All elements are fed with equal current. The two central elements are fed at 0° (reference), the rear element is fed at 90°, and the front element is fed at −90°. The direction of maximum signal is along the diagonal from the rear to the front element (an array always radiates in the direction of the element with the lagging current). Several feed methods can be developed. I have worked out two systems, the Lewallen method and the Collins method.

4.9.2.1. The Lewallen feed method

The 2 EL AND 4 EL VERTICAL ARRAYS module of the NEW LOW BAND SOFTWARE is a tutorial and engineering program that takes you step by step through the design of the 4-element square array. The results as displayed from that program will be slightly different from the results shown here, as in that program ideal feed lines (no

losses) are used. It is interesting, however, to compare the values from this paragraph with the values from the software program to assess the error caused by using lossless cables.

The impedances at the elements are:

Z1 = 61.7 + j59.4 Ω (−90° element)

Z2 = Z3 = 41 − j19 Ω

$Z4 = -0.4 - j15 \ \Omega$ (+90°element)

We cannot reach the center of the array with quarter-wave feed lines with a velocity factor of 0.66 (solid PE, eg, RG-213). Foam-type RG-8 will be required. For the calculations I used VF = 0.79 and attenuation = 0.3 dB/100 ft.

Using the software module COAX TRANSFORMER, voltages, currents and impedances at the end of the four lengths of coaxial cable can be calculated.
270° line from element 1 (90° current forcing line plus additional 180° phasing line):

$Z1' = 23.56 - j19.31 \ \Omega$

$E1' = 53.43$ V, $\underline{/-176.62°}$

90° lines from element 2 and from element 3:

$Z2' = Z3' = 50.37 + j22.45 \ \Omega$

$E2' = E3' = 50.73$ V, $\underline{/89.62°}$

90° line from element 4:

$Z4' = -6.24 + j166.44 \ \Omega$

$E4' = 50.00$ V, $\underline{/179.7°}$

The feed lines to element 1 and element 4 can be connected in parallel (the voltages are practically identical). The resulting impedance is

$Z_{total}(1,4) = 29.26 - j16.92 \ \Omega$

The elements of the Lewallen L network that provide the necessary 90° voltage phase shift can easily be calculated (see Section 3.3.3.):

In this case we have to apply the formula suitable for more than 2 elements:

$$X_S = \frac{Z_{coax}^2}{\Sigma(R)}$$

$$X_P = \frac{Z_{coax}^2}{[\Sigma(X) - \Sigma(R)]}$$

As the R and X values are the same:

$\Sigma(R) = 2 \times R2$ with $R2 = 41 \ \Omega$

$\Sigma(X) = 2 \times X2$ with $X2 = -j19 \ \Omega$

R2 and X2 are the real and the imaginary parts of the impedance of element 2 (the element with the lagging current)

$$X_S = \frac{Z_{coax}^2}{2 \times R2} = \frac{50^2}{2 \times 41} = 30.48 \ \Omega$$

$$X_P = \frac{Z_{coax}^2}{2(X2 - R2)} = \frac{50^2}{2 \times (-19 - 41)} = -j20.83 \ \Omega$$

Series element (inductor): 1.28 μH at 3.8 MHz

Parallel element (capacitor): 2012 pF

Calculating the total feed impedance of the array

The feed lines from elements 2 and 3 are in parallel at the output of the L network. The total impedance is:

$Z_{total}(2, 3) = 25.18 + j11.23 \ \Omega$

This impedance is first shunted by a capacitor with a reactance of $-20.83 \ \Omega$. Using the SHUNT IMPEDANCE NETWORK module we calculate the resulting impedance:

$Z(2, 3)' = 15.04 - j15.09 \ \Omega$

In series with this impedance we have the reactance of the coil from the L network, $j30.48 \ \Omega$. The net resulting impedance becomes:

$Z(2, 3)'' = 15.04 + j15.39 \ \Omega$

The input impedance of the array is the parallel connection of:

$Z(2, 3)'' = 15.04 + j15.39 \ \Omega$

$Z(1, 4)'' = 29.26 - j16.92 \ \Omega$

The result is $10.22 + j8.45 \ \Omega$.

The Lewallen feed method for this array is worked out in great detail in *The ARRL Antenna Book*, and L-network values are listed for a range of feed-line impedances and ground systems. The layout of the feed system is shown in **Fig 11-57A**.

Matching the $10.22 + j8.45 \ \Omega$ impedance to a 50-Ω feed line can be done in several ways eg, with an L network, as shown in Fig 11-57.

4.9.2.2. The Collins feed method

The feed-line configuration (length, type, etc) is identical to that for the Lewallen feed system. The impedance and voltage values at the end of the lines are calculated in Section 4.9.2.1.

In the Collins feed system we replace the L network with the hybrid coupler. In order to optimize the performance of the coupler we must determine the system impedance that best matches the impedances at the end of the feed lines.

270° cable from element 1 (90° current forcing cable + 180° extra phase shift):

$Z1' = 23.56 - j19.31 \ \Omega$

$E1' = 53.43$ V, $\underline{/-176.62°}$

90° cables from element 2 and element 3:

$Z2 = Z3 = 50.37 + j22.45 \ \Omega$

$E2' = E3' = 50.73$ V, $\underline{/89.62°}$

90° cable from element 4:

$Z4 = 6.24 + j166.44 \ \Omega$

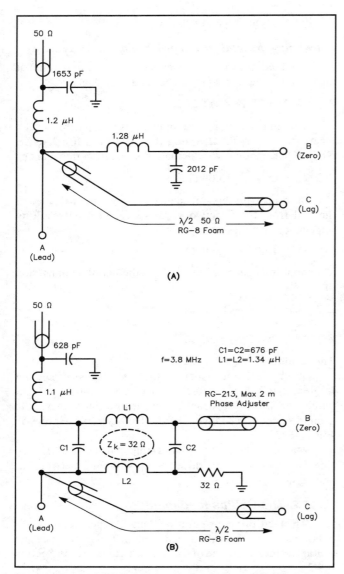

Fig 11-57—Lewallen feed system at A, and Collins feed system at B for the 4-square array with ¼-λ element spacing. The identification of the feed system units matches the direction switching system as shown in Fig 11-55.

$E4' = 50$ V, $\underline{/179.7°}$

Paralleling the two feed lines to elements 2 and 3 yields a total impedance of:

$Z(2, 3)' = 25.18 + j11.22 \ \Omega$

Paralleling the lines to elements 1 and 4 yields:

$Z(1, 4)' = 29.26 - j16.92 \ \Omega$

Running both impedances in the SWR ITERATION soft-ware program shows that the lowest SWR is obtained with a hybrid-coupler impedance of 32 Ω (SWR 1.7:1 and 1.6:1).

The 32-Ω hybrid coupler can be constructed as described in Section 3.3.4, with:

$X_{L1} = X_{L2} = j32 \ \Omega$

$X_{C1} = X_{C2} = -j64 \ \Omega$

Remember that the termination resistance (dummy load) must be 32 Ω.

For a design frequency of 3.65 MHz, the component values are:

$L = 1.29 \ \mu H$

$C = 704$ pF

For 160 meters these values can be doubled.

The hybrid coupler can be fed with a 50-Ω feed line via an appropriate L network.

The front-to-back ratio can be fine-tuned for ultimate rejection by inserting a variable length of feed line between port 2 and the paralleled feed lines to the −90° elements. This line should be between 1 and 9 electrical degrees long (1 to 6 feet on 80 meters).

Collins measured the power in the terminating resistor of the hybrid coupler for a 4-square array that was designed for 3.8 MHz. **Table 11-1** shows the power in the terminating resistor over a frequency range from 3.5 to 4 MHz. If you want to use the array over such a wide frequency range you will have to provide a terminating resistor that can take approximately 200 W if you run 1500 W into the array. Collins also measured the feed-current magnitude and phase over this wide frequency range. The results are shown in

Table 11-1

Power Reflected into Dummy Resistor at Port 4 of Collins Hybrid Coupler

Frequency (MHz)	Power Reflected into Load (dB)
3.5	−7.2
3.6	−10.0
3.7	−12.2
3.8	−18.0
3.9	−13.5
4.0	−10.5

Table 11-2

Measured Element Feed Current with Hybrid Coupler

Frequency (MHz)	Back Element Current (amps)	Angle (deg)	Side Elements Current (amps)	Angle (deg)	Front Element Current (amps)	Angle (deg)
3.5	7.7	0	7.5	83	7.8	173
3.8	7.9	0	8.5	90	7.9	180
4.0	6.3	0	8.2	90	6.5	185

Table 11-2. It is clear that these results will, to a very large extent, depend on the effective diameter of the radiator (Q-factor). If we calculate the directivity patterns using the current data from that table, we note a constant gain (within less than 0.1 dB from 3.5 to 4 MHz), and an F/B ratio of more than 22.5 dB from 3.5 to 3.8 MHz. On 4 MHz the F/B ratio has dropped to 16 dB. The layout of the feed system is shown in Fig 11-45.

It has been found in practice that with a hybrid coupler, having a design impedance of 50 Ω, it is generally better to use 75-Ω cable to feed the array elements. The transformation is such that the impedances presented at the hybrid coupler result in less power being dissipated in the dummy load resistor. In that case it is of course not necessary to use an L network between the 50-Ω hybrid coupler and the 50-Ω feed line. The terminating resistor can then also be a plain vanilla 50-Ω dummy load.

4.9.3. The $^1/_8$-λ spaced 4-square

Array data

Number of elements: 4

Placement of elements: In a square, spacing 0.125 λ per side

Elements 2 and 3 = 1 diagonal of the square

Feed currents: I1 = 1 A, $\underline{/-135°}$; I2 = 1 A, $\underline{/0°}$; I3 = 1 A, $\underline{/0°}$; I4 = 1 A, $\underline{/+135°}$

Gain: 4.7 dB over a single vertical

3-dB beamwidth: 90°

Radiation: Unidirectional

The feed-point impedances (including 2 Ω ground loss resistance) are:

Z1 = – 10.4 + j21.3 Ω

Z2 = Z3 = 20.2 – j7.6 Ω

Z4 = – 21.1 – j12.4 Ω

Fig 11-58 shows the radiation patterns over very good ground.

Note that, especially with close-spaced designs, driving-point impedances with a negative resistance part can be found. This means that the parasitic coupling supplies too much current (power) to this element, and that the element is then supplying this power back into the feed network.

The operating bandwidth over which this array shows a usable front-to-back ratio is expected to be rather narrow, which is not a great handicap on 160 meters.

4.9.3.1. The modified Lewallen method

As the $^1/_8$-λ-spaced square array does not use quadrature feed angles, the Lewallen feed method using a single L network cannot be used. We will use the modified method

having a shunt network plus a line stretcher, as explained in detail in Section 3.3.4.

We chose the two center elements as the elements to be fed directly with quarter-wave feed lines. This is the logical choice, as these are the only two elements that take power from the feed line; the other two supply power to the feed system by virtue of mutual coupling.

The networks between A and B and between A and C are designed as follows. The impedances at the elements are:

Z1 = –10.4 + j21.3 Ω (–135°)

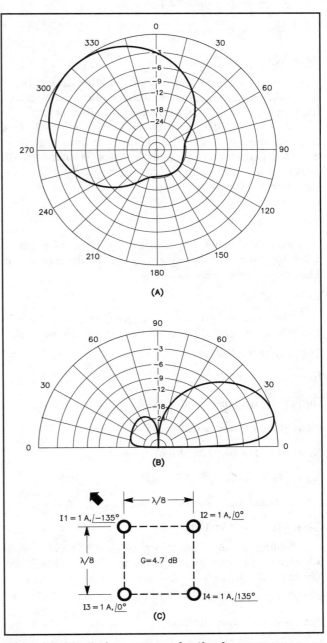

Fig 11-58—Radiation patterns for the 4-square array with $^1/_8$-λ element spacing.

$Z2 = Z3 = 20.2 - j7.6 \, \Omega$

$Z4 = -21.1 - j12.4 \, \Omega \, (135°)$

At the end of the 90° feed lines the impedances and voltages are:

$Z1' = -42.91 - j97.73 \, \Omega$

$E1' = 49.83 \, V, \, \underline{/-44.85°}$

$Z2' = Z3' = 105.86 + j37.89 \, \Omega$

$E2' = E3' = 50.36 \, V, \, \underline{/89.95°}$

$Z4' = -88.92 + j55.01 \, \Omega$

$E4' = 49.64 \, V, \, \underline{/-135.25°}$

Because we used "current-forcing" line lengths (90°), we see the same voltage magnitude at the end of all four feed lines (except for slight differences due to line losses).

We must now design a network to be inserted in the feed line to element 1 and another one for insertion at the end of line 3 to equalize the voltage phase angles. The network will consist of a shunt reactance (coil or capacitor) and a constant-impedance line stretcher.

Line to element 1:

$Z1' = -42.91 - j97.73 \, \Omega$

Adding a shunt impedance of $j117 \, \Omega$ (4.9 μH at 3.8 MHz) tunes out the imaginary part of the impedance. The new impedance is:

$Z1'' = -265 \, \Omega$

The minus sign indicates that we are still dealing with power being delivered from the element into the network.

The pi-network line stretcher is:

Impedance: −265 Ω

Output voltage phase angle: −44.5°

Input voltage phase angle: 89.85°

$X_P = j111 \, \Omega$ (4.67 μH at 3.8 MHz)

$X_S = -j189 \, \Omega$ (221 pF at 3.8 MHz)

Line to element 2:

$Z4' = -88.92 + j55.01 \, \Omega$

Adding a shunt impedance of $-j199 \, \Omega$ (211 pF at 3.8 MHz) tunes out the imaginary part of the impedance. The new impedance is:

$Z4'' = -123 \, \Omega$

The pi-network line stretcher is:

Impedance: −123 Ω

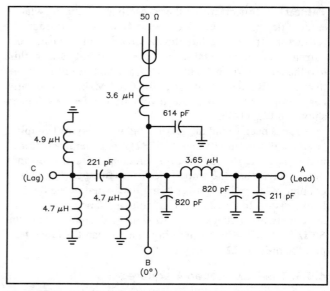

Fig 11-59—Current-forcing feed system for the four-square array with ⅛-λ spacing. The identification of the feed system units matches the direction-switching system shown in Fig 11-55. The two parallel connected coils of 4.9 μH and 4.7 μH can be replaced with a single coil of 2.4 μH, while the parallel capacitor of 820 and 211 pF can be combined in one unit of 1031 pF.

Output voltage phase angle: −135.22° = 224.78°
Input voltage phase angle: 89.85°

$X_P = -j51.1 \, \Omega$ (820 pF at 3.8 MHz)

$X_S = j87 \, \Omega$ (3.65 μH at 3.8 MHz)

The impedance at the T junction of the three branches is given by the parallel connection of:

$Z1'' = -265 \, \Omega$

$Z2' = Z3' = 105.86 + j37.89 \, \Omega$

$Z4'' = -123 \, \Omega$

$Z_{array} = 82.6 + j97.35 \, \Omega$

The layout of the modified Lewallen feed method is shown in **Fig 11-59**.

4.9.4. Optimized 4-square array

Jim Breakall, WA3FET, optimized the quarter-wave-spaced 4-square array to obtain a better F/B ratio. From Fig 11-56 we learn that the original 4-square exhibits a very major high-angle back lobe (−18 dB only). Changing the current magnitudes and angles of the front and the back elements changes the size and the shape of the back lobe. Full optimization is a compromise between optimization in the elevation and the azimuth planes. With Breakall's optimization, the gain of the array went up by 0.7 dB. He came up with the following design.

Array data

$^1/_4$-λ spaced elements

Feed currents: I1 = 0.969 A, $\underline{/-107°}$; I2 = 1 A, $\underline{/0°}$; I3 = 1 A, $\underline{/0°}$; I4 = 1.11 A, $\underline{/111°}$

Gain: 6.2 dB over a single vertical

3-dB beamwidth (at main wave angel): 84°

The feed-point impedances (including 2 Ω ground loss resistance) are

Z1 = 37.5 + *j*57.7 Ω

Z2 = Z3 = 30.8 – *j*7.0 Ω

Z4 = 60 – *j*3.4 Ω

Fig 11-60 shows the radiation patterns over very good ground.

While this optimized design shows what can be achieved by tweaking a design, it has a number of disadvantages:
• The elements are not quadrature fed; this means that neither the Lewallen nor the Collins feed system can be used.
• Because of the odd values of feed current, the current-forcing method with the modified Lewallen feed method is also out of the question.

I developed a modified Christman-type feed system. A Gehrke system would have been possible as well, although the network would have contained more components.

I first ran the VOLTAGES ALONG FEED LINES software module to see if there were points of identical voltage. I used 75-Ω feed lines (0.35 dB loss/100 ft at 3.8 MHz). The following points were selected:

At 119° on the line to element 1:

Z1′ = 25.9 – *j*23 Ω

E1′ = 41.8 V, $\underline{/9.5°}$

At 164° on the lines to elements 2 and 3:

Z2′ = Z3′ = 36.5 – *j*24.5 Ω

E2′ = E3′ = 41.8 V, $\underline{/138.7°}$

At 212° on the line to element 4:

Z4′ = 11.24 + *j*41.52 Ω

E4′ = 41.8 V, $\underline{/9.5°}$

The feed lines to elements 1 and 4 can be connected in parallel as the voltages are identical. The voltage magnitude at the input ends of the feed lines to elements 2 and 3 are identical. The phase angle needs adjusting, which can be done with a constant-impedance line stretcher (pi network).

The parallel impedance of the lines to elements 2 and 3 is

Z(2, 3)′ = 18.25 – *j*12.25 Ω

Using the SHUNT IMPEDANCE module we calculate

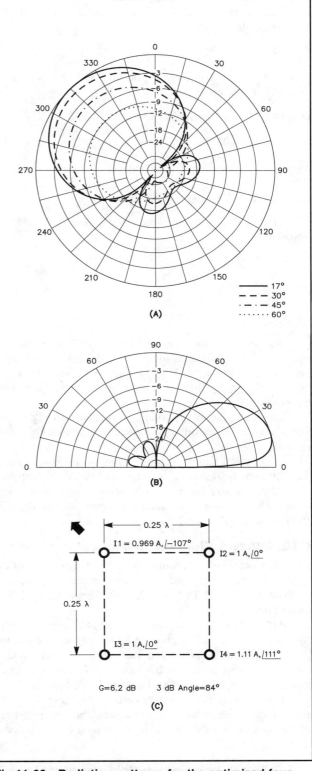

Fig 11-60—Radiation patterns for the optimized four-square array. The horizontal patterns at A are for a 17° wave angle (main wave angle), 30, 45 and 60°. The array was optimized in both the horizontal and the vertical planes in order to get the minimal total three-dimensional back lobe.

the shunt impedance that will tune the reactive component in this impedance: 39.5 Ω. A 1.7 μH coil (3.8 MHz) turns the above impedance into:

$$Z(2, 3)''' = 26.49 \text{ } \Omega$$

Now we design the pi-network line stretcher around a 26.49-Ω characteristic impedance:

Output voltage phase angle: 138.7°

Input voltage phase angle: 9.5°

$$X_S = -j20.5 \text{ } \Omega \text{ (2040 pF at 3.8 MHz)}$$

$$X_P = j12.6 \text{ } \Omega \text{ (0.53 μH at 3.8 MHz)}$$

The input impedance of the array is the parallel of three impedances:

$$Z1' = 25.9 - j23 \text{ } \Omega$$

$$Z2, 3''' = 26.49 \text{ } \Omega$$

$$Z4' = 11.24 + j41.52 \text{ } \Omega$$

The resulting impedance is: 15.25 + j0.76 Ω. Instead of matching this impedance to the 50-Ω feed line with an L network as shown in **Fig 11-61**, one could also use two parallel 1/4-λ 50-Ω lines as a quarter-wave matching transformer.

The feed system for this array is shown in Fig 11-61. There are, of course, other possible combinations for a feed system.

This design has the disadvantage of not using a "current-forcing" feed system, which means that the directivity, while theoretically better, will be more sensitive to variations in changes of element feed-point impedances when the array is switched around.

4.10. Other 4-Element Rectangular Arrays

I have analyzed a few more 4-element arrays. All of the following arrays are made up of two groups of 2-element cardioid arrays with different X and Y spacings and feed current phase angles.

Although the Gehrke, Christman and Collins feed methods may be valid alternatives, I have only calculated the (modified) Lewallen method for the following arrays. Where the array is quadrature fed, the Lewallen L network can be replaced by a hybrid coupler. For details see Section 3.3.4.

The schematics of the feed systems for these arrays do not include a direction-switching system, but the system in all cases is extremely simple; only a single DPDT relay is required.

4.10.1. Two 1/4-λ spaced cardioid arrays side by side, spaced 1/2 λ, fed in phase

The basic group is a 2-element cardioid array, 90° spacing, 90° phase shift (see Section 4.1.). The groups are spaced 1/2-λ apart, placed side by side, and fed in phase. This

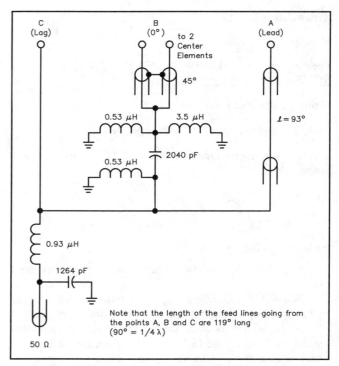

Fig 11-61—Modified Christman feed system for the optimized 4-square array of Fig 11-60. The identification of the feed system units matches the direction-switching system shown in Fig 11-55. This design is for 3.8 MHz. All coax is 75-Ω RG-11; the four feed lines are 119° in length.

array is also mentioned by W7EL in *The ARRL Antenna Book*.

Array data

Two 1/4-λ spaced cardioids, spaced 1/2 λ side by side

Feed currents: I1 = 1 A, /−90°; I2 = 1 A, /0°; I3 = 1 A, /−90°; I4 = 1 A, /0°

Gain: 6.8 dB over a single vertical

3-dB beamwidth: 62°

The feed-point impedances (including 2-Ω ground loss resistance) are

$$Z1 = Z3 = 54.6 - j5.1 \text{ } \Omega$$

$$Z2 = Z4 = 5.5 - j21.2 \text{ } \Omega$$

Fig 11-62 shows the radiation patterns over very good ground. Note the high gain, and also note the relative narrow beamwidth (which of course always go together!).

The element feed impedances are:

$$Z1 = Z3 = 54.6 - j5.1 \text{ } \Omega$$

$$Z2 = Z4 = 5.5 - j21.2 \text{ } \Omega$$

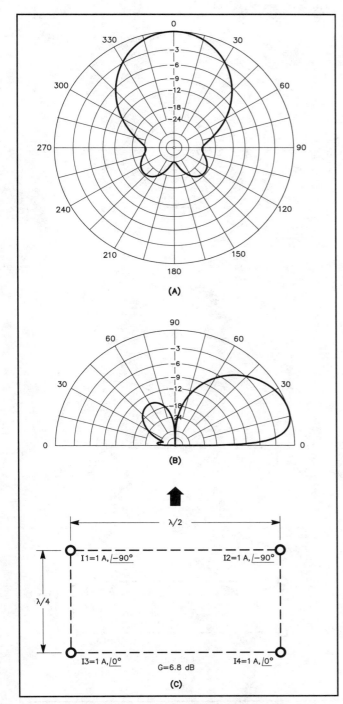

Fig 11-62—Radiation patterns for a rectangular array made of two 2-element cardioid arrays (90° spacing, 90° phase shift), spaced ½ λ.

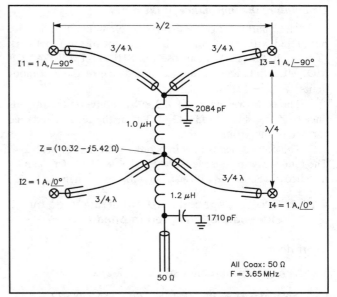

Fig 11-63—Lewallen feed system for the rectangular array shown in Fig 11-62.

Lewallen described a feed method where he runs four ³/₄-λ-long 50-Ω feed lines to the center of the array (see **Fig 11-63**). The data at the end of the ³/₄-λ-long lines are:

Z1′ = Z3′ = 45.84 + j3.86 Ω

E1′ = E3′ = 52.89 V, /179.71°

Z2′ = Z4′ = 42.7 + j103.1 Ω

E2′ = E4′ = 50.36 V, /− 91.24°

Calculating the L network:

The elements of the Lewallen L network that provide the necessary 90° voltage phase shift can easily be calculated (see Section 3.3.3.):

In this case we have to apply the formula suitable for more than 2 elements:

$$X_S = \frac{Z_{coax}^{2}}{\Sigma\left(R\right)}$$

$$X_P = \frac{Z_{coax}^{2}}{\left[\Sigma\left(X\right) - \Sigma\left(R\right)\right]}$$

As the R and X values are the same:

Σ(R) = 2 × R2 with R2 = 54.6 Ω

Σ(X) = 2 × X2 with X2 = −5.1 Ω

R2 and X2 are the real and the imaginary part of the impedance of the elements with the lagging current.

$$X_S = \frac{Z_{coax}^{2}}{2 \times R2} = \frac{50^2}{2 \times 54.6} = j22.89\ \Omega$$

$$X_P = \frac{Z_{coax}^{2}}{2\left(X2 - R2\right)} = \frac{50^2}{2 \times \left(-5.1 - 54.6\right)} = -\ j20.93\ \Omega$$

Series element (inductor): 1.00 μH at 3.65 MHz
Parallel element (capacitor): 2084 pF at 3.65 MHz

Calculating the impedance of the array

The quarter-wave feed lines to elements 1 and 3 are paralleled with, in addition, the capacitor of the L network ($Z = -j20.93\ \Omega$). We can use the SHUNT IMPEDANCE MODULE to calculate the net impedance of the 3 components: $Z_{tot1} = 11.33 - j11.54\ \Omega$

The series reactance ($+ j22.89\ \Omega$) converts the impedance to $Z1''' = 11.33 + j34.43\ \Omega$. This is the feed impedance for one of the groups.

This impedance is now in parallel with $Z'2$ and with $Z'4$. The total array impedance is: $= 15.31 - j8.41\ \Omega$. This can be matched to a 50-Ω line with an L network as shown in Fig 11-63.

4.10.2. Two $^1/_8$-λ spaced cardioid arrays side by side, spaced $^1/_2$ λ, fed in phase

Array data

Two $^1/_8$ λ spaced cardioids, spaced $^1/_2$ λ apart

Feed currents: I1 = 1 A, $\underline{/-135°}$; I2 = 1 A, $\underline{/0°}$; I3 = 1 A, $\underline{/-135°}$; I4 = 1 A, $\underline{/0°}$

Gain: 7.0 dB over a single vertical

3-dB beamwidth: 58°

The feed-point impedances (including 2-Ω ground loss resistance) are:

$Z1 = Z3 = 24.9 + j12.5\ \Omega$

$Z2 = Z4 = 3.1 - j19.2\ \Omega$

Fig 11-64 shows the radiation patterns over very good ground.

The feed system is a modified Lewallen feed system as the feed angles are not in quadrature. Because the two cardioids are $^1/_2$-λ spaced, we need $^3/_4$-λ-long feed lines to reach the center of the array.

At the end of the 270°-long feed lines the impedances and voltages are:

$Z1' = Z3' = 77.76 - j34.20\ \Omega$

$Z2' = Z4' = 37.9 + j119.42\ \Omega$

Because we used current-forcing line lengths (90°), we see the same voltage magnitude at the end of all four feed lines (except for slight differences due to line losses).

We will feed the array at the junction of the feed lines going to elements 1 and 3 (in principle we could have chosen the junction of lines 2 and 4 instead). The impedance at the junction of the feed lines 1 and 3 is

$Z(1, 3)' = 38.83 - j17.10\ \Omega$

We must now design a network to be inserted in the joined feed lines to elements 2 and 4 to equalize the voltage phase angles. The network will consist of a shunt reactance (coil or capacitor) and a constant-impedance line stretcher. The two feed lines connected in parallel give:

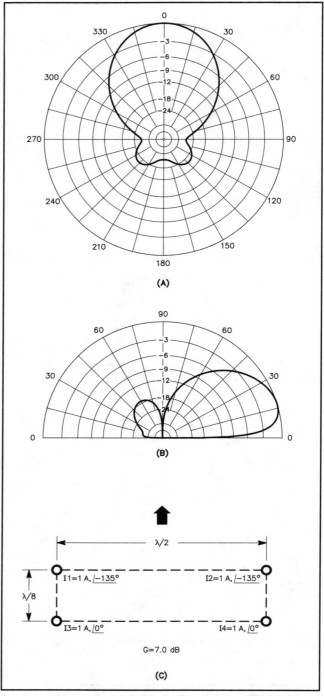

(A)

(B)

(C)

Fig 11-64—Radiation patterns for a rectangular array made of two 2-element cardioid arrays ($^1/_8$-λ spacing, 135° phase shift), spaced $^1/_2$ λ.

$Z(2, 4) = 18.95 + j59.71\ \Omega$

We will connect a shunt impedance to cancel the reactive part of the impedance (calculated using the SHUNT REACTANCE software module). A shunt capacitance of $-65.7\ \Omega$ (637 pF at 3.8 MHz) tunes out the imaginary part of the impedance. The new impedance is:

$Z(2, 4)' = 207.1 \ \Omega$

The pi-network line stretcher is

Impedance: 207.1 Ω

Output voltage phase angle: −91.13° = +268.87°

Input voltage phase angle: 135.72°

$X_P = j90 \ \Omega$ (3.76 μH at 3.8 MHz)

$X_S = -j151 \ \Omega$ (277 pF at 3.8 MHz)

The impedance at the T junction of the two branches is given by the parallel connection of:

$Z(2, 4)' = 207.\Omega$

$Z(1,3)' = 38.83 - j17.10 \ \Omega$

$Z_{array} = 22.54 - j12.07 \ \Omega$

The layout of the modified Lewallen feed method is shown in **Fig 11-65**.

4.10.3. Two ¹/₄-λ spaced cardioid arrays side by side, spaced ¹/₄ λ, fed in phase

This is the same physical layout as the famous quarter-wave-spaced 4-square. In Section 4.9. I referred to the possibility of making a square array that could be switched in eight directions. This is the array that would fill in the "other" four directions.

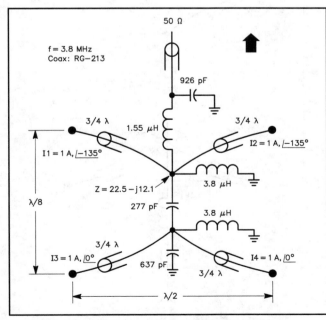

Fig 11-65—Modified Lewallen feed system for the rectangular array described in Fig 11-64.

Array data

Two ¹/₄-λ spaced cardioids, spaced ¹/₄ λ

Feed currents: I1 = 1 A, /−90°; I2 = 1 A, /0°; I3 = 1 A, /−90°; I4 = 1 A, /0°

Gain: 4.3 dB over a single vertical

3-dB beamwidth: 122°

The feed-point impedances (including 2-Ω ground loss resistance) are:

$Z1 = Z3 = 88.2 + j7.2 \ \Omega$

$Z2 = Z4 = 18.7 - j37.7 \ \Omega$

Fig 11-66 shows the radiation patterns over very good ground.

To reach the center of the array with ¹/₄-λ feed lines we must use coax with cellular PE insulation. I assumed a VF of 0.79 (RG-8 foam type), with 0.30 dB loss per 100 ft at 3.8 MHz.

The data at the end of the ¹/₄-λ-long lines are

$Z1' = Z3' = 28.7 - j2.3 \ \Omega$

$E1' = E3' = 51.4$ V, /0.12°

$Z2' = Z4' = 27.8 + j52.3 \ \Omega$

$E2' = E4' = 50.3$ V, /89.34°

Calculating the L network:

The elements of the Lewallen L network that provide the necessary 90° voltage phase shift can easily be calculated (see Section 3.3.3):

In this case we have to apply the formula suitable for more than 2 elements:

$$X_S = \frac{Z_{coax}^2}{\Sigma (R)}$$

$$X_P = \frac{Z_{coax}^2}{[\Sigma (X) - \Sigma (R)]}$$

As the R and X values are the same:

$\Sigma(R) = 2 \times R2$ with R2 = 88.2 Ω

$\Sigma(X) = 2 \times X2$ with X2 = 7.2 Ω

R2 and X2 are the real and the imaginary part of the impedance of the elements with the lagging current.

$$X_S = \frac{Z_{coax}^2}{2 \times R2} = \frac{50^2}{2 \times 88.2} = j14.2 \ \Omega$$

$$X_P = \frac{Z_{coax}^2}{2(X2 - R2)} = \frac{50^2}{2 \times (7.2 - 88.2)} = -j15.43 \ \Omega$$

Series element (inductor): 0.62 μH at 3.65 MHz

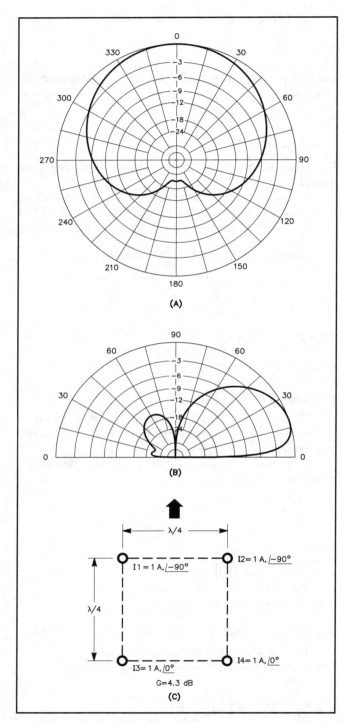

Fig 11-66—Radiation patterns for a rectangular array made of two 2-element cardioid arrays (¹/₄-λ spacing, 90° out of phase), spaced ¹/₄ λ.

Parallel element (capacitor): 2827 pF at 3.65 MHz

Calculating the impedance of the array

The 2827 pF capacitor is in parallel with the ends of the two ¹/₄-λ feed lines that go to elements 1 and 3. The resulting impedances is the parallel connection of three impedances:

$$Z1' = 28.7 - j2.3 \ \Omega$$

$$Z_{cap} = 0 - j15.43 \ \Omega$$

The sum (as calculated with the SHUNT IMPEDANCE MODULE) is Z = 7.11 − j7.22 Ω

Adding the series impedance of the inductance of the L network, the impedance at the other side of the inductance becomes:

$$Z_{t1} = (7.11 - j7.22) + (0 + j14.2) = 7.11 + j6.98 \ \Omega$$

At this point we connect the two feed lines going to the elements 2 and 4. Three parallel impedances need to be calculated:

$$Z2' = Z4' = 27.8 + j52.3 \ \Omega$$

$$Z_{t1} = 7.11 + j6.98 \ \Omega$$

The resulting impedance is: 4.95 +j5.66 Ω. This impedance can be matched to the 50-Ω line using an L network as shown in **Fig 11-67**.

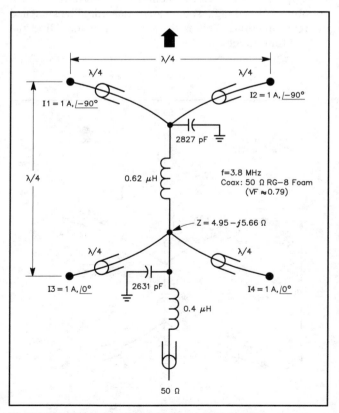

Fig 11-67—Lewallen-type feed system for the rectangular array shown in Fig 11-66.

4.10.4. Two $^1/_8$-λ-spaced cardioid arrays side by side, spaced $^1/_4$ λ, fed in phase

Array data

Two $^1/_8$-λ spaced cardioids, spaced $^1/_4$ λ

Feed currents: I1 = 1 A, $\underline{/-135°}$; I2 = I A, $\underline{/0°}$; I3 = 1 A, $\underline{/-135°}$; I4 = 1 A, $\underline{/0°}$

Gain: 4.9 dB over a single vertical

3-dB beamwidth: 100°

The feed-point impedances (including 2-Ω ground loss resistance) are

Z1 = Z3 = 38.2 + j28.5 Ω

Z2 = Z4 = 6.8 − j33.6 Ω

Fig 11-68 shows the radiation patterns over very good ground.

The feed system is a modified Lewallen feed system (the feed angles are not in quadrature). Feed lines that are 90° long will reach the center of the array. At the end of the 90° feed lines the impedances and voltages are:

Z1′ = Z3′ = 42.62 + j28.50 Ω

E1′ = E3′ = 50.57 V, $\underline{/-44.45°}$

Z2′ = Z4′ = 17.0 + j70.71 Ω

E2′ = E4′ = 50.13 V, $\underline{/89.34°}$

We will feed the array at the junction of the feed lines going to elements 1 and 3 (in principle we could have chosen the junction of lines to elements 2 and 4 instead).

The impedance at the junction of the feed lines 1 and 3 is:

Z(1, 3)′ = 21.31 − j14.25 Ω

We must now design a network to be inserted in the joined feed lines to elements 2 and 4 to equalize the voltage phase angles. The network will consist of a shunt reactance (coil or capacitor) and a constant-impedance line stretcher.

The two feed lines connected together give:

Z(2, 4) = 8.55 + j30.35 Ω

E(2, 4) = 50.13 V, $\underline{/89.34°}$

We will connect a shunt impedance to cancel the reactive part of the impedance (calculated using the SHUNT IMPEDANCE software module): a shunt inductance of − j32.8 Ω (1279 pF at 3.8 MHz) tunes out the imaginary part of the impedance. The new impedance is:

Z(2, 4)′ = 116.28 Ω

The pi-network line stretcher is:

Impedance: 116.28 Ω

Output voltage phase angle: − 44.45°

Input voltage phase angle: 89.34°

X_P = − j49.6 Ω (844 pF at 3.8 MHz)

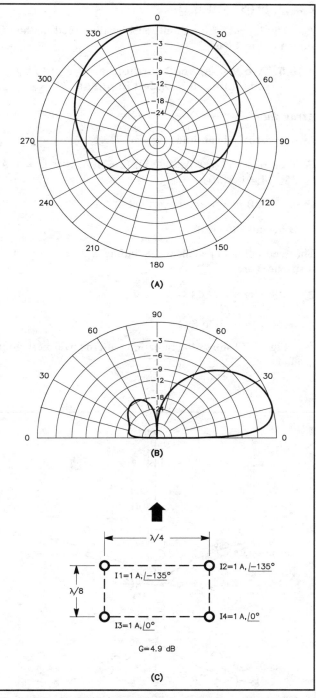

Fig 11-68—Radiation patterns for a rectangular array made of two 2-element cardioid arrays ($^1/_8$-λ spacing, 135° out of phase), spaced $^1/_4$ λ.

$X_S = j83.9 \, \Omega$ (3.5 µH at 3.8 MHz)

The impedance at the T junction of the two branches is given by the parallel connection of:

$Z(2, 4)' = 116.28 \, \Omega$

$Z(1, 3)' = 21.31 - j14.25 \, \Omega$

$Z_{array} = 19.05 - j10.07 \, \Omega$

The layout of the modified Lewallen feed method is shown in **Fig 11-69**.

4.10.5. Two ⅛-λ-spaced cardioid arrays side by side, spaced ⅛ λ, fed in phase

Array data

Two ⅛-λ spaced cardioids, spaced ⅛ λ

Feed currents: I1 = 1 A, /–135°; I2 = 1 A, /0°; I3 = 1 A, /–135°; I4 = 1 A, /0°

Gain: 4.2 dB over a single vertical

3-dB beamwidth: 128°

The feed-point impedances (including a 2-Ω ground loss resistance) are:

$Z1 = Z3 = 39.8 + j43.8 \, \Omega$

$Z2 = Z4 = 14.8 - j36.8 \, \Omega$

Fig 11-70 shows the radiation patterns over very good ground.

The feed system is again a modified Lewallen feed system (the feed angles are not in quadrature). At the end of the 90° feed lines the impedances and voltages are:

$Z1' = Z3' = 29.31 - j30.65 \, \Omega$

$E1' = E3' = 50.7 \, V, /\underline{-44.15°}$

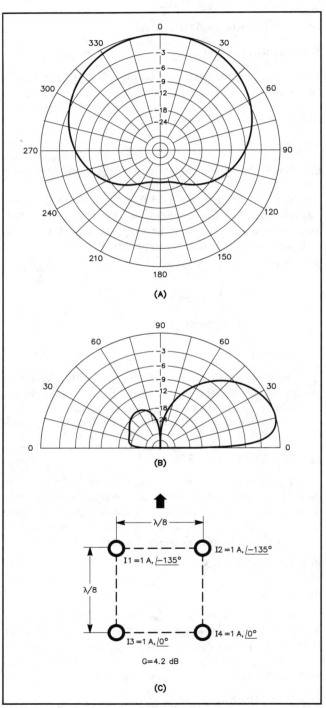

(A)

(B)

(C)

Fig 11-70—Radiation patterns for a square array made of two 2-element cardioid arrays (⅛-λ spacing, 135° out of phase), spaced ⅛ λ.

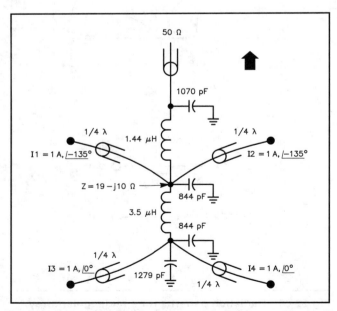

Fig 11-69—Feed system according to the modified Lewallen system for the array shown in Fig 11-68.

$Z2' = Z4' = 25.33 + j57.5 \, \Omega$

$E2' = E4' = 50.27 \, V, \, \underline{/89.28°}$

We will feed the array at the junction of the feed lines going to elements 1 and 3. The impedance at the junction of the feed lines 1 and 3 is:

$Z(1, 3)' = 14.65 - j15.32 \, \Omega$

We must now design a network to be inserted in the joined feed lines to elements 2 and 4 to equalize the voltage phase angles. The network will consist of a shunt reactance (coil or capacitor) and a constant-impedance line stretcher.

The two feed lines connected together give:

$Z(2, 4) = 12.66 + j28.75 \, \Omega$

$E(2, 4) = 50.27 \, V, \, \underline{/89.28°}$

We will connect a shunt impedance to cancel the reactive part of the impedance (calculated using the SHUNT IMPEDANCE software module). A shunt inductance of $-j34.3 \, \Omega$ (1220 pF at 3.8 MHz) tunes out the imaginary part of the impedance. The new impedance is:

$Z(2, 4)' = 78 \, \Omega$

The pi-network line stretcher is:

Impedance: 78 Ω

Output voltage phase angle: −44.15°

Input voltage phase angle: 89.28°

$X_P = -j33.6 \, \Omega$ (1247 pF at 3.8 MHz)

$X_S = j56.6 \, \Omega$ (2.37 µH at 3.8 MHz)

The impedance at the T junction of the two branches is given by the parallel connection of:

$Z(2, 4)' = 78 \, \Omega$

$Z(1, 3)' = 14.65 - j15.32 \, \Omega$

$Z_{array} = 14.08 - j10.57 \, \Omega$

The layout of the modified Lewallen feed method is shown in **Fig 11-71**.

4.11. Triangular Arrays

A last group of antennas consists of the triangular arrays. The gain is comparable to the gain of the 4-square arrays, and they also have a broad forward-lobe beamwidth. I have analyzed three different equilateral triangle arrays. All three can be fed in two different ways:
• Beaming off the top of the triangle. The top corner is fed with a current of lagging phase angle. The two elements at the bottom of the triangle are fed with the leading current.
• Beaming off the bottom of the triangle. Both bottom-corner elements are fed by the current with the lagging phase angle.

The top vertical is fed with the leading phase angle.
• In both cases the "solitary" top element is fed with twice the current magnitude when compared to the two elements at the corners of the bottom base line of the triangle.

Being a triangle, each array can be switched in three directions. The "alternative" feed method adds another three directions, which means that with a rather complex switching system a triangular array can be made switchable in six directions. The alternative array has the same gain (within 0.1 dB) and an almost identical radiation pattern.

If both alternatives are to be used in an array that can be switched in six directions, separate phasing networks will be required. I will leave it to your imagination as an array designer to develop your switching harness. The problem is that you will have to run two RG-213 cables to each of the element feed points, because there always will be either one of the elements that will require a feed current of "double magnitude," and hence the need for a 25-Ω feed line. When there is only one cable required (50 Ω), you will have to short the quarter-wave feed line. At the antenna element the short will look like an open.

As each of the directivity patterns has a 3-dB beamwidth of approximately 146°, it is questionable if the added complexity is worth the effort.

I will describe three different triangle arrays. The difference is in the element spacing and current phase angles.

4.11.1. The quadrature-fed triangular array

Atchley, W1CF, described a 3-element array where the

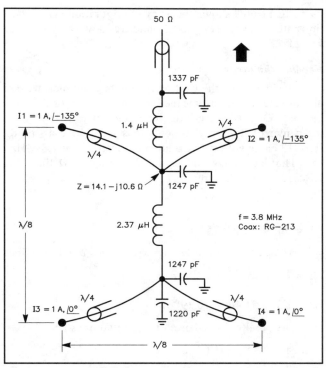

Fig 11-71—Modified Lewallen-type feed system for the square array shown in Fig 11-70.

verticals are positioned in an equilateral triangle with sides measuring 0.29 λ (Ref 939 and 941). The original version of the array used equal current magnitude in all elements. Later, Gehrke, K2BT, improved the array by feeding the two back elements with half the current of the front element. This improved the zero-wave-angle front-to-back ratio to almost infinity, with a back rejection of 20 dB or better at the main wave angle over approximately 60°.

4.11.1.1. First alternative

One element is fed with twice the current magnitude as the other two, and is leading the other two by 90°. Radiation is always off the element with the lagging current. Thus, it radiates broadside to the line connecting the two elements with the lagging current.

Array data

Side triangle: 0.29 λ

Feed currents: I1 = 2 A, $\underline{/0°}$; I2 = 1 A, $\underline{/-90°}$; I3 = 1 A, $\underline{/-90°}$

Gain: 3.9 dB over a single vertical

3-dB beamwidth: 140°

The feed-point impedances (including a 2-Ω ground loss resistance) are:

Z1 = 18.6 – j12.7 Ω

Z2 = Z3 = 84.6 + j9.7 Ω

Fig 11-60C shows the array configuration. The radiation patterns over very good ground are shown at A and B of **Fig 11-72**.

Feeding the array

Besides using the Collins-type hybrid coupler, we can also use the Lewallen-type feed system (see Section 3.3.3 for details). Quarter-wave-long feed lines will easily reach the center of the array. Calculations were done assuming RG-213 cable with 0.35 dB attenuation/100 ft at 3.8 MHz.

The data at the end of the ¹/₄-λ-long 25-Ω line (2 × RG-213 in parallel) are:

Z1′ = 23.03 + j15.40 Ω

E1′ = 50.32 V, $\underline{/89.75°}$

At the end of the 50-Ω lines to elements 2 and 3 we find:

Z2′ = Z3′ = 29.73 – j3.28 Ω

E2′ = E3′ = 51.46 V, $\underline{/0.19°}$

The parallel impedance of these two lines is:

Z(2, 3)′ = 14.86 – j1.64 Ω

Calculating the L network

The elements of the Lewallen L network that provide the necessary 90° voltage phase shift can easily be calculated (see Section 3.3.3). In this case we have to apply the formula suitable for more than 2 elements:

$$X_S = \frac{Z_{coax}^2}{\Sigma (R)}$$

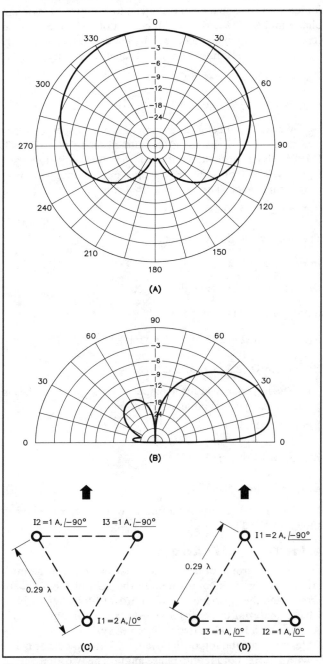

Fig 11-72—Radiation patterns for a triangular array with an element spacing of 0.29 λ (quadrature fed). The alternative feed method at D produces almost identical gain and radiation patterns.

$$X_P = \frac{Z_{coax}^2}{\left[\Sigma(X) - \Sigma(R)\right]}$$

As the R and X values are the same:

$\Sigma(R) = 2 \times R2$ with $R2 = 84.62 \ \Omega$

$\Sigma(X) = 2 \times X2$ with $X2 = j9.7 \ \Omega$

R2 and X2 are the real and the imaginary part of the imped-ance of the elements with the lagging current.

$$X_S = \frac{Z_{coax}^2}{2 \times R2} = \frac{50^2}{2 \times 84.6} = j14.8 \ \Omega$$

$$X_P = \frac{Z_{coax}^2}{2(X2 - R2)} = \frac{50^2}{2 \times (9.7 - 84.6)} = -j16.7 \ \Omega$$

Series element (inductor): 0.65 μH at 3.65 MHz

Parallel element (capacitor): 2613 pF at 3.65 MHz

Calculating the array impedance

The shunt impedance (−187.5 Ω) is in parallel with the two feed lines going to the elements 2 and 3. The total net impedance (calculated with the SHUNT IMPEDANCE MODULE) is $7.44 - j17.8 \ \Omega$

The series reactance (14.8 Ω) converts the impedance to $Z_{t1} = 7.44 - j \ 3 \ \Omega$.

At that point the two parallel feed lines to element 1 are connected in parallel. The impedance at the end of these 2 parallel feed lines is $Z1' = 23.03 + j15.40 \ \Omega$

The array impedance is the parallel connection of Z_{t1} and $Z1'$, which is $= 6.65 - j1.21 \ \Omega$. This impedance can be matched to a 50-Ω line using an L network as shown in **Fig 11-73A**.

4.11.1.2. Second alternative

In this configuration the radiation is off one of the "tips" of the triangle. The element at this tip is fed with twice the current magnitude when compared to the other ones, and with a current lagging the other elements by 90°. This configuration in combination with the configuration de-scribed in Section 4.10.1.1 makes it possible to make a triangular array that covers six directions. The impedances are different and the Lewallen L networks will be different.

Array data

Side triangle: 0.29 λ

Feed currents: I1 = 2 A, /−90°; I2= 1 A, /0°; I3 = 1A, /0°

Gain: 3.8 dB over a single vertical

3-dB beamwidth: 145°

The feed-point impedances (including 2-Ω ground-loss re-sistance) are:

Z1 = 54.2 − j13.5 Ω

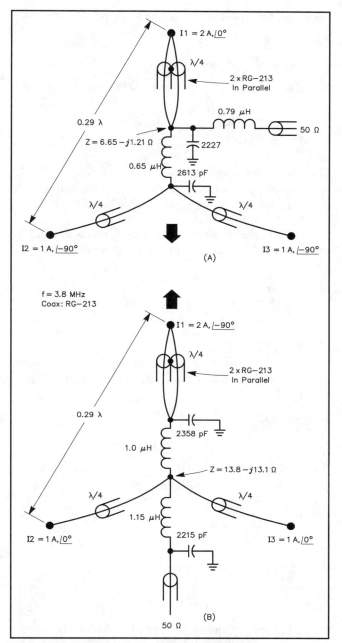

Fig 11-73—Lewallen-type feed systems for the triangular array described in Fig 11-72. The phase-shift L networks, as well as the array impedances, are different in the two different "firing positions."

Z2 = Z3 = 14.4 − j43.9 Ω

Fig 11-72D shows the array configuration. The radia-tion pattern is for all practical purposes identical to the pattern generated with the array described in Section 4.10.1.1.

Feeding the array

Here too, the Collins-type hybrid coupler will be the easy approach, especially if we want to make the array switchable in six directions. I shall describe the Lewallen-type feed system (see Section 3.3.3 for details). The data at the end of the $^1/_4$-λ-long 25-Ω line to element 1 are:

Z1$'$ = 11.3 + j2.68 Ω

E1$'$ = 50.93 V, $\underline{/-0.26°}$

At the ends of the two $^1/_4$-λ 50-Ω lines to elements 2 and 3:

Z2$'$ = Z3$'$ = 18.52 + j50.79 Ω

E2$'$ = E3$'$ = 50.26 V, $\underline{/89.14°}$

Calculating the L network

The elements of the Lewallen L network that provide the necessary 90° voltage phase shift can easily be calculated (see Section 3.3.3):

In this case we have to apply the formula suitable for more than 2 elements (we have two parallel elements):

$$X_S = \frac{Z_{coax}^{\ 2}}{\Sigma\,(R)}$$

$$X_P = \frac{Z_{coax}^{\ 2}}{\left[\Sigma\,(X) - \Sigma\,(R)\right]}$$

As we feed the same element with two coaxial cables in parallel, we have:

Σ(R) = 2 × R1 with R1 = 54.2 Ω

Σ(X) = 2 × X1 with X1 = – j13.5 Ω

R1 and X1 are the real and the imaginary part of the impedance of the element with the lagging current.

$$X_S = \frac{Z_{coax}^{\ 2}}{2 \times R2} = \frac{50^2}{2 \times 54.2} = j23.1\ \Omega$$

$$X_P = \frac{Z_{coax}^{\ 2}}{2\,(X2 - R2)} = \frac{50^2}{2 \times (-13.5 - 54.2)} = -\,j18.5\ \Omega$$

Series element (inductor): 1.0 µH at 3.65 MHz

Parallel element (capacitor): 2358 pF at 3.65 MHz

Calculating the array impedance

The shunt impedance (–18.5 Ω) is in parallel with the two feed lines going to the elements 2 and 3. The total net impedance (calculated with the SHUNT IMPEDANCE MODULE) is 23.8 – j36.2 Ω

The series reactance (j23.1 Ω) converts the impedance to Z$_{t1}$ = 13.8 – j13.1 Ω.

At that point the two parallel feed lines to element 2 and 3 are connected in parallel. The impedance at the end of these 2 feed lines is Z2$'$ = Z3$'$ = 18.52 + j50.79 Ω

The array impedance is the parallel connection of , Z2$'$ and Z3$'$, which is = 6.71 – j9.38 Ω. This impedance can be matched to a 50-Ω line using an L network as shown in Fig 11-73B.

4.11.2. Triangular array with improved phasing

This array was optimized to reduce the high angle back lobe. Instead there is a large back lobe at a 15° wave angle. I think it is more important to have a good high-angle front-to-back. It may even be advantageous to hear the DX (low angle) off the back a little better. At the same time the gain has increased by 0.6 dB, which is not negligible.

4.11.2.1. The first alternative

Array data

Side triangle: 0.29 λ

Feed currents: I1 = 2 A, $\underline{/0°}$; I2 = 1 A, $\underline{/-115°}$; I3 = 1 A, $\underline{/-115°}$

Gain: 4.5 dB over a single vertical

3-dB beamwidth: 120°

The feed-point impedances (including 2-Ω ground-loss resistance) are:

Z1 = 14.9 – j4.3 Ω

Z2 = Z3 = 70.6 + j21.5 Ω

Fig 11-74 shows the radiation patterns over very good ground and the array configuration at C.

This is not a quadrature-fed array, so the modified Lewallen-type feed system is indicated (see Section 3.3.4 for details). At the end of the 90° 25-Ω feed line (two parallel RG-213 cables) to element 1, the impedance and voltage are:

Z1$'$ = 38.47 + j10.88 Ω

E1$'$ = 50.26 V, $\underline{/89.92°}$

At the end of the 90°-long 50-Ω feed lines to elements 2 and 3 we find:

Z2$'$ = Z3$'$ = 32.93 – j9.65 Ω

E2$'$ = E3$'$ = 51.22 V, $\underline{/-24.59°}$

We must now design a network to be inserted in the joined feed lines to elements 2 and 3 to equalize the voltage phase angles. The network will consist of a shunt reactance (coil or capacitor) and a constant-impedance line stretcher. The two feed lines connected together give:

Z(2, 3) = 16.46 + j4.82 Ω

E(2, 3) = 51.22 V, $\underline{/-24.59°}$

We will connect a shunt impedance to cancel the reactive part of the impedance (calculated using the SHUNT IMPEDANCE software module).

A shunt reactance of – j61 Ω (686 pF at 3.8 MHz) tunes out the imaginary part of the impedance. The new impedance is:

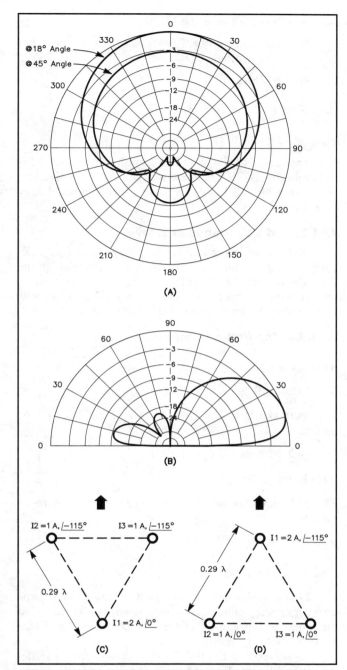

Fig 11-74—Radiation patterns for a triangular array with an element spacing of 0.29 λ and improved phase angle (115°). The alternative feed method at D produces almost identical gain and radiation patterns.

The pi-network line stretcher is:

Impedance: 17.8 Ω

Output voltage phase angle: −24.59°

Input voltage phase angle: 89.92°

$X_P = -j11.4 \ \Omega$ (3658 pF at 3.8 MHz)

$X_S = j16.2 \ \Omega$ (0.68 μH at 3.8 MHz)

The impedance at the T junction of the two branches is given by the parallel connection of:

$Z(2, 4)' = 17.8 \ \Omega$

$Z(1, 3)' = 38.47 + j10.88 \ \Omega$

$Z_{array} = 12.37 - j1.05 \ \Omega$

The layout of the modified Lewallen feed method is shown in **Fig 11-75**. There are many other possible alternative networks, all using this same network methodology (see Section 3.3.4).

4.11.2.2. The second alternative

Array data

Side triangle: 0.29 λ

Feed currents: I1 = 2 A, /−155°; I2 = 1 A, /0°; I3 = 1 A, /0°

Gain: 4.4 dB over a single vertical

3-dB beamwidth: 124°

The feed-point impedances (including 2-Ω ground-loss resistance) are:

$Z1 = 64.9 + j19.7 \ \Omega$

$Z2 = Z3 = 6.5 - j26.5 \ \Omega$

Fig 11-74D shows the array configuration. The radiation pattern is for all practical purposes identical to the pattern generated with the array described in Section 4.11.2.1.

This is not a quadrature-fed array, so the modified Lewallen-type feed system is indicated (see Section 3.3.4 for details). At the end of the 90° 25-Ω feed line (two parallel RG-213 cables) to element 1, the impedance and voltage are:

$Z1' = 9.01 - j2.66 \ \Omega$

$E1' = 51.12 \ V, \ /−24.62°$

At the end of the 90°-long 50-Ω feed lines to elements 2 and 3 we find:

$Z2' = Z3' = 25.18 - j87.56 \ \Omega$

$E2' = E3' = 50.12 \ V, \ /−24.62°$

We must now design a network to be inserted in the joined feed lines to elements 2 and 3 to equalize the voltage phase angles. The network will consist of a shunt reactance (coil or capacitor) and a constant-impedance line stretcher.

The two feed lines connected together give:

$Z(2, 3) = 12.59 - j43.78 \ \Omega$

We will connect a shunt impedance to cancel the reactive part of the impedance (calculated using the SHUNT IMPEDANCE software module).

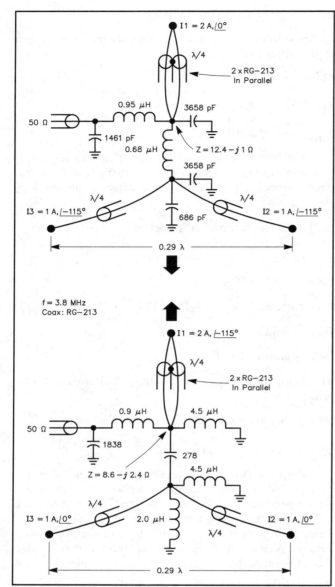

Fig 11-75—Lewallen-type feed systems for the triangular array described in Fig 11-74. The phase-shift L networks, as well as the array impedances, are different in the two different "firing positions."

A shunt inductance of $j47.4$ Ω (2.0 μH at 3.8 MHz) tunes out the imaginary part of the impedance. The new impedance is:

$Z(2, 3)' = 164.8$ Ω

The pi-network line stretcher is

Impedance: 164.8 Ω

Output voltage phase angle: 89.48°

Input voltage phase angle: −24.62°

$X_P = j106.8$ Ω (4.5 μH at 3.8 MHz)

$X_S = -j150$ Ω (278 pF at 3.8 MHz)

The impedance at the T junction of the two branches is given by the parallel connection of

$Z(2, 3)' = 164.8$ Ω

$Z(1)' = 9.01 - j2.66$ Ω

$Z_{array} = 8.58 - j2.39$ Ω

The layout of this modified Lewallen feed method is shown in Fig 11-75. There are many other alternative networks possible, all using this same network methodology (see Section 3.3.4).

4.11.3. Half-size triangular array

In this array I have adjusted the phase angles for a well-balanced front-to-back ratio. Above a wave angle of 30° the F/B is better than 20 dB. This looks like an attractive array for 160 meters. This array trades in only 0.2 dB of gain as compared to the array with twice the dimensions.

4.11.3.1. The first alternative

Array data

Side triangle: 0.145 λ

Feed currents: I1 = 2 A, $\underline{/0°}$; I2 = 1 A, $\underline{/-145°}$; I3 = 1 A, $\underline{/-145°}$

Gain: 4.3 dB over a single vertical

3-dB beamwidth: 122°

The feed-point impedances (including 2 Ω ground loss resistance) are

$Z1 = 9.1 - j13$ Ω

$Z2 = Z3 = 26.6 + j35.9$ Ω

Fig 11-76 shows the array layout and radiation patterns over very good ground. This is not a quadrature-fed array, so the modified-Lewallen-type feed system is indicated (see Section 3.3.4 for details).

At the end of the 90° 25-Ω feed line (two parallel RG-213 cables) to element 1 the impedance and voltage are:

$Z1' = 22.98 + j31.76$ Ω

$E1' = 50.16$ V, $\underline{/- 89.74°}$

At the end of the 90°-long 50-Ω feed lines to elements 2 and 3 we find:

$Z2' = Z3' = 34.46 - j43.93$ Ω

$E2' = E3' = 50.47$ V, $\underline{/- 54.3°}$

We must now design a network to be inserted in the joined feed lines to elements 2 and 3 to equalize the voltage phase angles. The network will consist of a shunt reactance

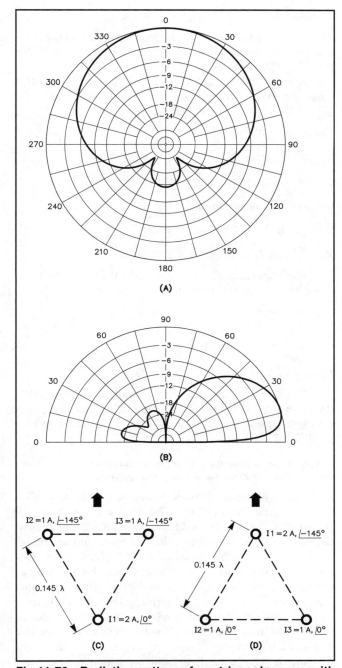

Fig 11-76—Radiation patterns for a triangular array with an element spacing of 0.145 λ (elements fed 145° out of phase). The alternative feed method at D produces almost identical gain and radiation patterns.

(coil or capacitor) and a constant-impedance line stretcher.

The feed lines to the elements 2 and 3 connected together give:

$$Z(2, 3) = 17.23 - j21.96 \ \Omega$$

We will connect a shunt impedance to cancel the reactive part of the impedance (calculated using the SHUNT

IMPEDANCE software module).

A shunt inductance of $j35.5 \ \Omega$ (1.5 µH at 3.8 MHz) tunes out the imaginary part of the impedance. The new impedance is:

$$Z(2, 3)' = 45.22 \ \Omega$$

The pi-network line stretcher is

Impedance: 45.22 Ω

Output voltage phase angle: −54.3°

Input voltage phase angle: 89.74°

$X_P = -j14.7 \ \Omega$ (2854 pF at 3.8 MHz)

$X_S = j26.6 \ \Omega$ (1.1 µH at 3.8 MHz)

The impedance at the T junction of the two branches is given by the parallel connection of $Z(2, 3)' = 45.22 \ \Omega$

$$Z(1)' = 22.98 + j31.76 \ \Omega$$

$$Z_{array} = 20.37 + j11.3 \ \Omega$$

The layout of this modified Lewallen feed method is shown in **Fig 11-77**. There are many other alternative networks possible, all using this same network methodology (see Section 3.3.4).

4.11.3.2. The second alternative

Array data

Side triangle: 0.145 λ

Feed currents: I1 = 2 A, $\underline{/- 145°}$; I2 = 1 A, $\underline{/0°}$; I3 = 1 A, $\underline{/0°}$

Gain: 4.2 dB over a single vertical

3-dB beamwidth: 124°

The feed-point impedances (including 2 Ω ground loss resistance) are:

$$Z1 = 17.2 + j20.3 \ \Omega$$

$$Z2 = Z3 = 10.3 - j31.5 \ \Omega$$

Fig 11-64D shows the array configuration. The radiation pattern is for all practical purposes identical to the pattern generated with the array described in 4.10.3.1. This is not a quadrature-fed array, so the modified Lewallen type of feed system is indicated (see Section 3.3.4 for details).

At the end of the 90° 25-Ω feed line (two parallel RG-213 cables) to element 1, the impedance and voltage are:

$$Z1' = 15.43 - j17.73 \ \Omega$$

$$E1' = 50.30 \ V, \ \underline{/- 54.60°}$$

At the end of the 90°-long 50-Ω feed lines to elements 2 and 3 we find:

$$Z2' = Z3' = 25.84 + j70.49 \ \Omega$$

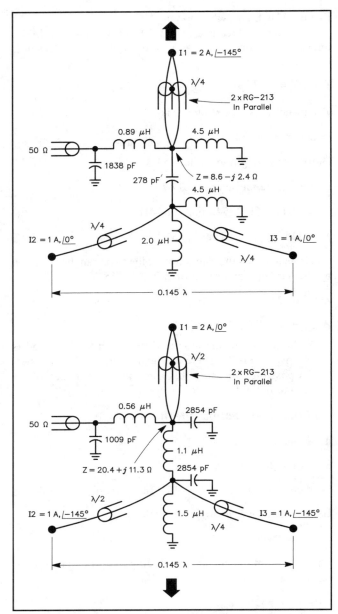

$Z1 = 15.43 - j17.73 \ \Omega$

We will connect a shunt impedance to cancel the reactive part of the impedance (calculated using the SHUNT IMPEDANCE software module).

A shunt inductance of $j31.2 \ \Omega$ (1.3 µH at 3.8 MHz) tunes out the imaginary part of the impedance. The new impedance is $Z(2,3)' = 35.8 \ \Omega$

The pi-network line stretcher is:

Impedance: 35.8 Ω

Output voltage phase angle: 89.38°

Input voltage phase angle: −54.6°

$X_P = j11.6 \ \Omega$ (0.49 µH at 3.8 MHz)

$X_S = -\ j21.1 \ \Omega$ (1989 pF at 3.8 MHz)

The impedance at the T junction of the two branches is given by the parallel connection of:

$Z(2, 3)' = 35.8 \ \Omega$

$Z(2, 3) = 12.92 + j35.24 \ \Omega$

$Z_{array} = 18.53 + j12.49 \ \Omega$

The layout of this modified Lewallen feed method is shown in Fig 11-77. There are many other alternative networks possible all using this same network methodology (see Section 3.3.4).

4.12. The Ultimate Vertical Array: The 9-Circle Array by WØUN

John Brosnahan, WØUN, developed what may seem like the ultimate vertical array for the low bands. It uses 9 elements, 8 of which are positioned on a circle, with the 9th element in the center of the circle. The radius of the circle is approx. 31 m for a frequency of 3.775 MHz, which makes this quite a sizable array! Including radials, this antenna has a footprint of 1 hectare (100 m by 100 m). But the performance data are just short of formidable. With its narrow forward angle (60°) this array has been made switchable in 8 directions. Its F/B at the main wave angle of 25° is better than 20 dB over an angle of not less than 230°, which is truly outstanding. The geometrical front to side ratio is 32 dB, and the geometrical front to back ratio is 33 dB. With an array like this, there definitely is no need for separate receiving antennas! **Fig 11-78** shows the layout of the antenna, with the feed currents.

Array data

Radius circle: 0.39 λ

Eight quarter-wave elements equispaced on the circle (one per 45°)

Feed currents: see Fig 11-78.

Gain: 7.7 dB over a single vertical (over average ground)

Fig 11-77—Lewallen-type feed systems for the triangular array described in Fig 11-76. The phase-shift L networks as well as the array impedances are different in the two different "firing positions."

$E2' = E3' = 50.19$ V, $/89.38°$

We must now design a network to be inserted in the joined feed lines to elements 2 and 3 to equalize the voltage phase angles. The network will consist of a shunt reactance (coil or capacitor) and a constant-impedance line stretcher. The two feed lines connected together give:

$Z(2, 3) = 12.92 + j35.24 \ \Omega$

At the end of the two parallel feed lines to element 1 we have:

3-dB beamwidth: 60°

The feed-point impedances (including 2-Ω ground-loss resistance) are:

$Z1 = 41.1 + j32.1\ \Omega$

$Z2 = 118.0 + j212\ \Omega$

$Z3 = Z4 = 87.8 + j30.1\ \Omega$

$Z5 = Z6 = 66 + j0.6\ \Omega$

$Z7 = Z8 = 15.7 + j16.1\ \Omega$

$Z9 = -32.3 + j28.7\ \Omega$

Fig 11-79 shows the radiation patterns over very good ground.

The feed currents

The nice thing about this array is that the elements are all fed in quadrature (increments of 90°), which greatly simplifies the design of the feed system. The current magnitudes may seem to be a little odd at first look, but they are really the result of some practical thinking.

We have learned before that with current-forcing feed methods (W7EL or W1FC) one can achieve different feed currents by using different coaxial cable impedances (for the $^1/_4$-λ current-forcing feed lines). The seemingly odd current values of the WØUN 9-circle array are obtained as follows:
- Center element (required 3-A feed current magnitude): three 50-Ω cables in parallel (total impedance = 16.7 Ω)
- Elements requiring 1-A feed current: single 50-Ω feed line
- Element requiring 1.66-A feed current: a 75 Ω and a 50-Ω cable in parallel (total impedance = 30 Ω).

The above indicated currents are, of course, relative (normalized) values.

Switching the array around

Each of the 8 elements on the circle are connected with two coaxial cables to the switching box, located in the center of the array. One cable is a 50-Ω cable (RG-213), the other one is a 75-Ω cable (RG-11). Both are exactly $^3/_4$-λ long (which is approx. 39 m). When an element needs to be fed with a current magnitude of 1 A, the 50-Ω feed line will be the actual feed line, the 75-Ω line will then be terminated at the switching box in a short circuit. When an element requires a feed current of 1.66 A, both coaxial cables will be used as parallel feed lines, totaling a net impedance of 30 Ω.

The switch box

The switch box has 20 coaxial cable terminals, two for each of the elements on the circle, 3 for the center element (three parallel 50-Ω lines) and one going to the trans-

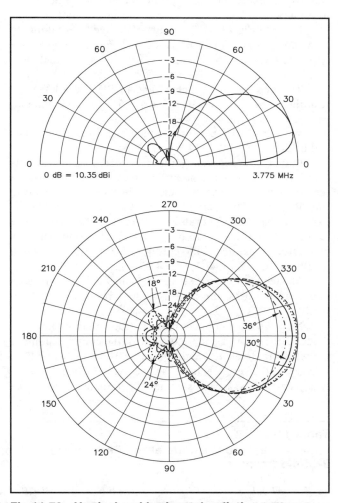

Fig 11-79—Vertical and horizontal radiation patterns for the WØUN 9-circle array. Horizontal patterns are shown for various wave elevation angles, showing the excellent F/B and F/S ratios for a wide range of elevation angles.

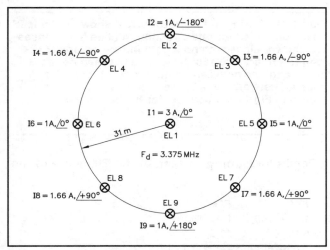

Fig 11-78—Element layout and associated feed currents for the WØUN 9-circle array.

mitter/receiver. The easiest way to construct the switching hardware is to use a matrix approach, whereby 4 feed bars (−90°, 0°, +90° and +180° which is the same as −180°) cross the 8 lines from the 8 elements. At the intersection we have a SPST relay that connects the X-lines to the Y-lines as appropriate (this is what we call the cross-bar system). At the same time there is another SPDT relay near each of the 2 × 8 coax terminals (to the circle elements). This relay will switch the end of the 75-Ω cable either in parallel with the 50-Ω line or short circuit its end. **Fig 11-80** shows a partial layout of such a cross-bar switching system.

Feeding the array

In **Table 11-3** I calculated the impedances and voltages at the end of the feed lines (at the switching box). For the cables I used a loss of 0.35 dB/100 ft. With loss-free cable the voltages at the ends of the cables should all be 50 V and the phase angle in increments of 90°. We note quite important deviations from this, which means that in reality we will *not* achieve the exact current amplitude and phase angles as we have in our model.

Refer to **Fig 11-81**. The ends of the feed lines going to elements 1, 5 and 6 will be paralleled. The net impedance will be: $Z(A) = Z(1, 5, 6) = 4.23 − j1.97 \, \Omega$, an extremely low value!

The feed lines to elements 3 and 4 will be paralleled, the resulting impedance is $Z(B) = Z(3, 4) = 5.3 − j1.56 \, \Omega$, another very low value!

We also need to parallel the feed lines to the elements 7 and 8: $Z(C) = Z(7, 8) = 14.75 − j13 \, \Omega$.

Point C (the common point of the feed lines of elements 7 and 8) is connected to point B (the common points of the feed lines to the elements 4 and 3) via a 180° long phasing line. Because of the very low impedances involved, I chose to use two parallel 50-Ω cables to do the transformation. The results are: $Z'(7, 8) = 15.6 − j12.4 \, \Omega$ This impedance is connected in parallel with $Z(3, 4) = 5.3 − j1.56 \, \Omega$, which results in $ZC = Z'(3, 4, 7, 8) = 4.09 − j2.58 \, \Omega$.

The feed lines to the elements 2 and 9 will be connected in parallel (−180° is the same as +180°), resulting in $Z(2, 9) = 5.6 − j10.1 \, \Omega$.

Now we must connect point D (the common points of the feed lines to el. 2 and 9) to point A (the common point of the feed lines to el. 1, 5 and 6) via a 180° long phasing line. Because of the very low impedances involved, I chose to use two parallel 50-Ω cables to do the transformation. The

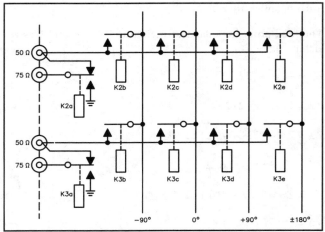

Fig 11-80—Possible partial layout of a matrix switching system for the WØUN 9-circle array.

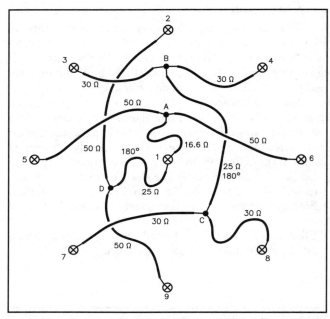

Fig 11-81—Wiring of the WØUN array, showing the feed lines connected for one direction. All line impedances are shown. Lines with no length indication are 270° long (there are two 180° lines, one between points B and C and one between points D and A). The 90° phasing network goes between points A and D as shown in Fig 11-82. See text for further details.

Table 11-3

Impedances and Currents at the Ends of the ³/₄-λ-long Feed Lines going to each of the Elements of the 9-Circle Array

	El 3,4	El 1	El 5,6	El 7,8	El 2	El 9
I_{ant}	1.66 A, /−90°	3 A, /0°	1 A, /0°	1.66 A, /90°	1 A, /−180°	1 A, /180°
Z_{cable}	30 Ω	16.66 Ω	50 Ω	30 Ω	50 Ω	50 Ω
Z_{end}	10.6−j3.03	5−j3.11	40−j0.6	29.5−j26	7.63−j8.89	−45.2−j35.1
E_{end}	57.5V, /−177°	56.7V, /−85°	53.4V, /−90°	51.3V, /1.5°	57.2V, /101°	51.7V, /91.6°

results are: $Z'(2, 9) = 6.6 - j9.9 \, \Omega$. This impedance goes in parallel with the impedance $Z(1, 5, 6)$. The resulting parallel impedance now becomes: $Z(A) = Z'(1, 5, 6, 2, 9) = 2.88 - j1.92 \, \Omega$. As we work along our way to define the final feed impedance of the array we get confronted with lower and lower impedances!

This still is *not* the impedance of the array. We will now insert the 90° phase shift network between point A and B.

How to calculate the phasing network

I used a Pi network to achieve the required phase shift. There are several alternatives. You can feed the array from point A or from point B (in one case the required shift is 90°, in the other –90°), and you can use a T or a Pi network. I chose to feed the array from point A and used a Pi-network.

A positive parallel reactance of $j9.1 \, \Omega$ tunes the impedance in B to pure resistive. It becomes 5.72 Ω. This is the characteristic impedance of the Pi network.

Z_{filter}: 5.72 Ω

Voltage input phase angle: – 90°

Voltage output phase angle: 0°

$X_S = -j5.72 \, \Omega$ (7.623 pF at 3.65 MHz)

$Z_P = 5.72 \, \Omega$ (0.25 µH at 3.65 MHz)

The final input impedance of the array (point A) is $2.88 - j1.92 \, \Omega$ in parallel with $5.72 + j\,0 \, \Omega = 2.1 - j\,0.81 \, \Omega$, which can be matched to a 50-Ω line with an L network as shown in **Fig 11-82**.

Conclusion

- This is quite an array, even on paper.
- It requires about 1 hectare (100 m by 100 m) of space.

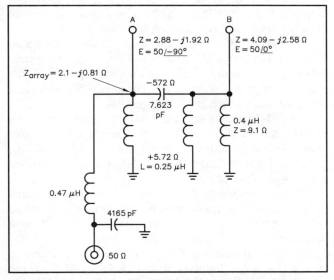

Fig 11-82—Pi-network 90° phase-shift network for the WØUN 9-circle array.

- It requires approximately 850 m of coaxial cable (yes, almost 1 kilometer!)
- The impedances involved are *very* low; not so much the feed-point impedances of the elements, but the impedances at the end of the feed lines (at the switch box)
- This also means there are very high SWR values on the feed lines, and the losses because of SWR tend to upset the correct magnitude and phase ratios.

If construction of such an array and switching system is anticipated, extreme good care must be taken of very low losses in the switching box. All connections must be made by wide copper straps, and contact resistances in the relays should be kept to a bare minimum. RF currents of up to 25 A are flowing in circuits with impedances as low as 2 Ω with a power of 2 kW! The relays are a very critical item in this antenna.

Once the array is working, look for lossy components by measuring the temperature. You can use the "touch" method (make sure you switch off the power first!), or you may use an infrared thermometer if you can borrow one.

■ 5. ELEMENT CONSTRUCTION

5.1. Mechanical Considerations

Self-supporting ¼-λ elements are easy to construct on 40 meters. On 80 meters it becomes more of a challenge, but self-supporting elements are feasible even with tubular elements when using the correct materials and element taper. Lattice-type constructions are more commonly used, with tapering-diameter aluminum tubing at the top. On Top Band, most vertical radiators will be guyed towers. As it is advisable to series feed the elements of an array, the elements must be insulated from ground, which poses extra mechanical constraint on the construction.

I used the ELEMENT STRESS ANALYSIS module of the YAGI DESIGN software (see chapter on software) to develop self-supporting elements for 40 and 80 meters that withstand high wind loads. As the element is vertical, there is no loading of the element by its own weight, which means that the same element in a vertical position will sustain a higher wind load than in a horizontal position. When using the ELEMENT STRESS ANALYSIS module, one can create this condition by entering a near-zero specific weight for the material used.

It will, however, be much easier if you plan to have at least one level where the vertical can be guyed. This will typically lower the material cost for constructing a wind-survival vertical by a factor of 3 or more. Finally, the element construction that is best for your project will be dictated to a large extent by material availability.

Needless to say, guying materials need to be electrically transparent guy wires (Kevlar, Phyllystran, Nylon, Dacron, etc) or metallic guy wires broken up into small nonresonant lengths by egg-type insulators. Refer to *The ARRL Antenna Book* (Chapter 22), which covers this aspect in great detail.

All the array data in this chapter are for ¼-λ full-size elements. It is not necessary, however, to use full-size elements. Top-loaded elements that are physically ⅔ full-size length can be used without much compromise. Make sure, however, that all elements in the array use the same amount of top-loading. If not, special precautions have to be taken (see Section 6.5). If guyed elements (eg, aluminum tubing) are used, the top set of guy wires can be used to load the element (see chapter on vertical antennas). If the array must cover 3.8 MHz as well as 3.5 MHz, a small inductance can be inserted at the base of each vertical (make sure the loading coils are identical!) to establish resonance for all elements at 3.5 MHz.

5.2. Shunt Versus Series Feeding

Shunt feeding the elements of an all-fed array is to be avoided in just about all cases. The matching system (gamma match, omega match, slant-wire match, etc) introduces additional phase shifts that are difficult to control. Such phase shifts will mess up the correct feed current in the antenna elements.

Only with arrays where all the feed impedances are identical could shunt feeding be applied successfully. The feed impedances of all elements of an array will be identical only when all the elements are fed in phase (or 180° out of phase). Shunt feeding may be considered for such arrays if the vertical elements as well as the matching systems are identical (including the values of any capacitors or inductors used in the matching system).

If you feel tempted to use your tower loaded with HF antennas as an element of an array, be aware that you might be trying to achieve the impossible.

• The loaded tower may be electrically "quite" long, which could very well be a hindrance to achieve the required directivity (see Section 2).
• You will be forced to use shunt feeding, which is just about uncontrollable, especially if all elements are not strictly identical (which will hardly ever be the case with "loaded" towers).

Loaded towers are just great for single verticals, but are more than a hassle in arrays.

5.3. A 4-Square with Wire Elements

5.3.1. The concept

An 80-meter 4-square takes a lot of room to put up. I have installed a somewhat special version of the 4-square around my full-size $^1/_4$-λ 160-meter vertical. This design has become very popular since it was first published. From the top of the vertical I run four 6-mm nylon ropes in 90° increments, to distant supports (poles). These nylon ropes serve as support cables from which I suspend the four verticals. A single radial is directed away from the center of the square (where the 160-meter vertical is located). See **Fig 11-83**.

In my particular case with the support being high enough, I managed full-size vertical elements, with the feed point and the radial 5 m above ground. In this setup the single radial serves three purposes:

1) It provides the necessary low-impedance connection for the feed line outer shield.
2) It helps to establish the resonance of the antenna (which is not the case with a large number of radials or buried

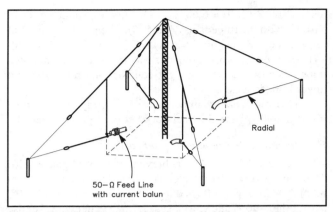

Radial

50–Ω Feed Line with current balun

Fig 11-83—If enough space is available, one can run four cables (insulated material) from the top of the support tower to four supports mounted in a square. These cables then support the verticals and their loading structures, if any. Sloping top-loading wires as shown exhibit no horizontal radiation component, provided the length is the same on both sides of the vertical.

radials, where the resonance is only determined by the length of the vertical member).

3) It provides some high-angle radiation. We can debate whether or not this is wanted, but in my particular case I wanted a fair amount of high-angle radiation as well, in order to be able to use the array successfully in contests, where shorter range contacts are also needed.

Just a single radial, without an extra ground screen (on or in the ground) will make you lose up to 6 dB of maximum achievable forward gain. I strongly advocate the use of extra radials or a ground screen under the verticals (see Chapter 9, Section 2.2.12). In my particular case, there are some 250 radials (20 to 60 m long) under the array, basically serving as the radial system for the 160-meter vertical that supports the array. With the extensive radial system, the array exhibits a low-angle F/B of 20 to 25 dB, and still a very reasonable amount of directivity at relatively high angles (20 dB F/B at 60°).

The gain of this array at low angles is very comparable (only 0.4 dB less) to the gain of a 4-square over a system using a perfect ground system. The slight drop is due to the power radiated at higher angles. The gain is 5.1 dB over a single element, which is very substantial.

Fig 11-84 shows the horizontal and vertical radiation patterns for the array as well as a single element over identical good ground (very good ground with 250 radials). Both the single vertical and the 4 elements of the 4-square use a single elevated radial.

The bottom ends of the four vertical wires are supported by steel masts that are located on the corners of a square measuring 20 m, with the 160-meter vertical (39 m tall) right in the center of the square. The masts can be folded over for easy access to the element feed point. The vertical elements are 19.5 m long. Together with a radial of 18.7 m, the elements are resonant at 3.75 MHz. The individual elements of the array were measured to have a feed-point resistance of

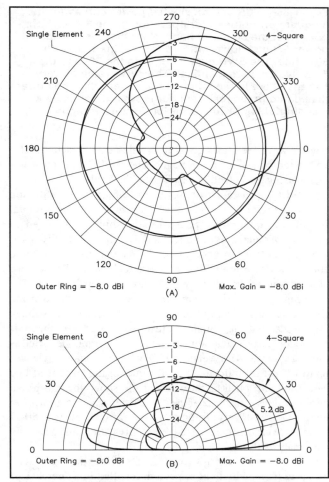

Fig 11-84—Horizontal and vertical radiation patterns of the 4-square array with one elevated radial. Also shown is the pattern of a single vertical element. Both are modeled with a single radial (per element), but over an extensive buried radial system, 5 m (17 ft) below the radial over very good ground. The buried radials are installed like spokes from the center of the square.

40 Ω at resonance (3.75 MHz). The impedance was measured over a frequency range going from 2 to 5 MHz using an HP network analyzer with a Smith Chart display. Mutual coupling to other antennas and surrounding structures shows up on the Smith Chart as a kink or a dip in the impedance chart of one or more elements at a specific frequency. It is important that the impedance curves be as near alike as possible over the frequency range of interest, if the impedance variations when switching antenna directions are to be kept at a minimum. Section 3.3.8.6 deals with the problem of eliminating unwanted mutual coupling.

One word of caution: if the central supporting tower is a base insulated tower, tuned to 160 meters, make sure that the tower is effectively grounded when you use the 4-square. If it would be left floating, this floating central element would act as a half-wave element on 80 meters and interfere heavily with the array. Grounding the central tower can be done in several ways as discussed in Chapter 7 on

special receiving antennas (Section 13). The grounded 160-meter resonant tower in the center does not influence the performance of the 4-square.

5.3.2. Loading the elements for CW operation

In order to make the antenna cover the CW end of the band as well, I use a stub, inserted in the radial at the feed point, to shift the resonance of the elements to 3.5 MHz. A small box is mounted on top of each mast. All connections (to the vertical element, radial, and feed lines) are made inside this box. The box also contains a relay that can switch the stub in and out of the circuit. The stub is supported by stand-off insulators along the metal support mast (see Fig 11-88).

The calculated reactance of the stub is 130 Ω. Using 3-mm-OD (AWG 9) copper wire with a spacing of 20 cm, the length of the stub turned out to be 2.25-m long to lower the resonant frequency to 3.505 MHz. The same stub, when shortened to 75 cm resonates the element at 3.65 MHz. A nice feature is that the resonant frequency can be changed anywhere between 3.5 and 3.75 MHz by using a movable shorting bar across the stub. This way, one can create different operating windows on 80 meters. A relay can be used to switch the 3.65-MHz shorting bar in and out of the circuit, making the window selection remotely controlled. The direction control box (see Fig 11-90) contains a 3-position lever switch, which selects the three band segments.

5.3.3. The ¼-λ feed lines

Each element is fed via an electrical ¼ λ of coaxial feed line with a current balun (50 stacked ferrite beads on a short length of small-diameter Teflon coax) at the feed point. The short length of coax together with the ferrite beads is covered by a heat-shrink (Raychem ATUM) tube to provide protection from the weather. The feed lines were cut to be ¼ λ at 3.75 MHz. If a perfect 90° phase shift is desired at 3.5 MHz, the feed lines can be lengthened by a 1-m long piece of coax (VF = 66%).

5.3.4. Wasted power

With 50-Ω feed lines, the combined feed-line impedances that load the ports of our 50-Ω hybrid coupler are quite low (22 + j19 Ω and 23 + j17 Ω), which results in up to 11% of the power being dissipated in the load resistor. With 75-Ω feed lines, these impedances are much higher (51 + j37 Ω and 48 + j43 Ω), which results in much less power being dissipated in the load resistor (4%). As already explained in Section 3.3.7, the main parameter that determines the operational bandwidth of an array fed with a hybrid coupler is the amount of power being dissipated in the load resistor.

Fig 11-85 shows both the dissipated power as well as the array input SWR for the array tuned to the high end of the band (3.7-3.8 MHz), for both the 50 Ω and the 75-Ω feed line impedance case.

5.3.5. Gain and directivity

Section 3.3.8 explains that the feed current in the elements can be assessed by measuring the voltage at the end of the

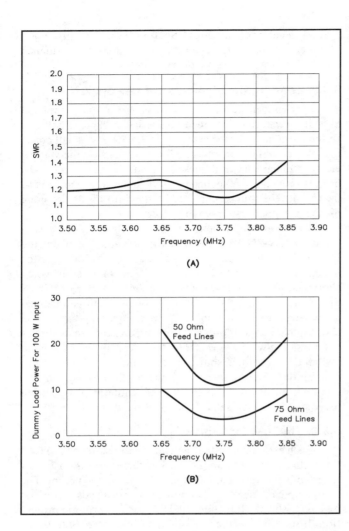

(A)

(B)

quarter-wave feed lines going to the elements. I used a vector voltmeter to measure the voltages. The results of the measurements using 50-Ω feed lines are listed in **Table 11-4**.

With the current-forcing method employed (see Section 3.3.3), the relative element feed-current requirement (equal magnitude, quadrature phase relationship) is reflected in voltages of equal magnitude (where $E = Z_k \times I$ or $E = 50$ V for a 50-Ω line) at the ends of the $1/4$-λ feed lines. The table shows the deviation from the theoretical values. From these voltage values the feed currents have been calculated. The resulting gain and F/B performance data as modeled using *ELNEC* are also listed in the table. As expected, the voltage magnitudes and phase angles were not exactly as in the theoretical model (perfect quadrature). The voltage magnitude varied as much as 1.7 dB (41 V versus 50 V), while the phase angle was up to 13° off from the theoretical value for the 50-Ω feed-line case. Table 11-4 lists all the data and also shows the transposed current values at the base of the verticals. In a pleasant surprise, even the relatively important deviations of the 50-Ω impedance case influenced the directivity pattern and gain only very marginally.

Later the 50-Ω $1/4$-λ (or $3/4$-λ) feed lines were replaced by 75-Ω lines, which, as expected, resulted in a decrease of wasted power (see Fig 11-85). A change from 11% to 4%

Table 11-4
Voltages at the Ends of the Quarter-Wavelength Feed Lines of the 4-Square Array with one Elevated Radial

50-Ω Feed Lines

		El #1	El #2	El #3	El #4
Voltage	Theoretical	50 V, $\underline{/0°}$	50 V, $\underline{/90°}$	50 V, $\underline{/90°}$	50 V, $\underline{/180°}$
	Measured	41 V, $\underline{/-13°}$	50 V, $\underline{/90°}$	50 V, $\underline{/90°}$	44.2 V, $\underline{/186°}$
Current	Theoretical	1 A, $\underline{/-90°}$	1 A, $\underline{/0°}$	1 A, $\underline{/0°}$	1 A, $\underline{/90°}$
	Calculated from Measurements	0.82 A, $\underline{/-103°}$	1 A, $\underline{/0°}$	1 A, $\underline{/0°}$	1 A, $\underline{/96°}$
Gain	Theoretical	8.13 dBi			
	Calculated from Measurements	8.07 dBi			
F/B	Theoretical	19-25 dBi			
	Calculated from Measurements	17-25 dBi			

wasted power represents a relative gain of 0.33 dB, which is respectable. **Fig 11-86** shows the superimposed vertical radiation patterns of the array with both the theoretical current values as well as the measured values.

Using 75-Ω ¼-λ (or ¾-λ) feed lines does not make this a 75-Ω system. In this particular case we are still using a hybrid coupler with a 50-Ω nominal design impedance. The 75-Ω cables are used only, because they transform the element feed-point impedances to more suitable values, resulting in less power dissipation in the dummy resistor.

5.3.6. Construction

The ComTek Systems 50-Ω hybrid coupler (with 180° phase shift) and the hybrid-coupler load resistor are located in a cabinet mounted at the base of the 160-meter vertical, which is in the center of the 4-square array. The ComTek unit

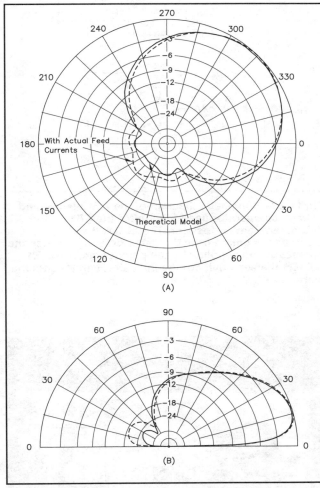

Fig 11-86—Vertical and horizontal radiation patterns (at a 20° wave angle) for the theoretical (ideal) currents and the actual currents at each element of the 4-square array (see Fig 11-85). Note that although there are some significant current deviations (phase angle and magnitude) from the theoretical values, the array suffers only very slightly from these differences. These patterns were calculated with *EZNEC*.

was removed from its normal housing and the PL-259 hardware was replaced by N connectors. The cabinet also contains the relay that switches the feed line between the 160-meter vertical and the 80-meter 4-square array (**Fig 11-87**). **Fig 11-88** shows one of the element supporting masts with the connection box at the top. Notice the 3.5 MHz stub running along the pole.

In order to know at all times how much power is being dissipated in the dummy load, I added a small RF detector to the dummy-load resistor and fed the dc voltage into the shack, where the relative power is displayed on a small moving-coil instrument that is mounted on the homemade direction-switching box. The box also contains the switch to select the subbands. In addition, a level-detector circuit is included, using an LM339 voltage comparator, which turns on a red LED if the dissipated power goes above a preset value. **Fig 11-89** shows the schematic of the system and **Fig 11-90** shows the actual switch box.

5.3.7. Using the Lewallen phasing-feed system

The logical alternative to prevent power from being wasted is to use the W7EL feed system. The drawback of the W7EL system is that the SWR and directivity bandwidth of the array are much narrower than with the hybrid coupler system. It is possible, however, to obtain a perfect quadrature feed at the design frequency, which, by the nature of the hybrid coupler, is impossible with an array presenting complex loads to the coupler. From an operating and performance point of view, the advantage of a "perfect" quadrature feed is quite unimportant as the deviations result in unnoticeably small variations in gain and F/B as compared to the theoretical "perfect" model.

A valid alternative would be to switch different LC

Fig 11-87—Cabinet located at the base of the 160-meter vertical, housing the (ComTek) hybrid coupler and directional switching-circuitry for the 4-square array. The original ComTek board is mounted on a new 20 × 20-m aluminum plate, equipped with N connectors. The four coaxial cable connectors on top of the cabinet are CATV type ½-inch 75-Ω connectors. The cabinet also contains the 50-Ω dummy load resistor.

Fig 11-88—A 10 × 10 × 3 cm plastic box is mounted on top of the 5 m support pole for the elevated verticals. Inside the box, the vertical wire and the single radial are connected to the feed line, which is equipped with a stack of 50 ferrite cores to remove any RF from flowing on the outside of the feed line. The box also houses the relay that switches the stub in and out of the circuit to lower the operating frequency to 3.5 MHz. The stub can be seen running along the steel mast.

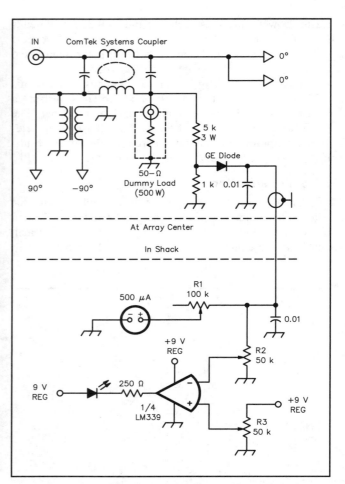

Fig 11-89—Schematic diagram of the RF detector and voltage comparator used to monitor the RF into the hybrid terminating resistor. The LED will switch on if the voltage coming from the detector is higher than the preset voltage supplied by the potentiometer R3. R1 adjusts the sensitivity of the indicator, and R2 sets the alarm level.

phase-shift networks in order to have optimum performance in all band segments.

5.3.8. Array performance

Assessing the array performance by measuring its SWR is totally meaningless (see Section 3.3.7). With a hybrid coupler, this array shows an SWR of less than 1.3:1 over the entire 80-meter band, wherever the resonance of the elements is.

Based on a sound wasted-power bandwidth criterion, the bandwidth of this array is 100 kHz. Fig 11-85 shows the wasted-power curve for the 4-square array with a single elevated radial. The steepness of the dissipated-power curve is determined by the Q factor of the array elements. In the case of this particular 4-square, the elements being made of wire, the Q is high and the bandwidth narrow. Also the fact that I use a single radial instead of a comprehensive (buried)

Fig 11-90—Array direction switch box, including the dummy-load relative RF power indicator and alarm circuit as shown in Fig 11-89. Note also the three position lever switch to select one of the three band segments (see Section 3.5.2).

radial system adds to the sharpness of the curve. While changing the frequency away from the design frequency, the single radial (just like the vertical element) will introduce reactance into the feed-point impedance, which would not be the case with a buried radial system.

Practically speaking, this array is by far the best antenna I have ever had on 80 meters. On-the-air tests continuously indicate that the signal strength on DX is ranging with the best signals from the continent. As far as directivity is concerned, it is clear that the array has a nice wide forward lobe, and that the relative loss half-way between two adjacent forward lobes is hardly noticeable (typically 2 dB). Long-haul DX very often reports, "You are S9 on the front and not copyable off the back." Even on high-angle European signals there is always a good deal of directivity with this array (typically 15 dB).

Array data

Feed currents: I1 = 1 A, *$\underline{/-90°}$*; I2 = 1 A, *$\underline{/0°}$*; I3 = 1 A, *$\underline{/0°}$*; I4 = 1 A, *$\underline{/+90°}$*

Gain: 5.25 dB over a single identical vertical with a single radial

3-dB beamwidth: 95°

Design frequency: 3.8 MHz

Length of verticals: 20 m (2-mm-OD wire)

Length of radials: 19.5 m

Height of feed point / radials: approx. 5 m

The calculated feed-point impedances are:

Z1 = 83 + j57 Ω

Z2 = Z3 = 40 – j0 Ω

Z4 = 8 + j13 Ω

These data were calculated using *EZNEC* (*NEC*-2 engine) over very good ground.

5.4. T-Loaded Vertical Elements

If the central tower is not high enough to support full-size quarter-wave verticals from the sloping support wires, these verticals can be top-loaded by a sloping top-wire. The top-loading wires can be part of the support system, as shown in Fig 11-83. The vertical elements are loaded with sloping top-wires in order to show resonance at 3.8 MHz. The sloping support wires have the property of not producing any horizontally polarized signal, provided the lengths on both sides of the vertical are the same.

As long as the vertical wire is not shorter than $^2/_3$ full size (approximately 15 m), the loaded verticals will produce the same results as the full-size verticals, with only some reduction in bandwidth.

■ 6. ARRAYS OF SLOPING VERTICALS

In the chapter on dipoles, I describe the vertical half-wave dipole as well as the sloping half-wave dipole and its evolution into a quarter-wave vertical with one radial. Sloping verticals are well suited for making a 4-square array from using a single, tall tower as a support. In all these arrays the elements should be arranged in such a way that the feed points are located on a square measuring $^1/_4$ λ on the side.

6.1. Four-Square Array with Sloping $^1/_2$-λ Dipoles

The 4-square array made of four slopers (sloping at 30° with respect to the support) requires a 36-m tower.

Array data

Feed currents: I1 = 1 A, *$\underline{/-90°}$*; I2 = 1 A, *$\underline{/0°}$*; I3 = 1 A, *$\underline{/0°}$*; I4 = 1 A, *$\underline{/90°}$*

Gain: 4.3 dB over a single sloping vertical

3-dB beamwidth: 108°

Design frequency: 3.65 MHz

Length of dipole: 40 m

The feed-point impedances calculated with a grounded 36-m support tower are:

Z1 = 205 + j172 Ω

Z2 = Z3 = 89 – j51 Ω

Z4 = 28 + j13 Ω

Fig 11-91 shows the radiation patterns. For comparison, the vertical pattern of a single element is included.

The array shows a fair amount of high-angle radiation, which is due to the horizontal radiation component originated by the sloping wires. The array can be fed with a Collins-type network, preferably designed for a nominal impedance of 75 Ω. The elements should be fed with 75-Ω quarter-wave-long feed lines to the hybrid network.

6.2. The K8UR Sloping-Dipoles Square Array

D. C. Mitchell, K8UR, described his 4-element sloping array (Ref 975). He uses half-wave sloping dipoles (sometimes called slopers) where the bottom half is sloped back toward the tower. This eliminates all the high-angle radiation, as the horizontal component is now canceled due to the folding of the elements.

Array data

Feed currents: I1 = 1 A, *$\underline{/-90°}$*; I2 = 1 A, *$\underline{/0°}$*; I3 = 1 A, *$\underline{/0°}$*; I4 = 1 A, *$\underline{/90°}$*

Gain: 4.0 dB over a single identical sloping vertical

3-dB beamwidth: 123°

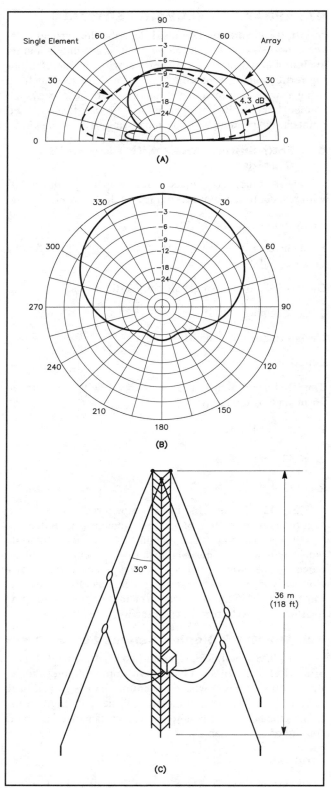

Fig 11-91—Configuration and radiation patterns of a 4-square array made of four ¹/₂-λ sloping verticals. For 80 meters a 35-m (118 ft) support is required. The antenna was modeled over very good ground. Note the high-angle radiation component. The vertical pattern for a single sloper is included for comparison.

Design frequency: 3.65 MHz

Length of sloping verticals 2×21.2 m

The feed-point impedances calculated with a grounded 36-m support tower are:

Z1 = 66 + j136 Ω

Z2 = Z3 = 90 − j2.5 Ω

Z4 = − 17 − j41 Ω

Fig 11-92 shows the radiation patterns. For comparison, the vertical pattern of a single element is included in B.

Because the elements are folded back toward the tower, the elements are very tightly coupled to the tower. There seems to be no influence on the radiation pattern. The feed-point impedance is much lower with the tower than without, however, which indicates heavy mutual coupling.

Mitchell uses the Collins-type feed system. In his design of the network, he has replaced the 180° phasing line with a hybrid-type network, taking care of the required 180° phase shift. These hybrid couplers are now commercialized by ComTek Systems.

Mike Greenway, K4PI, developed an innovative way for switching the dipoles of his K8UR-type array from the phone end of the band to the CW end of the band: see **Fig 11-93**.

6.3. The 4-Square Array with Sloping Quarter-Wave Verticals

Another variant I developed does not require such a high tower. I modeled the 4-square array using the sloping quarter-wave vertical with a single radial, as described in the chapter on dipoles (Section 6.3).

Because I had put up my full-size 40-meter vertical for 160 meters, I was looking for something better than a single vertical on 80, something I could support from the tall tower without too much coupling from the 160-meter antenna into the 80-meter one and vice versa. **Fig 11-94** shows the array that evolved. At 25 m height, two 8-m cross-arms are mounted in the tower. Each cross-arm tip (4 m from the tower) will support the top of the sloping vertical. The feed point of the vertical is 6 m above ground, and the four feed points are the corners of a square measuring 21.1 m (¹/₄ λ). This makes the feed point of each vertical approx. 14.5 m from the base of the tower. Each sloping vertical has a single sloping radial with the end connected to a stake 6 m above ground.

Array data

Feed currents: I1 = 1 A, /−90°; I2 = 1 A, /0°; I3 = 1 A, /0°; I4 = 1 A, /90°

Gain: 2.7 dB over a single sloping vertical

3-dB beamwidth: 112°

Design frequency: 3.65 MHz

Length sloping verticals and radials: 20.1 m

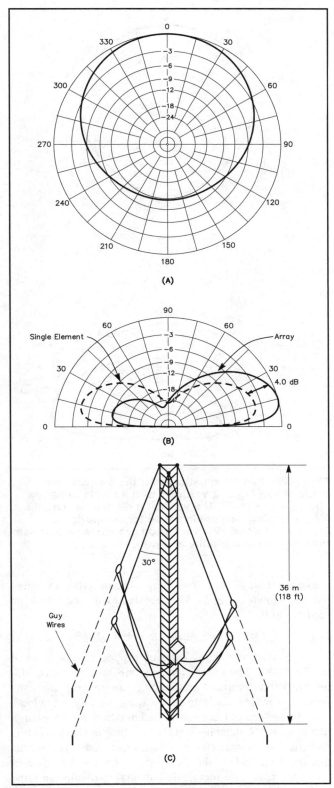

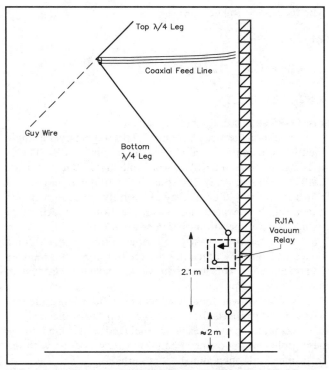

Fig 11-93—Method of switching a K8UR-style dipole from 3.8 to 3.5 MHz. The bottom-end of the dipole is lengthened by a piece of wire (which can be called capacitive loading) to lower its resonant frequency to 3.5 MHz. K4PI mentions that a 2.1-m long wire, spaced approx. 0.5 m from the tower does the job. The relay must be able to withstand the high voltage. Mike uses RJ1A vacuum relays for the job. Using this system, the wasted power in the dummy load is approx. 3 W (for 1500 W input) on both 3.8 as well as 3.5 MHz.

Fig 11-92—Configuration and radiation patterns of the 4-element K8UR array. The high-angle radiation component has been completely eliminated by folding the bottom half of the elements back to the tower. The vertical radiation pattern for a single element is shown for comparison.

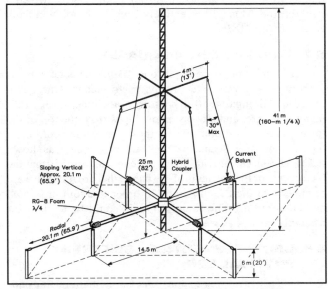

Fig 11-94—Four-square array made of four sloping verticals, each with one elevated radial. With arms extending 4 m outward from the tower, to support the top of the vertical wires, we have a slope angle of 30°. This is about the maximum slope angle we should tolerate.

The feed-point impedances are:

$$Z1 = 57 + j76 \ \Omega$$

$$Z2 = Z3 = 53.6 - j12 \ \Omega$$

$$Z4 = 10.2 - j22 \ \Omega$$

Fig 11-95 shows the radiation patterns.

Although the array has only 2.7 dB of gain over a single identical element, the gain over isotropic is 7.0 dBi. This is 0.2 dB better than the K8UR array, only 0.05 dB less than the half-wave sloper array described in Section 6.1, and 0.7 dB down from the 4-square array using fully vertical members and a single elevated radial (see Section 5.3). All these values are for operation over very good ground.

The array could easily be further optimized to exhibit a better F/B ratio by adjusting the feed-current phase angles and magnitudes. This would, however, make the feed system much more complicated.

The array was modeled including the 39-m grounded vertical that supports the array. The coupling of the array on the 160-meter band was also analyzed and found to be negligible (no impedance change). This was done with the feed lines both open and short circuited.

Using 2-mm OD (AWG 12) wire and a slope angle of 20°, the sloping wire length is 20.12 m to produce resonance at 3.79 MHz over a perfect mirror. The single elevated radial (at 6 m) that produces a resonant element on the same frequency, is 21.0 m long.

If space is available, the single radial can be replaced by two radials (in line). This will decrease the high-angle radiation, which is mainly coming from the single sloping radial (see Section 6.3 in the chapter on dipoles). If two radials (or 4, or 8) are used, install them two-by-two in a straight line. This will effectively cancel all radiation from each pair of radials.

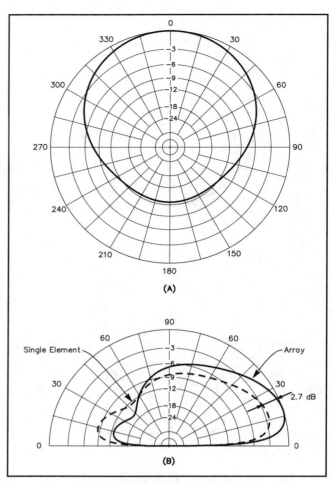

Fig 11-95—Radiation patterns for the 4-square array made of sloping $^1/_4$-λ verticals with a single radial, as shown in Fig 11-94. The vertical radiation pattern of a single sloping element is added for comparison. Modeling is done over very good ground, and includes the tower supporting the array.

■ 7. THREE-IN-ONE 4-SQUARES

It is quite possible to put one 4-square inside another without in any way affecting its performance. One could imagine building four 40-m tall towers as elements for a 160-meter 4-square, and slope support ropes from the top of the towers to a central point. From these ropes one could hang the elements of an 80 and even a 40-meter 4-square, as shown in **Fig 11-96**. The only thing we have to make sure is that the feed points of the verticals of the bands that are not in use are grounded (at their feed points) in order to avoid unwanted mutual coupling. The same principle could be applied for triangular arrays as well.

■ 8. USING ELEMENTS OF DIFFERENT PHYSICAL LENGTH

One practical way to construct a 3-element in-line end-fire array (see Section 4.8) would be to use a quarter-wave long tower in the center and slope long catenary ropes supporting the two outer elements. Let's work out the example of such an array with $^1/_8$-λ spacing between the

elements. Using three $^1/_4$-λ long elements, the excellent pattern, shown in Fig 11-53 is obtained with the following feed currents:

$$I1 = 1 \ A, \ \underline{/-149°}; \ I2 = 2 \ A, \ \underline{/0°}; \ I3 = 1 \ A, \ \underline{/146°}$$

A construction with a catenary support rope is sketched in **Fig 11-97**. The outer elements will be top-loaded elements, with the vertical section being approx. 60° long, the remainder will be made up eg, by a sloping T-type top load.

All the arrays I have described in this chapter are using identical length elements that are $^1/_4$-λ long or shorter. In this specific case we have elements of different sizes. This means that the magnitude of the element currents will be different (from the case with identical elements) to obtain the same pattern (F/B).

When confronted with such a design, it is best to model the array using a *MININEC*-based modeling program eg, *ELNEC*.

I have modeled an array with a central element measur-

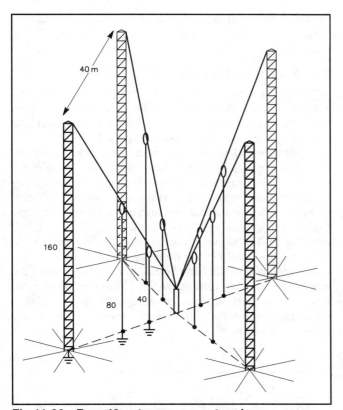

Fig 11-96—Four 40-m towers are set up in a 4-square array for 160 meters. From the top of these towers support ropes are run to the center of the square. These ropes support the elements of the 80 and the 40-meter 4-square arrays.

ing 38.5 m high (30 cm effective diameter), and two sloping T-wire loaded outer elements (#12 wires), each 27-m high, with an 8-m long sloping wire, which resonated these elements at 1.835 MHz.

The best F/B was obtained with the center element current being approximately 1.78 times the current in the outer elements. The feed-point impedances are very low:

$$Z1 = -12 + j11 \ \Omega$$

$$Z2 = 11 + j1 \ \Omega$$

$$Z3 = 5 - j13 \ \Omega$$

It's a good idea to use a very low-impedance feed line to minimize the SWR on the lines and keep losses under control. In addition we have to find a suitable ratio of feed impedances that will give the correct current magnitude ratio of 1.78/1.

The solution is to use three parallel 50-Ω feed lines to the center element (Z=16.66 Ω), and feed the outer elements with a parallel combination of a 50 Ω and a 75-Ω feed line (Z = 30 Ω). The current ratio will be 30/16.66 = 1.8. **Fig 11-98** shows the radiation pattern for both current ratios.

■ 9. RADIAL SYSTEMS FOR ARRAYS

9.1. Buried Radials

Radials of the elements of an array cross each other. It is standard procedure to install a bus (AWG 6 or 8 copper strip) half way between the elements and to connect the radials to this "bus." The radials can be any wire size, if many are used. The size will be dictated more by mechanical strength than current carrying capability. **Fig 11-99** shows

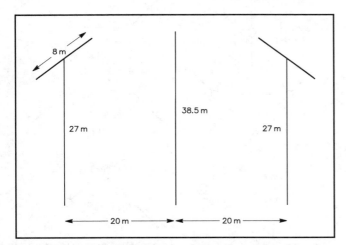

Fig 11-97—When elements of different physical length are used, the required current magnitude to achieve best F/B will be different from the case where all elements of the same length are used. A configuration is shown of a three-in-line end-fire array, where the center element is a tower, and where the outer elements are supported from catenary ropes. These elements are top-loaded by slant T-shaped wires.

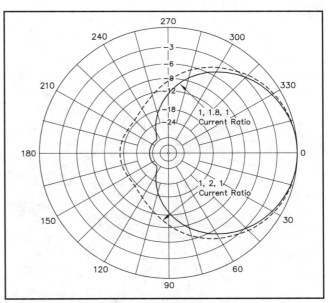

Fig 11-98—Horizontal radiation patterns for the array shown in Fig 11-97 with current ratios 1, 2, 1 and 1, 1.8, 1. The shorter the length of the outer elements, the lower the ratio for optimum directivity.

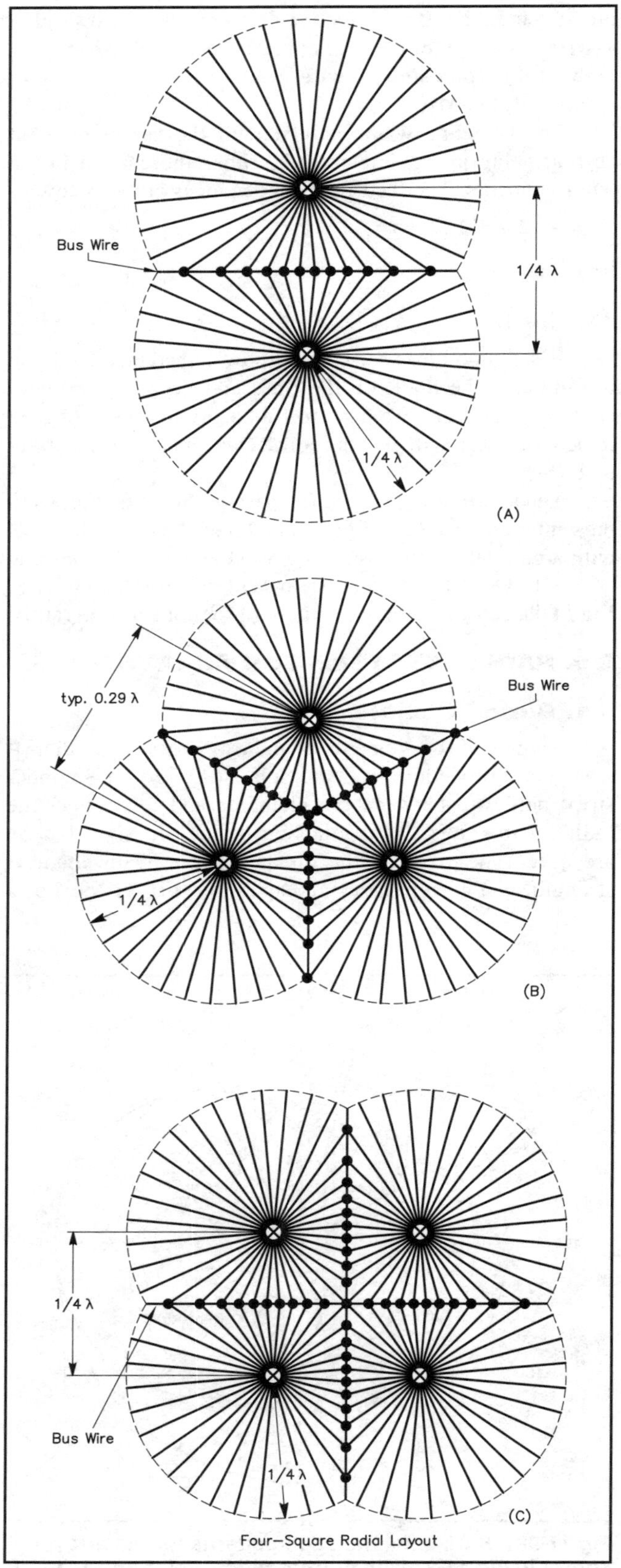

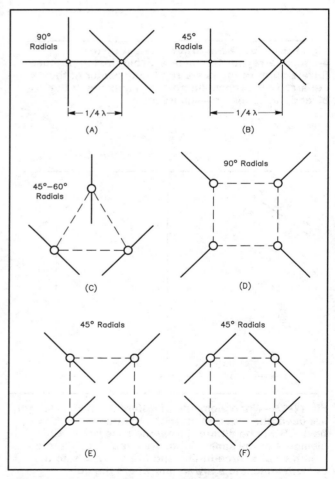

Fig 11-99—Recommended radial layouts for multi-element arrays.

various typical radial layouts eg, for a 2-element cardioid array, a 3-element triangular array and for the classic 4-square array.

9.2. Elevated Radials

When a high number of elevated radials are used on each of the elements of an array, these radials become non-resonant, and they can be connected to a bus system, in exactly the same way as shown in Fig 11-99.

When only a few radials are used (typically 1 to 4 radials), the situation is very different. In this case the radials, especially when ¼-λ long, can couple heavily with adjacent (especially parallel) radials from other elements and can upset the directivity of the array, and create uncontrolled and unwanted high-angle radiation from the elevated radials.

Fig 11-100 shows a few possible layouts that try to minimize the coupling. Short radials (45° to 60° long—see Chapter 9, Section 2.2.7 and Section 2.2.14) help reduce the mutual coupling between the radials. Note that the way the single (λ/4) radials are laid out in case (D), the coupling is effectively minimized. This is the layout I use on my 80-meter 4-square.

Fig 11-100—Recommended elevated radial layout schemes.

■ 10. CONCLUSION

Now that we have powerful modeling programs available (eg, *ELNEC*), I would like to encourage everyone to try to develop an array that fits his or her own requirements.

As far as array gain is concerned, always compare the array gain against an antenna made by a single element, modeled over the same ground. This will tell you exactly how much you will gain by going from a single vertical to an array. All other comparisons are pretty much meaningless. Make sure you use the same ground quality specifications in all circumstances. If you want to model the antenna for designing your own feed system, don't forget to include the equivalent loss resistance for the less than perfect ground radial system.

Let me warn the readers once again: antenna modeling is one thing; practical antenna design is a totally different thing. One can model complex arrays with staggering characteristics, but which may often be totally impossible to build. Don't forget you have to feed the array, by practical means, with feed lines. Feed lines have losses— losses that can be quite high when the SWR on the lines is high. The various computer programs, part of the NEW LOW BAND DXING software make it possible to calculate practical feed systems, with real feed lines, not with the ideal lossless lines.

Especially watch out for very low or very high impedances. These usually make it very difficult to achieve in real life what may seem like a very good design on a computer screen!

Modeling an antenna is one thing; building it and making it work as it says on paper (or the computer screen) is another thing. If you have to compromise a little in order to be able to use a much easier-to-make feed-system (eg, quadrature fed versus the exotic phase angles or current magnitude), I would advise the compromise, unless you have the required measuring setup.

Finally, measure. If you use a quadrature-fed array, there are some very simple tools available that will tell you exactly how well the array has been fed, and how well it will work. If you choose an array with more exotic current angles, try to obtain a quality oscilloscope or a vector voltmeter. I guarantee the results will be well worth the effort.

OTHER ARRAYS

12 OTHER ARRAYS

Chapter 11 on phased arrays only covered arrays made of vertical (omnidirectional) radiators. You can of course design phased arrays using elements that, by themselves, already exhibit some horizontal directivity, eg, horizontal dipoles.

Even at relatively low heights (0.3 λ), arrays made of horizontal elements (dipoles) can be quite attractive. Their intrinsic radiation angle is certainly higher than for an array made of vertical elements, but unless the electrical quality of the ground is good to excellent, the horizontal array may actually outperform the vertical array even at low angles.

The vertical radiation angle (wave angle) of arrays made with vertical elements (typical 1/4-λ long elements) depends only on the quality of the ground in the Fresnel zone. Radiation angles range typically from 15° to 25°.

The same is true for arrays made with horizontally polarized elements, but we have learned that reflection efficiency is better over bad ground with horizontal polarization than it is with vertical polarization (see Chapter 9, Section 1.1.2 and Chapter 8, Section 1.2.1.1).

The wave angle for antennas with horizontally polarized elements basically depends on the height of the antenna above ground. For low antennas (with resulting high wave angles), the quality of the ground right under the antenna (near field) will also play a role in determining the wave angle (see Chapter 8, Section 1.2.1.4). But as DXers, we are not interested in antennas producing wave angles that radiate almost at zenith.

Over good ground, a dipole at 1/4-λ height radiates its maximum energy at the zenith. Over average ground, the wave angle is 72°. The only way to drastically lower the radiation angle with an antenna at such low height is to add another element.

If we install a second dipole at close spacing (eg, 1/8 λ), and at the same height (1/4 λ), and feed this second dipole 180° out of phase with respect to the first dipole, we achieve two things:
• Approximately 2.5 dB of gain in a bidirectional pattern.
• A lowering of the wave angle from 72° to 37°!

At the zenith angle the radiation is a perfect null, whatever the quality of the ground is. This is because, at the zenith, the reflected wave from element no. 1 (reflected from the ground right under the antenna) will cancel the direct wave from element no. 2. The same applies to the reflected wave from element no. 1 and the direct 90° wave

from element no. 2. All the power that is subtracted from the high angles is now concentrated at lower angles. Of course there also is a narrowing of the horizontal forward lobe. Example: A 1/2-λ 80-m dipole at 25 m has a −3 dB forward-lobe beamwidth of 124° at a wave angle of 45°. The 2-element version, described above, has a −3 dB angle of 95° at the same 45° wave angle. The impedance of the two dipoles has dropped very significantly to approximately 8 Ω.

Fig 12-1 shows the vertical radiation angle (wave angle) for three types of antennas over average ground: a horizontal dipole, two half waves fed 180° out of phase (spaced 1/8 λ), and a 2-element Yagi. From this graph you can see that the only way to achieve a reasonably low radiation angle from a horizontally polarized antenna at low height (typically 1/3 λ or less) is to add a second element. The 180° out-of-phase element lowers the radiation angle at lower antenna heights (below 0.35 λ) significantly more than a Yagi or a 2-element all-fed array. It also has the distinct advantage to suppress all the high-angle radiation, which is not the case with the other types of arrays (Yagi or all-fed arrays).

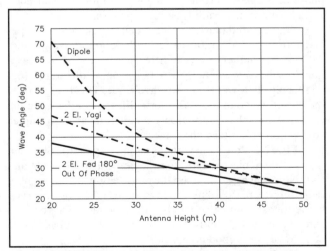

Fig 12-1—Vertical radiation angle (wave angle) for three types of antennas over average ground: a half-wave dipole, a 2-element parasitic array (Yagi) and two close-spaced half-wave dipoles fed 180° out of phase. Note the remarkable superiority of the last antenna at low heights. The graph is applicable for 80 meters.

■ 1. TWO-ELEMENT ARRAY SPACED ¹/₈ λ, FED 180° OUT OF PHASE

The vertical and the horizontal radiation patterns of the 2-element array are shown in **Fig 12-2**. As the antenna elements are fed with a 180° phase difference, the feeding is simple. The impedances at both elements are identical. **Fig 12-3** gives the feed-point impedance of the elements as a function of the spacing between the elements and the height. Within the shown limits, spacing has no influence on the gain or the directivity pattern. Very close spacings give very low impedances, which makes feeding more complicated and increases losses in the system. A minimum spacing of 0.15 λ is recommended.

Compared to a single dipole at the same height, this antenna has a gain of 3.5 dB at its main wave angle of 37°, and of 4.5 dB at a wave angle of 25° (see **Fig 12-4**).

Feeding the array is done by running a ¹/₄-λ feed line to one element, and a ³/₄-λ feed line to the other element. The feed point at the junction of the two feed lines is approximately 100 Ω for an element spacing of 0.125 λ. A ¹/₄-λ long 75-Ω cable will provide a perfect match to a 50-Ω feed line.

You will have a 5:1 SWR on the two feed lines, so be careful when running high power! Another feed solution that may be more appropriate for high power is to run two parallel 50-Ω feed lines to each element, giving a feed line impedance of 25 Ω. In this case the SWR will be a more acceptable 2.2:1 on the line. At the end of the feed lines (¹/₄ λ and ³/₄ λ) the impedances will be 54 Ω. The parallel combination will be 27 Ω, which can be matched to a 50-Ω line through a quarter-wave transformer of 37.5 Ω (two parallel 75-Ω cables) or via a suitable L network.

■ 2. UNIDIRECTIONAL 2-ELEMENT HORIZONTAL ARRAY

Starting from the above array, we can now alter the phase of the feed current to change the bidirectional horizontal pattern into a unidirectional pattern.

The required phase to obtain beneficial gain and especially front-to-back ratio varies with height above ground. At ¹/₂ λ and higher, a phase difference of 135° produces a good result. At lower heights, a larger phase difference (eg, 155°) helps to lower the main wave angle. This is logical, as the closer we go to the 180° phase difference, the more the effect of the phase radiation cancellation at high angles comes into effect (see above).

Fig 12-5 shows the vertical radiation patterns obtained with different phase angles for a 2-element array at ¹/₄ λ and ¹/₂ λ. Note that as we increase the phase angle, the high-angle radiation decreases, but the low angle F/B worsens. The higher phase angle also yields a little better gain. For both antenna heights (25 and 40 m or 0.3 and 0.4 λ), a phase angle of 145° seems a good compromise.

Feeding these arrays is not simple, as the feed-current phase angles are not in quadrature (phase angle differences in steps of 90°). For a discussion of feed methods see Chapter 11 on vertical arrays. Current forcing using a modified Lewallen feed system seems to be the best choice.

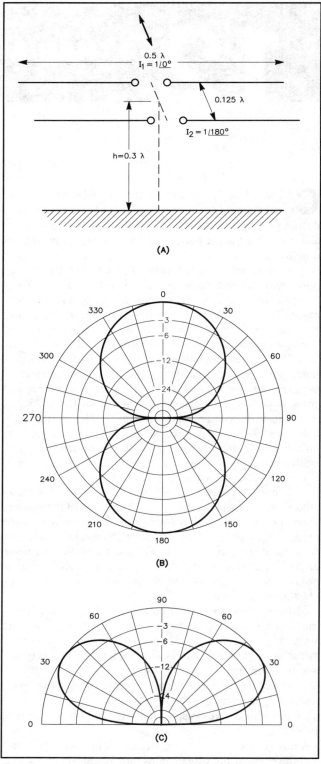

Fig 12-2—Configuration and radiation patterns of two close-spaced half-wave dipoles fed 180° out of phase, at a height of 0.3 λ above average ground. The azimuth pattern at B is taken for a wave angle of 36°. Note in the elevation pattern at C that all radiation at the zenith angle is effectively canceled (see text for details).

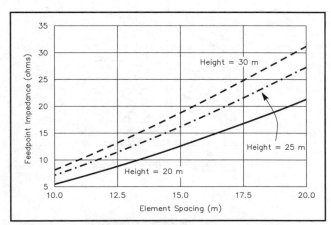

Fig 12-3—Feed-point impedance of the 2-element close-spaced array with elements fed 180° out of phase, as a function of spacing between the elements and height above ground. The design frequency is 3.75 MHz.

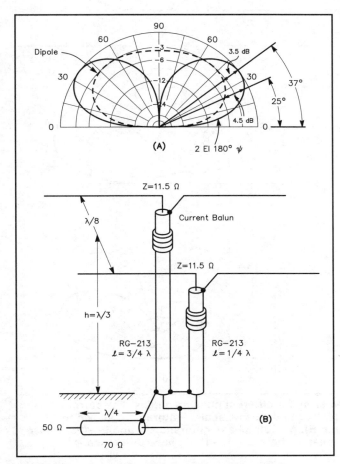

Fig 12-4—At A, vertical radiation pattern of the 2-element close-spaced array as compared to a single dipole at the same height of 0.3 λ (25 m for 3.8 MHz). The feed method for a spacing of ¹/₈ λ is shown at B. The feed-point impedance is approximately 100 Ω at the junction of the ¹/₄ and the ³/₄-λ 50-Ω feed lines. A ¹/₄-λ long 70-Ω feed line can be used to provide a perfect match to a 50-Ω feed line.

The question that comes to mind is, "Can we obtain similar gain and directivity with a parasitic array?" Let's see.

■ 3. TWO-ELEMENT PARASITIC ARRAY

Our modeling tools teach us that we can indeed obtain exactly the same results with a parasitic array. A 2-element director-type array produces the same gain and a front-to-back ratio that is even slightly superior.

As a practical 2-element parasitic-type wire array, I have developed a Yagi with 2 inverted-V-dipole elements. **Fig 12-6** shows the configuration as well as the radiation patterns obtained at a height of 25 m (0.3 λ). In order to make the array easily switchable, both wire elements are made equally long (39.94 m for a design frequency of 3.8 MHz). The inverted-V-dipole apex angle is 90°. A 25-meter high support (mast or tower) is required. At that height we need to install a 10-m long horizontal support (boom), from the end of which we can hang the inverted-V dipoles. The gain is 3.9 dB versus an inverted-V dipole at the same height, measured at the main wave angle of 45°.

A loading capacitor with a reactance of $-j\,60\;\Omega$ produces the right current phase in the director. The radiation resistance of the array is 24 Ω. In order to make the array easily switchable, we run two feed lines of equal length to the elements.

From here on there are two possibilities:
• We use a length of coax feed line to provide the required reactance of $-j65\;\Omega$ at the element.
• We use a variable capacitor at the end of a ¹/₂-λ feed line. The theoretical value of the capacitor is:

$$\frac{10^6}{2\pi \times 3.8 \times 65} = 644 \text{ pF}$$

Now we calculate the length of the open feed line that exhibits a capacitance of 644 pF on 3.8 MHz. The reactance at the end of an open feed line is given by:

$$X = -Z_C \times \tan(90 - L)$$

where
  Z_C = characteristic impedance of the line
  L = length of the line in degrees
  This can be rewritten as

$$L = 90 - \arctan\frac{X}{Z_C}$$

In our case we need X = –60 Ω. Thus,

$$L = 90 - \arctan\frac{60}{50} = 39.8°$$

The physical length of this line is given by

$$L_{meters} = \frac{833 \times Vf \times \ell}{1000 \times F_q}$$

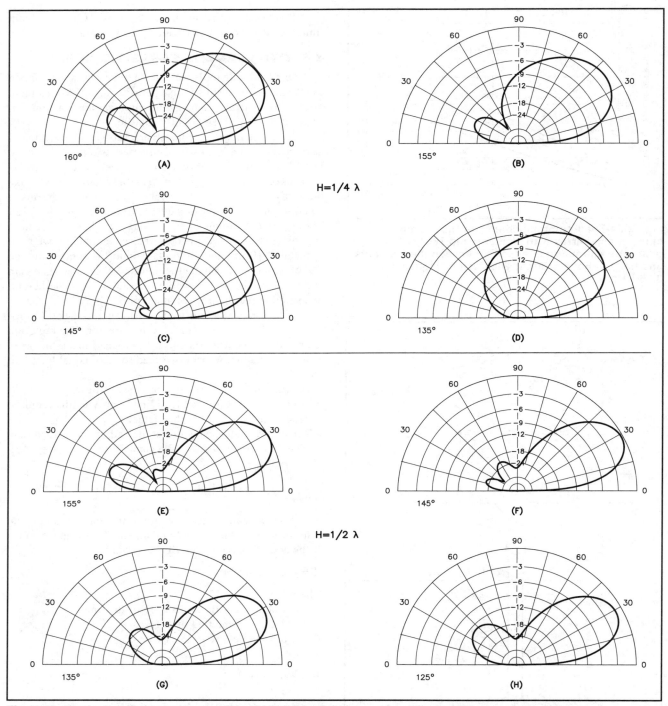

Fig 12-5—Vertical radiation patterns of the 2-element all-fed array for different phasing angles. The current magnitude is the same for both elements. All patterns are plotted to the same scale. Patterns are shown for antenna heights of ¼ λ (at A through D and ½ λ (at E through H). A—160° phase difference. B and E—155° phase difference. C and F—145° phase difference. D and G—135° phase difference. H—125° phase difference.

where
    Vf = velocity factor (0.66 for RG-213)
    F_q = design frequency
    ℓ = length in degrees

$$L_{meters} = \frac{833 \times 0.66 \times 39.8}{1000 \times 3.8} = 5.76 \text{ m}$$

Fig 12-7 shows the feed and switching arrangements according to the two above-mentioned systems.

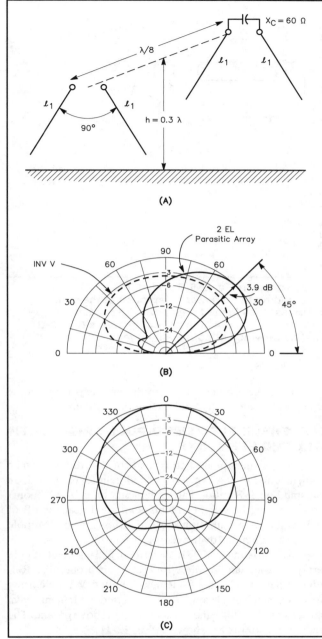

Fig 12-6—Configuration and calculated radiation patterns for the 2-element parasitic array using inverted-V-dipole elements. The array is installed with an apex angle of 90°, at a height of 0.3 λ (25 m for 3.8 MHz). Element spacing is ¹/₈ λ. The vertical pattern of a single inverted-V dipole is included at B for comparison. At C the azimuth pattern is shown for a wave angle of 45°. The gain at the main wave angle (45°) is 3.9 dB over the single inverted-V dipole.

■ 4. TWO-ELEMENT DELTA-LOOP ARRAY

Using the same support as described above (a 10-m long boom at 25 m), we can also design a 2-element delta-loop configuration. If the ground conductivity is excellent,

and if we can install radials (a ground screen), the 2-element delta-loop array should provide a lower angle of radiation and comparable gain as compared to the 2-element inverted-V-dipole array as described in Section 3.

4.1. Two-Element Delta Loop with Sloping Elements

As the low-impedance feed point of the vertically polarized delta loop is quite a distance from the apex, and as most of the radiation comes from the high-current areas of the antenna, we can consider using delta-loop elements that are sloping away from the tower. We could not do this with the inverted-V 2-element array, as the high-current points are right at the apex.

In our example I have provided a boom of 6 m length at the top of our support (25 m). From the tips of the boom we slope the two triangles so that the base lines are now 8 m away from the support and approximately 2.5 m over the ground.

Fig 12-8 shows the radiation pattern obtained with the array when the loops are fed with equal current magnitude and with a phase difference of 120°. Note the tremendous F/B at low angles (more than 45 dB!). Gain over a single-element loop is 3.5 dB. The wave angle is 18° over a very good ground. One of the problems is, of course, the feed system for an array that is not fed in quadrature.

Fig 12-9 shows the radiation patterns for the 2-element array with a parasitic reflector. The gain is the same as for the all-fed array and 3.4 dB over a single delta-loop element. The parasitic array shows a little less F/B at low angles, as compared to the all-fed array (see Fig 12-8), but the difference is slight.

As with the 2-element dipole array, my personal preference goes to the parasitic array, as the all-fed array is not fed in quadrature, which means that the feed arrangement is all but simple (it requires a modified Lewallen feed system). The obvious feed method for the 2-element parasitic array uses two equal-length feed lines to a common point mid-way between the two loops. A small support can house the switching and matching hardware.

As with the 2-element inverted-V array, we use two loops of identical length, and use a length of shorted feed line to provide the required inductive loading with the reflector element. The length of the feed line required to achieve the required 140-Ω inductive reactance is calculated as follows:

$$X_L = Z_C \times \tan \ell$$

where

X_L = required inductance
Z_C = cable impedance
ℓ = cable length in degrees

This can be rewritten as

$$\ell = \arctan \frac{X_L}{X_C}$$

or

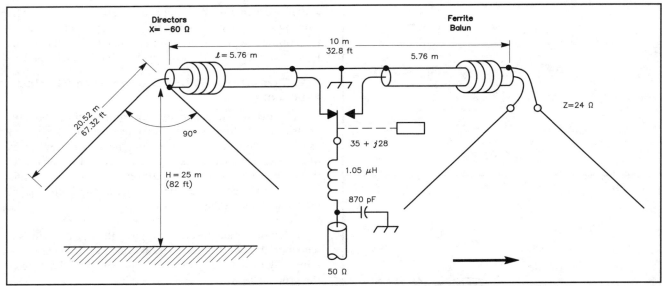

Fig 12-7—Feeding arrangement for the 2-element parasitic array shown in Fig 12-6. Two lengths of RG-213 run to a switch box in the center of the array. The coax feeding the director is left open at the end, producing a reactance of −j 65 Ω (equivalent of 644 pF at 3.8 MHz) at the element feed point. The radiation resistance of the 2-element array is 29 Ω. An L network can be provided to obtain a perfect match to the 50-Ω feed line. A current type balun (eg, stack of ferrite beads) *must* be provided at both element feed points.

$$\ell = \arctan \frac{140}{75} = 61.8°$$

The physical length is given by

$$L_{meters} = \frac{833 \times Vf \times \ell}{1000 \times F_q}$$

where

L_{meters} = length, meters
ℓ = length in degrees
Vf = velocity factor of the cable
F_q = design frequency, MHz

We use foam-type RG-11 (Vf = 0.81), because solid PE-type coax (Vf = 0.66) will be too short to reach the switch box.

$$L_{meters} = \frac{833 \times 0.81 \times 61.8}{1000 \times 3.8} = 10.98 \text{ m}$$

Fig 12-10 shows the feed line and the switching arrangement for the array. Note that the cable going to the reflector must be short-circuited. The two coaxial feed lines must be equipped with current-type baluns (a stack of ferrite beads).

The impedance of the array varies between 75 Ω and 150 Ω, depending on the ground quality. If necessary, the impedance can easily be matched to the 50-Ω feed line using a small L network. This array can be made switchable from the SSB end of the band to the CW end by applying the capacitive loading technique as described in Chapter 10.

Since this array was published in the second edition of this book, I have received numerous comments from people who have successfully constructed this antenna.

■ 5. THREE-ELEMENT DIPOLE ARRAY WITH ALL-FED ELEMENTS

A 3-element phased array made of $\frac{1}{2}$-λ dipoles can be dimensioned to achieve a very good gain together with an outstanding F/B ratio. Three elements on a $\frac{1}{4}$-λ "boom" ($\frac{1}{8}$-λ spacing between elements) can yield nearly 6 dB of gain at the major radiation angle of 38° over a single dipole at the same height (over average ground).

A. Christman, KB8I, described a 3-element dipole array with outstanding directional and gain properties. (Ref. 963.) I have modeled a 3-element inverted-V-dipole array using the same phase angles. The inverted-V elements have an apex angle of 90°, and the apex at 25 m above ground. The radiation patterns are shown in **Fig 12-11**.

The elements are fed with the following currents:

I1 = 1 $\underline{/-149°}$ A
I2 = 1 $\underline{/0°}$ A
I3 = 1 $\underline{/146°}$ A

With the antenna at 25 m above ground and elements that are 39.72-m long (F_{design} = 3.8 MHz), the element feed-point impedances are

Z1 = −36 + j24.5 Ω
Z2 = 12.3 + j25 Ω
Z3 = 7.6 − j12.2 Ω

If you are confused with the minus sign in front of the *real* part of the impedance, it just means that in this array, element no. 1 is actually *delivering* power into the feed system, rather than taking power from it. This is a very common situation with

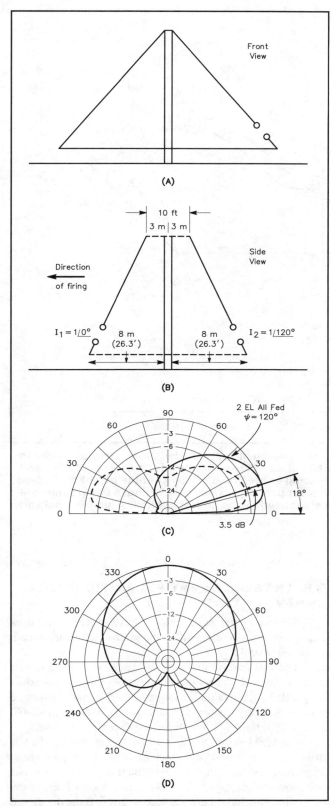

Fig 12-8—Configuration and radiation patterns of a 2-element delta-loop array, using sloping elements. The elements are fed with equal-magnitude current and with a phase difference of 120°. The horizontal pattern at D is taken for a wave angle of 18°.

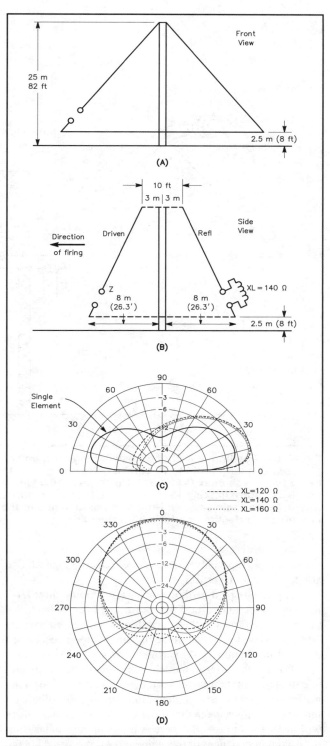

Fig 12-9—Radiation patterns for the 2-element delta-loop array having the same physical dimensions as the all-fed array of Fig 12-8, but with one element tuned as a reflector. In practice both triangles are made equally long, and the required loading inductance is inserted to achieve the required phase angle. Patterns shown are for different values of loading coils (X_C =120, 140 and 160 Ω). The feed-point impedance of the array will vary between 80 and 150 Ω, depending on the ground quality.

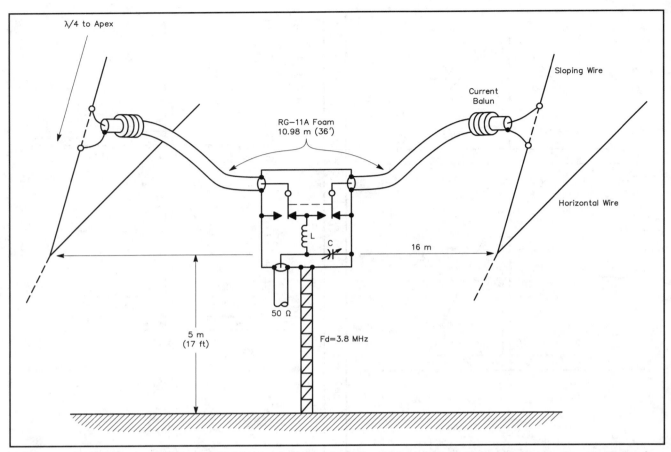

Fig 12-10—Feeding and direction-switching arrangement for the 2-element parasitic delta-loop array as shown in Fig 12-9. The length of the 75-Ω feed lines going from the feed points to the switch box is 61.8°. For 3.8 MHz, and using foam-type coax (Vf = 0.81), this equals 10.98 m. The spacing between the elements at the height of the feed points is approximately 15 m. The switch box is mounted on a support approximately 5 m above the ground, half way between the elements. Note that the feed line to the reflector needs to be short-circuited. A simple L network provides a perfect match for a 50-Ω feed line.

driven arrays, especially where "close" spacing is used. See also the chapter on vertical arrays.

A possible feed method consists of running three $^1/_4$-λ lines to a common point. Current forcing is employed: We use 50-Ω feed lines to the outer elements, and two parallel 50-Ω lines to the central element. The method is described in detail in Chapter 11 on vertical arrays.

It is much easier to model such a wonderful array and to calculate a matching network than to build and align the matching system. Slight deviations from the calculated impedance values mean that the network component values will be different as well. There is no method of measuring the driven impedances of the elements. All you can do in the way of measuring is use an HF vector voltmeter and measure the voltages at the end of the three feed lines. The voltage magnitudes should be identical, and the phase as indicated above (E1, E2 and E3). If they are not, the values of the networks can be tweaked in order to obtain the required phase angles. Good luck!

We have seen that we can just about match the performance of a 2-element all-fed array with a parasitic array. We will see that the same can be done with a 3-element array.

■ 6. THREE-ELEMENT PARASITIC DIPOLE ARRAY

The model that was developed has a gain of 4.5 dB over a single inverted V-element (at the same height) for its main wave angle of 43°.

The F/B ratio is just over 20 dB, as compared to just over 30 dB with the all-driven array. At the same antenna height (0.3 λ), the radiation angle of the 3-element parasitic was also slightly higher (43°) than for the 3-element all-fed array (38°), modeled over the same (average) ground.

Fig 12-11 shows the superimposed patterns for the all-driven and the parasitic 3-element array (for 80 meters at 25 m height). Note that the 3-element all-fed has a better rejection at high angles. This is because the currents in the outer elements have a greater phase shift (versus the driven element) than in the parasitic array. These phase shifts are

Reflector:
All-driven array: −149°
Parasitic array: −147°

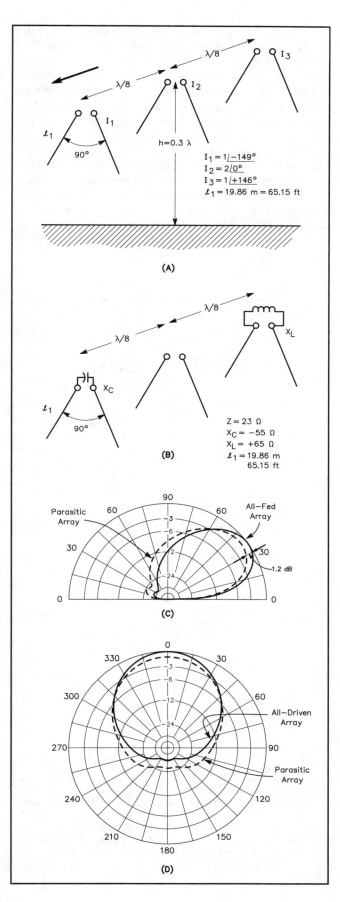

(A)

$\lambda/8$
I_3
$\lambda/8$
I_2
I_1
ℓ_1
$90°$
$h=0.3 \lambda$

$I_1 = 1\underline{/-149°}$
$I_2 = 2\underline{/0°}$
$I_3 = 1\underline{/+146°}$
$\ell_1 = 19.86\ m = 65.15\ ft$

(B)

$\lambda/8$
X_L
$\lambda/8$
X_C
ℓ_1
$90°$

$Z = 23\ \Omega$
$X_C = -55\ \Omega$
$X_L = +65\ \Omega$
$\ell_1 = 19.86\ m$
$65.15\ ft$

(C)

Parasitic Array — All-Fed Array — 1.2 dB

(D)

All-Driven Array — Parasitic Array

Director:
    All-driven array: +147°
    Parasitic array: +105°

This demonstrates again that, with an all-driven array, we have more control over all the parameters that determine the radiation pattern of the array.

Like the 2-element array described in Section 3, the 3-element array is also made using three elements identical in length. The required element reactances for the director and reflector are obtained by inserting the required inductance or capacitance in the center of the element. In practice we bring a feed line to the outer elements as well. The feed lines are used as stubs, which represent the required loading to turn the elements into a reflector or director.

The question is, which is the most appropriate type of feed line for the job, and what should be its impedance. **Table 12-1** shows the stub lengths obtained with various types of feed lines. The length of the open-ended stub serving to produce a negative reactance (for use as a director stub) is given by

$$\ell° = 90 - \arctan \frac{X_C}{Z_C}$$

For the short-circuited stub serving to produce a positive reactance (for the reflector), the formula is:

$$\ell° = \arctan \frac{X_L}{Z_C}$$

Table 12-1

Required Line Length for the Loading Stubs of the Parasitic Version of the 3-Element Array of Fig 12-11

Z_C, Ohms	VF	Length, Degrees	Length, Meters	Length, Feet
Director				
50	0.66	42.3	6.12	20.08
75	0.66	53.75	7.77	25.49
100	0.95	83.03	8.85	29.04
450	0.95	83.03	17.28	56.7
Reflector				
50	0.66	52.53	7.58	24.87
75	0.66	40.91	5.91	13.39
100	0.66	33.02	4.77	15.65
450	0.95	8.22	1.71	5.61

Other data:
Design frequency = 3.8 MHz, wavelength = 78.89 m
Director $X_C = -55$ ohms
Reflector $X_L = +65$ ohms

Fig 12-11—Configuration and radiation patterns for two types of 3-element inverted-V-dipole arrays with the apex at 0.3 λ. At both C and D, one pattern is for the all-fed array and the other for an array with a parasitic reflector and director. The all-fed array outperforms the Yagi-type array by approximately 1 dB in gain as well as 10 dB in F/B ratio.

- From Table 12-1 we learn the 450-Ω stub requires a very long length to produce the required negative reactance for the director (17.28 m).
- When made from 50-Ω or 75-Ω coax, we obtain attractive short lengths. The disadvantage is that you need to put a current balun at the end of the stubs to keep any current from flowing on the outside of the coax shield.
- A third solution is to use a 100-Ω shielded balanced line, made of two 50-Ω coax cables. The lengths are still very attractive, and you no longer require the current balun.
- A final solution is to use the 450-Ω transmission line for the reflector (1.71 m long) and to load the line with an extra capacitor to turn it into a capacitor. I assumed a velocity factor of 0.95 for the transmission line. You must check this in all cases (see Chapter 11 on vertical arrays). The capacitive reactance produced by an open-circuited line of 1.71 m length at 3.8 MHz is

$$X_L = 450 \tan (90° - 8.22°) = +j\, 3115\ \Omega$$

This represents a capacitance value of only

$$\frac{10^6}{2\pi \times 3.8 \times 3115} = 13.4\ \text{pF}$$

The required capacitive reactance was $-j\, 55\ \Omega$, which represents a capacitance value of

$$\frac{10^6}{2\pi \times 3.8 \times 95} = 762\ \text{pF}$$

This means we need to connect a capacitor with a value of 762 – 13.4 = 750 pF across the end of the open stub. This last solution seems to be the most flexible one. A parallel connection of two transmitting-type ceramic capacitors, 500 pF and 250 pF, will do the job perfectly. If you want even more flexibility you can use a 500-pF motor-driven variable in parallel with a 500-pF fixed capacitor. This will allow you to tune the array for best F/B.

The practical arrangement is shown in **Fig 12-12**. From each outer element we run a 1.71-m long piece of 450-Ω line to a small box which is mounted on the boom. The box can also be mounted right at the center of the inverted-V element, whereby the 1.71-m transmission line is shaped in a large 1-turn loop. The box houses a small relay which either shorts the stub (reflector) or opens, leaving the 750-pF capacitor across the line.

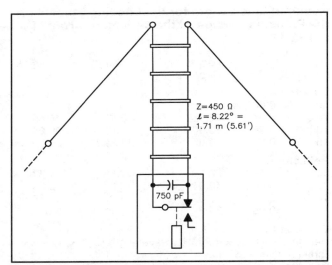

Fig 12-12—The 3-element parasitic type inverted-V dipole array is made with elements that have exactly the same length. The required element loading is obtained by inserting the required capacitance or inductance in the center of these elements. This is obtained by using stubs, as shown here. With a 450-Ω transmission line we require only a short 1.71-m long piece of short-circuited line to make a stub for the reflector. For the director we connect a 750-pF capacitor across the end of the open-circuit line. This can be switched with a single-pole relay, as explained in the text.

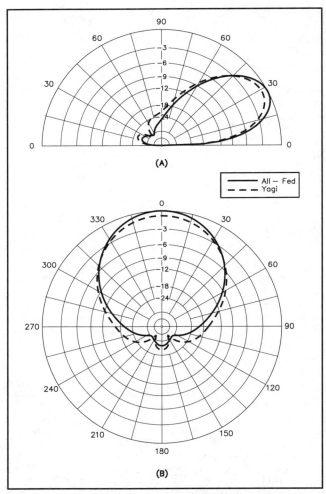

Fig 12-13—Radiation patterns of the 3-element inverted-V type array at a height of ½ λ. Note that the all-fed array still outperforms the Yagi-type array, but with a smaller margin than at a height of 0.3 λ (Fig 12-11). In order to produce an optimum radiation pattern, the values of the loading impedances were different than those for a height of 0.3 λ. See text for details.

Is the relative "inferiority" of the parasitic array due to the low height? In order to find out I modeled the same antennas at $1/2 \lambda$ height. **Fig 12-13** shows the vertical and the horizontal radiation patterns for the all-driven and parasitic-array versions of the 3-element inverted-V array at this height. Note that the all-driven array still has 0.9 dB better gain than the parasitic array. The F/B is still a little better as well, although the difference is less pronounced than at lower height. The optimum pattern was obtained when loading the director with a -50-Ω impedance and the reflector with a $+30$-Ω impedance. The gain of the all-fed array is 5.7 dB versus a dipole at the same height (at 28° wave angle). For the 3-element parasitic array, the gain is 4.8 dB versus the dipole at its main wave angle of 29°.

In looking at the vertical radiation pattern it is remarkable again that the all-driven array excels in F/B performance at high angles. Notice the "bulge" that is responsible for 5 to 10 dB less F/B in the 35°-50° wave-angle region.

It must be said that I did not try to further optimize the parasitic array by shifting the relative position of the elements. By doing this, further improvement could no doubt be made. This, of course, would make it impossible to switch directions, as the array would no longer be symmetrical.

Conclusion

All-fed arrays made of horizontal dipoles or inverted-V dipoles always outperform the parasitic-type equivalents in gain as well as F/B performance. As they are not fed in quadrature, it is elaborate or even "difficult" to feed them correctly.

The parasitic-type arrays lend themselves very well for remote tuning of the parasitic elements. Short stubs (open-ended to make a capacitor, or short-circuited to make an inductor) make ideal tuning systems for the parasitic elements. Switching from director to reflector can easily be done with a single-pole relay and a capacitor at the end of a short open-wire stub.

The same 3-element array made of fully horizontal (flat top) dipoles exhibits 1.0 dB more gain than the inverted-V version at the same apex height.

■ 7. DELTA LOOPS IN PHASE (COLLINEAR)

Two delta loops can be erected in the same plane and fed with in-phase currents to provide gain and directivity. In order to obtain maximum gain, the loops must be separated about $1/8 \lambda$, as shown in **Fig 12-14**. In this case the two loops, fed in phase exhibit a gain of almost 3.5 dB over a single loop! The array has a front-to-side directivity of at least 15 dB, not negligible. The impedance on a single loop is between 125 and 160 Ω. Each element can be fed via a 75-Ω $1/2$-λ feed line. At the point where they join the impedance will be 60 to 80 Ω. The radiation patterns and the configuration are shown in Fig 12-14.

This may be an interesting array if you happen to have two towers with the right separation and pointing in the right direction. As with all vertically polarized delta loops,

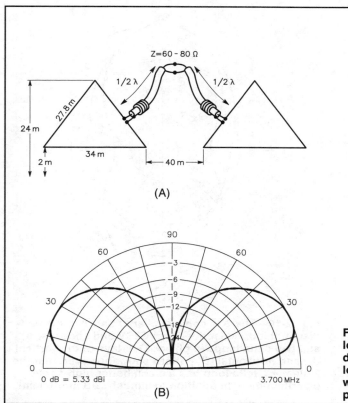

(A)

(B)

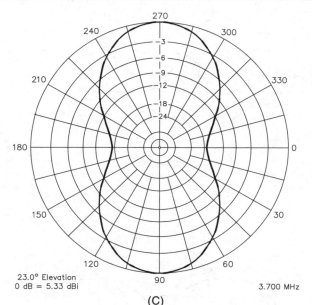

(C)

Fig 12-14—Configuration of the 2-element collinear delta-loop array with 10-m spacing between the tips of the deltas. This array has a gain of 3.0 dB over a single delta loop. The loops are fed $1/4 \lambda$ from the apex on the sloping wire in the center of the array (see text for details). The pattern at C is taken for a wave angle of 18°.

the ground quality is very important as to the efficiency and the low angle radiation of the array (see Chapter 10 on large loops).

Putting the loops closer together results in a spectacular drop in gain. loops with touching tips only exhibit approx. 1-dB gain over a single element, not worth the effort!

In one of his articles on elevated radials, John Belrose, VE2VC, mentioned the half-diamond loop, which has a significant resemblance to the delta loop (Ref 7824). I modeled this array and compared it to the 2-element delta loop shown in Fig 12-14. **Fig 12-15** shows both the horizontal and the vertical radiation pattern of both antennas in overlay. The 2-element delta has almost 0.7 dB more gain and has excellent high-angle rejection, while the half-diamond loop has some very strong high-angle response, which is of course due to the way the radials are laid out, resulting in zero high angle cancellation. The extra gain that was thought to be achieved by laying radials in one direction, is apparently more than wasted in high angle radiation. It seems that the two in-phase delta loops are still, by far, the best choice.

■ 8. ZL SPECIAL

The ZL Special, sometimes called the HB9CV, is a 2-element dipole array with the elements fed 135° out of phase. This configuration is described in Section 2. It is the equivalent of the vertical arrays described in Chapter 11, Section 4.2.

These well-known configurations make use of a specific feeding method. The feed points of the two elements are connected via an open-wire feed line that is crossed. The crossing introduces a 180° phase shift. The length of the line, with a spacing of $^1/_8$ λ between the elements, introduces an additional phase shift of approximately 45°. The net result is $180° + 45° = 225°$ phase shift, lagging. This is equivalent to $360° - 225° = 135°$ leading.

Different dimensions for this array have been printed in various publications. Correct dimensions for optimum performance will depend on the material used for the elements and the phasing lines. Jordan, WA6TKT, who designed the ZL Special entirely with 300-Ω twin lead (Ref 908), recommends that the director (driven element) be $447.3/f_{MHz}$ and the reflector be $475.7/f_{MHz}$, with an element spacing of approximately 0.12 λ.

Using air-spaced phasing line with a velocity factor of 0.97, the phasing-line length is $119.3/f_{MHz}$. This configuration of the ZL Special with practical dimensions for a design frequency of 3.8 MHz is given in **Fig 12-16**, along with radiation patterns. As it is rather unlikely that this antenna will be made rotatable on the low bands, I recommend the use of open-wire feeders to an antenna tuner. Alternatively, a coaxial feed line can be used via a balun.

■ 9. LAZY H

The Lazy-H antenna is an array that is often used by low-banders that have a bunch of tall towers, where they can support Lazy-Hs in between. **Fig 12-17** shows a typical

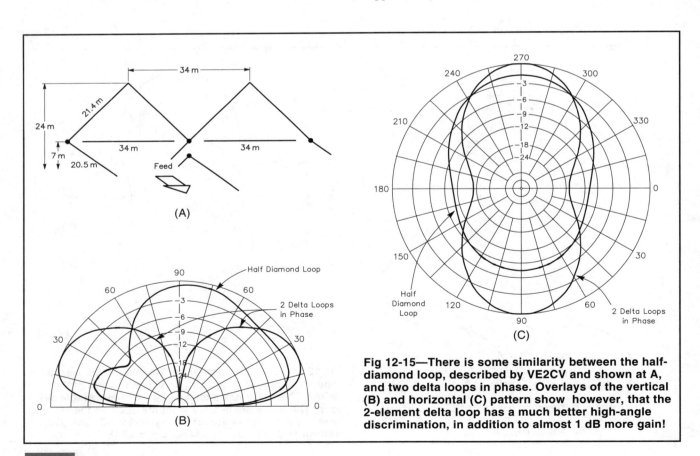

Fig 12-15—There is some similarity between the half-diamond loop, described by VE2CV and shown at A, and two delta loops in phase. Overlays of the vertical (B) and horizontal (C) pattern show however, that the 2-element delta loop has a much better high-angle discrimination, in addition to almost 1 dB more gain!

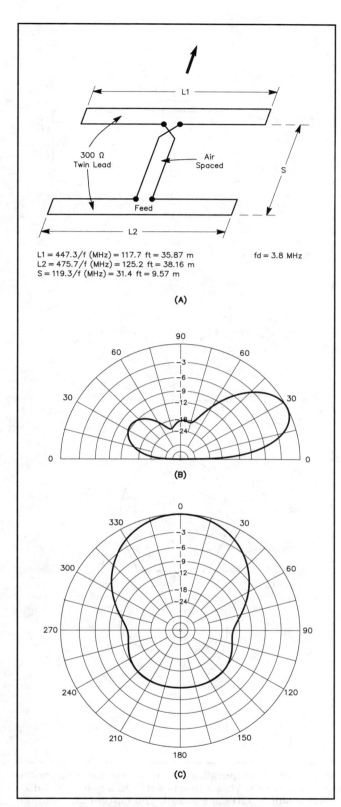

L1 = 447.3/f (MHz) = 117.7 ft = 35.87 m
L2 = 475.7/f (MHz) = 125.2 ft = 38.16 m
S = 119.3/f (MHz) = 31.4 ft = 9.57 m

fd = 3.8 MHz

(A)

(B)

(C)

Fig 12-16—The ZL Special (or HB9CV) antenna is a popular design that gives good gain and F/B for close spacing. Radiation patterns were calculated with *ELNEC* for the dimensions shown at A, for a height of ¹/₂ λ above average ground. The horizontal pattern at C is taken for a wave angle of 27°.

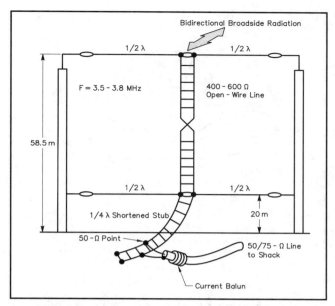

Fig 12-17—Typical Lazy-H configuration for 80 meters. The same array can obviously be made for 40 meters with all dimensions halved.

Lazy-H layout for use on 80 meters. Such a 4-element Lazy-H has a very respectable gain of approx. 11 dBi over average ground, as shown in **Fig. 12-18**. Its gain at 20° wave angle is nearly 4 dB above a flat-top dipole at the same height, and 1.7 dB over a collinear (two ¹/₂ waves) at the same height. The outstanding feature of the Lazy-H is however, that the 90° (zenith) radiation, which is very dominant with the dipole and the collinear is almost totally suppressed, which makes it a good DX-listening antenna as well!

The easiest way to feed the array is shown in Fig 12-18. A ¹/₄-λ open-wire line, shorted at its end, is probed to find the low-impedance point (50 or 75 Ω). Fine adjustment of the length of the line and the position of the tap make it possible to find a perfect resistive 50 or 75-Ω point. One of the popular antenna analyzers is a valuable tool to find the exact match. The same antenna can be used for both ends of the 80-meter band, all that is required is a different set of values for the length of the ¹/₄-λ stub and the position of the tap. This can be achieved with some rather simple relay switching.

■ **10. BOBTAIL CURTAIN**

The bobtail curtain consists of three phased ¹/₄-λ verticals, spaced ¹/₂ λ apart, where the center element is fed at the base, while the outer elements are fed via a horizontal wire section between the tips of the verticals. Through this feeding arrangement, the current magnitude in the outer verticals is half of the current in the center vertical. The current distribution in the top wire is such that all radiation from this horizontal section is effectively canceled. The configuration as well as the radiation patterns are shown in **Fig 12-19**.

The gain of this array over a single vertical is 4.4 dB. The –3 dB forward-lobe beamwidth is only 54°, which is quite

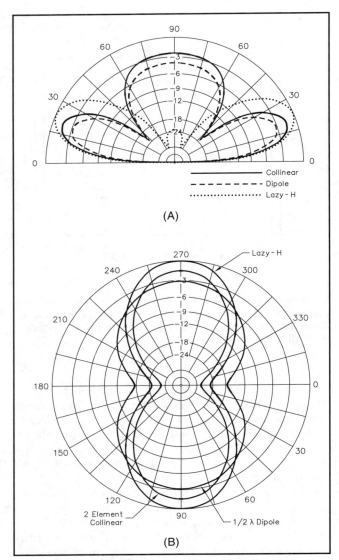

(A)

(B)

Fig 12-18—Vertical and horizontal radiation patterns of the 80-meter Lazy H shown in Fig 12-17 compared to the patterns of a flat-top dipole and a 2 × ¹/₂-λ collinear at the same height (over average ground).

narrow. This is because the radiation is bidirectional. K. Svensson, SM4CAN, who published an interesting little booklet on the bobtail array, recommends the following formulas for calculating the lengths of the elements of the array.

Vertical radiators: $\ell = 68.63/F$

Horizontal wire: $\ell = 143.82/F$

where

 F = design frequency

 ℓ = length, meters

  The antenna feed-point impedance is high (several thousand ohms). The array can be fed as shown in **Fig 12-20**. This is the same feed arrangement as for the voltage-fed T antenna, described in Chapter 9 on vertical antennas. In order to make the bobtail antenna cover both the CW as well

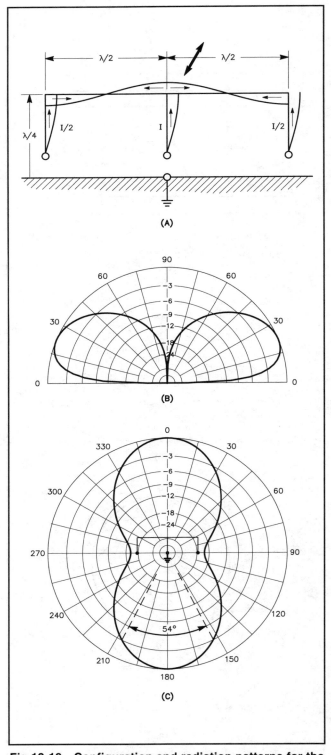

Fig 12-19—Configuration and radiation patterns for the bobtail curtain. This antenna exhibits a gain of 4.4 dB over a single vertical element. The current distribution, shown at A, reveals how the three vertical elements contribute to the low-angle broadside bidirectional radiation of the array. The horizontal section acts as a phasing and feed line and has no influence on the broadside radiation of the array. The horizontal pattern at C is taken for a wave angle of 22°.

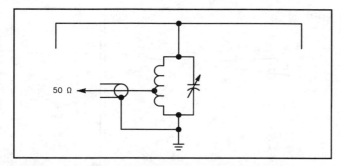

Fig 12-20—The bobtail curtain is fed at a high-impedance point. The antenna can best be fed via a parallel-tuned circuit, where the coax is tapped a few turns from the cold end of the coil. The array can be made to operate over a very large bandwidth by simply retuning the tuned circuit.

as the phone end of the band, it is sufficient to retune the parallel resonant circuit. This can be done by switching a little extra capacitor in parallel with the tuned circuit of the lower frequency, using a high-voltage relay.

The bottom ends of the three verticals are *very hot* with RF. You must take special precautions so that people and animals cannot touch the vertical conductors.

Do not be misled into thinking that the bobtail array does not require a good ground system because it is a voltage-fed antenna. As for all vertically polarized antennas, it is the electrical quality of the reflecting ground that will determine the efficiency and the low-angle radiation of the array

■ 11. HALF-SQUARE ANTENNA

The half-square antenna was first described by Vester, K3BC (Ref 1125). As its name implies, the half-square is half of a bi-square antenna (on its side), with the ground making up the other half of the antenna (see Chapter 10 on large loop antennas). It can also be seen as a bobtail with part of the antenna missing.

Fig 12-21 shows the antenna configuration and the radiation patterns. The feed-point impedance is very high (several thousand ohms), and the antenna is fed like the bobtail. The gain is somewhat less than 3.4 dB over a single $1/4$-λ vertical. The forward-lobe beamwidth is 68°, and the pattern is essentially bidirectional. There is some asymmetry in the pattern, which is caused by the asymmetry of the design: The current flowing in the two verticals is not identical. As far as the required ground system is concerned, the same remarks apply as for the bobtail antenna.

■ 12. A 14-ELEMENT YAGI FOR 80 METERS?

Both N7UA and K6UA use side-by-side Yagi arrays, made of two 3-element Yagis, and they are amongst the most successful 80-meter DXers on the West Coast. This is what inspired N7ML to make his giant 2 × 7-element Yagi array.

The 80-meter antenna, with probably the highest gain that's ever been erected and still is operational, was built by Mike Lamb, N7ML. **Fig 12-22** shows two, side-by-side

7-element Yagis (sloping at an angle of approx. 45°), which Mike hung from a catenary rope between two 54-m high towers, separated by 81 m. Note that these are *not* inverted-V dipoles; what may look like an inverted-V dipole are two collinear half-wave dipoles. The supporting towers are all guyed with (dielectric) Kevlar rope, to avoid any spurious

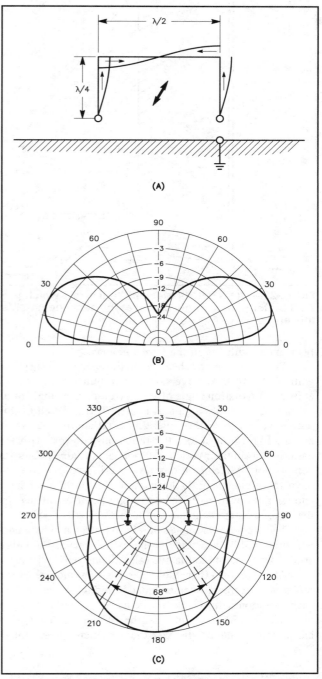

Fig 12-21—Configuration and radiation patterns of the half-square array, with a gain of 3.4 dB over a single vertical. The antenna pattern has a somewhat asymmetrical radiation pattern because the currents in the two vertical conductors are not identical. The azimuth pattern at C is for a wave angle of 22°.

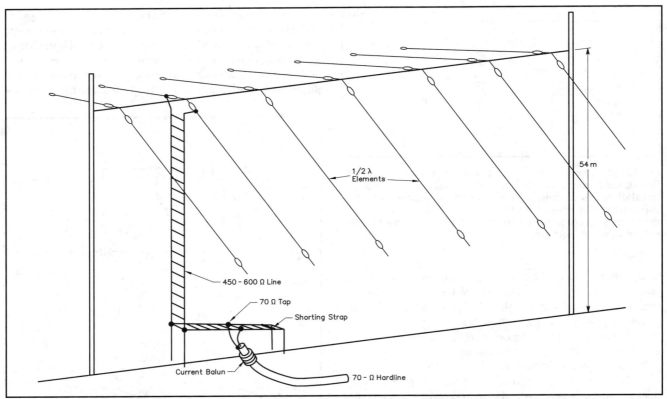

Fig 12-22—The N7ML 14-element Yagi array, which produces a very respectable gain of nearly 15 dBi, which is 7 dB better than a single dipole at the same height. It's not gain, but rather directivity that's the main virtue of this array.

resonances with any of the wire antennas.

The array, which is beamed on Europe, has a free space gain of 11 dBi, and an over-average ground gain of 14.9 dB at its main radiation angle of 20°. At a radiation angle of 5° the antenna still has a gain of nearly 7 dBi, and all of this with a worst case F/B of over 25 dB (at approx. 32° wave angle). **Fig 12-23** shows the radiation patterns, with the pattern of a single inverted-V dipole at the same height for comparison. Mike mentions he would have preferred to have the elements fully horizontal, but he did not have enough tall towers to do that (poor guy!). He estimates that the sloping elements made him lose 2 to 3 dB of gain.

The elements are made of #14 pre-stretched bare copper wire, and 350-lb seine twine, which makes the entire array barely visible to the eye (also to photographs!).

At Mike's QTH in Montana, this new monster-array replaces his former 5-element quad (three active elements in each direction, with a reflector in the center) with the top at 54 m as well. Mike claims that the 14-element Yagi array has a gain of about 5 dB over the 3-element quad, though

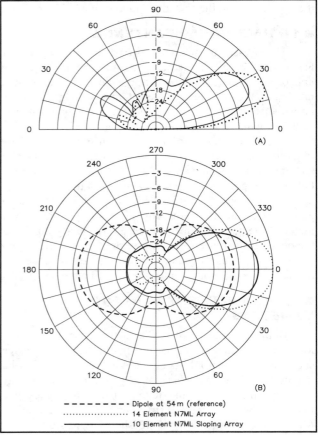

Fig 12-23—Vertical and horizontal radiation pattern (at 20° wave angle) of the N7ML 14-element Yagi array, and the 10-element sloping-Yagi array. For comparison the pattern of a half-wave dipole at the same height (54 m) is also included. The gain is 7.16 dB over the reference dipole.

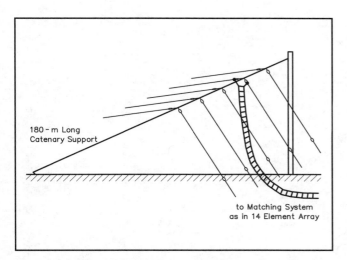

180 - m Long
Catenary Support

to Matching System
as in 14 Element Array

Fig 12-24—An array similar to the array shown in Fig 12-22 can be supported by a sloping catenary cable, with only little sacrifice in performance. N7ML uses this approach for his long path to Europe array.

computer modeling only predicts 3 dB! It goes to prove that the Yagi array is not as critical to tune as a quad array.

Mike uses a $^3/_4$-λ long 600-Ω open-wire line. He used the AEA/CIA HF Antenna Analyzer to determine the exact length of the line as well as the point where the impedance is 75 Ω. At that point he attached the 75-Ω Hardline using a stack of ferrite beads to make a good common-mode current balun.

The little brother of this monster array is shown in **Fig 12-24**. This 10-element array (two side by side 5-element antennas) is supported by a catenary cable that slopes from 54 m height to ground level some 200 m away (slope angle 12 to 15°). This is the antenna Mike uses for the long path to Europe. Its gain is 2.2 dB lower than for the 14-element array (see Fig 12-23), but the most striking difference is the "inferior" F/B as compared to the 14-element array. But the F/B is still a healthy 25 dB at 20° wave angle!

Mike points out that the most striking improvement of these arrays over his former quad array is a very significant reduction in noise, which is of course normal. Gain equals directivity, lots of gain equals lots of directivity, and that means eliminating areas where the noise comes from.

YAGIS and QUADS

13 YAGIS AND QUADS

THANKS K3LR

Tim Duffy, K3LR, whom you may know as "Mr Dayton Antenna Forum," or "Mr Super Professional Multi-Op Contest Station Owner and Operator," probably needs to be introduced to only a few readers. The way Tim runs his forum and his contest station tells a lot about this man. He is thorough, well-organized, punctual, and a super host on top of all that! No wonder there's a long line of operators that want to operate from K3LR in the big contests!

Tim is Director of Engineering for Sygnet Communications, Inc (a regional cellular telephone company based in Youngstown, Ohio). He has been with Sygnet for 14 years. Previous to Sygnet Tim spent several years in AM and FM radio broadcast engineering. Tim is a graduate of the Pennsylvania State University and has been a ham for over 26 years. He started as WN3SZX when he was 12 years old, and upgraded to Extra Class at age 14. He is currently building a full-size 80-meter Yagi, for installation at a height of 45 meters.

Tim took the time to review this chapter on Yagis and Quads, for which I am very thankful.

On the higher HF bands, almost all dedicated DXers use some type of rotatable directional antenna. Directional antennas produce gain to be better heard. They also show directivity, which is a help when listening. Yagi and cubical-quad antennas are certainly the most popular antennas on those bands.

On the low bands, rotatable directive antennas are big. Forty-meter Yagis and quads, even full-size, exist in great numbers these days. On 80 meters there are only a few full-size Yagis and quads, while reduced-size Yagis and quads are a little more common. They seem to come and go, and are rather difficult to keep in the air. On 160 meters, rotatable Yagis still belong to dreamland.

I had the chance to operate a 3-element full-size quad as well as a 3-element full-size Yagi on 80 meters, and I must admit that it is only when you have played with such monsters that you appreciate what you are missing without them. The same is even more true on 40, where full-size Yagis and quads appear in ever-growing numbers on the band. Until the day I had my own full-size 40-meter Yagi, I always considered 40 as my "worst" band. Now that I have the full-size Yagi, I think it has become my "best" band.

Much of the work presented in this chapter is the result of a number of major antenna projects that were realized with the help of R. Vermet, ON6WU, who has been a most assiduous supporter and advocate in all my antenna work.

I also had the pleasure of developing a somewhat novel design for a full-size 3-element 80-meter Yagi, featuring full 80-meter band coverage plus instantaneous direction reversal. The design methodology and some of the exclusive mechanical features of this Yagi are described in detail in this chapter.

In the past few years a number of novel designs of 40 and 80-meter Yagis have been made; some of the most interesting ones are reviewed in detail in this chapter.

Until recently, little had been published in the amateur literature covering the mechanical design of Yagis. I deal in great detail with this aspect when designing a full-size 40-meter Yagi.

■ 1. ARRAYS WITH PARASITIC ELEMENTS

In the chapter on vertical arrays I discuss groups of antennas (arrays) where each antenna element was fed via an individual feed line. During the analysis of these arrays we noticed that elements sometimes exhibit a negative impedance, which means that these elements do not draw power from the feed line, but actually deliver power into the feed system.

In such a case mutual coupling has already supplied enough (or too much) current into the element. Negative feed-point impedances are typical with close-spaced arrays where the coupling is more intensive than with wide-spaced arrays.

Parasitic arrays are arrays where (most often) only one element is fed, and where the other elements obtain their feed current only by mutual coupling with the various elements of the array. In order to obtain a desired radiation pattern and gain, feed-current magnitudes and phases need to be carefully adjusted. This is done by changing the relative positions of the elements and by changing the lengths of the elements. The exact length of the "driven" element (the fed element) will not influence the pattern nor the gain of the array; it will only influence its feed-point impedance.

Unlike with driven arrays, you cannot obtain any specific feed-current magnitude and angle. In driven arrays you "force" the antenna currents, which means you add (or subtract) feed current to the element current already obtained by mutual coupling. You can make a driven array with three elements in line where all elements have an identical feed current. You cannot make a parasitic array where the three elements have the same current (phase and magnitude).

Arrays with parasitic elements are limited as to the current distribution in the elements. The best-known configuration is the configuration used with Yagi (Yagi/Uda) and cubical-quad antennas.

In a 3-element Yagi or quad the two parasitic elements are adjusted (in length and position) to provide the required current with the lagging phase angle for the director, and with the required leading phase angle for the reflector.

■ 2. QUADS VERSUS YAGIS

It is not my intention to get into the debate of quads versus Yagis. Before I tackle both in more depth, let me clarify a few points and kill a few myths:

- For a given height above ground, the quad does *not* produce a markedly lower radiation angle than the Yagi. The vertical radiation angle of a horizontally polarized antenna in the first place depends on the height of the antenna above ground.
- There is a very slight difference (max. a few degrees depending on actual height) in favor of the quad, as there is some more bundling in the vertical plane due to the effect of the stacked two horizontal elements that make a horizontally polarized cubical quad (Ref 980).
- For a given boom length, a quad will produce slightly more gain than a Yagi. This is logical as the aperture (capture area) is larger. The principle is simple: Everything being optimized, the antenna with the largest capture area has the highest gain, or can show the highest directivity.
- Yagis as a rule are easier to build and maintain. Being two-dimensional, the problems involved with low-band antennas are simplified by an order of magnitude. Problems of wire breaking are nonexistent with Yagis. Large Yagis are also easier to handle and to install on a tower than large quads.
- There are other factors that will determine the eventual choice between a Yagi or a quad, eg, material availability, maximum turning radius (the quad takes less rotating space!), and of course, personal preference.

■ 3. YAGIS

There have been a number of good publications on Yagi antennas. Until about 10 years ago, before we all knew about the effect of tapered elements, the W6SAI Yagi book was in many circles considered the Yagi bible. I built my first Yagi based on information from Orr's work.

It was Dr J. Lawson, W2PV (SK), who wrote a very good series on Yagis back in the early 1980s. Later the ARRL published his work in the excellent book, *Yagi Antenna Design* (Ref 957). In his work, Lawson explains how he scientifically designed a winning contest station, based on high-level engineering work.

Lawson was the first in amateur circles to bring up and study the effect of tapered elements. He came up with a tapering algorithm, which is still widely referred to as the "W2PV algorithm." It calculates the exact electrical length of an element as a function of the length of the individual (in diameter) tapered sections.

3.1. Modeling Yagi Antennas

The most widely used antenna modeling code is still *MININEC*. Many of the modeling programs today, with all the bells and whistles have a *MININEC* engine. *MININEC*

was released to the public in the early 1980s, and finally every serious antenna builder was presented with a tool to model antenna performance. The W2PV taper algorithm together with *MININEC* have opened the eyes of many. I remember how I found out, many years ago, that my 5-element 20-meter Yagi peaked in both gain and front-to-back ratio around 14.45 MHz!

Since then progress has been spectacular. We now have very sophisticated modeling tools available, most of them based on the method of moments.

ELNEC is a very user-friendly version of *MININEC*, made available by R. Lewallen, W7EL, at a very attractive price. Other *MININEC*-based optimizing programs are around, eg, a *Yagimax* by Lew Gordon, K4VX. B. Beezley has made available *YO* (Yagi Optimizer), which is a Yagi (only) modeling program, and *AO* (Antenna Optimizer), which is a program that uses some simple optimizing algorithms. Both are also *MININEC*-based.

MININEC-based programs are just fine for developing Yagis, as long as they are at least $^1/_4$ λ above ground. In recent years *NEC*-based modeling programs have become available (eg, *EZNEC*). They will more accurately model antennas close to ground, but *NEC-2* based programs have problems with accurately modeling structures using connections of wires of largely varying diameters. *NEC-4*, which is not released to the general public (and extremely expensive) can handle these situations more accurately. Both however, are known to be producing rather optimistic gain figures for antennas very close to ground (see also Chapter 9, Sections 2.2.1 and 2.2.2).

If you consider modeling your own "classic configuration" Yagi for the low bands, stick to the following guidelines:

1. Make sure you know exactly what you want before you start: maximum boom length, maximum gain, maximum directivity, low Q (large SWR bandwidth) etc.
2. Always model the antenna first in free space.
3. Always model the antenna on a range of frequencies (eg, 7.0, 7.1 and 7.2 MHz), so you can immediately assess the bandwidth characteristics (SWR, gain, F/B) of the design. You can use one of the faster modeling programs (eg, *YO*) for your initial modeling, but make sure you always verify the design using a full-blown version of *MININEC* (eg, *ELNEC*).
4. Make sure the feed-point impedance is reasonable (it can be anything between 18 Ω and 30 Ω). Generally, Yagis with a higher feed-point impedance are very low Q, and do not achieve maximum gain. There are exceptions however.
5. When the array is optimized and meets your requirements in free space, you must repeat the exercise over real ground at the actual antenna height.
6. If the antenna is stacked with other antennas, include the other antennas in the model as well. This is especially so when considering stacking Yagis for the same band. F/B may be totally ruined due to stacking. Stacks need to be optimized as stacks!
7. If you consider making a Yagi with loaded elements, first model the full-size equivalent. When applying the loading

devices, don't forget to include the resistance losses (especially for loading coils!) and possible parasitic capacitances or inductances.

8. If you are about to model your own Yagi using loading devices such as linear-loading stubs or capacity-loading wires, you should be very careful. The best approach is to first model the antenna using all wires of the same diameter. This should prove the feasibility. Next, you could determine the resonant frequency of the individual elements, which is an excellent guiding parameter during the actual tune-up of the antenna. Further modeling issues are discussed while describing particular antenna designs in this chapter.

3.2. Mechanical Design

Making a perfect electrical design of a low-band Yagi is a piece of cake nowadays with all the magnificent modeling software available. The real challenge comes when you have to turn your model into a mechanical design. When building a mechanically sound 40-meter Yagi, there is no room for guesswork. Don't ever take anything for granted when you are building a 40-meter Yagi. If you want your beam to survive the winds and ice loading you expect, you *must* go through a fair bit of calculating (making sure). The same holds true for an 80-meter Yagi of course, but squared!

Physical Design of Yagi Antennas, by D. Leeson, W6QHS, (Ref 964) covers all the aspects of mechanical Yagi design. The book covers the theoretical aspects in detail. Leeson uses the "variable area" principle to assess the influence of wind on the Yagi. The book unfortunately does not give any design examples of practical full-size 40 or for 80-meter Yagis. The only low-band antenna covered is the Cushcraft 40-2CD, a shortened 2-element 40-meter Yagi. Leeson's modification to strengthen the Cushcraft 40-2CD has become a classic, and is a must for everyone who has this antenna and who does not want to see it ripped to pieces in a storm.

Over the years standards dealing with mechanical issues for towers and antennas have evolved. The well-known EIA RS-222 standard has evolved from 222-C through suffix D and eventually to the RS-222-E standard. While the earlier versions of Leeson's software that he supplies with his book were based on C, the latest versions are now based on E. The E-version (and also ASCE 74) treats wind statistics and force on cylindrical elements more physically than C and D, and the difference shows up in the question of forces on cylinders at an angle to the wind. This affects boom strength and rotating torque. The original article by K5IU (Ref 958) uses the E approach, as well as the ON4UN LOW BAND SOFTWARE modules dealing with boom strength and torque balancing.

More recently NI6W wrote an interesting software package that addresses all of the mechanical issues concerning antenna strength. *YS* is easy to use, has lots of data about materials and tubing in easy-to-access form (costs approximately $50 and is available from Kurt Andress, NI6W, tel. 702-267-5290, e-mail: ni6w@yagistress.minden.nv.us).

All of these tools deal with static wind-load models. The question of course is how reliable all these models are in a complex aerodynamic situation. As Leeson puts it, ". . . but we're not dealing with mathematical models when the wind is roaring through here at 134 mi/h. Either model (C or E) results in booms that break upward in the wind if you ignore vertical gusting..." In particular locations, such as hill-top QTHs, there may be vertical updraft winds that would break a boom unless three-way boom guys are used. But these are rather extreme conditions, and not the run-of-the mill situations.

The real proof of the pudding is in the building of the big antennas, and more so to keep them up, year after year.

The mathematics involved in calculating all the structural aspects of a low-band Yagi element are rather complex. It is a subject which is ideally suited for computer assistance. Together with my friend R. Vermet, ON6WU, I have written a comprehensive computer program, YAGI DESIGN, which was released in early 1988 and updated a few times since. In addition to the "traditional" electrical aspects, YAGI DESIGN tackles the mechanical-design aspects. This is especially of interest to the prospective builder of 40 and 80-meter Yagi antennas. While Yagis for the higher HF bands can be built "by feel," 40 and 80-meter Yagis require much closer attention if you want these antennas to stay up.

The different modules of the YAGI DESIGN software are reviewed in the chapter on low-band software.

This book is not a textbook on mechanical engineering, but a few definitions are needed in order to better understand some of the formulas I use in this chapter.

3.2.1. Terms and definitions

Stress

Stress is the force applied to a material per unit of cross-sectional area. Bending stress is the stress applied to a structure by a bending moment. Shearing stress is the stress applied to a structure by a shearing moment. The stress is expressed in units of force divided by units of area (usually expressed in kg/mm^2 or $lb/in.^2$).

Breaking Stress

The breaking stress is the stress at which the material breaks.

Yield Stress

Yield stress is the stress where a material suddenly becomes plastic (non-reversible deformation). The yield stress to breaking stress ratio differs from material to material. For aluminum the yield stress is usually close to the breaking stress. For most steel materials the yield stress is approximately 70% of the breaking stress. Never confuse breaking stress with yield stress, unless you want something to happen that you will never forget.

Elastic Deformation

Elastic deformation of a material is deformation that will revert to the original shape after removal of the external force causing the deformation.

Compression or Elongation Strain

Compression strain is the percentage change of dimension under the influence of a force applied to it. Being a ratio, strain is an abstract figure.

Shear Strain

Shear strain is the deformation of a material divided by the couple arm. It is a ratio and thus an abstract figure.

Shear Angle

This is the material displacement divided by the couple arm. As the angles involved are small, the ratio is a direct expression of the shear angle expressed in radians. To obtain degrees, multiply by $180/\pi$.

Elasticity Modulus

Elasticity modulus is the ratio stress/strain as applied to compression or elongation strain. This is a constant for every material. It determines how much a material will deform under a certain load. The elasticity modulus is the material constant that plays a role in determining the sag of a Yagi element. The elasticity modulus is expressed in units of force divided by the square of units of dimension (unit of area).

Rigidity Modulus

Rigidity modulus is the ratio shear-stress/strain as applied to shear strain. The rigidity modulus is the material constant that will determine how much a shaft (or tube) will twist under the influence of a torque moment (eg, the drive shaft between the antenna mast and the rotator). The rigidity modulus is expressed in units of force divided by units of area.

Bending Section Modulus

Each material structure (tube, shaft, plate T-profile, I-profile, etc) will resist a bending moment differently. The section modulus is determined by the shape as well as the cross-section of the structure. The section modulus determines how well a particular shape will resist a bending moment. The section modulus is proper to a shape and not to a material.

The bending section modulus for a tube is given by

$$S = \delta \times \frac{OD^4 - ID^4}{32 \times OD}$$

where
    OD = outer diameter of tube
    ID = inner diameter of tube
The bending section modulus is expressed in units of length to the third power.

Shear Section Modulus

Different shapes will also respond differently to shear stresses. The shear stress modulus determines how well a given shape will stand stress deformation. For a hollow tube the shear section modulus is given by

$$S = \delta \times \frac{OD^4 - ID^4}{16 \times OD}$$

where
    OD = outer diameter of tube
    ID = inner diameter of tube
The bending section modulus is expressed in units of length to the third power.

3.3. Computer-Designed 3-Element 40-Meter Yagi at ON4UN

Let us go through the design of a very strong 3-element full-size 40-meter Yagi. This is not meant to be a step-by-step description of a building project, but I will try to cover all the critical aspects of designing a sound and lasting 40-meter Yagi. The Yagi described also happens to be the Yagi I have been using successfully over the past several years on 40 meters (it has brought several new European records in major contests on 40 meters).

The design criteria for the Yagi are
- Low Q, good bandwidth, F/B optimized.
- Survival at wind speeds up to 140 km/h with the elements broadside to the wind.
- Maximum ice load 10 mm at 60 km/h wind.
- Lifetime greater than 20 years.
- Boom length 10.7 m maximum (only because I happened to have this boom)

3.3.1. Selecting an electrical design

Design no. 10 from the database of the YAGI DESIGN software program meets all the above specifications. **Fig 13-1** shows a copy of the screen with all the data (performance and generic dimensional data) for the chosen design.

While another design with up to 0.5 dB more gain could have been selected, the no. 10 design was selected because of its excellent F/B pattern and wide bandwidth (SWR, gain and F/B ratio).

The Yagi was to be mounted 5 m above the 20-meter Yagi (design no. 68 from the database), at 30 m above ground. The combination of both antennas was modeled once more over real ground at the final height using a *MININEC*-based modeling program, to see if there would be an important change in pattern and gain due to the presence of the second antenna. The performance figures (gain, F/B) and directivity pattern of the 40-meter Yagi changed very little at the 5-m stacking distance.

3.3.2. Principles of Mechanical Load and Strength Calculations for Yagi Antennas

D. Weber, K5IU, brought to our attention (Ref 958) that the "variable-area" method, commonly employed by most Yagi manufacturers, and used by many authors in their publications as well as software, has *no* basis in science, nor is there any experimental evidence of the method.

The variable-area method assumes that the direction of the force created by the wind on an element is always in line with the wind direction, and that the magnitude is proportional to the area of the element as projected onto a plane perpendicular to the wind direction (proportional to the sine of the wind angle).

The scientifically correct method of analyzing the wind-force behavior, called the "cross-flow" principle, says that the direction of the force due to the wind is *always* perpendicular to the plane in which the element is situated, and that its magnitude is proportional to the square of the sine of the wind angle.

Fig 13-2 shows both principles. It is easy to understand that the cross-flow principle is the correct one. The experiment described by D. Weber, K5IU, can be carried out by anyone, and should convince anyone who has doubts: "Take a 1-m long piece of aluminum tubing (approximately 25 mm in diameter) for a car ride. One person drives, while another sits in the passenger seat. The passenger holds the tube in his hand, and puts his arm out the window positioning the tube vertically. The tube is now perpendicular to the wind stream (wind angle = zero). It is easy to observe a force (drag force) which is *in line* with the wind (and at the same time perpen-

dicular to the axis of the tube). The passenger now rotates the tube approximately 45°, top end forward. The person holding the tube will now clearly feel a force which pushes the tube *backward* (drag force), but at the same time tries to *lift* (cross force) the tube. The resulting force of these two components (the drag and cross force) is a force which is *always* perpendicular to the direction of the tube. If the tube is inclined with the bottom end forward, the force will try to push the tube downward."

This means that the direction of the force developed by the wind on an object exposed to the wind is not necessarily the same as the wind direction. There are some specific conditions where the two directions are the same, such as the case where the (flat) object is broadside to the wind direction. If you put a plate (1 m²) on top of a tower, and have the wind hit the plate at a 45° angle, it will be clear that the "push" developed by the wind hitting the plate will not be developed

```
 7547                         SELECT DESIGN MODULE                    on4un/on6wu

   DESIGN #  10  ELEMENTS: 3        NAME: FREDA          BOOM: 0.249   WVL

          ──────── Performance data ────────
    FREQ.   GAIN    F/B      RESIST    REACT.    SWR    FOM
   -1.5%    7.4    20.5       28.8     -12.6     1.5    9.0          ┌──────────────────┐
   -1.0%    7.4    23.4       29.1      -9.0     1.3    9.4          │ ANT. Q  =    14  │
   -0.5%    7.4    23.9       28.8      -5.3     1.1    9.7          │                  │
    0.0%    7.5    24.4       28.0      -1.4     1.0    9.9          │ SWR BW  >   3  % │
   +0.5%    7.5    25.4       26.8       2.8     1.2    9.9          │                  │
   +1.0%    7.6    22.8       25.2       7.5     1.4    9.5          │ F/B BW  >   3  % │
   +1.5%    7.7    19.2       23.4      12.5     1.8    8.9          └──────────────────┘

   ──── Dimensions in wavelengths ────      ──── Physical Boomlength ────
  ELEMENT       LENGTH      POSITION      28 Mhz ->   2.63 m. OR    8.6 ft
  REFLECTOR    0.510217     -.105595      24 Mhz ->   2.99 m. OR    9.8 ft
  DRIV. EL.    0.483032     0.000000      21 Mhz ->   3.52 m. OR   11.5 ft
  DIR #  1     0.452359     0.143370      18 Mhz ->   4.12 m. OR   13.5 ft
  DIR #  2     0.000000     0.000000      14 Mhz ->   5.27 m. OR   17.3 ft
  DIR #  3     0.000000     0.000000      10 Mhz ->   7.37 m. OR   24.2 ft
  DIR #  4     0.000000     0.000000       7 Mhz ->  10.51 m. OR   34.5 ft

 El. lengths are for el. diam of .0010527 wavelengths (7/8 inch on 14.2 MHz).
     S = SELECT THIS DESIGN      C = CONTINUE        H = HELP     X = EXIT
```

```
 7547                         SELECT DESIGN MODULE                    on4un/on6wu

  DESIGN #  10        NAME: FREDA                          ELEMENTS =   3
  FREQ. =    7.10 MHz     WAVEL.:  42.22535  m.      BOOM: 10.51 m or  34.49 ft
  DRIVEN ELEMENT REACTANCE =   -1.4 ohm.

       ELEMENT/POSITION          CENTIMETERS   INCHES    WAVELENGTHS
       POSITION REFLECTOR           -445.9     -175.6     -0.105595
       LENGTH REFLECTOR             2149.1      846.1      0.508953
       POSITION DRIV. ELEM.            0.0        0.0      0.000000
       LENGTH DRIV. ELEM.           2046.9      805.9      0.484767
       POSITION DIR # 1              605.4      238.4      0.143370
       LENGTH DIR # 1               1930.2      759.9      0.457122

      THESE LENGTHS ARE FOR A CONSTANT DIAMETER OF 7/8 INCH OR 2.2225 CM.
    X=EXIT   S=SAVE    F=FREQ.CHANGE   O=OTHER DESIGN   C=CHANGE DR. EL   H=HELP
```

Fig 13-1—Dimensional and performance data for the 3-element Yagi design no. 10 from the YAGI DESIGN software program database, for building a 40-meter full-size Yagi. The element lengths are expressed in terms of wavelength, for a fixed element diameter.

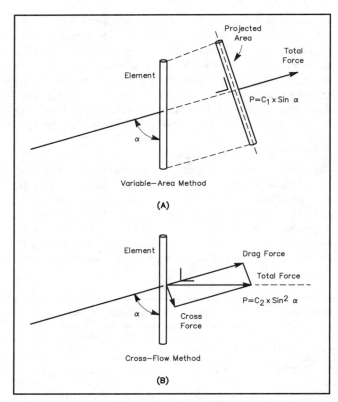

Fig 13-2—Most amateur literature uses the "variable area" method shown at A for calculating the effect of wind on an element. The principle says that the direction of the force created by the wind on an element is *always* in line with the direction of the wind, which is clearly incorrect. If this were correct, no plane would ever fly! The "cross-flow" principle, illustrated at B, states that the direction of the force is *always* perpendicular to the element, and is the resultant of two components, the drag force and the cross force (which is the lifting force in the case of an airplane wing). See text for details.

in the direction of the wind, but in the direction perpendicular to the plane of the flat plate. If you have any "feeling" for mechanics and physics, this should be fairly evident.

To remove any doubt from your mind, D. Weber states that Alexandre Eiffel, builder of the Paris Eiffel tower, used the cross-flow principle for calculating his tower. And it still stands there after more than 100 years.

Now comes a surprise: Take a Yagi, with the wind hitting the elements at a given wind angle (forget about the boom at this time). The direction of the force caused by the wind hitting the element at whatever wind angle, will always be perpendicular to the element. This means that the force will be in line with the boom. The force will not create any bending moment in the boom; it will merely be a compression or elongation force in the boom. All of this of course provided the element is fully symmetrical with respect to the boom.

This force in the boom should not be of any concern, as the boom will certainly be strong enough to cope with the bending moments caused by the broadside (to the boom) winds. These bending moments in the boom (at the mast

attachment plate) are caused only by the force created by the wind on the boom only (by the same "cross-flow" principle) or any other "components" which have an exposed wind area in line with the boom.

If the mast-to-boom plate is located in the center of the boom, the wind areas on both sides of the mast are identical, and the bending moments in the boom, on both sides of the mast (at the boom-to-mast plate) will be identical. This means there is no *mast torque*. If the areas are unequal, mast torque will result. This mast torque puts extra strain on the rotator, and should be avoided. Torque balancing can be done by adding a *boom dummy*, which is a (small) plate placed near the end of the shorter boom half, and which serves to reestablish the balance in bending moments between the left and the right side of the boom.

This may seem strange as intuitively one may have it difficult to accept that the extreme case, of a Yagi having one element sitting on one end of a boom, would not create any rotating torque in the mast, whatever the wind direction is. Surprisingly enough, this is the case. You cannot compare this situation with a weathervane, where the boom area at both sides of the rotating mast is vastly different. It is the vast difference in boom area that makes the weathervane turn into the wind. **Fig 13-3** shows what the situation is in theory, and what's likely to happen in a real world. At A and B the wind only sees the element (the boom is not visible), and if the element is fully symmetrical with respect to the boom, there will be no torque moment at the element-to-boom interface. Hence this is a fully stable situation. At C the situation where the boom is facing the wind is shown. The element is "invisible" now and as the boom is supposed to be wind-load balanced, the boom by itself creates no torque at the boom-mast interface. At D we see that the cross-flow principle (see above) only creates a force in-line with the boom. This means that this example still guarantees a well balanced situation, and the structure will not rotate in the wind.

But let's be practical. The wind blowing on the long flexible elements of a Yagi will make the elements bend slightly as shown in Fig 13-3E. In this case now it is clear that the pressure induced by the wind on side (a) of the element will be much greater than on side (b) as side (a) now faces the wind much more than side (b). In this case the antennas will tend to rotate in the sense indicated by the arrow.

Taking all of this into account it seems to be a good idea not only to try to achieve full boom (area) symmetry but full element (area) symmetry as well. Leeson came to the conclusion that he prefers to balance in the element plane by offsetting the element ensemble to eliminate the need for a torque balancing element, then using a vane (boom torque compensating plate) on the now unbalanced boom. If offsetting the element ensemble creates an important weight imbalance, this can always be compensated for by inserting some form of weight in the boom near one tip.

Not adding extra dummy elements seems to be a good idea, as in dynamic situations (wind turbulence) these may actually deteriorate the situation rather than improve it.

As, in principle, the Yagi elements do *not* contribute to the boom moments, and therefore not to the mast torque, it

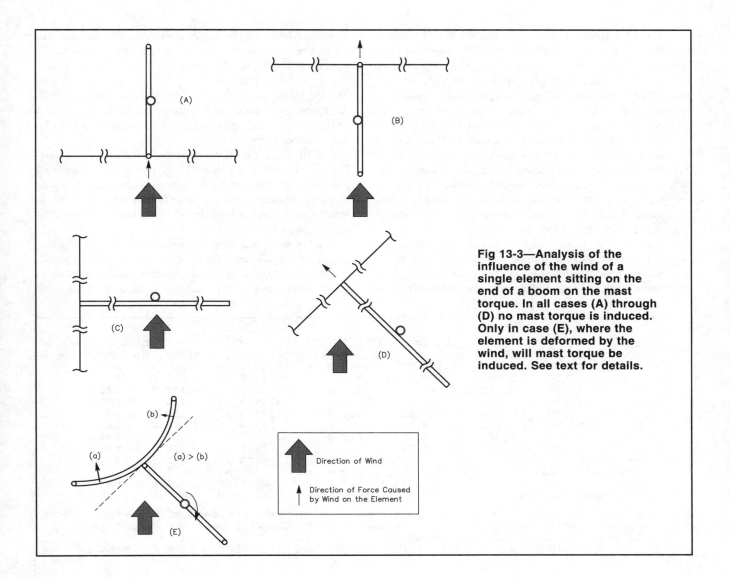

Fig 13-3—Analysis of the influence of the wind of a single element sitting on the end of a boom on the mast torque. In all cases (A) through (D) no mast torque is induced. Only in case (E), where the element is deformed by the wind, will mast torque be induced. See text for details.

makes no sense to create "dummy elements" to try to achieve a torque-balanced Yagi.

The MECHANICAL YAGI BALANCE module of the YAGI DESIGN software addresses all the issues as explained above, and uses the cross-flow principle. It uses latest data from the latest EIA/TIA-222-E specification, which is somewhat different from the older EIA standard RS-222-C. (See also Section 3.2.)

3.3.3. Element strength calculation

While it is standard procedure to correct the boom sag by using support cables, element sag must be controlled to a maximum degree by using the properly designed tapered sections for making the element. Guyed elements are normally only used with 80-meter Yagis, although full-size 80-meter self-supporting elements with negligible sag have been designed and demonstrated (see Section 3.6.1). Unguyed 40-meter full-size tubular elements (24 m) can be built to withstand very high wind speeds, as well as a substantial degree of ice loading.

D. Weber, K5IU, (Ref 966) wrote an excellent article concerning the structural behavior of Yagi elements. The mathematics involved are quite tedious, and a very good subject for a computer program. Leeson (Ref 964) addresses the issue in detail in his book, and he has made a spreadsheet type of program available for calculating elements. As the element strength analysis is always done with the wind blowing broadside to the elements, the issue of "variable area" or "cross-flow principle" does not have to be taken into consideration at this point.

The ELEMENT STRENGTH module of the YAGI DESIGN software is a dedicated software program that allows the user to calculate the structural behavior of Yagi elements with up to nine tapering elements. The ELEMENT STRENGTH module operates in the English measurement system as well as in the metric system (as do all other modules of the integrated YAGI DESIGN software). It is based on the latest data from the latest EIA/TIA-222-E specification. A drag factor of 1.2 is used for the element calculations (as opposed to 0.66 in the older RS-222-C standard).

Fig 13-4 shows a screen printout of the ELEMENT STRENGTH module of the YAGI DESIGN software. From the prompt line, each of the inputs can be easily changed whereby the impact on the performance is immediately displayed as in a spreadsheet. Input data that can be changed are section dimensions (length, diameter and wall thickness), wind speed, ice loading, material properties, etc.

The interactive designing of elements enables the user to achieve element sections that are equally loaded (ratio of actual bending moment to allowable bending moment). Many of the published designs show one section which is loaded to the limit, while other sections still exhibit a large safety margin. Such unbalanced designs are always inefficient as to weight, wind area (and load), as well as cost.

Each change (number of sections, section length, section diameter, wind speed, aluminum quality, ice load, etc) is immediately reflected in a change of the moment value at the interface of each taper section, as well as at the center of the

element. When a safe limit is exceeded, the unsafe value will blink. The screen also shows the weight of the element, the wind area, and the wind load for the specified wind speed.

It is obvious that the design in the first place will be dictated by the material available. Material quality, availability and economical lengths are discussed in Section 3.3.6 where Table 13-2 shows a range of aluminum tubing material commonly available in Europe.

A 40-meter Yagi reflector is approximately 23-m long. This is twice the length of a 20-meter element. Designing a good 40-meter element can be done by starting from a sound 20-meter element, which is then "lengthened" by more tapered sections toward the boom, calculating the bending stresses at each section drop.

When designing a Yagi element one must make sure that the actual bending moments (LM_t) at all the critical points match the maximum allowable bending moments (RM) as closely as possible. LM_v is the bending moment in the verti-

```
  8586                        ELEMENT STRESS ANALYSIS                    on4un/on6wu
  SEC#    OD(in)    WT(in)     L(in)    RM(in.lbs) LMt(lbs.in) LMv(lbs.in) CONDIT.
   1      2.500     0.125     144.00    18463.8     19389.3     3207.7     FAIL
   2      2.000     0.083      66.00     8051.7      6416.7      935.0     SAFE
   3      1.250     0.110      42.00     3617.3      3382.5      458.9     SAFE
   4      1.000     0.110      36.00     2164.2      2124.5      269.0     SAFE
   5      1.000     0.058      45.00     1337.8      1331.6      161.4     SAFE
   6      0.625     0.110      18.00      691.0       651.7       74.9     SAFE
   7      0.625     0.058      40.00      469.9       464.7       52.0     SAFE
   8      0.500     0.058      65.00      280.1       171.5       18.9     SAFE
  Velocity=  85.0 Mph                              Wind press.=  17.3 lb/sqft
  Material= 6061-T6         Tens. str. = 35000 psi  Ice thickn.= 0 inch
  Rope    = YES             Ele. weight= 45.2 lbs   El.windload= 200.6 lbs
  Pr. area= 1389 sq.in      Half el.lgt= 456.0 inch El. sag   = 56.4 inch

  If you intend to do a full physical design of a yagi, run each of the yagi
  elements and make a screen dump of the results. You will need the weight data
  as inputs to the BALANCE program.
    1=SPEED  2=GUSTFAC  3=MATER  4=ICE  5=DIM  6=NRUN  7=SECT   H=HELP   X=EXIT

  8586                        ELEMENT STRESS ANALYSIS                    on4un/on6wu
  SEC#    OD(in)    WT(in)     L(in)    RM(in.lbs) LMt(lbs.in) LMv(lbs.in) CONDIT.
   1      2.500     0.125     104.00    18463.8     14842.1     2384.1     SAFE
   2      2.000     0.083      66.00     8051.7      6416.7      935.0     SAFE
   3      1.250     0.110      42.00     3617.3      3382.5      458.9     SAFE
   4      1.000     0.110      36.00     2164.2      2124.5      269.0     SAFE
   5      1.000     0.058      45.00     1337.8      1331.6      161.4     SAFE
   6      0.625     0.110      18.00      691.0       651.7       74.9     SAFE
   7      0.625     0.058      40.00      469.9       464.7       52.0     SAFE
   8      0.500     0.058      65.00      280.1       171.5       18.9     SAFE
  Velocity=  85.0 Mph                              Wind press.=  17.3 lb/sqft
  Material= 6061-T6         Tens. str. = 35000 psi  Ice thickn.= 0 inch
  Rope    = YES             Ele. weight= 37.8 lbs   El.windload= 171.7 lbs
  Pr. area= 1189 sq.in      Half el.lgt= 416.0 inch El. sag   = 41.6 inch

  If you intend to do a full physical design of a yagi, run each of the yagi
  elements and make a screen dump of the results. You will need the weight data
  as inputs to the BALANCE program.
    1=SPEED  2=GUSTFAC  3=MATER  4=ICE  5=DIM  6=NRUN  7=SECT   H=HELP   X=EXIT
```

Fig 13-4—Design table for the reflector of a 40-meter Yagi. The total reflector length is 912 inches (23.16 m), which should be long enough for a reflector. With the YAGI DESIGN software you can work either in the English system (inches, ft, lb) or the metric system (m, cm, mm, kg). For each of the element sections the OD, ID, the maximum allowable moment, and the actual moment in the vertical and horizontal planes are displayed, together with the safety status in the last column. Any of the input data can be changed via the prompt line. See text for details.

```
7547                      ELEMENT STRESS ANALYSIS                  on4un/on6wu
SEC.#   OD(mm)   WT(mm)    L(cm)     RM(kgm)   LMt(kgm)   LMv(kgm)   CONDITION
  1     60.000   5.000     200.0     241.49    199.51     37.26      SAFE
  2     50.000   5.000     285.0     159.36    108.81     17.19      SAFE
  3     35.000   2.000      84.0      35.61     34.49      3.69      SAFE
  4     30.000   2.000     100.0      25.41     22.85      2.30      SAFE
  5     25.000   1.500     176.0      13.51     12.71      1.19      SAFE
  6     15.000   1.000      82.0       3.18      2.77      0.23      SAFE
  7     12.000   1.000     113.2       1.93      0.89      0.07      SAFE
Velocity= 140.0 Kph                            Wind press.=   88.7 kg/m²
Material= OTHER          Tens. str. = 22.0 kg/mm²  Ice thickn.= 0 mm
Rope     = YES           Ele. weight=  25.1 kg     El.windload=  83.4 kg
Pr. area=  7836 cm²      Half el.lgt=1040.2 cm     El. sag    =  83.6 cm



7547                      ELEMENT STRESS ANALYSIS                  on4un/on6wu
SEC.#   OD(mm)   WT(mm)    L(cm)     RM(kgm)   LMt(kgm)   LMv(kgm)   CONDITION
  1     60.000   4.000     300.0     203.28    256.02     48.88      FAIL
  2     50.000   5.000     285.0     159.36    108.81     17.19      SAFE
  3     35.000   2.000      84.0      35.61     34.49      3.69      SAFE
  4     30.000   2.000     100.0      25.41     22.85      2.30      SAFE
  5     25.000   1.500     176.0      13.51     12.71      1.19      SAFE
  6     15.000   1.000      82.0       3.18      2.77      0.23      SAFE
  7     12.000   1.000     113.2       1.93      0.89      0.07      SAFE
Velocity= 140.0 Kph                            Wind press.=   88.7 kg/m²
Material= OTHER          Tens. str. = 22.0 kg/mm²  Ice thickn.= 0 mm
Rope     = YES           Ele. weight=  27.2 kg     El.windload=  96.2 kg
Pr. area=  9036 cm²      Half el.lgt=1140.2 cm     El. sag    = 132.4 cm


           Element half length = 1140.2 cm (ref file: FREDA9.ANT)
  If you intend to do a full physical design of a yagi, run each of the yagi
  elements and make a screen dump of the results. You will need the weight data
  as inputs to the BALANCE program.
   1=SPEED   2=GUSTFAC   3=MATER   4=ICE   5=DIM   6=NRUN   7=SECT   H=HELP   X=EXIT
```

Fig 13-5—Design of a 40-meter reflector (23 m length) using metric-dimension aluminum available in Europe. The element is first modeled being 1 m short (top table). Then the center section is lengthened by 1 meter, which is the length of the steel insert. At full length we see that the aluminum tube fails marginally (259 kg-m versus 241 kg-m). This is of no concern, as the steel insert will strengthen the tube considerably in the center of the element. The moment at the tip of the steel insert is 201 kg-m, which is well below the allowable moment of 241 kg-m.

cal plane, created by the weight of the element. This is the moment that creates the sag of the element. LM_t is the sum of LM_v and the moment created by the wind (in the horizontal plane). Adding those together may seem to create some safety, although it can be argued that turbulent wind may in actual fact blow vertically in a downward direction.

Fig 13-5 shows the design of the reflector element using material of metric dimensions available in Europe. The design was done for a maximum average wind speed of 140 km/h, using F22 quality (Al Mg Si 0.5%) material. This material has a yield strength of 22 kg/mm² (31,225 lb/in.²). For material specifications see Section 3.3.6.

All calculations have been done in a static condition. Dynamic wind conditions can be significantly different, however. The highest bending moment is at the center of the element. Inserting a 2-m long steel tube (5 or 7-mm wall)

in the center of the center element will not only provide additional strength but also further reduce the sag.

Whether 140 km/h will be sufficient in your particular case depends on the following factors:

• The rating of the wind zone where the antenna is to be used. The latest EIA/TIA-222-E standard lists the recommended wind speed per county in the US.
• Whether modifiers or safety factors are recommended (see EIA/TIA-222-E standard).
• Whether you will expose the element to the wind, or put the boom into the wind (see Section 3.3.4.1).
• Whether you have your Yagi on a crank-up tower, so that you can nest it at protected heights during high wind storms.

Fig 13-6 shows the 3-element full-size 40-meter Yagi (4 m above a 5-element 20-m Yagi), with a similar taper design.

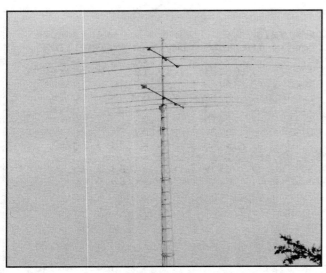

Fig 13-6—Three-element 40-meter Yagi at ON4UN. The Yagi is mounted 5 m above a 5-element 20-meter Yagi with a 15-m boom, at a height of 30 m. Note the very limited degree of element sag, which is proof of a good physical design.

Note the very limited sag on the elements. The telescopic fits are discussed in Section 3.3.7. **Figs 13-7** and **13-8** show the section layout of the 40-meter reflector element as calculated for both metric and US (inch) materials.

3.3.3.1. Element sag

Although element sag is not a primary design parameter, I have included the mathematics to calculate the sag of the element in the ELEMENT STRENGTH module of the YAGI DESIGN software. While designing, it is interesting to watch the total element sag. Minimal element sag is an excellent indicator of a good mechanical design.

A well-known manufacturer of Yagis sells a full-size 40-meter Yagi that exhibits an excessive sag of well over 200 cm. That is proof of rather poor engineering. Too much sag means there is somewhere along the element too much weight that does not contribute to the strength of the element. The sag of each of the sections of an element depends on:

- The section's own weight.
- The moment created by the section(s) beyond the section being investigated (toward the tip).
- The length of the section.
- The diameter of the section.
- The wall thickness of the section.
- The elasticity modulus of the material used.

The total sag of the element is the sum of the sag of each section.

The elasticity modulus is a measure of how much a material can be deformed (bent, stretched) without inducing permanent deformation. The elasticity modulus for all aluminum alloys is 700,000 kg/cm^2 (9,935,000 lb/in^2). This means that an element with a stronger alloy will exhibit the same sag as an element made with an alloy of lesser strength.

The 40-meter reflector, as designed above, has a calculated sag of 129.5 cm, not taking into account the influence of the steel insert (coupler). The steel coupler reduces the sag to approximately 91 cm. These are impressive figures for a 40-meter Yagi. With everything scaled down properly, the sag is comparable to that of most commercial 20-meter

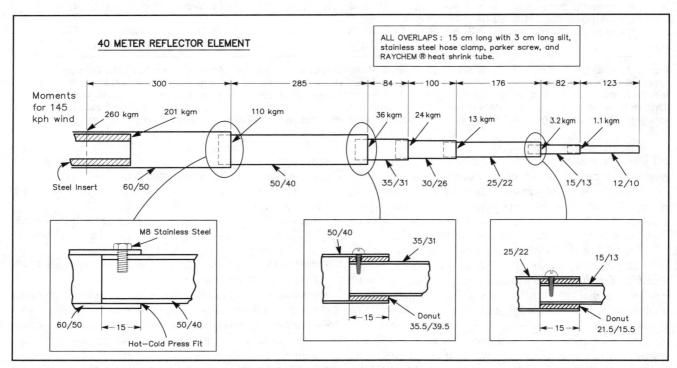

Fig 13-7—Mechanical layout of a 40-meter full-size reflector element using metric materials, as shown in the design sheet of Fig 13-5.

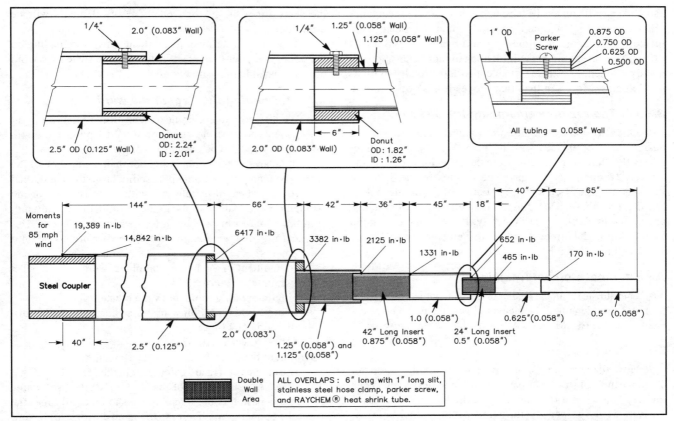

Fig 13-8—Layout of the 40-meter reflector using US materials (inch dimensions). The element was calculated in Fig 13-4.

Yagis. After mounting the element, the total element sag turned out to be exactly as calculated by the software.

3.3.3.2. Alternative Element Designs Using US Materials

The US design is made by starting from standard tubing lengths of 144 inches. Tables 13-3 and 13-4 (appearing later in this chapter) show a list of some of the standard dimensions commonly available in the US. The availability of aluminum tubes and pipes is discussed in Section 3.3.6.

For the two larger diameter tubes, I used aluminum pipe. The remaining sections are from the standard tubing series with 0.058-inch wall thickness. From the design table we see that for some sections I used a wall thickness of 0.11 inch, which means that we are using a tight-fit section of $^1/_8$-inch less diameter as an internal reinforcement.

The design table shows that the center sections would (marginally) fail at the 90-mi/h design wind speed. In reality this will not be a problem, as this design requires an internal coupler to join the two 144-inch center sections. This steel coupler must be strong enough to take the entire bending moment.

The section modulus of a tube is given by

$$S = \delta \times \frac{OD^4 - ID^4}{32 \times OD}$$ (Eq 13-1)

where

   S = section modulus
   OD = tube outer diameter
   ID = tube inner diameter
   The maximum moment a tube can take is given by

$$M_{max} = YS \times S$$

where

   YS = yield strength of the material
   S = section modulus as calculated above

or

$$M_{max} = YS \times \delta \times \frac{OD^4 - ID^4}{32 \times OD}$$ (Eq 13-2)

The yield strength varies to a very large degree (Ref 964 p. 7-3). For different steel alloys it can vary from 21 kg/mm^2 (29,800 lb/in^2) to 50 kg/mm^2 (71,000 lb/in^2).

A 2-inch OD steel insert (with aluminum shimming material) made of high tensile steel with a YS = 55,000 lb/in^2 would require a wall thickness of 0.15 inch to cope with the maximum moment of 19.622 in.-lb at the center of the 40-meter reflector element.

Note that the element sag (42.1 inches with 2×40-inch-long steel coupler) is very similar to the sag obtained with the metric design example. It is obvious that for an optimized Yagi element (and for a given survival wind speed), the element sag will always be the same, whatever the exact taper scheme may be. In other words, a good 40-meter Yagi reflector element, designed to withstand a 140 km/h (87 mi/h) wind should not exhibit a sag of more than approximately 40 inches (100 cm) when constructed totally of tubular elements. More sag than that proves it is a poor design.

3.3.3.3. The driven element and the director

Once we have designed the longest element, we can design the shorter ones with no pain. We should consider taking the "left over" lengths from the reflector for use in the director (economical design). The lengths of the different sections for the 3-element Yagi no. 10 from the YAGI DESIGN database, according to the metric and US systems, are shown in **Table 13-1**. Typically, if the reflector is good for 144 km/h, the director and the driven element will withstand 160 to 170 km/h.

3.3.3.4. Final element tweaking

Once the mechanical design of the element has been finalized, the exact length of the element tips will have to be calculated using the ELEMENT TAPER module of the software.

3.3.4. Boom design

Now that we have a sound element for the 40-meter Yagi, we must pay the necessary attention to the boom. When the wind blows at a right angle onto the boom, a maximum pressure is developed by the wind on the boom area. At the same time, the loading on the Yagi elements will be minimum.

There is no intermediate angle at which the loading on the boom is higher than at a 90° wind angle (when the wind blows broadside onto the boom).

3.3.4.1. Pointing the Yagi in the wind

We all know the often heard question, "Should I point the elements toward the wind, or should I point the boom toward the wind?" The answer is simple.

If the area of the boom is smaller than the area of all the elements, then put the boom perpendicular to the wind. And vice versa. Let me illustrate this with some figures for the 40-meter Yagi. Calculations are done for a 140 km/h wind, with the boom-to-mast plate in the center of the boom.

Zero-degree wind angle (wind blowing broadside onto the elements)
• Boom moment in the horizontal plane: Zero
• Thrust on tower/mast 323 kg
• Maximum bending moment in the elements
• 90° wind angle (wind blowing broadside onto the boom)
• Boom moment 114 kg-m
• Thrust on tower/mast: 87 kg
• Minimum bending moment in the elements

The above figures are calculated in the MECHANICAL YAGI BALANCE module of the YAGI DESIGN software.

In this case it is obvious that we should at all times try to put the boom perpendicular to the wind during a high wind storm. For calculating and designing the rotating mast and tower, it is recommended, however, to take into account the worst case wind pressure of 323 kg.

Relying on the exact direction of the Yagi as a function of wind direction to reduce safety design margins is a dangerous practice which should not be encouraged. This does not mean that in case of high winds one could not take advantage of the "best" wind angle to relieve load on the Yagi (or parts thereof), mast or tower, but what is gained by doing so should only be considered as extra safety margin only.

Remark: With long-boom Yagis on higher frequencies eg, a 6-element 10-meter Yagi, it is likely that putting the elements perpendicular to the wind will be the logical choice.

3.3.4.2. Weight balancing

In **Fig 13-9**, I have assumed that the mast is at the physical center of the boom. As the driven element is offset toward the reflector, the Yagi will not be weight balanced. A good physical design must result in a perfect weight balance, as it is extremely difficult to handle an unbalanced 40-meter monster on a tower when trying to mount it on the rotating mast. The obvious solution is to shift the mast attachment point in such a way that a perfect balance is achieved.

The MECHANICAL YAGI BALANCE module of the software will weight-balance the Yagi. **Fig 13-10** shows the screen print of the worksheet showing the weight-balanced Yagi. The software automatically calculates the area of the required boom dummy plate (see Section 3.3.4.2), to reestablish torque balance. Components taken into account for calculating the weight balance are:

Table 13-1

Element Design Data for the 3-Element 40-Meter Yagi Reflector, Driven Element and Director

Section	OD/Wall	Dir.	Dr. El.	Refl.
1	60/5	300	300	300
2	50/5	285	285	285
3	35/2	60	85	84
4	30/2	60	112	100
5	25/1.5	135	135	176
6	15/1	60	80	82
7	12/1	111	80	113
Total length (cm)		1011	1077	1150

Section	OD/Wall	Dir.	Dr. El.	Refl.
1	2.375/0.154	144	144	144
2	1.00/0.109	55	66	66
3	1.25/0.11	34	42	50
4	1.00/0.11	30	30	30
5	1.00/0.058	30	38	42
6	0.625/0.11	18	15	21
7	0.625/0.058	28	30	34
8	0.50/0.058	60	63	65
Total length (inches)		399	428	452

Note: This design assumes a boom diameter of 75 mm (3 inches) and U-type clamps to mount the element to the boom (L=300 mm, W=150 mm, H=70 mm). This design is only meant as an example. Availability of materials will be the first restriction when designing a Yagi antenna.

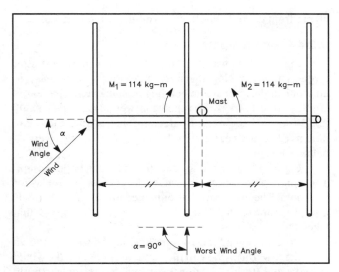

Fig 13-9—Boom moments in the horizontal plane as a result of the wind blowing onto the boom and the elements. The forces produced by the wind on the Yagi *elements* do not contribute to the boom moment; they only create a compression force in the boom (see text). The highest boom moments occur when the wind blows at a 90° angle (broadside to the boom).

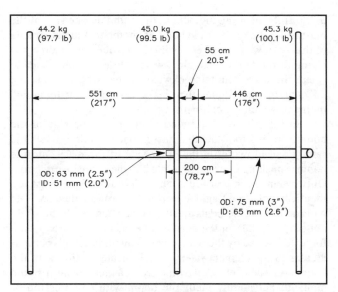

Fig 13-11—Weight-balanced layout of the 3-element 40-meter Yagi, showing the internal boom coupler. The net weight, without a match box (containing the gamma or omega matching capacitors) and without the boom-to-mast plate is 183 kg.

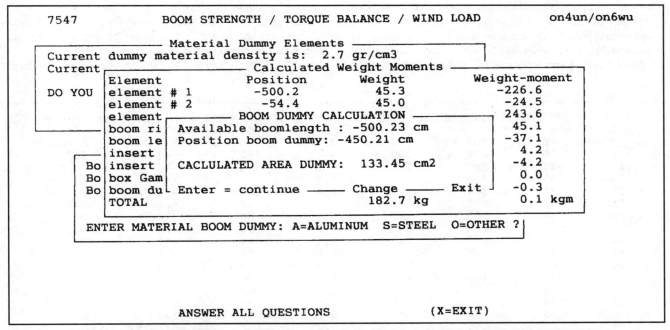

Fig 13-10—Screen print of the WEIGHT BALANCE screen from the YAGI MECHANICAL BALANCE software module. The mast attachment point has been moved 25 cm toward the reflector to restore balance in boom moments in order to achieve a zero mast torque.

- The Yagi elements.
- The boom.
- The boom coupler (if any).
- The boom dummy (see Section 3.3.4.3).
- The match box (box containing gamma/omega matching components).

Fig 13-11 shows the layout which produces perfect weight balance. In our example I have assumed no match box.

Slightly offsetting the driven element of the 3-element Yagi avoids the conflict between the location for the mast and for the driven element attach point.

3.3.4.3. Yagi torque balancing

The cause of mast torque has been explained in Section

3.3.2. If the bending moment in the boom on one side of the mast is not the same as the bending moment at the other side of the mast, we have a resultant mast torque. One moment is trying to rotate the mast clockwise, while the other tries to rotate the mast counterclockwise. If the two moments are unequal in magnitude, there is a resulting "differential" moment, which we call *mast torque*.

In other words, the wind blowing on one side of the boom is trying to rotate the mast in one direction, while the same wind blowing on the other side of the mast onto the boom is pushing the Yagi to rotate it in the opposite direction. Only when the boom areas on both sides of the mast are identical will the Yagi be perfectly torque-balanced. The wind area of the elements and their placement on the boom do *not* play any role in the mast torque, as the direction of the force developed by the wind on an element is always perpendicular to the element itself, which means in line with the boom. As such, element wind area cannot create a boom moment, but merely loads the boom with compression or elongation (see also Section 3.3.2).

It is the mast torque which makes an antenna "windmill" in high winds. A good mechanical design must be torque-free at all wind angles.

During our "weight-balancing" exercise we shifted the mast attachment point somewhat to reestablish weight balance. This has caused the boom moments on both sides of the mast to become different. In order to reestablish balance, a small *boom dummy plate* will be mounted near the end of the shorter boom half. Fig 13-10 shows the size of the dummy as calculated by the software: A small plate of 133 cm2 should be mounted

50 cm from the reflector to achieve full torque-balance.

3.3.4.4. Boom moments

Fig 13-12 shows the listing of the boom moments after torque-balancing. Note that a zero mast torque is obtained for all wind angles. The boom bending moments have increased slightly from 114 kg-m for the "non-weight-balanced Yagi" to 120 kg-m after weight balancing and adding the boom dummy. This is a negligible price to pay for having a weight-balanced Yagi.

Fig 13-13 shows all the data related to the boom design. The material stresses are shown for the coupler, as well as for the boom. The boom stress is only meaningful if the boom is not split in the center. With a split boom it is the coupler that takes the entire stress.

Note that even for a 140-km/h wind, the stresses shown are quite low. But as we will likely put the boom "in the wind" in high wind storms (Section 3.3.4.1), it is advisable to build in a lot of safety. Also, as mentioned before, the 140-km/h does not include any safety factors or modifiers, as may be prescribed in the standard EIA/TIA-222.

It is proof of *poor* engineering to design a boom which needs support guys in order to render it strong enough to withstand the forces from the wind and the bending moments caused by it. If guy wires are employed to provide the required strength, guying will have to be done in both the horizontal as well as the vertical plane. Guy wires can be used to eliminate boom sag. This will only be done for cosmetic rather than strength reasons.

Three way guying may be necessary where vertical

```
  7547              BOOM STRENGTH / TORQUE BALANCE / WIND LOAD        on4un/on6wu
  - 1 -      - 2 -       - 3 -       - 4 -       - 5 -       - 6 -       - 7 -       - 8 -
    0 deg       0 Kg    302 Kg      302 Kg     0.0 deg       0 Kgm       0 Kgm       0 Kgm
    5         -25       299         300        0.1           1          -1          -0
   10         -48       289         293        0.5           4          -4           0
   15         -67       274         282        1.2           8          -8           0
   20         -81       254         267        2.2          14         -14           0
   25         -91       232         249        3.7          21         -21          -0
   30         -94       208         228        5.6          30         -30          -0
   35         -92       183         205        8.2          39         -39          -0
   40         -86       160         181       11.7          49         -49           0
   45         -75       138         158       16.5          60         -60          -0
   50         -62       120         135       22.8          70         -70           0
   55         -47       106         116       31.1          80         -80           0
   60         -32        96         101       41.5          90         -90           0
   65         -18        89          91       53.6          98         -98           0
   70          -6        86          86       65.8         106        -106           0
   75           2        86          86       76.3         112        -112          -0
   80           6        87          87       84.0         116        -116           0
   85           5        89          89       88.5         119        -119          -0
   90           0        89          89       90.0         120        -120           0
  ─── Wind Speed: 140 kph ───────────────────────────────────────────────────────
   1 : Wind angle     2 : cross force       3 : drag force       4 : total force
   5 : load angle     6 : boom momt left    7 : boom momt right  8 : mast torque
  ──── H = HELP ───────── W = Change Wind Speed ───────── ENTER = MENU ────
```

Fig 13-12—Screen print from the YAGI MECHANICAL BALANCE software. Column 4 lists the wind load of the antenna. Column 5 represents the load angle. The load angle is usually not the same as the wind angle (see text). Columns 6 and 7 list the boom moments at the mast attachment point, and column 8 gives the resultant mast torque. The mast torque is zero if the boom moments are the same on both sides of the mast.

```
    7547         BOOM STRENGTH / TORQUE BALANCE / WIND LOAD      on4un/on6wu
             — BOOM DIMENSIONAL DATA —         ——— BOOM STRENGTH DATA ———
         Boomlength: 1051 cm              Boom stress left:    6.6 kg/mm2
    1.  Boom OD:  75.00 mm                Boom stress right:   6.6 kg/mm2
    2.  Boom Wall:  5.000 mm              Stress INSERT left:  8.5 kg/mm2
        Boom length LEFT  500 cm          Stress INSERT right: 8.5 kg/mm2
        Boom length RIGHT:  551 cm
        Area boom LEFT =   3752 cm2       6.  Wind speed =  140  Kph
        Area boom RIGHT =   4133 cm2
    3.  Boom Insert length:  200 cm       ——————— MATCH BOX DATA ———————
    4.  Boom Insert OD:  63  mm           7.   AREA ALONG BOOM:    0 cm2
    5.  Boom Insert wall:  6  mm          8.   AREA ALONG ELEMENTS:    0 cm2
        Spec. gravity insert:  7.87 kg/dm3  9.   POSITION ON BOOM:    0 cm
                                          0.   WEIGHT:  0.0 Kg

    ┌─────────────────────────────────────────────────────────────────┐
    │ BOOM STRESS: stress in the boom AT the mast attach point. If the yagi is │
    │ torque balanced the LEFT and the RIGHT values are equal.          │
    │ STRESS INSERT: the stress of a boom insert or boom coupler with dimensions │
    │ as specified, taking into account ONLY the coupler.              │
    └─────────────────────────────────────────────────────────────────┘

    ┌─────────────────────────────────────────────────────────────────┐
    │           Enter the number of the item to be changed             │
    │    H = HELP                                          ENTER = MENU │
    │ ─────────────────── Antenna File: FREDA9.ANT ─────────────────── │
    └─────────────────────────────────────────────────────────────────┘
```

Fig 13-13—The boom moment of 120 kg-m (see Fig 13-14) results in a material stress of 8.5 kg/mm² for a boom coupler of 63-mm OD, with a wall of 6 mm. If no boom coupler were used, the 75-mm OD boom, with a 5-mm wall would endure a stress of 6.6 kg/mm². In the final design of the 40-meter Yagi a steel boom coupler is used. This means that the entire moment is taken by the coupler. As explained in the text, the boom stress in the vertical plane due to weight loading is three times higher than the stress in the horizontal plane due to wind loading!

gusts can be expected (hill-top QTHs) to prevent the boom from dancing up and down due to vertical up drafts.

3.3.4.5. Boom sag

The boom as now designed will withstand 140-km/h winds, with a good safety factor. The same boom however, without any wind loading, will have to endure a fair bending moment in the vertical plane, caused by the weight of the elements and the boom itself.

Fig 13-14 shows the forces and dimensions that create these bending moments. The weight moments were obtained earlier when calculating the Yagi weight balance. (See Fig 13-12.)

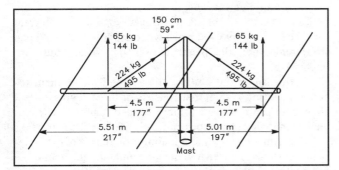

Fig 13-14—Layout of the boom-support cables with the forces and tensions involved. The boom-support cables are not installed to provide additional strength to the boom; they merely support the boom in order to compensate for the sag from weight loading of the boom.

Weight moments to the "left" of the mast:
Element no. 1: –226.6 kg-m
Element no. 2: –24.5 kg-m
Boom left: –37.1 kg-m
Boom insert left: –4.2 kg-m
Boom dummy: –0.3 kg-m
Total: –292.7 kg-m

Weight moments to the "right" of the mast:
Element no. 3: 243.6 kg-m
Boom right: 45.2 kg-m
Boom insert right: 4.2 kg-m
Total: 293 kg-m

The weight moment to the left of the mast is the same as to the right of the mast as the Yagi is weight-balanced.

Here comes another surprise: The boom is loaded almost *three times* as much by weight loading in the vertical plane (293 kg-m) than it is by wind loading at 140 km/h in the horizontal plane (120 kg-m).

The maximum allowable bending moment for the boom steel insert with a diameter of 63 mm and 6-mm wall is 619 kg-m as calculated with Eq 13-2 for a material yield strength of 20 kg/mm². This steel coupler has a safety factor of *two* as far as the loading in the vertical plane (weight-loading) is concerned. Boom stress by weight will often be the condition that will specify the size of the boom with large low-band Yagis using heavy elements.

The boom, using the above calculated coupler, does not require any guying for additional strength. However, the high weight loading of the very long elements sitting at the end of the boom halves will cause a very substantial sag in the boom. In the case of the 40-meter beam the sag amounts to nearly 65 cm, which is really excessive from a cosmetic point of view. A sag of 10 cm is due to the boom's own weight and 55 cm is due to the weight of the elements at the tips of the boom.

It is proof of good engineering to eliminate the sag by supporting the boom using slant support cables. The two boom halves are supported by two sets of two parallel guy wires attached on the boom at a point 4.5 m from the mast attachment point. The guy wires are supported from a 1.4-m high support mast made of a 35 mm OD stainless steel tube, which is welded to the boom-to-mast plate. See Fig 13-14.

The weight to be supported is given by the moment (calculated before) divided by the distance of the cable attachment point to the boom center (or mast attachment point).

$$\text{Vertical force} = \frac{293 \text{ kg} - \text{m}}{4.5 \text{ m}} = 65 \text{ kg}$$

Assuming the two boom halves are hinged at the mast, each support cable must support the total weight as shown above, divided by the sine of the angle the support cable makes with the boom.

$$\text{Force in the cable} = \frac{65 \text{ kg}}{\sin 17°} = \frac{65}{0.29} = 223 \text{ kg}$$

Remark: Leeson (Ref 964) covers the aspects of guyed booms in his publication. In the above case we are *not* guying the boom to give it additional strength, we do it only to eliminate the boom sag. Guying a boom is not a simple problem of moments, but a problem of a compressed column, where the slenderness of the boom and the compression force caused by the guy wire (usually in 3 directions) come into the picture. In our case these forces are so low that we can simplify the model as done above. In the above case we assume that the boom has enough lateral strength (which we had calculated). For solving the wire-support problem we assume that the boom is a "nonattached" cantilever. The fact that the boom is attached introduces an additional safety factor.

If a single steel cable is used, a 6-mm (0.24-inch) OD cable is required to safely support this weight. I use *two* cables of 4-mm (0.16-inch) OD Kevlar (also known as Phyllistran in the US). I use this because it was available at no cost, and it does not need to be broken up by egg insulators (Kevlar is a fully dielectric material which has the same breaking strength as steel and the same elongation under load). I recommend not using turnbuckles, as they may prove to be the weak link in the system. In addition, stainless-steel turnbuckles are very expensive. If two parallel cables are used, a tension equalizer must be used to ensure perfect equal stress in both cables. In the case of two support cables without equalization, one of the cables is likely to take most of the load.

Let me go into detail why I use two parallel support guys. **Fig 13-15** shows the top of the support mast, on which two triangular-shaped stainless-steel plates are mounted. These plates can pivot around their attachment point, which consists of a 1-cm diameter stainless-steel bolt. The two guy wires are connected with the correct hardware (very impor-

Fig 13-15—Detail of the tension-equalizing system at the top of the support mast, where the two boom-support guy-wires are attached. The triangular-shaped plate can rotate freely around the 10-mm bolt, which serves to equalize the tension in the two guy wires. See text for details.

tant—consult the supplier of the cable!) at the base of these triangular pivoting plates. The pivoting plates now serve a double purpose:

- To equalize the tension in the two guy wires.
- To serve as a visual indicator of the status of the guy wires.

If something goes wrong with one of the support wires, the triangular plate will pivot around its attachment point. At the same time the remaining support (if properly designed) will still support the boom, although with a greatly reduced safety factor.

In order to install the support cables and adjust the system for zero or minimum boom sag without the use of turnbuckles, place the beam on two strong supports near the end of the boom so as to induce some inverse sag in the boom. Lift the center of the boom to control the amount of inverse sag. Now adjust the position of the boom attachment hardware to obtain the required support behavior.

Make sure you properly terminate the cables with thimbles and all. The load involved is not small, and improper terminations will not last long. This is especially true when Kevlar rope is used.

3.3.5. Element-to-boom and boom-to-mast clamps

With an element weighing well over 40 kg, attaching such a mast at the end of a 5-m arm needs to be done with great care. The forces involved when we rotate the Yagi (start and stop) and when the beam swings in storm winds are impressive.

After an initial failure, I designed an element-to-boom mounting system that consists of three stainless-steel U-channel profiles (50-cm long) welded together. The element is mounted inside the central channel profile using four U bolts with 12-mm wide aluminum saddles. Four double-saddle systems are used to mount the unit onto the boom (see **Fig 13-16**). U bolts must be used together with saddles. You must use saddles on both sides. The bearing strength of U bolts is far too low to provide a durable attachment under extreme wind loads without saddles on both sides. Never use U bolts made of threaded stainless-steel rods directly on the

Fig 13-16—The element-to-boom mounting system as used on the 40-meter Yagi.

Fig 13-17—The omega matching system and plastic "drainpipe" box containing the two variable capacitors. Note also the boom-to-mast mounting plate made of 1-cm thick stainless steel. The boom is attached to the plate with eight U bolts and double saddles.

boom; if they can move but a hair, they become like perfect files which will machine a nice groove in the boom in no time.

At the center of the boom I have mounted a 60-cm wide 1-cm thick stainless-steel plate to which the 1.5-m long support mast for the boom guying is welded. The boom is bolted to the boom-to-mast plate using eight U bolts with saddles matching the 75-mm OD boom (see **Fig 13-17**). On the tower, this plate is bolted to an identical plate (welded to the rotating mast) using just four 18-mm OD stainless-steel bolts.

3.3.6. Materials

In the metric world (mainly Europe), aluminum tubes are usually available in 6-meter sections. **Table 13-2** lists dimensions and weights of a range of readily available tubes. Aluminum tubing in F22 quality (Al Mg Si 0.5%) is readily available in Europe in 6-meter lengths. The yield strength is 22 kg/mm^2.

Tables 13-3 and **13-4** show a range of material dimensions that are available in the US. *The ARRL Antenna Book* also lists a wide range of aluminum tubing sizes. Make sure

Table 13-2
Dimensions and Weight of Aluminum Tubing in F22 Quality

OD mm	Wall mm	Weight g/m	OD mm	Wall mm	Weight g/m
10	1	76	40	1.5	489
12	1	93	44	2	541
13	1	103	48	1.5	603
14	1	110	50	5	1923
16	1	127	50	2	820
19	1.5	227	52	1.5	654
20	1.5	235	57	2	940
22	2	339	60	5	2350
22	1.5	261	60	3	1460
25	2.5	477	62	2	1040
25	2	398	70	5	2757
25	1.5	298	70	3	1718
28	1.5	336	80	5	3181
30	3	687	80	4	2579
30	2	484	84	2	1385
32	1.5	387	90	5	3605
35	2	564	100	5	4029
36	1.5	438	100	2	1676
40	5	1495	110	5	4485
40	2	644			

Table 13-3
List of Currently Available Aluminum Tubing in the US

OD in.	Wall in.	Weight lb/ft	OD in.	Wall in.	Weight lb/ft
0.25	0.058	0.04	1.5	0.083	0.43
0.375	0.058	0.07	1.625	0.058	0.34
0.5	0.058	0.10	1.75	0.058	0.36
0.625	0.058	0.12	1.75	0.083	0.51
0.750	0.058	0.15	2.0	0.065	0.45
0.875	0.058	0.18	2.0	0.083	0.59
1.0	0.058	0.20	2.5	0.0	0.
1.125	0.058	0.23	2.5	0.0	0.
1.25	0.058	0.26	2.5	0.083	0.74
1.375	0.058	0.28	2.5	0.083	1.10
1.5	0.058	0.31	3.0	0.065	11.33
1.5	0.065	0.34			

Table 13-4
List of Currently Available Aluminum Pipe in the US

OD in.	Wall in.	OD in.	Wall in.
1.05	0.113	1.90	0.109
1.05	0.154	1.90	0.145
1.315	0.133	1.90	0.2
1.315	0.179	2.375	0.065
1.66	0.065	2.375	0.109
1.66	0.109	2.375	0.154
1.66	0.140	2.375	0.218
1.66	0.191	2.875	0.203
1.90	0.065	2.875	0.276

you know which alloy you are buying. The most common aluminum specifications in the US are:

6061-T6: Yield strength = 24.7 kg/mm^2
6063-T6: Yield strength = 17.6 kg/mm^2
6063-T832: Yield strength = 24.7 kg/mm^2
6063-T835: Yield strength = 28.2 kg/mm^2

Economical Lengths

When designing the Yagi elements, a maximum effort should be made to use full fractions of the 6-meter tubing lengths, in order to maximize the effective use of the material purchased. A proper section overlap is 15 cm (6 inches). The effective net lengths of fractions of a 600-cm tube are 285, 185, 135, 85 and 60 cm.

In the US, aluminum is available in 12-ft lengths. The effective economical cuts (excluding the 6-inch overlap) are 66, 42, 30, 22.8 inches, etc.

3.3.7. Telescopic Fits

Good-fit telescopic joints are made as follows: With a metal saw, make two slits of approximately 30-mm length into the tip of the larger section. To avoid corrosion, use plenty of Penetrox (available from Burndy) or other suitable contact grease when assembling the sections. A stainless-steel hose clamp will tighten the outer element closely onto the inner one (with the shimming material in between if necessary). A stainless-steel Parker screw will lock the sections lengthwise. For large diameters and heavy-wall sections, a stainless steel 6 or 8-mm bolt is preferred in a pre-threaded hole.

Metric tube sections do not provide a snug telescoping fit as do the US series with a 0.125-in.-diameter step and 0.058-in. wall thickness. At best there is a 1-mm difference between the OD of the smaller tube and the ID of the larger tube. A fairly good fit can be obtained, however, by using a piece of 0.3-mm-thick aluminum shimming material. US tubes with 0.125-inch diameter increase and 0.058-inch wall provide a very good fit. The slit, hose clamp, Parker screw and heat-shrink tube make this a reliable joint as well.

Sometimes sections must be used where the OD of the smaller section is the same as the ID of the larger section. In order to achieve a fit, make a slit approximately 5 cm (2 inches) in length in the smaller tube. Remove all burrs and then drive the smaller tube inside the larger to a depth of 3 times the slit length (eg, 15 cm). Do this after heating up the outer tube (use a flame torch) and cooling down the inner tube (use ice water). The heated-up outer section will expand, while the cooled-down inner section will shrink. Use a good-sized plastic hammer and enough force to drive the inner tube quickly inside the larger tube before the temperature-expansion effect disappears. A solid unbreakable press fit can be obtained. A good Parker screw or stainless-steel bolt (with pre-threaded hole) is all that's needed to secure the taper connection.

Under certain circumstances a very significant drop in element diameter is required. In this case a so-called donut is required. The donut is a 15-cm long piece of aluminum tubing that is machined to exhibit the right OD and ID to fill up the gap between the tubes to be fit. Often the donut can be made from short lengths of heavy-wall aluminum tubing.

I always cover each taper-joint area with a piece of heat-shrinkable tube that is coated with a hot melt on the inside (Raychem, type ATUM). This makes a perfect protection for the element joint and keeps the element perfectly watertight.

3.3.8. Material ratings and design conditions

All the above calculations are done in a static environment, assuming a wind blowing horizontally at a constant speed. Dynamic modeling is very complex and falls out of the scope of this book. If all the rules, the design methodology and the calculating methods as outlined above and as used in the mechanical design modules of the YAGI DESIGN software are closely followed, a Yagi will result that will withstand the forces of wind, even in a "normal" dynamic environment, as has been proved in practice.

The 40-meter Yagi was designed to be able to withstand wind speeds of 140 km/h, according to the EIA/TIA-222-E standard. The 140-km/h wind does *not* include any safety factors or other modifiers.

The most important contribution of all the above calculations is that the stresses in all critical points of the Yagi are kept at a similar level when loading. In other words, the mechanical design should be well balanced as the system will only be as strong as the weakest element in the system.

Make sure you know exactly the rating of the materials you are using. The yield stress for various types of steel and especially stainless steel can vary with a factor of 3! Do not go by assumptions. Make sure.

3.3.9. Element finishing

As a final touch I always paint my Yagi beams with three layers of transparent metal varnish. It keeps the aluminum nice and shiny for a long time.

3.3.10. Ice loading

Ice loading greatly reduces the wind survival speed. Fortunately, heavy ice loading is not often accompanied by very high winds, with an exception for the most harsh environments (near the poles).

Table 13-5
Ice Loading Performance of the 40-Meter Beam

Radial Ice		Max Wind Speed		Sag	
mm	inch	kph	mph	cm	inch
2.5	0.1	116	72	132	52
5.0	0.2	96	60	183	72
7.5	0.3	79	49	242	95
10	0.4	64	40	310	122
12.5	0.5	47	29	386	152
15	0.6	25	15	435	171
16	0.63	0	0	Break	

Note: As designed, the Yagi element will break with a 16-mm (0.63 inch) radial ice thickness at zero wind load, or at lower values of ice loading when combined with wind. The design was *not* optimized to resist ice loading. Optimized designs will use elements that are overall thicker, especially the tip elements.

Although we are almost never subject to ice loading here in Northern Belgium, it is interesting to evaluate what the performance of the Yagi would be under ice loading conditions.

Table 13-5 shows the maximum wind survival speed and element sag as a function of radial ice thickness. As the ice thickness increases, the sections that will first break are the tips. The reflector of our metric-design element will take up to 16 mm of radial ice before breaking. At that time the sag of the tips of the reflector element will have increased from 100 cm without ice to approximately 500 cm with the ice load. If the Yagi must be built with heavy ice loading in mind, you will have to start from heavier tubing at the tips. The ELEMENT STRENGTH module will help you design an element meeting your requirements in only a few minutes.

3.3.11. Material fatigue

It has often been observed that especially light elements (thin wall, low wind-survival designs) will oscillate (flutter) under mild wind conditions. Element tips can oscillate with an amplitude of well over 10 cm. Under such conditions a mechanical failure will be induced after a certain time. This failure mechanism is referred to as material fatigue.

Element vibrations can be prevented by designing elements consisting of strong heavy-wall sections. Avoid tip sections that are too light. Tip sections of a diameter of less than approximately 15 mm are not recommended, although difficult to avoid with a large 40-meter Yagi. Through the entire length of the element I run an 8-mm nylon rope which lies loosely in the element. This rope will dampen any self-oscillation that might start in the element.

At both ends, the rope is fastened at the element tips by injecting a good dose of silicone rubber into the tip of the element and onto the end of the rope. The tip is then covered with a heat-shrinkable plastic cable-head cover with internal hot melt.

At both ends of the element you must drill a small hole (3 mm) at the underside of the element about 5 cm from the tip of the element to allow the draining of any condensation water that may accumulate inside the element.

Make sure the cord lays loosely inside the element. The method is very effective, and not a single case of fatigue element failure has occurred when these guidelines were followed. A simple test consists of trying to hand excite the elements into a vibration mode. Without internal rope this can usually be done quite easily. You can get really frustrated in trying to get in an oscillation mode when the rope is present. Try for yourself!

3.3.12. Matching the Yagi

The only thing left to do is design a system that will match the antenna impedance (28 Ω) to the feed-line impedance (50 Ω). The choice of the omega match is obvious:
- No need for a split element (mechanical complications).
- No need to adjust the length of a gamma rod.
- Fully adjustable from the center of the antenna.

The two capacitors are mounted in a housing made of a 50-cm long piece of plastic drainpipe (15-cm OD), which is mounted below the boom near the driven element (Fig 13-17). This is a very flexible way of constructing boxes for housing gamma and omega capacitors. The drain pipes are available in a range of diameters, and the length can be adjusted by cutting to the required length. End caps are available that make professional-looking and perfectly watertight units.

The design of the omega match is described in detail in Section 3.10.2. **Fig 13-18** shows the SWR curve as shown on the screen of the Alpha Network Analyzer. The 1.5:1 SWR bandwidth turned out to be 210 kHz.

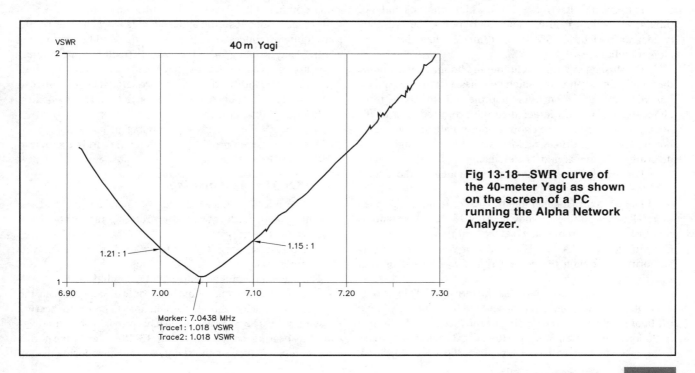

Fig 13-18—SWR curve of the 40-meter Yagi as shown on the screen of a PC running the Alpha Network Analyzer.

3.3.13. Tower, mast, mast bearings, drive shaft and rotator

If you want a long-lasting low-band Yagi system, the necessary attention should be paid to:
- The tower.
- The rotating mast.
- The mast bearings.
- The rotator.
- The drive shaft.

3.3.13.1. The tower

Your tower supplier or manufacturer will want to know the wind area of your antenna. Or maybe you have a tower that's good for 2 m² of top load. Will it be okay for the 40-meter antenna?

Specifying the wind area of a Yagi is an issue of great confusion. Wind thrust (force) is generated by the wind hitting a surface exposed to that wind. The thrust is the product of the dynamic wind pressure multiplied by the exposed area, and with a so-called drag coefficient, which is related to the *shape* of the body exposed to the wind. The "resistance" to wind of a flat-shaped body (panel) is obviously different (higher!) than the resistance of a ball-shaped or tubular-shaped body.

This means that if we specify or calculate the wind area of a Yagi, we must always specify if this is the equivalent wind area for a flat plate (which really should be the standard) or if the area is simply meant as the sum of the projected areas of all the elements (or the boom, whichever has the largest projected area; see Section 3.3.4.1).

In the former case we must use a drag coefficient of 2.0 (according to the latest EIA/TIA-222-E standard) to calculate the wind load, while for an assembly of (long and slender) tubes a coefficient of 1.2 is applicable.

This means that for a Yagi which consists only of tubular elements (Yagi elements and boom), the flat-plate wind area will be 66.6% lower (2.0/1.2) than the round-element wind area.

The 40-meter Yagi, excluding the boom-to-mast plate, the rotating mast and any match box, has a flat-plate equivalent wind area of 1.65 m². As the projected area of the three elements is 2.5 times larger than the projected area of the boom, the addition of the boom-to-mast plate and the match box will not change the wind load, which for this Yagi is only determined by the area of the elements.

The round-element equivalent wind area for the Yagi is 2.74 m².

The wind thrust generated by this Yagi at a wind speed of 140 km/h is 302 kg, as shown in Fig 13-12. This figure is for 140 km/h, without any safety margins or modifiers. Consult the EIA/TIA-222-E standard or your local building authorities to obtain the correct figure to be used in your specific case.

Let me make clear again that the thrust of 302 kg is only generated with the element broadside to the wind. If you put the boom in the wind, the loading on the tower will be limited to 90 kg. See Section 3.3.4.1. However, I would not advise using a tower that will take less than 300 kg of top load.

Consider the margin between the boom in the wind and the elements in the wind as a safety margin.

3.3.13.2. The rotating mast

Leeson (Ref 964) covered the issue of masts very well. Again, what you use will probably be dictated in the first place by what you can find. In any case, make sure you calculate the mast. My 3-element 40-meter beam sits on top of a 5-m long stainless-steel mast, measuring 10 cm in diameter with a wall thickness of 10 mm. This mast is good for a wind load of 579 kg at the top. I calculated the maximum wind load as 302 kg (see Fig 13-12). At the end of a 5-m cantilever the bending moment caused by the beam is 1510 kg-m. Knowing the yield strength of the tube we use, we can calculate the minimum required dimensions for our mast using Eq 13-2.

$$M_{max} = YS \times \delta \times \frac{10^4 - 8^4}{32 \times 10} = YS \times 58$$

where YS = yield strength

The stainless-steel tube I use has a yield strength of 50 kg/mm².

$$M_{max} = 5000 \times 58 = 290,000 \text{ kg-cm}$$

It appears that we have a safety factor of 75% versus the moment created by the Yagi (1510 kg-m). I have not included the wind load of the mast, but the safety margin is more than enough to cover the bending moment caused by the mast itself.

In my installation I have welded plates on the mast at the heights where the beam needs to be mounted. These plates are exact replicas of the stainless-steel plates mounted on the booms of the Yagis (the boom-to-mast coupling plates). When mounting the Yagi on the mast, you do not have to fool around with U bolts; the two plates are bolted together at the four corners with 18-mm-OD stainless-steel bolts.

One word of caution about stainless-steel hardware. Do not tighten stainless-steel bolts like you would do with steel bolts. They would grip and be very difficult to remove later. It is always wise to use a special grease before assembling stainless-steel hardware. Also, where safety is a concern, use one normal bolt, doubled up with a special safety self-locking bolt (with plastic insert).

Between the two plates a number of stairs have been welded in order to provide a convenient working situation when installing the antennas.

3.3.13.3. The mast bearings

The mast bearings are equally important parts of the antenna setup. Each tower with a rotating mast should use two types of bearings:
- The thrust bearing; it should take axial (weight) as well as radial load.
- The second bearing should only take radial load.

The thrust bearing should be capable of safely bearing the weight of the mast and all the antennas. The thrust-bearing assembly must be waterproof and have provisions for lubricating the bearing periodically. **Fig 13-19A** shows the thrust collar being welded on the stainless-steel mast inside the top

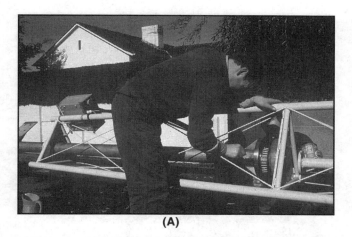

(A)

(B)

Fig 13-19—At A, the thrust bearing for the 100-mm OD-mast inside the top section of the 24-m tower at ON4UN. At B, base of the self-supporting 25-m tower (measuring 1.5 m across), with the prop-pitch motor installed 1 m above the ground. The drive shaft is coupled to the prop-pitch motor via a cardan axle from a heavy truck. Having the motor at ground level facilitates service, and takes torque load off the tower. In addition, the long drive shaft acts as a shock (momentum) absorber, greatly reducing strain on the motor.

tower section. Notice the stainless-steel housing of the thrust bearing. The bearing is a 120-mm ID, FAG model FAG30224A (T4FB120 according to DIN ISO 355). In my tower the thrust bearing is 2 meters below the top of the tower.

The second (radial) bearing is mounted right at the top of the tower and consists of a simple 10-cm long nylon bushing with approximately 1-mm clearance with the mast OD.

The thrust bearing can also be at the top with the radial bearing at the lower point. This does not make any difference. The choice is dictated by practical construction aspects.

The mast and antenna weight should never be carried by the rotator. In my towers I have the rotator sitting at ground level, with a long drive shaft in the center of the self-supporting tower. The drive shaft is supported by a thrust

bearing near the top of the tower. The fact that the heavy drive shaft "hangs" in the center of the tower adds to the stability of the tower. Replacing the rotator can be easily done. The coupling between the rotator and the drive shaft is done with a cardan axle as shown in Fig 13-19B.

3.3.13.4. The rotator

I would not dare to suggest using one of the commercially available rotators with antennas of this size. Use a prop-pitch or a large industrial-type worm-gear reduction with the appropriate reduction ratio and motor.

3.3.13.5. The drive shaft

The drive shaft is the tube connecting the rotating mast with the rotator. The drive shaft must meet the following specifications:
• It must have enough spring effect to act as a torque absorber when starting and stopping the motor. This effect can be witnessed when you start the rotator and the antenna starts moving but a second later. This action relieves a lot of stress from the rotator. Leeson (Ref 964) uses an automotive transmission damper as a torque spring.
• The drive shaft should not have too much spring effect so as to keep the antenna in the right direction in high winds. Also, if there is too much spring effect, the excessive swinging of the antenna could damage the antenna. The acceleration and the forces induced at the element-to-boom mounting hardware at the tips of the boom may induce failure at the element-to-boom mounting system.

The torque moment will deform (twist) the drive shaft (hollow tube). The angle over which the shaft is twisted is directly proportional to the length of the shaft. In practice, we should not allow for more than ±30° of rotation under the worst torque moment.

In an ideal world the Yagi is torque balanced, which means that even under high wind load there is no mast torque. In practice nothing is less true: Wind turbulence is the reason that the large wind capture area of the Yagi always creates a large amount of momentary torque moment during wind storms.

When rotation is initiated, the inertia of the Yagi induces twist in the drive shaft. This is witnessed by the fact that the antenna starts rotating some time after the rotator has been switched on. The same is true after stopping the rotator, when the antenna overshoots a certain degree before coming back to its stop position.

In practice you will have to make a judicious choice between the length of the drive shaft and the size of the shaft. Using a long drive shaft and the rotator at ground level has the following advantages (in a non-crank-up, self-supporting tower):
• No torque induced on the tower above the point where the rotator isolated.
• Motor at ground level facilitates maintenance and supervision.
• Long crank shaft works as torsion spring and takes torque load off the motor.
• The disadvantage is that you will need a sizable shaft to keep the swinging under control.

Calculating the drive shaft

Refer to Section 3.2.1 for some of the definitions used. It is difficult, if not impossible to calculate the torque moment caused by turbulent winds. I have estimated the momentary maximum torque moment to be 3 *times* as high as the *torque* moment *on one side of the boom*, as calculated before for a wind speed of 140 km/h. This is 360 kg-m. Taking this figure as a maximum momentary shaft torque, caused by highly turbulent winds, means that we consider that the wind momentarily causes the antenna to rotate in only one direction, and that we disregard the forces trying to rotate the antenna in the opposite direction. In addition I added a 200% safety factor. I use this figure as the maximum momentary torque moment to calculate the requirements for the drive shaft. I have not found any better approach yet, and it is my practical experience that, using this approach, a fair approximation is obtained of what can happen under worst circumstances with peak winds in a highly turbulent environment.

Assumed momentary maximum torque moment T = 360,000 kg-mm

Calculate the section shear modulus (Z):

$$Z = \delta \times \frac{D^4 - d^4}{16 \times D}$$

Assume the following:

D = 8 cm
d = 6.5 cm
Z = 56.7 cm^3

Calculate the shear stress (ST):

ST = T/Z

where

Z = modulus of section under shear stress
T = applied torque moment

ST = 36,000 kg-cm/56.7 cm^3 = 635 kg/cm^2
= 6.35 kg/mm^2

This is a very low figure, meaning the tube will certainly not break under the torque moment of 36,000 kg-cm.

Calculate the max. twist angle (TW):

The twist angle of the shaft is, of course, directly proportional to the shaft length. In my case the rotator is 21 m below the lower bearing, which makes the shaft 21-m long. The critical part of the whole setup is the shaft-twist angle under maximum mast torque.

$$TW = \frac{T \times L}{J \times G}$$

where

T = applied torque (360,000 kg-mm)
L = length of shaft (21,000 mm)
G = rigidity modulus of the material = 8000 kg/mm^2
J = section modulus × radius of tube = 56,700 mm^3 × 40 mm = 2,268,000 mm^4

$$TW = \frac{360,000 \text{ kg} - \text{mm} \times 21,000 \text{ mm}}{2,268,000 \text{ mm}^4 \times 8,000 \text{ kg}/\text{mm}^2}$$

TW = 0.44 radians = 25°

This means that our anticipated 360 kg-m torque moment,

Fig 13-20—The 40-meter 3-element full-size Yagi is lowered on top of the rotating mast at a height of 30 m with the use of a 48-m hydraulic crane.

applied to a 21-m-long drive shaft, with OD = 80-mm ID = 65 mm and a rigidity modulus of 8000 kg/mm^2, will produce a twist angle of 25°. This is an acceptable figure. The twist should in all cases be kept below 30°, in order to keep the antenna from excessively swinging back and forth in high winds.

It is clear that the same result could be obtained with a much lighter tube, provided it was a much shorter length.

3.3.14. Raising the antenna

A 3-element full-size 40-meter Yagi, built according to the guidelines outlined in the previous paragraphs, is a "monster." Including the massive boom-to-mast plate, it weighs nearly 250 kg and is huge.

A few years ago I met a man who has his own crane company. He has a whole fleet of hydraulic cranes that come in very handy for mounting large antennas on their tower.

Fig 13-20 shows the crane arm extended to a full 48 m, maneuvering the 40-meter Yagi on top of the 30-m self-supporting tower.

With the type of boom-to-mast plates shown in Fig 13-17, it takes but a few minutes to insert the four large bolts in the holes at the four corners of the plates and get the Yagi firmly mounted on the mast.

3.3.15. Conclusion

Long-lasting full-size low-band Yagis are certainly not the result of much improvisation. The 40-meter Yagi that has been described has been up for almost 10 years now, without ever having needed any repair.

Long lasting Yagis, especially for the low bands are the result of a serious design effort, which is 90% a mechanical engineering effort. Software is now available that will help design mechanically sound, large low-band Yagis. This makes it possible to build a reliable antenna system that will out-perform anything that is commercially available by a large margin. It also brings the joy of home-building back into our hobby, the joy and pride of having a no-compromise piece of equipment.

3.4. A Super-Performance Super-Lightweight 3-Element 40-Meter Yagi

Nathan Miller, NW3Z designed a very novel and at-tractive 3-element Yagi that was featured in *QST* (Ref 979). See **Fig 13-21**. It weighs only a tiny fraction of the battleship described in Section 3.3. This antenna can be mounted on a tower by the *armstrong method*, and turned with a run-of-the-mill good-quality rotator. The antenna is based on a similar 2-element design by Jim Breakall, WA3FET.

In addition this 3-element Yagi also uses the principle of instantaneous pattern reversal which I described about 7 years ago in a previous edition of this book (see also Section 3.5). Basically the 3-element Yagi uses two directors, sym-metrically located with respect of the driven element. By using small relays an inductor or short stub is inserted in the middle of the parasitic element to turn it into a reflector. This

makes instantaneous direction switching possible. You must have experienced this feature in order to fully appreciate it. I would say that it's a must for a serious contest station.

3.4.1. Electrical performance

I modeled the antenna using *EZNEC*, both in free space as well as over real ground. The dimensions shown in **Fig 13-22** are very close to those published by NW3Z.

In free space the antenna exhibits 7.34 dBi gain, with a feed impedance of 40.5 Ω, using a loading inductance with a reactance of 138 Ω as indicated in the *QST* article.

The Yagi has a gain of 7.1 to 7.2 dBi, all across the 40-meter band. The F/B performance in free space is illus-trated in **Fig 13-23**.

The design was made for a fairly low-Q antenna, yield-

Fig 13-21—The 3-element 40-meter Yagi is mounted on a 21-m crank-up tower at the Penn State University Dept of Electrical Engineering antenna research facility at Rock Springs.

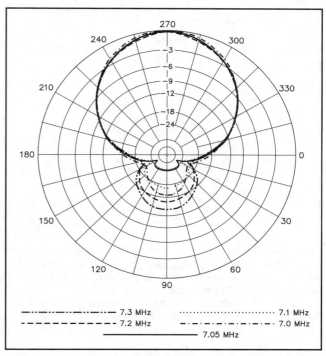

Fig 13-23—Horizontal radiation pattern in free space for the NW3Z/WA3FET 40-meter Yagi. The F/B is 20 dB or better from 7.0 to 7.1 MHz, and still a usable 17 dB at 7.2 MHz.

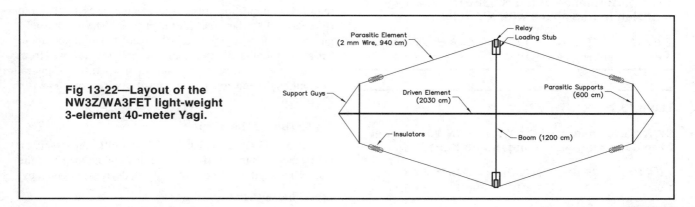

Fig 13-22—Layout of the NW3Z/WA3FET light-weight 3-element 40-meter Yagi.

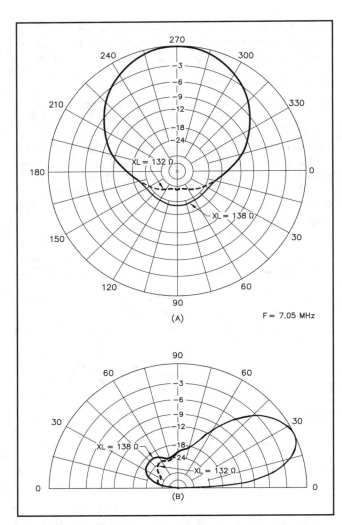

F = 7.05 MHz

(A)

(B)

Fig 13-24—At A horizontal-radiation pattern at 21-m height over average ground for the NW3Z/WA3FET 40-meter Yagi. The patterns are for 7.05 MHz. Reducing the value of the loading reactance from 138 to 132 Ω improves the F/B performance.

Table 13-6

SWR performance of the WA3FET/NW3Z Yagi as modeled in free space with XL=138 Ω.

7.0 MHz	7.05 MHz	7.1 MHz	7.2 MHz	7.3 MHz
1.4:1	1.2:1	1.1:1	1.3:1	1.7:1

Table 13-7

SWR performance of the WA3FET/NW3Z Yagi at 21 m over average ground, using XL=132 Ω.

7.0 MHz	7.05 MHz	7.1 MHz	7.2 MHz	7.3 MHz
1.6:1	1.3:1	1.2:1	1.3:1	1.8:1

ing a feed-point impedance of nearly 50 Ω, which means that the antenna is split fed without any type of matching network (a current balun was of course used). The computer SWR values are shown in **Table 13-6**.

I also modeled the antenna over real ground, at a height of (only) 21 m. We learned in Chapter 5, Section 1.1.1.2 that for most DX paths a wave angle between 10 and 15° seems to be optimum. if the wave angle is 15°, a Yagi at 0.6 λ will trade in about 1.5 dB vs. its brother at 1 λ, but it will have much better high angle rejection. The high antenna rejects a signal at a wave angle of 60° in the forward direction by approx. 8 dB. The antenna at 0.6 λ will reject the same signal, about 18 dB!

This antenna, which is within reach of many, can perform quite outstandingly at a 0.5 to 0.6-λ height. **Fig 13-24** shows the radiation patterns for 7.05 MHz. Note that in order to obtain best F/B at that frequency, the reactance of the loading coil should be changed from 138 Ω to 130 Ω. The SWR curve becomes a little steeper, especially on the low side, as shown in **Table 13-7**.

3.4.2. Mechanical design

The prototype was made using two types of aluminum tubing: The 2.5 and 2.25-inch-OD tubing is an extruded 6061-T6 alloy with 0.0125-inch walls. All other tubing is 6063-T832 with 0.058-inch wall. The parasitic elements are made of #10 aluminum plated steel wire. Copper-clad steel wire or bronze wire would also be appropriate.

The boom, for which the taper schedule is shown in **Fig 13-25** weights only 9 kg. The entire Yagi weighs well under 50 kg, which makes this really a super lightweight 3-element full-size 40-meter antenna!

3.4.3. The parasitic element supports

Miller uses an aluminum spreader (1.5-inch OD) tubing broken up by fiberglass rods in order to minimize loading of the director. Also, where the element supports are attached to the driven element, he uses a 30-cm long fiberglass rod, in order to keep the metal of the support far enough from the driven element. A valid alternative would of course be to use fiberglass poles along the entire length.

3.4.4. Truss wiring

Because of the additional cross-arm and parasitic-wire weight loading on the tips, the full-size driven element requires a supporting truss. The antenna uses Phyllystran (PVC coated Kevlar rope) for the purpose. The boom is also guyed. Both sets of guy wires are attached to a support about 2 m above the antenna. Horizontal support guys are used from the driven-element tip to the ends of the parasitic supports to counter the tension in the parasitic wires, as shown in **Fig 13-26**.

3.4.5. Tuning the Yagi

The loading reactance of 132 to 139 Ω represents an inductance of 3 to 3.1 μH. Let us calculate the length of the stub. The length of the closed stub (in degrees) is given by:

$$L(°) = \arctan (XL/Z)$$

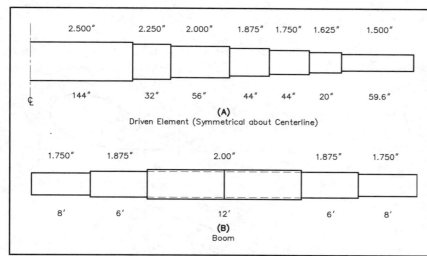

Fig 13-25—Taper schedule for the driven element (A) and boom (B) of the NW3Z/WA3FET Yagi. The driven-element taper schedule is quite different from an ordinary Yagi element taper, as the element supports the 6-m (20-ft) long cross bars at the end, which in turn support the ends of the parasitic wires. In this design the driven element is more like a boom, while the boom can be much lighter because it only supports the centers of the parasitic wires. The driven element weighs approx. 22 kg.

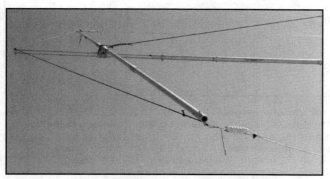

Fig 13-26—The parasitic-wire, cross support, parasitic elements, horizontal and vertical supports in place on the NW3Z/WA3FET Yagi.

where
   XL = reactance of the stub
   Z = characteristic impedance of the open-wire line.
Let us take Z = 450 Ω: L = arctan (132/450) = 16.3°
Assuming a velocity factor of 95%, the length of the stub is:

Length (meters) = (299.8 × 16.3) / (7.05 × 360) = 1.92 m

Make sure you measure the inductance!

The *NEC* model shows that the reflector is resonant on 7.7 MHz, and the director on 7.0 MHz! The best way to make sure that the parasitic elements are resonant on 7.7 MHz would be to feed the elements temporarily with ¹/₂-λ feed lines (make sure they are exactly ¹/₂-λ long) and cut them for zero reactance as measured on a network analyzer or an antenna analyzer. While doing this the driven element and the second parasitic element must be left "open." Adjusting the loading coil or stub can be done the same way: connect the feed line in series (not in parallel!) with the loading coil.

3.4.6. Feeding the Yagi

The original design uses a very simple split-element feed, as the antenna impedance is around 40 Ω. This requires the driven element to be split. A fiberglass rod used as a boom

insert can be used for the purpose. When direct feed is used, a current balun is required. Alternatively one could use any of the other matching systems described in Section 3.10.

3.4.7. Conclusion

Considering that this antenna only trades in 0.2 dB of forward gain vs the heavy-weight 3-element Yagi described in Section 3.3, and given its additional feature of instant direction reversal, this antenna is one of the most interesting designs that has been published for a long time, and deserves great popularity.

When will we see the first 80-meter version of this design?

3.5. A 3-Element Full-Size 80-Meter Yagi

One of the disadvantages of a rotatable Yagi is the fact that "switching directions" takes a while. Very large and heavy Yagi antennas should not be rotated at speeds of more than 0.5 to 1 rev/min maximum. In order to overcome this problem, it is possible to design a Yagi where, by means of relays, the director is instantly transformed into a reflector, and vice versa. This means that at 1 rev/min it would never take more than 15 seconds to get the antenna in any direction. On average, it would take 7.5 seconds.

Also, due to its high relative bandwidth (8.2% as compared to 2.4% for the 20-meter band), it is impossible to design a Yagi that will exhibit good gain, good F/B and an acceptable SWR at the high end (3.8 MHz) as well as the low end (3.5 MHz) of the band, without resorting to our bag of special tricks.

3.5.1. Antenna height for an 80-meter Yagi

Fig 13-27 shows the radiation patterns for a "standard" 3-element Yagi at heights ranging from ¹/₂ λ to 1 λ. Above ¹/₂ λ, an annoying high angle lobe appears, and a lot of RF is wasted at that angle. At a ¹/₂-λ height (Fig 13-27A), the radiation angle is approximately 25° to 30° (depending on the ground quality), with a reasonable broad lobe (29° at −3 dB).

Chapter 5, Section 1.1.1.2 showed us that wave angles

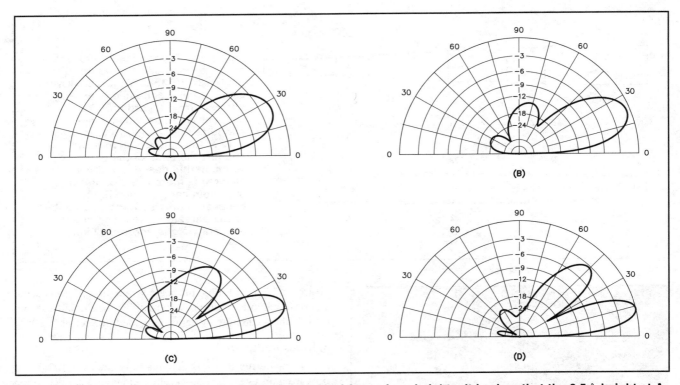

Fig 13-27—Vertical radiation patterns of a 3-element Yagi for various heights. It is clear that the 0.5-λ height at A is by far the most suitable height for general DXing on 80 meters. The high-angle secondary lobes and the narrow first lobe plus the minimum (dip) between the first and the second lobe make higher heights a bad choice for 80 meters, where the bulk of DX signals come in at wave angles between 25 and 50°. The patterns are calculated for flat ground with good ground conductivity. A—0.5 λ height B—0.6 λ height C—0.8 λ height D—1.0 λ height.

as low as 10° are not unusual, and that the bulk of DX happens at angles between 10 and 20°. This means that with horizontally polarized antennas you can't really get high enough: 40 m ($^1/_2$ λ) seems to be a bare minimum (radiation angle approx. 25°), but I know of 80-meter Yagis at 30 m that perform very well also.

If you are tempted to put the Yagi much higher, eg, at 1 λ (78 m), the main lobe is as low as 14° (I know that's way too low for almost all DX on 80 meters). The lobe will be quite narrow (only 14° at –3 dB) and you have a null at 30°, which happens to be the angle where you will have a lot of DX coming in. The second lobe is at 45°, which in turn is already too high for serious DX work. I know very high antennas are like a status symbol, but this time too high is no good! It is true that at 1-λ height the Yagi exhibits 1.0 dB more gain than at 0.5 λ, but what's the point of concentrating more energy at the wrong elevation angle?

The 3-element full-size Yagi described here has been developed to be installed at a height of 38 m over flat ground with good conductivity properties.

3.5.2. Electrical design

The Yagi has been developed to be physically "fully symmetrical." This means that the driven element is right at the center of the boom, with two parasitic elements of equal physical length (the director length at the highest operating frequency). The reflector is then loaded in the center (by an inductance) in order to lower its resonant frequency. This means that both parasitic elements need to be split (at the boom). By a set of relays it is possible to either short the split (turn the element into a director) or insert the required inductance (turn it into a reflector). This is the approach that was later adopted by Jim Breakall, WA3FET, and N. Miller, NW3Z, for designing their super-lightweight Yagis (Section 3.4).

I also set out to design a Yagi which should be switchable from the SSB to the CW portion of 80 meters without any compromise in performance (gain, F/B).

The constant-element-diameter design is shown in **Fig 13-28**. I used a constant diameter of 100 mm (4 inches) which (later) turned out to be the equivalent diameter of the tapering diameter element of our mechanical design (see Section 3.5.8). Using this element diameter and inserting a coil with $X_L = 65$ Ω in the reflector turns this into a Yagi with a very good gain and F/B ratio. Note that, with a smaller element diameter, the Q factor of the element would be higher, which in turn means that one would require more inductance to tune the element to the same frequency. For a constant diameter of 22.225 mm ($^7/_8$ inch), the required reactance would be 85 Ω.

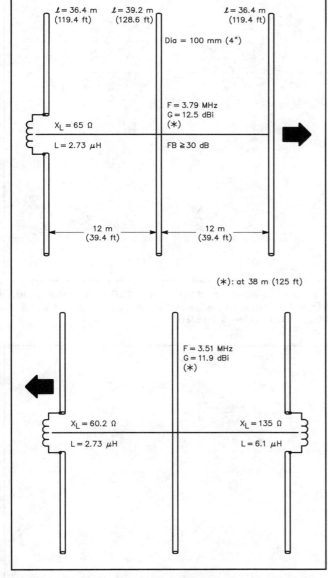

$\ell = 36.4$ m
(119.4 ft)

$\ell = 39.2$ m
(128.6 ft)

$\ell = 36.4$ m
(119.4 ft)

Dia = 100 mm (4")

$X_L = 65$ Ω

L = 2.73 μH

F = 3.79 MHz
G = 12.5 dBi
(✶)

FB ≧ 30 dB

12 m
(39.4 ft)

12 m
(39.4 ft)

(✶): at 38 m (125 ft)

F = 3.51 MHz
G = 11.9 dBi
(✶)

$X_L = 60.2$ Ω

L = 2.73 μH

$X_L = 135$ Ω

L = 6.1 μH

Fig 13-28—Design of the equally spaced 3-element 80-meter Yagi. The element lengths shown are for a constant element diameter (100 mm). The loading coils make this an excellent Yagi for 3.5 as well as 3.8 MHz. Note that the same coil (2.73 μH) is used as a loading element for the reflector on 3.8 MHz and for the director for 3.5 MHz.

To make the same Yagi work on 3.5 MHz, all that is required is a coil in the director element, and a second (larger) coil in the reflector. It turns out that on 80 meters, an element length that makes a perfect reflector for 3.8 MHz, is a perfect director on 3.5 MHz. In other words, the same coil that is used for loading the reflector on 3.8 MHz can be used as a loading coil for the director on 3.5 MHz.

In our example, the coil that has a reactance of 65 Ω on 3.79 MHz (2.73 μH) has a reactance of 65 × 3.51/3.79 = 60.2 Ω on 3.51 MHz. Together with a loading coil having a reactance of $+j135$ Ω at 3.51 MHz (6.1 μH), this value results

in a very good 3-element Yagi for the CW end of the 80-meter band. If the antenna is erected at a height of $\frac{1}{2}$ λ, the F/B ratio is between 25 and 30 dB at any wave angle between 0 and 90°, at both design frequencies (3.79 and 3.51 MHz).

The initial design was modeled with *ELNEC*. Modeling and optimizing of the Yagi for best gain and F/B was done over real (good) ground at a height of $\frac{1}{2}$ λ. This is the ideal height for such an antenna. Under these conditions the gain is calculated as 12.5 dBi at 3.79 MHz and 11.9 dBi at 3.51 MHz. The horizontal and vertical radiation patterns for the 3-element Yagi are shown in **Figs 13-29** and **13-30**.

3.5.3. Parasitic parallel capacitance with split elements

Split elements cannot be realized without introducing some parallel capacitance between the inside end of the half-element and the boom, or between the two element halves (in case you have no boom or have a dielectric boom). The ends of the insulated elements have a certain capacitance with the boom because of the mechanical construction of the insulating material and all the mounting hardware. If we were to use the loading coils as modeled above, without taking into account the "parasitic" capacitance, the loading effects could be way off.

The parasitic capacitance is the value of the series connection of the capacitances of each half element to the boom. In other words, the values shown in **Fig 13-31** are half the values as measured on one of the element legs. It is essential that this capacitance be measured. This can easily be done before the Yagi is raised. However, you cannot measure it on a finished element, because the self-capacitance (from one side to the other and also to ground) of the full element itself would upset the results.

I made a mockup of the center insulator consisting of the boom and the mounting hardware, but no element. Then I measured the capacitance at the Yagi operating frequency. The capacitance can range from just a few pF, if special care has been taken to reduce it, to several hundred pF.

The mechanical design shown in **Fig 13-32** turned out to have an extremely low parasitic capacitance of only 32 pF between the ends of the split elements (64 pF between each element half and the boom).

3.5.4. Modeling the Yagi including the parasitic parallel capacitance

Now that we know we have 32 pF across the split elements where the loading lines (hairpins) will be connected, we must model the Yagi using a parallel tuned circuit as a loading element, instead of just an inductor. The parallel capacitor of the tuned circuit is 32 pF. We must find the required inductance to achieve the desired loading as modeled before in our simplified model without parallel capacitance.

The following design methodology was used.

• The Yagi was first modeled and optimized without taking into account the parasitic capacitance.
• When the model was optimized, the resonant frequency of the director and the reflector was determined. This can be easily done as follows.

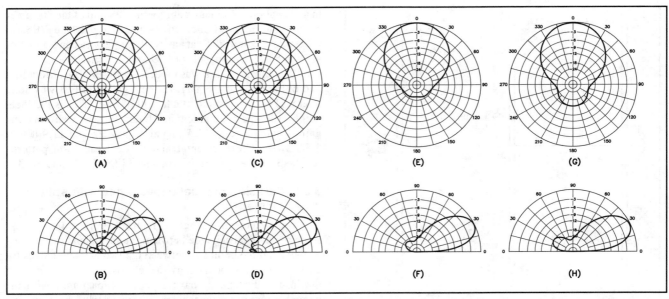

Fig 13-29—Horizontal and vertical radiation patterns for the 3-element 80-meter Yagi on the SSB end of the band. These patterns are for the Yagi design frequency of 3.79 MHz. All azimuth patterns are for a wave angle of 27°.
A and B—3.8 MHz   C and D—3.79 MHz   E and F—3.775 MHz   G and H—3.75 MHz

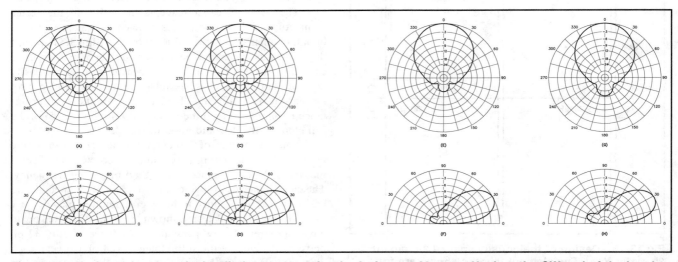

Fig 13-30—Horizontal and vertical radiation patterns for the 3-element 80-meter Yagi on the CW end of the band. These patterns are for the Yagi design frequency of 3.51 MHz. All azimuth patterns are for a wave angle of 28°.
A and B—3.5 MHz   C and D—3.51 MHz   E and F—3.53 MHz   G and H—3.55 MHz

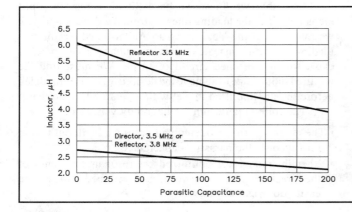

Fig 13-31—Value of the tuning coil for the 80-meter Yagi as a function of parallel capacitance. Center-insulated elements always suffer from parasitic parallel capacitance. This means that in fact the loading coils are part of the parallel circuits. The values must be adjusted in order to obtain the desired reactance. This chart shows the required reactance (in μH) as a function of the total parasitic capacitance (capacitance per half element ÷ 2), for the two loading coils. See text for details.

Fig 13-32—Detail of the mounting of the parasitic element on the boom. The boom is made of a steel lattice-tower type of construction (solid bars, no tubes!). The mounting and insulation method has been developed by DJ6JC, and ensures extremely low parasitic capacitance as well as simplicity in mounting and top mechanical strength. In this picture the boom is straight up (vertical). In this setup the boom supports the hairpin loading elements during the element resonant-frequency measuring session. DJ4PT (owner of the 3-element full-size 80-meter quad described later) is in the foreground, with DJ6JC looking on in the background.

1. Delete all elements from the model, except the element whose resonant frequency we want to know.

2. Keep the loading device (if any), and excite the center of the element. The loading device can be simply in series with the excitation.

3. Change the resonant frequency until you find a feed-point impedance where the reactive part is zero (this is the definition of resonance). In our Yagi the director for the SSB design (F_{design} = 3.79 MHz) is resonant at 4.005 MHz; the reflector is resonant at 3.745 MHz. The CW design (F_{design} = 3.51 MHz) has a director that is resonant at 3.745 MHz, and a reflector that is resonant at 3.465 MHz.

• Now the loading inductors are replaced in the modeling program by a parallel tuned circuit ($C_{parallel}$ = 32 pF), and the inductance values are found that produce the same resonant frequencies as found in our simplified (no parallel capacitance) model.

The 3.745-MHz element turns out to require a loading inductance of 2.6 μH (in parallel with the 32 pF of parallel capacitance). This is + j 62 Ω at 3.79 MHz, or 57.3 Ω at 3.51 MHz. Compare these values with the values of 65 Ω and 60.2 Ω (L = 2.73 μH) as required when there is no parasitic parallel capacitance.

The 3.465-MHz CW-band reflector requires a loading coil of 5.6 μH (in parallel with 32 pF). This represents a reactance of + j 123 Ω at 3.51 MHz. Without the parallel capacitance the required loading inductance was 6.1 μH.

Fig 13-31 shows the adapted values of inductive reactance as a function of the parasitic capacitance. This chart is only valid for the Yagi with a given Q factor. In our design

case, this is for a Yagi with an equivalent constant element diameter of 100 mm (4 inches). A Yagi with smaller diameter elements will require more loading inductance and vice versa. A similar chart can be constructed easily for any element Q factor by modeling the combinations using, eg, *ELNEC*.

3.5.5. The loading elements

The loading coils can be made preferably in a hairpin configuration. I used a transmission line made of 8 mm (0.3 inch) OD aluminum rods, spaced 10 cm (4 inches), which gives a feed-line impedance of 389 Ω (Z = 276 × log (2S/D).

The length of the hairpin is given by:

$$\ell° = \arctan (X_L/Z)$$

where

X_L = reactance of the stub

Z = characteristic impedance of the open-wire line.

1) The 2.6-μH hairpin (reflector loading in the SSB band, director loading in the CW band):

$$X_L = 62 \ \Omega$$

$$L° = \arctan (62/389) = 9.1°$$

Assuming a velocity factor of 95%, the length of the stub is:

Length (meters) = (299.8 × 9.1) / (3.78 × 360) = 1.92 m

2) The 5.6-μH hairpin (reflector loading in CW band):

$$X_L = 123 \ \Omega$$

$$L° = \arctan (123/389) = 17.5°$$

Assuming a velocity factor of 95%, the length of the stub is:

Length (meters) = (299.8 × 17.5) / (3.52 × 360) = 4.14 m

Fig 13-33 shows the loading and switching layout, which is identical for both parasitic elements. The switching has two purposes:
1. Switching from SSB to CW band.
2. Instantaneous direction reversal.

3.5.6. Remote tuning for optimum F/B

The radiation patterns are shown in Figs 13-29 and 13-30. Note that the F/B deteriorates quite rapidly in the SSB band below 3.76 MHz.

We can tune the Yagi for a high F/B ratio over quite a wide spectrum by connecting a capacitor in parallel with the hairpin at the center of the reflector element. In practice, this will not be needed on the CW band, where an excellent F/B is obtained from 3.5 to 3.53 MHz. In the phone band, however, adding a variable capacitor across the hairpin of the reflector will allow us to tune the Yagi for an F/B ratio of better than 25 dB at any frequency between 3.68 and 3.8 MHz!

Without the extra capacitance, the F/B is better than 22 dB from 3.76 to 3.8 MHz. With 100 pF in parallel, the F/B is better than 23 dB from 3.73 to 3.78 MHz, and with 200 pF in parallel, an F/B of better than 24 dB can be

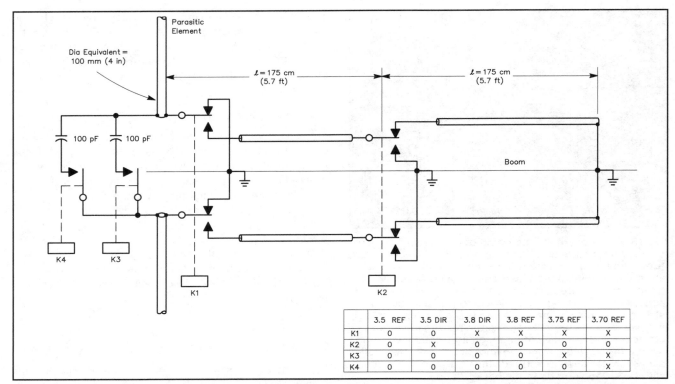

	3.5 REF	3.5 DIR	3.8 DIR	3.8 REF	3.75 REF	3.70 REF
K1	0	0	X	X	X	X
K2	0	X	0	0	0	0
K3	0	0	0	0	X	X
K4	0	0	0	0	0	X

Fig 13-33—Switching harness for the parasitic elements. To make a director at 3.8 MHz the two element halves are stripped. As a reflector element on 3.8 MHz (and as a director element on 3.5 MHz) the short (175 cm) hairpin is used, which has an inductance of 2.6 μH. When operating as a reflector element on 3.5 MHz the extra length of hairpin (making it in total 350 cm long) is switched into the circuit, resulting in a hairpin with an inductance of 5.6 μH.

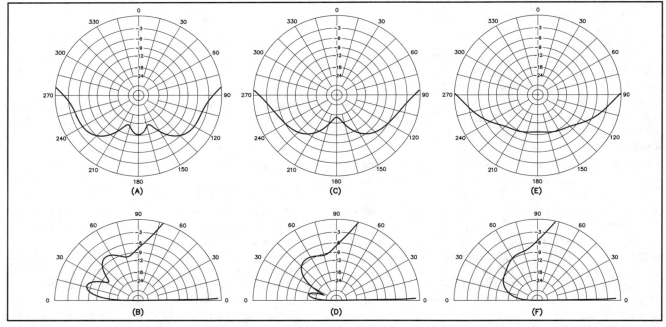

Fig 13-34—Expanded radiation patterns showing the front-to-back ratio that can be obtained by adjusting the capacitor value across the hairpin of the reflector element. Note that the outer ring is at −20 dB from the maximum response (front lobe) of the Yagi. In our example we obtain better than 30 dB anywhere between 3.7 and 3.8 MHz. All azimuth patterns are for a wave angle of 28°.

A and B—Patterns for 3.79 MHz, without parallel capacitor.

C and D—Patterns for 3.75 MHz, with 100 pF across the reflector stub.

E and F—Patterns for 3.7 MHz, with 200 pF across the reflector stub.

achieved between 3.69 and 3.73 MHz. These are worst-case F/B values over the entire 90° wave angle in the back of the Yagi. **Fig 13-34** shows the back patterns of the Yagi (on a very much stretched scale—outer ring equals –20 dB referenced to the maximum response) when tuned for maximum F/B using the variable capacitor across the reflector element.

In practice we can mount two transmitting-type 100-pF ceramic capacitors right at the center of the reflector element and switch these capacitors in parallel with the loading hairpin with vacuum relays. Fig 13-33 shows the switching and loading arrangement which must be provided at both parasitic elements.

This arrangement has been successfully implemented by OZ8BV in a 3-element full-size 80-meter Yagi, built along these principles. OZ8BV reports 30 dB F/B over a wide frequency range, tunable by the parallel capacitor across the tuning stub. See **Fig 13.35**.

3.5.7. Feeding the Yagi

3.5.7.1. Omega match

In our model, which has a constant element diameter of 100 mm (4 in.), the reactive part of the impedances at 3.79 MHz and at 3.51 MHz differ approximately 75 Ω. I dimensioned the driven-element length so that on 3.79 MHz the reactance would be approximately +27 Ω, and –49 Ω at 3.5 MHz. The element was then isolated from its parasitic elements and its resonant frequency was found to be 3.745 MHz. It is interesting to see that our driven element has the same resonant frequency as our reflector for 3.79 MHz and our director for 3.51 MHz.

There is no special meaning to this; it only illustrates that the driven-element length does not contribute to the exact radiation pattern of the Yagi. Its exact length will only

Fig 13-35—OZ8BV's 3-element 80-meter Yagi showing the loading hairpin mounted in a fiberglass boom extension. This boom extension makes part of the element rigging (see Fig 13-45). The vacuum variable capacitors used to remotely tune the Yagi for best F/B are mounted in the PVC pipe, which is mounted alongside the vertical mast, serving as a truss support for the element.

play a role in designing a matching system for the Yagi. In general the driven element will be between 0.47 and 0.50 λ long.

Fig 13-36 shows the matching that was obtained with the GAMMA/OMEGA OPTIMIZER module of the YAGI DESIGN program.

In an effort to cover both band ends with a smooth and well balanced SWR curve, I decided to use two sets of omega/gamma capacitors, with a single omega rod length. Two vacuum relays are used to select either set. The schematic with the component values is given in **Fig 13-37**, as well as the resulting SWR curves, together with the worst-case F/B value. This shows that the Yagi exhibits a well matched SWR and F/B performance over both the DX CW section (3.5 to 3.55 MHz) as well as the DX SSB section (3.73 to 3.8 MHz), with peak performances centered on 3.79 and 3.51 MHz.

The Yagi can also be fed via a direct feed, if the driven element is split and insulated from the boom. Section 3.10.4 describes 2 very attractive alternatives in detail.

3.5.8. Mechanical design of the elements

The half-element lengths for our theoretical model, with a constant diameter of 100 mm are:
- Director/reflector: 18.2 m
- Driven element: 19.60 m

In terms of wavelengths (f = 3.79 MHz) the dimensions of the Yagi are:
- Director/reflector: 0.46 λ
- Driven element: 0.4955 λ
- Spacing: 0.1517 λ

The final elements' lengths, when using a tapering schedule, are much longer than for the constant reference diameter, and can be calculated using the ELEMENT TAPER module of the YAGI DESIGN software. Depending on the exact taper configuration, a full-size reflector will be approximately 42-m long. There are two practical approaches for constructing elements that are that long:

- All tubular construction.
- Tubular tips and lattice construction for central section.

The full-size Yagis that were built by OZ8BV, I5NPH and W6MBK use the first approach. The elements are top and side braced to ensure the required structural strength.

Using the ENTER YOUR OWN DESIGN module of the YAGI DESIGN software, the ELEMENT TAPER module, and the ELEMENT STRENGTH module, we can make the physical design of tubular elements.

Above a certain diameter the use of a lattice construction may be more economical than the use of a tubular construction. The lattice construction creates less wind load and weighs much less. OH6RM (**Fig 13-38**) uses this approach in his full-size 80-meter design. **Fig 13-39** shows two possible designs where a central lattice section is extended by a 10-m long tapered-diameter section.

To keep the weight of the element within practical limits, and in order to obtain the required strength, it may be necessary to guy the central part of the elements toward the

```
        DIAMETER DRIVEN ELEMENT =    400        GAMMA WIRE LENGTH = 300.0 mm
                                                              11.81 inch
        DIAMETER GAMMA ROD =    30               GAMMA CAPACITOR =    252 pF (C1)
        SPACING CENTER TO CENTER =   500         OMEGA CAPACITOR = 280.9 pF (C2)
                                                 ROD LENGTH = 450.0 cm
        DESIGN FREQUENCY =   3.78 MHz                   177.2 inches

        TUNING RATE = 5  %                       FEEDLINE-Z = 50  ohm
```

———————————— Press F/C to change tuning rate ————————————

ANTENNA IMPEDANCE				SWR		MATCHED IMPEDANCE		
FREQ	Real	/	Imag			Real	/	Imag
3.820	17.2	/	37.3	1.99		31.0	/	20.0
3.800	18.6	/	30.1	1.30		39.9	/	6.0
3.790	19.2	/	26.7	1.11		45.0	/	-0.7
3.780	20.4	/	20.1	1.32		58.3	/	-12.4
3.750	23.8	/	14.8	1.62		74.4	/	-17.0
3.740	24.3	/	11.8	1.81		82.8	/	-20.9
3.730	24.7	/	9.0	2.00		91.5	/	-24.0

```
1=DECR.C1  2=INCR.C1   3=DECR.C2   4=INCR.C2   5=DECR.ROD   6=INCR.ROD   Z=CABLE-Z
7=DECR.WIRE  8=INCR.WIRE   C=COURSE   F=FINE   T=TUBE DIMS   S=SAVE   X=EXIT   H=HELP
```

```
        DIAMETER DRIVEN ELEMENT =    400        GAMMA WIRE LENGTH = 300.0 mm
                                                              11.81 inch
        DIAMETER GAMMA ROD =    30               GAMMA CAPACITOR =    176 pF (C1)
        SPACING CENTER TO CENTER =   500         OMEGA CAPACITOR = 262.5 pF (C2)
                                                 ROD LENGTH = 450.0 cm
        DESIGN FREQUENCY =   3.55 MHz                   177.2 inches

        TUNING RATE = 5  %                       FEEDLINE-Z = 50  ohm
```

———————————— Press F/C to change tuning rate ————————————

ANTENNA IMPEDANCE				SWR		MATCHED IMPEDANCE		
FREQ	Real	/	Imag			Real	/	Imag
3.500	25.3	/	-52.0	1.18		44.1	/	-5.3
3.510	24.8	/	-49.4	1.02		49.8	/	0.8
3.525	23.7	/	-45.6	1.30		59.7	/	10.9
3.550	21.3	/	-38.4	2.04		87.4	/	30.2
3.575	18.7	/	-30.0	3.18		145.1	/	42.3
3.600	16.1	/	-20.8	4.82		240.3	/	-13.5

```
1=DECR.C1  2=INCR.C1   3=DECR.C2   4=INCR.C2   5=DECR.ROD   6=INCR.ROD   Z=CABLE-Z
7=DECR.WIRE  8=INCR.WIRE   C=COURSE   F=FINE   T=TUBE DIMS   S=SAVE   X=EXIT   H=HELP
```

Fig 13-36—Design of the omega match for the 80-meter Yagi. The top table shows the settings for the CW end of the band. The gamma was tuned for 1:1 SWR on 3.51 MHz. The bottom worksheet shows the values for the SSB part of the band. Adjustment was made to cover 3.73 to 3.82 MHz with an SWR of 2:1 or better. The SWR bandwidth (2:1 points) is approx. 100 kHz.

boom. In this construction you can calculate the inner section (= the guyed section) using the approach as outlined by Leeson (Ref 964) for guyed booms. The unguyed section can be calculated using the YAGI DESIGN software.

3.6. Building a Full-Size 80-Meter Yagi

Doing a fancy design on paper (on screen, really) is one thing, doing the physical design, constructing it, and keeping it up in the air is another thing!

3.6.1. The DJ6JC mechanical design

The 3-element 80-meter Yagi described here has been built by H. Lumpe, DJ6JC, who has his own company (WIBI) making commercial radio towers and antenna systems.

DJ6JC designed and constructed an element which is 36.42-m long, and which is entirely made as a square tapering lattice construction. At the center (near the boom), the lattice construction measures 42 cm and it tapers to approximately 5 cm at the tips. The lattice-type sections are made of alumi-

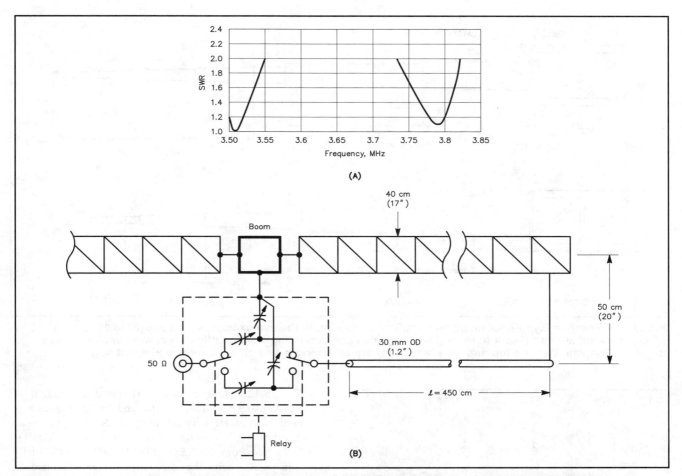

(A)

(B)

Fig 13-37—At A, SWR curves for the 3-element full-size 80-meter Yagi (optimized for 3.51 and 3.79 MHz), obtained with the omega matching system as shown at B. A single omega rod-length was used. Switching from the CW to the SSB portion of the band consists of switching from one set of capacitors (series and parallel) to a second set. This can be done by using vacuum relays.

Fig 13-38—Four-element full-size 80-meter Yagi on a 30-m boom. The antenna has been operational at the PJ9M contest QTH on Curacao. This extremely lightweight Yagi is suitable only for use during short periods of contests, and will certainly not withstand high winds.

5 cm at the tips. The lattice-type sections are made of aluminum bars. The sections use bars with an OD of 20 mm, 15 mm and 10 mm for the four horizontal members, while bars with an OD of 10 mm, 8 mm and 6 mm are used for the oblique members. The weight of the full-size element is only 114 kg. This is extremely low, if you compare it to the full-size elements for the 40-meter Yagi described in Section 3.3.3, which weigh nearly half that much.

This design has an incredible lack of sag over its total length, only 30 cm, without any support cables! **Fig 13-40** clearly shows that the sag of the full-size 80-meter element is hardly visible to the naked eye.

It is clear that such an element should be okay for all but maybe the strongest hurricane winds. The dimensions and construction of the DJ6JC-built Yagi are shown in **Fig 13-41**.

The lattice-type elements are connected to the boom by four 2.5 cm thick fiberglass-reinforced plates, in a construction which keeps the parasitic capacitance (to the boom) as low as 64 pF! Fig 13-32 shows how the (horizontal) split element is mounted on the boom (vertical). The fiberglass-

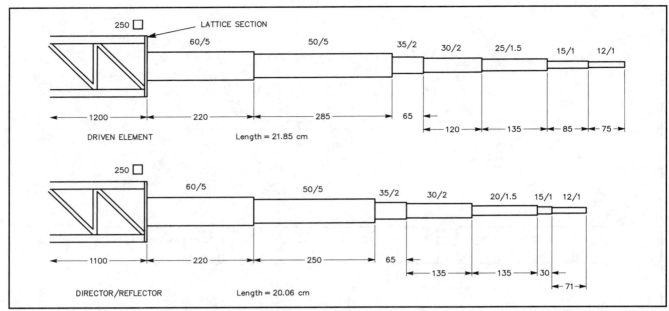

Fig 13-39—Possible layouts for an 80-meter full-size element. In the two examples we see a 10 to 11-m long tubular tapered section (like a full-size 40-m Yagi element) extending from a lattice tower section. The lattice tower section will likely be top and side braced for added strength as well as to reduce element sag.

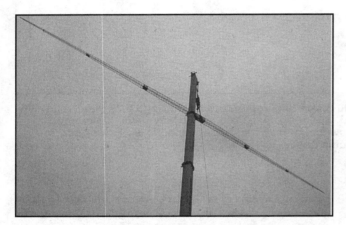

Fig 13-40—The amazing full-size 80-meter element, weighing only 114 kg and good for 160 km/h winds. The element is lifted ½ λ in the air to make resonant-frequency measurements. If you are not sure, no, there are no support guys. The totally self-supporting element has a sag of only 30 cm. The mechanical design is by H. Lumpe, DJ6JC.

reinforced plates are clearly visible.

As there is little, if any information available about the electrical behavior of elements with such an important (and continuous) taper, ON6WU and I decided to check the electrical element length on a test setup.

From earlier modeling, we knew that the director for the SSB band needed to resonate on 4.005 MHz. In this configuration the parasitic element would have no loading coil (hairpin), but the insulated halves would simply be strapped through the contacts of a vacuum relay (see **Fig 13-42**).

The 36.42-m long element was raised to a height of 38 m (½ λ) using a hydraulic crane, and the impedance was measured using an HP network analyzer. After we compensated for the 32 pF capacitance (using the SHUNT IMPEDANCE module from the NEW LOW BAND SOFTWARE) that is in parallel with the impedance bridge, the element turned out to be resonant at 4.090 MHz. Tubular tips with a length of 65 cm and with a diameter of 3-cm OD were added to the lattice construction to bring the resonant frequency down to 4.050 MHz. **Fig 13-43** shows the tubular tip being adjusted to bring the element resonance to 4.005 MHz. Note once again how straight the element is. It is interesting to note that the impedance at resonance was 67 Ω, which corresponds perfectly with the value obtained using *ELNEC* when modeling an element with an equivalent constant diameter of 100 mm at the same height (1 λ) above good ground.

Next we installed a hairpin to bring the resonance down to 3.745 MHz, as required. We had calculated the required hairpin to be 196 cm, but it turned out that a 175-cm long line did the trick. The 10% difference is probably due to the inductance of the straps connecting the element to the hairpin.

To bring the resonant frequency of the reflector for CW operation down to 3.465 MHz, we installed a hairpin with a length of 350 cm.

The driven element was then brought to resonance on 3.745 MHz by extending the lattice section (2 × 18.21 m) by 3.5 m of aluminum tubing (3.2-cm OD). The 3.745-MHz resonant frequency provides a good balance between the positive reactance (at 3.79 MHz) and the negative reactance (at 3.51 MHz), as explained in Section 3.5.7.

This procedure of individually checking the resonant frequency of the elements of the Yagi avoids guesswork when trying to optimize for best performance. It is obvious

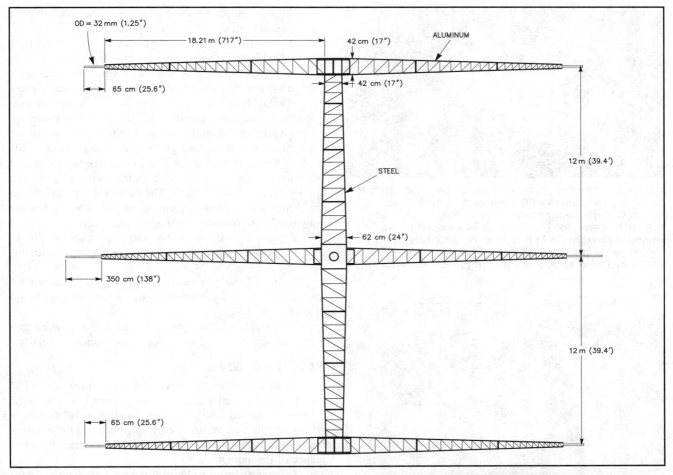

Fig 13-41—Element dimensions for the 80-meter Yagi elements, as used by DJ6JC. The entire element, with the exception of its tubular tips, is made of a continuously tapering square-section lattice construction. At the boom the lattice measures 42 cm. This tapers to 5 cm at the tip (18.2 m). Tubular outer tips (32-mm OD) are used to adjust the final element length. The lattice construction is made of solid aluminum bars, and weighs approx. 57 kg for a half-element. The total sag is an unbelievable 30 cm.

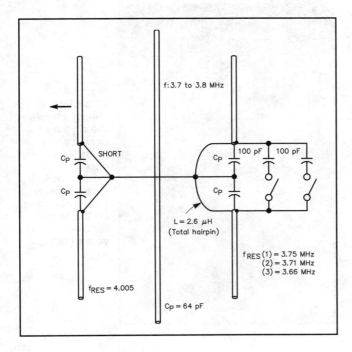

that checking the resonant frequencies should be done one element at a time; this means without any mutual coupling between elements. This is also how we determined the resonant frequencies of the elements in the model (first, model the Yagi, then isolate the elements to determine their electrical lengths).

Fig 13-44 shows the full-size 80-meter element 38 m up in the air supported by a hydraulic crane above the place of business of DJ6JC. It is important that the electrical tests are done at the same height at which the Yagi will be installed, as well as over the same terrain. It was very instructive to see the resonant frequency of the 36.42-m long element change from 3.5 MHz just above ground to over

Fig 13-42—Adding a capacitor across the center insulator of the reflector makes it possible to obtain better than 25 dB F/B ratio at any frequency between 3.68 and 3.8 MHz. Having 100 pF in parallel optimizes the F/B between 3.73 and 3.76 MHz, while 200 pF takes care of the range between 3.68 and 3.73 MHz. See text for details.

Fig 13-43—The length of the tubular tip is adjusted to bring the resonance of the element to 4.005 MHz, which is the resonant frequency of the director for 3.8-MHz operation. The element was raised to the operating height ($\frac{1}{2}$ λ) to do the resonant-frequency measure-ments. Note the sag of the element, which is really invisible even when looking along the element.

Fig 13-44—Eighty-meter full-size dipole element supported 38-m high above the place of business of H. Lumpe, DJ6JC, who is responsible for the outstanding mechanical design of the 80-meter full-size Yagi.

4.1 MHz at 38 m. Between a height of 20 m and 38 m, the resonant frequency shifted as much as 200 kHz.

The boom

The boom for such a large antenna is made of square lattice-type sections, capable of handling the very high boom moment. The moment at the center of the boom (at the mast attachment point) is approximately 7000 kg-m in the vertical plane (weight loading without top-guying) and 4000 kg-m in the vertical plane. This is 25 to 30 times the moments that we encountered with our 40-meter full-size Yagi!

Here again there are two options. You can design the boom of adequate strength without any guying, or you can rely on guy cables for additional strength. I would suggest only to use the guy wires to compensate for the boom sag.

The DJ6JC-designed boom consists of a (square-section) steel lattice construction. Fig 13-32 shows the boom being attached to one of the elements. While the elements are made of aluminum (for weight reasons), each boom half is made of four 3-meter long sections of square-section lattice construction. High-yield-strength steel (solid) bars (not tubes!) are used for the construction. The boom measures 62 cm at the center and tapers to 42 cm at the tips. The total boom weight is approximately 1500 kg.

Top guying will be used on the boom only in order to reduce its sag to zero. Side bracing is not necessary with this design.

Including the element-to-boom coupling "satellites," the weight of the 3-element full-size 80-meter Yagi is approximately 1850 kg.

Whereas DJ6JC had planned to have this antenna up a long time ago, the local authorities have decided otherwise.

3.6.2. The OZ8BV design

Dr Ben Moeller, OZ8BV, first used an original KLM 3-element shortened Yagi, which was blown to pieces in the first gale wind to hit the Danish coast. Ben subsequently reinforced the design, using extensive side and top bracing, but he still was not fully happy with the electrical perfor-mance of the antenna.

This caused him to build his own 3-element full-size Yagi, according to the design outlined in Section 3.5, includ-ing the F/B tuning (see Section 3.5.6).

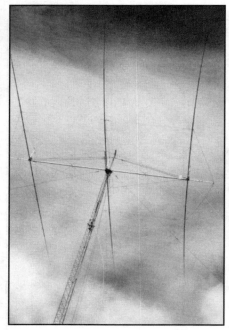

Fig 13-45—This picture clearly shows the rigging used by OZ8BV to keep his 3-element full-size 80-meter Yagi intact and on the tower for over 5 years (1998).

Fig13-46—OZ8BV's 3-element full-size 80-meter Yagi atop his 48-m tower on Denmark's Baltic sea shore. Yes, that is all salt water you see in the picture. The antenna uses the remote director and reflector tuning as described in Section 3.5.6.

OZ8BV used extensive guying and bracing to achieve a 3-element full-size Yagi weighing approx. 700 kg (only! See **Fig 13-45**). The antenna has been up 5 years now without any damage, which speaks for its design. As can be seen from the picture in **Fig 13-46**, the antenna is mounted right near the Baltic sea, and has gone through wind storms with registered wind speeds of 160 km/h.

Remote fine tuning of the F/B is achieved through the used of vacuum variables, fine tuning the hair-pins that resonate the parasitic elements, as explained in Section 3.4.6. Instantaneous F/B direction reversal is also incorporated. This facility is extremely useful for measuring F/B. Using calibrated test equipment, Ben measures 30 dB F/B at any frequency in the 80-meter band. In addition it has not been uncommon for him to receive reports indicating 40 dB and more F/B.

The antenna is made out of all aluminum tubular elements. The taper schedule is shown in **Table A**. The elements are split, with a fiberglass bar (not rod), measuring 83-mm OD inserted in the center.

The boom is 18-m long and has a diameter of 112.5 mm with 6-mm wall thickness. The boom is extended at both ends with 2.7-m long fiberglass tubes, measuring 60-mm OD,

with 4-mm wall. These extensions are required to do the side-trussing of the elements, as is clearly visible from Fig 13-45. The boom is also side trussed with 10-mm Phillystran, and vertically trussed with 15-mm Phillystran.

The 38-m long elements are guyed 4 m and 7.82 m out, using 6-mm OD and 4-mm OD Phillystran (Kevlar) cable. The element with this dimension is resonant on 4.035 MHz. The resonant frequency of the elements were measured, at the final operating height, before calculating the required loading inductances (see also Section 3.6.1).

If you play around with a modeling program, you will soon find out that various values of loading will produce slightly different back patterns. Ben points out that he found it best to optimize the pattern to show the highest F/B at higher angle lobes (30° to 60°) rather than optimizing for best F/B at a very low angle. This makes sense, as the (strong) signals likely to cause most QRM will always come from more or less local stations, coming in at "higher" angles than the DX. Actually Ben prefers zero F/B on the DX signals and infinite F/B on nearer signals!

A shorted transmission line, made out of 10-mm diameter tubes, spaced 15 cm (Z_k = 360 Ω), was calculated to 176 cm (but ended up with 155 cm and a vacuum relay controlled shortening point at 118 cm) is used to do the initial loading of the two parasitic elements. This line has an inductance of 2.16 μH on 3.79 MHz. As explained in Section 3.6.1 this inductance must be tuned by a parallel capacitor to give the required loading for acting as either a director or a reflector. A capacitor value of 427 pF was eventually used for the director and 665 pF for the reflector. At each of the two parasitic elements, Ben installed two motor-driven vacuum capacitors, that each can be paralleled with the loading stub by a small vacuum relay. The system has a capacitance value readout in the shack with an accuracy of 1 pF. Tuning of the vacuum variables is done by stepper motors. This configuration allows Ben to accurately set these values for any value giving the best F/B on any desired frequency. This design is slightly different from what is shown in Fig 13-33, where both stub length and capacitor values are changed, when changing from director to reflector.

The driven element is brought to resonance on the design frequency by using a short section of linear-loading. In this particular case a 82-cm long line (10-mm OD conductor, 13.5-cm spacing), spaced 14.5 cm from the driven element is attached to each side of the driven element. The exact length of these stubs can be adjusted to tune the driven element to the design frequency. The feed impedance is

Table A
Element taper schedule for the OZ8BV full-size 80-meter Yagi. The taper was calculated using the ELEMENT STRENGTH module of the ON4UN YAGI DESIGN program. The element weight is approx 62 kg.

	Sec #1	Sec #2	Sec #3	Sec #4	Sec #5	Sec #6	Sec #7	Sec #8	Sec #9
OD (mm)	90	75	60	35	30	25	18	14	10
Wall (mm)	3	3	3	2	2	1.5	1.5	1	1

approx. 14 Ω. This impedance can easily be matched to a 50-Ω line using a 25-Ω quarter-wave transformer, consisting of two parallel 50-Ω transmissions lines, coiled up to make a current (common-mode) balun ($15 \times 50 \approx 25^2$). The coil diameter should be approx. 30 cm. Such a coil will represent approx. $+j1500$ Ω reactance on 3.8 MHz, effectively decoupling common-mode currents. (See also Fig 13-88 where a 37.5-Ω quarter-wave transformer is used, as the Yagi impedance is a little higher.)

3.6.3. Conclusion of full-size 80-meter Yagis

A project such as the construction of a full-size 3-element Yagi for 80 meters is not a simple task. Very few of the full-size 80-meter Yagis built so far have had a long life. Depending on what wind speed you want the "monster" to be able to survive, an 80-meter Yagi weighs between 700 and 2000 kg. The material cost is substantial, not to talk about the many hundreds of hours of labor that will go into such a project.

The design of the Yagi described in Section 3.5 and realized by OZ8BV and DJ6JC is thought to be novel in a few ways:
• Instantaneous 180° directional switching with no compromises.
• Instantaneous SSB to CW switching with no compromises at DJ6JC (OZ8BV does not work CW).
• Optimum F/B ratio over a wide bandwidth by capacitor-controlled compensation.

The mechanical designs of both OZ8BV and DJ6JC have their own outstanding merits:
• The full-size self-supporting elements of the DJ6JC design are continuously tapered lattice sections made of aluminum, showing extreme strength and an unbelievably small sag of only 30 cm for a total element length of nearly 40 m.
• The OZ8BV design is done with classic tubular aluminum materials, and its outstanding longevity is the proof of sound mechanical engineering.

3.7. Loaded 80-meter Yagi designs

Full-size 80-meter Yagis are not for everyone. The investment is very important, and they are, let it be said, hard to keep up. If carefully designed and well made, Yagis with shortened elements can perform almost as well as a full-size Yagi. A 3-element Yagi with shortened elements can be made to have just as good a directivity as its full-sized counterpart. It may, however, give up a dB (or less) in gain. Let me review some interesting designs that have appeared recently in literature or on the Internet.

3.7.1. The W6KW 3-element Yagi using high-Q coil loading

Peter Dalton, W6KW (ex-W6NLZ) published a design (Ref 977) which makes us think again about loading devices for Yagis with shortened elements (see **Fig 13-47**). Until now it has been common understanding and acceptance that linear-loading devices, such as used for many years by KLM and M-Squared, were the best solution for low loss loading of shortened low-band Yagis, because they were low loss.

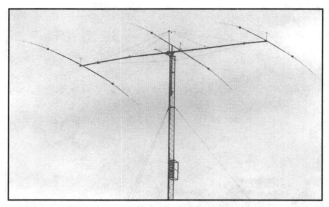

Fig 13-47—Three-element 80-meter Yagi using shortened elements at W6KW. The Yagi uses the high-Q loading coils developed by W6ANR. See text for details.

Fig 13-48—W6KW holding two of the coils for his new 80-meter Yagi, as developed by W6ANR. These loading coils measure 7 inches in diameter and exhibit a Q of nearly 700!

That was the general assumption.

Practical experiments have indicated that this design, probably due to some uncontrolled radiation off the sloping loading wires, does not achieve the kind of directivity one can achieve using perfectly horizontal elements. This leaves us with linear loading devices along the elements (like the KLM 40-meter antennas) or with capacitive or inductive loading. Until now, however, coil-loading has generally been assumed to be a lossy affair. Does it really have to be that way?

David Padrick, W6ANR (ex-KC7LU and WB6IIS), decided to analyze loading coils in detail and found out that most commercial coils were exhibiting poor Qs. But David also came to the conclusion that it was not all that hard to make coils with Qs of 650 or even more. However, the dimensions are critical, and high-Q coils able to handle high power ended up being quite large! (See **Fig 13-48**.) A high-Q coil is not enough by itself; it also is of equal importance to make very low loss RF connections. This point is of utmost importance, and that's where a lot of commercial designs have failed in the past. Good RF connections mean connections that are wide with respect to their length! The same holds true for RF conductors as well. Also the location of the coils on the elements is important (see Chapter 7, Section 2.3.3.2). With very high-Q coils we can afford putting them out further on the elements, which increases the radiation resistance without introducing additional losses, provided the Q remains high.

David found that the commercially made Yagis using elements loaded with sloping stubs exhibit two sorts of problems: inferior F/B (because of radiation off the sloping loading stubs), and accumulated resistive losses due to the way they have been built. He points out that the length of the linear-loading wire is 2 to 3 times longer than the wire or tubing in a high-Q coil providing the same degree of loading. In addition the gauge of a loading stub is usually only a fraction of what's used for high-Q coils. Poor connections to the element and the relays and jumpers used to switch band-segments add to these problems, and all of these critical items have been found to deteriorate over time.

Going from a linear-loading stub to a coil means lower radiation resistance (both inserted in the same point), assuming the losses are identical. There is no two ways about it: the coil does not radiate, the linear-loading device does radiate a little (the currents in both wires don't completely cancel, as shown in Chapter 9, Section 3.1.5). However the marginal increase in radiation resistance may not outweigh the possible increase in equivalent loss resistance. This is what W6ANR found. Also, it may be easier to maintain the low losses of a well designed high-Q coil than for a loading stub.

W6KW's new Yagi uses elements that are approximately $2/3$ full-size, and he has incorporated the loading coils 6.85 m from the center of the elements. In this particular design it appeared that loading coils with an inductance of 17 μH were required to resonate the elements. The final fine tuning of the parasitic elements was done by varying the length of the element tips. A coil of 17 μH represents a reactance of $2 \times \pi \times 3.8 \times 17 = 405$ Ω. A Q of 650, as quoted by W6ANR means a series loss resistance of 405/650 = 0.624 Ω per coil.

I calculated the influence of the Q factor on the gain and directivity. The influence on gain can be seen in **Table 13-8**.

In a typical 3-element coil-loaded Yagi, with the reflector spaced closer than the director (tuned for best F/B), the current magnitude in the reflector is almost the same as for the driven element, while the director only carries 15 or 20% that much current. As a consequence, the influence of coil losses on antenna gain is the same for driven element and

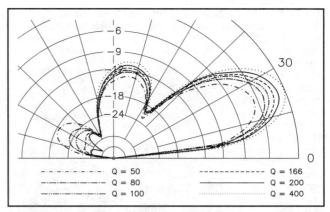

Fig 13-49—Vertical directivity of a 3-element 80-meter Yagi (design similar to W6KW) for various loading coil Q factors.

reflector, but much less for the director. In other words, if you have a lossy coil, better put it in the director element.

For a range of "reasonable" Q factors, it appears that the Q factor has very little impact on directivity. An antenna which has been optimized for an excellent F/B with no coil losses, may actually show even marginally better directivity when some losses (1 or 2 Ω) are introduced! (See **Fig 13-49**.) Too high losses reduce the mutual coupling to a degree that the proper current magnitude and phase can no longer be set up in the elements to achieve a good F/B. As far as directivity is concerned, in the model we used, it appeared that it was possible to achieve a little deeper null with Q factors of 200 as compared to 400, which is quite irrelevant and only says something about the model, not about what can be achieved with the antenna. In general one can say that, as far as directivity is concerned, there is very little to be gained by going for Q factors above 150 or 200, *but* from a total antenna gain point of view it is clear that the higher the Q, the higher the gain. The difference in antenna gain between a Q of 700 and a Q of 175 is about 1 dB, which is certainly not negligible—that is like losing 30% of your power!

How to make high-Q coils

The picture of the loading coils, made by W6ANR (see **Figs 13-50** and **13-51**) gives you part of the answer. Heavy gauge copper tubing conductor (6 mm or $1/4$ inch) and good low-loss contacts. But there is more. There are not only series losses involved, but also parallel losses. A leakage current between adjacent turns of the coil causes parallel losses. If we keep the Q high (800), it means that we have an equivalent series resistance (for a coil measuring = 400 Ω) of 400/800 = 0.5 Ω. This is the equivalent of a parallel loss resistance of 400 × 800 = 320 kΩ. If we allow dirt, smoke deposits, etc, to accumulate on the surface of the bare copper tube of an unprotected coil, we can expect parallel losses to drop well below 320 kΩ, especially when it gets wet. Therefore the coil must be properly protected. One neat way would be to protect the copper tubing used to wind the coil, with a heat shrink tube of a material that has a good dielectric for HF and is UV-resistant (consult

Fig 13-50—The W6ANR loading coil, wound using ¼ inch copper tubing on a grooved ABS coil form, measuring 10 inches long and 7 inches in diameter. Note the husky, large area contact clamp used to connect the coil to the element. The ¼ inch copper tubing is covered by a plastic heat shrink tube to protect it from surface contamination and surface leakage.

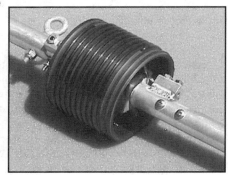

Fig 13-51—Forty-meter very high-Q loading coil developed by W6ANR. Note the contact blocks where the heavy gauge special enameled copper wire (AWG #8 or 3.3-mm OD) attaches to the element. The contact block is actually welded onto the element to minimize loss resistance.

Table 13-8

Influence of Coil Losses on Antenna Gain

The gain at 0 Ω coil losses is calculated over real ground for a reference

Loss R	0 Ω	0.5 Ω	1.0 Ω	1.5 Ω	2.0 Ω	2.5 Ω	3.0 Ω
Q factor	–	800	400	333	200	160	133
Gain in dB	0 dB	−0.34	−0.66	−0.97	−1.27	−1.55	−1.82

Ref 12.3 dBi

Raychem for such material). Other ways of achieving the same goal are left to the imagination of the builder. In summary, it does not pay to reduce the serial losses to almost zero, if we don't take care of the parallel losses to the same order of magnitude.

Conclusion

The publication by W6KW and the work by W6ANR have tremendous merit in bringing to our attention the importance of low-loss loading devices, as well as the importance of the geometrical configuration of loading devices. Losses do cause the gain of an array to drop significantly, and high losses make it impossible to achieve very deep nulls *in parasitic arrays*. Wires used as loading devices do radiate,

and unless they are inline with the elements of the Yagi, or arranged in a way that radiation from these wires is completely canceled in the far field, can upset directivity and reduce F/B and F/S by a substantial amount.

I have written before that all claims of F/B and F/S of more than 25 or 30 dB are paper dB by all means. You may indeed model an array with even a 50 dB deep null (or more), but move the frequency 1 kHz and you will be down to more reasonable values. In other words, I will never quote F/B or F/S figures over 25 or 30 dB as they are totally meaningless. When we look at the radiation patterns in Fig 13-49, what is the message? The message is *not* that a Q of 200 is better than a Q of 400. It shows that way on paper, but the small differences in currents (phase and magnitude) causing these slight differences on paper, are totally out of control in real life anyhow! Let's keep both feet on the ground. The figures tells us that high Qs (150 and up) achieve good F/B, and lower Qs don't. We have to understand figures, not just read and write them. It sure helps to understand the physics behind all this modeling! And don't forget that going from a Q of 100 to a Q of 800 makes you gain exactly 2 dB in signal strength! That is the same as increasing your power by 60%.

If you have any particular interest in a low-band antenna using these Hi-Q coils, contact W6ANR (**w6anr@jps.net**). New antennas and conversion kits for KLM and M-Squared Yagis are commercially available from David's new company Ionic Wave Corporation for the various low bands.

3.7.2. W7CY 2-element 80-meter Yagi

Rod Mack, W7CY, developed an interesting 2-element capacitively end-loaded 2-element Yagi for 80 meters, of which he published pictures and some raw dimensions on the Internet Web pages (**http://www.ulio.com/ants.html**; also see **Fig 13-52**). He claims in excess of 20 dB F/B, which for a 2-element Yagi is quite a lot.

I set out to "reconstruct" his array using *EZNEC*, and noticed some very interesting things.

Fig 13-52—Rod, W7CY, stands proudly on the boom of his excellent 2-element capacitively loaded 80-meter Yagi.

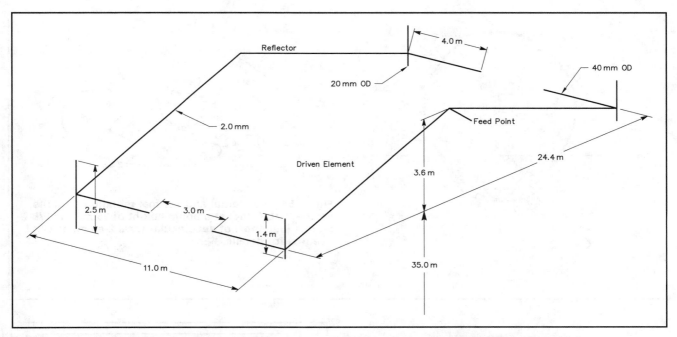

Fig 13-53—Approximate dimensions of the W7CY-like 2-element 80-meter capacitively loaded Yagi. See text for details.

The layout of the array is shown in **Fig 13-53**. The array uses a 24.4-m long boom. At both ends of this boom he mounted 11-m long spreaders, which serve as capacitive loading devices for the sloping elements. The center of the 2 elements is supported by a 11-m boom, which is 3.6 m above the boom. This means that the central part of the 12 elements is in an inverted-V fashion.

I didn't model the array even though Rod sent me the modeling input file. Rather, I set out to develop my own version, hoping to understand more about his design. The process was very revealing indeed.

Rod mentioned that the (capacitive loading) spreaders are, of course, insulated from the boom as they carry very high voltages at their ends. This is where he "tunes" the array by varying their length. It appears soon that the distance between the tips of the loading devices (near the boom) is a very critical item! By varying this distance one can control the coupling between the 2 elements to a fine degree. For an element spacing of 11 m, the ideal current relationship in the two elements is 1:1 for current magnitude and 125° for phase shift. You can model this easily by "planting" two verticals spaced 11 m (on 3.8 MHz) and feeding them that way. If the tips are too close together you will have too much current in the reflector. With parts near the element tips so close together we can create a great deal of capacitive coupling (see also Chapter 10, Section 3.1.2). The exact position of these loading devices with respect to one another is an important issue in this design.

I developed a system by which I loaded the sloping elements in two different ways (both capacitively): by changing the length of the horizontal loading wires (the support structure), and by adding some vertical aluminum tubing at

the same point. By judiciously weighing the ratio of these two capacitive loading devices, I arrived to a point where the required current ratios in both elements were obtained. At that point the F/B was over 24 dB (that is just short of spectacular for a 2-element Yagi) together with a feed-point impedance of very close to 50 Ω.

I could *not* obtain these results by loading the elements with only the horizontal "spreader" loading tubes; they gave me too much coupling between the 2 elements, as their tips came too close together. I found out by inserting a resistor in the reflector, which reduced the current magnitude, of course at a gain sacrifice. The same result was obtained by spacing the tips of the horizontal "loading wires" further apart!

It is obvious that the same result can be obtained in many different ways. You should note that by using close coupled capacitive loading like this, you can control the amount of coupling and achieve a very high F/B which you could never achieve by using a 2-element Yagi with elements that are full size and perfectly parallel along their entire length.

The array has a very nice bandwidth as well. **Fig 13-54** shows the radiation patterns for the array over a span of 40 kHz, without retuning the reflector.

It is of course possible to "tune" the reflector by installing a variable capacitor in the center of the element and changing its capacitance as you move around on the band. By doing so the same F/B can be achieved (20 to 25 dB) *anywhere* in the band (see Section 3.6.3).

Without doing any retuning, this array has an SWR of less than 2:1 over more than 150 kHz. **Fig 13-55** shows some essential array data for a frequency range of 3750 to 3890 kHz.

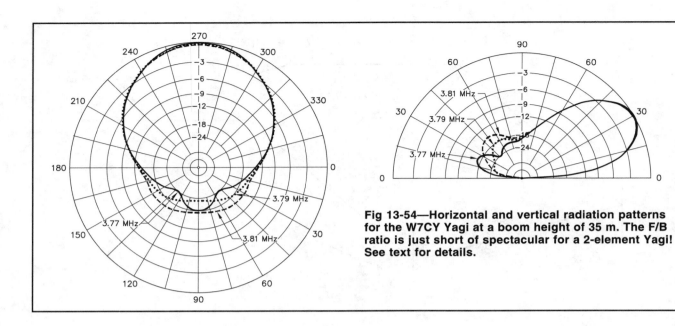

Fig 13-54—Horizontal and vertical radiation patterns for the W7CY Yagi at a boom height of 35 m. The F/B ratio is just short of spectacular for a 2-element Yagi! See text for details.

Duplicating the antenna

First of all it is important to know that the dimensions shown in Fig 13-53 are "ballpark" figures. These are by no way "build and forget" dimensions.

This antenna was modeled with several modeling programs such as *AO* (*MININEC* based), with *ELNEC* (*MININEC*), with *EZNEC2* (*NEC2* based) and *EZNEC-PRO* (*NEC4.1* based), and although with all of these modeling programs achieved essentially the same radiation patterns after fine tuning, these results were all obtained with slightly different dimensions. The main reason for this is the inability of some of these programs to handle wires with vastly different diameters. The problem lies in modeling the capacitive hat, which is made out of aluminum tubing, while the rest of the element is made of a much thinner wire. As an example, with the dimensions optimized for 3.79 MHz using *NEC-2*, the frequency shifted down approx. 80 kHz using *NEC4-1*. The dimensions shown in Fig 13-53 are the results of modeling with the *NEC4-1* engine, which is supposed to give the most accurate results in this case.

Whereas these models may not give us the exact lengths for a precise operating frequency, they give us a good idea of what can be achieved as far as directivity is concerned. Further, it is very important to model in order to determine the resonant frequency of the parasitic reflector and the driven element. We can use this information to tune the array in real life.

The models tell us that for an array optimized for 3.79 MHz, the reflector is resonant on 3.80 MHz, and the driven element, by itself is resonant on 3.94 MHz! This clearly shows what mutual coupling does! How did we find this information? Model the array using, eg, *EZNEC* or *ELNEC*. Determine the resonant frequency of the driven element and the director. To find the resonant frequencies of an individual element, decouple the other element (insert a

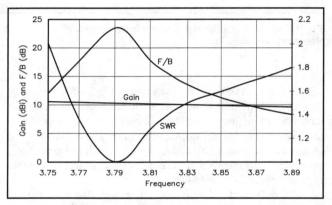

Fig 13-55—Gain (over average ground), F/B and SWR for the W7CY 2-element shortened 80-meter Yagi as a function of frequency.

load of R = 9999 Ω and X = 9999 Ω in the center of the element or simply open it up), and feed the first element. Change the frequency until you find resonance (reactive part of impedance = zero).

Armed with this information, here is how we can tune this array:

1. Build the array, according to the dimensions of your model. Be prepared however to change the dimensions of the loading devices.

2. Cut a feed line that is $^1/_2 \lambda$ on the resonant frequency of the reflector (in our case 3.8 MHz). Using RG-213 cable the length is $(0.66 \times 299.8) / 3.8 \times 2) = 26.03$ m. If you cannot reach the end of the feed line, you can use a full wavelength feed line as well (52.06 m).

3. Connect this feed line to the reflector, raise the antenna to final height, decouple the driven element (leave the center open) and now adjust the loading devices symmetrically on both sides until you get resonance on 3.8 MHz. That

takes care of tuning the reflector. Remove the feed line and close the reflector.

4. Now connect the feed line to the driven element, and prune the length of the loading devices for minimum SWR on the design frequency (3.79 MHz).

3.7.3. The K6UA 2-element 80-meter Yagi

Dale Hoppe, K6UA, must have been around 80-meter DXing almost as long as the band has been there. Some old-timers may remember Dale as "W6 Very Strong Signal." With his beautiful hilltop QTH on an avocado plantation, not only avocados grow well, but also antennas!

Although from this way-above-average QTH almost any antenna would work, Dale has been an avid antenna experimenter and builder. The latest of his designs is a 2-element shortened 80-meter Yagi, which he described in *CQ* Magazine (Ref 978).

It is clear that the tower and the boom, which I have seen at Dale's place for ages, were what set him on his way to develop the array (see **Fig 13-56**).

The boom of the array is a 22-m long triangular tower, 30-cm wide. Fiberglass vaulting poles, measuring 4.5-m long were mounted at both ends of the boom, providing the 9-m spacing between the driven element and the reflector. The 22-m long horizontal elements are loaded at both ends by loading stubs, as shown in **Fig 13-57**. The linear-loading stubs also serve as vertical bracing for the vaulting poles.

Dale reports raising and lowering the antenna about 5 times and cutting the length of the vertical trim wires to tune the array (see Ref 978). Tuning the reflector is quite critical, and changes of a few cm can make an important difference in antenna Q. If the reflector is too short, the feed-point impedance will be much lower than 50 Ω and the bandwidth will be very narrow. When properly tuned, the array exhibits a gain, F/B and SWR pattern as shown in **Fig 13-58**. The antenna has a fairly high bandwidth above its design frequency, and the gain remains fairly constant as well. Depending on the wave angle considered the F/B is >20 dB over approx. 50 kHz. This is quite a good figure for a 2-element Yagi.

This array is very similar to the W7CY array (Section 3.7.2), the only difference being a little shorter elements, closer spacing and partial inductive loading of the elements. As with the W7CY array the position of the tips of the loading elements, facing one another (the tips of the stubs), determines the degree of coupling of the 2 elements in the array. The amount of coupling is quite critical to obtain maximum F/B ratio. In the model I used, a physical spacing of approx. 4 m between the tips of the loading stubs gave the best results.

Instead of using the aluminum tubing capacity hats, which cause a modeling problem due to the vast difference in diameter between the wires in the antenna (2 mm) and the tubes, I decided to keep the original K6UA approach and use "dangling" wires to tune the array. The length of these wires can be trimmed to change the frequency of the elements. To keep them more or less in place in the breeze, one could hang small weights at the end of those wires.

The dimensions given in Fig 13-57 were obtained by modeling through *EZNEC* (*NEC-2* engine).

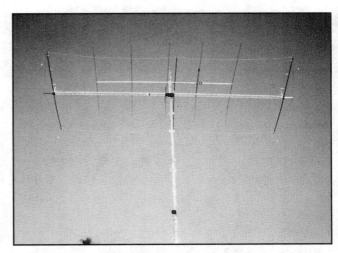

Fig 13-56—View of the K6UA 2-element 80-meter array mounted on a Telrex rotating pole, just under a 5-element 20-meter Yagi.

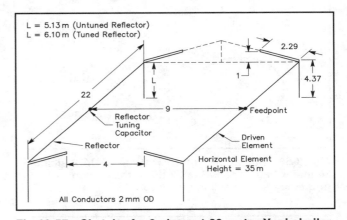

Fig 13-57—Sketch of a 2-element 80-meter Yagi similar to K6UA's design. The dimensions are approximate, and were used to calculate the patterns of the array. The wires of the loading stubs in this model are separated 20 cm.

One way of tuning the reflector is to watch the SWR about 30 kHz below the design frequency and adjust the reflector (shorten it) until the SWR is about 2:1 (see **Fig 13-59**). Tuning of the driven element to obtain lowest SWR at the design frequency is the last step in tuning the array. The antenna can be fed directly with a 50-Ω feed line via a current balun.

Alternative tuning method

Raising the Yagi repeatedly in order to tune the antenna for best F/B may not be the most attractive job. There is an alternative way though, that brings you an additional advantage. I intentionally lengthened the reflector a substantial degree, and made the vertical tuning wires approx. 1 m longer (6.1 vs 5.13 m). Now that we have a reflector that is way too long, we can electrically tune it to where we want it, by simply inserting a capacitor in the center of the element. In the

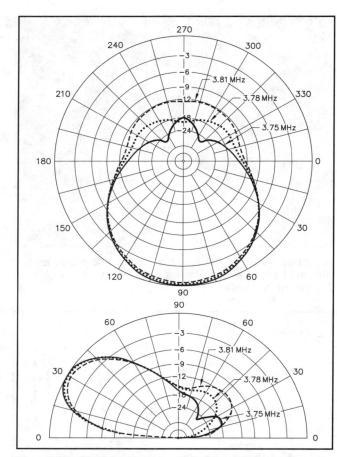

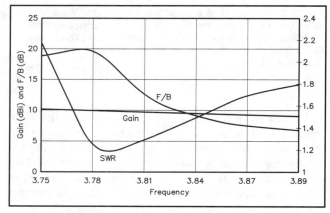

Fig 13-58—Radiation patterns for the K6UA 2-element 80-meter Yagi. The F/B is 23 dB at the design frequency for a wave angle of 30°.

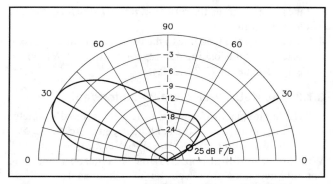

Fig 13-60—Vertical radiation pattern for the K6UA 2-element 80-meter Yagi, showing the directivity that can be obtained across 150 kHz of bandwidth by remotely tuning the reflector with a variable capacitor. See text for details.

Table 13-9

Values of the capacitive reactance (X_C) and corresponding capacitance (pF) to tune the K6UA for best F/B across a wide spectrum. Curve SWR (1) is for the array tuned for best F/B at >23 dB (R_{rad} approx. 37 Ω). SWR (2) is for the array adjusted for a 50-Ω impedance (F/B approx. 15 dB).

Freq (MHz)	3.75	3.78	3.81	3.84	3.87	3.9
X_C (Ω)	−64	−77	−90	−103	−116	−129
C (pF)	663	547	464	403	354	317
SWR (1)	2.0	1.5	1.3	1.5	1.9	>2.0
SWR (2)	1.4	1.1	1.2	1.6	2.0	>2.0

Fig 13-59—Gain, F/B and SWR for the 2-element 80-meter loaded Yagi modeled after a design by K6UA. The gain figure is valid for average ground.

model case I achieved 23 dB F/B on *any* frequency between 3.75 MHz and 3.87 MHz, simply by adjusting the value of the capacitor from 663 pF on 3.75 MHz and 354 pF on 3.87 MHz (see **Table 13-9**)! **Fig 13-60** shows the vertical radiation patterns obtained at various frequencies in that range.

The same approach for remotely tuning the reflector can of course be used on the W7CY 2-element Yagi.

In conclusion, this design is a good example of what can be achieved using locally available materials and a good deal of knowledge, insight and imagination. Listen for Dale's big signal on 80 meters!

3.8. Horizontal Wire Yagis

Yagis require a lot of space and electrical height in order to perform well. Excellent results have been obtained with fixed wire Yagis strung between high apartment buildings, or as inverted-V shaped or sloping elements from catenary cables strung between towers. There are few circumstances, however, where supports at the right height and in a favorable direction are available. When using wire elements, it is easy to determine the correct length of the elements using a *MININEC* or a *NEC*-derived modeling program (eg, *EZNEC*).

Wire Yagis have been described in detail in Chapter 12 (Other Arrays), Sections 3.6 and 12.

3.9. Vertical Arrays with Parasitic Elements (Vertical Yagis)

Do vertical arrays with parasitic elements work on the low bands? If you are a Top Bander, look in your log for

Fig 13-61—Bird's eye view from the driven element of the 160-meter array at KØHA. The first director, on the left of the picture is hiding the second director. On the right a line of elements for the 80-meter array aims at Europe. It also appears that Bill is enjoying some of the best ground conductivity around. No wonder he's loud!

KØHA. He's either there long time before anyone else from his area, or he's there all by himself, or he's there much stronger than anyone else. Bill Hohnstein, KØHA, swears by them. His farm grows vertical parasitic arrays in all sorts and sizes (see **Fig 13-61**).

There is no need to use full-size elements for putting together effective and efficient arrays. Bill uses a shunt-fed 32-m tower as the driven element, while his parasitic elements are approx. 26-m high, and top loaded.

It is obvious that in a parasitic array, neither the feed method nor the exact electrical length affect the performance of the array. Shunt or series feeding can be used without preference. The elements should, however, not be much longer than $^1/_4$ λ.

I will take you on a little tour of some of the classic parasitic arrays, and point out what you should watch for if you want to build one. A modeling program seems to be essential, as you probably will be using existing towers as part of the antenna, and you will need to do some specific modeling. Watch out that you understand what the modeling program tells you, and be aware of what it does *not* tell you.

3.9.1. Turning your tower guy wires into parasitic elements

Several good articles have been published on this subject (Refs 981, 982 and 983). I recommend reading those if you plan to try one of these antennas.

3.9.1.1. One sloping wire

It seems logical to think of a sloping (guy) wire as a reflector or a director. But, you can also use the sloper as a driven element, and use the tower as a parasitic element! This last case may not be so practical, as in many cases it will probably not be possible to tune the tower to the exact

required length. The tower could be tuned by changing its length, or by tuning it, eg, at its insulated bottom by inserting a coil or capacitor to ground.

I analyzed the case of a 40-m tower (25 cm equivalent effective diameter) with a sloping 2-mm-OD wire measuring 40.5-m long, sloping from the top of the tower (via a length of insulated rope) to the ground point, 27 m away from the tower base. This appears to be the distance for the guy cable anchor points for a 40-m high tower. The tip of the sloping wire is approx 4.7 m from the tower. These dimensions are valid only for conductors of the diameter indicated. The resonant frequency of the vertical conductor by itself is 1.78 MHz, and the sloping wire is resonant at 1.822 MHz. These data make it possible to duplicate the array with conductors of different diameters. All one has to do is to dip the "wires" to the listed frequencies.

With the tower fed, the wire acts as a perfect reflector, giving 3.4 dB gain over the tower by itself and a useful 17 dB of F/B on 1.83 MHz. With the tower grounded, and feeding the sloping wire, the array now shoots in the opposite

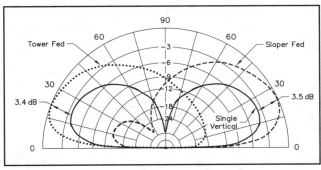

Fig 13-62—Vertical radiation patterns for the tower and 1 sloping wire. See text for details.

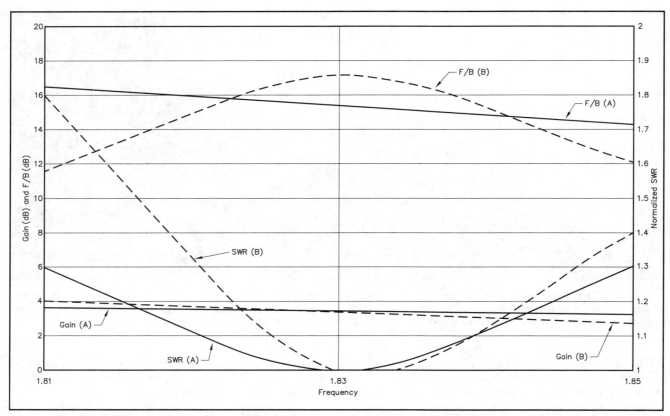

Fig 13-63—Gain, F/B ratio and normalized SWR for the vertical and sloping wire array. Case A is the array with the sloper fed, and the tower acting as director, and Case B is the tower fed and the sloper acting as a reflector. See text for details.

direction. The tower now acted as a director, and the gain was about the same (3.5 dB), with a F/B of 15 dB on 1.83 MHz. **Fig 13-62** shows the vertical patterns of these arrays, as compared to the tower by itself. Modeling was done over average ground, and a perfect radial system was assumed for both conductors.

Fig 13-63 shows the main performance data for both configurations. In most practical cases, however, one would probably try to use the sloping wire as either a director or as a reflector, in which case the sloping wire would need to be tuned by an appropriate reactive element (L or C; see **Fig 13-64**). This specific case is interesting as it demonstrates that prefect directivity and pattern reversal can be obtained without need of a capacitor or inductor. The feed-point imped-ance, in the case of the fed tower, is $43 + j62 \, \Omega$ on 1.83 MHz. When the sloping wire is fed, its feed impedance on 1.83 MHz is $43 + j62 \, \Omega$. In both cases a small L-network (or just a series capacitor, if you can live with about 1.3:1 SWR at the design frequency) should be used to match the antenna to a 50-Ω feed line.

Using the same sloping wire (dimensions, placement) I tuned it by a series capacitor ($X_C = -j50 \, \Omega$) for best perfor-mance. With less than 4 dB of F/B the gain was 2.9 dB over the vertical by itself. Not a very spectacular result. This was

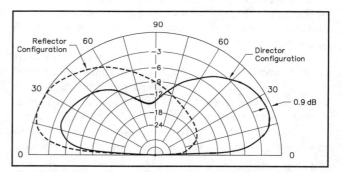

Fig 13-64—The same sloping wire tuned as a reflector and as a director. The director configuration yields poor F/B and mediocre gain. See text for details.

also reported by J. Stanley, K4ERO (Ref 982). The gain obtained is also almost 1-dB less than for the reflector case. Note that the results of this configuration are far inferior to those obtained when feeding the sloping wire and using the tower as a director (see Fig 13-62).

Most of the 40-m tall towers used as 160-meter $1/4$-λ verticals are probably guyed in 4 directions. That means that we probably can hang 4 sloping wires from the top. This is the next case I investigated.

3.9.1.2. A reflector and a director

Continuing with the same physical configuration (the guy cables being anchored approx. 29 m from the base of the 40-m tower), the combination of using both a director as well as reflector was obvious. This combination can typically boost the gain another 1 dB, but has the disadvantage of reducing the F/B by about 6 dB. Tuning the director with a series reactance of $-j70$ Ω is a compromise situation that does not yield maximum gain, but still yields a more or less acceptable F/B ratio. This compromise was described in detail by Christman (Ref 389). **Fig 13-65** shows the resulting radiation patterns. The resonant frequencies of the parasitic elements, when fully decoupled from one another and from the driven element are: director: 1.952 MHz, reflector: 1.822 MHz.

It is clear that for all of the above arrays it is important to have a good ground radial system, not only for the driven element but also for the parasitic elements. Fig **13-66** shows the gain of the array as a function of the equivalent ground loss resistance. Case A is for a radiator with a perfect ground radial system ($R_{loss}= 1$ Ω) but for varying ground loss resistances at the parasitic elements. Whereas 1 Ω ground systems yield a gain of 5.6 dBi for the array, the gain drops to 3.67 dBi if all elements have an equivalent ground loss of 8 Ω. If we have the 1 Ω loss resistance for the radiator, but a rather mediocre loss resistance of 8 Ω for the parasitic elements, the gain drops to 4.13 dBi. This shows that there really is very little room for a poor ground system under the parasitic elements.

It is obvious that it makes no difference whatsoever how the driven element is fed, series or parallel.

3.9.2. Three element vertical parasitic array

The arrays I analyzed in Section 3.9.1 all showed rather

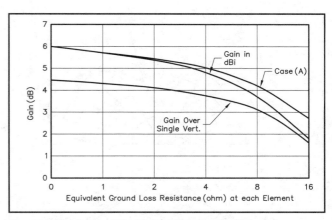

Fig 13-66—Gain of the 3-element guy-wire array as a function of the equivalent ground loss resistance. Case A is for a driven element with a fixed 1-Ω loss resistance.

substantial high-angle radiation which is caused by the horizontal component of the sloping parasitic wires. To improve on that situation we can try to bring the sloping elements as vertical as possible. If you do not use the parasitic-element wires as guy wires for the tower, you can consider a configuration as shown in **Fig 13-67**.

Four cross arms, which support four sloping wires, are mounted at the top of the tower (the quarter-wave vertical). The bases of the parasitic elements are 0.125 λ away from the driven element, which is much closer spacing than with the arrays from Section 3.9.1 (0.17 λ). All four sloping wires are dimensioned to act as directors. When used as a reflector, a parasitic element is loaded with a coil at the bottom. The two sloping elements off the side are left floating. This array has

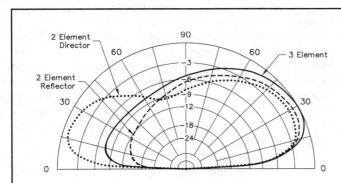

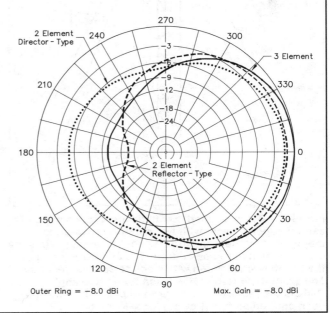

Fig 13-65—Vertical and horizontal radiation patterns (over perfect ground) of 3 types of slant-wire parasitic arrays: reflector parasitic, director parasitic and reflector plus director parasitic. These patterns are valid for parasitic elements spaced 0.18 λ from the driven element (at their base), which appears to be a typical situation for guy cables of a 40-m guyed tower. The horizontal patterns are for a wave angle of 20°.

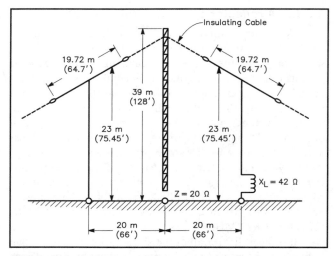

Fig 13-68—This 160-meter 3-element parasitic array produces 4.8 dB gain over a single vertical, and better than 25 dB F/B ratio over 30 kHz of the band. With such an array there is no need for Beverage receiving antennas! The drawing shows only two of the four parasitic elements. The two other elements are left floating.

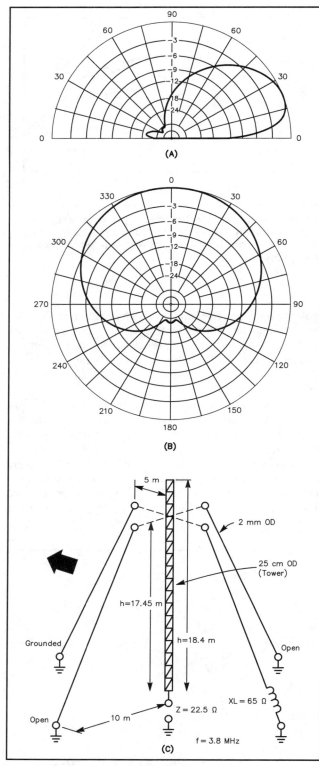

Fig 13-67—Three-element parasitic array, consisting of a central support tower with two support cross-arms mounted at 90° near the top. Two of the sloping wires are left floating, a third one is grounded as a director, and the fourth one is loaded with a coil to act as a reflector. The azimuth pattern at B is taken for a wave angle of 22°. Radials are required at all 5 ground points but they have been omitted on this drawing for sake of clarity.

a very respectable gain of 4.5 dB over a single vertical. At the major wave angle the F/B ratio is an impressive 30 dB, as can be seen from the patterns in Fig 13-67.

You can "dip" the sloping wires. Make sure the driven element as well as the other three sloping wires are left floating when dipping a parasitic element. The resonant frequency should be 4.055 MHz. (f_{design} = 3.8 MHz). You can, of course also dip the wires with the loading coil in place in order to find the resonant frequency of the reflector. Again, all other elements must be fully decoupled when dipping the element. The resonant frequency for the reflector element is 3.745 MHz (f_{design} = 3.8 MHz).

If you want to totally eliminate the horizontal high angle radiation component, there is the solution where you hang top-loaded elements from catenary cables, as shown in **Fig 13-68**. In this 160-meter (f_{design} = 1.832 MHz) version the parasitic elements are top loaded with sloping T-shaped wires. These wires can be supported along the catenary support cable or may be an intrinsic part of the support structure. Such slanted top-loading wires do not produce any far-field horizontal radiation because they are perfectly symmetrical with respect to the vertical wire.

To make this an array that can be switched in 4 directions, you should slope 4 catenary cables at 90° increments from the top of the driven-element tower. With the appropriate hardware you can connect the parasitic elements directly to ground (director), to ground via a loading coil (reflector) or leave the element floating (the unused elements off the side of the bearing direction). It's a good idea to provide a position in your switching system to have all elements floating, in which case you will have an omnidirectional antenna. This may be of interest for testing the array, or for taking a quick look around in all directions.

The example I analyzed uses 23-m-long vertical para-

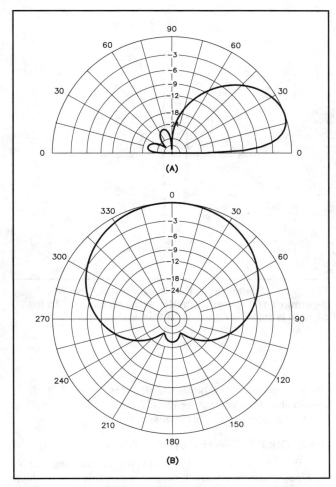

Fig 13-69—Horizontal and vertical radiation patterns of a 3-element vertical parasitic array for 160 meters. The azimuth pattern at B is taken for a wave angle of 20°. Note the near perfect radiation pattern and the excellent F/B of the array.

Fig 13-70—Feed point of the K3LR vertical Yagi for 160 meters. The aluminum box contains an L network to transform the array impedance (approx. 20 Ω) to the 75-Ω Hardline impedance. The coaxial cable coil is a short-circuited ¼ λ long (on 160 meters) stub, which serves as static drain, and also provides attenuation of the 80-meter harmonic (which is very important at a multioperator contest station). Note the 120 radials connected to the annular ring.

sitic elements, which are top-loaded with a 19.72-m long sloping top wire, which is part of the support cable. As the length of the top-loading wire is the same on both sides of the loaded vertical member, there is no horizontally polarized radiation from the top-loading structure.

The four parasitic elements are dimensioned to be resonant at 1.935 MHz. The same procedure as explained above for the 80-meter array can be used to "tune" the parasitic elements. When used as a reflector, the elements are tuned to be resonant at 1.778 MHz. This can be done by installing an inductance of 3.65 µH (reactance = 42 Ω at 1.832 MHz) between the bottom end of the parasitic element and ground. The radiation resistance of this array is around 20 Ω, and it has a gain of 4.8 dB over the single full-size vertical. **Fig 13-69** shows the radiation patterns of this array. The bandwidth behavior is excellent. The array shows a constant gain over more than 50 kHz and better than 25 dB F/B over more

than 30 kHz. When tuned for a 1:1 SWR at 1.832 MHz, the SWR will be less than 1.2:1 from 1.820 to 1.850 kHz. This really is a winner antenna, and it requires only one full-size quarter-wave element, plus a lot of real estate to run the sloping support wires and the necessary radials.

The same principle with the sloping support wires and the top-loaded parasitic elements could, of course, be used with the 80-meter version of the 3-element vertical parasitic array.

Tim Duffy, K3LR, made an almost exact copy of this array, after initially having used inverted Ls for the parasitic elements. Tim recognized that these inverted-L elements introduce a fair amount of horizontally polarized high-angle radiation, however. For a single-element vertical, this may be of very little importance, but for a parasitic element of an array, this will greatly reduce the directivity of the array, especially at high angles This can be important if the array is also used for receiving. Tim reported changing from inverted-L shaped elements to the sloping-T shaped elements and reported that the T-shaped elements work much better.

At K3LR the parasitic directors were resonated at 1.903 kHz, and the loading coils were 4.0 µH with a vertical length of 19.58 m and a sloping-T-shaped top hat of 17.78 m (all made of #12 copperweld wire). The array is matched to a 75-Ω coaxial feed line by an L network (see **Fig 13-70**). The measured SWR is 1.3:1 on 1.8 MHz, 1:1 on 1.83 MHz and 1.3:1 on 1.85 MHz.

Tim has an omnidirectional mode (floating all parasitic elements—see **Fig 13-71**) and reports approx. 5 dB of gain and 30 dB of F/B at the design frequency (1.83 MHz). At 1.82 MHz the measured F/B is still 25 dB and at 1.84 MHz Tim measured 15 dB.

Tim, who cannot run many Beverage antennas (just one

Fig 13-71—A plastic food box contains the Air Dux loading coil and a small relay to switch the coil in and out the circuit at each of the 4 parasitic elements. The box is covered by an inverted plastic dust bin (trash container). The base of each parasitic element is also equipped with a ¹/₄-λ coaxial stub and complemented with not less than 120 radials.

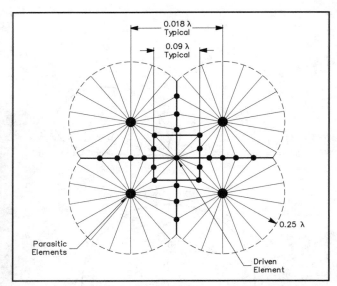

Fig 13-72—Radial layout as used by K3LR on his 3-element 160-meter parasitic array. A total of 15 km of wire is used for radials in this array.

1200 footer on Europe) sure appreciates this excellent directivity for a 160-meter array. Don't forget that in order to make such an array work correctly, you need an impressive radial system under each of the elements. K3LR uses not less than 15 km of radials in this system. A recommended radial layout, which K3LR uses, is shown in **Fig 13-72**. Note that radials are even more important than for all driven phased arrays. With parasitic arrays the gain seems to suffer even more quickly from poor ground systems, so a good radial system is mandatory. Incidentally, computer modeling indicates that elevated radials do not work well with parasitic arrays. No matter what kind of elevated radial system I modeled (different numbers of radials, varied lengths, and different orientations), the result was a badly distorted pattern. This is logical in view of the influence of the capacitive coupling of the raised resonant radials. Compare this situation with what was experienced with the top-loaded 80-meter Yagis designed by W7CY and K6UA. With phased arrays

using current-forcing methods, the feed method itself is responsible for overcoming the effects of mutual coupling due to the proximity of the wires.

3.9.3. The K1VR/W1FV Spitfire Array

Fred Hopengarten, K1VR, and John Kaufmann, W1FV, designed a somewhat novel 3-element parasitic array, which they called the "Spitfire" array. **Fig 13-73** shows the layout of the antenna. John, W1FV, described the array as a 3-element parasitic array with a vertical (tower) driven element and 2 sloping wire parasitics (director and reflector). He adds that the parasitic elements are *not* grounded and do *not* require a separate radial system of their own. Rather the parasitic wires are folded around to achieve the required

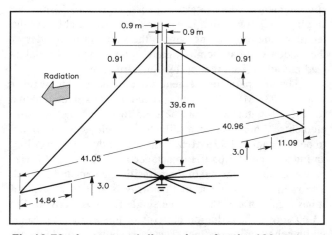

Fig 13-73—Layout and dimensions for the 160-meter Spitfire array.

≈1/2 wave resonances. The tower driven element does have its own radial system.

The antenna was modeled with *EZNEC*, using the *MININEC* ground analysis method (the *NEC* ground analysis method cannot be used because the driven element is a grounded element). Over average ground, the model shows a gain of 4.8 dB over a single vertical at a wave angle of 23° (wave angle of single vertical over average ground).

The antenna has a substantial high amount of high angle radiation and its pattern resembles that of a EWE antenna (see **Fig 13-74**). The high angle radiation is mainly caused by the radiation of the bottom half of the sloping half-wave parasitic elements. You can consider the bottom half of each parasitic element as a single radiating radial, that is bent backward toward the driven element.

To change directions, it is "simply" a matter of using a relay to switch in or out an additional wire segment on the lower horizontal portion of each parasitic to change from director to reflector operation. It is important to point out that this switching happens at high voltage points, and a well insulated vacuum relay is certainly no luxury.

A model for 4 switching directions can be constructed by adding an identical set of parasitic wires (for a total of 4 wires) oriented at right angles to the original 2. Only 2 wires at a time are "active." The other 2 are detuned so as not to couple. Simply grounding them appears to accomplish this function.

Critical analysis

If you compare this array with the classic 3-element parasitic array as described in Section 3.9.2, you will notice that the main difference is that this Spitfire array claims *not to require radials* for the parasitic elements. We learned in Chapter 9 (Section 1.3.3 and Section 2) about ground losses and radial systems for vertical antennas. I also pointed out that the Spitfire array has been modeled with a *MININEC*-based modeling program, which means that a perfectly conducting ground is assumed in the near field of the antenna. This is certainly not true in real life. The bottom half of the sloping parasitic elements are very close to ground, and undoubtedly will cause a great deal of near-field absorption losses in the lossy ground, unless the ground is hidden from these low wires by an effective ground screen. The model used to develop this antenna does not take any of this into account. What does that mean? It does *not* mean that the antenna will not be able to give good directivity. But it means that the quoted gain figures are probably several dB higher than what can be accomplished in real life, if no radial or ground screen system is used that effectively screens the lossy ground under the array from the antenna. This could be accomplished by using extra long radials on the driven element that extend at least $1/8$ λ beyond the tips of the parasitic elements. This would mean radials that are at least 60-m long, with their tips separated not more than 0.015 λ (see Chapter 11, Section 2). This means that 157, 60-m long radials fulfill this requirement. Only under these circumstances will we achieve the same gain as with ground-mounted quarter-wave parasitic elements, each using their own elaborate radial system (a la K3LR).

I modeled the same antenna, but using $1/4$-λ grounded parasitic elements, and maintained the same average spacing from the tower. The sloping parasitic elements are both 38.7-m long, spaced 9.2 m from the tower at the top and 29 m from the tower at the bottom. The tops of the parasitic elements are at 33 m, which is 6 m below the top of the 39-m tower. The directors can be tuned to become reflectors by loading them with a coil with an inductance of 2.85 μH. This classic antenna has 0.7 dB more gain than the Spitfire

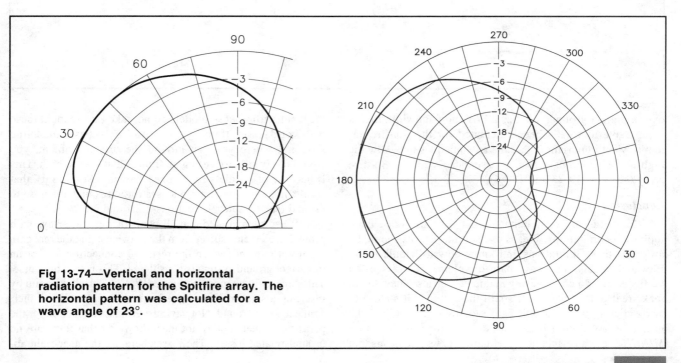

Fig 13-74—Vertical and horizontal radiation pattern for the Spitfire array. The horizontal pattern was calculated for a wave angle of 23°.

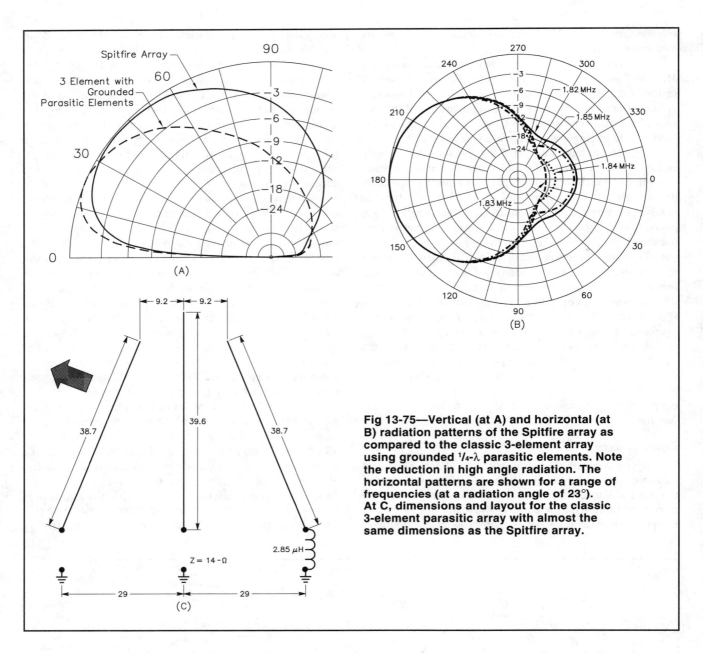

Fig 13-75—Vertical (at A) and horizontal (at B) radiation patterns of the Spitfire array as compared to the classic 3-element array using grounded ¹/₄-λ parasitic elements. Note the reduction in high angle radiation. The horizontal patterns are shown for a range of frequencies (at a radiation angle of 23°). At C, dimensions and layout for the classic 3-element parasitic array with almost the same dimensions as the Spitfire array.

over a prefect ground, and has a F/B and bandwidth that is comparable to what's been calculated for the Spitfire antenna. Most important is that the antenna does not show the high-angle radiation, which is associated with the Spitfire array (see **Figs 13-75** and **13-76**).

Conclusion

Modeling tools are fantastic, but they are tools. Each tool has its limitations. We, as users of these tools should be aware of these limitations and know how to handle them. Elevated radials, half-wave parasitic elements, voltage-fed verticals, etc, do not have any magic properties. They do not make real ground vanish. It's still there, and if it's close to any radiating wires, it will cause losses, what we call the near-field absorption losses. Modeling programs based on *MININEC* use a perfect near-field ground, which means that

the results from those models do not take into account these real-world losses. If you would use a *NEC-4* based modeling program, where you can ground the driven element, you would likely still arrive at gain figures that are too high. This is because of a widely recognized flaw in *NEC* (so far), that results in too low near-field losses for wires that are close to ground (see Chapter 9, Sections 2.2.1 and 2.2.2).

It has been argued that all parasitic vertical arrays with grounded elements suffer from the drawback that the real gain rapidly diminishes when the resistive connection loss of an imperfect ground system is considered. This is true of course. This can easily be modeled on a *MININEC*-based program by inserting a small resistor in series with the elements at their connection to ground. Not having a ground connection for the parasitic elements does not mean however that there are no ground-related losses. The losses here are the near-field absorption losses, associated with low to the ground wires, and these *cannot* be properly modeled with today's modeling tools. This does not detract from the fact that they are there, and can account for several dB of signal loss!

The Spitfire is an array that has its merits. The extra high-angle radiation may be an asset under certain circumstances, like in contesting where some extra local presence is welcome. Potential builders should know that a good ground screen is as essential with this antenna as it is for an array using grounded near-quarter-wave parasitic elements. Sorry, but again, there is simply no free lunch!

3.10. Yagi matching systems

The matching systems for Yagis I describe in this section are not only valid for the low bands. The concept, design and realization of various popular matching systems is covered in greater detail than in any other book I know of.

I describe four of the most popular matching systems:
- Gamma match.
- Omega match.
- Hairpin match.
- Direct feed.

The YAGI DESIGN software contains modules which make it possible to design these matching systems with no guesswork.

3.10.1. The gamma match

In the past, gamma-match systems have often been described in an over-simplifying way. A number of home-builders must have gone half-crazy trying to match one of W2PV's 3-element Yagis with a gamma match. The reason for that is the low radiation resistance, and the fact that the driven-element lengths, as published, are "too long" (positive reactance). The driven element of the 3-element 20-meter W2PV Yagi (which has a high Q factor of 58) has

a radiation resistance of only 13 Ω, and an inductive reactance of +18 Ω at the design frequency, with the published radiator dimensions of 0.489661 λ (Ref 957). Yagis with such low radiation resistance, together with a positive reactance (element too long!) cannot be matched with a gamma (or omega) match, unless the driven element is first shortened to introduce the required negative (capacitive) reactance in the feed impedance!

Yagis with a relatively high radiation resistance (25 Ω) or with some amount of negative reactance (slightly short elements), typically −10 Ω, will easily be matched with a whole range of gamma-match element combinations.

Fig 13-77 shows the electrical equivalent of the gamma match. Z_g is the element impedance to be matched. The gamma match will have to match half of the element impedance ($Z_g = Z_g/2$) to the feed-line impedance (50 Ω).

The step-up ratio of a gamma match depends on the dimensions of the physical elements (element diameter, rod diameter and spacing) making up the matching section.

Fig 13-78 shows the step-up ratio as a function of the driven-element diameter, the gamma rod diameter and spacing between the two.

The procedure to calculate the elements of a gamma match is as follows:

1. Calculate the step-up ratio (**Fig 13-78**).
2. Multiply Z_h (half of the radiation resistance) with the step-up ratio (Z_i).
3. Calculate the (inductive) reactance (X_r) of the (shorted) transmission-line length made up by the gamma rod and the driven element. The gamma shorting bar is the shorted end of the transmission line, while the open end of the transmission line is in parallel with half of the feed-point impedance of the Yagi.

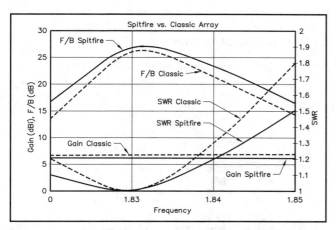

Fig 13-76—Performance characteristics of the Spitfire array compared to a classic 3-element parasitic array of essentially the same dimensions (Fig 13-75). Note that the gain shown for the Spitfire array can only be achieved with an extensive radial or ground screen which is mandatory to prevent several dB of near-field ground-absorption losses of the half-wave elements that are very close to ground.

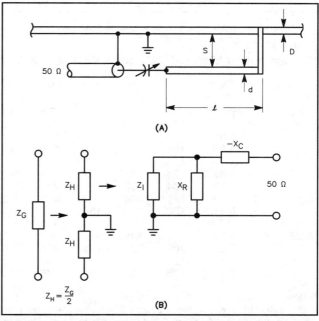

Fig 13-77—Layout and electrical equivalent of the gamma match.

4. Calculate the parallel impedance (Z_p) made up by paralleling Z_i and X_r. If the gamma rod has the correct length, then the resistive part of the impedance Z_p will equal 50 Ω. If not, lengthen or shorten the gamma rod until 50 Ω is reached.

5. A capacitor giving a negative reactance of $-X_C$ will have to be connected in series with the gamma rod to tune out the inductive part of the impedance Z_p. This capacitor is the so-called gamma capacitor.

The calculation involves a fair bit of complex mathematics, but software tools have been made available from different sources to solve the gamma match problem. The YAGI DESIGN software addresses the problem in one of its modules (MATCHING SYSTEMS).

In order to illustrate the matching problems evoked above, I have listed the gamma-match element variables in **Table 13-10** for a Yagi with $R_{rad} = 25$ Ω, and in **Table 13-11** for a Yagi with $R_{rad} = 15$ Ω.

From Table 13-10 it is clear that a Yagi with a radiation resistance of 25 Ω can easily be matched with a wide range of gamma-match parameters, while the exact length of the driven element is not at all critical. It is clear that "short" elements (negative reactance) require a shorter gamma rod and a slightly smaller value of gamma capacitor.

Table 13-11 tells the story of a high-Q Yagi with a radiation resistance of 15 Ω (similar to the 3-element W2PV or W6SAI Yagis). If such a Yagi is in addition using a "long"

Table 13-10
Gamma-Match Element Data for a Yagi with a Radiation Resistance of 25 Ohms

Rod d	S	Step up Rat.	−20 ohms L	C	−15 ohms L	C	−10 ohms L	C	−5 ohms L	C	0 ohms L	C	+5 ohms L	C
0.50	5.0	5.28	118	350	123	502	138	614	171	734	231	396	317	734
	4.0	5.42	131	342	135	488	151	592	184	700	255	376	331	700
	3.0	5.65	152	332	155	468	172	562	207	656	267	349	351	654
	2.5	5.83	169	324	172	452	189	540	224	634	285	332	369	624
0.38	5.0	5.87	119	322	121	450	133	536	158	618	203	328	269	618
	4.0	6.08	132	314	133	434	145	514	171	588	216	311	281	584
	3.0	6.43	153	302	154	412	165	482	192	548	238	288	302	558
	2.5	6.71	170	292	169	396	188	462	208	520	255	273	319	520
0.25	5.0	6.75	120	290	119	394	128	458	147	516	181	270	230	516
	4.0	7.07	133	282	131	374	140	430	158	482	192	251	239	482
	3.0	7.62	154	268	151	356	160	408	179	452	213	236	262	452
	2.5	8.06	172	258	167	340	175	398	195	428	230	223	278	428

Design parameters: D = 1.0; Z_{ant} = 25 ohms; Z_{cable} = 50 ohms. The element diameter is normalized as 1. Values are shown for a design frequency of 7.1 MHz. L is the length of the gamma rod in cm, C is the value of the series capacitor in pF. The length of the gamma rod can be converted to inches by dividing the values shown by 2.54.

Table 13-11
Gamma Match Element Data for a Yagi with a Radiation Resistance of 15 Ohms

Rod d	S	Step up Rat.	−20 ohms L	C	−15 ohms L	C	−10 ohms L	C	−5 ohms L	C	0 ohms L	C	+5 ohms L	C
0.50	5.0	5.28	93	410	92	586	116	1180	—	—	—	—	—	—
	4.0	5.42	103	400	102	566	121	1074	—	—	—	—	—	—
	3.0	5.65	120	386	117	538	136	948	—	—	—	—	—	—
	2.5	5.83	134	376	131	518	123	874	—	—	—	—	—	—
0.28	5.0	5.87	94	372	91	514	130	860	—	—	—	—	—	—
	4.0	6.08	104	362	101	494	113	996	206	3906	—	—	—	—
	3.0	6.43	121	346	117	466	128	716	208	1680	—	—	—	—
	2.5	6.71	136	334	130	446	140	666	210	1306	—	—	—	—
0.25	5.0	6.75	96	334	91	442	99	660	147	1268			—	—
	4.0	7.07	106	322	101	424	107	614	152	1060	376	1268	—	—
	3.0	7.62	123	304	117	396	122	556	161	864	309	1188	—	—
	2.5	8.06	138	292	131	376	135	518	172	766	295	982	—	—

Design parameters: D = 1.0; Z_{ant} = 15 ohms; Z_{cable} = 50 ohms. The element diameter is normalized as 1. The frequency is 7.1 MHz. C is expressed in pF, L in cm (divide by 2.54 to obtain inches). Note there is a whole range where no match can be obtained. If sufficient negative reactance is provided in the driven element impedance (with element shortening), there will be no problem in matching Yagis even with a low radiation resistance.

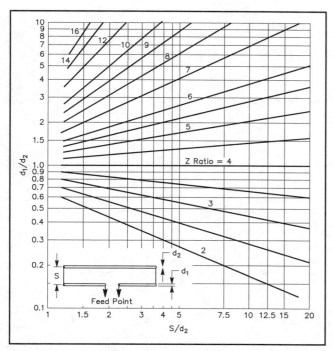

Fig 13-78—Step-up ratio for the gamma and omega match as a function of element diameter (d2), rod diameter (d1) and spacing (S). (*After ARRL Antenna Book*)

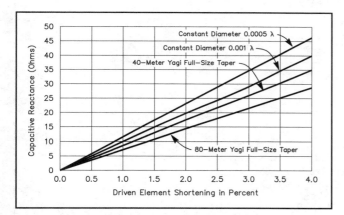

Fig 13-79—Capacitive reactance obtained by various percentages of driven-element shortening. The 40-meter full-size taper is the taper described in Table 13-1; the 80-meter taper is shown in Fig 13-41.

driven element (eg, +18 Ω reactance), a match cannot be achieved, not even with a step-up ratio of 15:1. With this type of gamma (step-up = 15), the highest positive reactance that can be accommodated with a radiation resistance of 13 Ω is approximately +12 Ω. In other words, it is simply impossible to match the 13 + j18 Ω impedance of the W2PV 3-element 20-meter Yagi with a gamma match without first reducing the length of the driven element.

The first thing to do when matching a Yagi with a relatively low radiation resistance is to decrease the element length as to introduce, eg, −15 Ω of capacitive reactance in the

driven element impedance. How much shortening is needed (in terms of element length) can be derived from **Fig 13-79**. From Table 13-11 we learn that an impedance of 15 − j15 Ω can be easily matched with step-up ratios ranging from 5 to 8.

Several Yagis have been built and matched with gamma systems, calculated as explained above. When the reactance of the driven element at the design frequency was exactly known, the computed rod length as calculated with the above procedure was always right on. In some cases the series capacitor value turned out to be smaller than calculated. This is caused by the stray inductance of the wire connecting the end of the gamma rod with the plastic box containing the gamma capacitor, and the wire between the series capacitor and the coaxial feed line (receptacle). The inductance of such a wire is not at all negligible, especially on the higher frequencies. With a pure coaxial construction, this should not occur.

A coaxial gamma rod is made of two concentric tubes, whereby the inner tube is covered by a proper dielectric material (eg, polyethylene tube). The length of the inner tube, as well as the material dielectric and thickness, determine the capacitance of this coaxial capacitor. Make sure to properly seal both ends of the coaxial gamma rod to prevent moisture penetration.

Feeding a symmetric element with an asymmetric feed system has a slight impact on the radiation pattern of the Yagi. The forward pattern is skewed slightly toward the side where the gamma match is attached, but only a few degrees, which is of no practical concern. The more elements the Yagi has, the less the effect is noticeable.

The voltage across the series capacitor is quite small even with high power, but the current rating must be sufficient to carry the current in the feed line without warming up.

For a power of 1500 W, the current through the series capacitor is 5.5 A (in a 50-Ω system). The voltage, depending on the value of the capacitor, will vary between 200 and 400 V in most cases. This means that moderate-spacing air-variable capacitors can be used, although it is advisable to over-rate the capacitors as slight corrosion of the capacitor plates normally caused by the humidity in the enclosure will derate the voltage handling of the capacitor.

3.10.1.1. Designing the gamma match with the YAGI DESIGN software

Fig 13-80 shows a screen print of the MATCHING SYSTEMS module. From the prompt line, any of the input data can be easily changed. This is immediately reflected in a different gamma-rod length and gamma capacitor value.

In the first iteration the Yagi feed-point impedance was purely resistive (no reactance). Introducing 20 Ω of negative reactance (shortening the driven element about 2%) results in a shorter gamma rod and a smaller gamma capacitor (second solution).

3.10.2. The omega match

The omega match is a sophisticated gamma match that uses two capacitors. Tuning of the matching system can be done by adjusting the two capacitors, without having to

```
,-------------------------------- Gamma match Design --------------------------------.
| DESIGN FREQUENCY  :    7.10 MHz                                                      |
| FEEDLINE IMPEDANCE :    50 ohm.                                                      |
| ANTENNA POWER  :  1500 WATT.             .--------------------------------------.    |
| Z-ANT RESISTIVE PART :   28.00 ohm.      | -->> STEP UP RATIO =  6.60 <<--      |    |
| Z-ANT REACTIVE PART :    0.00 ohm.       '--------------------------------------'    |
| ELEMENT DIAMETER :   6                                                               |
| GAMMA ROD DIAMETER :   2                                                             |
| SPACING (CENTER TO CENTER) :   20                                                    |
'-------------------------------------------------------------------------------------'
```

```
,------------------------------------ Results ---------------------------------------.
| GAMMA ROD LENGTH : 216.8 cm  OR  85.4 inch.                                         |
| SERIES CAPACITOR :   487 pF.                                                        |
| VOLTAGE ACROSS SERIES CAPACITOR :   252 Volts.                                      |
| CURRENT THROUGH SERIES CAPACITOR :  5.5 Amp.                                        |
'------------------------------------------------------------------------------------'
```

1=SAVE   2=FREQ   3=Z-CABLE   4=DIMENS   5=MATCH SYS   6=Z-ANT   7=PWR H=HELP X=EXIT

7547   YAGI IMPEDANCE MATCHING   on4un/on6wu

```
,-------------------------------- Gamma match Design --------------------------------.
| DESIGN FREQUENCY  :    7.10 MHz                                                      |
| FEEDLINE IMPEDANCE :    50 ohm.                                                      |
| ANTENNA POWER  :  1500 WATT.             .--------------------------------------.    |
| Z-ANT RESISTIVE PART :   28.00 ohm.      | -->> STEP UP RATIO =  6.60 <<--      |    |
| Z-ANT REACTIVE PART : -20.00 ohm.        '--------------------------------------'    |
| ELEMENT DIAMETER :   6                                                               |
| GAMMA ROD DIAMETER :   2                                                             |
| SPACING (CENTER TO CENTER) :   20                                                    |
'-------------------------------------------------------------------------------------'
```

```
,------------------------------------ Results ---------------------------------------.
| GAMMA ROD LENGTH : 149.9 cm  OR  59.0 inch.                                         |
| SERIES CAPACITOR :   335 pF.                                                        |
| VOLTAGE ACROSS SERIES CAPACITOR :   367 Volts.                                      |
| CURRENT THROUGH SERIES CAPACITOR :  5.5 Amp.                                        |
'------------------------------------------------------------------------------------'
```

1=SAVE   2=FREQ   3=Z-CABLE   4=DIMENS   5=MATCH SYS   6=Z-ANT   7=PWR H=HELP X=EXIT

Fig 13-80—Screen dump of the design of a gamma match with the YAGI DESIGN software. In the first example I did not shorten the driven element (zero-Ω reactance). In the second alternative I introduced −20 Ω, which resulted in a much shorter gamma rod and a series capacitor of smaller value.

adjust the rod length.

Fig 13-81 shows the omega match and its electrical equivalent. Comparing it with the gamma electrical equivalent (Fig 13-77) reveals that the extra parallel capacitor, together with the series capacitor, now is part of an L network that follows the original gamma match.

The steps in calculating an omega match are:

- Calculate the step-up ratio (use Fig 13-78).
- Multiply Z_h (half of the radiation resistance) with the step-up ratio (Z_i).
- Decide which omega rod length you will use. Do not use too short a rod, because very high currents will circulate in the low-impedance elements associated with a very short omega rod. As a rule of thumb, use a rod which is $2/3$ to

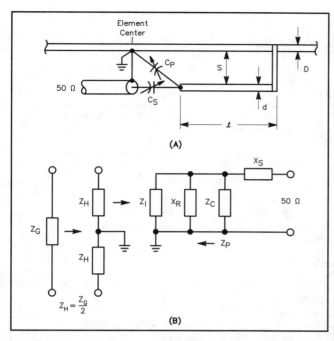

Fig 13-81—Layout and electrical equivalent of the omega match.

$^3/_4$ the length of the equivalent gamma rod.

- Calculate the (inductive) reactance (X_r) of the (shorted) transmission line length made up by the omega rod and the driven element. The omega shorting bar is the shorted end of the transmission line, while the impedance at the open end of the transmission line is in parallel with half of the feed-point impedance of the Yagi.
- Calculate the parallel impedance (Z_p) made up by paralleling Z_i and X_r. If the omega rod is not too long, then the resistive part of the impedance (Z_p) will be less than 50 Ω.
- Calculate the impedance resulting from connecting the omega capacitor (Z_C) in parallel with Z_p. Change the value of this capacitor (tune the omega capacitor) until the resistive part of Z_p becomes 50 Ω. For 14.2 MHz and a rod of $^2/_3$ to $^3/_4$ the full-size gamma rod length, the value of this capacitor will usually be between 50 and 150 pF.
- Calculate the value of the (omega) series capacitor required to tune out the inductive reactance of the impedance Z_p (Z_p being 50 + X_s).

Again, the mathematics involved are complex, but the MATCHING section of the YAGI DESIGN software will do the job in a second.

From a practical point of view the omega match is really unbeatable. The ultimate setup consists of a box containing the two capacitors, together with dc motors and gear-reductions. **Fig 13-82** shows the interior of such a unit using surplus capacitors and dc motors from a flea market. This system makes the adjustment very easy from the ground, and is the only practical solution when the driven element is located away from the center of the antenna.

Fig 13-82—Motor-driven omega matching unit. The two capacitors with the dc motors and gear box are mounted in line on a piece of insulating substrate material. This can then be slid inside the housing, which is made of stock lengths of PVC water drainage pipe. The PVC pipe is available in a range of diameters, and the pipe can be easily cut to the desired length. The round shape of the housing also has an advantage as far as wind loading is concerned.

The remarks given for the gamma capacitors as to the required current and voltage rating are valid for the omega match as well. The voltage across the omega capacitor is of the same magnitude as the voltage across the gamma capacitor, usually varying between 300 and 400 V, with a current of 2 to 4 A for a power of 1500 W.

3.10.2.1. Designing the omega match with the YAGI DESIGN software

Fig 13-83 shows the screen print from the MATCHING module for an omega-match design. With the omega match you must also enter the rod length. The program calculates the values of both the parallel (omega) and the series (gamma)

```
┌──────────────────── Omega match Design ─────────────────────────┐
│  DESIGN FREQUENCY :    7.10 MHz                                  │
│  FEEDLINE IMPEDANCE :  50 ohm.                                   │
│  ANTENNA POWER :   1500 WATT.        ┌─────────────────────────┐ │
│  Z-ANT RESISTIVE PART :  28.00 ohm.  │ -->> STEP UP RATIO = 6.60 <<-- │
│  Z-ANT REACTIVE PART : -20.00 ohm.   └─────────────────────────┘ │
│  ELEMENT DIAMETER :  6                                           │
│  OMEGA ROD DIAMETER :  2                                         │
│  SPACING (CENTER TO CENTER) :  20                                │
└─────────────────────────────────────────────────────────────────┘

┌──────────────────────────── Results ────────────────────────────┐
│  OMEGA ROD LENGTH : 120.0 cm  OR  47.2 inch.                     │
│  SERIES CAPACITOR :   335 pF.                                    │
│  VOLTAGE ACROSS SERIES CAPACITOR :   367 Volts.                 │
│  CURRENT THROUGH SERIES CAPACITOR :  5.5 Amp.                   │
│  PARALLEL CAPACITOR :   96 pF                                    │
│  VOLTAGE ACROSS PARALLEL CAPACITOR :   458 Volts.              │
│  CURRENT THROUGH PARALLEL CAPACITOR :  1.95 Amp.               │
└─────────────────────────────────────────────────────────────────┘

  1=SAVE   2=FREQ   3=Z-CABLE   4=DIMENS   5=MATCH SYS   6=Z-ANT   7=PWR H=HELP X=EXIT
```

Fig 13-83—Screen dump for the omega-match design for a 40-meter Yagi with a 28-Ω feed-point impedance. Compare the results with those obtained for the gamma match (Fig 13-80).

capacitors, including voltages and currents for a given power.

3.10.2.2. Tuning the omega / gamma matching systems

The GAMMA/OMEGA OPTIMIZER module of the YAGI DESIGN software is a very interesting software modeling tool. It allows you to change all the parameters of the matching system, while observing the effects of varying these parameters on up to 7 different frequencies. **Fig 13-84** shows the design of an omega match for one of the designs from the YAGI DESIGN database (no. 10). Note that in the model I have included the parasitic reactance of the wire connecting the rod with the capacitors in the box (in this example the wire was 164 mm (6.5 inches) long. I have adjusted the omega for a 1:1 SWR not in the center of the passband, but 0.5% above the center frequency.

3.10.3. The hairpin match

The (center) feed impedance of the driven element of a Yagi (in the neighborhood of 0.5-λ long) consists of a resistive part (the radiation resistance) in series with a reactance. The reactance is positive if the element is "long" and negative if the element is "short." "Long" means longer than the resonant length. A dipole is resonant when the (center fed) feed-point impedance shows zero reactance. In practice, resonance never occurs at a physical length of exactly 0.5 λ, but always at a shorter length (see further). With a hairpin matching system we deliberately make the element "short," meaning that the feed-point impedance will be capacitive.

Fig 13-85 shows the electrical equivalent of the hairpin matching system.

If we connect an inductor (coil) across the terminals of a "short" driven element, we can now consider the capacitor (the element that is responsible for the capacitance part in the short element feed-point impedance) and the parallel inductor to be the two arms of an L network. This L network can be dimensioned to give a 50-Ω output impedance. In other words, a perfect match can be obtained by shortening the driven element to produce the required capacitive reactance so that, through the parallel combination of the reactive feed-point impedance (negative reactance) and the positive reactance of the inductor, a 50-Ω impedance is obtained. The parallel inductor is normally replaced with a short length of short-circuited open-wire feed line, having the shape of a hairpin, and hence the matching system's name.

The resistive part of the impedance of the driven element (radiation resistance) changes only slightly as a function of length if the element length is varied plus or minus 5% around the resonant length. The change in the reactance, however, is quite significant. The rate of change will be greatest with elements having the smallest diameter (see Fig 13-79).

The required hairpin reactance is given by

$$X_{hairpin} = 50 \times \sqrt{\frac{R_{rad}}{50 - R_{rad}}} \qquad \text{(Eq 13-3)}$$

The formula for calculating the size of the shunt reac-

```
7547              GAMMA / OMEGA OPTIMIZER PROGRAM                 on4un/on6wu

                                         GAMMA WIRE LENGTH = 164.5 mm
      DIAMETER DRIVEN ELEMENT =  6                          6.48 inch
      DIAMETER GAMMA ROD =  2            GAMMA CAPACITOR =   442.6 pF (C1)
      SPACING CENTER TO CENTER =  20     OMEGA CAPACITOR = 134.3 pF (C2)
                                         ROD LENGTH = 150.0 cm
      DESIGN FREQUENCY =  7.05 MHz                   59.1 inches

      TUNING RATE = 1  %                 FEEDLINE-Z = 50  ohm

                         Press F/C to change tuning rate

    ANTENNA IMPEDANCE                     SWR        MATCHED IMPEDANCE
  FREQ         Real / Imag                             Real  /   Imag
  6.944        28.8 / -12.6               1.67         83.2  /    4.5
  6.979        29.1 /  -9.0               1.49         74.4  /    2.7
  7.015        28.8 /  -5.3               1.33         66.2  /    1.1
  7.050        28.0 /  -1.4               1.16         58.2  /    0.1
  7.085        26.8 /   2.8               1.00         50.1  /    0.0
  7.121        25.2 /   7.5               1.20         41.8  /    1.0
  7.156        23.4 /  12.5               1.48         34.1  /    3.3

1=DECR.C1 2=INCR.C1  3=DECR.C2  4=INCR.C2  5=DECR.ROD  6=INCR.ROD   Z=CABLE-Z
7=DECR.WIRE  8=INCR.WIRE  C=COURSE  F=FINE T=TUBE DIMS  S=SAVE  X=EXIT  H=HELP
```

Fig 13-84—Screen-print of the omega match as optimized using the OPTIMIZE GAMMA/OMEGA module of the YAGI DESIGN software. Note the prompt line at the bottom of the screen from where *all* parameters can easily be changed. This is really tweaking the matching system on the computer. "Playing" with the module is most instructive to understand the behavior of gamma and omega matching systems.

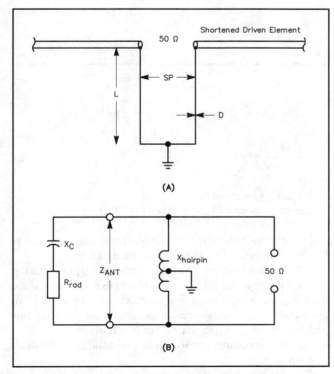

Fig 13-85—Layout and electrical equivalent of the hairpin match.

tance depends on the shape of the inductor. There are two common types:

- The hairpin inductor.
- The beta-match inductor.

The hairpin inductor is a short piece of open-wire transmission line. The boom is basically outside the field of the transmission line. This means that the boom is not near the transmission line. In practice the separation between the line and the boom should be at least equal to twice the spacing between the conductors of the transmission line.

The characteristic impedance of such a transmission line is given by

$$Z_C = 276 \times \log \frac{2 \times SP}{D} \quad \text{(Eq 13-4)}$$

where

SP = spacing between wires

D = diameter of the wires

In the so-called beta-match, the transmission line is made of two parallel conductors with the boom in between. This is the system used by Hy-Gain.

The characteristic impedance of such a transmission line is given by

$$Z_C = 553 \times \log \frac{2 \times S}{\sqrt{D \times DB}} \quad \text{(Eq 13-5)}$$

where

    S = spacing of wire to boom center
    D = diameter of wire
    DB = diameter of boom

    Z_C is the characteristic impedance of the open wire line made by the two parallel conductors of the hairpin or beta-match. The length of the hairpin or beta-match is given by

$$\ell^\circ = \arctan \frac{X_{hairpin}}{Z_C} \qquad \text{(Eq 13-6)}$$

where

    ℓ° is the length of the hairpin expressed in degrees
    arctan is the inverse tangent

    To convert to real dimensions (assuming a velocity factor of 0.98 for a transmission line with air dielectric)

$$L_{cm} = \ell^\circ \times \frac{81.6}{f} \qquad \text{(Eq 13-7)}$$

where f = design frequency, MHz

    The required driven element reactance is given by

$$X_C = -\frac{R_{rad} \times 50}{X_{hairpin}} \qquad \text{(Eq 13-8)}$$

    Table 13-12 shows the required values of capacitive reactance in the driven element as well as the required reactance for a range of radiation resistances. (For a hairpin with S = 10D as in the table, Z = 359 Ω.) The question now is how long the driven element must be to represent the required amount of negative reactance ($-X_C$). Fig 13-79 lists the reactance values obtained with several degrees of element shortening. Although the exact reactance differs for each one of the listed element diameter configurations, one can derive the following formula from the data in Fig 13-79.

$$X_C = -Sh \times A \qquad \text{(Eq 13-9)}$$

where

    Sh = shortening in % versus the resonant length.
    X_C= reactance of the element in Ω
    A = 8.75 (for a 40-meter full-size Yagi) or 7.35 (for an 80-meter full-size Yagi)

    This formula is valid for shortening factors of up to 5%. The figures are "typical" and depend on the effective diameter of the element.

3.10.3.1. Design guidelines for a hairpin system

    Most HF Yagis have a radiation resistance between 20 Ω and 30 Ω. For these Yagis the following rule of thumb applies:

- The required element reactance to obtain a 50-Ω match with a hairpin is approximately −25 Ω (Table 13-12).
- This almost constant reactance value can be translated to an element shortening of approximately 2.8% as compared to the resonant element length for a 40-meter Yagi, and 3.5% for an 80-meter Yagi.
- The value of reactance of the hairpin inductor is equal to 2 times the radiation resistance.

Table 13-12
Required Capacitive Reactance in Driven Element Impedance and in Hairpin Inductance

R_{rad} ohms	Antenna Reactance (ohms)	Inductance Hairpin (ohms)	Length hairpin (cm) (SS = 10D) 3.65 MHz	7.1 MHz
10.0	−20.0	25.0	89	46
12.5	−21.6	28.9	103	53
15.0	−22.9	32.7	117	60
17.5	−23.8	36.7	130	67
20.0	−24.5	40.8	144	74
22.5	−24.9	45.2	160	82
25.0	−25.0	50.0	177	91
27.5	−24.9	55.3	194	100
30.0	−24.5	61.2	216	111

Note: The feed-point impedance is 50 ohms. To obtain the hairpin length in inches, divide values shown by 2.54.

Table 13-13
Hairpin Line Impedance as a Function of Spacing-to-Diameter Ratio

S/D Ratio	Impedance, Ohms
5	193
7.5	325
10	359
15	408
20	442
25	469
30	491
35	510
40	525
45	539
50	552

    The length of the hairpin is given by

$$\ell = \frac{9286 \times R_{rad}}{f \times Z} \qquad \text{(Eq 13-10)}$$

where

    f = design frequency
    Z = impedance of hairpin line (Eq 13-4).
    ℓ = length in cm

    The impedance of the hairpin line for a range of spacing-to-wire diameter ratios is shown in **Table 13-13**.

    The real area of concern in designing a hairpin matching system is to have the correct element length that will produce the required amount of capacitive reactance. As we have an open (split) element, we can theoretically measure the impedance, but this is impractical for two reasons:

- The impedance measurement must be done at final installation height.
- The average ham does not have access to measuring equipment that can measure the impedance with the re-

Table 13-14
Values of Transformed Impedance and SWR for a Range of Driven-Element Impedances

Driven Element Impedance	Hairpin Inductance	Resulting Impedance	SWR (vs 50 ohms)
$20 - j\,20$	40.8 ohms	$40 - j\,0.81$	1.25
$20 - j\,24.5$	40.8	$50 +$	1
$20 - j\,25$	40.8	$51.2 + j\,0.26$	1.02
$20 - j\,30$	40.8	$64.5 + j\,5.92$	1.32

Table 13-15
Capacitive Reactance Obtained by Various Percentages of Driven-Element Shortening

Shorten Element	Diameter in wavelengths 0.0010527	0.0004736	Light Taper	Heavy Taper
0%	0 ohms	0 ohms	0 ohms	0 ohms
0.5	−4.8	−5.5	−4.6	−4.8
1.0	−9.6	−11.1	−9.1	−9.7
1.5	−14.3	−16.5	−13.6	−14.3
2.0	−19.1	−22.2	−18.2	−19.2
2.5	−23.8	−27.5	−22.7	−23.7
3.0	−28.6	−32.8	−27.2	−28.6
3.5	−33.5	−38.2	−31.7	−33.5
4.0	−38.5	−43.5	−36.2	−38.3

quired degree of accuracy. The run-of-the-mill noise bridge will not suffice, and a professional impedance bridge or network analyzer is required.

Let us examine the impact of a driven element that does not have the required degree of capacitive reactance. **Table 13-14** shows the values of the transformed impedance and the resulting (minimum) SWR if the reactance of the driven element was off $+5\ \Omega$ and $-5\ \Omega$ versus the theoretically required value, for an R_{rad} of 20 Ω. An error in reactance of 5 Ω either way is equivalent to an error length of 0.5% (see **Table 13-15**). In other words, an inaccuracy of 0.5% in element length will deteriorate the minimum SWR value from 1:1 to 1.25 or 1.3:1.

The mounting hardware for a split element will always introduce a certain amount of shunt capacitance at the driven-element feed point. This must be taken into account when designing a hairpin- or beta-match system (see example in Section 3.8.3.3).

3.10.3.2. Element loading and a hairpin

Getting the correct element length that will yield a perfect 50-Ω match in principle requires a very accurate element length. Adjusting the element length while the antenna is in the air is usually not practical.

The length of the driven element that produces zero reactance (at the design frequency) is called the resonant length. This length also depends on the element diameter (in terms of wavelength). If any taper is employed for the

construction of the element, the degree of taper will have its influence as well. Finally, the resonant length will differ with every Yagi design. This is caused by the effect of mutual coupling between the elements of the Yagi. For elements with a constant diameter of approximately 0.001 λ, the resonant-frequency length will usually be between 0.477 λ and 0.487 λ. The exact value for a given design can be obtained by modeling the Yagi or by obtaining it from a reliable database.

If the exact resonant length is not known, then it is better to make the element somewhat too short (too much negative reactance), whereby the element can be electrically lengthened by loading it in the center with a short piece of transmission line. Adjusting the amount of loading (as a function of how much the element was too short) can usually be done more easily than adjusting the element length (element tips).

The loading is done by using a short length of open-wire line. The short length of line can have the same impedance and configuration (wire diameter and spacing) as used for the hairpin (usually between 300 Ω and 450 Ω). The layout and the electrical equivalent of this approach is shown in **Fig 13-86**. The transmission line acts as a loading device between the element feed point and the 50-Ω tap, and as the matching inductor beyond the 50-Ω tap (the hairpin). Another method of changing the electrical length of the driven element is described in Section 3.10.3.5, where a parallel capacitor is used to shorten the electrical length of the driven element.

Let us examine the impedance of the antenna feed point along a short high-impedance (200-Ω to 450-Ω) transmission line:

• The value of the resistive part will remain almost constant (change negligible).
• The value of the capacitive reactance will decrease by X ohms per degree, where X is given by

$$\frac{X}{deg\,rees} = Z \times 0.017 \qquad \text{(Eq 13-11)}$$

The change in reactance per unit of length is:

$$\frac{X}{cm} = \frac{Z \times f \times 0.204}{1000} \qquad \text{(Eq 13-12)}$$

where
  f = frequency, MHz
  Z = characteristic impedance of the line made by the two parallel wires of the hairpin or beta-match (see Eq 13-4 and Table 13-13)
  Eqs 13-9 and 13-10 are valid for line lengths of 4° maximum.
  The line length required to achieve a given reactance shift X is given by:

$$L_{cm} = \frac{X \times 4900}{Z \times f} \qquad \text{(Eq 13-13)}$$

This formula is valid for values of X of 25 Ω maximum.
Example: Let us assume that we start from a

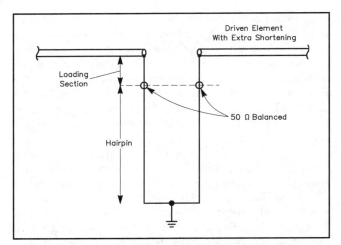

Fig 13-86—Layout of hairpin match combined with element loading, whereby the 50-Ω point is to be found on the hairpin at some distance from the element.

(20 – j30) Ω impedance, and we need to electrically lengthen the driven element to yield an impedance of (20 – j24.5) Ω (see Table 13-12). The design frequency is 7.1 MHz.

The required reactance difference is X = (30 – 24.5) = 5.5 Ω. The required 359-Ω-line length is

$$L = \frac{5.5 \times 1929}{359 \times 7.1} = 10.6 \text{ cm}$$

The length of the hairpin section can be determined from Table 13-12 as 111 cm. This means that we can electrically adjust (load) the element to the required length by adding an extra piece of (hairpin) line. The length of this line will be only a few inches long. In this case the 50-Ω tap will not be at the element but at a short distance on the hairpin line. The length of the hairpin matching inductor will remain the same, but the total transmission-line length will be slightly longer than the matching hairpin itself.

To adjust the entire system, look for the 50-Ω point on the line (move the balun attachment point) while at the same time adjusting the total length of the hairpin. The end of the hairpin (shorting bar) is usually grounded to the boom.

Design Rule of Thumb: The transmission-line loading device can be seen as part of the driven element folded back in the shape of the transmission line. For a 359-Ω transmission line (spacing = 10 × diameter), the length of the loading line will be exactly as long as the length that the element has been shortened. In other words, for every inch of total element length you shorten the driven element, you must add an equivalent inch in loading line. This rule is applicable only for 359-Ω lines and for a maximum length of 406/f cm, where f = design frequency. For other line impedances the calculation as shown above should be followed.

Example

A driven element is resonant at 7.1 MHz with a length

of 2200 cm. We want to shorten the total element length by 25 cm, and restore resonance by inserting a 359-Ω loading line in the center. The length of the loading line will be approximately 25 cm.

3.10.3.3. Hairpin match design with parasitic element-to-boom capacitance

As explained in Section 3.4.3, it is virtually impossible to construct a split element without any capacitive coupling to the boom. With tubular elements a "coaxial" construction is often employed, which results in an important parasitic capacitance.

The Hy-Gain Yagis, which use a form of "coaxial" insulating technique to provide a split element for their Yagis, exhibit the following parallel capacitances.

- 205BA: 27 pF
- 105BA and 155BA: 10 pF

Let's work out an example for a Yagi designed at 7.1 MHz. We model the driven element to be resonant (zero reactance) at the design frequency (7.1 MHz). Assume the resonant length is 1985 cm.

The capacitance introduced by the split-element mounting hardware is 300 pF per side (I have measured this with a digital capacitance meter). Do not forget to measure the capacitance without the full element attached (see Section 4.3.3).The reactance of this capacitor is

$$X_C = \frac{10^6}{2\delta \times 7.1 \times 300} = 75 \text{ Ω}$$

The capacitance across the feed point is 150 pF (2 times 300 in series):

$$X_C = 150 \text{ Ω}$$

Using the SHUNT NETWORK module of the NEW LOW BAND SOFTWARE, we calculate the resulting impedance of this capacitor in parallel with 28 Ω impedance at resonance as

$$Z = 27.1 – j5 \text{ Ω}$$

The required inductance of the hairpin (using Eq 13-3) will be

$$X_{hairpin} = 50 \times \sqrt{\frac{R_{rad}}{50 - R_{rad}}} = \sqrt{\frac{27.1}{50 - 27.1}} = 54.30 \text{ Ω}$$

Assume we are using a hairpin with two conductors with spacing = 10 × diameter. The impedance of the line is given by Eq 13-4 as

$$Z_C = 276 \times \log\frac{2 \times SP}{D} = 276 \times \log(20) = 359 \text{ Ω}$$

The length of the hairpin is given by Eq 13-6 as

$$\ell° = \arctan\frac{X_{hairpin}}{Z_C} = \arctan\frac{54.3}{359} = 8.6°$$

The length in cm is given by Eq 13-7 as

$$\ell_{cm} = \ell° \times \frac{81.6}{f} = 8.6 \times \frac{81.6}{7.1} = 98.5 \text{ cm}$$

For an impedance of 27.1 Ω we need an impedance

reactance (using Eq 13-8) of

$$X_C = -\frac{R_{rad} \times 50}{X_{hairpin}} = -\frac{27.1 \times 50}{54.3} = -24.95 \ \Omega$$

This means we have to add another 19.95 Ω of negative reactance to our driven element. This can be done by shortening the element approximately 2.2% (see Fig 13-79), which amounts to $1985 \times \dfrac{2.2}{100} = 44$ cm or 22 cm on each side.

Instead of shortening the element 20 cm on each side, we can, eg, shorten it 40 cm on each side, which will now give us some range to fine tune the matching system. The 20 cm we have shortened the driven element on each side will be replaced with an extra 20-cm length of transmission line at the feed point. The 50-Ω point will now be located some 20 cm from the split driven element. The hairpin will extend another 98.5 cm beyond this point. Tuning the matching system consists of changing the position of the 50-Ω point on the hairpin as well as changing the length of the hairpin.

If you use "wires" to connect the split driven element to the matching system, you must take the inductance of this short transmission line into account as well.

3.10.3.4. Designing the hairpin match with the YAGI DESIGN software

Fig 13-87 shows the screen print from the MATCHING SYSTEMS module with the data of the example used in Section 3.3.1. Note exactly the same results in the bottom part of the screen as those calculated above. From the prompt line you can change any of the input data, which will be immediately reflected in the dimensions of the matching system.

The value of the "parasitic" parallel capacitance (see Section 3.10.3.3) can be specified, and is accounted for during the calculation of the matching system.

3.10.3.5. Using a parallel capacitor to fine-tune a hairpin matching system

A parallel capacitance (of reasonable value) across the split element only slightly lowers the resistive part of the impedance, while it introduces an appreciable amount of negative reactance.

Example

A capacitor of 150 pF in parallel with an impedance of 28 + $j0$ Ω (at 7.1 MHz) lowers this impedance to 27.1 $- j5 \ \Omega$ (see example in Section 3.10.3.3). This means that instead of fine tuning the matching system by accurately shortening the driven element in order to obtain the required negative reactance (see Table 13-12), you can use a variable capacitor across the driven element in order to electrically shorten the element. This is a very elegant way of tuning the hairpin matching system "on the nose." The only drawback is that it requires another (vulnerable) component. This method is an alternative fine-tuning method to the configuration where the length of the driven element is altered by using a short length of transmission line as described in Section 3.10.3.2.

3.10.4. Direct feed

If the driven element is split (not grounded to the boom), one can also envisage a direct feed. A 3-element Yagi will never present a 50-Ω feed-point impedance, unless it would be designed with an extremely low Q and trade in a lot of

```
 7547                    YAGI IMPEDANCE MATCHING                on4un/on6wu

                         ─────── Hairpin match Design ───────
    DESIGN FREQUENCY :    7.10 MHz     PARALLEL CAPACITANCE:    150   pF
    FEEDLINE IMPEDANCE :   50 ohm.
    Z-ANT RESISTIVE PART :   27.5 ohm.  This Impedance now includes the effect
    Z-ANT REACTIVE PART : -4.0 ohm.     of the parasitic parallel capacitance
    INITIAL DR. ELEMENT HALF LENGTH :  1023.5 cm  OR  402.9 Inches.
    HAIRPIN SPACING :   10
    HAIRPIN WIRE DIAMETER :   1



                         ─────── Results ───────
    THE DR. EL. HALF LENGTH MUST BE SHORTENED  21.7 cm  OR    8.5 Inches.
    LENGTH OF HAIRPIN : 100.1 cm  OR  39.4 Inches.
    THE INDUCTANCE OF THE HAIRPIN IS 1.24 µH



    1=SAVE  2=FREQ  3=Z-CABLE  4=DIMENS  5=SYST  6=SWR  7=PAR CAP  H=HELP  X=EXIT
```

Fig 13-87—Screen dump of the hairpin matching system as designed with the YAGI DESIGN software.

forward gain. Three-element Yagis will typically show feed-point impedances varying between 18 Ω and 30 Ω. This means we have to use some kind of system to match the Yagi impedance to the feed-line impedance. In addition, if we want to use a feed system for an 80-meter Yagi, which has to cover both the CW and the SSB end of the band, we also will have to deal with the reactances involved.

3.10.4.1. Split element direct feed system with series compensation

We can often use a quarter-wave transformer to achieve a reasonable match between the Yagi impedance (at resonance) and a 50-Ω feed line. There are two solutions. For a feed-point impedance lower than 25 Ω one can use a $^1/_4$-λ length of line with an impedance of 30 Ω. This can be made by paralleling a 50-Ω $^1/_4$-λ cable with a 75-Ω $^1/_4$-λ cable. The cable can be coiled up in coil measuring approx. 30 cm in diameter, which serves as common-mode current balun.

For impedances between 25 and 30 Ω the impedance of the quarter-wave transformer is 37.5 Ω, made by paralleling two 75-Ω $^1/_4$-λ cables. An example of such a matching system is given in **Fig 13-88**.

Let us analyze the case of a direct feed for the 80-meter Yagi described in Section 3.3. The real part of the impedance of the Yagi is around 28 Ω, which is easy to match to a 50-Ω feed line via a 37.5-Ω quarter-wave transformer, made by two parallel 75-Ω cables.

There are different approaches to handling the inductive part of the impedance at the opposite end of the band. One could dimension the driven element to be resonant on 3.8 MHz and tune out the capacitive reactance (approx. 75 Ω) by using a coil in series with the coaxial feed line. The other alternative is to dimension the driven element for resonance in the CW band and then tune out the inductive reactance on, eg, 3.75 MHz by using a series capacitor. In an example I dimensioned the driven element for resonance on 3.55 MHz, and calculated the value of the series capacitor to achieve resonance on 3.75 MHz. The capacitor value is 900 pF. This matching system is extremely simple, and will guarantee maximum bandwidth as well. The quarter-wave 37.5 Ω transformer can be coiled up and serve as a current balun.

Fig 13-89 shows the SWR curves for this feed arrangement. Note that with such an arrangement it is impossible to have a good SWR in the middle of the band.

3.10.4.2. Split element direct feed system with parallel compensation

The direct feed system with parallel compensation does not require a quarter-wave transformer as it provides a good match directly to a 50-Ω impedance. In the case of the driven element of an 80-meter Yagi, one could dimension the driven element for resonance in the middle of the band (3.65 MHz). In that case the reactance of a typical 3-element Yagi such as the antenna developed in Section 3.3, exhibits approx. $-j35$ Ω on 3.5 MHz and $+j35$ Ω on 3.8 MHz. The reactance can be tuned out by a parallel coil or capacitor (see **Fig 13-90**). A coil with an inductance of 2.2 μH will tune the driven element to resonance on 3.55 MHz and yield an impedance of 50 Ω (lucky coinci-

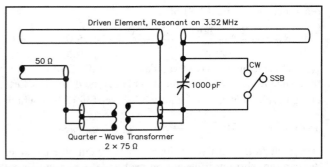

Fig 13-88—Split-element matching system for the 80-meter Yagi. The driven element is tuned to resonance in the CW end of the band. On phone (3.8 MHz) the inductive reactance is tuned out by a simple series of capacitor of approx. 560 pF. A simple relay can short circuit the capacitor on CW. A quarter-wave transformer (Z=37.5 Ω) is made of two parallel 75-Ω cables that may be coiled up as a current balun, and represents a 50-Ω impedance at its end.

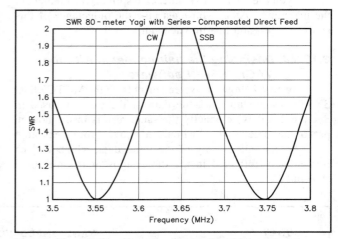

Fig 13-89—SWR curves for the split-element feed method with series compensation and a 37.5-Ω quarter-wave transformer.

dence!). Likewise, a capacitor with a value of 700 pF will tune the element to resonance on 3.8 MHz, also with an impedance of 50 Ω.

The value of these components can easily be calculated using the SHUNT IMPEDANCE NETWORK module of the NEW LOW BAND SOFTWARE. With a simple relay one can switch either the coil or the capacitor in parallel with the feed point, and obtain a fine matching system for either the CW or the SSB end of the band. **Fig 13-91** shows the SWR curves obtained with such an arrangement.

3.10.5. Selecting a Yagi matching system

Hairpin or beta-matching systems are generally used by manufacturers of commercial Yagis because they are simple and cost effective to reproduce in volume. For the home-builder, things are not quite the same.

Hairpin and beta match

Advantages:

- No capacitor required, no box, one less (vulnerable) component.
- Fully symmetric feed system.
- The split element makes it possible to measure the feed-point impedance before matching.

Disadvantages:

- Length adjustment of driven element required.
- Difficult to adjust at full height unless the modified hairpin system (with the extra loading line) is used.
- No way to adjust if driven element is at a distance from the mast.
- A balun is required.

Gamma and omega match

Advantages:

- No length adjustment of driven element required.
- Motor driven capacitors (omega) make remote tuning extremely flexible.
- No split element required (all plumber's delight).
- No risk for insulator breakdown.
- In principle mechanically stronger.

Disadvantages:

- Requires box and one or two capacitors.
- Causes very slight asymmetric pattern.

Direct feed

Advantages:

- Very simple.
- Easily adaptable to wide band coverage using series or parallel compensation.
- Switching between CW and SSB can be done with a single SPDT relay and one component.
- Split element allows direct measurements.

Disadvantages:

- Split element required.

I use omega matching systems on all my Yagi antennas from 10 through 40 meters, but would use a split feed with parallel compensation if I would build an 80-meter Yagi, because of its simplicity and ease to switch from the CW to the SSB end of the band.

■ 4. QUADS

4.1. Modeling Quad Antennas

MININEC-Based programs

Modeling quad antennas with *MININEC*-based programs requires very special attention. To obtain proper results the number of wire segments should be carefully chosen. Near the corners of the loop, the segments must be short enough not to introduce a significant error in the results. Segments as short as 20 cm must be used on an 80-meter quad to obtain reliable impedance results on multielement loop antennas.

Taper technique

If you would break up the entire loop conductor into

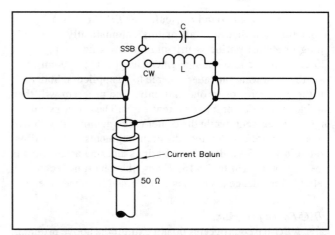

Fig 13-90—Direct feed system for the driven element of an 80-meter array with parallel compensation, which makes it possible to obtain a good SWR in both the CW as well as the SSB section of the band. See text for details.

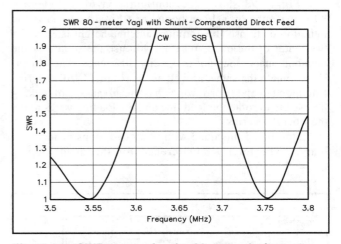

Fig 13-91—SWR curves for the 80-meter 3-element Yagi using parallel compensation. The driven element, initially tuned to resonance on 3.65 MHz was tuned to resonance on 3.55 MHz using a parallel inductor of 2.2 μH. Likewise the same element was tuned to resonance on 3.75 MHz using a parallel capacitor of 700 pF. The resulting SWR curves have an outstanding bandwidth.

20-cm segment lengths, this would mean a very high number of segments, and hence a very long computing time or more pulses than can be handled by the software. To avoid this problem, a conductor can be broken up into a number of (in-line) wires of varying lengths. In our example, the wire closest to the junction is the shortest one (30 cm), and has 1 segment. The next wire has twice that length and also has 1 segment, the third wire again has double the length of the previous wire, etc, until we come to a point where we don't want to make the wires any longer because of accuracy limitations. This last wire can be a "long" one, broken up into different segments of equal length.

This may all sound difficult, but *ELNEC* has a special taper function that does all of this automatically. The only thing one must define is the minimum segment length (eg, 20 cm) and the maximum segment length (eg, 200 cm), and whether you want the taper scheme to be applied to both ends of the conductor or to one end only. The program will then divide the conductor into different wires. The tip wires will all have one segment, while the center (or remaining) length will be a long wire divided into different segments. This is illustrated in **Fig 13-92**. Results of various tapering arrangements are shown in **Table 13-16**. Using the tapering technique, reliable impedance results can be obtained for quad antennas.

NEC-based programs

NEC-based programs do not exhibit the above problem, and no special precautions have to be taken to obtain correct results.

4.2. Two-element full-size 80-meter quad with a parasitic reflector

Fig 13-93 shows the configuration of a 2-element 80-meter quad on a 12-m boom, and **Fig 13-94** shows the radiation patterns. The optimum antenna height is 35 m for the center of the quad. Whether you use the square or the diamond shape does not make any difference. The dimensions remain the same, as well as the results. I will describe a diamond-shaped quad, which has the advantage of making it possible to route the feed line and the loading wires along the fiberglass arms.

I designed this quad with two quad loops of identical length. The total circumference for the quad loop is

Table 13-16

Influence of the Number of Sections on the Impedance of a Quad Loop

Taper Arrangement	Calculated Impedance
Nontapered, 4 × 5 segments	123 − j20 ohms
Nontapered, 4 × 10 segments	130 + j44
Nontapered, 4 × 20 segments	133 + j78
Nontapered, 4 × 40 segments	135 + j95
Nontapered, 4 × 50 segments	135 + j97
Nontapered, 4 × 60 segments	135 + j98
32 sections, tapering from 1.0 to 5.0 m	131 + j85
56 sections, tapering from 0.4 to 2 m	134 + j102
64 sections, tapering from 0.2 to 2 m*	135 + j104
104 sections, tapering from 0.2 to 1 m	135 + j104

Note: See Fig 13-92 regarding the taper procedure.
*Taper arrangement illustrated in Fig 13-92.

1.0033 λ (for a 2-mm-OD conductor or #12 wire). The parasitic element is loaded with a coil (or stub) having an inductive reactance of +j150 Ω. The gain is 3.7 dB over a single loop at the same height over the same ground.

In the model I used 3.775 kHz as a central design frequency. This is because the SWR curve rises more sharply on the low side of the design frequency than it does on the high side.

We can optimize the quad by changing the reactance of the loading stub as we change the operating frequency.

Figs 13-95 and **13-96** show the gain, F/B and SWR for the 2-element quad with a fixed loading stub (150 Ω) as well

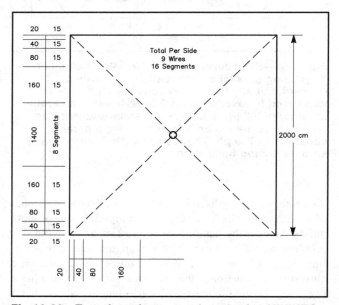

Fig 13-92—Tapering of segment lengths for *MININEC* analysis. See Table 13-16 for the results with different tapering arrangements. With the segment length taper procedure shown here, the result with a total of just 56 tapered segments is as good as for 240 segments of identical length.

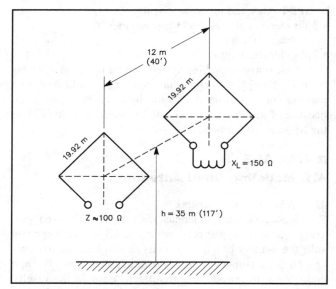

Fig 13-93—Configuration of a 2-element cubical quad antenna designed for 80-meter SSB. Radiation patterns are shown in Fig 13-94. By using a remote tuning system for adjusting the loading of the reflector, the quad can be made to exhibit an F/B of better than 22 dB over the entire operating range. See text for details.

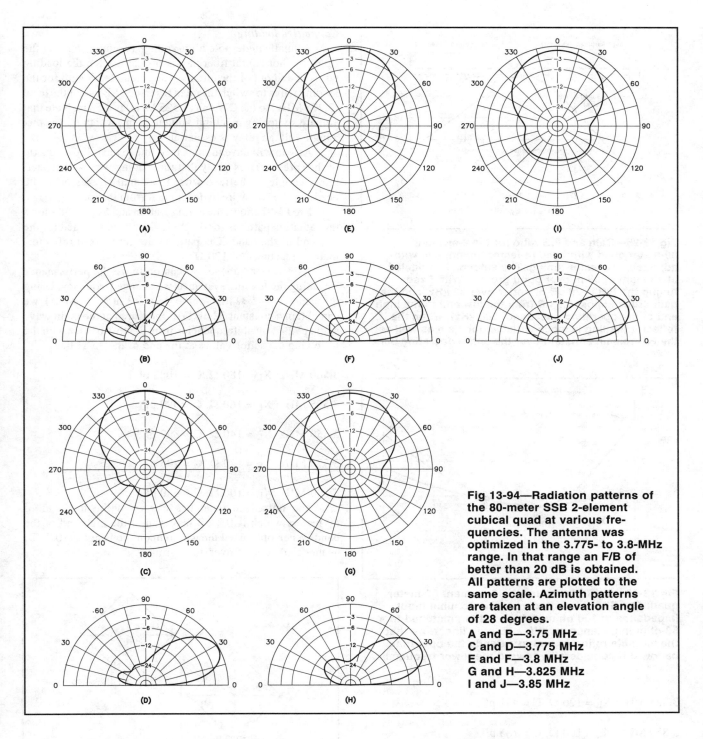

Fig 13-94—Radiation patterns of the 80-meter SSB 2-element cubical quad at various frequencies. The antenna was optimized in the 3.775- to 3.8-MHz range. In that range an F/B of better than 20 dB is obtained. All patterns are plotted to the same scale. Azimuth patterns are taken at an elevation angle of 28 degrees.

A and B—3.75 MHz
C and D—3.775 MHz
E and F—3.8 MHz
G and H—3.825 MHz
I and J—3.85 MHz

as for a design where the loading stub reactance is varied.

To make the antenna instantly reversible in direction, we can run two ¼-λ 75-Ω lines, one to each element. Using the COAX TRANSFORMER/SMITH CHART module from the NEW LOW BAND SOFTWARE we see that a $+j160 \, \Omega$ impedance at the end of a ¼-λ-long 75-Ω transmission line (at 3.775 MHz) looks like a $-j35 \, \Omega$ impedance. This means that a ¼-λ 75-Ω (RG-11) line terminated in a capacitor having a reactance of $-35 \, \Omega$ is all that we need to tune the parasitic element into a reflector. A switch box mounted at the center of the boom houses the necessary relay switching harness and the required variable capacitor to do the job.

The required optimal loading impedances can be obtained as follows:

3.750 MHz: $X_L = 180 \, \Omega$, C = 1322 pF

3.775 MHz: $X_L = 160 \, \Omega$, C = 1205 pF

3.800 MHz: $X_L = 150 \, \Omega$, C = 1148 pF

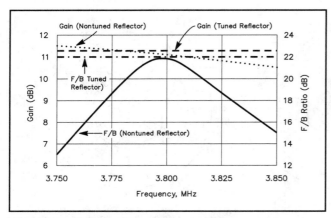

Fig 13-95—Gain and F/B ratio for the 2-element 80-meter quad with fixed reflector tuning, and with adjustable reflector tuning. The antenna is modeled at a height of 35 m over good ground. With fixed tuning the F/B is 20 dB or better over 30 kHz, and the gain drops almost 0.5 dB from the low end to the high end of the operating passband (100 kHz). When the reflector loading is made variable, the gain as well as the F/B remain constant over the entire operating band.

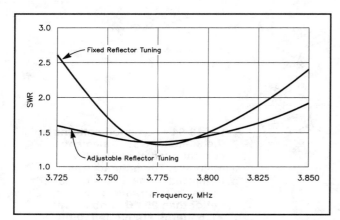

Fig 13-96—SWR curves for the 2-element 80-meter quad. The SWR is plotted versus a nominal input impedance of 100 ohms, which is then matched to a 50-ohm impedance by a $^1/_4$-λ 75-ohm line. Note that the variable reflector tuning extends the operating bandwidth considerably toward the lower frequencies.

3.825 MHz: $X_L = 120 \ \Omega$, C = 931 pF

3.850 MHz: $X_L = 100 \ \Omega$, C = 785 pF

If we tune the reflector for optimum value we will obtain better than 22 dB F/B ratio at all frequencies from 3.75 to 3.85 MHz, and the SWR curve will be much flatter than without the tuned reflector (see Figs 13-95 and 13-96).

The quad can also be made switchable from the SSB to the CW end of 80 meters. There are two methods of loading the elements, inductive loading and capacitive loading (see also the chapter on Large Loops). The capacitive method, which I will describe here, is the most simple to realize.

Capacitive loading

A small single-pole high-voltage (vacuum) relay at the tip of the horizontal fiberglass arms can switch the loading wires in and out of the circuit. The calculated length for the loading wires to switch the quad from the SSB end of the band (3.775 MHz) to the CW end (3.525 MHz) is 4.58 m. Note that you are switching at a high-voltage point, which means that a high-voltage relay is essential.

As you will have to run a feed line to the relay on the tip of the spreader, it is likely that the loading wire will capacitively couple to the feed wire. Use small chokes or ferrite beads on the feed wire to decouple it from the loading wires.

Fig 13-97 shows the configuration and **Fig 13-98** shows the radiation patterns for the 2-element quad as tuned for the low end of the band. The patterns are for a fixed reflector-loading reactance of 170 Ω.

As described above, we can optimize the performance by tuning the loading system as we change frequency. Using the same $^1/_4$-λ 75-Ω line (cut to be a $^1/_4$ λ at 3.775 MHz), we can obtain a constant 22 dB F/B (measured at the main wave angle of 29°) on all frequencies from 3.5 to 3.6 MHz with the following capacitor values at the end of the 75-Ω line:

3.500 MHz: $X_L = 180 \ \Omega$, C = 1083 pF

3.525 MHz: $X_L = 160 \ \Omega$, C = 999 pF

3.550 MHz: $X_L = 140 \ \Omega$, C = 897 pF

3.575 MHz: $X_L = 120 \ \Omega$, C = 795 pF

3.600 MHz: $X_L = 100 \ \Omega$, C = 660 pF

The optimized quad has a gain at the low end of 80 meters which is 0.3 dB less than at the high end of the band. When optimized the gain remains constant at 10.8 dBi as modeled at 35 m over good ground.

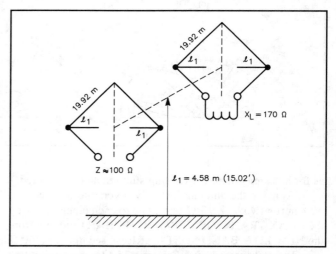

Fig 13-97—Configuration of the 2-element 80-meter cubical quad of Fig 13-93 when loaded to operate in the CW portion of the band. Radiation patterns are shown in Fig 13-98. See text and Fig 13-100 for information on relay switching between SSB and CW.

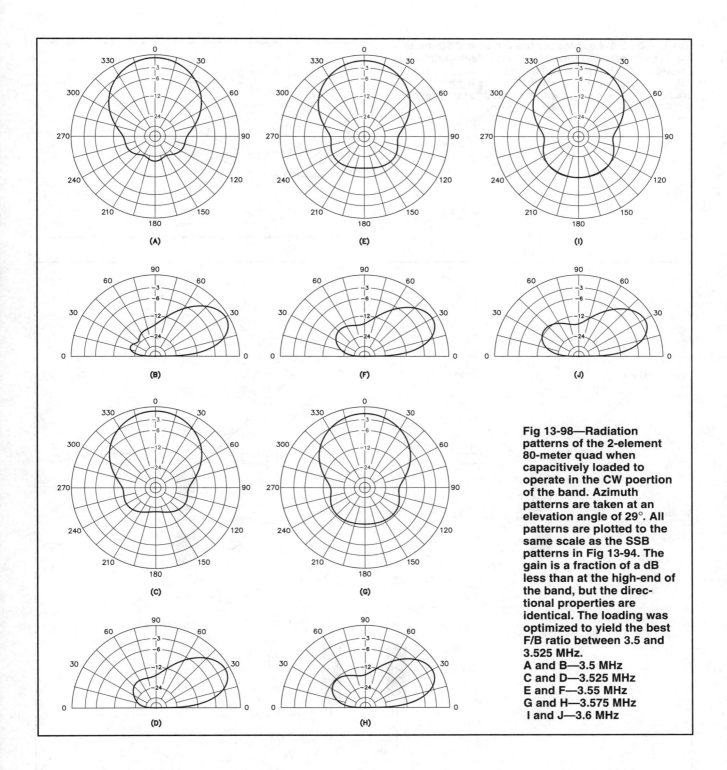

Fig 13-98—Radiation patterns of the 2-element 80-meter quad when capacitively loaded to operate in the CW poertion of the band. Azimuth patterns are taken at an elevation angle of 29°. All patterns are plotted to the same scale as the SSB patterns in Fig 13-94. The gain is a fraction of a dB less than at the high-end of the band, but the directional properties are identical. The loading was optimized to yield the best F/B ratio between 3.5 and 3.525 MHz.
A and B—3.5 MHz
C and D—3.525 MHz
E and F—3.55 MHz
G and H—3.575 MHz
I and J—3.6 MHz

Fig 13-99 shows the SWR curve of the quad at the CW end of the band, with both a fixed reflector loading ($X_L = 170$ Ω) and a variable setup as explained above. The switching harness for the 2-element quad is shown in **Fig 13-100**. The tuning capacitor at the end of the 75-Ω line going to the reflector can be made of a 500-pF fixed capacitor in parallel with a 100 to 1000-pF variable capacitor. Note that you need a current balun at both 75-Ω feed lines reaching the loops (a stack of ferrite beads).

It is also possible to design a 2-element quad array with both elements fed. With the dimensions used in the above design, a phase delay of 135° (identical feed-current magnitude) yields a gain that is very similar to what is obtained with the parasitic reflector. The F/B may be a little better than with the parasitic array. As the array is not fed in quadrature, the feed arrangement is certainly not simpler than for the parasitic array. The parasitic array is simpler to tune, as the reflector stub (the capacitor value) can be simply adjusted for best F/B.

Fig 13-99—SWR curves for the 2-element 80-meter quad as referred to a nominal 100-Ω feed-point impedance. The tuned reflector does not significantly improve the SWR on the high-frequency. The design was adjusted for the best SWR in the 3.5 to 3.525-MHz region.

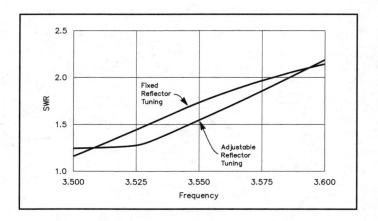

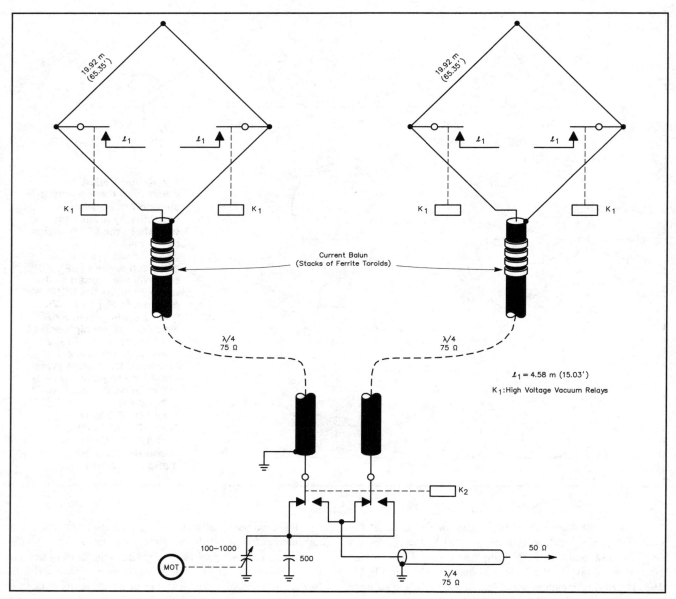

Fig 13-100—Feeding and switching method for the 2-element 80-meter quad array. The four high-voltage vacuum relays connect the loading wires to the high-voltage points of the quad, to load the elements to resonance in the CW band. Relay K2 switches directions. The motor-driven variable capacitor (50-1000 pF) is used to tune the reflector for maximum F/B at any part in the CW or phone band. See text for details.

4.3. Two-element reduced-size quad

D. Courtier-Dutton, G3FPQ, built a reasonably sized 2-element rotatable quad that performs extremely well. The quad side dimensions are 15 m, and the elements are loaded as shown in **Fig 13-101**. The single loop showed a radiation resistance of 50 Ω. Adding a reflector 12 m away from the driven element (0.14 λ), dropped the radiation resistance to approximately 30 Ω. The loading wires are spaced 110 cm from the vertical loop wires, and are almost as long as the vertical loop wires. The loading wires are trimmed to adjust the resonant frequency of the element. G3FPQ reports a 90-kHz bandwidth from the 2-element quad with the apex at 40 m. The middle 7 m of the spreaders are made of aluminum tubing, and 3.6-m long tips are made of fiberglass. A front-to-back ratio of up to 30 dB has been reported.

G3FPQ indicates that the length of the reflector element is exactly the same as the length of the driven element, for obtaining the best F/B ratio. This may seem odd, and is certainly not the case for a full-size quad. The same has been found with some 2-element Yagi arrays (Section 3.6.2).

4.4. Three-element 80-meter quad

Fig 13-102 shows the 3-element full-size 80-meter quad at DJ4PT. The boom is 26-m long, and the boom height is 30 m. Interlaced on the same boom are 5 elements for a 40-meter quad.

The greatest challenge in building a quad antenna of such proportions is mechanical in nature. The mechanical design was done by H. Lumpe, DJ6JC, who is a well-known professional tower manufacturer in Germany. The center parts of the quad spreaders are made of aluminum lattice sections that are insulated from the boom and broken up at given intervals as well. The tubular sections are made of fiberglass. The driven element is mounted less than 1 m from

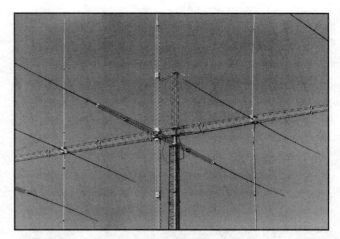

Fig 13-102—This impressive 3-element full-size 80-meter quad, with an interlaced 5-element 40-meter quad on a 26-m (87-ft) long boom sits on top of a self-supporting 30-m (100-ft) tower at DJ4PT. The antenna was built by DJ6JC.

the center. This makes it possible to reach the feed point from the tower. In order to be able to reach the lower tips of the two parasitic elements for tuning, a 26-m tower was installed exactly 13 m from the tower. On top of this "small" tower a special platform was installed from where one can easily tune the parasitic elements.

The weight of the quad is approximately 2000 kg (4400 lb). The monster quad is mounted on top of a 30-m self-supporting steel tower, also built by DJ6JC.

The rotator is placed at the bottom of the tower, and a 20-cm (8-in.) OD rotating pipe with a 10-mm (0.4-in.) wall takes care of the rotating job.

4.5. The W6YA 40-m quad

Jim McCook, W6YA, lives in a fairly typical suburban QTH, and has his neighbors and family accustomed to one crank-up tower (Tri-EX LM-470), on which he must put all of his antennas. Jim has 4-element monoband Yagis for 10, 15 and 20 and a WARC triband dipole. McCook set out to make it work on 9 bands.

For 40 meters, Jim has extended the 12-m boom of the

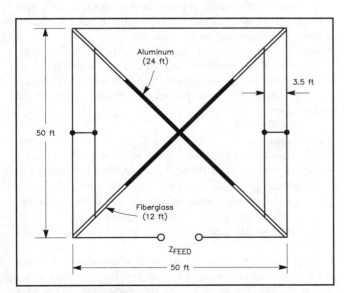

Fig 13-101—Reduced-size 2-element quad designed by D. Courtier-Dutton, G3FPQ. The elements are capacitively loaded as explained in detail in the chapter on large loop antennas.

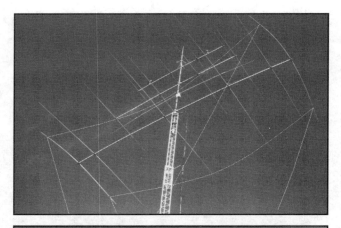

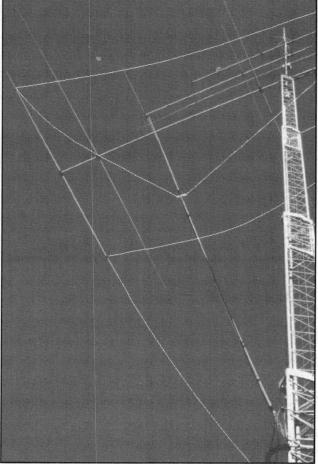

Fig 13-103—Two-element inverted delta-loop array at W6YA. The top of the loop is approx. 21-m high. See text for details.

supported only by the loop wires. The assembly pivots around the tower during rotation. The top horizontal sections are made to sag slightly (about 2 m!) to minimize interaction with the 20-meter Yagi. This arrangement has been up for 18 years, and has helped Jim to work all but 3 countries on 40 meters! Jim reports a 2:1 SWR bandwidth of 200 kHz and a F/B of 15-20 dB.

The driven element loop measures 14.78-m "flat top," 14.32 sloping length on one side and 4.63 m on the other side. This offset is to keep the bottom fiberglass-pole assembly free from the tower. The reflector measures 14.78 m, 15.04 m and 15.34 m respectively. The above lengths are for peak performance on 7.020 MHz.

This quad arrangement has very low wind load. Jim also uses this arrangement on 80 and even on 160 meters.

On 80 and 160, Jim straps the feed point of the driven loop, and feeds the loops with its feed line, at ground level via appropriate matching networks. If you feel tempted to try this combination, I would advise you to use an antenna analyzer to measure the feed-point impedance on both 80 and 160, and design an appropriate network. The feed-point impedance on 80 meters is approx. $90 + j366\ \Omega$, and on 3.8 MHz is $120 + j460\ \Omega$. On 1.83 MHz the impedance, including an estimated series equivalent ground loss resistance of $10\ \Omega$ is $25 - j72\ \Omega$. The appropriate matching networks for the different frequencies are shown in **Fig 13-104**.

It goes without saying that a good ground radial system is essential for this antenna.

Jim compliments his 9-bands-on-one-tower antenna system with a modified 30-meter rotary dipole, which he center loads for 80 meters. He anticipated adding a second set of loading coils to use the same short loaded dipole on 160 as well (see Chapter 7, Section 25).

4.6. Quad or Yagi

I must admit I have very little first-hand experience with quad antennas. But I can think of a few disadvantages of quad antennas as compared to Yagi antennas:
• Much better rejections (F/B) can be obtained with Yagis.
• Quads are three-dimensional; you can't assemble the quad on the ground, and then pick it up with a crane and put it on the tower. You must do a lot of assembly work with the boom in the air.
• Wires will break.
• It is a non-efficient material user: all the metalwork you put up is not part of the antenna; it is just a support structure.
• As far as electrical performance is concerned, a well-tuned quad antenna should (marginally) outperform a Yagi with the same boom length, at least as far as gain is concerned. The difference, being in order of a fraction of a dB to maximum 1 dB, is more of an academic than of a practical nature.

In order to prevent ice build up on the quad wire, one can feed a current (ac or dc) through the loops. The voltage should be adjusted so as to raise the temperature in the wire just enough to prevent ice loading.

The fact is that the great majority of rotatable arrays on the low bands are Yagis. This seems to indicate that the

20-meter Yagi to 14.5 m. At the tips of the boom he mounted fiberglass quad poles, which support two inverted delta loops, separated 6 m from each other. One loop is tuned as a reflector (3% longer). The driven element is fed through a $^{1}/_{4}$-λ section of RG-11 (75-Ω) cable. The inverted delta loops are kept taut by supporting two more abutted quad poles at the bottom. This quad-pole assembly hangs freely,

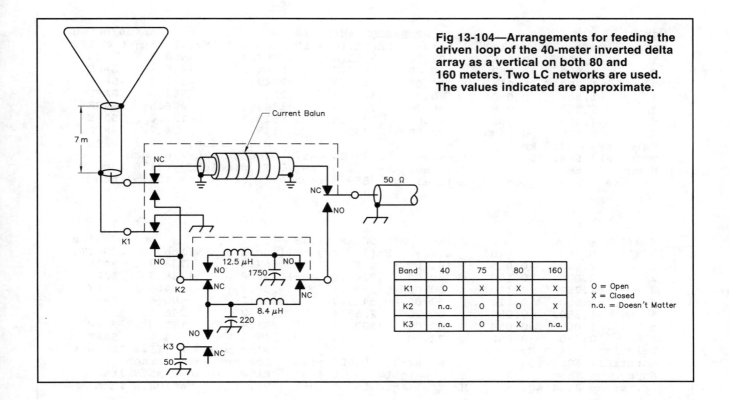

Fig 13-104—Arrangements for feeding the driven loop of the 40-meter inverted delta array as a vertical on both 80 and 160 meters. Two LC networks are used. The values indicated are approximate.

Band	40	75	80	160
K1	O	X	X	X
K2	n.a.	O	O	X
K3	n.a.	O	X	n.a.

O = Open
X = Closed
n.a. = Doesn't Matter

mechanical issues are probably harder to solve with quads than with Yagis.

■ 5. BUYING A COMMERCIAL LOW-BAND YAGI ANTENNA

5.1. 80-Meter Yagis

The linear loading approach

To my knowledge, there is no manufacturer who is currently offering a full-size 80-meter rotatable Yagi antenna. The 2 and 3-element shortened Yagis with linear-loaded elements, as originally developed by Mike Staal, K6MYC, for KLM more than 20 years ago, have held up over the years. This is seen in the example to follow for both home builders as well as commercial manufacturers. Today not only KLM, but also M-Square, Mike Staal's new company, and Force-12 sell 80-meter linear-loaded Yagis based on his original design.

The merits of this design have long been established. The only inherent design compromise seems to be the sacrifice in top-notch directivity which is caused by some radiation from the slant loading wires (see Section 3.7.1).

Mechanical issues

When deciding to buy one of these antennas check the mechanical issues closely. This is what makes an 80-meter Yagi last or not. I must admit I am really scared when I see how some of these antennas are made. I see 80-meter Yagis using booms with a wall thickness that is less than half of the wall on my 40-meter Yagi boom. I see how a simple aluminum plate of a few

mm thickness, is connected by a few simple rivets to the boom, and this is supposed to hold the 25-m long element in a lasting way. I have doubts these antennas can ever stay up in windy areas. In my QTH they would not last one winter! Mechanical issues are the real issues for a lasting 80-meter antenna. So, if you decide to spend a lot of money, spend good money and take a very close look at the mechanics. An antenna built to withstand high winds and lots of ice loading will inevitably use more aluminum than a flimsy antenna that won't withstand a 90 km/h breeze. And aluminum costs money. There is a price for a good mechanical design, there are no two ways about it.

KLM, as an example specifies an antenna weight of 122 kg for their 3-element 80-meter Yagi and Force-12 lists a weight of 160 kg for their 3-element 80-meter Yagi. Compare this with the weight of the earlier described 3-element 40-meter beam which weighs more than twice as much, and you will understand that these are antennas out of two different worlds. The manufacturer's documentation (80M-3, REV 9-9-92) claims that the elements are built to withstand winds in excess of 160 km/h.

I know it takes approximately 45 kg of 6061-T6 aluminum to make a full-size 40-meter element that will withstand 160 km/h winds (+30% gusts). I have my doubts that an 80-meter 3-element Yagi, with elements that are 20% longer than for a 40-meter reflector, can be built for a total weight of 120 kg.

I modeled the elements of the KLM 80M-3 Yagi to assess its wind survival speed. The element mechanical data were taken from the assembly manual of the 80M-3 antenna. The safe wind survival speed turned out to be 90 km/h, ex-

```
   7547                        ELEMENT STRESS ANALYSIS                      on4un/on6wu
   SEC#     OD(in)    WT(in)    L(in)   RM(in.lbs) LMt(lbs.in) LMv(lbs.in) CONDIT.
   1        3.000     0.065    191.00    15065.6     17187.7      4400.3    FAIL
   2        2.000     0.058     60.00     5843.7      5754.3      1448.7    SAFE
   3        1.750     0.058     60.00     4418.3      3749.1       940.0    SAFE
   4        1.500     0.058     60.00     3192.2      2253.7       562.0    SAFE
   5        1.250     0.058     60.00     2165.4      1199.8       296.8    SAFE
   6        1.000     0.058     60.00     1337.8       519.3       127.0    SAFE
   7        0.750     0.058     72.00      709.4       143.9        34.7    SAFE
   Velocity=  53.0 Mph                                   Wind press.=    6.7 lb/sqft
   Material= 6061-T6         Tens. str. = 35000 psi     Ice thickn.= 0 inch
   Rope     = YES            Ele. weight=  42.5 lbs      El.windload= 121.0 lbs
   Pr. area= 2154 sq.in      Half el.lgt= 563.0 inch     El. sag    = 103.5 inch


   7547                        ELEMENT STRESS ANALYSIS                      on4un/on6wu
   SEC#     OD(in)    WT(in)    L(in)   RM(in.lbs) LMt(lbs.in) LMv(lbs.in) CONDIT.
   1        3.000     0.065    191.00    15065.6     30667.3      5738.9    FAIL
   2        2.000     0.110     60.00    10241.8     10239.8      1846.3    SAFE
   3        1.750     0.110     60.00     7655.9      6580.7      1104.6    SAFE
   4        1.500     0.110     60.00     5447.7      3898.2       600.4    SAFE
   5        1.250     0.058     60.00     2165.4      2057.2       296.8    SAFE
   6        1.000     0.058     60.00     1337.8       891.8       127.0    SAFE
   7        0.750     0.058     72.00      709.4       247.6        34.7    SAFE
   Velocity=  74.0 Mph                                   Wind press.=   13.1 lb/sqft
   Material= 6061-T6         Tens. str. = 35000 psi     Ice thickn.= 0 inch
   Rope     = YES            Ele. weight=  52.6 lbs      El.windload= 235.9 lbs
   Pr. area= 2154 sq.in      Half el.lgt= 563.0 inch     El. sag    = 102.8 inch



   If you intend to do a full physical design of a yagi, run each of the yagi
   elements and make a screen dump of the results. You will need the weight data
   as inputs to the BALANCE program.
    1=SPEED  2=GUSTFAC  3=MATER  4=ICE  5=DIM  6=NRUN  7=SECT   H=HELP   X=EXIT
```

Fig 13-105—Wind-survival analysis for the element of a KLM 80-meter Yagi. The unmodified survival wind speed is 54 mi/h (80 km/h) *provided* the inner 3-inch section is side braced. Without side bracing the maximum wind speed is even less. Doubling the wall thickness of sections 2, 3 and 4 increases the maximum wind speed to 74 mi/h (112 km/h), together with side bracing the 3-inch (OD) inner section of the elements. Calculations are done according to EIA/TIA-222-E standard, and exclude any safety factors or modifiers that may be applicable.

cluding 30% higher gust factor. **Fig 13-105** shows the element stress analysis results, which indicate a very unbalanced design: While sections 2 (2-inch OD) and 3 (1.75-inch OD) are loaded to the limit, the 3 next sections are only loaded to about 60% of their possibilities. The tip section is only loaded 25%. This does not necessarily mean that the element will disintegrate at 90 km/h, as this assumes that the wind blows at a right angle with respect to the elements. Putting the boom in the wind (perpendicular to the wind direction) will take all the stress off the elements, and provided the side bracing of the boom is well done, it is likely that the Yagi (boom) will survive wind speeds above 90 km/h. Using the guidelines as explained by Leeson in his book (Ref 964), the sections 2, 3 and 4 can be reinforced by the double-walling technique to increase the wind survival speed to 123 km/h. In any case, one should add side guying

of the central 3-inch section of the elements. Short boom extensions will be required to do this.

In **Fig 13-106** we see the 80-meter KLM Yagi mounted on a telescoping tower at K7EG. Using a motorized telescoping mast and nesting the antenna at minimum height is one way to protect the antenna from high wind loads that could break the elements in high wind storms.

Ben Moeller, OZ8BV, has done a lot of experiments with his 3-element KLM Yagi, and he concluded that each storm at his Baltic coast QTH blew the antenna to pieces. Eventually Ben built himself a 3-element full-size Yagi, of which the elements were designed using the ELEMENT STRENGTH module of the YAGI DESIGN SOFTWARE. The antenna has now been up almost 10 years and has survived many Baltic coast storms without any ill effects (see Figs 13-45 and Fig 13-46).

Fig 13-106—Three-element linear-loaded KLM 80-meter Yagi topped by a 4-element linear-loaded 40-meter Yagi at K7EG. The telescoping tower is shown at its full height (top photo), and nests down to a safe height to protect the antennas from high wind during storms (bottom photo).

Low losses

A very interesting point was brought up by W6ANR. Often the weak point of a design is lack of long-lasting low-resistance electrical contacts. Lossy contacts will ruin the gain, and the pattern, of any array. Invest in some good contact grease, Parker screws and heat shrink tube for assembling the Yagi. Make sure all is done to prevent corrosion in the linear-loading wires.

Fig 13-107—Close-up view of part of the KLM 80-meter 4-element array. The linear-loading lines that also serve as truss wires for the central part of the elements are clearly visible.

High-Q coil loaded Yagis

We also have learned that Yagis using high-Q coil loading can achieve superior directivity as compared to elements using (inevitably) radiating slant loading stubs. Ionic Wave is the company that now sells Yagis with high-Q traps as described in Section 3.7.1.

Creative Design Co, Ltd, is, to my knowledge the only other commercial manufacturer using a coil-loaded shortened 80-meter array. The 3-element array has both elements driven in a ZL-Special configuration (135° out of phase). The element spacing is $1/8$ λ (9 m). The elements are 24 meters long (or approx. 62% of full size), and the loading is done with high-Q coils and a small capacitance hat about $2/3$ out on the elements. The elements are also loaded at the center with hairpin loading coils, which allows precise matching to the phasing line and the coaxial feed line. The array weighs "only" 80 kg. This is a very popular 80-meter antenna in Japan. Judging from its weight, it is probably not an antenna to put up where I live, though!

Gain figures

Be very careful when comparing published gain figures. The only thing that really makes sense are free space dBi gain figures, but the sales and marketing guys like to inflate these low figures and add ground reflection gain, which could be anything up to approx. 6 dB, depending on ground quality

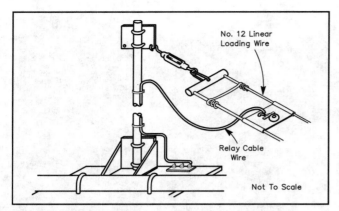

No. 12 Linear Loading Wire

Relay Cable Wire

Not To Scale

Fig 13-108—The linear-loading stub on the KLM shortened 80-meter Yagi also serves to support the element. Near the end of the stub, a small vacuum relay mounted on a printed board is used to shorten the stub for changing the antenna from the CW to the Phone end of the band. Bad contacts and wire breakage has been reported with this system, and it is a good idea to mechanically re-engineer this system, if long lasting top performance is desired.

and antenna height. Don't let these guys fool you!

I have withheld from publishing a list of commercial low-band antennas as I fear that I might not list all, also these listings might seem to indicate some kind of endorsement on my part. If you plan to buy a commercial low-band antenna, I suggest you get a reference list from the manufacturer, and contact some of the customers, or better yet, ask around on the Internet.

5.2. 40-Meter Yagis

It is amazing that none of the major antenna manufacturers advertise 2 or 3-element full-size 40-meter Yagis! That fortunately leaves a place for the real hams, the home builders to excel!

From the poll I did with over 200 active and successful low-band operators, it appears that 50% use a rotatable Yagi of some kind. Of those that listed the make they're using the breakdown is:

Cushcraft 40-2CD	32%
Hy-Gain	16%
KLM	9%
Force-12	7%
Mosley	6%

Undoubtedly the most popular commercial low band Yagi is the 40-meter Cushcraft 40-2CD Yagi. It appears to be the best value (excellent performance/price ratio) on the 40-meter shortened Yagi market.

Leeson, W6QHS, calculated the wind survival speed of an unmodified 40-2CD as 108 km/h. (Ref 967). The referenced article describes how to increase the wind survival speed to 150 km/h by using internal boom and element reinforcements.

LOW BAND DXING FROM A SMALL LOT

14. LOW BAND DXING FROM A SMALL LOT

THANKS K2UO

The story to follow is undoubtedly the story of many, and it could be the story of even more, provided they tried. If you don't have a large piece of property or a farm, you'll want to read it. It's the story on 160 of a very good friend of mine, George Oliva, K2UO.

Having been an avid DXer for many years and having achieved "Number One" Honor Roll status on CW, SSB and MIXED, 5BDXCC, etc, I was in search of a new challenge. Some of the locals had started on 160 meters but I assumed that I didn't have the space for the antennas needed to work Top Band on a half-acre lot. My amplifier didn't cover 160 and my tower was a crank-up type so a shunt feed wouldn't work very well. Eventually in 1985, I grew bored with the WARC bands and took on the challenge! I put up what has since become known as my "stealth" dipole, a full quarter wave on 160, not in a straight line and not very high in the air. I worked 75 countries over the next 36 months with 100 watts and no special receiving antennas. Although most were relatively non-exciting, I did manage to snag 3B8CF, D44BC and even VK7BC.

I next picked up a linear which did cover 160 meters. Now I began to see the need for special receiving antennas. I could now work everything I could hear but knew from the locals and packet clusters that I was not hearing a lot. I asked my "friendly" neighbor if I could run a wire up the back end of his property line and I was now in business with a 550+ foot single wire, terminated Beverage antenna pointed to about 65 degrees. This antenna is truly amazing. I could now hear stations that I couldn't even imagine hearing on the "stealth" dipole.

Although I am not the first to get through, I usually make it in the pileups. I have worked Bouvet, Peter I, Heard Island and South Sandwich, and now have 203 countries worked on 160 meters, almost all on CW of course.

When other hams visit my station and look at my Top Band antennas, they are amazed at the results I have achieved. The bottom line of all this is that you do not need a super station to work a lot of DX on Top Band. What you do need is a little imagination, ingenuity and perseverance to succeed and have a lot of fun.

What better introduction could I have than the above testimony of a dedicated Top Band DXer, who's not frustrated living in a (beautiful I must say) but fairly typical suburban house on a $1/2$ acre lot. George did not use his QTH handicap as an excuse. No, for him it was just another challenge, another hurdle to overcome.

So don't lament if you don't have a million dollar QTH. You can work DX on the low bands as well. Maybe you will not be the first in the pileups, but you will get even more satisfaction from succeeding, as you did have to make the extra hurdle!

My good friend George Oliva, K2UO, holds BSEE and MSEE degrees and is an Associate Director at the US Army's Communications and Electronics Command's Research, Development and Engineering Center at Fort Monmouth, New Jersey. He is responsible for research and development programs involving information technology. He got his first amateur license in 1961 and has operated from a few exotic locations such as Lord Howe Island, Guernsey, Turkey and even Belgium. He is a Senior Member of the IEEE and holds several patents.

George not only volunteered the above striking testimony, but he also volunteered to review this chapter of the book, for which I am very grateful.

■ 1. THE PROBLEM

If you have decided to read on, this is not going to be new for you. But let me nevertheless describe the typical suburban antenna syndrome.

You have this wonderful house, in this wonderful looking neighborhood, at the right driving distance from your work. A dream, however, may not be a ham's dream. There really isn't enough space for the three towers and the four-square you would like to put up, and the neighbors would rather see trees growing than antennas. And your spouse won't really tell it to your face, but thinks one multiband vertical is more than enough. At very best one tower is what you can obtain your spouse's permission for.

If you really want to compete with the big guns on the HF bands, you need Yagis. Not a simple tribander, but monoband Yagis. On the low bands though, you can be relatively competitive with rather simple antennas! This is good news! Read on. . .

■ 2. SET YOURSELF A GOAL

Maybe you should set yourself a goal that is realistic for your circumstances. You can get satisfaction that way as well. Compete with your equals.

But there are nevertheless "fantastic" stories from average suburban QTHs. Here is another testimony of perseverance, or maybe addiction: "I was a young engineer working for IBM, just emigrated from Europe and lived until 1986 in a Toronto suburb, on a 46×120-foot city lot surrounded by houses, TVI, power line noise and nasty neighbors. First I had a home-brewed 65-foot TV tiltover tower with used TH6 and 402BA and inverted Vs ($350). Later I thought I struck gold when I found a second-hand Telrex Big Bertha monopole with the antennas for $1200. I designed and built my own antennas (about $200 in material from junkyards). The rig was a used Drake B-line + R4C (about $500). All the rest of the station, the amplifier and the gadgets were home-brewed. I realized that I had a hard time beating the M/M stations in the contests, so I specialized in single-band operation. This netted me about 16 world records and all Canadian monoband records from 160 through 10 meters in CQWW and WPX contests. . . ."

All that from a 24×36-m city lot! Whaw! This was Yuri Blanarovich, VE3BMV, ex-OK3BU, now K3BU. But you are not that addicted? Keep on reading. . .

This book has explained propagation and focused on various types of antenna configurations for both receiving and transmitting. Factors such as gain, polarization, radiation angle, incoming signal direction and angle, soil conductivity and the many other factors affecting receive and transmit performance. It is up to *you* as an individual to assess your own situation, set your own goals and use the information in this book in conjunction with basic engineering judgment to experiment in the true amateur spirit.

Every QTH has its own limitations, and you must apply your own skill to optimize your station based on your individual goals. Let's have a look at some simple but very effective antennas that might help overcome some of the limitations.

■ 3. THE FLAGPOLE VERTICAL ANTENNA

A $1/4$-λ vertical for 7 MHz measures 10 m, about the size of a really good patriot's flagpole. There you have a wonderful full-size 40-meter vertical. If the pole is a metal pole, make sure there is a good electrical contact between the different sections. If you are using a wooden flagpole, you will have to run a wire along the pole. It is best to use small stand-off insulators, so that the wire does not make contact with the wood. If your neighbor is curious about the wire, tell him it's part of a lightning protection system. Being a vertical antenna, the flagpole requires radials, but you can hide these in the ground, so nobody should object. You should of course insulate the flagpole from the ground. If the flagpole is exactly resonant on 40 meters, you can probably feed it directly with a 50-Ω feed line. Chances are the flagpole may be a little shorter, so you can load it at the bottom with a coil. An L network, as shown in **Fig 14-2** will load and match the antenna at the same time. For a flagpole measuring 8 m, typical component values (assuming a 5 Ω

Fig 14-1—Showing a stealth antenna is easy—you show the sky. Rather than just the sky, here's the sky at K2UO's QTH showing his low-profile then his 10/15/20 m quad and his beautiful home in a wooded residential area in New Jersey. With his invisible 160-meter stealth dipole, George worked 200 countries on top band!

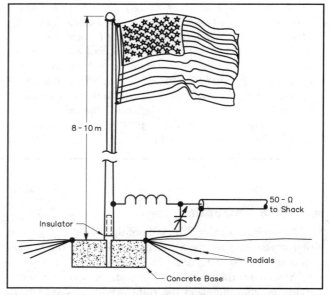

Fig 14-2—Forty-meter flagpole antenna. Any metal flagpole between 8 and 10 m will do. Use an L-network to match to the 50-Ω feed line.

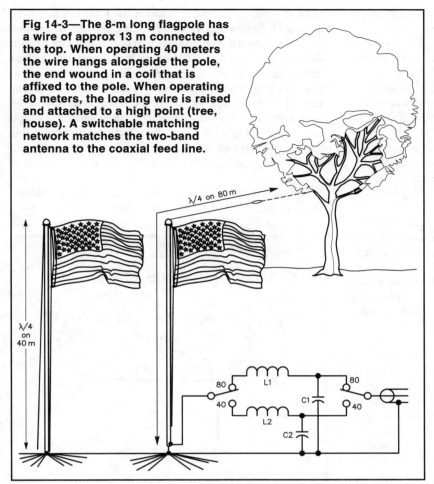

Fig 14-3—The 8-m long flagpole has a wire of approx 13 m connected to the top. When operating 40 meters the wire hangs alongside the pole, the end wound in a coil that is affixed to the pole. When operating 80 meters, the loading wire is raised and attached to a high point (tree, house). A switchable matching network matches the two-band antenna to the coaxial feed line.

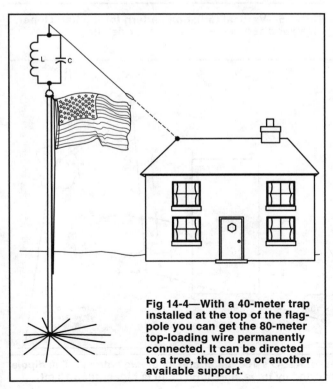

Fig 14-4—With a 40-meter trap installed at the top of the flagpole you can get the 80-meter top-loading wire permanently connected. It can be directed to a tree, the house or another available support.

equivalent ground loss resistance) are: C = 500 pF and L = 2.8 μH. With a 10-m long flagpole, no matching network will be required on 40 meters.

How about 80 meters? You can transform the 8/10-m tall 40-meter vertical into an efficient inverted L at night, if it has to be a super stealth antenna. See **Fig 14-3**. Connect the top loading wire to the top of the metal flagpole. When you operate 40 meters, or during daytime, hang the top wire along the flagpole (coil up the bottom end so that it does not touch the ground). When you want to operate 80, raise the wire with an invisible nylon fishing line and stretch it toward the house or a tree. The top loading wire can be any thin wire, as it hardly carries any current (all the current is at the base of the flagpole). Now you're all set on 80 meters .

For this 40/80-meter flagpole antenna (using an 8 m long flagpole) the typical L-network component value for 80 meters is: L1 = 1.1 μH and C1 = 1100 pF.

If your spouse or the neighbors won't object to a permanent tiny wire running from the top of the flagpole (maybe they haven't even seen it), install an 40-meter trap at the top of the flagpole. Disguise it using your fantasy. Appropriate traps are described in Chapter 9, Sections 6.5.1 and 6.7.1.

For a more or less efficient 160-meter vertical antenna you need at least 15 m of vertical conductor. Have you looked at the trees in that corner of your lot? They should do as supports. Maybe you need to exercise a bit with bow and arrow, but if you can shoot a nylon wire over the trees, you're probably set for a good 160-meter antenna. If you use a BC-Trapper (Chapter 9, Section 6.7.1) inverted L or T antenna, you can use the tree-supported vertical on 40, 80 and 160 meters. And your neighbors will hardly see it! Don't forget that this antenna requires a good radial system. But you can put those down during the night. . .

Don't forget that the open ends of an antenna are always at very high voltage. If you run the outer ends of these wire antennas through the foliage toward a tree, it's a good idea to use Teflon-insulated wire. This will help prevent setting your tree on fire. And, by the way, all these wires don't have to be perfectly horizontal or perfectly vertical. Slopes of up to 20° will not noticeably upset the antenna performance!

You could of course also buy a commercial antenna, and spend lots of money for lots of loss. Use your imagination instead, and put your brains to work instead of your wallet!

■ 4. LOADING YOUR EXISTING TOWER WITH THE HF ANTENNAS ON 80 OR 160 METERS

If you have a tower with one or more HF or VHF Yagis, you can probably turn it into an efficient vertical on 80 or 160 meters. A tower of about 15 m with a simple tribander will give you the right amount of loading to turn it into an excellent 80-meter vertical.

For 160 meters you will need a little higher tower, but starting about 18 m with a reasonably sized tribander antenna will get you approx 70° electrical length on 160. See Chapter 9, Section 6.8 for details on how to shunt-feed these antennas.

If the tower is guyed, make sure the guy wires are broken up in "short" sections. Short means approx ¼ λ. Better still, use dielectric guy rope, such as Phyllistran (Kevlar).

If you use a crank-up tower, you will do better running a solid copper cable along the sections (an old coaxial cable will be fine), as the electrical contact between the sections may not be all that good. In case of doubt, climb your tower and measure the resistance.

It is imperative that you run the cables *inside the tower* all the way down to ground level, and run them underground to the house; otherwise it will be extremely difficult to decouple these cables.

Don't forget these shunt-loaded towers do require a good ground system. Run as many radials as you can in as many directions. Don't overly worry if the tower is next to the house—you may lose a couple of dBs in that direction, that's all.

■ 5. HALF SLOPERS

Half slopers are covered in Chapter 10, Section 6. These antennas are popular with those who have a tower with a rotary antenna, and who want to get it working on 80 meters. A minimum height of approx 13 m (depending on the loading on top of the tower) is required to make a good vertical radiator on 80 meters. For a 160-meter sloper to work well, you would need a tower that's about twice that high. Don't forget that it is not the sloping wire that does most of the radiating, it is the vertical tower. The sloping wire merely serves as a kind of resonating counterpoise for the feed line to push against. As with all vertical antennas the efficiency of a half sloper will depend primarily on the radial system used.

Don't feel tempted to use sloping wires in various (switchable) directions. As the sloping wire only radiates a small part of the total field, this effort would be in vain.

As with shunt-fed towers, all cables that run to the top of the tower should run inside the tower, and run underground to the shack to maximize RF decoupling.

■ 6. HALF LOOPS

Half loops are covered in Chapter 10, Section 5. The half sloper, with the bottom of the sloping wire fed, seems to be an attractive antenna where space is limited. A 15-m high tree could support the vertical wire, and from the top a slant wire can run to the shack or any other convenient place. If we use a 26-m long sloping wire, the antenna will be resonant around 3.5 MHz, and have a feed-point impedance of 60 to 75 Ω, good for a direct feed to the transmitter. To

make it work on 3.8 MHz, shorten the total length of the antenna by approx 3 m, or simply feed it through an antenna tuner or L network. This antenna will also work quite well on 160. Its feed impedance will be very high, however. The best feed system is to use a parallel-tuned circuit as shown in Chapter 12, Fig 12-19. Needless to say, the feed point is at very high RF voltage, and the necessary precautions should be taken to prevent accidental touching of the antenna at this high voltage point. **Fig 14-5** shows the radiation patterns for this half-loop for 80 and 160 meters. On 80 meters the antenna shows some directivity, about 4 dB in favor of the direction of the sloping wire.

Again, a good ground system is required for this antenna, at both ground connection points.

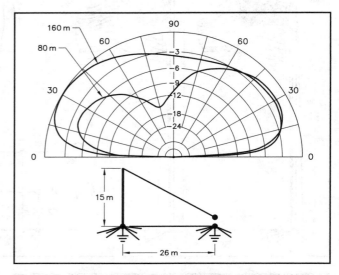

Fig 14-5—Vertical radiation pattern for the half sloper on 80 and 160 meters. See text for details.

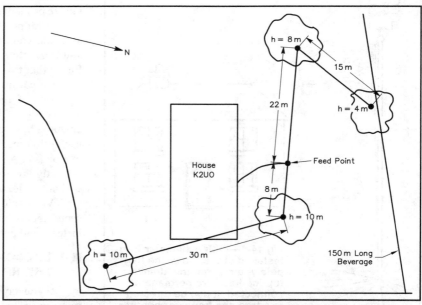

Fig 14-6—Top view of K2UO's 160-meter stealth dipole antenna. The dipole is completely supported by trees, and at no point is higher than 10 m!

■ 7. VERY LOW TOP BAND DIPOLES

The saying is that very low dipoles (10 m up) are only good as receiving antennas. Is that so?

Fig 14-6 shows the layout of K2UO's Zig-Zag dipole for 160. When you walk in his lot, you can hardly see the wire. It really is a stealth antenna, but it has given George 200 countries on Top Band. And that's not only "heard" countries, but those worked and confirmed!

In this book I have described high dipoles as efficient low-angle radiators. In order to be competitive with vertical antennas at really low angles, a dipole must be at least 0.5 λ high. I think we will hardly ever find such high antennas on a typical suburban lot, though! But low dipoles can still function quite well on the low bands. The antenna at K2UO is a standing testimony for such low dipoles.

Fig 14-7 shows the vertical radiation patterns of low λ/2 dipoles, compared to a 15 m long vertical (R_{rad} = 17 Ω) using a fairly decent radial system (R_{loss} = 5 Ω). A 160-meter dipole between 10 and 15 m high produces the same signal as our reference vertical (±1 dB) at a wave angle of 30°, which may come as a surprise. At very low angles, (10°), the vertical will be 13 dB better than the 10-m high dipole. **Fig 14-8** shows the gain of the various antennas for wave angles of 10, 20, 30 and 40°.

Looking at the patterns in Fig 14-7 we see that the big difference is in the high angles. The low dipole will be much better than the vertical for local coverage, but that means also that the signals from "local" stations will be much stronger than they would be on a vertical. Although the dipole may have the big advantage of reducing man-made noise (which is generally vertically polarized), it has the disadvantage of producing very strong signals received at high wave angles.

What may come as an even bigger surprise is that we have learned that not all (though most) of the DX on Top Band comes in at very low angles. We know that, especially on 160 meters, the gray line enhancement at sunrise or sunset often coincides with an optimum angle of radiation that is rather high, and that would definitely give the advantage to the low dipole. So, you might even beat the big gun with his super low-angle antenna, using a K2UO-style dipole!

As a rule I'd like to stress that it is important that you keep the center of the antenna as clear and as high as possible. The ends are just "capacitance hats"—they don't really radiate a lot, so you can bend and hide them as appropriate without hurting the antenna's performance a lot. If you don't have room for a straight 80-m long dipole (who has?), rather than loading it with coils, or using a W3DZZ-type dipole, just bend the ends. That's much better, and will introduce less loss than the usual lossy coils.

What holds for 160 meters is of course applicable to 80 as well.

K2UO is certainly not the only one who's been successful with low dipoles. Recently I read a similar testimony from Ivo, 5B4ADA (ex-HH2AW): "My 160-meter antenna is $^1/_{10}$-λ high (apex of inverted V is 16.5 m above ground, wire ends are 1.5 m above ground). Theoretically, it radiates up most of the RF, but I still have fun working USA, JA, VK, breaking XW30 pileup, etc. I had 57-meter long wire in Haiti on a bamboo pole 10 m above around. Worked many USA and EU stations on 160. Don't be scared with too much theory, get on the air . . ."

I would not necessarily agree with the "theory" part of Ivo's statement, as the theory does predict that low dipoles are a viable alternative. . . to nothing at all.

■ 8. WHY NOT A GAIN ANTENNA FROM YOUR SMALL LOT?

8.1. An Almost Invisible 40-Meter Half-Square Array

I am convinced that on 40 meters you can get up this almost invisible gain antenna. You need to be able to run a

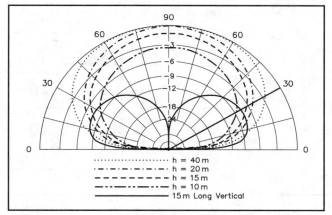

Fig 14-7—Vertical radiation patterns of dipoles at various heights as compared to a short 15-meter long vertical with a 5 Ω equivalent ground loss resistance.

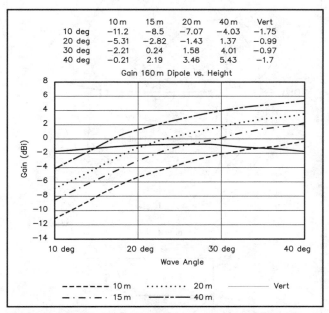

	10 m	15 m	20 m	40 m	Vert
10 deg	−11.2	−8.5	−7.07	−4.03	−1.75
20 deg	−5.31	−2.82	−1.43	1.37	−0.99
30 deg	−2.21	0.24	1.58	4.01	−0.97
40 deg	−0.21	2.19	3.46	5.43	−1.7

Fig 14-8—Gain of low dipoles as compared to the reference 15-meter long vertical.

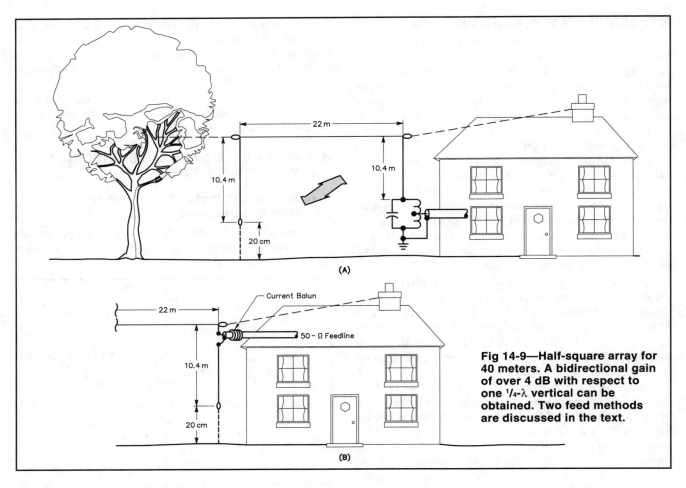

Fig 14-9—Half-square array for 40 meters. A bidirectional gain of over 4 dB with respect to one ¹/₄-λ vertical can be obtained. Two feed methods are discussed in the text.

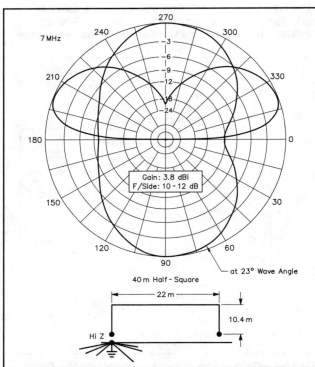

Fig 14-10—Radiation pattern for the 40-meter half-square.

horizontal wire about 10 m up, and 20 m long. Perhaps from the chimney of the house to a tree in the corner of the lot. **Fig 14-9** shows a 40-meter half-square array that can be squeezed in many small lots. Gain is approx 3.4 dB over a single full-size λ/4 vertical. The ends of the vertical wires are also at very high RF potential, and precautions should be taken to prevent accidental touching. The half-square is fed via a parallel-tuned circuit as shown in Chapter 12, Fig 12-19.

You can also feed the half-square in one of the top-corners. This may be a good idea if one element is close to the house as shown in Fig 14-9 (B). The feed impedance, when fed in a corner, is about 52 Ω, a perfect match for a 50-Ω feed line. Do not forget to install a current balun on the coaxial feed line.

8.2. Using the 40-Meter Half-Square on 80 Meters

What about using the 40-meter half-square on 80 meters? A bit of magic turns the antenna into two close-spaced in-phase fed end-fire arrays with top loading. The only thing one needs is to short the base of the second element to ground, and feed the array at the other element at ground level (**Fig 14-11**). This 2-element array has a gain of 1.6 dB over a single full-size (20 m high) vertical and provides excellent low-angle radiation. The antenna has

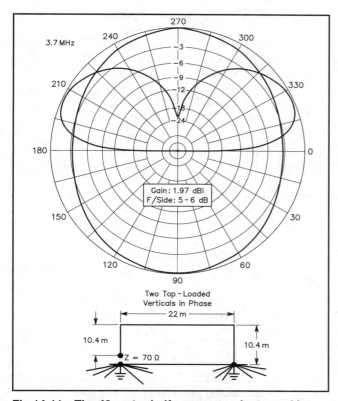

Fig 14-11—The 40-meter half-square can be turned into a 2-element close-spaced top-loaded array for 80 meters, where both elements are also fed in phase. Both vertical as well as horizontal radiation pattern (at a 30° wave angle) are shown.

about 4 dB front to side ratio. Its feed-point impedance is approx. 70 Ω ground loss resistance at each vertical element. This antenna requires a good ground radial system at the base of both elements.

With some ingenuity one could easily brew a switching system that grounds/ungrounds one element, and either feed the other element directly with coax on 80 meters, or feed it via a parallel-tuned network on 40 meters.

8.3. 40-Meter Wire-Type End-Fire Array

Maybe the half-square doesn't suit your most wanted direction. You can also turn this into a 2-element parasitic array as shown in **Fig 14-12**. I worked out the example of an array where a maximum height of 8 m was available as the catenary wire. The elements were top-loaded as shown in Fig 14-10 and 14-11. The array has a very good F/B and gain, and a feed-point impedance of approx 25 Ω. Matching can be done through a $1/4$-λ, 35-Ω line, consisting of two parallel 75-Ω coaxial cables (each measuring 7.03 m for RG-11 or RG-59 (solid PE insulated coax with VF=0.66).

It is important to install as good a radial system as possible on this array. Where radials from the two elements meet they can be connected to a bus wire, as shown in Chapter 11, Section 9, Fig 11-79.

8.4. And an 80-Meter End-Fire Array

Maybe you have two high trees in the back that could help you support a 2-element array for 80 meters. I would recommend a minimum height of the elements of approx.

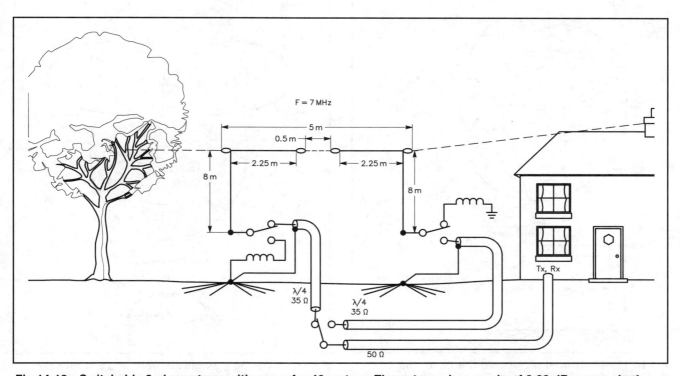

Fig 14-12—Switchable 2-element parasitic array for 40 meters. The antenna has a gain of 3.08 dB over a single vertical.

13 m; the remainder can be top loaded if necessary. **Fig 14-14** shows two T-loaded 13-m high verticals, suspended from a single catenary rope, eg between two tall trees.

This array has an excellent F/B ratio and gain, and will certainly put you in the front seat in a pileup if you care to install a good ground system. When properly adjusted, the array impedance, assuming about 5 Ω equivalent ground loss resistance, is approx 20 Ω. The array can be fed via a 37.5 Ω, 1/4-λ transformer (two parallel 75-Ω cables) as shown in Fig 14-12, or via an un-un transformer (20-50 Ω).

It is important that the top-loading wires are as shown (points facing one another). If you are forced to try another configuration, I would advise you to model the array exactly as in reality. Needless to say, if you have some really tall trees on your property, this antenna can be scaled up for 160 meters.

8.5. The Half-Diamond Array

Maybe you don't have the two supports required to put up the box-shaped arrays I described above. With just one support, a few good arrays can be created as well. You will require one high support (15 m), eg, a tree. We reshaped the half-square to become a half-diamond. We lose about 1 dB gain, but the pattern remains unchanged (see **Fig 14-15**).

8.6. The Half-Diamond Array on 80 Meters

The same inverted-V-shaped half-diamond 40-meter array can be used on 80 meters as well. As with the half-square (see Section 8.2) all you must do is ground the bottom

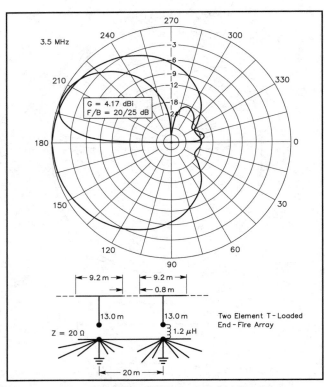

Fig 14-14—Horizontal radiation patterns for the 2-element T-loaded parasitic array for 80 meters. The gain is over 4 dBi , and the F/B is 20 to 25 dB.

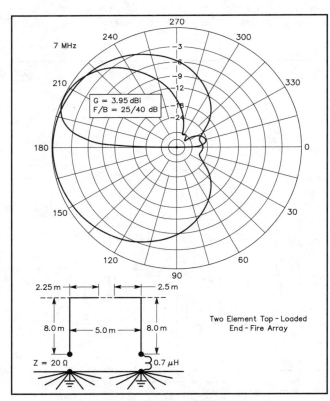

Fig 14-13—Horizontal radiation patterns for the 2-element parasitic array from Fig 14-10. The gain is 3.95 dBi, and the F/B is an impressive 25 to 40 dB.

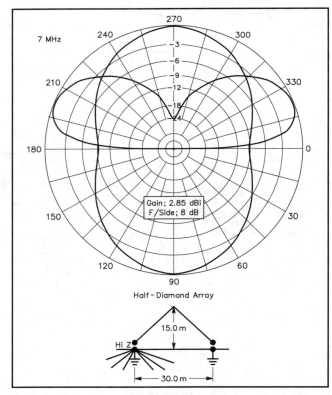

Fig 14-15—Using a single support, one can turn the half-square array into a half-diamond array, at the sacrifice of approx 1 dB of gain.

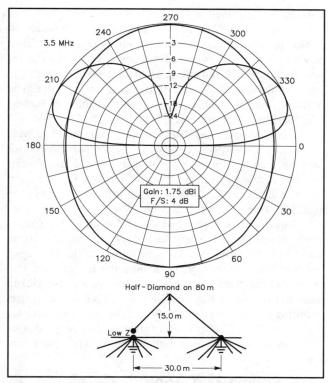

Fig 14-16—Radiation patterns of a capacitively loaded inverted-V array for 80 meters, which requires only one 12-m high support, and a 28-m-long base line.

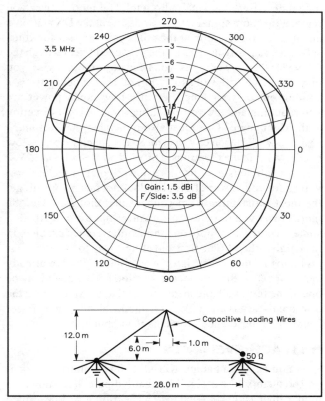

Fig 14-17—This 80-meter array requires only 12 m of height. The array is capacitively loaded at the top, using two wires in a V shape.

end of the array at the side opposite to the feed point. The array has an impedance of approx 75 Ω. The exact resonant frequency can be tuned by simply changing the total length of the antenna. For use on 40 meters the exact length is not so critical, as the array can easily be tuned for low SWR anywhere in the band using the resonant tuning circuit.

8.7. Capacitively Loaded Diamond Array for 80 Meters

Maybe you can't quite get the 15 m height. One solution is to top load the two sloping verticals with a common capacity wire, hanging right down as shown in **Fig 14-17**. Two wires are connected to the apex of the V. Their length and the angle between the wires is varied to tune the antenna to the required frequency. The gain of this antenna is still about 1.5 dBi, which is more than 1 dB better than a single full-size (20 m high) vertical. The feed-point impedance, including about 10 Ω loss resistance, is approx 50 Ω, a perfect match for the feed line!

8.8. A Midget Capacitively Loaded Delta Loop for 80 meters

The half diamond antennas (Sections 7.7 and 7.8) looks very much like a delta loop with its bottom wire laying on the ground. Let us raise the wire, and turn it into a real delta loop. The model shown in **Fig 14-18** has similar dimensions to the antenna from Fig 14-17, and yields the same gain, the

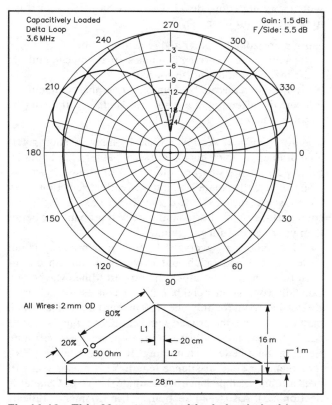

Fig 14-18—This 80-meter capacitively loaded midget delta loop array requires only a 16-m high support. With one 14-m support, and sloping the antenna about 30° the radiation pattern will not be overly affected.

same front to side ratio, and even the same feed-point impedance. Needless to say, this is once more proof that the delta loop is nothing else than two sloping verticals, fed in phase (see Chapter 10, Section 2).

In this example I used capacitive loading in a little different way. This delta loop can be tuned anywhere from 3.8 to 3.5 MHz by just changing the length of the bottom capacity wire. L1 is 11.0 m, and L2 is 6 m for f = 3.5 MHz. For f = 3.8 MHz the bottom loading wire can be eliminated altogether.

There is nothing magical about these dimensions. Just keep in mind that the capacitive loading wires should be attached at the high-voltage points, and that they carry very high voltage indeed. Where crossing each other, the loading wires should be kept about 20 cm from each other.

Do not fool yourself and think that this antenna does not require radials. The radiation is affected just as much by near-field absorption losses under the antenna as in the case of the grounded verticals. In other words, delta loops require a ground screen, just as is the case with all antennas that do have radiating elements close to ground!

■ 9. SPECIAL RECEIVING ANTENNAS

Typical suburban QTHs mean rather dense housing, which in turn means a lot of man-made noise. Now that you have used your imagination, and squeezed an efficient vertical—or even a couple—onto your lot, you're faced with a very high noise level. There are basically four ways to tackle this problem:

• use a horizontally polarized receiving antenna
• use a directive receiving antenna so that you can null out the main offending noise source
• use a noise reduction system based on phased antennas
• locate the offending noise sources, and . . . kill them (the noise sources)

Jim McCook, W6YA, swears by his very short rotatable dipole on top of his tower (see Chapter 7, Section 25, Fig 7-83). Such a small dipole would fit almost every lot. You can actually rotate it as it has excellent rejection when its ends are turned toward the directions of the noise source. But any low horizontal wire will probably be better than your vertical for receiving.

How about a Beverage? I know your property is not quite like a Texas ranch, but maybe you can run one or even two "short" ones along the property line. Maybe you can talk your good neighbor into a concession. George, K2UO, has room for only one (150 m long) Beverage, which partly runs along the property lines of his neighbors, but he admits this antenna really was an eye-opener. Even $1/2$-λ long Beverages are better than nothing! If you're after best F/B, tune them for cone-of-silence length. This means you will need a complex impedance to terminate them (see Chapter 7, Section 6.3). Also, read Section 19.1 in Chapter 7. It deals with Short Beverages and Slinkies. If your neighbor does not care to see the tiny Beverage wire along his property line, maybe you should try a B.O.G. (Beverage on Ground)

antenna, as described in Chapter 7, Section 19.2.

How about EWEs, K9AYs and the like? These antennas are mostly vertically polarized, and as such are not ideal in very (man-made) noisy environments. This is not true, of course, if you make yourself a rotatable version, where you can turn the back right at the offending noise source. The Titanex rotatable receiving array (see Chapter 7, Section 23.4) has the same drawback.

Also, do not forget that these antennas should be clear from any transmitting antennas. I would recommend $1/10$ λ as a very strict minimum for distance between a special receiving antenna and your transmit antenna. If you cannot achieve that, you can always detune the transmit antenna on receiving (see Chapter 7, Section 13).

I have been extremely successful in eliminating man-made noise from a particular source (a chemical plant 10 km away!) by using a so-called noise eliminator. This is nothing but a circuit in which you combine the inputs from two antennas, the main receiving antenna and a noise pickup antenna) in such a way that they are added out of phase, resulting in complete cancellation of the noise (see Chapter 7, Section12.6). The MFJ 1206 unit has been a very valuable asset for me when dealing with noises from one particular source.

■ 10. POWER AND MODE

Power and mode both factor into the results you will receive in terms of working DX as well as "working" your neighbor's TV/telephone/radio, etc! The more power you radiate, the more signal will be available at the DX station's antenna. However, look at the goals you have set for yourself. Do you really need to be the *first* one to get through the pileup? If the answer is yes, you need power; otherwise, you may not. Again the question is not how much power is coming out of your amplifier; it's how much power are you radiating and are you radiating it in the right direction at the right angle? One must also remember that it is *much* easier to work DX on CW than on SSB, especially on the low bands. With few exceptions, DX on Top Band is on CW.

It is obvious that the more power you run the easier it will be for you to work DX. That does not mean you will get the most satisfaction from your results, though. Also, the more power you run, the more chances there are of creating some kind of TVI / BCI / telephone interference problem in the neighborhood. The first and easiest way to avoid similar problems is not to run high power. But, if you are already handicapped with your pocket-size lot, some power may be one of the few available means to get that evasive DX on the low bands after all. CW as a mode also creates fewer "audible" interference problems than phone.

■ 11. ACHIEVEMENTS

You have read about K2UO's 200+ countries worked on 160 meters with a Zig-Zag Stealth dipole that's nowhere higher than 10 m. In Chapter 13 (Section 4.6) I described W6YA's 21 m tower, on a "typical" suburban lot, which carries antennas for all nine bands! Don't let space restrictions scare you away from the low bands. Be sure that, if you work the evasive 3B7 on Top Band from your small lot QTH, you will get triple the satisfaction of the big gun who maybe got through a couple minutes earlier.

Recently I read a very applicable statement: "...I was always complaining about my shoes, until I saw the man without legs." Let's have fun with what we have and do our best under the circumstances. Be convinced you're not the only one who's not living in ham's paradise.

There are many others in the same boat.

■ 12. THE ULTIMATE ALTERNATIVE

Don't put up any antennas, don't get on the air from your pocket-sized QTH, but save your money and energy, and go on a DXpedition once or even twice a year, and provide us hard-core Low Banders with the new countries we need to be able to prove we're the best!

FROM LOW-BAND DXING TO CONTESTING

THANKS

When I wrote the 2nd edition of the Low-Band DXing book, almost 12 years ago, I was a very active, omnipresent, low-band DXer. We in Belgium had just been "given" 160 meters, and really lived and slept on the low bands. Some of you may have wondered why, more recently, ON4UN has been so absent on the bands. Here's why.

It was my neighbor and friend Peter, ON6TT (from Peter I and Heard Island fame, among others) who stirred up my interest in contesting. Peter achieved "mission almost impossible" by helping us (in Belgium) to obtain a high-power permission from the PTT to operate during international contests. At last we could compete with equal means; at last we no longer had to "lie."

Building a competitive contest station has been a unique experience: getting all the help from friends and club members, and building a team of excellent operators. It not only widened my technical horizons, but also, and not at the least, my social horizons: working with people, enjoying Amateur Radio as a group, and making new friends from all over the world.

I would in particular like to mention two good friends, whom I met through contesting, and who played an important role in my becoming an avid contester.

I first saw Harry Booklan, RA3AUU, at the Clipperton Club Convention in Bordeaux (France) in 1993. Young Harry was 23 years old, and won both the CW as well as the phone pileup 'test in Bordeaux. That was my first introduction to Harry. Harry quickly became my friend as well as one of the fixed assets of our OTxT contest station. As ON9CIB, Harry and I represented Belgium at WRTC in San Francisco in 1997.

Frank Grossmann, DL2CC (ex-DL1SBR), is another highly esteemed operator and friend of the house. Frank is a young computer programming professional who runs his own company in Stuttgart, and between jobs, makes the 750 km trip (one-way) several times per year to operate contests from here.

Both Harry and Frank are superb CW operators. Frank was Germany's high speed champion some years ago. They both turned me into a CW addict. What a wonderful addiction, I am glad and proud to admit.

Frank, DL2CC, agreed to be my helper, counselor and critic for this chapter. Thank you, Frank!

■ 1. IN SEARCH OF EXCELLENCE

Amateur Radio is all about satisfaction, self-fulfillment.

My Elmer was ON4GV. He was also my uncle. At his home I saw my very first Amateur Radio station. I was not quite 10 years old. That was around 1950.

Fifty years ago Amateur Radio, in my eyes, was all about adventure, discovery. It was *adventureland* and *wonderland*, all in one. When I was a young boy, in my eyes telecommunication *was* Amateur Radio. You must realize that in the early years after WW II, out on the countryside, where I lived with my parents, we still had hand-cranked telephones with manually operated telephone exchanges! These exchanges closed down at 10 PM. No chance to telephone anyone during the night. If you wanted to communicate with someone across the ocean, you wrote a letter. If you wanted to travel across the water, you took the boat. But for all I knew, if you wanted to talk to someone anywhere in the world, you needed to be a radio amateur. Being a ham made you an explorer, a discoverer. You could expose yourself to worlds others had hardly ever heard about.

It was this magic, this thrill of radio that lured me into this hobby.

I will never forget the oh-so-typical

Fig 15-1—A 1965 picture showing ON4GV, the author's Elmer, ON4UN and his XYL to be. In the background we recognize a Drake 2B, an SB-1000 and an NCL-2000.

smell of bakelite, wax and tar-filled capacitors and transformers that was very typical for the early-day radios. And the white filament glow of early tubes. Some of the early-day triodes were so "brilliant" that you could literally read a book by them at night. My very first hands-on experiences with electronics (not that it was called that, in those days; we called it simply "radio") were in building small audio amplifiers, using directly heated triodes, such as the E, A416. My father's wooden cigar boxes served as chassis for building three-stage audio amplifiers using heavy 3:1 interstage transformers. This was all about discovering an amazing and intriguing world, the world of radio.

It was technology (a modern word for these early-day sensations) that hooked me to Amateur Radio.

For a while my discovery trips were somewhat curtailed and it was not until I was 20 years of age, in high school, that I finally got my license. The challenge then was to prepare for the license, and get it on the first try. The sense of fulfillment, once you got it, was enormous, as was the first antenna you built, the first QSO you made. I was doing something not every one else could do! Now I was part of them. . .

In the early '60s I stumbled across some guys working DX on 80 meters. I remember a few calls: GI6TK, GI3CDF, GW3AX and G3FPQ. Only David, G3FPQ, is still there and still a very active low-band DXer. This really seemed like something else—working across the pond and into New Zealand on frequencies where the others would work stations in a 500-km range. What a challenge! This really put you in a separate class among hams.

Working the elusive DX on the low bands was my next challenge, and it became my passion.

This time I found out that, in order to be part of those low-band DXers, you needed to be successful. You needed to have a good signal. That meant you needed to have the know-how to do it. Amateur Radio was no longer a communicating hobby for me, but became an experimenter's hobby: building new, better and bigger antennas, experimenting with and learning about propagation, becoming a better technician and becoming a better operator.

DXing on the low bands is all about *overcoming difficult hurdles*: everybody can work DX on 10, 15 and 20 meters. There is not much sense of satisfaction involved. In 1987 my last iron curtain was lifted. We finally got 160 meters in Belgium. The last frontier. A vast terrain for chasing the difficult game. And yes, top band certainly is where the DXer can get the ultimate sense of satisfaction. Technical knowledge and technical achievements are undoubtedly great assets in achieving success in low-band DXing. But, even with a modest station, provided dedication, patience and operating experience, one can be a successful DXer.

If so, what can provide you with the ultimate sense of achievement, of technical excellence, in Amateur Radio?

Throughout history, *competition* has been one of the important leverages for progress in many fields. So is contesting to Amateur Radio. To be a very successful low-band DXer you need to be a very good operator, know propagation, be patient, be persevering and have a "decent" antenna system and station. You can determine yourself whether or not you are successful. You can set your goals as a function of your possibilities, and if you have worked 300 countries on 80 and 200 countries on 160 from an average

urban-lot QTH, then you are, by all standards, a very successful DXer.

To be successful in "big game" contesting, you cannot compromise with yourself. You need to have the best antennas, the best station, the best operators, nothing but the best if you want to score high in world ranking. The best multi-operator contest stations are all built, improved, maintained and run by engineers. This is no coincidence.

Undoubtedly international contesting is the ultimate challenge—it provides the truth by excellence. It is truly the Formula 1 competition in Amateur Radio. This is what attracted me to this radio-sport.

Jim Reid, KH7M, who was an operator at Stanford University, W6YX, in the years when SSB techniques were worked out there by Art Collins (yes, later from Collins Radio) in the '50s. He was a witness to how Amateur Radio contributed to important advancements in communication technology, now almost half a century ago.

He made an interesting comparison between the world of contesting and the world of car racing.

"Today, about the entire globe, the most sophisticated, and elaborate HF band stations are owned by contesters such as CT1BOH, WB9Z, IR4T, PI4COM, IY4FGM, GW4BLE, JA3ZOH, PY5EG, W3LPL, K3LR, VE3EJ, and so on (ON4UN was also mentioned—thanks Jim!). Each of these stations has elaborate and multiple rig setups, multiple antenna installations, many computers with each operating position networked with logging programs, band mapping programs, propagation monitoring radios, and so on.

These Amateur Radio stations are all Contest/DX stations. They each have the most sophisticated and up to date technology possible. Each has invested into it what would be comparably invested into a stable of Ferrari racing automobiles; and I have seen both, especially in Southern California!!

These guys have pushed and pushed at manufacturers, at antenna designers, at software writers, and continue to do so. The station owners themselves are all first class operators and technicians. They spend virtually all of there time "tinkering" and pushing the state of the HF art, in every way feasible. Every station I listed represents thousands of man-hours of work, every year to maintain and remain in the top ranks of competition in the sport of DX contesting, which of course has no more purpose than the sport of auto racing: fun to have and maintain and win with the BEST.

The owners of these stations have pushed the state of the art of Amateur Radio every bit as directly as the owners of Indianapolis racing machines have pushed the state of the art of tires, lubricants, engines, brakes, frame design, and so on."

In the highly competitive sport of international contesting, it does not suffice to have the best car or engine; you also need the best drivers, the best mechanics, the best engineers. Contesting is indeed very much like car racing.

In DXing you can get the fulfillment of working all countries on 40 or 80 or even 160. Once it's done, the game is over. Not that the game of working all countries on top band will ever be over, I guess. But in contesting, there is a new competition calendar every year. Every year you can measure the station's performance, you can measure your improvements, plan your progress and enjoy the fulfillment of your victories, over and over.

That is why international contesting is the ultimate playground of ever advancing, competitive and self fulfilling Amateur Radio.

■ 2. WHY CONTESTING?

Now and then I read on the Internet how a proud antenna builder tells us all about the wonderful performance of his new antenna. He is trying to convince the world by telling us all about the rare DX he worked with his new antenna. What does that prove? Very little, really. Working DX is in no way a proof of technical performance of an antenna. Not in the strictly technical sense, in any case.

Working rare DX can be the proof of outstanding operating, dedication and perseverance, when it's done from a modest QTH with small antennas.

If you want to prove the technical capabilities of the station, there are really only two ways. Number one consists of elaborate full scale field testing and measuring in a precisely controlled environment. This is beyond the reach of almost every ham.

The second possibility consists of testing your weapons, not in a shooting range or in a lab, but on the battlefields. These battlefields are the major international contests. Contests are possibly as close as we can get to a controlled environment, simply because it is extremely nonconditioned. In major international contests you are competing against all the best engineered and equipped stations, under a variety of continuously changing conditions, which really makes it a fair and equal battle and test.

This is why, after having been an "occasional" contester for almost 40 years, I decided to get into some serious contesting, thereby putting emphasis on the low frequency capabilities of my station.

Fig 15-2—The author, 33 years later (1998), operating a phone contest (rare exception) from his two-station's contest shack. Note the two FT-1000MP transceivers, side by side. Each one is topped by a PC monitor screen.

■ 3. WHAT IS THE CATEGORY— MULTI-MULTI OR MULTI-SINGLE?

In order to convert my station into a successful contesting station the first decision was—"we want to win—but in which category?" In other words, what is the appropriate battleground for the weapons we have?

The biggest and probably best-known stations are the *multi-multi* stations.

The most successful of them have 2 stations per band; that means 12 fully equipped stations. These stations are on each of the 6 bands, 24 hours per day, and they must catch every single opening. They need to have access to a wide variety of antennas with different wave angles. Therefore, they are generally equipped with various stacked Yagis for the HF bands, even including 40 meters. The second station, whose task it is to look for multipliers, generally has access to a simpler antenna setup. It is located as far away as possible from the running station's antennas, in order to minimize interference, although eliminating same-band interference is quite impossible.

Interstation interference is the most challenging technical challenge in multi-transmitter station design. With multi-multi contest stations, though, each of the band stations can be completely (galvanically) separated from each other, which certainly helps prevent leakage paths for unwanted coupling between stations. In this respect a multi-multi is simpler to design and make than a so-called small multi-single, where all of the antennas have to be accessible by both stations. This makes eliminating leakage paths much more difficult.

There are a number of *multi-single* stations which, as far as station design is concerned, really fall in the category of multi-multi stations. I call them big multi-single stations. They have six well-separated stations, one of which "runs" while the five other stations are manned and are checking each of the five other bands simultaneously. In this configuration as well, it is possible to achieve better isolation between bands because there are six completely separate stations. These big multi-single stations normally also have antennas for each band on separate towers. To build a really top-notch multi-multi station you probably require at least 2.5 to 5 acres (1 to 2 hectares or 10,000 to 20,000 m²) of land—and that does not include what you need for Beverage antennas.

For a big multi-multi setup, in addition to the financial limitations, there is simply not enough space for putting up additional towers in my backyard.

This is why we decided on going for the category of multi-single, or—as I call it—small multi-single.

Small multi-single stations can be built much smaller than multi-multi stations. A small multi-single is a station with only two operating positions, one for the "run" station, and one for the "multiplier" station. The operator of the multiplier station has to scan all bands for multipliers. In general both the run as well as the multiplier station will have access to all antennas, which means that a fairly complex antenna switching system is part of the setup. Such switching systems increase the potential for unwanted coupling between the two stations. This is what makes designing a small multi-single station technically more difficult than a large multi-single or a multi-multi station. It goes without saying that a station designed for small multi-single is also well suited for single-op two-transmitter contesting. While it is imperative for a multi-multi station to catch every single band opening, and therefore needs antennas to match all possible wave angles, this is not necessary for a multi-single station. The run station will run on the bands at the

Fig 15-3—The author's contest station is located on a 4000 m² (2 acre) terrain about 10 km from the nearby major city of Ghent. On this terrain are all antennas, except for the Beverages, of course. In the foreground you see part of the 30,000 m² (3 hectares) that are used in winter for putting up 12 Beverages, one per 30°. See text for further details.

times the wave angle of his antenna matches propagation best. In other words, the height of the antennas should be such as to accommodate the average wave angle, the wave angle that produces most QSOs for the longest period of time. This means antenna heights between 18 and 30 m for 10 though 40 meters. The multiplier station may have to call a multiplier with an antenna that is not at the ideal height. He may not get through on the first call, but this is not as important for the multiplier station.

But there is not only multi-operator all-band HF contesting. Any station that is successful in DXing could be a candidate for *single-operator* contesting. And if the station is not equipped for all bands, a *single band* effort can be contemplated. Also, if you are not 20 any more, single band contesting is attractive. If you operate the low bands, you have all day to rest. You can still prove the technical excellence of your station on the band of your choice!

■ 4. ANTENNAS

The ON4UN/OTxT contest station was designed as a multi-single and a single-op two-transmitter station. **Fig 15-3** shows the QTH and some of the antennas. One tower supports the 40 meter (at 30 m height) and the 20 meter Yagi (at 25 m). As they are both on the same tower, they cannot be rotated independently. A similar combination exists on tower no. 2, where a 6-element 15-meter Yagi (at 24 m) tops a 6-element 10-meter Yagi (at 19 m). The third tower is quarter-wave 160-meter antenna, which also serves as a support tower for the 80-meter four-square. About 100 m behind the house is a fourth tower (18 m) with a KT34-XA triband Yagi. This is what we call the multiplier-

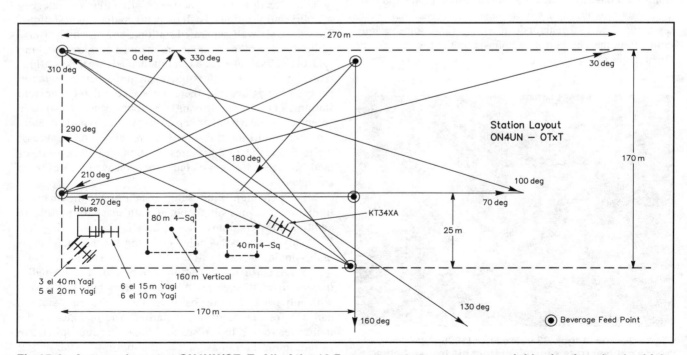

Fig 15-4—Antenna layout at ON4UN/OTxT. All of the 12 Beverage antennas run over neighboring farm land, which is accessible for this purpose between the end of October and late April. The shortest Beverages are about 165 m, the longer ones 300 m.

antenna. There are two more multiplier antennas, a 40-meter four-square and an 80-meter low inverted V. These multiplier antennas are used whenever the main antenna is not available for the multiplier station; eg, when the run station runs on 20 with the big Yagi, the 40-meter Yagi is not available, and the 40-meter four-square must be used to work multipliers on that band.

With this variety of antennas it is of utmost importance to design a foolproof antenna switching system. At ON4UN/ OTxT the switching system meets the following requirements:

- The transceiver is inhibited when an improper antenna is selected. To do this, the band info from the FT-1000MP (band data output) is compared to the band info available from the antenna switch boxes. If they do not match, an inhibit signal is sent to the transceiver, and a warning light flashes.
- The run station has total priority in selecting antennas. If the multiplier station selects the same antenna, it is automatically switched to a dummy load.
- It is impossible for the two stations to get on the air on the same bands on 80 and 40 meters. Tests have proven that antennas are so close together (all on a 4000 m² lot) that simultaneous operation on these bands results in burned-out receiver front ends.
- In addition, a protective circuit is installed in the switch boxes for the 40 and the 80-meter four-squares to prevent accidental switching while transmitting (see Chapter 11, Section 4.1.3).

■ 5. THE OPERATOR'S STATION

While for a "normal" or everyday DXing station there is practically no rule on how things have to work in the shack, for a contest station equipped for a team effort it is different. Things have to be simple and ergonomic. A contest operator is usually someone who doesn't read manuals. He or she wants to sit down and start operating right away. Out of this we found you need to have the following:

5.1. Antenna Switching

A foolproof antenna switching system—no way to transmit on the wrong antenna. Visual indicators of what is switched where plus error lamps (see **Fig 15-5**).

5.2. Antenna Directions

On the receiving side the visual direction indication of the Beverage antenna selector proved to be very helpful.

Point and Forget rotators: In the heat of the battle you don't want to sit and press that "turn left" or "turn right" button—just select your direction and press one button. On commercial rotator controls these buttons are too small.

5.3. Radio-Computer Interface

Radios connected to the contesting software. Now many might think of this as a useless feature—but when it comes to having very marginal point differences to your competitors, any mistakes are especially painful. Operators sometimes tend to forget to switch bands in

Fig 15-5—Control "tower" at ON4UN's contest station. The top box selects the antennas for the second station. The second box from the top is the 80-meter four-square control box. underneath is the main (run) station antenna selector. The smaller box on the right is the 40-meter four-square control box. On the bottom of the stack are the antenna direction control units (40+20 meters and 10+15 meters), with digital read-out and large 8 cm diameter set and forget direction pointer knobs. The smaller box behind the keyboard is the Beverage antenna selector unit. A large handy knob is used, with 12 LEDs in a circle, giving unambiguous indication of the selected direction.

the logging software. This connection forces the right frequency band information into the software.

5.4. Easy to Tune Power Amplifiers

In a typical multi-single or single-op two-transmitter contest setup one transmitter is tuned to the main band, where a pileup is worked. This is usually called the "run station." With the second transceiver the other bands are scanned, and multipliers are picked up between QSOs on the running band. In order to be able to concentrate fully on the operating aspects, band-switching should be automated as much as possible.

Tuning the second linear between bands often has been a problem, because:

- contest operators often don't know how to properly tune a linear,
- it takes them too long. The minimum to have is labels stuck to the linear front-panel with all settings for all antennas and modes. Even that sometimes seems to be too difficult for the operators trying to concentrate on moving this very rare and weak KH8 station from one band to another.

An amplifier that automatically switches bands, and automatically tunes to preset values of band-switch, load C and tune C, or better yet, performs a fully automatic tune-up, is the answer to that problem.

During 1998 the ON4UN station was equipped with an ACOM 2000A linear. The use of this new full auto-tune linear proved to be very helpful for working multipliers by quickly changing bands and antennas.

The ACOM 2000A (see **Fig 15-6**) is an auto-tune nominal 1500 W output amplifier (max. 2000 W) using two Russian-made 4CX800A (or GU74B) tetrodes (replacement cost in Europe typically $50 to $60). This is the first real auto-tune amateur HF-amplifier I have seen. By pressing a simple button on the remote control panel it fully automatically tunes itself completely within 0.5 seconds. The auto-tune function is not limited to recalling preset values; it actually tunes for a conjugate match for whatever load within the 2:1 SWR circle (on some bands up to 3:1), and does that fully automatically, with no human intervention at all. The amplifier has an absolutely blank front panel

Fig 15-7—Peter, ON7PC, operating at the OTxT contest station using his Alpha 87A amplifier, which like the ACOM 2000A is ideally suited for contest operating.

(except for an ac on-off switch), which makes it possible to "hide" the amplifier in any convenient place. All control and monitoring functions are grouped on a remote small control box, which can easily be positioned next to the keyboard during operation. The ACOM amplifier can be connected via an RS-232 connector to a PC for either remote control or testing. It even has a built-in processor that keeps track of all the important data (currents, voltages, temperatures). In case of a fault, you can send the information stored in the INFO BOX for the recent 12 faults to the dealer or the factory by means of Baudot code on the telephone—simply put the microphone close to the tiny loudspeaker on the RCU rear, or by means of a personal computer and its inherent communications channels (internet, modem, etc). Needless to say, the use of this amplifier has greatly increased flexibility and efficiency at the OTxT contest station.

Alpha/Power's well known Alpha 87A amplifier (see **Fig 15-7**) has very similar, but not identical, characteristics (also 1500 W nominal output, 2000 W max.), and is certainly a valid candidate for a top notch contesting station linear as well. The Alpha 87A uses two 3CX800A7 (triode) tubes. The Alpha 87A has an RS-232 interface port, which allows the linear to be controlled remotely. In addition, key parameter measurement values can be monitored remotely. Over 1000 units have been sold in eight years. This amplifier has the reputation of being an excellent auto-tune amplifier, with an excellent reliability record. Not only the 87A, but other models, including the Alpha 89, and the Alpha 91b are very popular with contesters as well as DXers and DXpeditioners. In many cases, these amplifiers made it possible for us to hear the DXpedition's signals on 160 meters.

Technically speaking, the main difference between the Alpha 87A and the ACOM 2000A amplifier is in how they program memory load-C / tune-C values, and how many presets are available on each band.

The Alpha 87A amplifier has "default" tune positions for a nominal 50-Ω load impedance (SWR 1:1), which generally works well for loads having an SWR below 1.5:1. The ampli-

Fig 15-6—Frank, DL2CC, at the two-station contesting setup at OT8T, checking out the capabilities of the new ACOM 2000A auto-tune linear (in the foreground). Two FT1000MPs are used. Two computers linked in a network run the contest logging program, and two separate keyboards are provided to ensure maximum flexibility. Separate keyers are used for both transceivers as well. When permitted by contest rules, a third computer is connected to the network to run a program that filters data from different packet clusters and feeds the DX spots into the network.

fier has memorized band-switch/tune-C/load-C positions for five band sections on each band, including the 12, 17 and 30-meter bands. One such band section is 100 kHz wide on all bands between 3.5 and 21 MHz, 400 kHz wide on 28 MHz and 40 kHz wide on 1.8 MHz. In addition, if the SWR is higher than 1.5:1, or if peak performance is wanted at lower SWR values, one can *manually* pre-tune the linear and load these "home-tuned" preset values into a memory bank. This means that in total 10 presets are available per band, 5 for the 50-Ω default and 5 for "home-tuned" values.

The ACOM 2000A amplifier automatically tunes the linear (change band switch, loading-C and tuning-C) for any frequency between 1.8 MHz and 29.8 MHz, and can memorize nine different sets of values per band segment. There is a possibility to tune manually as well. Segment width depends on frequency: 1.8-2.2 MHz: 25 kHz, 2.2-5 MHz: 50 kHz, 5-15 MHz: 100 kHz, 15 to 21.9 MHz: 150 kHz, 21.9 to 22.5 MHz: 200 kHz, 22.5 to 26.5 MHz 250 kHz and above 26.5 MHz: 300 kHz. Having segments of 50 kHz wide on 80 meters could be an advantage over having 100 kHz wide sections, as the impedance of an antenna can vary significantly over a 100 kHz span on 80 meters. The same holds true for 160 meters, of course. In the ACOM 2000A you can load into memory presets for up to 10 different antennas (per band segment), each one having slightly different impedances, and that for each of the band segments. The correct preset values, matching a particular selected antenna can be recalled from the remote control panel. Alternatively one can use a computer program to control this selection via the RS232 connector. The same program could also control the antenna selection, making matching of antenna selection and amplifier tuning foolproof.

What is similar with both units is that they both recall previously memorized values for bandswitch, tune-C and load-C. The Alpha 87A amplifier has one set of customer pre-programmable values per band segment plus one set of default 50-Ω settings, while the ACOM 2000A has not less than 10 customer-preprogrammable sets per band segment. These presets must be manually tuned in the Alpha (unless the 50-Ω default is used). In the ACOM amplifier, tuning of the presets is done fully automatically. Manually fine tuning the Alpha for peak output on 10 meters takes me about 30 seconds. The ACOM does it automatically in typically less than 1 second.

Both amplifiers automatically sense the driver RF signal, and switch to the appropriate band segment after a band change on the exciter. To change bands all you really need to do is send a short dot, or say "a" ("b" is fine as well) on phone, and the amplifiers will tune up in a fraction of a second.

It looks like the available sales channels, the long term reliability and the service will decide whether or not the new ACOM amplifier will be a tough competitor for the Alpha 87A.

■ 6. THE STATION AS SEEN BY THE TECHNICAL PERSON

It is clear that the technical requirements for a performing contest station are far superior to what's needed for casual or even serious DXing. Think of harmonic suppression. Stations built for multi-transmitter operation must be technically superior in order to be successful. These stations transmit the cleanest signals, and their harmonics are suppressed far in excess of what's the standard. Another issue is to keep the contest station "up and running" all year long. It takes good mechanical engineering to keep the antennas up.

6.1. The Operating Table

After having been housed in too small a shack (3 by 3.5 m) for years, from where two multi-single first places in Europe were scored (we were literally sitting on each other's laps . . .), one wall was taken out, which made it possible to install a single 7 m long × 1 m wide operating table. (No, the house did not collapse!)

The new shack layout was conceived with contesting in mind. To provide the best possible RF and safety ground, the underside of the 7 × 1 m table was entirely covered by a 1-mm thick aluminum sheath. This sheet provides maximum capacity to the equipment standing on the table, and minimum resistance and especially inductance for good RF grounding. Forty ac outlets are mounted on the aluminum sheath, providing the shortest possible safety ground return for the outlets. Short and wide straps are connected to the sheath and are available on the back side of the table to ground various equipment. The table is separated from the wall by approx 15 cm, which allows wires to pass and for ventilation as well. The aluminum ground sheath is grounded with a short strap to an excellent RF ground just outside the shack, with a 40-cm heavy-gauge cable going right through the wall.

The table is equipped with three separate mains distribution circuits, each equipped with a professional-grade mains filter. Circuit one powers all the run station equipment, circuit two all the multiplier station equipment and circuit three all the VHF/UHF (PacketCluster, and so on) equipment.

6.2. A Monitor 'Scope at Each Station

In the heat of a battle, operators sometimes have the tendency, on phone, to crank up the audio gain, resulting in poor and distorted audio, unnecessary splatter, and so on. I always have a monitor 'scope connected to the output of each of the stations. I use a second-hand commercial 20-MHz 'scope, and tap off a little RF using a resistive voltage divider across the output of the linear. This way the operator always has the pattern of the transmitted signal right in view.

6.3. The Problem of Interband Interference

In a two or more station setup, interband interference is the number one technical problem. But as the saying goes, every problem is an opportunity. In this field lies the opportunity to excel. Here also lies the opportunity for equipment manufacturers to improve their equipment.

Interference can be minimized by using the following techniques:

- separate the antennas as much as possible
- use vertical and horizontal polarization to take advantage of the additional attenuation of unlike polarization
- use band-pass filters between the exciter and the amplifier
- use amplifiers with Pi-L networks, not simple Pi networks
- avoid common-mode currents on the feed lines
- galvanically separate the feed lines of the separate bands
- use band-reject filters between the amplifier and the antenna
- push the equipment manufacturers to produce transmitters with much lower in-band noise output.

It is obvious that interference will be heard on the harmonic frequencies. This poses much more of a problem on CW than on phone. The CW band segments are all in the low end of the bands, and the harmonics of 3.503 will be 7.006, 14.012, 21.018 and 28.024 MHz—all right in the CW window. On phone, if you operate on 3.775 kHz, the harmonics will be on 7.550, 15.100, and so on, all outside the band. There is no real problem with the direct harmonic frequencies when operating phone.

Unfortunately our present-day transmitters do not only transmit just the wanted signal; they also transmit a lot of noise around the transmit frequency. This noise can often make it difficult to copy, even many kHz away from the exact harmonic of the transmit frequency, unless effective filtering is applied. And even then, the final improvement will have to come from the designers and manufacturers of our transceivers, putting out equipment producing less in-band noise.

6.3.1. Medium power band-pass filters

There are a few commercial sources for medium-power band-pass filters that are widely used in multi-station contest setups as well as during DXpeditions. I have experience with the ICE and the Dunestar units. The ICE units are rated 200 W, and if the SWR is low they will indeed cope with 200 W. The Dunestar filters are rated at 100 W. I have been using both and I did some comprehensive measuring some time ago. Both units have insertion losses of between 0.3 and 1.0 dB, depending on band. Due to the circuitry used, the Dunestar filters have significantly steeper shape factors. **Table 15-1** lists some of the major characteristics I measured for 160, 80 and 40 meters.

It is important that the filters be operated at a low SWR. If not, you will likely blow the capacitors. It is important, when driving a linear amplifier through one of these filters, that the linear is switched to the right band. If not, a high input SWR

may result. If, in that case, the exciter is equipped with a built-in tuner, it may try to get the full power into the filter, at a very bad mismatch, which guarantees fried components. Therefore, never switch the automatic tuner on when operating with band-pass filters. Also, it is a good idea to control the selection of the right filter right from the transceiver's band data output, so you do not dump RF of the wrong band into the filter. There must be countless contesters and DXpeditioners who have done that. I am sure replacement capacitors for these units must be a top selling item!

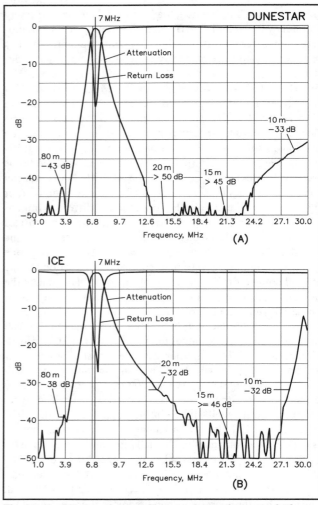

Fig 15-8—Attenuation and return loss characteristics for the Dunestar and the ICE 40-meter band-pass filters as measured by the author.

Table 15-1

Attenuation Characteristics for the ICE and Dunestar Band-Pass Filters (160, 80 and 40 M)

	1.8 MHz	3.5 MHz	7 MHz	14 MHz	21 MHz	28 MHz
ICE 160 m	0.4 dB	15 dB	27 dB	40 dB	>45 dB	>45 dB
Dunestar 160 m	1.0 dB	28 dB	>45 dB	>45 dB	>45 dB	>45 dB
ICE 80 m	25 dB	0.34 dB	17 dB	30 dB	>40 dB	>45 dB
Dunestar 80 m	42 dB	0.64 dB	37 dB	>45 dB	>45 dB	>45 dB
ICE 40 m	>45 dB	38 dB	0.8 dB	32 dB	>45 dB	32 dB
Dunestar 40 m	>45 dB	43 dB	0.6 dB	>50 dB	>45 dB	33 dB

6.3.2. High power filters

If you run power, it really is a must to run filters beyond the amplifier as well, because the amplifier generates harmonic power as well. In addition, these filters should not only be designed to suppress harmonics, they should attenuate signals on all bands, also on frequencies *below* the transmit frequency. It is not uncommon for signals from one of the stations of a multi-operator station to mix in the linear with other signals (BC or from another amateur band) and create unwanted mixing products. The ultimate filter is indeed a filter that attenuates all "other" bands, but gives the highest attenuation on the second harmonic.

The most common way of achieving out-of-band attenuation is by using band-reject filters. These can be made with discrete components or by using coaxial cable.

6.3.2.1. Using discrete components

High-power filters using discrete components can be made much smaller than those using coaxial cable, but the components are hard to come by (high-power, high-voltage, high-current capacitors) and the design requires some expertise and the use of a quality network analyzer. The design of such filters is beyond the scope of this book. I have designed a series of such filters, which perform very well. **Fig 15-9** shows a 10-meter band-reject filter that will take 3-kW continuous-duty power. It is built in a box measuring $25 \times 6 \times 6$ cm. The box is made of double-sided glass-epoxy printed board material, which is ideal for the application.

With single-pole series-tuned circuits for each band, an attenuation of 38 to 46 dB was obtained on all 5 bands,

with an insertion loss of approx 0.1 dB.

The principle for designing such band-reject filters is really quite simple. You design five series-tuned circuits, each one tuned to the frequency you want to suppress and simply connect all these traps in parallel. For the 10-meter filters, all these tuned circuits will exhibit an inductive reactance on 10 meters. You can easily calculate this value: calculate the impedance of all the coils and all the capacitors (five of each) used in this filter. As they are connected in series (for each band), you can simply add the values, taking the sign (+ for a coil, − for a capacitor) into account. Then calculate the parallel value of all of these, just as you calculate parallel resistors. Now we can "tune" out this positive reactance by using a parallel capacitor, which resonates the whole thing on 28 MHz. It really is that simple.

For other bands, series-tuned circuits below the design

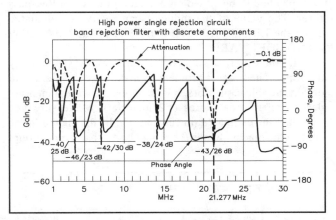

Fig 15-10—Rejection curve for the home-made high-power 10-meter filter. The rejection figures quoted are for band-center and band edges. Example: The rejection in the center of the 15 meter band is 43 dB, and 26 dB on the band edges. This measured plot was generated using the Alpha/Power network analyzer.

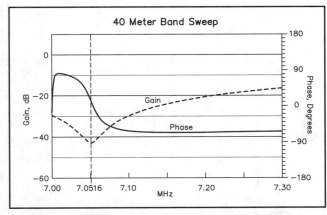

Fig 15-11—With the Alpha/Power network analyzer one can also do a preprogrammed sweep of any of the amateur bands, which is really handy. In this example we see a detailed analysis of the performance of the filter on 40 meters.

Fig 15-9—Ten-meter high-power band-reject filter. This simple unit uses one series-tuned trap for each of the five bands to be rejected.

frequency will show as inductors on the design frequency, and as capacitors above the design frequency. By judiciously choosing the LC ratio of the series-tuned traps, one can now design filters where the positive reactance of a group of traps will cancel the negative reactance of another group, which means there will be no need for a parallel capacitor or inductor to tune the filter to a 1:1 SWR on the operating frequency.

Fig 15-12 shows a high-performance 80-meter filter using a pair of 40-meter traps for improved rejection. The basic configuration is a π low-pass section. The effect of the low-pass section can clearly be seen at the overall shape of the rejection curve. Filters like this can easily be modeled using the *ARRL Radio Designer Software*, which is an ideal tool for this purpose. In this case I designed a symmetrical low-pass π filter, and arranged the traps on both sides of the inductor to obtain the same capacitance value. A capacitor (900 pF) had to be added on one side to tune the low-pass filter. The value of the inductor can easily be calculated using the LINE STRETCHER module of the NEW LOW BAND SOFTWARE.

If you want to design your own filters, the sky is really the limit. The biggest problem in making such filters is to obtain suitable capacitors. Inductors can be wound on powdered-iron toroidal cores (#2 material). Make sure you calculate the estimated power that will be dissipated in the cores. On adjacent bands there may be a substantial amount of heating in the cores, and 2-inch cores may be required in some circumstances.

It is beyond the scope of this book to deal with the concept, design and construction of such filters, but they are necessary to make a multi-transmitter station fully competitive.

6.3.2.2. Using coaxial cable stubs

Let's work out a situation where we want to operate an 80-meter station and a 40-meter station simultaneously in the CW contest.

A single quarter-wave long shorted stub, made of RG-213, cut for 80 meters, will provide approx. 26 dB attenuation on 7 MHz, 24 dB on 14 MHz, 23 dB on 21 MHz and 22 dB on 80 MHz (**see Fig 15-13**). The attenuation on 80 meters will be less than 0.1 dB.

A quarter-wave RG-213 stub cut for 20 meters typically shows an attenuation of 37 dB. The same stub with RG-58 shows about 25 dB of attenuation. A 10-m stub made of RG-213 can achieve 40 dB of attenuation.

We can use two identical stubs to almost double the attenuation, but *not* by merely connecting them in parallel! Connecting a short across a short can in the best case, when the two shorts are equally "good" or "bad," brings you 6 dB additional attenuation. There is one way, however, to obtain much more.

Look from the linear amplifier into the feed line. With a well designed and built amplifier, the pi-L filter will provide a good deal of attenuation on 40 meters. But there is some 40-meter power at the linear output. Assume it is 50 dB down from the 80-meter fundamental. At the output of the linear we assume a low Z for the second harmonic, an acceptable assumption at the output terminal of the pi-L filter. If we now put the stub (which is a short on 40) right at the output of the transmitter, we are putting a short across a short, which is not very effective. If we insert a quarter-wave coaxial line between the output of the amplifier and the stub, we have transformed the very low impedance point (on 40 meters) to a very high impedance point. If we now connect the stub at that point, we will have the most effect of the short that the stub represents. All of this holds true

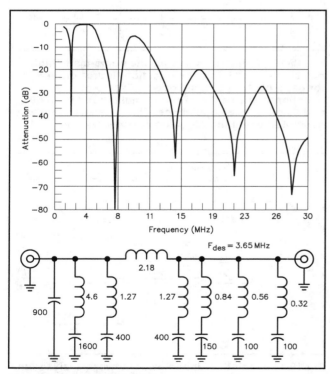

Fig 15-12—High performance 80-meter filter, using discrete components and two 40-meter traps. This filter has a rejection of 80 dB on 40 meters, and between 60 and 75 dB on 20, 15 and 10 meters. The insertion loss is less than 0.1 dB.

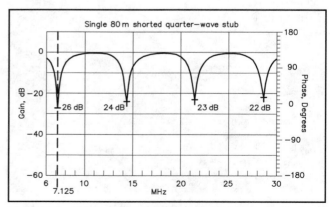

Fig 15-13—Attenuation "dips" are obtained on all of the harmonically related frequencies. Plot generated with the Alpha/Power Analyzer.

only if the output of the amplifier is representing a low Z for the second harmonic (40 meters). In practice it is a good idea to experiment: install the stub right at the output of the amplifier and check the attenuation on 40. Then insert the quarter-wave line between the linear and the stub. If the attenuation is better (which is likely), leave it there.

Fig 15-13 shows the attenuation of a single 80-meter shorted stub (between 6 and 30 MHz).

But you wanted more than 25 dB. You can install another quarter-wave "isolation" line between the first and the second stub. I call it an isolation line because it effectively isolates the two stubs. The reasoning is the same as explained above. Two stubs isolated by a quarter-wave coax (on 40 meters) now exhibit 56 dB of attenuation on 7 MHz, and 50 dB on 21 MHz, but "only" 31 dB on 20 meters and 30 dB on 10 meters. This is logical as the "isolation" line , to do its up-transformation job properly, must be an odd number of quarter-waves long on the reject frequency. This is true on 7 and on 21 MHz only. On 20 and 10 meters we only get the predicted 6-dB improvement.

Stubs can also be used as elements in a low-pass configuration. They can also be used in combination with discrete components. The example in **Fig 15-15** is a combination of a very simple 160-meter low-pass filter with four stubs. The attenuation pattern is amazingly clean, and gives better than 70 dB on all bands.

The impedance of the coaxial cable used for making stubs is irrelevant. The cable loss is important, however. An 80-meter stub made with RG-58 will yield approx. 15-17 dB attenuation, RG-213 gives 25 dB and $^7/_8$" Hardline will give 40 dB!

In a contest station setup, one can also install fixed stubs at the feed points of single-band antennas. An example is given in Chapter 13 (**Fig 13-43/1**). In this example a short-circuited quarter-wave long (on 160 meters) stub, which serves as static drain and also provides attenuation of the 80-meter harmonic, is installed at the feed point of the 160-meter vertical antenna.

Top Ten devices makes a multi-band switched stub system that can be driven from the band-data output of most transceivers (for all info consult: **http://www.qth.com/** topten/ or contact **w2vjn@rosenet.net**).

■ 7. COMPUTERS AND SOFTWARE

For years, computers have taken an important place in the average ham's shack. In a contest station, the computer, or very often computers, are like the heart and the brains of the station.

During a typical multi-single operation we have six computers running at OTxT. At one time we had as many as 10, but many functions have now been integrated into one PC. There is a PC at each of the two positions, with the screen on top of the FT-1000MPs. We call these the RUN and the MULT computers, installed at the Run and Mult operating positions. A third PC, also running the same contest software (either *CT* or *Writelog*), is the "MC" computer. MC stands for Master of Ceremony. The Master of Ceremony is a third operator who makes sure that we get all the PacketCluster info that is available. He interprets the info, and will decide where to "run" and where to work "multipliers." He is really in charge. We have been using this technique quite satisfactorily for many years. A fourth PC runs *QWin*, the packet radio program. All these computers are connected in a network.

7.1 Connecting Computers

For first multi-single as OT3T in 1993, we had linked a variety of PCs ranging from 286s to a very modern 386 66-MHz machine with copper-wire serial cables. Despite pounds of ferrite rods and toroids, we certainly did not

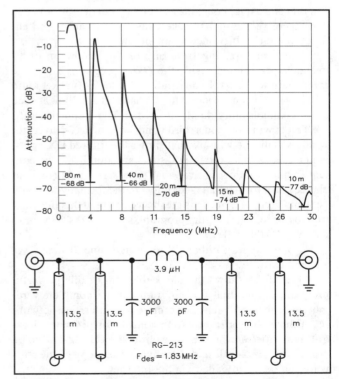

Fig 15-15—Attenuation curves for stubs used in conjuction with discrete components to form a low-pass filter. See text for details.

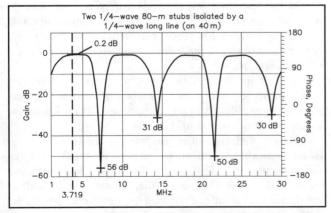

Fig 15-14—Attenuation curve for two stubs separated by a $^1/_4$-λ feed line (see text). This measured plot was generated with the Alpha/Power network analyzer.

achieve a totally RF-free situation. Often a computer would hang, without apparent reason. Several times during the contests logs had to be merged and computers started up again. All of this was certainly far from ideal.

In a second phase we replaced all the copper links with fiber-optic links. This certainly was an improvement, but still not 100% bulletproof.

David Robbins, K1TTT, and Wayne Wright, W5XD, provided the ultimate solution to the problem. K1TTT wrote the required drivers to run *CT* via Ethernet, and the Windows-based contest logging software by W5XT evidently supports Ethernet linking. Since we switched to Ethernet (RG-58 coax-based) linking of the computers, I have never seen any computer act up, or any computer lose a single QSO.

7.2. Connecting to the PacketCluster

In our contest station, we use a dedicated PC to take care of collecting the PacketCluster info during multi-op competitions or single-op assisted efforts. We use the latest version of *QWin Packeteer*, which is a packet radio plus Internet Telnet program that has a lot of interesting features for contest operations. This present version is still a 16-bit version, which means it does run on older computers using Windows 3.x. In the future, however, a 32-bit version will be made available.

On packet radio you can choose between host mode or terminal mode. In terminal mode you can connect to one PacketCluster at a time. This is what most DXers would use for normal PacketCluster operating. In host mode, however, you can connect up to 10 PacketClusters simultaneously.

In addition you can connect to a PacketCluster via the Internet and a telephone line as well (Telnet operation). You can be connected to 10 PacketClusters via packet and another one via Telnet once again simultaneously.

All spots arriving from any of the maximum of 11 channels are filtered inside the *QWin* program. There are three main windows. In one, called Packeteer (packet radio window), you can select up to 10 channels and look at each of the channels individually, if needed. The second is called the Telnet window, and a third window is the DX-Navigator window. This the window that shows the filtered DX-spots, which means that no two DX-spots are identical. The content of this window is available on a serial port (of your choice), in a format that is identical to the format you would get from your modem. This means we can feed this information to our regular logging program, such as *DX4WIN*, or to *CT*, *NA*, *N6TR* or *Writelog*.

QWin supports either CW or Voice announcement via a sound card.

Fig 15-16 shows the block diagram of this powerful program. Notice that the program can be completely remotely controlled by another remote computer via a network (eg, Ethernet). For information consult: **http://www.fietz-online.de/qw/** or via e-mail: **webmaster@fietz-online.de** or **mfietz@fietz-online.de**. An evaluation copy of the software can be downloaded via the Internet.

7.3. Computer Noise

In all the years we have been using various computers,

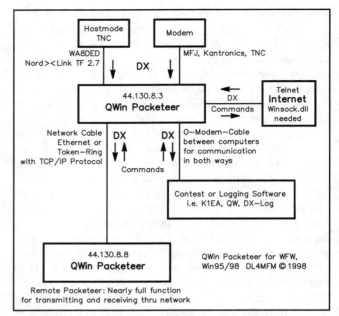

Fig 15-16—The *QWin* integrated PacketCluster program provides utmost flexibility at our contest station by making unattended operation possible. You remain connected with up to 10 PacketClusters via packet radio plus one connection via Telnet/Internet.

I have never had a problem with direct radiation from a computer. Of course, if you use an antenna inside the shack, very close to the computer, you will pick up all kinds of hash. It is very important that your antennas are at a sufficient distance from the computers, and that the feed lines are well shielded and well grounded, and that the coax connector makes perfect contact with the receptacle.

Computer screens can be very noisy, though. If you buy a new one, make sure you can return it if it is radiating too much. You may also want to add extra ferrite cores on the cable between the PC and the monitor.

■ 8. THE RESULTS

Remember why we participate in contests. There are of course the operators, the Formula 1 drivers, who want to win, full bore. But there is also the technical guy, who wants to see the fruit of his labor, his engine, his station, his antenna win in an international competition. This is what it was all about for me.

But at the same time I met a lot of good drivers (operators) and technicians (the always available helpers from the local radio club). This is undoubtedly the important social aspect of contesting.

Since 1993 my station has been tested in 47 international contests (organized by either the ARRL or CQ), which resulted in 32 first places (Europe or worldwide), and six second places. This proves that the antennas as well as the station are capable of winning top-notch contests. In the ARRL DX contests the station's 80-meter capabilities were tested in not less than 9 single-band 80-meter operations (5 on CW, 4 on phone). All 9 resulted in first places in

Table 15-2

Low Band Results During 6 Consecutive Entries in the CQWW DX CW Contest at OTxT

CQWW Contest	160 DXCC	160 Zones	80 DXCC	80 Zones	Place
1993 CW—single band 80			114	35	1st World
1994 CW—multi-single	83	17	107	33	2nd Europe
1995 CW—single band 80			118	35	1st World
1996 CW—multi-single	**102**	32	**116**	32	1st Europe
1997 CW—multi-single	87	21	112	33	2nd Europe

Table 15-3

Nine Entries in the CQ 160-M CW Contest Resulted in 9 First Places in Europe

Contest	States	DXCC	Place
1987 CQ 160 CW	—	—	1st Europe
1988 CQ 160 CW	33	64	1st World
1991 CQ 160 CW	37	57	1st World
1992 CQ 160 CW	27	62	1st Europe
1993 CQ 160 CW	30	70	1st Europe
1994 CQ 160 CW	32	80	1st Europe
1995 CQ 160 CW	47	82	1st World
1996 CQ 160 CW	47	79	1st Europe
1997 CQ 160 CW	49	76	1st Europe

Europe (or even worldwide). This is real battlefield testing for the four-square array with one radial, as described in Chapter 11 (Section 5.3). The results are further confirmed by the 80-meter country and zone totals during CQWW CW contests during the same period ('93-'97), as shown in **Table 15-2.** During the 1994 and the 1996 CQWW CW we scored 5-band DXCC in one weekend, 10 through 80 in 1994, and 15 through 160 in 1996. The 100 plus countries during a single weekend on 160 meters was a first, I believe.

Nine entries in the CQWW 160-meter contests over the past 12 years have resulted in 9 first places (either Europe or worldwide), and during most of these contests we scored the highest number of country multipliers, worldwide (see **Table 15-3**).

The results on Top Band and 80 meters not only speak for the performance of the transmit antennas (four-square on 80 and single quarter-wave vertical on 160) but also, and even more importantly, of the receiving capabilities of the station.

■ **9. FURTHER IMPROVEMENTS**

To stay competitive in international contesting one must improve the station year after year. Our competitors do the same. This is the driving force that leads to technological and conceptual improvements. Really, it is almost the opposite from DXing. The more successful you are in DXing, the more countries you have worked, the less there are left for you to work, the less pressure there is, the more you can relax. The more successful you are in DXing, the easier your call will be recognized in the pileups (that helps, too). No need to add another couple of dBs for those last two or three countries.

With contesting it is just the opposite. Competition grows and improves steadily, and if you don't match their efforts, you'll be at the losing end.

Planned areas of improvement are:
- installation of a second auto-tune amplifier
- use more phased Beverage antennas
- use Titanex rotatable 2-el RX antenna for 80 and 160 (for multiplier station)
- reduce interference level through more and better high-power band-reject filters
- add stacks of fixed 10-meter Yagis to one tower (beamed to USA).

16 LITERATURE REVIEW

This literature review lists over 1000 Reference works, catalogued by subject. The Ref.erence numbers are those used in the different chapters of the book. Copies may be obtained directly from the magazines listed at the following addresses:

CQ/Ham Radio: 25 Newbridge Rd, Hicksville, NY 11801.

CQ-DL: DARC, Postfach 1155, D34216 Baunatal 1, Germany.

QST: ARRL HQ, 225 Main St, Newington, CT 06111.

Radio Communication: RSGB HQ, Lambda House, Cranborne Rd, Potters Bar, Herts EN6 3JE, England.

For other magazines, contact the author directly: J. Devoldere, ON4UN, 215 Poelstraat, B9820 Merelbeke, Belgium.

■ 1. PROPAGATION

Ref. 100: K. J. Hortenbach et al, "Propagation of Short Waves Over Long Distances: Predictions and Observations," *Telecommunications Journal*, Jun 1979, p 320.

Ref. 101: R. D. Straw - N6BV - ed., *The ARRL Handbook for Radio Amateurs*, ARRL, Newington, CT.

Ref. 102: John Devoldere - ON4UN, "80-Meter DXing," *Communications Technology*, 1978.

Ref. 103: George Jacobs - W3ASK - et al, *The New Short-wave Propagation Handbook*.

Ref. 104: Peter Saveskie - W4LGF, *Radio Propagation Handbook*

Ref. 105: William Orr - W6SAI, *Radio Handbook*, 1991.

Ref. 106: R. D. Straw - N6BV - ed., *The ARRL Antenna Book*, ARRL - Newington CT, 1997.

Ref. 107: Wayne Overbeck - N6NB - et al, "Computer Programs for Amateur Radio"

Ref. 108: Dale Hoppe - K6UA - et al, "The Grayline Method of DXing," *CQ*, Sep 1975, p 27.

Ref. 109: Rod Linkous - W7OM, "Navigating to 80 Meter DX," *CQ*, Jan 1978, p 16.

Ref. 110: Yuri Blanarovich - VE3BMV, "Electromagnetic Wave Propagation By Conduction," *CQ*, Jun 1980, p 44.

Ref. 111: Guenter Schwarzbeck - DL1BU, "Bedeuting des Vertikalen Abstralwinkels von KW-Antennen," *CQ-DL*, Mar 1985, p 130.

Ref. 112: Guenter Schwarzbeck - DL1BU, "Bedeutung des Vertikalen Abstrahlwinkels von KW-Antennen (Part 2)," *CQ-DL*, Apr 1985, p 184.

Ref. 113: Henry Elwell - N4UH, "Calculator-Aided Propagation Predictions," *Ham Radio,* Apr 1979, p 26.

Ref. 114: Donald C. Mead - K4DE, "How to Determine True North for Antenna Orientation," *Ham Radio,* Oct 1980, p 38.

Ref. 115: Henry Elwell - N4UH, "Antenna Geometry for Optimum Performance," *Ham Radio,* May 1982, p 60.

Ref. 116: Stan Gibilisco - W1GV/4, "Radiation of Radio Signals," *Ham Radio,* Jun 1982, p 26.

Ref. 117: Garth Stonehocker - KØRYW, "Forecasting by Computer," *Ham Radio,* Aug 1982, p 80.

Ref. 118: Bradley Wells - KR7L, "Fundamentals of Grayline Propagation," *Ham Radio,* Aug 1984, p 77.

Ref. 119: Van Brollini - NS6N - et al, "DXing by Computer," *Ham Radio,* Aug 1984, p 81.

Ref. 120: Calvin R. Graf - W5LFM - et al, "High-Frequency Atmospheric Noise - Part 2," *QST*, Feb 1972, p 16.

Ref. 121: Jim Kennedy - K6MIO - et al, "D-Layer Absorption During a Solar Eclipse," *QST*, Jul 1972, p 40.

Ref. 122: Edward P. Tilton - W1HDQ, "The DXer's Crystal Ball," *QST*, Jun 1975, p 23.

Ref. 123: Edward P. Tilton - W1HDQ, "The DXer's Crystal Ball - Part II," *QST*, Aug 1975, p 40.

Ref. 124: Paul Argo - et al, "Radio Propagation and Solar Activity," *QST*, Feb 1977, p 24.

Ref. 125: Kenneth Johnston - W7LIX - et al, "An Eclipse Study on 80 Meters," *QST*, Jul 1979, p 14.

Ref. 126: V. Kanevsky - UL7GW, "Ionospheric Ducting at HF," *QST*, Sep 1979, p 20.

Ref. 127: Robert B. Rose - K6GKU, "MINIMUF: Simplified MUF-Prediction Program for Microcomputers," *QST*, Dec 1982, p 36.

Ref. 128: Tom Frenaye - K1KI, "The KI Edge," *QST*, Jun 1984, p 54.

Ref. 129: Richard Miller - VE3CIE, "Radio Aurora," *QST*, Jan 1985, p 14.

Ref. 130: Pat Hawker - G3VA, "Technical Topics: Trans-equatorial Supermode Theories," *Radio Communication*, Feb 1972, p 94.

Ref. 131: Pat Hawker - G3VA, "Technical Topics: Whispering Galleries," *Radio Communication*, May 1972, p 306.

Ref. 132: Pat Hawker - G3VA, "Technical Topics: Low Angles for Chordal Hops and TEP," *Radio Communication*, Nov 1972, p 746.

Ref. 133: A. P. A. Ashton - G3XAP, "160M DX from Suburban Sites," *Radio Communication*, Dec 1973, p 842.

Ref. 134: Pat Hawker - G3VA, "Technical Topics: Path Deviations - One-Way Propagation - HF Tropo and LDEs," *Radio Communication*, Oct 1974, p 686.

Ref. 135: Pat Hawker - G3VA, "Fading and the Ionosphere," *Radio Communication*, Mar 1978, p 217.

Ref. 136: Pat Hawker - G3VA, "Technical Topics: When Long-Path is Better," *Radio Communication*, Sep 1979, p 831.

Ref. 137: V. Kanevsky - UL7GW, "DX QSOs," *Radio Communication*, Sep 1979, p 835.

Ref. 138: Pat Hawker - G3VA, "Technical Topics: Chordal Hop and Ionospheric Focusing," *Radio Communication*, Apr 1984, p 315.

Ref. 139: Dr. Alexsandr V. Gurevich and Dr. Elena E. Tsedilina : "Long Distance Propagation of HF Radio Waves," ISBN 3-540-15139-7 and ISBN 0-387-15139-7.

Ref. 140: Bob Brown, NM7M, "160 Meter DXing, part 1," *The DX Magazine*, May/Jun 1996, p 22.

Ref. 141: Bob Brown, NM7M, "160 Meter DXing, part 2," *The DX Magazine*, Jul/Aug 1996, p 10.

Ref. 142: Cary Oler and Ted Cohen, N4XX, "The 160 Meter Band: An Enigma Shrouded in Mystery," *CQ* Mag, Part 1—Mar 1998, p 9; Part 2—Apr 1998, p 11.

Ref. 143: Carl Luetzelschwab - K9LA, "Working VKØIR on 160 m"; *The DX Magazine*, Jul/Aug 1997, p 49.

Ref. 144: Carl Luetzelschwab - K9LA, "Popular Great circle paths"; *The Low Band Monitor*, Oct 1997.

Ref. 145: Carl Luetzelschwab - K9LA, "A General Summary of Propagation from VKØIR to the US and VE," *QST*, Oct 1997, p 88.

Ref. 146: Carl Luetzelschwab - K9LA, "Stratwarms and their Effect on HF Propagation," *QST*, Technical Correspondence, Feb 1997, p 76.

Ref. 147: Robert Brown - NM7M, "The Low Band Logs of VKØIR," *Communications Quarterly*, Fall 1997, p 45.

Ref. 148: Bob Brown, NM7M, "Heard & Unheard Islands"; *The DX Magazine*, May/Jun 1997, p 9.

Ref. 149: "Boulder A Index - Propagation Predictor," *The Top Band Monitor*, Aug 1994.

Ref. 150: KØCS: "Boulder A-Index Daily Variations"; *The Top Band Monitor*, Dec 1993.

Ref. 151: NM7M - "An Argument for 160-meter Ducting," *The Low Band Monitor*, Aug 1997.

■ 2. RECEIVERS

Ref. 200: R. D. Straw - N6BV - ed, *The ARRL Handbook for Radio Amateurs*, ARRL, Newington, CT.

Ref. 201: ON4EG, "Sensibilite des Recepteurs," *CQ-QSO*, Jan 1985.

Ref. 202: ON4EG, "Sensibilite des Recepteurs," *CQ-QSO*, Feb 1985.

Ref. 203: Michael Martin - DJ7JV, "Emfangereingangsteil mit Grossem Dynamik Bereich," *DL-QTC*, Jun 1975.

Ref. 204: R. D. Straw - N6BV - ed, *"The ARRL Handbook for Radio Amateurs,"* ARRL, Newington, CT.

Ref. 205: John Devoldere - ON4UN, "80-Meter DX-ing," *Communications Technology*, Jun 1978.

Ref. 206: William Orr - W6SAI, *Radio Handbook*, Howard W. Sams & Co. Inc.

Ref. 207: Robert Sternowsky, "Using Preselectors to Improve HF Performance," *Communications International*, May 1980, p 34.

Ref. 208: John Devoldere - ON4UN, "Improved Performance from the Drake R-4B and T4X-B," *CQ,* Mar 1976, p 37.

Ref. 209: Michael Martin - DJ7VY, "Rauscharmer Oszillator fur Empfaenger mit grossem Dynamikbereich," *CQ-DL*, Dec 1976, p 418.

Ref. 210: Wes Hayward - W7ZOI, "Der Dynamische Bereich eines Empfaengers," *CQ-DL*, Mar 1977, p 93.

Ref. 211: Richard Waxweller - DJ7VD, "Hochfrequenz-Zweitongenerator," *CQ-DL*, Sep 1980, p 412.

Ref. 212: Guenter Schwarzbeck - DL1BU, "Testbericht: NF-Filter Datong FL2," *CQ-DL*, Feb 1981, p 56.

Ref. 213: Guenter Schwarzbeck - DL1BU, "Grosssignalverhalten von Kurzwellenempfaengern," *CQ-DL*, Mar 1981, p 117.

Ref. 214: Walter Flor - OE1LO, "KW-Eingangsteile: Eingangsfilter," *CQ-DL*, Aug 1981, p 373.

Ref. 215: Walter Flor - OE1LO, "IM-feste Verstaerker fuer den KW-bereich," *CQ-DL*, Aug 1981, p 473.

Ref. 216: Walter Flor - OE1LO, "KW-Eingangsteile: Extrem IM-feste selektive Vorverstaerker," *CQ-DL*, Aug 1981, p 376.

Ref. 217: Guenter Schwarzbeck - DL1BU, "Geraeteeigenschaften: Besonderheiten zwischen Testbericht und Praxis," *CQ-DL*, Sep 1982, p 424.

Ref. 218: Erich Vogelsang - DJ2IM, "Grundrmauschen und Dynamiekbereich bei Kurzwellenempfaengern," *CQ-DL*, Sep 1982, p 432.

Ref. 219: Michael Martin - DJ7YV, "Intermodulationsfster Preselector fur 1.5 - 30 MHz," *CQ-DL*, Jul 1984, p 320.

Ref. 220: Ray Moore, "Designing Communication Receivers for Good Strong-Signal Performance," *Ham Radio*, Feb 1973, p 6.

Ref. 221: Wes Hayward - W7ZOI, "Bandpass Filters for Receiver Preselectors," *Ham Radio*, Feb 1975, p 18.

Ref. 222: Ulrich Rohde - DJ2LR, "High Dynamic Range Receiver Input Stages," *Ham Radio,* Oct 1975, p 26.

Ref. 223: James Fisk - W1DTY, "Receiver Noise Figure Sensitivity and Dynamic Range, What The Numbers Mean," *Ham Radio,* Oct 1975, p 8.

Ref. 224: Marvin Gonsior - W6FR, "Improved Selectivity for Collins S-line Receivers," *Ham Radio*, Jun 1976, p 36.

Ref. 225: Howard Berlin - K3NEZ, "Increased Flexibility for MFJ CW Filters," *Ham Radio*, Dec 1976, p 58.

Ref. 226: Ulrich Rohde - DJ2LR, "I-F Amplifier Design," *Ham Radio*, Mar 1977, p 10.

Ref. 227: Alex Burwasser - WB4ZNV, "Reducing Intermodulation Distortion in High-Frequency Receivers," *Ham Radio,* Mar 1977, p 26.

Ref. 228: Wayne C. Ryder - W6URM, "General Coverage Communications Receiver," *Ham Radio,* Nov 1977, p 10.

Ref. 229: R. Sherwood - WBØJGP - et al, "Receivers: Some Problems and Cures," *Ham Radio,* Dec 1977, p 10.

Ref. 230: R. Sherwood - WBØJGP - et al, "New Product Detector for R-4C," *Ham Radio,* Oct 1978, p 94.

Ref. 231: R. Sherwood - WBØJGP - et al, "Audio Amplifier for the Drake R-4C," *Ham Radio,* Apr 1979, p 48.

Ref. 233: James M. Rohler - NØDE, "Biquad Bandpass Filter," *Ham Radio,* Jun 1979, p 70.

Ref. 234: Sidney Kaiser - WB6CTW, "Measuring Receiver Dynamic Range," *Ham Radio,* Nov 1979, p 56.

Ref. 235: Ulrich Rohde - DJ2LR, "Recent Developments in Circuits and Techniques for High Frequency Communications Receivers," *Ham Radio,* Apr 1980, p 20.

Ref. 236: Edward Wetherhold - W3NQN, "High Performance CW Filter," *Ham Radio,* Apr 1981, p 18.

Ref. 237: D. A. Tong - G4GMQ, "Add-On Selectivity for Communication Receivers," *Ham Radio,* Nov 1981, p 41.

Ref. 238: Ulrich Rohde - DJ2LR, "Communication Receivers for the Year 2000: Part 1," *Ham Radio,* Nov 1981, p 12.

Ref. 239: Jan K. Moller - K6FM, "Understanding Performance Data of High-Frequency Receivers," *Ham Radio,* Nov 1981, p 30.

Ref. 240: Ulrich Rohde - DJ2LR, "Communication Receivers for the Year 2000 - Part 2," *Ham Radio,* Dec 1981, p 6.

Ref. 241: Ulrich Rohde - DJ2LR, "Performance Capability of Active Mixers - Part 1," *Ham Radio,* Mar 1982, p 30.

Ref. 242: Ulrich Rohde - DJ2LR, "Performance Capability of Active Mixers - Part 2," *Ham Radio,* Apr 1982, p 38.

Ref. 243: R. W. Johnson - W6MUR, "Bridged T-Filters for Amateur Use," *Ham Radio,* Oct 1982, p 51.

Ref. 244: Cornell Drentea - WB3JZO, "Designing a Modern Receiver," *Ham Radio,* Nov 1983, p 3.

Ref. 245: Edward Wetherhold - W3NQN, "Elliptic Lowpass Audio Filter Design," *Ham Radio,* Jan 1984, p 20.

Ref. 246: J. A. Dyer - G4OBU, "High Frequency Receiver Performance," *Ham Radio,* Feb 1984, p 33.

Ref. 247: E. A. Andrade - WØDAN, "Recent Trends in Receiver Front-End Design," *QST,* Jun 1962, p 17.

Ref. 248: William K. Squires - W2PUL, "A New Approach to Receiver Front-End Design," *QST,* Sep 1963, p 31.

Ref. 249: Byron Goodman - W1DX, "Some Thoughts on Home Receiver Design," *QST,* May 1965, p 11.

Ref. 250: E. H. Conklin - K6KA, "Front-End Receiving Filters," *QST,* Aug 1967, p 14.

Ref. 251: Doug DeMaw - W1CER, "Rejecting Interference from Broadcast Stations," *QST,* Dec 1967, p 35.

Ref. 252: Rudolf Fisher - DL6WD, "An Engineer's Solid-State Ham-Band Receiver," *QST,* Mar 1970, p 11.

Ref. 253: Douglas A. Blakeslee - W1KLK, "An Experimental Receiver for 75-Meter DX Work," *QST,* Feb 1972, p 41.

Ref. 254: Wes Hayward - W7ZOI, "Defining and Measuring Receiver Dynamic Range," *QST,* Jul 1975, p 15.

Ref. 255: Doug DeMaw - W1FB, "His Eminence: the Receiver," *QST,* Jun 1976, p 27.

Ref. 256: Wes Hayward - W7ZOI, "CER-Verters," *QST,* Jun 1976, p 31.

Ref. 257: Doug DeMaw - W1FB, "Build this Quickie Preamp," *QST,* Apr 1977, p 43.

Ref. 258: Wes Hayward - W7ZOI, "More Thoughts on Receiver Performance Specification," *QST,* Nov 1979, p 48.

Ref. 259: Ulrich Rohde - DJ2LR, "Increasing Receiver Dynamic Range," *QST,* May 1980, p 16.

Ref. 260: Edward Wetherhold - W3NQN, "Modern Design of a CW Filter Using 88- and 44-mH Surplus Inductors," *QST,* Dec 1980, p 14.

Ref. 261: Doug DeMaw - et al, "Modern Receiver Mixers for High Dynamic Range," *QST,* Jan 1981, p 19.

Ref. 262: Wes Hayward - W7ZOI, "A Progressive Communications Receiver," *QST,* Nov 1981, p 11.

Ref. 263: Robert E. Lee - K2TWK, "Build an Audio Filter with Pizzazz," *QST,* Feb 1982, p 18.

Ref. 264: Harold Mitchell - NØARQ, "88-mH Inductors: A Trap," *QST,* Jan 1983, p 38.

Ref. 265: Gerald B. Hull - AK4L/VE1CER, "Filter Systems for Multi-transmitter Amateur Stations," *QST,* Jul 1983, p 28.

Ref. 266: John K. Webb - W1ETC, "High-Pass Filters for Receiving Applications," *QST,* Oct 1983, p 17.

Ref. 267: Doug DeMaw - W1FB, "Receiver Preamps and How to Use Them," *QST,* Apr 1984, p 19.

Ref. 268: Pat Hawker - G3VA, "Trends in H.F. Receiver Front-ends," *Radio Communication,* Sep 1963, p 161.

Ref. 269: D. A. Tong - G4GMQ, "Audio Filters as an Aid to Reception," *Radio Communication,* Feb 1978, p 114.

Ref. 270: John Bazley - G3HCT, "The Datong Multi-Mode Filter FL-2," *Radio Communication,* Aug 1980, p 783.

Ref. 271: Pat Hawker - G3VA, "Technical Topics: More Thoughts On 'Ideal' HF Receivers," *Radio Communication,* Oct 1982, p 861.

Ref. 272: Edward Wetherhold - W3NQN, "Simplified Elliptic Lowpass Filter Design Using Surplus 88-mH Inductors," *Radio Communication,* Apr 1983, p 318.

Ref. 273: P. E. Chadwick - G3RZP, "Dynamic Range; Intermodulation and Phase Noise," *Radio Communication,* Mar 1984, p 223.

Ref. 274: Pat Hawker - G3VA, "Technical Topics: Comparing Receiver Front-Ends," *Radio Communication,* May 1984, p 400.

Ref. 275: Pat Hawker - G3VA, "Technical Topics: Receivers: Numbers Right or Wrong?," *Radio Communication,* Aug 1984, p 677.

Ref. 276: Pat Hawker - G3VA, "Technical Topics: Receivers of Top Performance," *Radio Communication*, Oct 1984, p 858.

Ref. 277: Edward Wetherhold - W3NQN, "A CW Filter for Radio Amateurs Newcomer," *Radio Communication*, Jan 1985, p 26.

Ref. 278: Pat Hawker - G3VA, "Technical Topics: Whither Experimentation?," *Radio Communication*, Mar 1985, p 189.

Ref. 279: Ian White - G3SEK, "Modern VHF/UHF Front End Design—Part 1," *Radio Communication*, Apr 1985, p 264.

Ref. 280: Ian White - G3SEK, "Modern VHF/UHF Front End Design—Part 2," *Radio Communication*, May 1985, p 367.

Ref. 281: Ian White - G3SEK, "Modern VHF/UHF Front End Design—Part 3," *Radio Communication*, Jun 1985, p 445.

Ref. 282: Pat Hawker - G3VA, "Technical Topics: Weak Signal Reception," *Radio Communication*, Jul 1985, p 540.

Ref. 283: D. H. G. Fritsch - GØCKZ, "Active Elliptic Audio Filter Design Using Op-Amps (Part 1)," *Radio Communication*, Feb 1986, p 98.

Ref. 284: Ulrich L. Rohde - KA2WEU/DJ2LR, "Designing a State-of-the-Art Receiver," *Ham Radio*, Nov 1987, p 17.

Ref. 285: Robert J. Zavrel - W7SX, "Tomorrow's Receivers," *Ham Radio*, Nov 1987, p 8.

Ref. 286: John Grebenkemper - KI6WX, "Phase Noise and its Effects on Amateur Communications," *QST*, Mar 1988, p 14.

Ref. 287: Zack Lau - KH6CP, "Eliminating AM-Broadcast interference on 160 Meters," *QST*, Apr 1992, p 75.

Ref. 288: Gary Nichols - KD9SV, "Bandpass Filters for 80 and 160 Meters," *QST*, Feb 1989, p 42.

Ref. 289: John Grebenkemper - KI6WX, "Phase Noise and its Effects on Amateur Communications," *QST*, Apr 1988, p 22.

Ref. 290: Dave Hershberger - W9GR, "Low Cost Digital Signal Processing for Radio Amateurs," *QST*, Sep 1992, p 43.

Ref. 291: Bruce C. Hale - KB1MW/7, "An Introduction to Digital Signal Processing," *QST*, Jul 1991, p 35.

Ref. 292: D. DeMaw - W1FB, "Receiver Preamps and How to Use Them," *QST*, Apr 1984, p 19.

Ref. 293: J. Kearman - KR1S, "Audio Filter Roundup," *QST*, Oct 1991, p 36.

Ref. 294: Wes Hayward - W7ZOI, "The Double-tuned Circuit.: An Experimenter's Tutorial," *QST*, Dec 1991, p 29.

Ref. 295: Lew Gordon - K4VX, "Band-Pass filters for HF transceivers," *QST*, Sep 1988, p 17.

Ref. 296: D. DeMaw - W1FB, "A Diode-switched Bandpass Filter," *QST*, Jan 1991, p 24.

Ref. 297: G. E. Myers - K9GZB, "Diode-switched-filter (corrections and amplifications)," *QST*, Aug 1991, p 41.

Ref. 298: KØCS - "A Classic BCI Filter for 160 Meters"; *The Low Band Monitor*, May 1995.

■ 3. TRANSMITTERS

Ref. 300: R. D. Straw - N6BV - ed, *"The ARRL Handbook for Radio Amateurs,"* ARRL, Newington, CT.

Ref. 301: Bob Heil - K9EID, "Equalise That Microphone," *CQ-DL*, Apr 1985, p 27.

Ref. 302: John Devoldere - ON4UN, "80-Meter DX-ing," *Communications Technology Inc*, Apr 1978.

Ref. 303: William Orr - W6SAI, *Radio Handbook*, Howard W. Sams & Co.- Inc.

Ref. 304: John Devoldere - ON4UN, "Improved Performance from the Drake R-4B and T4X-B," *CQ*, Mar 1976, p 37.

Ref. 305: John Schultz - W4FA, "An Optimum Speech Filter," *CQ*, Oct 1978, p 22.

Ref. 307: L. McCoy - W1ICP, "The Design Electronics QSK-1500," *CQ*, Apr 1985, p 40.

Ref. 308: Guenter Schwarzbeck - DL1BU, "Geraeteeigenschaften: esonderheiten zwischen Testbericht und Praxis," *CQ-DL*, Sep 1982, p 424.

Ref. 309: Leslie Moxon - G6XN, "Performance of RF Speech Clippers," *Ham Radio*, Nov 1972, p 26.

Ref. 310: Charles Bird - K6HTM, "RF Speech Clipper for SSB," *Ham Radio*, Feb 1973, p 18.

Ref. 311: Henry Elwell - W2MB, "RF Speech Processor," *Ham Radio*, Sep 1973, p 18.

Ref. 312: Barry Kirkwood - ZL1BN, "Principles of Speech Processing," *Ham Radio*, Feb 1975, p 28.

Ref. 313: Timothy Carr - W6IVI, "Speech Processor for the Heath SB-102," *Ham Radio*, Jun 1975, p 38.

Ref. 314: Jim Fisk - W1DTY, "New Audio Speech Processing Technique," *Ham Radio*, Jun 1976, p 30.

Ref. 315: Frank C. Getz - K3PDW, "Logarithmic Speech Processor," *Ham Radio*, Aug 1977, p 48.

Ref. 316: Michael James - W1CBY, "Electronic Bias Switching for the Henry 2K4 and 3KA Linear Amplifiers," *Ham Radio*, Aug 1978, p 75.

Ref. 317: Wesley D. Stewart - N7WS, "Split-Band Speech Processor," *Ham Radio*, Sep 1979, p 12.

Ref. 318: J. R. Sheller - KN8Z, "High Power RF Switching With Pin Diodes," *Ham Radio*, Jan 1985, p 82.

Ref. 319: William Sabin - WØIYH, "R.F. Clippers for S.S.B.," *QST*, Jul 1967, p 13.

Ref. 320: J. A. Bryant - W4UX, "Electronic Bias Switching for RF Power Amplifiers," *QST*, May 1974, p 36.

Ref. 321: Robert Myers - W1FBY, "Quasi-Logarithmic Analog Amplitude Limiter," *QST*, Jul 1974, p 22.

Ref. 322: Hal Collins - W6JES, "SSB Speech Processing Revisited," *QST*, Aug 1976, p 38.

Ref. 323: Bob Heil - K9EID, "Equalize Your Microphone and Be Heard!," *QST*, Jul 1982, p 11.

Ref. 324: R. C. V. Macario - G4ADL - et al, "An Assured Speech Processor," *Radio Communication*, Apr 1978, p 310.

Ref. 325: H. Leerning - G3LLL, "Improving the FT-101," *Radio Communication*, Jun 1979, p 516.

Ref. 326: L. McCoy - W1ICP, "The Design Electronics QSK-1500," *CQ,* Apr 1985, p 40.

Ref. 327: Alfred Trossen - DL6YP, "Die AMTOR-II einheit nach G3PLX," *CQ-DL*, Jul 1983, p 316.

Ref. 328: Alfred Trossen - DL6YP, "Die AMTOR-II einheit nach G3PLX," *CQ-DL*, Aug 1983, p 368.

Ref. 329: J. R. Sheller - KN8Z, "High Power RF Switching With Pin Diodes," *Ham Radio,* Jan 1985, p 82.

Ref. 330: Peter Martinez - G3PLX, "AMTOR: an Improved Error-Free RTTY System," *QST*, Jun 1981, p 25.

Ref. 331: Paul Newland - AD7I, "Z-AMTOR: An Advanced AMTOR Code Converter," *QST*, Feb 1984, p 25.

Ref. 332: Peter Martinez - G3PLX, "AMTOR: an Improved Radio Teleprinter System Using a Microprocessor," *Radio Communication*, Aug 1979, p 714.

Ref. 333: Peter Martinez - G3PLX, "AMTOR the Easy Way," *Radio Communication*, Jun 1980, p 610.

Ref. 334: Jon Towle - WB1DNL, "QSK 1500 High-Power RF Switch," *QST*, Sep 1985, p 39.

Ref. 335: Dr. J. R. Sheller - KN8Z, "What Does 'QSK' Really Mean?," *QST*, Jul 1985, p 31.

Ref. 336: Paul Newland - AD7I, "A User's Guide to AMTOR Operation," *QST*, Oct 1985, p 31.

Ref. 337: W. J. Byron - W7DHD, "Designing an amplifier around the 3CX1200A7," *Ham Radio,* Dec 1978, p 33.

Ref. 338: Richard L. Measures - AG6K, "QSK Modification of the Trio Kenwood TL-922 Amplifier," *Ham Radio,* Mar 1989, p 35.

Ref. 339: Paul A. Johnson - W7KBE, "Homebrewing Equipment - From Parts to Metal Work," *Ham Radio,* Mar 1988, p 26.

Ref. 340: Richard L. Measures - AG6K, "Adding 160-meter Coverage to HF Amplifiers," *QST*, Jan 1989, p 23.

Ref. 341: Safford M. North - KG2M, "Putting the Heath SB-200 on 160 Meters," *QST*, Nov 1987, p 33.

Ref. 342: W.J. Byron - W7DHD, "Design Program for the Grounded-grid 3-500Z," *Ham Radio,* Jun 1988, p 8.

■ 4. EQUIPMENT REVIEW

Ref. 400: Guenter Schwarzbeck - DL1BU, "Testbericht: Transceiver TS820," *CQ-DL*, Apr 1977, p 130.

Ref. 401: Guenter Schwarzbeck - DL1BU, "Testbericht und Beschreibung TS-520S," *CQ-DL*, Feb 1978, p 50.

Ref. 402: Guenter Schwarzbeck - DL1BU, "Testbericht un Messdaten FT-901 DM (receiver section)," *CQ-DL*, Feb 1978, p 438.

Ref. 403: Guenter Schwarzbeck - DL1BU, "Testbericht und Messdaten FT-901 DM (transmitter section)," *CQ-DL*, Nov 1978, p 500.

Ref. 404: Guenter Schwarzbeck - DL1BU, "KW-Empfaenger Drake R-4C mit usatzfiltern," *CQ-DL*, Feb 1979, p 56.

Ref. 405: Guenter Schwarzbeck - DL1BU, "KW Transceiver ICOM IC-701," *CQ-DL*, Feb 1979, p 65.

Ref. 406: Guenter Schwarzbeck - DL1BU, "Testbericht: Vorausbericht FT-ONE," *CQ-DL*, Jan 1982, p 11.

Ref. 407: Guenter Schwarzbeck - DL1BU, "Testbericht und Messwerte IC-730," *CQ-DL*, Mar 1982, p 117.

Ref. 408: Guenter Schwarzbeck - DL1BU, "Testbericht und Messwerte FT-102," *CQ-DL*, Aug 1982, p 387.

Ref. 409: Guenter Schwarzbeck - DL1BU, "Testbericht und Messwerte TS930S," *CQ-DL*, Oct 1982, p 484.

Ref. 410: Peter Hart - G3SJX, "The Trio TS-830 HF Transceiver," *Radio Communication*, Jul 1982, p 576.

Ref. 411: Peter Hart - G3SJX, "The Yaesu Musen FT102 HF Transceiver," *Radio Communication*, Jan 1983, p 32.

Ref. 412: Peter Hart - G3SJX, "The ICOM IC740 HF Transceiver," *Radio Communication*, Nov 1983, p 985.

Ref. 413: Peter Hart - G3SJX, "The Yaesu FT77 HF Transceiver," *Radio Communication*, Jun 1984, p 482.

Ref. 414: Peter Hart - G3SJX, "The Yaesu Musen FT980 HF Transceiver," *Radio Communication*, Sep 1984, p 761.

Ref. 415: Peter Hart - G3SJX, "The Ten-Tec Corsair HF Transceiver," *Radio Communication*, Nov 1984, p 957.

Ref. 416: Peter Hart - G3SJX, "The Yaesu Musen FT757GX HF Transceiver," *Radio Communication*, May 1985, p 351.

Ref. 417: Peter Hart - G3SJX, "The Trio TS430S HF Transceiver," *Radio Communication*, Jun 1985, p 441.

Ref. 418: G. Schwarzbeck - DL1BU, "Yaesu FT-1000 Test Review—Part 1," *CQ-DL*, Mar 1991, p 91.

Ref. 419: G. Schwarzbeck - DL1BU, "Yaesu FT-1000 Test Review—Part 2," *CQ-DL*, Apr 1991, p 215.

Ref. 420: G. Schwarzbeck - DL1BU, "Yaesu FT-1000 Test Review—Part 3," *CQ-DL*, May 1991, p 273.

Ref. 421: G. Shwarzbeck - DL1BU, "TS-850S Test Review," *CQ-DL*, Feb 1991, p 79.

Ref. 422: G. Shwarzbeck - DL1BU, "TS-950 SD Test Review," *CQ-DL*, Dec 1989, p 750.

Ref. 423: G. Shwarzbeck - DL1BU, "ICOM IC-761 Test Review," *CQ-DL*, Aug 1988, p 479.

Ref. 424: Peter Hart - G3SJX, "FT-1000 Review," *Radio Communication*, Jun 1991, p 49.

Ref. 425: Peter Hart - G3SJX, "The Yaesu Musen FT767GX HF Transceiver," *Radio Communication*, Jul 1987, p 490.

Ref. 426: Peter Hart - G3SJX, "Kenwood TS-860S Transceiver," *Radio Communication*, Mar 1989, p 47.

Ref. 427: Peter Hart - G3SJX, "Yaesu Musen FT-747GX HF Transceiver," *Radio Communication*, May 1989, p 47.

Ref. 428: Peter Hart - G3SJX, "ICOM IC-725 HF Transceiver," *Radio Communication*, Sep 1989, p 56.

Ref. 429: Peter Hart - G3SJX, "Kenwood TS-950S Transceiver Review," *Radio Communication*, Apr 1990, p 35.

Ref. 430: Peter Hart - G3SJX, "ICOM-781 HF Transceiver," *Radio Communication*, Jul 1980, p 52.

Ref. 431: G. Shwarzbeck - DL1BU, "Ten-Tec Paragon 585 Test Review," *CQ-DL*, May 1988, p 277.

■ 5. OPERATING

Ref. 500: R. D. Straw - N6BV - ed, *The ARRL Handbook for Radio Amateurs*, ARRL, Newington, CT.

Ref. 501: Erik - SMØAGD, "Split Channel Operation," *CQ-DL*, Apr 1985, p 10.

Ref. 502: John Devoldere - ON4UN, "80-Meter DX-ing," *Communications Technology*, Apr 1987.

Ref. 503: Wayne Overbeck - N6NB, *Computer Programs for Amateur Radio*,

Ref. 504: Rod Linkous - W7OM, "Navigating to 80 Meter DX," *CQ*, Jan 1978, p 16.

Ref. 505: Larry Brockman, "The DX-list Net. What a Mess," *CQ*, May 1979, p 48.

Ref. 506: Wolfgang Roberts - DL7RT, "Wie werde ich DX-er?," *CQ-DL*, Oct 1981, p 493.

Ref. 507: John Lindholm - W1XX, "Is 160 Your Top Band?" *QST*, Aug 1985, p 45.

Ref. 508: V. Kanevsky - UL7GW, "DX QSOs," *Radio Communication*, Sep 1979, p 835.

Ref. 509: Top Band operators survey, *The Low Band Monitor*, Jun 1996.

Ref. 510: Top band operators survey, *The Low Band Monitor*, Jun 1995.

Ref. 511: Jeff Briggs, K1ZM, *DXing on the Edge, The Thrill of 160 Meters*, ARRL, Newington, CT.

■ **6. ANTENNAS: GENERAL**

Ref. 600: R. D. Straw - N6BV - ed, *The ARRL Handbook for Radio Amateurs*, ARRL, Newington CT.

Ref. 601: J. J. Wiseman, "How Long is a Piece of Wire," *Electronics and Wireless World*, Apr 1985, p 24.

Ref. 602: William Orr - W6SAI - et al, *Antenna Handbook*, Howard W. Sams & Co, Inc,

Ref. 603: Gerald Hall - K1TD - Ed., *"The ARRL Antenna Compendium, Volume 1,"* ARRL, Newington, CT,

Ref. 604: Keith Henney, *Radio Engineering Handbook*, 5th Edition.

Ref. 605: *Ref.erence Data for Radio Engineers*, Howard W. Sams, 5th Edition.

Ref. 606: John Kraus - W8JK, *Antennas*.

Ref. 607: Joseph Boyer - W6UYH, "The Multi-Band Trap Antenna - Part I," *CQ*, Feb 1977, p 26.

Ref. 608: Joseph Boyer - W6UYH, "The Multi-Band Trap Antenna - Part II," *CQ*, Mar 1977, p 51.

Ref. 609: Bill Salerno - W2ONV, "The W2ONV Delta/Slope Antenna," *CQ*, Aug 1978, p 52.

Ref. 610: Cornelio Nouel - KG5B, "Exploring the Vagaries of Traps," *CQ*, Aug 1984, p 32.

Ref. 611: K. H. Kleine - DL3CI, "Der Verkuerzte Dipol," *CQ-DL*, Jun 1977, p 230.

Ref. 612: Hans Wuertz - DL2FA, "Bis zu Einer S-Stufe mehr auf 80 Meter," *CQ-DL*, Dec 1977, p 475.

Ref. 613: Hans Wuertz - DL2FA, "DX-Antennen mit spiegelenden Flaechen," *CQ-DL*, Aug 1979, p 353.

Ref. 614: Hans Wuertz - DL2FA, "DX-Antennen mit spiegelenden Flaechen," *CQ-DL*, Jan 1980, p 18.

Ref. 615: Willi Nitschke - DJ5DW - et al, "Richtungskarakteristik Fusspunktwiederstand etc von Einelementantennen," *CQ-DL*, Nov 1982, p 535.

Ref. 616: Guenter Schwarzbeck - DL1BU, "Bedeuting des Vertikalen Abstralwinkels von KW-Antennen," *CQ-DL*, Mar 1985, p 130.

Ref. 617: Guenter Schwarzbeck - DL1BU, "Bedeutung des Vertikalen Abstrahlwinkels von KW-antennas (part 2)," *CQ-DL*, Apr 1985, p 184.

Ref. 618: E. Vogelsang - DJ2IM, "Vertikaldiagramme typische Kurzwellenantenne," *CQ-DL*, Jun 1985, p 300.

Ref. 619: John Schultz - W2EEY, "Stub-Switched Vertical Antennas," *Ham Radio*, Jul 1969, p 50.

Ref. 620: Malcolm P. Keown - W5RUB, "Simple Antennas for 80 and 40 Meters," *Ham Radio*, Dec 1972, p 16.

Ref. 621: Earl Whyman - W2HB, "Standing-Wave Ratios," *Ham Radio*, Jul 1973, p 26.

Ref. 622: Robert Baird - W7CSD, "Nonresonant Antenna Impedance Measurements," *Ham Radio*, Apr 1974, p 46.

Ref. 623: H. Glenn Bogel - WA9RQY, "Vertical Radiation Patterns," *Ham Radio*, May 1974, p 58.

Ref. 624: Robert Leo - W7LR, "Optimum Height for Horizontal Antennas," *Ham Radio*, Jun 1974, p 40.

Ref. 625: Bob Fitz - K4JC, "High Performance 80-Meter Antenna," *Ham Radio*, May 1977, p 56.

Ref. 626: Everett S. Brown - K4EF, "New Multiband Longwire Antenna Design," *Ham Radio*, May 1977, p 10.

Ref. 627: William A. Wildenhein - W8YFB, "Solution to the Low-Band Antenna Problem," *Ham Radio*, Jan 1978, p 46.

Ref. 628: John Becker - K9MM, "Lightning Protection," *Ham Radio*, Dec 1978, p 18.

Ref. 629: James Lawson - W2PV, "Part V *Yagi Antenna Design*: Ground or Earth Effects," *Ham Radio*, Oct 1980, p 29.

Ref. 630: Henry G. Elwell - N4UH, "Antenna Geometry for Optimum Performance," *Ham Radio*, May 1982, p 60.

Ref. 631: Randy Rhea - N4HI, "Dipole Antenna over Sloping Ground," *Ham Radio*, May 1982, p 18.

Ref. 632: Bradley Wells - KR7L, "Lightning and Electrical Transient Protection," *Ham Radio*, Dec 1983, p 73.

Ref. 633: R. C. Marshall - G3SBA, "An End-Fed Multiband 8JK," *Ham Radio*, May 1984, p 81.

Ref. 634: David Atkins - W6VX, "Capacitively Loaded High-Performance Dipole," *Ham Radio*, May 1984, p 33.

Ref. 635: David Courtier-Dutton - G3FPQ, "Some Notes on a 7-MHz Linear Loaded Quad," *QST*, Feb 1972, p 14.

Ref. 636: John Kaufmann - WA1CQW - et al, "A Convenient Stub-Tuning System for Quad Antennas," *QST*, May 1975, p 18.

Ref. 637: Hardy Lankskov - W7KAR, "Pattern Factors for Elevated Horizontal Antennas Over Real Earth," *QST*, Nov 1975, p 19.

Ref. 638: Robert Dome - W2WAM, "Impedance of Short Horizontal Dipoles," *QST*, Jan 1976, p 32.

Ref. 639: Donald Belcher - WA4JVE - et al, "Loops vs Dipole Analysis and Discussion," *QST*, Aug 1976, p 34.

Ref. 640: Roger Sparks - W7WKB, "Build this C-T Quad Beam for Reduced Size," *QST*, Apr 1977, p 29.

Ref. 641: Ronald K. Gorski - W9KYZ, "Efficient Short Radiators," *QST*, Apr 1977, p 37.

Ref. 642: Byron Goodman - W1DX, "My Feed Line Tunes My Antenna," *QST*, Apr 1977, p 40.

Ref. 643: Doug DeMaw - W1FB, "The Gentlemen's Band: 160 Meters," *QST*, Oct 1977, p 33.

Ref. 644: David S. Hollander - N7RK, "A Big Signal from a Small Lot," *QST*, Apr 1979, p 32.

Ref. 645: Dana Atchley - W1CF, "Putting the Quarter Wave Sloper to Work on 160," *QST*, Jul 1979, p 19.

Ref. 646: Stan Gibilisco - W1GV, "The Imperfect Antenna System and How it Works," *QST*, Jul 1979, p 24.

Ref. 647: John Belrose - VE2CV, "The Half Sloper," *QST*, May 1980, p 31.

Ref. 648: Larry May - KE6H, "Antenna Modeling Program for the TRS-80," *QST*, Feb 1981, p 15.

Ref. 649: Colin Dickman - ZS6U, "The ZS6U Minishack Special," *QST*, Apr 1981, p 32.

Ref. 650: Doug DeMaw - W1FB, "More Thoughts on the 'Confounded' Half Sloper," *QST*, Oct 1981, p 31.

Ref. 651: John S. Belrose - VE2CV, "The Effect of Supporting Structures on Simple Wire Antennas," *QST*, Dec 1982, p 32.

Ref. 652: Gerald Hall - K1TD, "A Simple Approach to Antenna Impedances," *QST*, Mar 1983, p 16.

Ref. 653: Jerry Hall - K1TD, "The Search for a Simple Broadband 80-Meter Dipole," *QST*, Apr 1983, p 22.

Ref. 654: Charles L. Hutchinson - K8CH, "Getting the Most out of Your Antenna," *QST*, Jul 1983, p 34.

Ref. 655: Doug DeMaw - W1FB, "Building and Using 30 Meter Antennas," *QST*, Oct 1983, p 27.

Ref. 656: James Rautio - AJ3K, "The Effects of Real Ground on Antennas - Part 1," *QST*, Feb 1984, p 15.

Ref. 657: James Rautio - AJ3K, "The Effects of Real Ground on Antennas - Part 2," *QST*, Apr 1984, p 34.

Ref. 658: James Rautio - AJ3K, "The Effect of Real Ground on Antennas - Part 3," *QST*, Jun 1984, p 30.

Ref. 659: Doug DeMaw - W1FB, "Trap for Shunt-Fed Towers," *QST*, Jun 1984, p 40.

Ref. 660: James Rautio - AJ3K, "The Effects of Real Ground on Antennas - Part 4," *QST*, Aug 1984, p 31.

Ref. 661: James Rautio - AJ3K, "The Effects of Real Ground on Antennas - Part 5," *QST*, Nov 1984, p 35.

Ref. 662: Robert C. Sommer - N4UU, "Optimizing Coaxial-Cable Traps," *QST*, Dec 1984, p 37.

Ref. 663: Bob Schetgen - KU7G, "Technical Correspondence" *QST*, Apr 1985, p 51.

Ref. 664: Pat Hawker - G3VA, "Technical Topics: Low Angle Operation," *Radio Communication*, Apr 1971, p 262.

Ref. 665: Pat Hawker - G3VA, "Technical Topics: Low-Angle Radiation and Sloping-Ground Sites," *Radio Communication*, May 1972, p 306.

Ref. 666: Pat Hawker - G3VA, "Technical Topics: All Band Terminated Long-Wire," *Radio Communication*, Nov 1972, p 745.

Ref. 667: A.P.A. Ashton - G3XAP, "160M DX from Suburban Sites," *Radio Communication*, Dec 1973, p 842.

Ref. 668: L. A. Moxon - G6XN, "Gains and Losses in HF Aerials - Part 1," *Radio Communication*, Dec 1973, p 834.

Ref. 669: A. Moxon - G6XN, "Gains and Losses in HF Aerials - Part 2," *Radio Communication*, Jan 1974, p 16.

Ref. 670: Pat Hawker - G3VA, "Technical Topics: Thoughts on Inverted-Vs," *Radio Communication*, Sep 1976, p 676.

Ref. 671: A. P. A. Ashton - G3XAP, "The G3XAP Directional Antenna for the Lower Frequencies," *Radio Communication*, Nov 1977, p 858.

Ref. 672: S. J. M. Whitfield - G3IMW, "3.5 MHz DX Antennas for a Town Garden," *Radio Communication*, Aug 1980, p 772.

Ref. 673: Pat Hawker - G3VA, "Technical Topics: Low Profile 1.8 and 3.5 MHz Antennas," *Radio Communication*, Aug 1980, p 792.

Ref. 674: Pat Hawker - G3VA, "Technical Topics: Half Delta Loop - Sloping One-Mast Yagi," *Radio Communication*, Oct 1983, p 892.

Ref. 675: R. Rosen - K2RR, "Secrets of Successful Low Band Operation - Part 1," *Ham Radio,* May 1986, p 16.

Ref. 676: R. Rosen - K2RR, "Secrets of Successful Low Band Operation - Part 2," *Ham Radio,* Jun 1986.

Ref. 677: J. Dietrich - WA0RDX, "Loops and Dipoles: A Comparative Analysis," *QST*, Sep 1985, p 24.

Ref. 678: Roy Lewallen - W7EL, *MININEC*: The Other Edge of the Sword," *QST*, Feb 1991, p 18.

Ref. 679: Rich Rosen - K2RR, "Secrets of Successful Low Band Operation," *Ham Radio,* May 1986, p 16

Ref. 680: Yardley Beers - W0JF, "Designing Trap Antennas: a New Approach," *Ham Radio,* Aug 1987, p 60.

Ref. 681: Guenter Schwarzbeck - DL1BU, "Die Antennw und ihre Umgebung," *CQ-DL*, Jan 1988, p 5.

Ref. 682: Maurice C. Hately - GM3HAT, "A No-compromise Multiband Low SWR Dipole," *Ham Radio,* Mar 1987, p 69.

Ref. 683: R. P. Haviland - W4MB, "Design Data for Pipe Masts," *Ham Radio,* Jul 1989, p 38.

Ref. 684: Gary E. O'Neil - N3GO, "Trapping the Mysteries of Trapped Antennas," *Ham Radio,* Oct 1981, p 10.

Ref. 685: J. Belrose VE2CV and P. Bouliane - VE3KLO, "The Off-center-fed Dipole Revisited," *QST*, Aug 1990, p 28.

Ref. 686: John J. Reh - K7KGP, "An Extended Double Zepp Antenna for 12 Meters," *QST*, Dec 1987, p 25.

Ref. 687: James W. Healy - NJ2L, "Feeding Dipole Antennas," *QST*, Jul 1991, p 22.

Ref. 688: Bill Orr, W6SAI, "Antenna Gain," *Ham Radio*, Jan 1990, p 30.

Ref. 689: B. H. Johns - W3JIP, "Coaxial Cable Antenna Traps," *QST*, May 1981, p 15.

Ref. 690: Doug DeMaw, W1FB, "The Paragon Technology NEC-Win Antenna Analysis Software" *CQ* mag, Nov 1996, p 28.

Ref. 691: N. Mullani, K0NM, "The Bent Dipole," *QST*, May 1997, p 56.

Ref. 692: Larry East, W1HUE, "Antenna Trap Design using a Home Computer," *ARRL Antenna Compendium Volume 2*, p 100 (ISBN 0-87259-254-5).

Ref. 693: Moxon, G6XN, *HF Antennas for All Locations*, published by the RSGB,

■ 7. VERTICAL ANTENNAS

Ref. 701: Paul Lee - K6TS, *Vertical Antenna Handbook*, CQ Publishing Inc.

Ref. 702: Wait and Pope, "Input Resistance of LF Unipole Aerials," *Wireless Engineer*, May 1955, p 131.

Ref. 703: J. J. Wiseman, "How Long is a Piece of Wire," *Electronics and Wireless World*, Apr 1985, p 24.

Ref. 704: Carl C. Drumeller - W5JJ, "Using Your Tower as an Antenna," *CQ*, Dec 1977, p 75.

Ref. 705: Karl T. Thurber - W8FX, "HF Verticals - Plain And Simple," *CQ*, Sep 1980, p 22.

Ref. 706: John E. Magnusson - WØAGD, "Improving Antenna Performance," *CQ*, Jun 1981, p 32.

Ref. 707: Larry Strain - N7DF, "A 3.5 to 30 MHz Discage Antenna," *CQ*, Apr 1984, p 18.

Ref. 708: Karl Hille - DL1VU, "Optimierte T-Antenne," *CQ-DL*, Jun 1978, p 246.

Ref. 709: Rolf Schick - DL3AO, "Loop - Dipol und Vertikalantennen - Vergleiche und Erfahrungen," *CQ-DL*, Mar 1979, p 115.

Ref. 710: Guenter Schwarzbeck - DL1BU, "DX Antennen fuer 80 und 160 Meter," *CQ-DL*, Apr 1979, p 150.

Ref. 711: Hans Wurtz - DL2FA, "DX-Antennen mit spiegelenden Flaechen," *CQ-DL*, Aug 1979, p 353.

Ref. 712: Hans Wurtz - DL2FA, "DX-Antennen mit spiegelenden Flaechen," *CQ-DL*, Sep 1979, p 400.

Ref. 713: Hans Wurtz - DL2FA, "DX-Antennen mit spiegelenden Flaechen," *CQ-DL*, Jan 1980, p 18.

Ref. 714: Hans Wurtz - DL2FA, "DX-Antennen mit spiegelenden Flaechen," *CQ-DL*, Jun 1980, p 272.

Ref. 715: Hans Wurtz - DL2FA, "DX-Antennen mit spiegelenden Flaechen," *CQ-DL*, Jul 1980, p 311.

Ref. 716: Hans Wurtz - DL2FA, "DX-Antennen mit spiegelenden Flaechen," *CQ-DL*, Feb 1981, p 61.

Ref. 717: Hans Wurtz - DL2FA, "DX-Antennen mit spiegelenden Flaechen," *CQ-DL*, Jul 1981, p 330.

Ref. 718: Guenter Schwarzbeck - DL1BU, "Groundplane und Vertikalantennenf," *CQ-DL*, Sep 1981, p 420.

Ref. 719: Hans Wurtz - DL2FA, "DX-Antennen mit spiegelenden Flaechen," *CQ-DL*, Apr 1983, p 170.

Ref. 720: Hans Wurtz - DL2FA, "DX-Antennen mit spiegelenden Flaechen," *CQ-DL*, May 1983, p 224.

Ref. 721: Hans Wurtz - DL2FA, "Antennen mit spiegelenden Flaechen," *CQ-DL*, Jun 1983, p 278.

Ref. 722: Hans Adolf Rohrbacher - DJ2NN, "Basic Programm zu Berechnungh von Vertikalen Antennen," *CQ-DL*, Jun 1983, p 275.

Ref. 723: Hans Wurtz - DL2FA, "DX-Antennen mit spiegelenden Flaechen," *CQ-DL*, Jul 1983, p 326.

Ref. 724: John Schultz - W2EEY, "Stub-Switched Vertical Antennas," *Ham Radio*, Jul 1969, p 50.

Ref. 725: John True - W4OQ, "The Vertical Radiator," *Ham Radio*, Apr 1973, p 16.

Ref. 726: John True - W4OQ, "Vertical-Tower Antenna System," *Ham Radio*, May 1973, p 56.

Ref. 727: George Smith - W4AEO, "80- and 40-Meter Log Periodic Antennas," *Ham Radio*, Sep 1973, p 44.

Ref. 728: Robert Leo - W7LR, "Vertical Antenna Characteristics," *Ham Radio*, Mar 1974, p 34.

Ref. 729: Robert Leo - W7LR, "Vertical Antenna Radiation Patterns," *Ham Radio*, Apr 1974, p 50.

Ref. 730: Raymond Griese - K6FD, "Improving Vertical Antennas," *Ham Radio*, Dec 1974, p 54.

Ref. 731: Harry Hyder - W7IV, "Large Vertical Antennas," *Ham Radio*, May 1975, p 8.

Ref. 732: John True - W4OQ, "Shunt-Fed Vertical Antennas," *Ham Radio*, May 1975, p 34.

Ref. 733: H. H. Hunter - W8TYX, "Short Vertical for 7 MHz," *Ham Radio*, Jun 1977, p 60.

Ref. 734: Laidacker M. Seaberg - WØNCU, "Multiband Vertical Antenna System," *Ham Radio*, May 1978, p 28.

Ref. 735: Joseph D. Liga - K2INA, "80-Meter Ground Plane Antennas," *Ham Radio*, May 1978, p 48.

Ref. 736: John M. Haerle - WB5IIR, "Folded Umbrella Antenna," *Ham Radio*, May 1979, p 38.

Ref. 737: Paul A. Scholz - W6PYK, "Vertical Antenna for 40 and 75 Meters," *Ham Radio*, Sep 1979, p 44.

Ref. 738: Ed Marriner - W6XM, "Base-Loaded Vertical for 160 Meters," *Ham Radio*, Aug 1980, p 64.

Ref. 739: John S. Belrose - VE2CV, "The Half-Wave Vertical," *Ham Radio*, Sep 1981, p 36.

Ref. 740: Stan Gibilisco - W1GV/4, "Efficiency of Short Antennas," *Ham Radio*, Sep 1982, p 18.

Ref. 741: John S. Belrose - VE2CV, "Top-Loaded Folded Umbrella Vertical Antenna," *Ham Radio*, Sep 1982, p 12.

Ref. 742: W. J. Byron - W7DHD, "Short Vertical Antennas for the Low Bands - Part 1," *Ham Radio*, May 1983, p 36.

Ref. 743: Forrest Gehrke - K2BT, "Vertical Phased Arrays - Part 1," *Ham Radio*, May 1983, p 18.

Ref. 744: John Belrose - VE2CV, "The Grounded Monopole with Elevated Feed," *Ham Radio*, May 1983, p 87.

Ref. 745: Forrest Gehrke - K2BT, "Vertical Phased Arrays - Part 2," *Ham Radio*, Jun 1983, p 24.

Ref. 746: W. J. Byron - W7DHD, "Short Vertical Antennas for the Low Bands - Part 2," *Ham Radio*, Jun 1983, p 17.

Ref. 747: Forrest Gehrke - K2BT, "Vertical Phased Arrays - Part 3," *Ham Radio*, Jul 1983, p 26.

Ref. 748: Forrest Gehrke - K2BT, "Vertical Phased Arrays - Part 4," *Ham Radio*, Oct 1983, p 34.

Ref. 749: Forrest Gehrke - K2BT, "Vertical Phased Arrays - Part 5," *Ham Radio*, Dec 1983, p 59.

Ref. 750: Marc Bacon - WB9VWA, "Verticals Over REAL Ground," *Ham Radio*, Jan 1984, p 35.

Ref. 751: Robert Leo - W7LR, "Remote Controlled 40 - 80 - and 160 Meter Vertical," *Ham Radio*, May 1984, p 38.

Ref. 752: Harry Hyder - W7IV, "Build a Simple Wire Plow," *Ham Radio,* May 1984, p 107.

Ref. 753: Gene Hubbell - W9ERU, "Feeding Grounded Towers as Radiators," *QST*, Jun 1960, p 33.

Ref. 754: Eugene E. Baldwin - WØRUG, "Some Notes on the Care and Feeding of Grounded Verticals," *QST*, Oct 1963, p 45.

Ref. 755: N. H. Davidson - K5JVF, "Flagpole Without a Flag," *QST*, Nov 1964, p 36.

Ref. 756: Jerry Sevick - W2FMI, "The Ground-Image Vertical Antenna," *QST*, Jul 1971, p 16.

Ref. 757: Jerry Sevick - W2FMI, "The W2FMI 20-Meter Vertical Beam," *QST*, Jun 1972, p 14.

Ref. 758: Jerry Sevick - W2FMI, "The W2FMI Ground Mounted Short Vertical Antenna," *QST*, Mar 1973, p 13.

Ref. 759: Jerry Sevick - W2FMI, "A High Performance 20, 40 and 80 Meter Vertical System," *QST*, Dec 1973, p 30.

Ref. 760: Jerry Sevick - W2FMI, "The Constant Impedance Trap Vertical," *QST*, Mar 1974, p 29.

Ref. 761: Barry A. Boothe - W9UCW, "The Minooka Special," *QST*, Dec 1974, p 15.

Ref. 762: Willi Richartz - HB9ADQ, "A Stacked Multiband Vertical for 80-10 Meters," *QST*, Feb 1975, p 44.

Ref. 763: John S. Belrose - VE2CV, "The HF Discone Antenna," *QST*, Jul 1975, p 11.

Ref. 764: Earl Cunningham - W5RTQ, "Shunt Feeding Towers for Operating on the Low Amateur Frequencies," *QST*, Oct 1975, p 22.

Ref. 765: Dennis Kozakoff - W4AZW, "Designing Small Vertical Antennas," *QST*, Aug 1976, p 24.

Ref. 766: Ronald Gorski - W9KYZ, "Efficient Short Radiators," *QST*, Apr 1977, p 37.

Ref. 767: Richard Lodwig - W2KK, "The Inverted-L Antenna," *QST*, Apr 1977, p 32.

Ref. 768: Asa Collins - K6VV, "A Multiband Vertical Radiator," *QST*, Apr 1977, p 22.

Ref. 769: Walter Schultz - K3OQF, "Slant-Wire Feed for Grounded Towers," *QST*, May 1977, p 23.

Ref. 770: Yardley Beers - WØJF, "Optimizing Vertical Antenna Performance," *QST*, Oct 1977, p 15.

Ref. 771: Walter Schultz - K3OQF, "Designing a Vertical Antenna," *QST*, Sep 1978, p 19.

Ref. 772: John S. Belrose - VE2CV, "A Kite Supported 160M (or 80M) Antenna for Portable Application," *QST*, Mar 1981, p 40.

Ref. 773: Wayne Sandford - K3EQ, "A Modest 45 Foot Tall DX Vertical for 160 - 80 - 40 and 30 Meters," *QST*, Sep 1981, p 27.

Ref. 774: Doug DeMaw - W1FB, "Shunt Fed Towers - Some Practical Aspects," *QST*, Oct 1982, p 21.

Ref. 775: John F. Lindholm - W1XX, "The Inverted L Revisited," *QST*, Jan 1983, p 20.

Ref. 776: Carl Eichenauer - W2QIP, "A Top Fed Vertical Antenna for 1.8 MHz. Plus 3," *QST*, Sep 1983, p 25.

Ref. 777: Robert Snyder - KE2S, "Modified Butternut Vertical for 80-Meter Operation," *QST*, Apr 1985, p 50.

Ref. 778: Doug DeMaw - W1FB, "A Remotely Switched Inverted-L Antenna," *QST*, May 1985, p 37.

Ref. 779: Pat Hawker - G3VA, "Technical Topics: Low Angle Operation," *Radio Communication*, Apr 1971, p 262.

Ref. 780: Pat Hawker - G3VA, "Technical Topics: Improving the T Antenna," *Radio Communication*, Sep 1978, p 770.

Ref. 781: J. Bazley - G3HCT, "A 7 MHz Vertical Antenna," *Radio Communication*, Jan 1979, p 26.

Ref. 782: P. J. Horwood - G3FRB, "Feed Impedance of Loaded $\lambda/4$ Vertical Antennas and the Effects of Earth Systems," *Radio Communication*, Oct 1981, p 911.

Ref. 783: Pat Hawker - G3VA, "Technical Topics: The Inverted Groundplane Family," *Radio Communication*, May 1983, p 424.

Ref. 784: Pat Hawker - G3VA, "Technical Topics: More on Groundplanes," *Radio Communication*, Sep 1983, p 798.

Ref. 785: Pat Hawker - G3VA, "Technical Topics: Sloping One-Mast Yagi," *Radio Communication*, Oct 1983, p 892.

Ref. 786: V. C. Lear - G3TKN, "Gamma Matching Towers and Masts at Lower Frequencies," *Radio Communication*, Mar 1986, p 176.

Ref. 787: B. Wermager - KØEOU, "A Truly Broadband Antenna for 80/75 Meters," *QST*, Apr 1986, p 23.

Ref. 788: Pat Hawker - G3VA, "Technical Topics: Elements of Non-Uniform Cross Section," *Radio Communication*, Jan 1986, p 36.

Ref. 789: Andy Bourassa - WA1LJJ, "Build a Top-hat Vertical Antenna for 80/75 Meters," *CQ*, Aug 1990, p 18.

Ref. 790: Carl C. Drumeller - W5JJ, "Using Your Tower as an Antenna," *CQ*, Dec 1977, p 75.

Ref. 791: Carl Huether - KM1H, "Build a High-performance Extended Bandwidth 160 Meter Vertical," *CQ*, Dec 1986, p 38.

Ref. 792: John Belrose - VE2CV, "More on the Half Sloper," *QST*, Feb 1991, p 39.

Ref. 793: Pat Hawker - G3VA, "The Folded Dipole and Monopole," *Radio Communication*, Jul 1987, p 496.

Ref. 794: C. J. Michaels - W7XC, "Evolution of the Short Top-loaded Vertical," *QST*, Mar 1990, p 26.

Ref. 795: Walter J. Schulz - K3OQF, "Calculating the Input Impedance of a Tapered Vertical," *Ham Radio,* Aug 1985, p 24.

Ref. 796: C. J. Michaels - W7XC, "Some Ref.lections on Vertical Antennas," *QST*, Jul 1987, p 15.

Ref. 797: C.J. Michaels - W7XC, "Loading Coils for 160 Meter Antennas," *QST*, Apr 1990, p 28.

Ref. 798: D. DeMaw - W1FB, "The 160-meter Antenna Dilemma," *QST*, Nov 1990, p 30.

Ref. 799: Al Christman - KB8I, "Elevated Vertical Antenna Systems Q&A," *QST*, May 1989, p 50.

Ref. 7991: H. Hille - DL1VU, "Vortschritte in der Entwicklung von Vertikalantennen," *CQ-DL*, Oct 1989, p 631.

Ref. 7992: Doug DeMaw - W1FB, "Trap for Shunt-fed Towers," *QST*, Jun 1984, p 40.

Ref. 7993: W. J. Byron - W7DHD, "Ground Mounted Vertical Antennas," *Ham Radio,* Jun 1990, p 11.

Ref. 7994: Dennis N. Monticelli - AE6C, "A Simple Effective Dual-Band Inverted-L Antenna," *QST*, Jul 1991, p 38.

Ref. 7995: Walter J. Schultz - K3OQF, "The Folded Wire Fed Top Loaded Grounded Vertical," *Ham Radio,* May 1989, p 32.

Ref. 7996: Gary Nichols - KD9SV and Lynn Gerig - WA9FGR, "Low Band Verticals and How to Feed Them," *CQ,* Aug 1990, p 46.

Ref. 7997: "The phase and magnitude of earth currents near transmitting antennas," *Proceedings IRE*, Vol 23, no. 2, Feb 1935, p 168.

Ref. 7998: Bruce Clark, KO1F, "How to Mount a Tower to use as a Low-Band Vertical Antenna"; *CQ* mag, Nov 1994, p 52.

Ref. 7999: Carl Moreschi, N4PY, "A DX Antenna for 160, 80, 40 and 30 Meters"; *CQ* mag, Apr 1995, p 36.

Ref. 7811: Paul Carr, N4PC, "A Short Two Band Vertical for 160 and 80 Meters"; *CQ* mag, Apr 1997, p 20.

Ref. 7812: W1XT "Gladiator Antennas - To Heard Island," *The Low Band Monitor*, Oct 1996.

Ref. 7813; N4KG, "Elevated Radial 80-Meter Reverse Feed," *The Low Band Monitor*, Sep 1996.

Ref. 7814 ; K3ND "The K3ND Low-Band Vertical," *The Low Band Monitor*, Jul 1996.

Ref. 7815: W0CD "The Battle Creek Special," *The Low Band Monitor*, Sep 1993.

Ref. 7816: K1VW "The S92SS 160 M Antenna"; *The Low Band Monitor*, Mar 1997.

Ref. 7817: K0JN, "The UNI-HAT CTSVR Antenna," *The Low Band Monitor*, Jan 1996.

Ref. 7818: Gerd Janzen, "Kuze Antennen," published by Franckh'sche Verlaghandlun, Stuttgart, Germany, ISBN 3-440-05469-1 (This book is only available directly from the author, DF6SJ, Hochvogelstrasse 29, D087435 Kempten, Germany).

Ref. 7819: Guy Hamblen, AA7ZQ/2, "A 75/80 Meter Full-size 1/4-λ Vertical," *ARRL Antenna Compendium, Volume 5* (ISBN 0-87259-562-5).

Ref. 7820: Al Christman, KB8I, "Elevated Vertical Antennas for the Low Bnds, Varying the Height and Number of Radials," *ARRL Antenna Compendium, Volume 5* (ISBN 0-87259-562-5).

Ref. 7821: John Belrose, VE2CV, "Elevated Radial Wire System for Vertically Polarized Ground-Plane Type Antennas, Part 1" *Communications Quarterly*, Winter 1998, p 29.

Ref. 7822: Dick Weber, K5IU, "Optimal Elevated Rdaial Vertical Antennas," *Communications Quarterly*, Spring 1997, p 20.

Ref. 7823: Dick Weber, K5IU, "Comments on Belrose's Article in *Comm.Q*. Winter 1998," *Communications Quarterly*, Spring '98, p 5.

Ref. 7824: John Belrose, VE2CV, "Elevated Radial Wire Systems for Vertically Polarized Ground-Plane Type Antennas, Part 2," *Communications Quarterly*, Spring 1998, p 45.

Ref. 7825: Al Christman, KB8I, "Elevated Vertical Antennas for the Low Bands: Varying the Height and Number of Radials," *ARRL Antenna Compendium, Volume 5*, ISBN 0-87259-562-5.

Ref. 7826: Carl. J. Moreschi, N4PY, "A DX Antenna for 160, 80, 40 and 30 Meters," *CQ* mag, Apr 1995, p 36.

Ref. 7827: Bill Orr, W6SAI, "The S92SS Limited Space Antenna for 160 Meters," Radio Fundamentals, *CQ* mag, Aug 1997, p 85.

Ref. 7828: Paul Carr, N4PC, "A Short, Two-Band Vertical for 160 and 80 Meters," *CQ* mag, Apr 1987, p 20.

Ref. 7829: Gale Stewart, K3ND, "The K3ND Low-Band Vertical (160/80)," *The Low Band Monitor*, Jul 1996, p 2.

Ref. 7830: "The S92SS 160 m vertical with 4 short, tuned radials," *The Low Band Monitor*, Mar 1997

Ref. 7831: Dave Bowker, W0RJU, "Low-Band Tri-Bander (Shunt-Fed Tower System)," *The DX Magazine*, Jan/Feb 1994, p 9.

Ref. 7832: Thomas Russell, N4KG, "Simple, Effective, Elevated Ground Plane Antenna," *QST*, Jun 1994, p 45.

Ref. 7833: L. Moxon, G6XN, "Ground Planes, Radial Systems and Asymmetric Dipoles" *ARRL Antenna Compendium Vol 3*, p 19 (ISBN 0-87259-401-7).

Ref. 7834: "Using elevated radials with ground-mounted towers" *IEEE Transactions on Broadcasting*, vol. 37. no. 3, Sep 1991, pp 77-82.

Ref. 7835: J. Devoldere, ON4UN "Radials made Clear," *CQ Contest*, Sep 1996, p 6-9.

■ 8. ANTENNAS: GROUND SYSTEMS

Ref. 800: Jager, "Effect of Earth's Surface on Antenna Patterns in the Short Wave Range," *Int. Elek. Rundshau*, Aug 1970, p 101.

Ref. 801: G. H. Brown - et al, "Ground Systems as a Factor in Antenna Efficiency," *Proceedings IRE*, Jun 1937, p 753.

Ref. 802: Abbott, "Design of Optimum Buried RF Ground Systems," *Proceedings IRE*, Jul 1952, p 846.

Ref. 803: Monteath, "The Effect of Ground Constants of an Earth System on Vertical Aerials," *Proceedings IRE*, Jan 1958, p 292.

Ref. 804: Caid, "Earth Resistivity and Geological Structure," *Electrical Engineering*, Nov 1935, p 1153.

Ref. 805: "Calculated Pattern of a Vertical Antenna With a Finite Radial-Wire Ground System," *Radio Science*, Jan 1973, p 81.

Ref. 806: John E. Magnusson - W0AGD, "Improving Antenna Performance," *CQ*, Jun 1981, p 32.

Ref. 807: "Improving Vertical Antenna Efficiency: A Study of Radial Wire Ground Systems," *CQ*, Apr 1984, p 24.

Ref. 808: Robert Leo - W7LR, "Vertical Antenna Ground System," *Ham Radio,* May 1974, p 30.

Ref. 809: Robert Sherwood - WBØJGP, "Ground Screen - Alternative to Radials," *Ham Radio,* May 1977, p 22.

Ref. 810: Alan M. Christman - WD8CBJ, "Ground Systems for Vertical Antennas," *Ham Radio,* Aug 1979, p 31.

Ref. 811: H. Vance Mosser - K3ZAP, "Installing Radials for Vertical Antennas," *Ham Radio,* Oct 1980, p 56.

Ref. 813: Bradley Wells - KR7L, "Installing Effective Ground Systems," *Ham Radio,* Sep 1983, p 67.

Ref. 814: Marc Bacon - WB9VWA, "Verticals Over REAL Ground," *Ham Radio,* Jan 1984, p 35.

Ref. 815: Harry Hyder - W7IV, "Build a Simple Wire Plow," *Ham Radio,* May 1984, p 107.

Ref. 816: John Stanley - K4ERO/HC1, "Optimum Ground System for Vertical Antennas," *QST,* Dec 1976, p 13.

Ref. 817: Roger Hoestenbach - W5EGS, "Improving Earth Ground Characteristics," *QST,* Dec 1976, p 16.

Ref. 818: Jerry Sevick - W2FMI, "Short Ground Radial Systems for Short Verticals," *QST,* Apr 1978, p 30.

Ref. 819: Jerry Sevick - W2FMI, "Measuring Soil Conductivity," *QST,* Mar 1981, p 38.

Ref. 820: Archibald C. Doty - K8CFU - et al, "Efficient Ground Systems for Vertical Antennas," *QST,* Feb 1983, p 20.

Ref. 821: Brian Edward - N2MF, "Radial Systems for Ground-Mounted Vertical Antennas," *QST,* Jun 1985, p 28.

Ref. 822: Pat Hawker - G3VA, "Technical Topics: Vertical Polarization and Large Earth Screens," *Radio Communication,* Dec 1978, p 1023.

Ref. 823: P. J. Horwood - G3FRB, "Feed Impedance of Loaded ¼ Wave Vertical Antenna Systems," *Radio Communication,* Oct 1981, p 911.

Ref. 824: J. A. Frey - W3ESU, "The Minipoise," *CQ,* Aug 1985, p 30.

Ref. 825: Al Christman - KB8I, "Elevated Vertical Antenna Systems," *QST,* Aug 1988, p 35.

Ref. 826 : Bill Orr, W6SAI "The S92SS Limited Space Antenna for 160 Meters"; *CQ* mag, Aug 1997, p 85.

Ref. 827: KØCS, "Elevated Radials, Try Two," *Low Band Monitor,* Feb 96.

Ref. 828: KØCS, "Real World Elevated Radials," *Low Band Monitor,* Jul 1994.

9. ANTENNA ARRAYS

Ref. 900: Bill Guimont - W7KW, "Liftoff on 80 Meters," *CQ,* Oct 1979, p 38.

Ref. 901: Guenter Schwarzbeck - DL1BU, "HB9CV Antenna," *CQ-DL,* Jan 1983, p 10.

Ref. 902: Rudolf Fisher - DL6WD, "Das Monster - eine 2 Element Delta-Loop fuer 3.5 MHz," *CQ-DL,* Jul 1983, p 331.

Ref. 903: G. E. Smith - W4AEO, "Log-Periodic Antennas for 40 Meters," *Ham Radio,* May 1973, p 16.

Ref. 904: G. E. Smith - W4AEO, "80- and 40-Meter Log-Periodic Antennas," *Ham Radio,* Sep 1973, p 44.

Ref. 905: G. E. Smith - W4AEO, "Log-Periodic Antenna Design," *Ham Radio,* May 1975, p 14.

Ref. 906: Jerry Swank - W8HXR, "Phased Vertical Array," *Ham Radio,* May 1975, p 24.

Ref. 907: Henry Keen - W5TRS, "Electrically-Steered Phased Array," *Ham Radio,* May 1975, p 52.

Ref. 908: Gary Jordan - WA6TKT, "Understanding the ZL Special Antenna," *Ham Radio,* May 1976, p 38.

Ref. 909: William Tucker - W4FXE, "Fine Tuning the Phased Vertical Array," *Ham Radio,* May 1977, p 46.

Ref. 910: Paul Kiesel - K7CW, "Seven-Element 40-Meter Quad," *Ham Radio,* Aug 1978, p 30.

Ref. 911: Eugene B. Fuller - W2LU, "Sloping 80-meter Array," *Ham Radio,* May 1979, p 70.

Ref. 912: Harold F. Tolles - W7ITB, "Scaling Antenna Elements," *Ham Radio,* Jul 1979, p 58.

Ref. 913: P. A. Scholz - W6PYK - et al, "Log-Periodic Antenna Design," *Ham Radio,* Dec 1979, p 34.

Ref. 914: James Lawson - W2PV, "Yagi Antenna Design: Experiments Confirm Computer Analysis," *Ham Radio,* Feb 1980, p 19.

Ref. 915: James Lawson - W2PV, "Yagi Antenna Design: Multi-Element Simplistic Beams," *Ham Radio,* Jun 1980, p 33.

Ref. 916: Paul Scholtz - W6PYK - et al, "Log Periodic Fixed-Wire Beams for 75-Meter DX," *Ham Radio,* Mar 1980, p 40.

Ref. 917: George E. Smith - W4AEO, "Log Periodic Fixed-Wire Beams for 40 Meters," *Ham Radio,* Apr 1980, p 26.

Ref. 918: Ed Marriner - W6XM, "Phased Vertical Antenna for 21 MHz," *Ham Radio,* Jun 1980, p 42.

Ref. 919: William M. Kelsey - N8ET, "Three-Element Switchable Quad for 40 Meters," *Ham Radio,* Oct 1980, p 26.

Ref. 920: Patrick McGuire - WB5HGR, "Pattern Calculation for Phased Vertical Arrays," *Ham Radio,* May 1981, p 40.

Ref. 921: Forrest Gehrke - K2BT, "Vertical Phased Arrays - Part 1," *Ham Radio,* May 1983, p 18.

Ref. 922: Forrest Gehrke - K2BT, "Vertical Phased Arrays - Part 2," *Ham Radio,* Jun 1983, p 24.

Ref. 923: Forrest Gehrke - K2BT, "Vertical Phased Arrays - Part 3," *Ham Radio,* Jul 1983, p 26.

Ref. 924: Forrest Gehrke - K2BT, "Vertical Phased Arrays - Part 4," *Ham Radio,* Oct 1983, p 34.

Ref. 925: Forrest Gehrke - K2BT, "Vertical Phased Arrays - Part 5," *Ham Radio,* Dec 1983, p 59.

Ref. 926: R. C. Marshall - G3SBA, "An End Fed Multiband 8JK," *Ham Radio,* May 1984, p 81.

Ref. 927: Forrest Gehrke - K2BT, "Vertical Phased Arrays - Part 6," *Ham Radio,* May 1984, p 45.

Ref. 928: R. R. Schellenbach - W1JF, "The End Fed 8JK - A Switchable Vertical Array," *Ham Radio,* May 1985, p 53.

Ref. 929: Al Christman - KB8I, "Feeding Phased Arrays - an Alternate Method," *Ham Radio,* May 1985, p 58.

Ref. 930: Dana Atchley - W1HKK, "A Switchable Four Element 80-Meter Phased Array," *QST,* Mar 1965, p 48.

Ref. 931: James Lawson - W2PV, "A 75/80-Meter Vertical Antenna Square Array," *QST,* Mar 1971, p 18.

Ref. 932: James Lawson - W2PV, "Simple Arrays of Vertical Antenna Elements," *QST,* May 1971, p 22.

Ref. 933: Gary Elliott - KH6HCM/W7UXP, "Phased Verticals for 40 Meters," *QST,* Apr 1972, p 18.

Ref. 934: Jerry Sevick - W2FMI, "The W2FMI 20-meter Vertical Beam," *QST,* Jun 1972, p 14.

Ref. 935: Robert Myers - W1FBY - et al, "Phased Verticals in a 40-Meter Beam Switching Array," *QST,* Aug 1972, p 36.

Ref. 936: Robert Jones - KH6AD, "A 7-MHz Parasitic Array," *QST,* Nov 1973, p 39.

Ref. 937: J. G. Botts - K4EQJ, "A Four-Element Vertical Beam for 40/15 Meters," *QST,* Jun 1975, p 30.

Ref. 938: Jarda Dvoracek - OK1ATP, "160-Meter DX with a Two-Element Beam," *QST,* Oct 1975, p 20.

Ref. 939: Dana Atchley - W1CF - et al, "360 Degree - Steerable Vertical Phased Arrays," *QST,* Apr 1976, p 27.

Ref. 940: Richard Fenwick - K5RR - et al, "Broadband Steerable Phased Arrays," *QST,* Apr 1977, p 18.

Ref. 941: Dana Atchley - W1CF, "Updating Phased Array Technology," *QST,* Aug 1978, p 22.

Ref. 942: Bob Hickman - WB6ZZJ, "The Poly Tower Phased Array," *QST,* Jan 1981, p 30.

Ref. 943: W. B. Bachelor - AC3K, "Combined Vertical Directivity," *QST,* Feb 1981, p 19.

Ref. 944: Walter J. Schultz - K3OQF, "Vertical Array Analysis," *QST,* Feb 1981, p 22.

Ref. 945: Edward Peter Swynar - VE3CUI, "40 Meters with a Phased Delta Loop," *QST,* May 1984, p 20.

Ref. 946: Riki Kline - 4X4NJ, "Build a 4X Array for 160 Meters," *QST,* Feb 1985, p 21.

Ref. 947: Trygve Tondering - OZ1TD, "Phased Verticals," *Radio Communication,* May 1972, p 294.

Ref. 948: Pat Hawker - G3VA, "Technical Topics: The Half Square Aerial," *Radio Communication,* Jun 1974, p 380.

Ref. 949: J. L. Lawson - W2PV, "Yagi Antenna Design," *Ham Radio,* Jan 1980, p 22.

Ref. 950: J. L. Lawson - W2PV, "Yagi Antenna Design," *Ham Radio,* Feb 1980, p 19.

Ref. 951: J. L. Lawson - W2PV, "Yagi Antenna Design," *Ham Radio,* May 1980, p 18.

Ref. 952: J. L. Lawson - W2PV, "Yagi Antenna Design," *Ham Radio,* Jun 1980, p 33.

Ref. 953: J. L. Lawson - W2PV, "Yagi Antenna Design," *Ham Radio,* Jul 1980, p 18.

Ref. 953: J. L. Lawson - W2PV, "Yagi Antenna Design," *Ham Radio,* Sep 1980, p 37.

Ref. 954: J. L. Lawson - W2PV, "Yagi Antenna Design," *Ham Radio,* Oct 1980, p 29.

Ref. 955: J. L. Lawson - W2PV, "Yagi Antenna Design," *Ham Radio,* Nov 1980, p 22.

Ref. 956: J. L. Lawson - W2PV, "Yagi Antenna Design," *Ham Radio,* Dec 1980, p 30.

Ref. 957: J. L. Lawson - W2PV, *Yagi Antenna Design,* ARRL, Dec 1986

Ref. 958: Dick Weber - K5IU, "Determination of Yagi Wind Loads Using the Cross-flow Principle," *Communications Quarterly,* Apr 1993.

Ref. 959: Roy Lewallen - W7EL, "The Impact of Current Distribution on Array Patterns," *QST,* Jul 1990, p 39.

Ref. 960: Robert H. Mitchell - N5RM, "The Forty Meter Flame Thrower," *CQ,* Dec 1987, p 36.

Ref. 961: Ralph Fowler - N6YC, "The W8JK Antenna," *Ham Radio,* May 1988, p 9.

Ref. 962: Jurgen A Weigl - OE5CWL, "A Shortened 40-meter Four-element Sloping Dipole Array," *Ham Radio,* May 1988, p 74.

Ref. 963: Al Christman - KB8I, "Phase Driven Arrays for the Low Bands," *QST,* May 1992, p 49.

Ref. 964: D. Leeson - W6QHS, *Physical Design of Yagi Antennas,* ARRL, Newington, CT.

Ref. 965: Dick Weber - K5IU, "Vibration Induced Yagi Fatigue Failures," *Ham Radio,* Aug 1989, p 9.

Ref. 966: Dick Weber - K5IU, "Structural Evaluation of Yagi Elements," *Ham Radio,* Dec 1988, p 29.

Ref. 967: Dave Leeson - W6QHS, "Strengthening the Cushcraft 40-2CD," *QST,* Nov 1991, p 36.

Ref. 968: R. Lahlum - W1MK, "Technical Correspondence," *QST,* Mar 1991, p 39.

Ref. 969: B. Orr, W6SAI , "The HGW Beam for 80 and 160 m," *CQ* mag, Nov 1994, p 120.

Ref. 970: Timothy Hulick, W9QQ, "The Box Antenna"; *CQ* mag, Jan 1997, p 9.

Ref. 971: JH1GNU, "Fishing Pole Low Band Yagis," *Low Band Monitor*, Feb 1996.

Ref. 972: JH1GNU, "Fishing Pole Yagi"; *Low Band Monitor*, Apr 1997.

Ref. 973: KF4HK, "The Comtec Systems ACB-4," *Low Band Monitor*, May 1994.

Ref. 974: KYØA: "Six for the Price of Three, A Novel 160 Meter Array"; *The Top Band Monitor*, Apr 1994.

Ref. 975: D. C. Mitchell, K8UR, "The K8UR Low-band Vertical Array," *CQ* mag, Dec 1989, p 42.

Ref. 976: John Bazley, G3HCT, "Phased Arrays for 7 MHz:, *RadCom,* Mar 1998, p 25.

Ref. 977: Peter Dalton, W6KW: "Designing and Building a 3-element 80-meter Yagi," *CQ* mag, Jul 1998, p 42.

Ref. 978: Dale Hoppe, K6UA, "How to Build a Relatively Small 2-Element 80-Meter Yagi," *CQ* mag, Jun 1998, p 24.

Ref. 979: Nathan A. Miller, NW3Z, and James K. Breakall, WA3FET, "The V-Yagi, A Light Weight Antenna for 40 Meters," *QST,* May 1988, p 38.

Ref. 980: Carl Luetzelschwab, K9LA, "Quad Versus Yagi at Low Heights*," The ARRL Compendium, Vol 4,* p 74 (ISBN 0-87259-491-2).

Ref. 981: Al Christman, KB8I, "The Slant-Wire Special," *The Antenna Compendium Vol. 4*, p 1, (ISBN 0-87259-491-2).

Ref. 982: John Stanley, K4ERO, "The Tuned Guy-Wire, Gain for (Almost) Free," *The ARRL Antenna Compendium Vol. 4*, ARRL, Newington, CT, p 27 (ISBN 0-87259-491-2).

Ref. 983: Al Christman, K3LC (ex KB8I), "Modifying the Slant-Wire Special for More Gain," Technical Correspondence, *QST*, May 1997, p 74.

■ 10. BROADBAND ANTENNAS

Ref. 1000: Larry Strain - N7DF, "3.5 to 30 MHz Discage Antenna," *CQ*, Apr 1984, p 18.

Ref. 1001: F. J. Bauer - W6FPO, "Low SWR Dipole Pairs for 1.8 through 3.5 MHz," *Ham Radio*, Oct 1972, p 42.

Ref. 1002: M. Walter Maxwell - W2DU, "A Revealing Analysis of the Coaxial Dipole Antenna," *Ham Radio*, Jul 1976, p 46.

Ref. 1003: Terry Conboy - N6RY, "Broadband 80-meter Antennas," *Ham Radio*, May 1979, p 44.

Ref. 1004: Mason Logan - K4MT, "Stagger Tuned Dipoles Increase Bandwidth," *Ham Radio*, May 1983, p 22.

Ref. 1005: C. C. Whysall, "The Double Bazooka," *QST*, Jul 1968, p 38.

Ref. 1006: John S. Belrose - VE2CV, "The Discone HF Antenna," *QST*, Jul 1975, p 11.

Ref. 1007: Allen Harbach - WA4DRU, "Broadband 80 Meter Antenna," *QST*, Dec 1980, p 36.

Ref. 1008: Jerry Hall - K1TD, "The Search for a Simple Broadband 80-Meter Dipole," *QST*, Apr 1983, p 22.

Ref. 1009: John Grebenkemper - KA3BLO, "Multiband Trap and Parallel HF Dipoles - a Comparison," *QST*, May 1985, p 26.

Ref. 1010: N. H. Sedgwick - G8WV, "Broadband Cage Aerials," *Radio Communication*, May 1965, p 287.

Ref. 1011: Pat Hawker - G3VA, "Technical Topics: Broadband Bazooka Dipole," *Radio Communication*, Aug 1976, p 601.

Ref. 1012: Frank Witt - AI1H, "The Coaxial Resonator Match and the Broadband Dipole," *QST*, Apr 1989, p 22.

Ref. 1013: Frank Witt - AI1H, "Match Bandwidth of Resonant Antenna Systems," *QST*, Oct 1991, p 21.

Ref. 1014: Rudy Severns, N6LF, "Wideband 80-Meter Dipole," *QST*, Jul 1995, p 27.

■ 11. LOOP ANTENNAS

Ref. 1100: William Orr - W6SAI, *All About Cubical Quads*, Radio Publications Inc.

Ref. 1101: Bill Salerno - W2ONV, "The W2ONV Delta/Sloper Antenna," *CQ*, Aug 1978, p 86.

Ref. 1102: Roy A. Neste - W0WFO, "Dissecting Loop Antennas To Find Out What Makes Them Tick," *CQ*, Aug 1984, p 36.

Ref. 1103: Rolf Schick - DL3AO, "Loop - Dipol und Vertikalantennen - Vergleiche und Erfahrungen," *CQ-DL*, Mar 1979, p 115.

Ref. 1104: Guenter Schwarzbeck - DL1BU, "DX Antennen fuer 80 und 160 Meter," *CQ-DL*, Apr 1979, p 150.

Ref. 1105: Gunter Steppert - DK8NG, "Zweielement Delta Loop mit einem Mast," *CQ-DL*, Aug 1980, p 370.

Ref. 1106: Hans Wurtz - DL2FA, "DX-Antennen mit spiegelenden Flaechen," *CQ-DL*, Apr 1981, p 162.

Ref. 1107: Hans Wurtz - DL2FA, "DX-Antennen mit spiegelenden Flaechen," *CQ-DL*, Dec 1981, p 583.

Ref. 1108: Dieter Pelz - DF3IK et al, "Rahmenantenne — keine Wunderantenne," *CQ-DL*, Sep 1982, p 435.

Ref. 1109: Willi Nitschke - DJ5DW - et al, "Richtkarakteristik - Fusspunktwiderstand etc von Einelementantennen," *CQ-DL*, Dec 1982, p 580.

Ref. 1110: Hans Wurtz - DL2FA, "DX-Antennen mit spiegelenden Flaechen," *CQ-DL*, Feb 1983, p 64.

Ref. 1111: Rudolf Fisher - DL6WD, "Das Monster - eine 2-Element Delta-Loop fuer 3.5 MHz," *CQ-DL*, Jul 1983, p 331.

Ref. 1112: John True - W4OQ, "Low Frequency Loop Antennas," *Ham Radio*, Dec 1976, p 18.

Ref. 1113: Paul Kiesel - K7CW, "7-Element 40-Meter Quad," *Ham Radio*, Jul 1978, p 30.

Ref. 1114: Glenn Williman - N2GW, "Delta Loop Array," *Ham Radio*, Jul 1978, p 16.

Ref. 1115: Frank J. Witt - W1DTV, "Top Loaded Delta Loop Antenna," *Ham Radio*, Dec 1978, p 57.

Ref. 1116: George Badger - W6TC, "Compact Loop Antenna for 40 and 80-Meter DX," *Ham Radio*, Oct 1979, p 24.

Ref. 1117: William M. Kesley - N8ET, "Three Element Switchable Quad for 40 Meters," *Ham Radio*, Oct 1980, p 26.

Ref. 1118: Jerrold Swank - W8HXR, "Two Delta Loops Fed in Phase," *Ham Radio*, Aug 1981, p 50.

Ref. 1119: Hasan Schiers - N0AN, "The Half Square Antenna," *Ham Radio*, Dec 1981, p 48.

Ref. 1120: John S. Belrose - VE2CV, "The Half Delta Loop," *Ham Radio*, May 1982, p 37.

Ref. 1121: V.C. Lear - G3TKN, "Reduced Size, Full Performance Corner Fed Delta Loop," *Ham Radio*, Jan 1985, p 67.

Ref. 1122: Lewis Mc Coy - W1ICP, "The Army Loop in Ham Communication," *QST*, Mar 1968, p 17.

Ref. 1123: J. Wessendorp - HB9AGK, "Loop Measurements," *QST*, Nov 1968, p 46.

Ref. 1124: F.N. Van Zant - W2EGH, "160, 75 and 40 Meter Inverted Dipole Delta Loop," *QST*, Jan 1973, p 37.

Ref. 1125: Ben Venster - K3DC, "The Half Square Antenna," *QST*, Mar 1974, p 11.

Ref. 1126: John Kaufmann - WA1CQW et al, "A Convenient Stub Tuning System for Quad Antennas," *QST*, May 1975, p 18.

Ref. 1127: Robert Edlund - W5DS, "The W5DS Hula-Hoop Loop," *QST*, Oct 1975, p 16.

Ref. 1128: Donald Belcher - WA4JVE et al, "Loops vs Dipole Analysis and Discussion," *QST*, Aug 1976, p 34.

Ref. 1129: Roger Sparks - W7WKB, "Build this C-T Quad Beam for Reduced Size," *QST*, Apr 1977, p 29.

Ref. 1130: John S. Belrose - VE2CV, "The Half-Delta Loop: a Critical Analysis and Practical Deployment," *QST*, Sep 1982, p 28.

Ref. 1131: Richard Gray - W9JJV, "The Two Band Delta Loop Antenna," *QST*, Mar 1983, p 36.

Ref. 1132: Edward Peter Swynar - VE3CUI, "40 Meters with a Phased Delta Loop," *QST*, May 1984, p 20.

Ref. 1133: Doug DeMaw - W1FB et al, "The Full-Wave Delta Loop at Low Height," *QST*, Oct 1984, p 24.

Ref. 1134: Pat Hawker - G3VA, "Technical Topics: Another Look at Transmitting Loops," *Radio Communication*, Jun 1971, p 392.

Ref. 1135: Pat Hawker - G3VA, "Technical Topics: Vertically Polarized Loop Elements," *Radio Communication*, Jun 1973, p 404.

Ref. 1136: Laury Mayhead - G3AQC, "Loop Aerials Close to Ground," *Radio Communication*, May 1974, p 298.

Ref. 1137: Pat Hawker - G3VA, "Technical Topics: The Half Square Aerial," *Radio Communication*, Jun 1974, p 380.

Ref. 1138: Pat Hawker - G3VA, "Miniaturized Quad Elements," *Radio Communication*, Mar 1976, p 206.

Ref. 1139: Pat Hawker - G3VA, "Technical Topics: Radiation Resistance of Medium Loops," *Radio Communication*, Feb 1979, p 131.

Ref. 1140: F. Rasvall - SM5AGM, "The Gain of the Quad," *Radio Communication*, Aug 1980, p 784.

Ref. 1141: Pat Hawker - G3VA, "Technical Topics: Polygonal Loop Antennas," *Radio Communication*, Feb 1981.

Ref. 1142: B. Myers - K1GQ, "The W2PV 80-Meter Quad," *Ham Radio*, May 1986, p 56.

Ref. 1143: Bill Orr - W6SAI, "A Two-Band Loop Antenna," *Ham Radio*, Sep 1987, p 57.

Ref. 1144: R. P. Haviland - W4MB, "The Quad Antenna - Part 1," *Ham Radio*, May 1988.

Ref. 1145: R. P. Haviland - W4MB, "The Quad Antenna - Part 2," *Ham Radio*, Jun 1988, p 54.

Ref. 1146: R. P. Haviland W4MB, "The Quad Antenna - Part 3," *Ham Radio*, Aug 1988, p 34.

Ref. 1147: C. Drayton Cooper - N4LBJ, "The Bi-Square array," *Ham Radio*, May 1990, p 42.

Ref. 1148: Bill Myers - K1GQ, "Analyzing 80 Meter Delta Loop Arrays," *Ham Radio*, Sep 1986, p 10.

Ref. 1149: Dave Donnelly - K2SS, "Yagi vs. Quad - Part 1," *Ham Radio*, May 1988, p 68.

Ref. 1150: D. DeMaw - W1FB, "A Closer Look at Horizontal Loop Antennas," *QST*, May 1990, p 28.

■ 12. RECEIVING ANTENNAS

Ref. 1200: I. Herliz, "Analysis of Action of Wave Antennas," *AIEE Transactions*, Vol 42, May 1942, p 260.

Ref. 1201: "Diversity Receiving System of RCA Communications for Radio Telegraphy," *AIEE Transactions*, Vol 42, May 1923, p 215.

Ref. 1202: Peterson Beverage, "The Wave Antenna: a New Type of Highly Directive Antenna," *Proc IRE*.

Ref. 1203: Dean Baily, "Receiving System for Long-Wave Transatlantic Radio Telephone," *Proc IRE*, Dec 1928.

Ref. 1204: *Antennas for Reception of Standard Broadcast Signals*, FCC Report, Apr 1958.

Ref. 1205: R. D. Straw - N6BV - ed, *The ARRL Handbook for Radio Amateurs*, ARRL, Newington, CT.

Ref. 1206: Victor Misek - W1WCR, *The Beverage Antenna Handbook*,

Ref. 1207: William Orr - W6SAI, *Radio Handbook*, Howard W. Sams & Co. Inc.

Ref. 1210: Bob Clarke - N1RC, "Six Antennas from Three Wires," 73, Oct 1983, p 10.

Ref. 1211: Davis Harold - W8MTI, "The Wave Antenna," *CQ*, May 1978, p 24.

Ref. 1212: Ulrich Rohde - DJ2LR, "Active Antennas," *CQ*, Dec 1982, p 20.

Ref. 1213: Karl Hille - 9A1VU, "Vom Trafo zum Kurzwellenempgaenger," *CQ-DL*, Mar 1977, p 99.

Ref. 1214: Dieter Pelz - DF3IK et al, "Rahmenantenne - keine Wunderantenne-," *CQ-DL*, Sep 1982, p 435.

Ref. 1215: Hans Wurtz - DL2FA, "Magnetische Antennen," *CQ-DL*, Feb 1983, p 64.

Ref. 1216: Hans Wurtz - DL2FA, "Magnetische Antennen," *CQ-DL*, Apr 1983, p 170.

Ref. 1217: Hans Wurtz - DL2FA, "Elektrisch Magnetische Beam Antennen (EMBA)," *CQ-DL*, Jul 1983, p 326.

Ref. 1218: Guenter Shwarzenbeck - DL1BU, "Rhamen- und Ringantennen," *CQ-DL*, May 1984, p 226.

Ref. 1219: Charles Bird - K6HTM, "160-Meter Loop for Receiving," *Ham Radio*, May 1974, p 46.

Ref. 1220: Ken Cornell - W2IMB, "Loop Antenna Receiving Aid," *Ham Radio*, May 1975, p 66.

Ref. 1221: John True - W4OQ, "Loop Antennas," *Ham Radio*, Dec 1976.

Ref. 1222: Henry Keen - W5TRS, "Selective Receiving Antennas: a Progress Report," *Ham Radio*, May 1978, p 20.

Ref. 1223: Byrd H. Brunemeier - KG6RT, "40-Meter Beverage Antenna," *Ham Radio*, Jul 1979, p 40.

Ref. 1224: David Atkins - W6VX, "Capacitively Loaded Dipole," *Ham Radio*, May 1984, p 33.

Ref. 1225: H. H. Beverage, "A Wave Antenna for 200 Meter Reception," *QST*, Nov 1922, p 7.

Ref. 1226: John Isaacs - W6PZV, "Transmitter Hunting on 75 Meters," *QST*, Jun 1958, p 38.

Ref. 1227: Lewis McCoy - W1ICP, "The Army Loop in Ham Communication," *QST*, Mar 1968, p 17.

Ref. 1228: J. Wessendorp - HB9AGK, "Loop Measurements," *QST*, Nov 1968, p 46.

Ref. 1229: Katashi Nose - KH6IJ, "A 160 Meter Receiving Loop," *QST*, Apr 1975, p 40.

Ref. 1230: Tony Dorbuck - W1YNC, "Radio Direction Finding Techniques," *QST*, Aug 1975, p 30.

Ref. 1231: Robert Edlund - W5DS, "The W5DS Hula-Hoop Loop," *QST*, Oct 1975, p 16.

Ref. 1232: Doug DeMaw - W1FB, "Build this 'Quickie' Preamp," *QST*, Apr 1977, p 43.

Ref. 1233: Larry Boothe - W9UCW, "Weak Signal Reception on 160. Some Antenna Notes," *QST*, Jun 1977, p 35.

Ref. 1234: Doug DeMaw - W1FB, "Low-Noise Receiving Antennas," *QST*, Dec 1977, p 36.

Ref. 1235: Doug DeMaw - W1FB, "Maverick Trackdown," *QST*, Jul 1980, p 22.

Ref. 1236: John Belrose - VE2CE, "Beverage Antennas for Amateur Communications," *QST*, Sep 1981, p 51.

Ref. 1237: H. H. Beverage, "H. H. Beverage on Beverage Antennas," *QST*, Dec 1981, p 55.

Ref. 1238: Doug DeMaw - W1FB et al, "The Classic Beverage Antenna Revisited," *QST*, Jan 1982, p 11.

Ref. 1239: John Webb - W1ETC, "Electrical Null Steering," *QST*, Oct 1982, p 28.

Ref. 1240: John F. Belrose - VE2CV et al, "The Beverage Antenna for Amateur Communications," *QST*, Jan 1983, p 22.

Ref. 1241: Doug DeMaw - W1FB, "Receiver Preamps and How to Use Them," *QST*, Apr 1984, p 19.

Ref. 1242: Pat Hawker - G3VA, "Technical Topics: Beverage Aerials," *Radio Communication*, Oct 1970, p 684.

Ref. 1243: Pat Hawker - G3VA, "Technical Topics: All Band Terminated Longwire," *Radio Communication*, Nov 1972, p 745.

Ref. 1244: Pat Hawker - G3VA, "Technical Topics: A 1.8 MHz Active Frame Aerial," *Radio Communication*, Aug 1976, p 601.

Ref. 1245: Pat Hawker - G3VA, "Technical Topics: Low Noise 1.8/3.5 MHz Receiving Antennas," *Radio Communication*, Apr 1978, p 325.

Ref. 1246: Pat Hawker - G3VA, "Technical Topics: Radiation Resistance of Medium Loops," *Radio Communication*, Feb 1979, p 131.

Ref. 1247: J.A. Lambert - G3FNZ, "A Directional Active Loop Receiving Antenna System," *Radio Communication*, Nov 1982, p 944.

Ref. 1248: R.C. Fenwick - K5RR, "A Loop Array for 160 Meters," *CQ*, Apr 1986, p 25.

Ref. 1249: Mike Crabtree - AB0X, "Some Thoughts on 160 Meter Receiving Antennas for City Lots," *CQ*, Jan 1988, p 26.

Ref. 1250: J. Wollweber - DF5PY, "Die Magnetisch Antenne - eine Wunderantenne," *CQ-DL*, Mar 1987, p 149.

Ref. 1251: Gary R. Nichols - KD9SV, "Variable gain 160-meter preamp," *Ham Radio*, Oct 1989, p 46.

Ref. 1252: Robert H. Johns - W3JIP, "How to Build an Indoor Transmitting Loop," *CQ*, Dec 1991, p 30.

Ref. 1253: Robert H. Johns - W3JIP, "How to Build an Indoor Transmitting Loop," *CQ*, Jan 1992, p 42.

Ref. 1254: D. DeMaw - W1FB, "On-Ground Low-Noise Receiving Antennas," *QST*, Apr 1988, p 30.

Ref. 1255: Juergen Schaefer - DL7PE, "Die Rahmantenne - eine Befehlsantenne zum Selbstbau," *CQ-DL*, Jan 1990, p 21.

Ref. 1256: D. DeMaw - W1FB, "Preamplifier for 80 and 160 M Loop and Beverage Antennas," *QST*, Aug 1988, p 22.

Ref. 1257: A. G. Lyner, "The Beverage Antenna: a practical way of assessing the radiation pattern using off-air signals," *BBC RD*, 1991/92.

Ref. 1258: B. H. Brunemeier - KG6RT, "Short Beverage for 40 meters," *HR* Jul 1979, p 40.

Ref. 1259 K9FD "The K9FD Receiving Loop," *The Low Band Monitor*, Nov 1995.

Ref. 1260: KY0A and G3SZA, "All Grounds are not equal—optimizing Beverage terminations," *The Low Band Monitor*, Apr 1995.

Ref. 1261: K0CS, "Real World Beverages"; *The Top Band Monitor*, Dec 1993.

■ 13. TRANSMISSION LINES

Ref. 1300: Wilkinson, "An N-Way Hybrid Power Divider," *IRE Transactions on Microwave*, Jan 1960.

Ref. 1301: R. D. Straw - N6BV, *The ARRL Handbook for Radio Amateurs*, ARRL, Newington, CT.

Ref. 1302: Pat Hawker - G3VA, *Amateur Radio Techniques*, RSGB.

Ref. 1303: Keith Henney, *Radio Engineering Handbook - 5th ed.*

Ref. 1304: *Ref.erence Data for Radio Engineers*, 5th ed, Howard W. Sams & Co. Inc.

Ref. 1305: John Devoldere - ON4UN, "80-Meter DXing," *Communications Technology Inc*, Jan 1979.

Ref. 1306: William Orr - W6SAI, *Radio Handbook*, Howard W. Sams & Co. Inc.

Ref. 1307: R. D. Straw - N6BV ed, *The ARRL Antenna Book*, ARRL, Newington, CT.

Ref. 1308: Walter Maxwell - W2DU, "Niedriges SWR aus falschem Grund (Teil 1)," *CQ-DL*, Jan 1976, p 3.

Ref. 1309: Walter Maxwell - W2DU, "Niedriges SWR aus falschem Grund (Teil 2)," *CQ-DL*, Feb 1976, p 47.

Ref. 1310: Walter Maxwell - W2DU, "Niedriges SWR aus falschem Grund (Teil 3)," *CQ-DL*, Jun 1976, p 202.

Ref. 1311: Walter Maxwell - W2DU, "Niedriges SWR aus falschem Grund (Teil 5)," *CQ-DL*, Sep 1976, p 272.

Ref. 1312: Jim Fisk - W1HR, "The Smith Chart," *Ham Radio*, Nov 1970, p 16.

Ref. 1313: Richard Taylor - W1DAX, "N-Way Power Dividers and 3-dB Hybrids," *Ham Radio*, Aug 1972, p 30.

Ref. 1314: Joseph Boyer - W6UYH, "Antenna-Transmission Line Analogy," *Ham Radio*, Apr 1977, p 52.

Ref. 1315: Joseph Boyer - W6UYH, "Antenna-Transmission Line Analogy - Part 2," *Ham Radio*, May 1977, p 29.

Ref. 1316: J. Reisert - W1JR, "Simple and Efficient Broadband Balun," *Ham Radio*, Sep 1978, p 12.

Ref. 1317: Jack M. Schulman - W6EBY, "T-Network Impedance Matching to Coaxial Feedlines," *Ham Radio*, Sep 1978, p 22.

Ref. 1318: Charlie J. Carroll - K1XX, "Matching 75-Ohm CATV Hardline to 50-Ohm Systems," *Ham Radio*, Sep 1978, p 31.

Ref. 1319: John Battle - N4OE, "What is Your SWR?," *Ham Radio,* Nov 1979, p 48.

Ref. 1320: Henry Elwell - N4UH, "Long Transmission Lines for Optimum Antenna Location," *Ham Radio,* Oct 1980, p 12.

Ref. 1321: John W. Frank - WB9TQG, "Measuring Coax Cable Loss with an SWR Meter," *Ham Radio,* May 1981, p 34.

Ref. 1322: Stan Gibilisco - W1GV/4, "How Important is Low SWR?" *Ham Radio,* Aug 1981, p 33.

Ref. 1323: Lewis T. Fitch - W4VRV, "Matching 75-Ohm Hardline to 50-Ohm Systems," *Ham Radio,* Oct 1982, p 43.

Ref. 1324: L. A. Cholewski - K6CRT, "Some Amateur Applications of the Smith Chart," *QST,* Jan 1960, p 28.

Ref. 1325: Walter Maxwell - W2DU, "Another Look at Ref.lections - Part 1," *QST,* Apr 1973, p 35.

Ref. 1326: Walter Maxwell - W2DU, "Another Look at Ref.lections - Part 2," *QST,* Jun 1973, p 20.

Ref. 1327: Walter Maxwell - W2DU, "Another Look at Ref.lections - Part 3," *QST,* Aug 1973, p 36.

Ref. 1328: Walter Maxwell - W2DU, "Another Look at Ref.lections - Part 4," *QST,* Oct 1973, p 22.

Ref. 1329: Walter Maxwell - W2DU, "Another Look at Ref.lections - Part 5," *QST,* Apr 1974, p 26.

Ref. 1330: Walter Maxwell - W2DU, "Another Look at Ref.lections - Part 6," *QST,* Dec 1974, p 11.

Ref. 1331: Gerald Hall - K1PLP, "Transmission-Line Losses," *QST,* Dec 1975, p 48.

Ref. 1332: Walter Maxwell - W2DU, "Another Look at Ref.lections - Part 7," *QST,* Aug 1976, p 16.

Ref. 1333: Charles Brainard - WA1ZRS, "Coaxial Cable: The Neglected Link," *QST,* Apr 1981, p 28.

Ref. 1334: Crawford MacKeand - WA3ZKZ, "The Smith Chart in BASIC," *QST,* Nov 1984, p 28.

Ref. 1336: R. C. Hills - G3HRH, "Some Ref.lections on Standing Waves," *Radio Communication,* Jan 1964, p 15.

Ref. 1337: Pat Hawker - G3VA, "Technical Topics: Another Look at SWR," *Radio Communication,* Jun 1974, p 377.

Ref. 1338: Garside - G3MYT, "More on the Smith Chart," *Radio Communication,* Dec 1977, p 934.

Ref. 1339: Pat Hawker - G3VA, "Technical Topics: SWR How Important?" *Radio Communication,* Jan 1982, p 41.

Ref. 1340: Kenneth Parker - G3PKR, "Ref.lected Power Does Not Mean Lost Power," *Radio Communication,* Jul 1982, p 581.

Ref. 1341: Pat Hawker - G3VA, "Technical Topics: Ref.lected Power is Real Power," *Radio Communication,* Jun 1983, p 517.

Ref. 1342: H. Ashcroft - G4CCM, "Critical Study of the SWR Meter," *Radio Communication,* Mar 1985, p 186.

Ref. 1343: Walter Maxwell - W2DU, "Niedriges SWR aus falschem Grund (Teil 4)," *CQ-DL,* Aug 1976, p 238.

Ref. 1344: Lew McCoy - W1ICP, "Coax as Tuned Feeders: Is It Practical?," *CQ,* Aug 1989, p 13.

Ref. 1345: Chet Smith - K1CCL, "Simple Coaxial Cable Measurements," *QST,* Sep 1990, p 25.

Ref. 1346: J. Althouse - K6NY, "Using a Noise Bridge to Measure Coaxial Cable Impedance," *QST,* May 1991, p 45.

Ref. 1347: A.E. Popodi - OE2APM/AA3K, "Measuring Transmission Line Parameters," *Ham Radio,* Sep 1988, p 22.

Ref. 1348: Byron Goodman - W1DX, "My Feed Line Tunes My Antenna," *QST,* Nov 1991, p 33.

Ref. 1349: Gary E. Myers - K9CZB, "Solving Transmission Line Problems on Your C64," *Ham Radio,* May 1986, p 74.

Ref. 1350: Fred Bonavita - W5QJM, "Working with Balanced Line, part I"; *CQ* mag, Jan 1994, p 56.

Ref. 1351: Fred Bonavita - W5QJM, "Working with Balanced Line, part II"; *CQ* mag, Feb 1994, p 26.

Ref. 1352: Warren Bruene, W5OLY, "Ref.lected Power Stays in the Coax," *CQ* mag, Jan 1995, p 13.

■ 14. MATCHING

Ref. 1400: R. D. Straw - N6BV - ed, *The ARRL Handbook for Radio Amateurs,* ARRL, Newington, CT.

Ref. 1401: Harold Tolles - W7ITB, "Gamma-Match Design," *Ham Radio,* May 1973, p 46.

Ref. 1402: Robert Baird - W7CSD, "Antenna Matching Systems," *Ham Radio,* Jul 1973, p 58.

Ref. 1403: Earl Whyman - W2HB, "Standing-Wave Ratios," *Ham Radio,* Jul 1973, p 26.

Ref. 1404: Robert Leo - W7LR, "Designing L-networks," *Ham Radio,* Feb 1974, p 26.

Ref. 1405: G. E. Smith - W4AEO, "Log-Periodic Feed Systems," *Ham Radio,* Oct 1974, p 30.

Ref. 1406: John True - W4OQ, "Shunt-Fed Vertical Antennas," *Ham Radio,* May 1975, p 34.

Ref. 1407: I. L. McNally - K6WX - et al, "Impedance Matching by Graphical Solution," *Ham Radio,* Mar 1978, p 82.

Ref. 1408: Jim Fisk - W1HR, "Transmission-Line Calculations with the Smith Chart," *Ham Radio,* Mar 1978, p 92.

Ref. 1409: Jack M. Schulman - W6EBY, "T-Network Impedance Matching to Coaxial Feed Lines," *Ham Radio,* Sep 1978, p 22.

Ref. 1410: Ernie Franke - WA2EWT, "Appreciating the L Matching Network," *Ham Radio,* Sep 1980, p 27.

Ref. 1411: Robert Leo - W7LR, "Remote-Controlled 40 - 80 - and 160-Meter Vertical," *Ham Radio,* May 1984, p 38.

Ref. 1412: James Sanford - WB4GCS, "Easy Antenna Matching," *Ham Radio,* May 1984, p 67.

Ref. 1413: Chris Bowick - WD4C, "Impedance Matching: A Brief Review," *Ham Radio,* Jun 1984, p 49.

Ref. 1414: Richard Nelson - WBØIKN, "Basic Gamma Matching," *Ham Radio,* Jan 1985, p 29.

Ref. 1415: L. A. Cholewski - K6CRT, "Some Amateur Applications of the Smith Chart," *QST,* Jan 1960, p 28.

Ref. 1416: Gene Hubbell - W9ERU, "Feeding Grounded Towers as Radiators," *QST,* Jun 1960, p 33.

Ref. 1417: J. D. Gooch - W9YRV - et al, "The Hairpin Match," *QST*, Apr 1962, p 11.

Ref. 1418: Eugene E. Baldwin - WØRUG, "Some Notes on the Care and Feeding of Grounded Verticals," *QST*, Oct 1963, p 45.

Ref. 1419: N. H. Davidson - K5JVF, "Flagpole Without a Flag," *QST*, Nov 1964, p 36.

Ref. 1420: Robert Leo - K7KOK, "An Impedance Matching Method," *QST*, Dec 1968, p 24.

Ref. 1421: D. J. Healey - W3PG, "An Examination of the Gamma Match," *QST*, Apr 1969, p 11.

Ref. 1422: Donald Belcher - WA4JVE, "RF Matching Techniques - Design and Example," *QST*, Oct 1972, p 24.

Ref. 1423: James McAlister - WA5EKA, "Simplified Impedance Matching and the Mac Chart," *QST*, Dec 1972, p 33.

Ref. 1424: John Kaufmann - WA1CQW - et al, "A Convenient Stub Tuning System for Quad Antennas," *QST*, May 1975, p 18.

Ref. 1425: Earl Cunningham - W5RTQ, "Shunt Feeding Towers for Operating on the Low Amateur Frequencies," *QST*, Oct 1975, p 22.

Ref. 1426: Doug DeMaw - W1CER, "Another Method of Shunt Feeding Your Tower," *QST*, Oct 1975, p 25.

Ref. 1427: Jerry Sevick - W2FMI, "Simple Broadband Matching Networks," *QST*, Jan 1976, p 20.

Ref. 1428: Walter Schulz - K3OQF, "Slant-Wire Feed for Grounded Towers," *QST*, May 1977, p 23.

Ref. 1429: Bob Pattison - N6RP, "A Graphical Look at the L Network," *QST*, Mar 1979, p 24.

Ref. 1430: Tony Dorbuck - K1FM, "Matching-Network Design," *QST*, Mar 1979, p 26.

Ref. 1431: Herbert Drake Jr. - N6QE, "A Remotely Controlled Antenna-Matching Network," *QST*, Jan 1980, p 32.

Ref. 1432: Doug DeMaw - W1FB, "Ultimate Transmatch Improved," *QST*, Jul 1980, p 39.

Ref. 1433: Colin Dickman - ZS6U, "The ZS6U Minishack Special," *QST*, Apr 1981, p 32.

Ref. 1434: Claude L. Frantz - F5FC/DJØOT, "A New More Versatile Transmatch," *QST*, Jul 1982, p 31.

Ref. 1435: Doug DeMaw - W1FB, "A 'Multipedance' Broadband Transformer," *QST*, Aug 1982, p 39.

Ref. 1436: Don Johnson - W6AAQ, "Mobile Antenna Matching. Automatically!" *QST*, Oct 1982, p 15.

Ref. 1437: Doug DeMaw - W1FB, "Shunt-Fed Towers: Some Practical Aspects," *QST*, Oct 1982, p 21.

Ref. 1438: Crawford MacKaend - WA3ZKZ, "The Smith Chart in BASIC," *QST*, Nov 1984, p 28.

Ref. 1439: J. A. Ewen - G3HGM, "Design of L-Networks for Matching Antennas to Transmitters," *Radio Communication*, Aug 1984, p 663.

Ref. 1440: H. Ashcroft - G4CCM, "Critical Study of the SWR Meter," *Radio Communication*, Mar 1985, p 186.

Ref. 1441: Robert C Cheek - W3VT, "The L-Match," *Ham Radio*, Feb 1989, p 29.

Ref. 1442: W. Orr - W6SAI, "Antenna Matching Systems," *Ham Radio*, Nov 1989, p 54.

Ref. 1443: Will Herzog - K2LB, "Broadband RF Transformers," *Ham Radio*, Jan 1986, p 75.

Ref. 1444: Richard A. Gardner - N1AYW, "Computerizing Smith Chart Network Analysis," *Ham Radio*, Oct 1989, p 10.

Ref. 1445: Joe Carr - K4IPV, "Impedance Matching," *Ham Radio*, Aug 1988, p 27.

Ref. 1446: I. L. McNally - W1CNK, "Graphic Solution of impedance matching problem," *HR*, May 78, p 82.

■ 15. BALUNS

Ref. 1500: J. Schultz - W4FA, "A Selection of Baluns from Palomar Engineers," *CQ*, Apr 1985, p 32.

Ref. 1501: William Orr - W6SAI, "Broadband Antenna Baluns," *Ham Radio*, Jun 1968, p 6.

Ref. 1502: J. Reisert - W1JR, "Simple and Efficient Broadband Balun," *Ham Radio*, Sep 1978, p 12.

Ref. 1503: John J. Nagle - K4KJ, "High Performance Broadband Balun," *Ham Radio*, Feb 1980, p 28.

Ref. 1504: George Badger - W6TC, "A New Class of Coaxial-Line Transformers," *Ham Radio*, Feb 1980, p 12.

Ref. 1505: George Badger - W6TC, "Coaxial-Line Transformers," *Ham Radio*, Mar 1980, p 18.

Ref. 1506: John Nagle - K4KJ, "The Half-Wave Balun: Theory and Application," *Ham Radio*, Sep 1980, p 32.

Ref. 1507: Roy N. Lehner - WA2SON, "A Coreless Balun," *Ham Radio*, May 1981, p 62.

Ref. 1508: John Nagle - K4KJ, "Testing Baluns," Ham Radio, Aug 1983, p 30.

Ref. 1509: Lewis McCoy - W1ICP, "Is a Balun Required?" *QST*, Dec 1968, p 28.

Ref. 1510: R. H. Turrin - W2IMU, "Application of Broadband Balun Transformers," *QST*, Apr 1969, p 42.

Ref. 1511: Bruce Eggers - WA9NEW, "An Analysis of the Balun," *QST*, Apr 1980, p 19.

Ref. 1512: William Fanckboner - W9INN, "Using Baluns in Transmatches with High-Impedance Lines," *QST*, Apr 1981, p 51.

Ref. 1513: Doug DeMaw - W1FB, "A 'Multipedance' Broadband Transformer," *QST*, Aug 1982, p 39.

Ref. 1514: R. G. Titterington - G3ORY, "The FerriteCored Balun Transformer," *Radio Communication*, Mar 1982, p 216.

Ref. 1515: John S. Belrose - VE2CV, "Transforming the Balun," *QST*, Jun 1991, p 30.

Ref. 1516: D. DeMaw - W1FB, "How to Build and Use Balun Transformers," *QST*, Mar 1987, p 34.

Ref. 1517: J. Sevick - W2FMI, "The 2:1 Unun Matching Transformer," *CQ*, Aug 1992, p 13.

Ref. 1518: J. Sevick - W2FMI, "The 1.5:1 and the 1.33:1 Unun Transformer," *CQ*, Nov 1992, p 26.

Ref. 1519: W. Maxwell - W2DU, "Some Aspects of the Balun Problem," *QST*, Mar 1983, p 38.

Ref. 1520: R. Lewallen - W7EL, "Baluns: What they do and how they do it," *ARRL Antenna Compendium, Vol 1*, p 157.

Ref. 1521: Jerry Sevick, W2FMI, "The Ultimate Multimatch Unun," *CQ* mag, Aug 1993, p 15.

Ref. 1522: Jerry Sevick, "A Multimatch Unun," *CQ* mag, Apr 1993, p 28.

Ref. 1523: Jerry Sevick, "Dual-Ratio Ununs," *CQ* mag, Mar 1993 p 54.

Ref. 1524: Bill Orr, W5SAI "The Coaxial Balun," *CQ* mag, Nov 1993, p 60.

Ref. 1525: Jerry Sevick, "Ununs for Beverage Antenna," *CQ* mag, Dec 1993, p 62.

Ref. 1526: Jerry Sevick, "A Balun Essay," *CQ* mag, Jun 1993, p 50.

Ref. 1527: Jerry Sevick, "More On the 1:1 Balun," *CQ* mag, Apr 1994, p 26.

Ref. 1528: Jerry Sevick: "A Subsequent Look at 4:1 Baluns"; *CQ* mag, Feb 1994, p 28.

Ref. 1829: Jerry Sevick, "The 4:1 UNUN," *CQ* mag, Jan 1993, p 30+.

Ref. 1830: Jerry Sevick, *Transmission Line Transformers*, Noble Publishing, Atlanta, Ga.

■ 16. ANTENNA MEASURING

Ref. 1600: R. D. Straw - N6BV - ed, *The ARRL Handbook for Radio Amateurs*, ARRL, Newington, CT.

Ref. 1601: John Schultz - W4FA, "The MFJ-202B R.F. Noise Bridge," *CQ,* Aug 1984, p 50.

Ref. 1602: Michael Martin - DJ7YV, "Die HF Stromzange," *CQ-DL*, Dec 1983, p 581.

Ref. 1603: Reginald Brearley - VE2AYU, "Phase-Angle Meter," *Ham Radio,* Apr 1973, p 28.

Ref. 1604: Robert Baird - W7CSD, "Nonresonant Antenna Impedance Measurements," *Ham Radio,* Apr 1974, p 46.

Ref. 1606: John True - W4OQ, "Antenna Measurements," *Ham Radio,* May 1974, p 36.

Ref. 1607: Frank Doting - W6NKU - et al, "Improvements to the RX Noise Bridge," *Ham Radio,* Feb 1977, p 10.

Ref. 1608: Charles Miller - W3WLX, "Gate-Dip Meter," *Ham Radio,* Jun 1977, p 42.

Ref. 1609: Kenneth F. Carr - WØKUS, "Ground Current Measuring on 160 Meters," *Ham Radio,* Jun 1979, p 46.

Ref. 1610: Forrest Gehrke - K2BT, "A Modern Noise Bridge," *Ham Radio,* Mar 1983, p 50.

Ref. 1611: William Vissers - K4KI, "A Sensitive Field Strength Meter," *Ham Radio,* Jan 1985, p 51.

Ref. 1612: Jerry Sevick - W2FMI, "Simple RF Bridges," *QST*, Apr 1975, p 11.

Ref. 1613: David Geiser - WA2ANU, "The Impedance Match Indicator," *QST*, Jul 1980, p 11.

Ref. 1614: Jerry Sevick - W2FMI, "Measuring Soil Conductivity," *QST*, Mar 1981, p 38.

Ref. 1615: Jack Priedigkeit - W6ZGN, "Measuring Impedance with a Reflection-Coefficient Bridge," *QST*, Mar 1983, p 30.

Ref. 1616: Doug DeMaw - W1FB, "Learning to Use Field Strength Meters," *QST*, Mar 1985, p 26.

Ref. 1617: Al Bry - W2MEL, "Beam Antenna Pattern Measurement," *QST*, Mar 1985, p 31.

Ref. 1618: M. R. Irving - G3ZHY, "The 'Peg Antennameter'," *Radio Communication*, May 1972, p 297.

Ref. 1619: Pat Hawker - G3VA, "Technical Topics: Current Probe for Open-Wire Feeders - Wire Antennas - etc," *Radio Communication*, Nov 1984, p 962.

Ref. 1620: D. DeMaw - W1FB, "A Laboratory Style RX Noise Bridge," *QST*, Dec 1987, p 32.

Ref. 1621: John Belrose - VE2VC, "RX Noise Bridges," *QST*, May 1988, p 34.

Ref. 1622: Andrew S. Griffith - W4ULD, "A Dipper Amplifier for Impedance Bridges," *QST*, Sep 1988, p 24.

Ref. 1623: John Grebenkemper - KI6WX, "Improving and Using R-X Noise Bridges," *QST*, Aug 1989, p 27.

Ref. 1624: D. DeMaw - W1FB, "MFJ-931 Artificial RF Ground," *QST*, Apr 1988, p 40.

Ref. 1625: Jim Smith, "How to Build a Snap On RF Current Probe," *CQ* mag, Aug 1996, p 22.

Ref. 1626: Gerd Janzen, DF6SJ/VK2BJZ "HF-Messungen mit einem aktiven Stehwellen-Messgeraet," ISBN 3-88006-170-X (This book is only available directly from the author, DF6SJ, Hochvogelstrasse 29, D087435 Kempten, Germany).

About the American Radio Relay League

The seed for Amateur Radio was planted in the 1890s, when Guglielmo Marconi began his experiments in wireless telegraphy. Soon he was joined by dozens, then hundreds, of others who were enthusiastic about sending and receiving messages through the air—some with a commercial interest, but others solely out of a love for this new communications medium. The United States government began licensing Amateur Radio operators in 1912.

By 1914, there were thousands of Amateur Radio operators—hams—in the United States. Hiram Percy Maxim, a leading Hartford, Connecticut, inventor and industrialist saw the need for an organization to band together this fledgling group of radio experimenters. In May 1914 he founded the American Radio Relay League (ARRL) to meet that need.

Today ARRL, with approximately 170,000 members, is the largest organization of radio amateurs in the United States. The League is a not-for-profit organization that:

• promotes interest in Amateur Radio communications and experimentation
• represents US radio amateurs in legislative matters, and
• maintains fraternalism and a high standard of conduct among Amateur Radio operators.

At League headquarters in the Hartford suburb of Newington, the staff helps serve the needs of members. ARRL is also International Secretariat for the International Amateur Radio Union, which is made up of similar societies in more than 150 countries around the world.

ARRL publishes the monthly journal *QST*, as well as newsletters and many publications covering all aspects of Amateur Radio. Its headquarters station, W1AW, transmits Morse-code practice sessions and bulletins of interest to radio amateurs. The League also coordinates an extensive field organization, which provides technical and other support for radio amateurs as well as communications for public service activities. ARRL also represents US amateurs with the Federal Communications Commission and other government agencies in the US and abroad.

Membership in ARRL means much more than receiving *QST* each month. In addition to the services already described, ARRL offers membership services on a personal level, such as the ARRL Volunteer Examiner Coordinator Program and a QSL bureau.

Full ARRL membership (available only to licensed radio amateurs) gives you a voice in how the affairs of the organization are governed. League policy is set by a Board of Directors (one from each of 15 Divisions). Each year, half of the ARRL Board of Directors stands for election by the full members they represent. The day-to-day operation of ARRL HQ is managed by an Executive Vice President and a Chief Financial Officer.

No matter what aspect of Amateur Radio attracts you, ARRL membership is relevant and important. There would be no Amateur Radio as we know it today were it not for the ARRL. We would be happy to welcome you as a member! (An Amateur Radio license is not required for Associate Membership.) For more information about ARRL and answers to any questions you may have about Amateur Radio, write or call:

ARRL
225 MAIN STREET
NEWINGTON CT 06111-1494
(860) 594-0200

Prospective new amateurs call:
800-32-NEW HAM (800-326-3942)

You can also contact us via e-mail:
newham@arrl.org

or check out our World Wide Web site:
http://www.arrl.org/

INDEX

581

ON4UN NEW LOW BAND SOFTWARE

Price: $50 + $5 for shipping and handling worldwide. MS-DOS format only.

If you have the original version of the LOW BAND SOFTWARE, you are eligible for a $10 price reduction, provided a copy of the registration form that came with the software is sent with your order.

NAME: _____ CALL: _____

ADDRESS: _____

CITY: _____ STATE: _____ ZIP: _____

Send this Order Form with payment to:

George Oliva, K2UO, 5 Windsor Dr, Eatontown NJ 07724, USA, or

John Devoldere, ON4UN, Poelstraat 215, B9820 Merelbeke, Belgium

PAYMENT BY: ☐ Personal check   ☐ Postal order

Make payable to George Oliva (for orders to K2UO) or John Devoldere (for orders to ON4UN)

ON4UN YAGI DESIGN SOFTWARE

Price: $65 + $5 for shipping and handling worldwide. MS-DOS format only.

NAME: _____ CALL: _____

ADDRESS: _____

CITY: _____ STATE: _____ ZIP: _____

Send this Order Form with payment to:

G. Oliva, K2UO, 5 Windsor Dr, Eatontown NJ 07724, USA, or

J. Devoldere, ON4UN, Poelstraat 215, B9820 Merelbeke, Belgium

PAYMENT BY: ☐ Personal check   ☐ Postal order

Make payable to: George Oliva (for orders to K2UO) or John Devoldere (for orders to ON4UN)

FEEDBACK

Please use this form to give us your comments on this book and what you'd like to see in future editions, or e-mail us at **pubsfdbk@arrl.org** (publications feedback). If you use e-mail, please include your name, call, e-mail address and the book title, edition and printing in the body of your message. Also indicate whether or not you are an ARRL member.

Where did you purchase this book?
□ From ARRL directly   □ From an ARRL dealer

Is there a dealer who carries ARRL publications within:
□ 5 miles   □ 15 miles   □ 30 miles of your location?   □ Not sure.

License class:
□ Novice   □ Technician   □ Technician Plus   □ General   □ Advanced   □ Amateur Extra

Name _____

Daytime Phone (   ) _____

Address _____

City, State/Province, ZIP/Postal Code _____

If licensed, how long? _____

Other hobbies _____

_____

Occupation _____

ARRL member? □ Yes □ No

Call Sign _____

Age _____

E-mail address _____

From _____

_____

_____

EDITOR, LOW-BAND DXING
AMERICAN RADIO RELAY LEAGUE
225 MAIN STREET
NEWINGTON CT 06111-1494

please fold and tape